D1694836

COLLOQUIA MATHEMATICA
SOCIETATIS JÁNOS BOLYAI, 56.

DIFFERENTIAL GEOMETRY
AND ITS APPLICATIONS

Edited by

J. SZENTHE and L. TAMÁSSY

NORTH-HOLLAND PUBLISHING COMPANY
AMSTERDAM – OXFORD – NEW YORK

© BOLYAI JÁNOS MATEMATIKAI TÁRSULAT

Budapest, Hungary, 1992

ISBN North-Holland: 0444 98759 2
ISBN Bolyai: 963 8022 59 0
ISSN Bolyai: 0139–3383

Joint edition published by

JÁNOS BOLYAI MATHEMATICAL SOCIETY

and

ELSEVIER SCIENCE PUBLISHERS B.V.
Saraburgerhartstraat 25, P.O. Box 103
1000 AC, Amsterdam, The Netherlands

In the U.S.A. and Canada:

ELSEVIER SCIENCE PUBLISHING COMPANY INC.
655 Avenue of Americas
New York, N.Y. 10010
U.S.A.

Assistant editor: L. VERHÓCZKI

Film transfer by ITEX Laser- and Computertechnics Ltd.
Printed in Hungary
Franklin Nyomda
Budapest

Contents

Preface

In succession to our former meetings on differential geometry a Colloquium took place in Eger from August 20 to August 25, 1989. The Colloquium was organized by the János Bolyai Mathematical Society with the assistance of the following committee: S. Bácsó, E. Molnár, P. T. Nagy, A. Rapcsák (honorary chairman), Gy. Soós, J. Szenthe (chairman), L. Tamássy (chairman), L. Verhóczki (secretary).

It was pleasant to meet again participants of our former colloquia and to see the considerable number of those who came for the first time. The Colloquium had 73 participants from 18 foreign countries and 22 from Hungary. The programme included 10 plenary lectures and 69 lectures in four sections. There were also some ad hoc evening sections and of course many informal discussions too. The historical sites of the town Eger and its famous surroundings offered ideal occasions to acquaint Hungarian cultural traditions and for evening programmes.

The present volume contains beside the papers submitted for publication the list of participants. The containt of the papers covers a wide range of topics in differential geometry so that it is impossible to give an exact survey of it in short. However, there were some frequented subjects, such as Riemannian geometry, Finsler geometry, the theory of submanifolds and applications of differential geometry in mathematical physics.

It is a pleasant duty for us to express our thanks to our guests who contributed to the success of the Colloquium and especially to those who offered us their manuscripts for publication.

The Editors

COLLOQUIA MATHEMATICA SOCIETATIS JÁNOS BOLYAI
56. DIFFERENTIAL GEOMETRY, EGER (HUNGARY), 1989

Riemannian Manifolds of Cohomogeneity One[*]

D. V. ALEKSEEVSKY

1. Introduction

There are many papers on the geometry of homogeneous spaces. Powerful technique of Lie groups can be used to obtain many deep results about homogeneous spaces which have various applications. However, some geometrical questions require the investigation of more general but still tractable class of spaces. For example it is well known that there are not Ricci-flat Riemannian homogeneous manifolds that are not flat. Hence, to construct Ricci-flat manifolds (and in particular Riemannian manifolds with the exceptional holonomy groups) one must deal with non homogeneous manifolds.

An immediate generalization of homogeneous spaces are the G-manifolds of cohomogeneity one, that is manifolds with a given connected Lie group of transformations G which has an orbit of codimension one. We shall assume also that G acts on M properly. This condition is equivalent to the existence on M a G-invariant complete Riemannian metric g [1]. The manifold M equipped with such metric g is called Riemannian cohomogeneity one (or for brevity $C1$) manifold.

---

[*] This paper is in final form and no version of it will be submitted for publication elsewhere.

The systematic study of Riemannian $C1$ manifolds was initiated by
L. Bérard–Bergery [2], who obtained many interesting results. Here we
continue his investigations. There are four types of $C1$ manifolds M with
the orbit space $\Omega = M/_G = \mathbb{R}$, S^1, $\mathbb{R}^+$, $[0, \pi]$ respectively. We give group
theoretic descriptions of $C1$ manifolds M of all these types. We describe
smooth G-invariant metrics on M and study its normal geodesics. It allows
us to compute and to study the curvature of an invariant metric g near a
singular orbit and to obtain some results on the existence and non-existence
of an invariant metric of some kinds of $C1$ manifold (for example a metric
with positive or negative Ricci curvature). The last questions will be treated
elsewhere.

We use the following notations:
$N_H(K)$ and $C_H(K)$ are the normalizer and the centralizer of a subgroup K
of a Lie group H, H_0 denotes the connected component of the identity of
the group H, G_x is the stabilizer of a point x in the group G.

2. The behaviour of normal geodesics

2.1. Let (M, g) be a Riemannian $C1$ manifold.
We say that a geodesic $\gamma = \{y_t\}$ on M is normal if it is orthogonal to
the orbit $P = Gy_0$. A normal geodesic is orthogonal to any orbit of the
group G since for a Killing vector X the function $g(\dot\gamma, X)$ is constant along
γ. The group G acts on the set of normal geodesics transitively. Hence a
normal geodesic intersects each orbit and its arc length defines a metric on
the orbit space $\Omega = M/_G$. The metric space Ω can be identified with one
of the following standard metric spaces [2]

$$\text{i)} \ \ \Omega = \mathbb{R}, \qquad \text{ii)} \ \ \Omega = S^1 = \mathbb{R}/_{2l\mathbb{Z}},$$
$$\text{iii)} \ \ \Omega = \mathbb{R}^+ = [0, \infty], \qquad \text{iv)} \ \ \Omega = [0, l].$$

Rescaling the metric g we may suppose that $l = \pi$. In the case i), ii)
all the orbits are principal. We have one singular (i.e. non-principal)
orbit $P_0 = \kappa^{-1}(0)$ in the case iii) and two singular orbits $P_0 = \kappa^{-1}(0)$,
$P_\pi = \kappa^{-1}(\pi)$ in the case iv), $\kappa : M \to \Omega = M/G$ is the natural projection.

Remark
2.1. It is clear that only in the case ii) a principal orbit does not separate
the manifold M into two parts.

2.2. Let P be an orbit of the group G in a Riemannian $C1$-manifold (M, g). Denote by $T^\perp P$ the normal vector bundle of P and by $SP = \{v \in T^\perp P,\ g(v, v) = 1\}$ its spherical subbundle. We have the following exponential mappings:

$$\exp : T^\perp P \to M, \qquad v \mapsto \exp(v)$$
$$\exp_t : SP \to M, \qquad v \mapsto \exp(tv), \quad t \geq 0.$$

Proposition 2.1.

i) Let P be an orbit. For any $t \geq 0$ the image $P_t = \exp_t SP$ is an equidistant set of the submanifold P and a connected component of P_t is an orbit of G.

ii) Suppose that P is a singular orbit. Then SP is connected and for any $t \geq 0$ P_t is an orbit. If P_t is a principal orbit, then $\exp_t : SP \to P_t$ is a G-equivariant diffeomorphism.

iii) Suppose that P is a principal orbit. Then SP has two connected components $S^\pm P$ and the mapping $\exp_t : S^\pm P \to \exp_t(S^\pm P) = P_t^\pm$ is a G-equivariant diffeomorphism if the orbit $P_t^\pm$ is principal.

Corollary 2.1. The set of focal points of a principal orbit P is the union of all singular orbits of codimension $m > 1$. In particular the submanifold P has no focal points if the codimension of any orbit is one.

Corollary 2.2. If the manifold (M, g) has one singular orbit $P_0 = G/H$ then the mapping $\exp : T^\perp P_0 \to M$ is a G-diffeomorphism.

3. $C1$ manifolds without singular orbits

3.1 Proposition 2.1 implies the following description of $C1$ manifold M with $\Omega = \mathbb{R}$ and invariant metrics on M.

Proposition 3.1. *Let (M, g) be a Riemannian C1 manifold with $\Omega = \mathbb{R}$. Fix an orbit $P = Gx = G/K$ in M and a unit normal vector $v_0 \in SP$ in the point x_0. Then the mapping*

$$\mathbb{R} \times P \to M, \qquad (t, ax_0) \mapsto a \exp(tv_0), \qquad a \in G,$$

is a G-equivariant diffeomorphism. Hence, we have the identification $\mathbb{R} \times P = M$.

The metric g on $M = \mathbb{R} \times P$ can be written as

$$g = dt^2 + g_t$$

where g_t is a one-parameter family of invariant Riemannian metrics on $P = G/K$. Any invariant Riemannian metric g on $M = \mathbb{R} \times P$ has the form

$$g' = f(t)\, dt^2 + 2\, dt \cdot \omega_t + g_t'$$

where $f(t) > 0$ is a function, ω_t (resp., g_t') is a family of invariant 1-forms (resp., metrics) on $P = G/K$ such that $\det g_t' > 0$ for any t. Following A. L. Onischik we say that a homogeneous space G/K is asystatic if the coset group $N_G(K)/K$ is discrete. On such space there aren't invariant 1-forms different from zero. Hence we have

Corollary 3.1. *Let M be a C1 manifold with $\Omega = \mathbb{R}$ and an asystatic orbit $P = G/K$. Then all invariant metrics on M have the same normal (non-parametrized) geodesics.*

3.2. Now we consider a Riemannian C1 manifold (M, g) with the orbit space $\Omega = S^1 = \mathbb{R}/2\pi\mathbb{Z}$. In this case, the natural projection $\kappa : M \to S^1$ is a fibration. Let $\gamma = \{y_t\}$ be a normal geodesic. We may suppose that $\kappa(y_0) = \kappa(y_\pi) = [0] \in S^1$. Denote by $K = G_y$ the subgroup which preserves the geodesic γ. There is an element $w \in N_G(K)$ (defined up to a factor from K) such that $wy_0 = y_{2\pi}$. The normal exponential mapping $\exp_t : P_0 = \kappa^{-1}(0) \to P_t = \kappa^{-1}(t)$ is a G-equivariant diffeomorphism. Hence, the manifold M can be constructed from the direct product $[0, 2\pi] \times P$, where $P = G/K$ is an orbit, by the following identification of the boundaries:

$$(0, aK) \leftrightarrow (2\pi, awK) \qquad aK \in G/K.$$

More precisely, we can identify the G-manifold M with the coset space $M_w = \mathbb{R} \times P/\mathbb{Z}$ of the G-equivariant covering $\nu : \mathbb{R} \times P \to M_w$ which is defined by the free action of the group $\mathbb{Z}$ on $\mathbb{R} \times P$ generated by the transformation

$$(3.1) \qquad\qquad \alpha_w : (t, aK) \mapsto (t + 2\pi, awK).$$

We have

Proposition 3.2.

i) *Any $C1$ manifold with an orbit $P = G/K$ and the orbit space $\Omega = S^1$ is determined by an element $w \in N_G(K)/K = W$ and has the form $M_w = \mathbb{R} \times P/\mathbb{Z}$.*

ii) *Two manifolds M_w, $M_{w'}$ for w, $w' \in W$ are G-diffeomorphic iff there is a smooth curve $w : \mathbb{R} \to W$ such that*

$$w' = w_t w w_{t+2\pi}^{-1}, \qquad t \in \mathbb{R}.$$

Example. Let T be a maximal torus of a compact semisimple Lie group G. Then we have natural $1 - 1$ correspondence between $C1$ manifold with the orbit $P = G/T$ and orbit space $\Omega = S^1$ and conjugated classes of elements of the Weyl group $W = N_G(T)/T$.

Proposition 3.3. *Let (M, g) be a Riemannian $C1$ manifold with an orbit $P = G/K$ and the orbit space $\Omega = S^1$. Then there is an element $w \in W = N_G(K)/K$ (defined up to conjugation in W) such that the manifold (M, g) is isometric to the manifold (M_w, g_w), where $M_w = \mathbb{R} \times P/\mathbb{Z}$ is the coset space defined by the transformation α_w (3.1) and the metric g_w on M_w (raised on the cover space $\mathbb{R} \times P$) has the form*

$$\nu^*(g_w) = dt^2 + g_t, \qquad t \in \mathbb{R},$$

where g_t is a one-parameter family of G-invariant Riemannian metric on $P = G/K$ such that $R_w^* g_t = g_{(t+2\pi)}$. Here $R_w : aK \to awK$ is the right translation by the element w in $P = G/K$.

A normal geodesic γ intersects each orbit P in k (=order w) points. In particular, if order $w < \infty$ a normal geodesic is closed.

Any invariant metric on M_w (raised on $\mathbb{R} \times P$) has the form

$$\nu^*(g') = f(t) \, dt^2 + 2 \, dt \cdot \omega_t + g_t',$$

where $f(t) > 0$ is a 2π-periodic function and ω_t, g_t are one-parameter families of invariant 1-forms and metrics on $P = G/K$ with $R_w^* \omega_t = \omega_{t+2\pi}$, $R_w^* g_t' = g_{t+2\pi}'$.

Since an invariant Riemannian metric g_t on $P = G/K$ is defined by an $\operatorname{Ad} K$-invariant Euclidean metric $g(t)$ on the vector space $\mathfrak{m}$ where $\mathfrak{g} = \mathfrak{k} + \mathfrak{m}$ is the reductive decomposition, we have

Corollary 3.2. *Any Riemannian $C1$ manifold with $\Omega = S^1$ and an orbit $P = G/K$ is defined by a pair $(w, g(t))$ where $w \in N_G(K)/K$ and $g(t)$, $t \in [0, 2\pi]$ is a smooth family of $\operatorname{Ad} K$-invariant Euclidean metrics of $\mathfrak{m}$ with*

$$(\operatorname{Ad}_w)^* j_0^\infty (g(t)) = j_{2\pi}^\infty (g(t)).$$

Here $j_{t_0}^\infty (g(t))$ denotes the infinite jet of the curve $g(t)$ in the point t_0. The pair $w, g(t))$ defines the manifold $(M_w = \mathbb{R} \times P/_{\mathbb{Z}}, g)$ where $\nu^*(g) = dt^2 + \hat{g}(t)$ and $\hat{g}(t)$ is the invariant metric on $P = G/K$ defined by $g(t)$.

4. $C1$ Manifolds with One Singular Orbit

4.1. Let (M, g) be a $C1$ manifold with the orbit space $\Omega = \mathbb{R}^+$ and one singular orbit $P_0 = \kappa^{-1}(0)$. According to Proposition 2.1, the G-manifold M can be identified with the normal bundle $T^\perp P_0$ by means of the exponential mapping

$$\exp : T^\perp P_0 \to M.$$

Now we fix an element $y_0 \in P_0$. Denote by $\rho : H \to GL(V)$ the isotropy representation of the stabilizer $H = G_{y_0}$ in the normal space $V = T_{y_0}^\perp P_0 \approx \mathbb{R}^m$. It is clear that $\rho(H)$ acts transitively on the unit sphere $S^{m-1} \subset V$ and the normal bundle $T^\perp P_0$ can be identified with the homogeneous vector bundle $M_\rho = G \times_{\rho(H)} V$ over P_0. We have

Proposition 4.1.

(i) *Any $C1$ manifold M with one singular orbit $P_0 = G/H$ can be identified with the total space $M_\rho = G \times_{\rho(H)} V$ of the homogeneous vector bundle $\nu : M_\rho \to P_0$ associated with a linear representation $\rho : H \to GL(V)$ of the stabilizer H in a vector space $V = \mathbb{R}^m$ which acts transitively on the sphere $S^{m-1} \subset V$. Principal orbits of the group G on $M = M_\rho$*

have the form G/K where $K \subset H$ is the stabilizer of a point $v \in S^{m-1}$ in the group H.

ii) $C1$ manifolds M_ρ and $M_{\rho'}$, associated with two representation ρ, ρ' : $H \to GL(V)$ are G-diffeomorphic iff there is an element $w \in N_G(H)$ such that the representations $h \mapsto \rho(h)$ and $h \mapsto \rho'(whw^{-1})$ are conjugated in $GL(V)$.

The identification $M = M_\rho$ is defined by an invariant metric g on M. The associated fibration $\nu : M = M_\rho \to P_0 = G/H$ will be called normal fibration and its fiber $V(x_0) = \exp(T_{x_0}^\perp P_0)$, $x_0 \in P_0$ is called a normal submanifold.

By Proposition 4.1 the classification of $C1$ manifolds with one singular orbit $P_0 = G/H$ reduces to the classification of linear representations $\rho :$ $H \to GL(V)$ of the compact Lie groups H which act transitively on the sphere $S^{m-1} \subset V$. Such classification is well known. We enumerate all connected compact linear groups H which act transitively on the sphere S^n in Table 1 on the following paper. Also we indicate the stabilizer K of a point $y_1 \in S^n$, the centre $Z(H)$, the normalizers $N_H(K)$ and $N_{0_{n+1}}(H)$ and the decomposition of the tangent space $\mathfrak{m} = T_{y_1} S^n$ into the sum of irreducible and respectively the trivial K-modules of the dimension K.

4.2. Now we give the description of invariant metrics on a Riemannian $C1$ manifold (M, g) with one singular orbit $P_0 = G/H$. We identify M with $M_\rho = G \times_{\rho(H)} V$. Fix a unit vector $v_0 \in V = T_{x_0}^\perp P_0$, $x_0 = [eH, 0] \in M_\rho$ and denote by $\gamma = \{y_t\} = \{\exp_{x_0} tv_0\}$ the normal geodesic starting from x_0 in the direction v_0 and by K the subgroup of G which preserves γ pointwise.

Lemma 4.1.

i) *The group K coincides with the stabilizer G_{y_t} of any point $y_t \in \gamma$, $t \neq 0$.*

ii) *The orbit $P_t = Gy_t$ of any point $y_t \in \gamma$, $t \neq 0$, intersects the geodesic γ in two points y_t, y_{-t}. There is a transformation $\sigma \in N_H(K)$ (defined up to an element from K) such that*

$$\sigma y_t = y_{-t}.$$

Consider the reductive decomposition

$$\mathfrak{G} = \mathfrak{k} + \mathfrak{m} + \mathfrak{p}, \qquad \mathfrak{k} + \mathfrak{m} = \mathfrak{h},$$

For Ad K-invariant Euclidean metric $g_\mathfrak{m}(t)$ on $\mathfrak{m} \simeq T_{y_t} S_t(y_0) = T_{y_t} H y_t$ we denote by $\tilde{g}_m(t)$ the H-invariant metric on the normal sphere $S_t(y_0) = H y_t \subset V$. We note that the Cartesian coordinates in $\mathfrak{m}$ near the origin can be considered as local coordinates on the sphere $S_t(y_0) = H/_K$ via the exponential map $\exp : \mathfrak{m} \to H \to H/_K$.

Proposition 4.3. *An admissible right family*

$$g(t) = g_\mathfrak{m}(t) \oplus g_\mathfrak{p}(t), \qquad t \in \mathbb{R},$$

defines a smooth invariant Riemannian metric $g = dt + g(t)$ iff

$$g^V = dt^2 + g_\mathfrak{m}(t), \qquad t > 0$$

is a smooth $\rho(H)$-invariant Riemannian metric in the vector space V in the polar coordinates.

Corollary. *Any invariant right metric g on M is defined by a $\rho(H)$-invariant Riemannian metric g^V in the vector space V and a smooth family $g_\mathfrak{p}(t)$, $t \geq 0$, of Ad K-invariant Euclidean metrics in $\mathfrak{p}$ starting with Ad H-invariant metric $g_\mathfrak{p}(0)$. In particular if $g_\mathfrak{p}(t) = g_\mathfrak{p} = \text{const}$ then the normal fibration of the corresponding metric g on M is a Riemannian submersion with totally geodesic fibers.*

Example. Let $g_\mathfrak{p}(t) = g_\mathfrak{p} = \text{const}$, $g_\mathfrak{m}(t) = f(t)^2 \times g_{\text{can}}$ where g_{can} is the canonical metric of the sphere S^{m-1} and $f(t) = \text{sh}(t), t, \sin(t)$. Then for the associated metric g on M, a fiber of the normal fibration is a totally geodesic submanifold of constant negative, zero or positive curvature. In the last case the metric is defined only for $0 \leq t \leq \pi$ and it isn't complete.

5. $C1$ manifolds with two singular orbits

5.1 In this paragraph we study a Riemannian $C1$ manifold (M, g) with $\Omega = [0, \pi]$ and singular orbits $P_0 = G/H$ and $P_\pi = G/H$. We assume that all the normal geodesics are closed and have length 2π, furthermore, we fixed a normal geodesic $\gamma = \{y_t\}$ starting with $y_0 \in P_0$ in the direction of a unit vector $v_0 \in V = Y_{y_0}^1 P_0$.

$$G_{y_0} = H, \qquad G_{y_\pi} = H', \qquad G_{y_{\pi/2}} = K.$$

Since any element from a stabilizer G_{y_t} $(t \notin \mathbb{Z}\pi)$ preserves the geodesic γ pointwisely, we have $G_{y_t} = K \subset H \cap H'$.

Denote by ρ (resp., ρ') the isotropy representation of the stabilizer H (resp., H') in the normal space $V = T_{y_0}^\perp P_0$ (resp., $V' = T_{y_\pi}^\perp P_\pi$). The linear groups $\rho(H)$ and $\rho'(H')$ act transitively on the unit spheres $S^{m-1} \subset V$ and $S^{m'-1} \subset V'$. The stabilizer of the points $v_0 = \dot{y}_0 \in S^{m-1}$ and $v_\pi = \dot{y}_\pi \in S^{m'-1}$ are equal to K.

Denote by σ an element from H which transforms the point $v_0 \in S^{m-1} = H/K$ into the antipodal point $-v_0$. It belongs to the normalizer $N_H(K)$ and is defined up to a factor from K. We set $\sigma \left(H/K, v_0 \right) = \sigma \cdot K$. If v' is another point from $S^{m-1} = H/K$ with $H_{v'} = K$, then $v' = \rho(h)v$ for some $h \in N_H(K)$ and

$$\sigma \left(H/K, v' \right) = h\sigma \left(H/K, v_0 \right) h^{-1}.$$

Hence the set

$$\sigma \left(H/K \right) = \mathrm{Ad}_{N_H(K)}\, \sigma \left(H/K, v_0 \right) =$$
$$\left\{ n\sigma n^{-1}k, \quad n \in N_H(K), \quad k \in K \right\} \subset H$$

depends only on H and K.

It is clear that

$$\sigma \left(H/K, v_0 \right) = \sigma \left(H'/K, v_\pi \right) \subset N_H(K) \cap N_{H'}(K).$$

Hence

$$\sigma \left(H/K \right) \cap \sigma \left(H'/K \right) \neq 0.$$

Definition. 1) We say that three compact subgroups of Lie group G compose an admissible triple (H, K, H') if

i) $K \subset H \cap H'$ ii) $H/K = S^{m-1}$, $\quad H'/K = S^{m'-1}$

are spheres,

iii) $\sigma(H/K) \cap \sigma(H'/K) \neq 0.$

2) Two admissible triples (H, K, H') and $(\overline{H}, \overline{K}, \overline{H}')$ are called equivalent if there are elements $b \in G$ and $a \in \left(N_G(K)|_K \right)_0$ such that

$$H = b\overline{H}b^{-1}, \qquad K = b\overline{K}b^{-1}, \qquad H' = ab\overline{H}'b^{-1}a^{-1}.$$

Proposition 5.1. *There is a natural $1-1$ correspondence between Riemannian C1 G-manifolds M with two singular orbits and closed normal geodesics of length 2π (considered up to a G-diffeomorphism) and admissible triples (considered up to an equivalence).*

Example 5.1. Let H be a subgroup of the group G which acts transitively on a sphere $S^{m-1} = H/K \subset V^m$. Then (H, K, H) is an admissible triple. The associated $C1$ manifold is the total spaces $M_{\tilde{\rho}}$ of the homogeneous sphere bundle

$$M_{\tilde{\rho}} = G \times_{\tilde{\rho}(H)} S^m.$$

Here $\tilde{\rho}$ is the action of H on the sphere $S^m = V^m \cup \{\infty\}$ associated with the representation $\rho : H \to GL(V^m)$ which acts transitively on the sphere. We say that $M_{\tilde{\rho}}$ is the standard compactification of the manifold M_ρ (or standard $C1$ manifold).

We have immediately

Proposition 5.2. *Any Riemannian C1 manifold with two singular orbits of codimension one and closed normal geodesics of length 2π is a standard C1 manifold M associated with a triple $(H = \mathbb{Z}_2 \cdot K, K, H)$.*

Remark. Suppose that the $C1$ manifold M admit a complete invariant Lorentzian metric g and that all orbits up to a countable number are spacelike. Then all normal geodesics are timelike and the codimension of any orbit is one. Changing the sign of g in the direction of normal geodesics we obtain an invariant Riemannian metrics g on M. Thus we have $1-1$ correspondence between invariant Lorentz metrics on $C1$ manifolds and invariant Riemannian metrics on $C1$ manifolds with all orbits of codimension one. The Propositions 3.1, 3.3, 4.3, 5.2 give a description of all invariant Lorentzian metrics on $C1$ manifolds.

5.2. Now we describe an invariant Riemannian metric on a $C1$ manifold M associated with a triple (H, K, H'). Let $\mathfrak{h}, \mathfrak{k}, \mathfrak{h}'$ be the Lie algebras of the groups H, K, H' and let

(5.1) $\mathfrak{G} = \mathfrak{k} + \mathfrak{n} + \mathfrak{m}_1 + \mathfrak{m}_1' + \mathfrak{b}$

be an Ad K-invariant decomposition of $\mathfrak{G}$ where

$$\mathfrak{h} \cap \mathfrak{h}' = \mathfrak{k} + \mathfrak{n}, \qquad \mathfrak{h} = \mathfrak{k} + \mathfrak{n} + \mathfrak{m}_1, \qquad \mathfrak{h}' = \mathfrak{k} + \mathfrak{n} + \mathfrak{m}_1'$$

and the subspace $\mathfrak{p} = \mathfrak{m}_1' + \mathfrak{b}$ (resp., $\mathfrak{p}' = \mathfrak{m}_1 + \mathfrak{b}$) is Ad H-invariant (resp., Ad H'-invariant).

Chose a normal geodesic $\gamma = \{y_t\}$ such that $G_{y_0} = H$, $G_{y_{\pi/2}} = K$, $G_{y_\pi} = H'$.

The tangent space $T_{y_t} P_t$ of an orbit $P_t = Gy_t$ can be identified with a subspace as follows:

$$T_{y_t} P_t = \begin{cases} \mathfrak{p} = \mathfrak{m}_1' + \mathfrak{b} & \text{for} \quad t = 0, \\ \mathfrak{q} = \mathfrak{n} + \mathfrak{m}_1 + \mathfrak{m}_1' + \mathfrak{b} & \text{for} \quad t \neq \mathbb{Z}\pi, \\ \mathfrak{p}' = \mathfrak{m}_1 + \mathfrak{b} & \text{for} \quad t = \pi. \end{cases}$$

Hence an invariant Riemannian metric g defines a 1-parameter 2π-periodic Ad K-invariant family $g(t) \in \mathrm{Bil}\,(\mathfrak{q})^K$ of symmetric 2-forms with the following properties

i) $\mathrm{Ker}\, g(0) = \mathfrak{n} + \mathfrak{m}$ and $g(0)$ is an Ad H-invariant Euclidean metric;

ii) $g(t)|_{\mathfrak{q}} > 0$ for $t \notin \mathbb{Z} \cdot \pi$,

iii) $\mathrm{Ker}\, g(\pi) = \mathfrak{n} + \mathfrak{m}_1'$ and $g(\pi)|_{\mathfrak{p}'}$ is an Ad H'-invariant metric,

iv) $(\mathrm{Ad}\,\sigma)^* g(t) = g(-t)$

where $\sigma \in \sigma\left(H/K\right) \cap \sigma\left(H'/K\right)$.

We call such a family admissible. Any admissible family $g(t)$ defines a Riemannian invariant metric

$$(5.2) \qquad\qquad g = dt^2 + \hat{g}(t)$$

on M, but this metric can be nonsmooth near the singular orbits P_0, P_π. An admissible family $g(t)$ is called right if the decomposition (5.1) is orthogonal with respect to any bilinear form $g(t)$, $t \in [0, 2\pi]$.

Denote by $\tilde{g}(t)_{\mathfrak{p}}$ and $\tilde{g}(\pi - t)_{\mathfrak{p}'}$ for small t the invariant metrics on the spheres $S_t(y_0) = Hy_0$ and $S_{\pi-t}(y_\pi) = H'y_{\pi-t}$ that are defined by the Ad K-invariant Euclidean metrics $g(t)|_{\mathfrak{p}}$ on $\mathfrak{p} \simeq T_{y_t}(S_t(y_0))$ and $g(\pi - t)|_{\mathfrak{p}'}$ on $\mathfrak{p}' \simeq Y_{y_{\pi-t}}(S_{\pi-t}(y_\pi))$.

Proposition 5.3. *Let $g(t)$ be a right admissible family of symmetric 2-forms on $\mathfrak{q} = \mathfrak{n} + \mathfrak{m}_1 + \mathfrak{m}_1' + \mathfrak{b}$.*

The associated metric

$$g = dt^2 + \hat{g}(t)$$

on M is smooth iff

$$g^V = dt^2 + \tilde{g}(t)_{\mathbf{p}} \quad \text{and} \quad g^{V'} = dt^2 + \tilde{g}(\pi - t)_{\mathbf{p}'}$$

define smooth metrics in neighbourhoods of the points y_0 and respectively y_π in the normal submanifolds

$$V = \exp_{y_0}(T_{y_0}^{\perp} P_0) \quad \text{and} \quad V' = \exp_{y_\pi}(T_{y_\pi}^{\perp} P_\pi).$$

References

[1] D. V. ALEKSEEVSKY, On proper actions of Lie groups, *Uspechi Math. Nayk.* **34** (1979), 219–220.

[2] L. BERARD BERGERY, Sur de nouvelles variétés riemanniennes d'Einstein, Publications de l'Institut E. Cartan **4** (1982), 1–60.

[3] A. L. BESSE, Einstein manifolds, Springer Verlag, Berlin, 1986.

[4] W. ZILLER, Homogeneous Einstein metrics on spheres and projective spaces, *Math. Ann.* **259** (1982), 351–358.

[5] J. SZENTHE, Orthogonally transversal submanifolds and the generalizations of the Weyl group. *Periodica Math. Hungarica* **15** (1984), 281–299.

D. V. Alekseevsky

ul. Antonova, d. 2
117279 Moscow
USSR

COLLOQUIA MATHEMATICA SOCIETATIS JÁNOS BOLYAI
56. DIFFERENTIAL GEOMETRY, EGER (HUNGARY), 1989

Cross-section Submanifolds in Vector Bundles[*]

M. ANASTASIEI

1. Introduction

Let (E, π, M) be a vector bundle of finite rank where π is a projection over a smooth n-dimensional manifold M. Denote by π^T the tangent map to π and let $VE := \ker \pi^T \to E$ be the vertical bundle over E, a subbundle of the tangent bundle to E. We shall take (x^i) $(i = \overline{1, n})$ as local coordinates on M and (x^i, y^a) $(a = \overline{1, m})$ as local coordinates on E. We shall put $\partial_i := \partial/\partial x^i$, $\partial_a := \partial/\partial y^a$.

Let $s : M \to E$ be a smooth cross-section of (E, π, M). It defines an embedding of M in E and we shall assume that $s(M)$ is closed in E. Locally s is given by

$$(1.1) \qquad x^i = x^i, \qquad y^a = s^a(x^1, \ldots, x^n).$$

We shall denote by B the tangent map s^T to s and we shall put

$$(1.2) \qquad B_i := B(\partial_i) = \partial_i + \partial_i(s^a)\partial_a.$$

It immediately follows that B_i and ∂_a are linearly independent so that

$$(1.3) \qquad TE|_{s(M)} = B(TM) + VE \quad \text{(direct sum)}.$$

---

[*] This paper is in final form and no version of it will be submitted for publication elsewhere.

We shall study the embedding s by using this decomposition i.e. taking VE as "normal" bundle endowing successively E with certain geometric structures.

2. Gauss and Weingarten formulae. Gauss and Codazzi equations

2.A. Suppose that (E, π, M) has a *nonlinear connection* N i.e. there exists on E a distribution $u \to H_u E$ $(u \in E)$ such that $T_u E = H_u E + V_u E$ (direct sum). We shall consider a basis (δ_i, ∂_a) adapted to this decomposition, where $\delta_i := \partial_i - N_i^a(x, y)\partial_a$. The set of functions $N_i^a(x, y)$ obeys a law of transformation which assures that every ∂_i $(i = \overline{1, n})$ is similar to a vector field on M. It completely determines the nonlinear connection N. We note that

$$(2.1) \qquad\qquad B_i = \partial_i + B_i^a(x, s(x))\partial_a$$

where

$$(2.2) \qquad\qquad B_i^a(x, (s(x)) = \partial_i(s^a) + N_i^a(x, s(x)).$$

In order to give a geometrical meaning to B_i^a we recall that a nonlinear connection defines a *covariant derivative* $\overset{\circ}{D} : \Xi(M) \times \Gamma(E) \to \Gamma(E)$, where $\Xi(M)$ is the $\mathcal{F}(M)$-module of vector fields on M, $\Gamma(E)$ is the $\mathcal{F}(M)$-module of sections of (E, π, M) and $\mathcal{F}(M)$ denotes the ring of real functions on M. It locally is as follows:

$$(2.3) \qquad\qquad \overset{\circ}{D}_X A = [X^i(\partial_i(A^a) + N_i^a(x, A(x)))]e_a$$

where $X = X^i\partial_i$, $A = A^a(x)e_a$ and (e_a) is a local basis of the fiber E_x $(x \in M)$. Particularly, $\overset{\circ}{D}_X s = X^i(\partial_i(s^a) + N_i^a(x, s(x))e_a = X^i \overset{\circ}{D}_i(s^a)e_a$. Hence $B_i^a = \overset{\circ}{D}_i s^a$ where we have put $\overset{\circ}{D}_{\partial_i} := \overset{\circ}{D}_i$.

A cross-section A of E is called *parallel* (with respect to N) if $\overset{\circ}{D}_X A = 0$ for every $X \in \Xi(M)$.

We note that $B_i^a = 0$ for the parallel cross-sections.

We denote by h and v the *projectors* defined by the horizontal and the vertical distribution, respectively. For every $X = X^i \partial_i \in \Xi(M)$ we have

$$(2.4) \qquad BX = X^i \partial_i + X^i B_i^a \partial_a.$$

But $X^i \partial_i = X^h$ (the horizontal lift of X) and $X^i B_i^a = \overset{\circ}{D}_X s^a$ so we obtain

$$(2.5) \qquad X^h = BX - (\overset{\circ}{D}_X s^a) \partial_a$$

There are some reasons to call (2.5) the *Gauss formula associated* to N and s. From (2.5) it results:

Theorem 2.1. X^h *is tangent to* $s(M)$ *iff* s *is parallel with respect to* N. ∎

To any nonlinear connection an *almost product structure* P is associated by

$$(2.6) \qquad P(\delta_i) := \delta_i; \qquad P(\partial_a) = -\partial_a.$$

From $P(B_i) = B_i - 2B_i^a \partial_a$ it follows:

Theorem 2.2. *The submanifold* $s(M)$ *is invariant by* P *iff* s *is parallel with respect to* N. ∎

Since N can be identified with an injective morphism $\Xi(M) \to \Xi(E)$, $X \to X^h$ the following *vector-valued* 2-form can be called the *curvature* of N:

$$(2.7) \qquad \Omega(X^h, Y^h) = [X,Y]^h - [X^h, Y^h].$$

This 2-form is "normal"-valued, hence the *Gauss equation* is an *identity*. The *Codazzi equation* is as follows:

$$(2.8) \qquad \Omega(X^h, Y^h) = \left(BX(\overset{\circ}{D}_Y s^a) - BY(\overset{\circ}{D}_X s^a) - \overset{\circ}{D}_{[X,Y]} s^a \right) \partial_a.$$

Here we have used $B[X,Y] = [BX, BY]$; $X, Y \in \Xi(M)$.

2.B. Let us assume that E is endowed with a *linear connection* D which preserves by parallel displacements the horizontal and vertical distributions. Such a connection will be called a distinguished connection, shortly *d*-connection [1]. Locally, D is completely determined by a set of functions on E

$$(2.9) \qquad \Gamma D = (L^i_{jk}(x,y), L^a_{bk}(x,y), C^i_{ja}(x,y), C^a_{bc}(x,y))$$

given by

$$(2.10) \quad D_{\delta k}\delta_j = L^i_{jk}\delta_i; \ \ D_{\delta k}\partial_b = L^a_{bk}\partial_a; \ \ D_{\partial a}\delta_j = C^i_{ja}\delta_i; \ \ D_{\partial c}\partial_b = C^a_{bc}\partial_a,$$

where L^i_{jk} and L^a_{bk} change like the connection coefficients on M and in (E, π, M), respectively, and C^i_{ja}, C^a_{bc} change like the components of a tensor field on E.

If BX and BY, $X, Y \in \Xi(M)$ are two vector fields tangent to $s(M)$, denoting by X', Y' their smooth extensions to E, then by a computation in local coordinates one obtains that $D_{X'}Y'|_{s(M)}$ is independent of X', Y' and denoting it by $D_{BX}BY$ we can write the *Gauss formula*

$$(2.11) \qquad D_{BX}BY = B(\nabla_X Y) + H(BX, BY)$$

denoting the *tangent* and *"normal" components* by the two terms of the right hand side. Assuming that D satisfies the axioms of a linear connection, we find that ∇ is a linear connection on M and H is a "normal"-valued $\mathcal{F}(M)$-bilinear form.

If T denotes the torsion of D and $\overset{\triangledown}{T}$ denotes the torsion of ∇ we have

$$(2.12) \qquad H(BX, BY) - H(BY, BX) = T(BX, BY) - B(\overset{\triangledown}{T}(X,Y)).$$

Hence H is not symmetric.

Putting $H(B_i, B_j) = H^a_{ij}\partial_a$ one obtains m bilinear scalar forms called the *second fundamental forms of the embedding s*.

Expressing (2.12) in local coordinates and equating the horizontal and vertical parts, one obtains

$$(2.13) \qquad \overset{\triangledown}{T}{}^k_{ji} = T^k_{ji} + B^a_i C^k_{ja} - B^a_j C^k_{ia}$$

$$(2.14) \qquad H_{ij}^a - H_{ji}^a = R_{ji}^a + B_i^c P_{jc}^a - B_j^c P_{ic}^a + B_i^c B_j^d S_{cd}^a - B_k^a \overset{\triangledown}{T}{}_{ji}^{\,k}$$

where

$$(2.15) \qquad \begin{aligned} T_{ji}^k &= L_{ji}^k - L_{ij}^k, & R_{ji}^a &= \delta_j N_i^a - \delta_i N_j^a \\ P_{jc}^a &= \delta_c N_j^a - L_{cj}^a, & S_{cd}^a &= C_{cd}^a - C_{dc}^a. \end{aligned}$$

Remark: If s is parallel with respect to N, then $\overset{\triangledown}{T}{}_{ji}^{\,k} = T_{ji}^k$ and $H_{ij}^a - H_{ji}^a = R_{ji}^a$. Hence in this case the symmetry of H is equivalent to the integrability of the horizontal distribution.

In local coordinates Gauss' formula looks as follows:

$$(2.16) \qquad D_{B_k} B_j = (L_{jk}^i + B_k^a C_{ja}^i) B_i + H_{kj}^a \partial_a$$

where

$$(2.17) \qquad H_{kj}^a = B_{j|k}^a + B_k^b (B_{j|b}^a).$$

(The long bar and the short bar denote the v- and h-covariant derivatives with respect to D.)

If we say that $s(M)$ is totally geodesic if $H_{jk}^a = 0$, then we get:

Theorem 2.3. *The cross-section submanifold $s(M)$ is totally geodesic if B_i^a is v- and h-covariant constant.* ∎

Now, if X' extends BX ($X \in \Xi(M)$) to the whole E then $D_{X'} \partial_a$ does not depend on X' and denoting it by $D_{BX} \partial_a$ we get the *Weingarten formula*:

$$(2.18) \qquad D_{BX} \partial_a = X^i (L_{ai}^c + B_i^b C_{ab}^c) \partial_c.$$

The local coefficients $\tilde{L}_{ai}^c := L_{ai}^c + B_i^b C_{ab}^c$ define a linear connection in the vertical bundle. We shall denote it by $\tilde{\nabla}$.

By a direct calculation one gets

$$(2.19) \quad \begin{aligned} R(BX, BY)BZ = {} & B(\overset{\triangledown}{R}(X,Y)Z) + H(BX, B(\nabla_Y Z)) - \\ & - H(BY, B(\nabla_X Z)) - H(B[X,Y], BZ) + \\ & + D_{BX} H(BY, BZ) - D_{BY} H(BX, BZ), \end{aligned}$$

where R and $\overset{\triangledown}{R}$ are the curvature tensor fields of D and ∇, respectively. Taking in (2.19) the tangential and "normal" components indexed by $\top$ and $\perp$ respectively, one obtains

$$
\begin{aligned}
(a) \quad & (R(BX, BY)BZ)^{\top} = B(\overset{\triangledown}{R}(X,Y)Z); \\
(2.20) \quad (b) \quad & (R(BX, BY)BZ)^{\perp} = H(BX, B(\nabla_Y Z)) - H(BY, B(\nabla_X Z)) - \\
& \qquad - H(B[X,Y], BZ) + D_{BX} H(BY, BZ) - \\
& \qquad - D_{BY} H(BX, BZ).
\end{aligned}
$$

Analogously, we get

$$
(2.21) \quad
\begin{aligned}
(a) \quad & (R(BX, BY)\partial_a)^{\top} = 0 \\
(b) \quad & (R(BX, BY)\partial_a)^{\perp} = [D_{BX}, D_{BY}]\partial_a - D_{[BX,BY]}\,\partial_a.
\end{aligned}
$$

The equations (2.20) and (2.21) are the *equations of Gauss and Codazzi for the embedding s.*

3. Induced metrical structure

In this section we assume that E is endowed with a metrical structure G. Then there exists a canonical nonlinear connection in (E, π, M) defined by the orthocomplement of VE with respect to G. Considering this nonlinear connection, we have

$$
(3.1) \qquad G = g_{ij}(x,y)dx^i \otimes dx^j + h_{ab}\delta y^a \otimes \delta y^b
$$

where

$$
(3.2) \qquad \delta y^a = dy^a + N_i^a(x,y)dx^i.
$$

A d-connection D on E is metrical if $DG = 0$. This is equivalent to the following conditions:

$$
(3.3) \qquad g_{ij|k} = 0, \quad g_{ij|a} = 0, \quad h_{ab|i} = 0; \quad h_{ab|c} = 0.
$$

There exists a unique metrical d-connection with $T^i_{jk} = 0$, $S^a_{bc} = 0$. Its local components are given by

(3.4)
$$
\begin{aligned}
L^i_{jk} &= \frac{1}{2}g^{il}(\delta_j g_{lk} + \delta_k g_{lj} - \delta_l g_{jk}) \\
L^a_{bi} &= \partial_b N^a_i + \frac{1}{2}h^{ac}(\partial_i h_{bc} - \partial_b N^d_i h_{dc} - \partial_c N^d_i h_{db}) \\
C^i_{jc} &= \frac{1}{2}g^{ik}\partial_c g_{jk} \\
C^a_{bc} &= \frac{1}{2}h^{ad}(\partial_c h_{db} + \partial_b h_{dc} - \partial_d h_{bc}).
\end{aligned}
$$

We continue to study the embedding s by using (B_i, ∂_a). The metrical structure G on E induces a metrical structure $\tilde{g}$ on M:

(3.5)
$$
\begin{aligned}
\tilde{g}(X) &= \tilde{g}_{ij}(x)dx^i \otimes dx^j, \\
\tilde{g}_{ij}(x) &= g_{ij}(x, s(x))+ \\
&\quad + B^a_i(x, s(x))B^b_j(x.s(x))h_{ab}(x, s(x)).
\end{aligned}
$$

The induced connection ∇ on M has the torsion

(3.6)
$$
\overset{\triangledown}{T}{}^k_{\ ji} = B^a_i C^k_{ja} - B^a_j C^k_{ia}.
$$

We also have

(3.7)
$$
H^a_{ij} - H^a_{ji} = R^a_{ji} + (B^c_i P^a_{jc} - B^c_j P^a_{ic}) - B^a_k(B^c_i C^k_{jc} - B^c_j C^k_{ic}).
$$

The *induced connection* ∇ is *not metrical* since if we set $\nabla_{\partial_k} := \nabla_k$, then

(3.8)
$$
\tilde{\nabla}_k g_{ij} = \nabla_k g_{ij} + 2\nabla_k(B^a_j)B^b_i h_{ab} + \partial_k(h_{ab})B^a_i B^b_j \neq 0.
$$

On the other hand, it is easy to see that

(3.9)
$$
\tilde{\nabla}_{\partial_i} h_{ab} = h_{ab|i} + B^c_i h_{ab|c} = 0
$$

i.e. $\tilde{\nabla}$ is a *metrical connection* in the vertical bundle.

Note that the equations of Gauss and Codazzi cannot be simplified since the spaces spanned by (B_i) and (∂_a) are not orthogonal, except if s is parallel with respect to N.

If we suppose (E, π, M) endowed with a nonlinear connection (which always can exist if M is paracompact), a metrical structure G on E can be obtained under the following assumptions:

(a) M has a metrical structure $g_{ij}(x)$ and there exists a *regular* Lagrangian $L : E \to R$. "Regular" means that *the matrix*

$$(h_{ab}) = \left(\frac{\partial^2 L(x, y)}{\partial y^a \partial y^b} \right)$$

is *nondegenerate*.
In this case we can put

$$(3.10) \qquad G = g_{ij}(x)dx^i \otimes dx^j + h_{ab}(x, y)\delta y^a \otimes \delta y^b$$

and we obtain a metrical structure on E. The unique metrical connection with respect to this G satisfies $C^i_{jc} = 0$, hence the *torsion* of the *induced connection* ∇ *vanishes* and its curvature is

$$\overset{\nabla}{R}{}_k{}^h{}_{ji} = R_k{}^h{}_{ji}(x, s(x)).$$

(b) M has a metrical structure $g_{ij}(x)$ as above and there exists a metric structure in the bundle (E, π, M) i.e. every fibre is endowed with a symmetric bilinear form of maximal rank, of matrix $(h_{ab}(x))$ depending differentiably on x. The metrical structure G has the same form as in the case (a). The unique metrical connection with respect to G satisfies $C^i_{ja} = 0$ and $C^a_{bc} = 0$.
It results that *the induced connection ∇ coincides with the Levi-Civita connection of G* (not of $\tilde{g}$), and $\tilde{\nabla}$ is metrical with respect to h and $\tilde{L}^a_{bi} = L^a_{bi}$.

References

[1] R. MIRON, Techniques of Finsler geometry in the theory of vector bundles. *Acta Sci. Math.* **49** (1985), 119–129.

[2] K. YANO, S. ISHIHARA, Tangent and cotangent bundle. M. Dekker, 1973.

Mihai Anastasiei

University "Al.I.Cuza" IASI
Faculty of Mathematics
6600 IASI, Romania

In this article we give differential geometric descriptions of the string class and its generalization (higher dimensional string classes) as follows: Let $\xi(\{g_{ij}\}$ be an ΩG-bundle over M, $\{\theta_i\}$ and $\{\Theta_i\}$ a connection and its curvature of ξ. Then, for any $p \geq 1$, there is a 1-cochain of $2p$-forms $\{\Psi_i\}$ such that

$$\int_0^1 \mathrm{tr}(\Theta_i^p g_{ij}{}' g_{ij}{}^{-1})\, dt = \Psi_j - \Psi_i.$$

Here, t is the loop variable, $g' = dg/dt$ and $\Theta^p = \overbrace{\Theta \wedge \ldots \wedge \Theta}^{p}$. By using this $\{\Psi_i\}$, we obtain

$$\int_0^1 \mathrm{tr}(\Theta_i{}^p \wedge \theta_i{}') - d\Psi_i = \int_0^1 \mathrm{tr}(\Theta_j{}^p \wedge \theta_j{}')dt - d\Psi_j,$$

on $U_i \cap U_j$. This form is closed and its de Rham class $\widetilde{c}^p(\xi)$ is determined by ξ. Especially, $\widetilde{c}^1(\xi)$ is the original string class and it vanishes if and only if ξ has an $\widetilde{\Omega}\mathfrak{g}$-valued connection. Here $\widetilde{\Omega}\mathfrak{g}$ means the basic central extension of $\Omega\mathfrak{g}$, the based loop algebra over $\mathfrak{g}$.

An ΩG-bundle ξ over M induces a G-bundle ξ^b over $M \times S^1$ by the correspondence

$$\xi = \{g_{ij}\} \to \xi^b = \{g_{ij}{}^b\}, \qquad g_{ij}{}^b(x,t) = (g_{ij}(x))(t).$$

On the other hand, a G-bundle ξ over M induces an LG-bundle ξ^L over ΩM by the correspondence

$$\xi = \{g_{ij}\} \to \xi^L = \{g_{ij}{}^L\}, \qquad (g_{ij}{}^L(\gamma))(t) = g_{ij}(\gamma(t)).$$

Here LG means the free loop group over G. It is shown that ξ^L is equivalent to an ΩG-bundle. By definition we get

$$(\xi^L)^b = ev^*(\xi).$$

Here $ev : \Omega M \times S^1 \to M$, $ev(\gamma, t) = \gamma(t)$ is the evaluation map ([9]). We also define the Gysin map
$\gamma : H^0(M, G_d)/\exp(H^0(M, \mathfrak{g}_d)) \to H^1(M \times S^1, G_d)$ by

$$\gamma(g) = (B_0(g))^b.$$

Since the inverse of the transgression $\tau^{-1} : H^{q+1}(M, \mathbb{C}) \to H^q(\Omega M$
the composition of the evaluation map and integration along S^1 ([

COLLOQUIA MATHEMATICA SOCIETATIS JÁNOS BOLYAI
56. DIFFERENTIAL GEOMETRY, EGER (HUNGARY), 1989

Characteristic Classes of Loop Group Bundles and Generalized String Classes

A. ASADA

§0. Introduction

In a talk at the last colloquium the author treated the differential geometric and non-abelian cohomological meanings of the logarithm of (complex) matrix valued functions ([5]). In the continuation of that research we get the following bijection:

$$B_0 : H^0(M, G_d)/\exp(H^0(M, \mathfrak{g}_d)) \cong H^1(M, \Omega G_d).$$

Here, M is a smooth (Hilbert) manifold, $G = GL(n, \mathbb{C})$ (or $U(n)$), $\mathfrak{g}$ its Lie algebra, ΩG the (based) loop group over G. G_d, etc., mean the sheaves of germs of smooth G, etc., valued functions over M. The bijection B_0 gives natural meaning and examples of loop group bundles. Another important example of a loop group bundle is the tangent bundle of the (based) loop space ΩM over M (cf. [11], [15], [22], [27]). On ΩM the Dirac–Ramond operator (loop space version of the Dirac operator) is defined if and only if the structure group of the tangent bundle of ΩM is lifted on $\widetilde{\Omega}G$, the basic central extension of ΩG ([2], [15], [22]). The obstruction for this lifting was named string class ([22], cf. [11], [15]). Its free part belongs to $H^3(\Omega M, \mathbb{C})$ and is mapped to the first (rational) Pontrjagin class of M by transgression. (The torsion part needs a more delicate discussion, cf. [22], [25], [26].)

together with the properties of the Gysin map ([14], cf. [3], [4]), we obtain a *trinity of β-classes (Chern–Simons classes), string classes and transgressed Chern classes.* All of these results are formulated in terms of non-abelian de Rham theory ([5], [6]). The use of non-abelian de Rham theory is essential in these studies. For example, geometric studies of integrable forms are mostly devoted to their monodromies. We can define the Gysin map for integrable forms with non-trivial monodromies. Their images are not G-bundles, but belong to $H^1(M \times S^1, \mathcal{M}^1)$, the two-dimensional non-abelian de Rham set of $M \times S^1$. On the other hand, as we have pointed out in our talk at the last Colloquium, one-dimensional non-abelian de Rham theory treats global properties of the equation

$$d^e f = df + \sum_{n=1}^{\infty} \frac{(-1)^n}{(n+1)!} (\operatorname{ad} f)^n (df) = \theta,$$

$$d\theta + \theta \wedge \theta = 0, \qquad (\operatorname{ad} f)(\zeta) = f\zeta - \zeta f.$$

The local properties of this equation and of the equation

$$dg = g\theta \qquad (g = e^f),$$

are the same, but global properties differ. Differences are measured by (integral) β-classes (g^*)-images of generators of $H^*(G, \mathbb{Z})$) and β-classes are the origins of Chern classes via transgression (cf. [7]). These results together with the Grassmannian model of the loop group ([23]) suggest that we may identify Q.F.T. on M with the Chern–Simons actions and Q.F.T. on $M \times S^1$ with the topological actions or Q.F.T. on ΩM with the (stable) Yang–Mills actions on the one hand, and Q.F.T. on M with the topological actions and Q.F.T. on $M \times S^1$ with the (stable) Chern–Simons actions, or Q.F.T. on ΩM with the (stable) Chern–Simons actions on the other hand (cf. [8], [13], [24], [24]', [28]). It also suggests that complex representations of the group of paths $[\Omega M]$ ([7], [18], [19], [21]) divides two-classes, one defines Q.F.T. on M with the topological actions and the other defines Q.F.T. on M with the Chern–Simons actions (cf. [19]).

This article is outlined as follows. In Sect. 1 we study $\Omega\mathfrak{g}$ and $\widetilde{\Omega}\mathfrak{g}$-valued integrable forms and cohomologies with coefficients in $\mathcal{M}^1{}_{\Omega\mathfrak{g}}$ and $\mathcal{M}^1{}_{\widetilde{\Omega}\mathfrak{g}}$, the sheaves of germs of $\Omega\mathfrak{g}$ and $\widetilde{\Omega}\mathfrak{g}$-valued integrable forms. In Sect. 2 we define B_0 and the Gysin map. Connections and $\widetilde{\Omega}\mathfrak{g}$-valued connections of ΩG-bundles are defined in Sect. 3. Properties of $\widetilde{\Omega}\mathfrak{g}$-valued connections and their curvatures give a prototype of the definitions of general string classes which are also defined in Sect. 3. It seems that similar discussions may be possible for the basic abelian extension valued connections of $\mathrm{Map}(s^{2n-1}, G)$-bundles by using the results in [20]. (cf. [20]') In Sect. 4, we show that string classes of ξ^L are the inverse-images of transgression of the Chern characters of ξ. The equivalence of the β-classes of g and the string classes of $B^0(g)$ and the existence of the Bott map

$$\widetilde{B} : H^0(M, \mathcal{M}^1)_0/d^e(h^0(M, \mathfrak{g}_d)) \to H^2(M, \mathcal{M}^1),$$

where $\mathcal{M}^1$ is the sheaf of germs of $\mathfrak{g}$-valued integrable forms and $H^0(M, \mathcal{M}^1)_0$ is a suitable subset of $H^0(M, \mathcal{M}^1)$, are shown in Sect. 5.

Acknowledgement. In an interesting study of the geometry of curves in a space, Prof. K. Abe (Dep. Math., General Education, Sinsyu Univ.) obtained a non-linear equation with a parameter ([1], cf. [7]). In the study of this equation, we recognized that it is more natural to consider this equation to be a solvable loop algebra valued integrable form. It was one of the starting points of this research. Another starting point is the relation of geometry of loop spaces and non-abelian de Rham theory which was suggested by S. I. Andersson and A. Connes. I would like to thank them. Mensky's book [19] gave many suggestions in this research. I would like to thank Dr. Terazawa (Dep. Phys., Fac. Sci., Sinsyu Univ.) who taught the author Mensky's book.

§1. Non-abelian de Rham theory with respect to loop groups

1. Let $G = G_n$ be $GL(n, \mathbb{C})$ and $\mathfrak{g} = \mathfrak{g}_n$ its Lie algebra. $LG = LG_n$, $\Omega G = \Omega G_n$, $L\mathfrak{g} = L\mathfrak{g}_n$ and $\Omega\mathfrak{g} = \Omega\mathfrak{g}_n$ are free and based loop groups and loop algebras over G and $\mathfrak{g}$. The basic central extensions of LG, ΩG, $L\mathfrak{g}$ and $\Omega\mathfrak{g}$ are denoted by $\widetilde{LG}$, $\widetilde{\Omega G}$, $\widetilde{L\mathfrak{g}}$ and $\widetilde{\Omega\mathfrak{g}}$, respectively.

If g (or ζ) is an LG-valued function (or an $L\mathfrak{g}$-valued form) on a smooth Hilbert manifold M, we define a G-valued function g^b (or a $\mathfrak{g}$-valued form ζ^b) on $M \times S^1$ by

$$(1) \qquad g^b(x,t) = (g(x))(t), \quad \zeta^b(x,t) = (\zeta(x))(t), \quad x \in M, \quad t \in S^1.$$

Convention. *We call g (or ζ) to be smooth on M if and only if g^b (or ζ^b) is smooth on $M \times S^1$.*

dg/dt and $d\zeta/dt$ are often denoted by g' and ζ'. They are smooth if g and ζ are smooth. An $\widetilde{L}G$-valued function $\tilde{g}$ induces an LG-valued function g. $\tilde{g}$ is said to be smooth if $\tilde{g}$ is a smooth map to $\widetilde{L}G$ and g is smooth in the above sense. An $\widetilde{L}\mathfrak{g}$-valued differential form $\tilde{\zeta}$ is written

$$\tilde{\zeta} = (\zeta, \beta), \quad \zeta \text{ is an } L\mathfrak{g}\text{-valued form, } \beta \text{ is a usual form.}$$

We call $\tilde{\zeta}$ to be smooth if ζ is smooth in the above sense and β is smooth.

A smooth $L\mathfrak{g}$-valued 1-form θ is said to be integrable (or flat) if it satisfies

(2) $$d\theta + \theta \wedge \theta = 0.$$

A smooth $\widetilde{L}\mathfrak{g}$-valued 1-form $\tilde{\theta} = (\theta, \beta)$ is said to be integrable (or flat) if it satisfies

(2)' $$d\tilde{\theta} + \frac{1}{2}[\tilde{\theta}, \tilde{\theta}] = 0, \qquad \text{i.e. } d\theta + \theta \wedge \theta = 0,$$

$$d\beta + \frac{1}{2} \int_0^1 \mathrm{tr}(\theta \wedge \theta') dt = 0.$$

Lemma 1. (i) *Let ζ be an $L\mathfrak{g}$-valued 1-form on M, then*

(3) $$\int_0^1 \mathrm{tr}(\zeta^{2p} \wedge \zeta') dt = 0, \quad p \geq 0, \quad \zeta^q = \overbrace{\zeta \wedge \ldots \wedge \zeta}^{q}.$$

(ii) *If θ is an integrable $L\mathfrak{g}$-valued 1-form, then*

(4) $$d\left(\int_0^1 \mathrm{tr}(\theta^{2p+1} \wedge \theta') dt\right) = 0, \qquad p \geq 0.$$

Proof. Since $\int_0^1 \mathrm{tr}(\zeta^{2p} \wedge \zeta') dt = 1/(2p+1) \int_0^1 (\mathrm{tr}(\zeta^{2p+1}))' dt$, we have (3). If θ is integrable, then $d\theta = -\theta \wedge \theta$ and $d(\theta') = (d\theta)'$ by Convention, so we get

$$d\left(\int_0^1 \mathrm{tr}(\theta^{2p+1} \wedge \theta')\right) dt = -\int_0^1 \mathrm{tr}\left(\theta^{2p+1} \wedge (d\theta)'\right) dt$$

$$= \int_0^1 \mathrm{tr}\left((\theta^{2p+1})' \wedge d\theta\right) dt$$

$$= \frac{2p+1}{2p+3} \int_0^1 (\mathrm{tr}(\theta^{2p+3}))' dt = 0. \qquad \blacksquare$$

Definition 1. *Let θ be an integrable L-valued 1-form on M. Then we set*

$$\alpha^p(\theta) = \text{the de Rham class of } \int_0^1 \text{tr}(\theta^{2p-1} \wedge \theta')dt \in H^{2p}(M, \mathbb{C}).$$

By definition and $(2)'$, we obtain

Proposition 1. (i) *If $\theta = g^{-1}dg$ on M, we have*

(5) $$\alpha^p(\theta) = c_p g^*(\widehat{e}_p).$$

Here $\widehat{e}_p$ is the $2p$-dimensional generator of $H^*(\Omega G, \mathbb{C})$ (cf. [23]) and c_p is a non-zero constant determined by the choice of $\widehat{e}_p$.

(ii) *If $\theta = g^{-1}dg$ and $g = e^f$ on M, then $\alpha^p(\theta) = 0$ for all p.*

(iii) *$\alpha^1(\theta) = 0$ if and only if there exists an $\widetilde{L}\mathfrak{g}$-valued integrable 1-form $\widetilde{\theta}$ such that $\widetilde{\theta} = (\theta, \zeta)$.*

2. Let $\zeta = \sum_i \zeta_i dx_i$ be an $L\mathfrak{g}$-valued 1-form on a (starlike) neighborhood U of the origin of a (separable) Hilbert space. Then we set

$$(I\zeta(x))(t) = \int_0^1 \sum_i sx_i(\zeta_i(sx))(t)ds,$$

$$P_\zeta(f) = \sum_{n=0}^\infty I_\zeta{}^n(f), \quad I_\zeta{}^0(f) = f, \quad I_\zeta(g) = I(\zeta g).$$

If ζ is integrable, $df = 0$ and $(f(x))(0) = ((x))(1)$, we get

$$dP_\zeta(f) = \zeta P_\zeta(f), \qquad (p_\zeta(f)(x))(0) = (P_\zeta(f)(x))(1).$$

Hence an integrable $L\mathfrak{g}$-valued 1-form θ is locally integrable, that is locally written as $\theta = g^{-1}dg$.

For a (scalar valued) 2-form $\zeta = \sum_{ij} \zeta_{ij} dx_i \wedge dx_j$, $I\zeta$ is defined by

$$I\zeta = \sum_i I\zeta_i dx_i, \qquad I\zeta_i = \int_0^1 \sum_j s^2 x_i \zeta_{ij}(sx)ds.$$

So we can define canonical local integration β_0 of $-1/2 \int_0^1 \text{tr}(\theta \wedge \theta')dt$ on U if θ is an integrable $L\mathfrak{g}$-valued 1-form. Hence, if $\widetilde{\theta} = (\theta, \beta)$ is an integrable $\widetilde{L}\mathfrak{g}$-valued 1-form on U, we can set

$$(\theta, \beta) = (g^{-1}dg, c^{-1}dc + \beta_0), \qquad c \text{ is a smooth } \mathbb{C}^*\text{-valued function.}$$

Therefore we can associate a smooth $\widetilde{L}G$-valued function (g, c), $g^{-1}dg = \theta$, to $\widetilde{\theta}$. We note that since $g(U)$ is contractible, $p^{-1}(g(U)) = g(U) \times \mathbb{C}^*$, where $p : \widetilde{L}G \to LG$ is the projection. We set

(6) $\rho(g) = g^{-1}dg,$ g is an LG-valued (or a G-valued) function,

$$\widetilde{\rho}_I((g, C)) = \left(g^{-1}dg, c^{-1}dc - \frac{1}{2}I\left(\int_0^1 \mathrm{tr}\left(g^{-1}dg \wedge (g^{-1}dg)'\right)dt\right)\right).$$

By definition, ρ is defined globally, but $\widetilde{\rho}_I$ is defined only locally.

In the sequel, we use the following notations.

$\mathbb{C}^{*t}$, LG_t and $\widetilde{L}G_t$: the sheaves of germs of constant $\mathbb{C}^*$, LG and $\widetilde{L}G$ valued
functions over M.

$\mathbb{C}^*{}_d$, LG_d and $\widetilde{L}G_d$: the sheaves of germs of smooth $\mathbb{C}^*$, LG and $\widetilde{L}G$ valued
functions over M.

Φ^p: the sheaf of germs of closed p-forms over M.

$\mathcal{M}^1{}_{L\mathfrak{g}}$ and $\mathcal{M}^1{}_{\widetilde{L}\mathfrak{g}}$: the sheaves of germs of integrable $L\mathfrak{g}$ and $\widetilde{L}\mathfrak{g}$ valued
1-forms over M.

Stalks of these sheaves at x are denoted by $\mathbb{C}^*{}_{t,x}$, etc. Then we have the following commutative diagram with exact lines and columns.

$$
\begin{array}{ccccccccc}
 & & 0 & & 0 & & 0 & & \\
 & & \uparrow & & \uparrow & & \uparrow & & \\
0 & \longrightarrow & \Phi^1 & \xrightarrow{i} & \mathcal{M}^1{}_{\widetilde{L}\mathfrak{g}} & \xrightarrow{j} & \mathcal{M}^1{}_{L\mathfrak{g}} & \longrightarrow & 0 \\
 & & \uparrow{\scriptstyle(\widetilde{\rho}_i)} & & \rho\uparrow & & \rho\uparrow & & \\
0 & \longrightarrow & \mathbb{C}^*{}_d & \xrightarrow{i} & \widetilde{L}G_d & \xrightarrow{j} & LG_d & \longrightarrow & 0 \\
 & & \uparrow & & \uparrow & & \uparrow & & \\
0 & \longrightarrow & \mathbb{C}^*{}_t & \xrightarrow{i} & \widetilde{L}G_t & \xrightarrow{j} & LG_t & \longrightarrow & 0 \\
 & & \uparrow & & \uparrow & & \uparrow & & \\
 & & 0 & & 0 & & 0 & &
\end{array}
$$

Here $\widetilde{\rho}_I$ is not a sheaf map (not continuous). But we have

$$\widetilde{\rho}_I(i(G)) = i(\rho(g)) = (0, g^{-1}dg),$$
$$\rho(j(g, c)) = j(\widetilde{\rho}_I(g, c)) = g^{-1}dg.$$

ρ and $\widetilde{\rho}_I$ are right logarithmic derivations of g and (g, c). Corresponding left logarithmic derivations ρ_L and $\widetilde{\rho}_{I,L}$ are given by

(6)$_L$ $\rho_L(g) = (dg)g^{-1},$

$$\widetilde{\rho}_{I,L}((g, c)) = \left(\rho_L(g), c^{-1}dc + \frac{1}{2}I\left(\int_0^1 \mathrm{tr}\left(\rho_L(g) \wedge (\rho_L(g))'\right)dt\right)\right).$$

3. **Definition 2.** Let $\widetilde{\zeta} = (\zeta, \beta)$ be an $\widetilde{L\mathfrak{g}}$-valued differential form and g is an LG-valued function. Then we define the adjoint action $\widetilde{\zeta}^g$ of g on $\widetilde{\zeta}$ by

(7) $$\widetilde{\zeta}^g = \left(\zeta^g, \beta + \int_0^1 \mathrm{tr}(\zeta g' g^{-1}) dt \right), \qquad \zeta^g = g^{-1}\zeta g.$$

We also define the left adjoint action $^g\widetilde{\zeta}$ by

(7)$_L$ $$^g\widetilde{\zeta} = \widetilde{\zeta}^{g^{-1}}.$$

By definition, if $\widetilde{\theta}$ is right integrable and locally takes the form $\widetilde{\theta} = (g^{-1}dg, \beta)$, then $^g\widetilde{\theta}$ is left integrable, that is, we have

$$d(^g\widetilde{\theta}) - \frac{1}{2}[^g\widetilde{\theta}, {}^g\widetilde{\theta}] = 0.$$

Let $\mathfrak{U} = \{U_i\}$ be a locally finite open covering of M, $C^p(\mathfrak{U}, \mathcal{M}^1{}_{L\mathfrak{g}})$ and $C^p(\mathfrak{U}, \mathcal{M}^1{}_{\widetilde{L\mathfrak{g}}})$ the sets of p-cochains with coefficients in $\mathcal{M}^1{}_{L\mathfrak{g}}$ and $\mathcal{M}^1{}_{\widetilde{L\mathfrak{g}}}$. If $\{\omega_{ij}\}$ belongs to $C^1(\mathfrak{U}, \mathcal{M}^1{}_{L\mathfrak{g}})$, then we define

$$\delta\omega_{ijk} = \omega_{jk} - \omega_{ik} + \omega_{ij}{}^{g_{jk}}, \qquad \omega_{ij} = g_{ij}{}^{-1}dg_{ij}.$$

Similarly, if $\{\widetilde{\omega}_{ij}\}$ belongs to $C^1(\mathfrak{U}, \mathcal{M}^1{}_{\widetilde{L\mathfrak{g}}})$, then we define

$$\delta\widetilde{\omega}_{ijk} = \widetilde{\omega}_{jk} - \widetilde{\omega}_{ik} + \widetilde{\omega}_{ij}{}^{g_{jk}}, \qquad \widetilde{\omega}_{ij} = (g_{ij}^{-1}dg_{ij}, \beta_{ij}).$$

Then we can define the cohomology sets $H^1(M, \mathcal{M}^1{}_{L\mathfrak{g}})$ and $H^1(M, \mathcal{M}^1{}_{\widetilde{L\mathfrak{g}}})$ (cf. [5], [6]).

Lemma 2. If $\{\omega_{ij}\} \in C^1(\mathfrak{U}, \mathcal{M}^1{}_{L\mathfrak{g}})$ is a cocycle, that is $\delta\omega_{ijk} = 0$, then

(8) $$\frac{1}{2}\int_0^1 \mathrm{tr}\left(\omega_{jk} \wedge \omega_{jk}{}' - \omega_{ik} \wedge \omega_{ik}{}' + \omega_{ij}{}^{g_{jk}} \wedge (\omega_{ij}{}^{g_{jk}})' \right) dt$$

$$= d\left(\int_0^1 \mathrm{tr}(\omega_{ij}g_{jk}{}'g_{jk}{}^{-1}) dt \right),$$

$$\omega_{ij} = g_{ij}{}^{-1}dg_{ij}, \qquad g_{ij}g_{jk}g_{ki} = c_{ijk}, \qquad \text{a constant } (\in LG).$$

Proof. Since $\delta\omega_{ijk} = 0$, we have

$$\mathrm{tr}\left(\omega_{jk} \wedge \omega_{jk}{}' - \omega_{ik} \wedge \omega_{ik}{}' + \omega_{ij}{}^{g_{jk}} \wedge (\omega_{ij}{}^{g_{jk}})' \right)$$

$$= 2\,\mathrm{tr}(\omega_{ij}{}^{g_{jk}}(g_{jk}{}^{-1}g_{jk}{}'\omega_{ik} - g_{jk}{}^{-1}d(g_{jk}{}'))).$$

Then, since $g_{jk}{}^{-1}g_{ij}{}^{-1} = g_{ki}c_{ikj}$, we get

$$\mathrm{tr}(\omega_{ij}{}^{g_{jk}}(g_{jk}{}^{-1}g_{jk}{}'\omega_{ik} - g_{jk}{}^{-1}d(g_{jk}{}')))$$
$$= -\mathrm{tr}(c_{ikj}(dg_{ij}d(g_{jk}{}'g_{ki}))) = d(\mathrm{tr}(\omega_{ij}{}^{g_{jk}}g_{jk}{}^{-1}g_{jk}{}')). \qquad \blacksquare$$

Corollary. *If $\delta\omega_{ijk} = 0$ and a 1-cocycle of scalar 1-forms $\{\alpha_{ij}\}$ satisfies*

$$d\alpha_{ij} + \frac{1}{2}\int_0^1 \mathrm{tr}(\alpha_{ij} \wedge \omega_{ij}{}')dt = 0,$$

then the 2-cochain $\{\beta_{ijk}\}$ given by

(9) $$\beta_{ijk} = \alpha_{jk} - \alpha_{ik} + \alpha_{ij} + \int_0^1 \mathrm{tr}(\omega_{ij}g_{jk}{}'g_{jk}{}^{-1})dt,$$

belongs to $\mathbb{Z}^2(\mathfrak{U}, \Phi^1)$, that is we have

(10) $$d\beta_{ijk} = 0, \qquad \beta_{ijk\ell} = \beta_{jk\ell} - \beta_{ik\ell} + \beta_{ij\ell} - \beta_{ijk} = 0,$$

provided $c_{ijk}{}' = 0$.

By this Corollary we can define the coboundary map δ : $H^1(M, \mathcal{M}^1{}_{L\mathfrak{g}}) \to H^2(M, \Phi^1)$ by $\delta((\langle\{\omega_{ij}\}\rangle)) = \langle\{\beta_{ijk}\}\rangle$. Here $\langle\langle\{\zeta\}\rangle\rangle$ means the cohomology class of $\{\zeta\}$. Then we obtain

Proposition 2. *The following diagram is commutative and each line is exact:*

$$0 \longrightarrow H^0(M, \Phi^1) \xrightarrow{i} H^0(M, \mathcal{M}^1{}_{\widetilde{L\mathfrak{g}}}) \xrightarrow{j} H^0(M, \mathcal{M}^1{}_{L\mathfrak{g}}) \xrightarrow{\delta} H^1(M, \Phi^1)$$

$$\uparrow \qquad\qquad\qquad \rho\uparrow \qquad\qquad\qquad \uparrow$$

$$0 \longrightarrow H^0(M, \mathbb{C}^*{}_d) \xrightarrow{i} H^0(M, \widetilde{L G}_d) \xrightarrow{j} H^0(M, L G_d) \xrightarrow{\delta} H^1(M, \mathbb{C}^*{}_d)$$

$$\xrightarrow{i^*} H^1(M, \mathcal{M}^1{}_{\widetilde{L\mathfrak{g}}}) \xrightarrow{j^*} H^1(M, \mathcal{M}^1{}_{L\mathfrak{g}}) \xrightarrow{\delta} H^2(M, \Phi^1) \cong H^3(M, \mathbb{C})$$

$$\uparrow \qquad\qquad\qquad \rho\uparrow \qquad\qquad\qquad \uparrow$$

$$\xrightarrow{i^*} H^1(M, \widetilde{L G}_d) \xrightarrow{j^*} H^1(M, L G_d) \xrightarrow{\delta} H^2(M, \mathbb{C}^*{}_d) \cong H^3(M, \mathbb{Z})$$

$$\uparrow \qquad\qquad\qquad \uparrow$$

$$\mathrm{Hom}(\pi_1(M), LG) \cong H^1(M, L G_t) \xrightarrow{\delta} H^2(M, G_t).$$

Note. We get the same commutative diagram with exact lines replacing LG, $L\mathfrak{g}$, etc., by ΩG, $\Omega\mathfrak{g}$, etc.

§2. Geometric meanings of loop group bundles and Gysin map in non-abelian de Rham theory

4. For a complex matrix A, we define linear maps $F_A : \mathfrak{g} \to \mathfrak{g}$ and $G_A : \mathfrak{g} \to \mathfrak{g}$ by

$$F_A(X) = X + \sum_{n=2}^{\infty} \frac{1}{n!} \left(\sum_{s=0}^{n-1} A^s X A^{n-s-1} \right),$$

$$G_A(X) = X + \sum_{n=1}^{\infty} \frac{(-1)^n}{(n+1)!} (\operatorname{ad} A)^n (X),$$

$$(\operatorname{ad} A)(X) = [A, X] = AX - XA.$$

In [5] we showed

Lemma 3. (i) $F_A(X)$ *is equal to* $e^A G_A(X)$. *It is also shown that*

$$F_A(X) = G_{A,L}(X)e^A, \quad G_{A,L}(X) = X + \sum_{n=1}^{\infty} \frac{1}{(n+1)!} (\operatorname{ad} A)^n (X).$$

(ii) *The Jacobian of* $\exp : \mathfrak{g} \to G$ *at* A *is* F_A.

Corollary. (i) *The Jacobian of* $\exp : \mathfrak{g} \to G$ *is non-degenerate at* A *if and only if* A *satisfies the following condition:*

(∗) *If* λ_i *and* λ_j *are distinct proper values of* A, *then* $\frac{1}{2\pi\sqrt{-1}}(\lambda_i - \lambda_j)$ *is not integer.*

(ii) *A smooth* G-*valued function* g *on* M *is locally written as* $g = e^f$, *where* f *is smooth* $\mathfrak{g}$-*valued function on some open set of* M.

We denote by $\mathfrak{g}_d$ and G_d the sheaves of germs of smooth $\mathfrak{g}$ and G-valued functions over M. Then $\exp : \mathfrak{g} \to G$ induces a sheaf map $\exp : \mathfrak{g}_d \to G_d$. Its kernel sheaf is denotes by $N_{\mathfrak{g},d}$. In [5], we defined the first cohomology set of M with coefficients in $N_{\mathfrak{g},d}$ as follows: Let

$$\delta_P n_{ijk} = n_{jk} - n_{ik} + n_{ij}{}^{P_{jk}}, \qquad \{n_{ij}\} \in C^1(\mathfrak{U}, N_{\mathfrak{g},d}),$$
$$\{P_{ij}\} \in C^1(\mathfrak{U}, G_d).$$

By using this coboundary map, we can define $H^1(M, N_{\mathfrak{g},d})$. Then we have the following exact sequence:

$$0 \longrightarrow H^0(M, N_{\mathfrak{g},d}) \longrightarrow H^0(M, \mathfrak{g}_d) \xrightarrow{\exp} H^0(M, G_d) \xrightarrow{\delta} H^1(M, N_{\mathfrak{g},d}).$$

If $\delta_P n_{ijk} = 0$, the relation

$$U_i \times N_{\mathfrak{g}} \ni (x, n(x)) \sim \left(x, P_{ij}(x)n(x)P_{ij}(x)^{-1} + n_{ij}(x)\right) \in U_j \times N_{\mathfrak{g}},$$

is an equivalence relation. The quotient space of $\bigcup U_i \times N_{\mathfrak{g}}$ by this relation is a fibre bundle over M with the fibre $N_{\mathfrak{g}}$. Hence we have

Lemma 4. *If M is contractible, then* $\exp : H^0(M, \mathfrak{g}_d) \to H^0(M, G_d)$ *is onto.*

Corollary 1. *Let $i : M \to E$ be a smooth imbedding of M into a contractible space E. Then we have*

$$(10) \qquad \exp(H^0(M, \mathfrak{g}_d)) = i^*(H^0(E, G_d)).$$

Corollary 2. $\exp(H^0(M, \mathfrak{g}_d))$ *is a normal subgroup of* $H^0(M, G_d) = \mathrm{Map}(M, G)$.

Note. In general $\exp(H^0(M, \mathfrak{g}_d))$ is *not* a closed subgroup of $\mathrm{Map}(M, G)$ (cf. [23]).

5. Let g be a smooth G-valued function on M. Then by Corollary of Lemma 3, there is a locally finite open covering $\mathfrak{U} = \{U_i\}$ of M such that $g(x) = \exp\left(2\pi\sqrt{-1}f_i(x)\right)$, f_i is a smooth $\mathfrak{g}$-valued function on U_i.

On $(U_i \cap U_j) \times \mathbb{C}^*$, we set

$$g_{ij}(x, z) = e^{f_i(x) \log z} e^{-f_j(x) \log z}, \quad x \in U_i \cap U_j, \quad z \in \mathbb{C}^*.$$

By definition $g_{ij}(x, z)$ is single-valued and $g_{ij}(x, 1) = I$, the unit matrix. Hence we can define a smooth ΩG-valued function $g_{ij}{}^{\Omega}$ on $U_i \cap U_j$ by

$$(g_{ij}{}^{\Omega}(x))(t) = g_{ij}\left(x, e^{2\pi\sqrt{-1}t}\right).$$

If $\exp\left(2\pi\sqrt{-1}f_{i,1}(x)\right) = \exp\left(2\pi\sqrt{-1}f_{i,2}(x)\right)$, then we have

$$g_{ij,2}(x) = h_i(x)g_{ij,1}(x)h_j(x)^{-1},$$
$$(h_i(x))(t) = e^{2\pi\sqrt{-1}f_{i,2}(x)t} e^{-2\pi\sqrt{-1}f_{i,1}(x)t}.$$

Hence $\{g_{ij}\}$ defines an ΩG-bundle $B_0(g)$ over M and its equivalence class as an ΩG-bundle is determined by g. By definition, $B_0(g)$ is trivial if $g = e^f$, f a smooth $\mathfrak{g}$-valued function on M.

On the other hand, if g is an ΩG-valued function, we define a $\mathrm{Map}(\mathbb{R}, G)$-valued function $\widetilde{g}$ by

$$(\widetilde{g}(x))(t) = (g(x))\left(e^{2\pi\sqrt{-1}t}\right), \qquad t \in \mathbb{R}.$$

We note that $(\widetilde{g}(x))(0) = I$ for all x. If $\xi = \{g_{ij}\}$ is an ΩG-bundle, then we define a $\mathrm{Map}(\mathbb{R}, G)$-bundle $\widetilde{\xi}$ by $\{\widetilde{g}_{ij}(x)\}$. Then, since $\mathrm{Map}(\mathbb{R}, G)$ is a contractible group, we can set

$$\widetilde{g}_{ij}(x) = \widetilde{h}_i(x)\widetilde{h}_j(x)^{-1}, \qquad \widetilde{h}_i(x) \text{ is a smooth } \mathrm{Map}(\mathbb{R}, G)\text{-valued}$$
$$\text{function on } U_i.$$

By definition we have $(\widetilde{g}_{ij}(x))(t) = (\widetilde{g}_{ij}(x))(t+1)$. Hence we have

$$\left(\widetilde{h}_i(x)(t)\right)^{-1}\left(\widetilde{h}_i(x)(t+1)\right) = \left(\left(\widetilde{h}_j(x)(t)\right)^{-1}\left(\widetilde{h}_j(x)(t+1)\right)\right).$$

Therefore we can define a smooth G-valued function g on M by

$$g(x) = \left(\widetilde{h}_i(x)(0)\right)^{-1}\left(\widetilde{h}_i(x)(1)\right) = \widetilde{h}_i(x)(1), \qquad x \in U_i.$$

By the definition of B_0 we get $B_0(g) = \xi$. Hence we obtain

Theorem 1. *There is a bijection*

$$B_0 \cdot H^0(M, G_d)/\exp(H^0(M, \mathfrak{g}_d)) \cong H^1(M, \Omega G_d).$$

Note. If $g(x) = \exp\left(2\pi\sqrt{-1}f(x)\right)$, we have $g(x)\exp(f(x)\log z) = \exp(f(x)\log z)g(x)$. Hence

$$g_{ij,L}(x, z) = e^{-f_i(x)\log z}\, e^{f_j(x)\log z}$$

is also a single valued smooth function on $(U_1 \cap U_j) \times C^*$. By using this $\{g_{ij,L}\}$, we can define an alternative bijection

$$B_{0L} : H^0(M, G_d)/\exp(H^0(M, \mathfrak{g}_d)) \cong H^1(M, G_d).$$

The relation between B_0 and B_{0L} is given by

(11) $$B_0(g^{-1}) = B_{0L}(g) \quad (= (B_0(g))^{-1}).$$

We denote the connected component of the identity of $H^0(M, G_{n,d}) = H^0(M, G_d)$ by $H^0(M, G_{n,d})_0$. It contains $\exp(H^0(M, \mathfrak{g}_{n,d}))$ and there is a map $k : H^0(M, G_{n,d})/H^0(M, G_{n,d})_0 \to K^1(M) \ (= K^{-1}(M))$. k is an isomorphism if n is sufficiently large. Hence there is a homomorphism

$$k^1 : H^1(M, G_{n,d}) \to K^1(M),$$

such that $\ker k^1 = B^0(H^0(M, G_{n,d})_0/\exp(H^0(M, \mathfrak{g}_{n,d}))$. k^1 is onto if n is sufficiently large.

6. We denote by $\mathcal{M}^1$ and $M^1{}_L$ the sheaves of germs of right and left integrable 1-forms. We have

$$\mathcal{M}^1 = \rho(G_d) = d^e(\mathfrak{g}_d), \qquad \mathcal{M}^1{}_L = \rho_L(G_d) = d^e{}_L(\mathfrak{g}_d).$$

Here $\rho(g) = g^{-1}dg$, $\rho_L(g) = (dg)g^{-1}$. d^e and $d^e{}_L$ are given by

$$d^e f = e^{-f} d(e^f) = df + \sum_{n=1}^{\infty} \frac{(-1)^n}{(n+1)!}(\operatorname{ad} f)^n(df),$$

$$d^e{}_L f = (d(e^f))e^{-f} = df + \sum_{n=1}^{\infty} \frac{1}{(n+1)!}(\operatorname{ad} f)^n(df).$$

If θ belongs to $H^0(M, \mathcal{M}^1)$, $\pi^*(\theta)$ is integrated on $\widetilde{M}$, the universal covering space of M. Here $\pi : \widetilde{M} \to M$ is the projection.

Definition 3. *We set*

$$H^0(M, \mathcal{M}^1)_0 = \{\theta | \pi^*(\theta) = \rho(g), \ \rho_L(g) \in \pi^*(H^0(M, \mathcal{M}^1{}_L))\}.$$

By definition θ belongs to $H^0(M, \mathcal{M}^1)_0$ if and only if $\pi^*(\theta) = \rho(g)$ and g satisfies

$$(12) \qquad g^\sigma = \chi_\sigma g = g\chi_\sigma, \qquad \sigma \in \pi_1(M), \quad \chi \in \operatorname{Hom}(\pi_1(M), G).$$

Theorem 1'. *There is a map* $\widetilde{B}_0 : H^0(M, \mathcal{M}^1)_0/d^e(H^0(H^0, \mathfrak{g}_d)) \to H^1(M, \mathcal{M}^1{}_{\Omega\mathfrak{g}})$ *such that the following diagram becomes commutative:*

$$
\begin{array}{ccc}
H^0(M, \mathcal{M}^1)_0/d^e(H^0(M, \mathfrak{g}_d)) & \xrightarrow{\widetilde{B}_0} & H^1(M, \mathcal{M}^1{}_{\Omega\mathfrak{g}}) \\
\rho \uparrow & & \rho \uparrow \\
H^0(M, G_d)_0/\exp(H^0(M, \mathfrak{g}_d)) & \xrightarrow{B_0} & H^1(M, \Omega G_d).
\end{array}
$$

Proof. By (1) we can take local integrations $\{g_i\}$ of $\theta \in H^0(M, \mathcal{M}^1)_0$ to be

$(12)'$ $$g_i = c_{ij}g_j = g_j c_{ij}, \qquad \text{on } U_i \cap U_j.$$

We assume $g_i = \exp\left(2\pi\sqrt{-1}f_i\right)$ on U_i and set

$$h_{ij}(x, z) = e^{f_i(x)\log z}\, e^{-f_j(x)\log z}\,.$$

Then by $(12)'$ we have

$$h_{ij}(x, z)h_{jk}(x, z)h_{ki}(x, z) = c_{ij}c_{jk}c_{ki}.$$

Hence if we define $\rho_x(h_{ij})^\Omega$ by

$$\left(\rho_x(h_{ij})^\Omega(x)\right)(t) = \rho_x\left(h_{ij}\left(x, e^{2\pi\sqrt{-1}t}\right)\right),$$
$$\rho_x(f(x, t)) = ((x, t)^{-1}d_x f(x, t),$$

$\{\rho_x(h_{ij})\}$ becomes a cocycle. Hence we can define $\widetilde{B}_0$ by

$$\widetilde{B}_0([\theta]) = \left\{\rho_x(h_{ij})^\Omega\right\} \quad (= \{\rho(h_{ij}{}^\Omega)\}).$$

Then we have the Theorem by the definitions of B_0 and $\widetilde{B}_0$. $\blacksquare$

$\widetilde{B}_0$ is not onto. We set $H^1(M, \mathcal{M}^1{}_{\Omega\mathfrak{g}})_0$ to be the subset of $H^1(M, \mathcal{M}^1{}_{\Omega\mathfrak{g}})$ whose representing cocycle $\{\omega_{ij}\}$ satisfies

$$\omega_{ij} = g_{ij}{}^{-1}dg_{ij}, \qquad \frac{d}{dt}(g_{ij}(x)g_{jk}(x)g_{ki}(x)) = 0.$$

Then we have

$$\widetilde{B}^0(H^0(M, \mathcal{M}^1)_0/d^e(H^0(M, \mathfrak{g}_d)) \subset H^1(M, \mathcal{M}^1{}_{\Omega\mathfrak{g}})_0.$$

If g is a smooth LG-valued function on M, we define a smooth G-valued function g^b on $M \times S^1$ by

(13) $$g^b(x, t) = (g(x))(t).$$

Since $d(g^b)$ contains derivation in t, $(g^b)^{-1}d(g^b)$ is not determined by $g^{-1}dg$. If $\{\omega_{ij}\}$ is a representing cocycle of an element of $H^1(M, \mathcal{M}^1{}_{\Omega\mathfrak{g}})_0$ and $\{g_{ij,1}\}$ and $\{g_{ij,2}\}$ are integrations of $\{\omega_{ij}\}$ such that

$$\frac{d}{dt}(g_{ij,1}(x)g_{jk,1}(x)g_{ki,1}(x)) = \frac{d}{dt}(g_{ij,2}(x)g_{jk,2}(x)g_{ki,2}(x)) = 0,$$

the difference between $\left\{(g_{ij,1}{}^b)^{-1}dg_{ij,1}{}^b\right\}$ and $\left\{(g_{ij,2}{}^b)^{-1}dg_{ij,2}{}^b\right\}$ comes from a representing cocycle of an element of $H^1(S^1, \mathcal{M}^1)$. Since an element of $H^1(S^1, \mathcal{M}^1)$ is determined by a representation of (the universal covering group of) $[\Omega S^1{}_e]$, the group of zero homotopic paths of S^1 ([7], cf. [18], [19], [21]), $H^1(S^1, \mathcal{M}^1)$ vanishes. Hence we can define the map $\natural : H^1(M, \mathcal{M}^1{}_{\Omega\mathfrak{g}})_0 \to H^1(M \times S^1, \mathcal{M}^1)$ by

$$\langle\{\omega_{ij}\}\rangle^\natural = \left\langle\left\{(g_{ij}{}^b)^{-1}d(g_{ij}{}^b)\right\}\right\rangle.$$

Here $\{g_{ij}\}$ is assumed to be $(g_{ij}g_{jk}g_{ki})' = 0$. On the other hand, we define $b : H^1(M, \Omega G_d) \to H^1(M \times S^1, G_d)$ by $\{g_{ij}\}^b = \{g_{ij}{}^b\}$. Then we have the commutative diagram

$$
\begin{array}{ccc}
H^1(M, \mathcal{M}^1{}_{\Omega\mathfrak{g}})_0 & \xrightarrow{\natural} & H^1(M \times S^1, \mathcal{M}^1) \\
{\scriptstyle\rho}\uparrow & & {\scriptstyle\rho}\uparrow \\
H^1(M, \Omega G_d) & \xrightarrow{b} & H^1(M \times S^1, G_d).
\end{array}
$$

Definition 4. *We define Gysin maps* $\gamma : H^0(M, G_d)/\exp(H^0(M, \mathfrak{g}_d)) \to H^1(M \times S^1, G_d)$ *and* $\widetilde{\gamma} : H^0(M, \mathcal{M}^1)_0/d^e(H^0(M, \mathfrak{g}_d)) \to H^1(M \times F^1, \mathcal{M}^1)$ *by*

$$\gamma([g]) = (B_0([g]))^b, \qquad \widetilde{\gamma}(\langle\omega\rangle) = (\widetilde{B}_0(\langle\omega\rangle))^\natural.$$

By the definitions the following diagram is commutative:

$$
\begin{array}{ccc}
H^0(M, \mathcal{M}^1)_0/d^e(H^0(M, \mathfrak{g}_d)) & \xrightarrow{\widetilde{\gamma}} & H^1(M \times S^1, \mathcal{M}^1) \\
{\scriptstyle\rho}\uparrow & & {\scriptstyle\rho}\uparrow \\
H^0(M, G_d)/\exp(H^0(M, \mathfrak{g}_d)) & \xrightarrow{\gamma} & H^1(M \times S^1, G_d).
\end{array}
$$

§3. Connections of loop group bundles and string classes

7. We can define connections and curvatures of loop group bundles and elements of $H^1(M, \mathcal{M}^1{}_{\Omega\mathfrak{g}})$ similarly as connections and curvatures of G-bundles and elements of $H^1(M, \mathcal{M}^1)$ (cf. [5], [6]). If M has a smooth partition of unity subordinate to any locally finite open covering of M, connections always exist. Next we define the connection form $\{\widetilde{\theta}_i\}$ of an element of $H^1(M, \mathcal{M}^1{}_{\widetilde{\Omega\mathfrak{g}}})$ whose representing cocycle is $\{\widetilde{\omega}_{ij}\}$ by the relation

$$(14) \qquad \widetilde{\omega}_{ij} = \widetilde{\theta}_j - \widetilde{\theta}_i{}^{g_{ij}}, \qquad \widetilde{\omega}_{ij} = (g_{ij}{}^{-1}dg_{ij}, \beta_{ij}).$$

Definition 5. *Let* $\widetilde{\xi}$ *be an* $\widetilde{\Omega}G$-*bundle and* $\langle\{\widetilde{\omega}\}\rangle$ *an element of* $H^1(M, \mathcal{M}^1{}_{\widetilde{\Omega\mathfrak{g}}})$ *such that*

$$j^*(\langle\{\widetilde{\omega}\}\rangle) = \rho^*(j^*(\widetilde{\xi})).$$

Then we say a connection of $\langle\widetilde{\omega}\rangle$ *to be a connection of* $\widetilde{\omega}$.

Note. Connections of elements of $H^1(M, \mathcal{M}^1{}_{\widetilde{L\mathfrak{g}}})$ and $\widetilde{L}G$-bundles are similarly defined.

The curvature $\{\widetilde{\Theta}_i\}$ of $\{\widetilde{\theta}_0\}$ is defined by

$$(15) \qquad\qquad \widetilde{\Theta}_i = d\widetilde{\theta}_i + \frac{1}{2}[\widetilde{\theta}_i, \widetilde{\theta}_i].$$

We set $\widetilde{\theta}_i = (\theta_i, \psi_i)$ and $\widetilde{\Theta}_i = (\Theta_i, \Psi_i)$. Then (14) and (15) mean

$$(14)' \qquad \omega_{ij} = \theta_j - g_{ij}{}^{-1}\theta_i g_{ij},$$

$$\beta_{ij} = \psi_j - \left(psi_i + \int_0^1 \mathrm{tr}(\theta_i g_{ij}{}' g_{ij}{}^{-1})dt\right),$$

$$(15)' \qquad \Theta_i = d\theta_i + \theta_i \wedge \theta_i,$$

$$\Psi_i = d\psi_i + \frac{1}{2}\int_0^1 \mathrm{tr}(\theta_1 \wedge \theta_i{}')dt.$$

Proposition 3. (i) *If* $\{\widetilde{\omega}_{ij}\}$ *is a representing cocycle of an element of* $H^1(M, \mathcal{M}^1{}_{\widetilde{\Omega\mathfrak{g}}})$, *then it has a connection.*

(ii) If $\{\widetilde{\theta}_i\} = \{(\theta_i, \psi_i)\}$ and $\{\widetilde{\theta}_{i,1}\} = \{(\theta_{i,1}, \psi_{i,1})\}$ are connections of $\{\widetilde{\omega}_{ij}\}$, then

(16)
$$\theta_{i,1} = \theta_i + \eta_i, \quad \eta_j = g_{ij}^{-1}\eta_i g_{ij},$$
$$\psi_{i,1} = \psi_i + \phi, \quad \phi \text{ is a global 1-form on } M.$$

(iii) If a collection of $\widetilde{L\mathfrak{g}}$-valued differential forms $\{\widetilde{\phi}_i\} = \{(\Phi_i, \zeta_i)\}$ satisfies $\widetilde{\phi}_i{}^{g_{ij}} = \widetilde{\phi}_j$, then we have

(17)
$$d\widetilde{\phi}_j + [\widetilde{\theta}_j, \widetilde{\theta}_j] = (d\widetilde{\phi}_i + [\widetilde{\theta}_i, \widetilde{\phi}_i])^{g_{ij}}.$$

Proof. Let $\{e\}$ be a smooth partition of unity subordinate to $\{U_i\}$. Then if we set $\widetilde{\omega}_{ki} = (\omega_{ki}, \beta_{ki})$ and define

$$\theta_i = \sum_{U_i \cap U_k \neq \emptyset} e_k \omega_{ki}, \qquad \psi_i = \sum_{U_i \cap U_k \neq \emptyset} e_k \beta_{ki},$$

we get $\omega_{ij} = \theta_j - g_{ij}^{-1}\theta_i g_{ij}$ and $\beta_{ij} = \psi_j - \psi_i + \sum e_k \int_0^1 \mathrm{tr}(\omega_{ki} g_{ij}' g_{ij}^{-1})dt = \psi_j - (\psi_i + \int_0^1 \mathrm{tr}(\theta_i g_{ij}' g_{ij}^{-1})dt$. Hence we have (i). (ii) follows from $(14)'$.

Since $\{\theta_i\}$ is a connection of $\{\omega_{ij}\}$, we have $d\phi_j + [\theta_j, \phi_j] = g_{ij}^{-1}(d\phi_i + [\theta_i, \phi_i])g_{ij}$. Since $\theta_j = g_{ij}^{-1}\theta_i g_{ij} + g_{ij}^{-1}dg_{ij}$, we get

$$d\left(\int_0^1 \mathrm{tr}(\phi_i g_{ij}' g_{ij}^{-1})dt\right) - \int_0^1 \mathrm{tr}(\theta_j' \phi_j)dt +$$
$$+ \int_0^1 \mathrm{tr}(\theta_i' \phi_i)dt - \int_0^1 \mathrm{tr}(d\phi_i g_{ij}' g_{ij}^{-1})dt$$
$$= \int_0^1 \mathrm{tr}([\theta_i, \phi_i]g_{ij}' g_{ij}^{-1})dt.$$

Hence we have

$$d\zeta_j + \int_0^1 \mathrm{tr}(\theta_j \phi_j')dt$$
$$= d\zeta_i + d\left(\int_0^1 \mathrm{tr}(\phi_i g_{ij}' g_{ij}^{-1})dt\right) + \int_0^1 \mathrm{tr}(\theta_j \phi_j')dt$$
$$= d\zeta_i + \int_0^1 \mathrm{tr}(\theta_i \phi_i')dt + \int_0^1 \mathrm{tr}((d\phi_i + [\theta_i, \phi_i])g_{ij}' g_{ij}^{-1})dt.$$

This shows (iii).

8. By straightforward calculations we obtain

Lemma 5. *Let* $\{\theta_i\}$ *be a connection form of* $\{\omega_{ij}\} = g_{ij}{}^{-1}dg_{ij}$, *a representing cocycle of an element of* $H^1(M, \mathcal{M}^1{}_{L\mathfrak{g}})$, *then*

$$(18) \quad \int_0^1 \operatorname{tr}\left(\theta_i{}^{g_{ij}} \wedge \omega_{ij}{}' + \omega_{ij} \wedge (\theta_i{}')^{g_{ij}} - \omega_{ij} \wedge [g_{ij}{}^{-1}g_{ij}{}', \theta_i{}^{g_{ij}}] - \right.$$
$$\left. -2\theta_i d(g_{ij}{}')g_{ij}{}^{-1} + 2\theta_i g_{ij}{}' g_{ij}{}^{-1}(dg_{ij})g_{ij}{}^{-1}\right) dt = 0.$$

Proposition 4. *The coordinate transformation law* $\widetilde{\Theta}_i{}^{g_{ij}} = \widetilde{\Theta}_j$ *and Bianchi identity* $d\widetilde{\Theta}_i + [\tilde{\theta}_i, \widetilde{\Theta}_i] =$ *hold for the* $\widetilde{L}\mathfrak{g}$-*valued curvature form* $\{\widetilde{\Theta}_i\}$. *That is, we have*

$$(19) \quad \Theta_j = g_{ij}{}^{-1}\Theta_i g_{ij}, \quad \Psi_j = \Psi_i + \int_0^1 \operatorname{tr}(\Theta_i g_{ij}{}' g_{ij}{}^{-1})dt,$$

$$(20) \quad d\Theta_i + [\theta_i, \Theta_i] = 0, \quad d\Psi_i + \int_0^1 \operatorname{tr}(\theta_i \wedge \Theta_i{}')dt = 0.$$

Proof. We need only to show the second equalities of (19) and (20). Since we obtain

$$d(\psi_j - \psi_i)$$
$$= -\frac{1}{2}\int_0^1 \operatorname{tr}\omega_{ij} \wedge \omega_{ij}{}')dt + \int_0^1 \operatorname{tr}(d\theta_i g_{ij}{}' g_{ij}{}^{-1} - \theta_i d(g_{ij}{}')g_{ij}{}^{-1} +$$
$$+ \theta_i g_{ij}{}' g_{ij}{}^{-1}(gd_{ij})g_{ij}{}^{-1})dt,$$

we have

$$2(\Psi_j - \Psi_i)$$
$$= \int_0^1 \operatorname{tr}(\theta_j \wedge \theta_j{}' - \theta_i \wedge \theta_i{}' - \omega_{ij}\omega_{ij}{}')dt + 2\int_0^1 \operatorname{tr}(d\theta_i g_{ij}{}' g_{ij}{}^{-1} -$$
$$- \theta_i d(g_{ij})g_{ij}{}^{-1} + \theta_i g_{ij}{}' g_{ij}{}^{-1}(fg_{ij})g_{ij}{}^{-1})dt.$$

Since $\theta_j = g_{ij}{}^{-1}\theta_i g_{ij} + \omega_{ij}$, we get the second equality of (19) by this equality and Lemma 5.

By the definition of Ψ_i we have $d\Psi_i = \int_0^1 \operatorname{tr}(d\theta_i \wedge \theta_i{}')dt$. Hence by Lemma 1, we obtain the second equality of (20). ∎

The second equalities of (19) and (20) are rewritten as

$$(19)' \quad \int_0^1 \operatorname{tr}(\Theta_i g_{ij}{}' g_{ij}{}^{-1})dt = \Psi_j - \Psi_i,$$

$$(20)' \quad d\Psi_i = \int_0^1 \operatorname{tr}(\Theta_i \wedge \theta_i{}')dt.$$

We generalize $(19)'$ and $(20)'$ as follows: Let $\{\Theta_i\}$ be a curvature form of $\{\omega_{ij}\} = \{g_{ij}^{-1}dg_{ij}\}$. Then we set

$$\phi_{p,ij} = \int_0^1 \mathrm{tr}(\Theta_i{}^p g_{ij}' g_{ij}^{-1})dt, \qquad \Theta^p = \overbrace{\Theta \wedge \ldots \wedge \Theta}^{p}.$$

$\{\phi_{p,ij}\}$ is a 1-cochain of $2p$-forms, and we have by straightforward calculations

Lemma 6. *As a 1-cochain of $2p$-forms, we obtain*

$$\delta\phi_{p,ijk} = \int_0^1 \mathrm{tr}(\Theta_k{}^p c_{kij}' c_{kij}^{-1})dt, \qquad c_{ijk} = g_{ij}g_{jk}g_{ki}.$$

Corollary. *If $\{\omega_{ij}\}$ is a representing cocycle of an element of $H^1(M, \mathcal{M}^1{}_{\Omega\mathfrak{g}})_0$, then there exists a 0-chain of $2p$-forms $\{\Psi_{p,i}\}$ such that*

$$(21) \qquad \int_0^1 \mathrm{tr}(\Theta_i{}^p g_{ij}' g_{ij}^{-1})dt = \Psi_{p,j} - \Psi_{p,i}.$$

Lemma 7. *Let $\{\theta_i\}$ be a connection form of $\{\omega_{ij}\} = \{g_{ij}^{-1}dg_{ij}\}$, $\{\Theta_i\}$ the curvature form of $\{\theta_i\}$. Then we have*

$$(22) \qquad d\left(\int_0^1 \mathrm{tr}(\Theta_i{}^p g_{ij}' g_{ij}^{-1})dt\right)$$

$$= \int_0^1 \mathrm{tr}(\Theta_j^p \wedge \theta_j')dt - \int_0^1 \mathrm{tr}(\Theta_i{}^p \wedge \theta_i')dt.$$

Proof. By the Bianchi identity we get $d(\Theta_i{}^p) + [\theta_i, \Theta_i{}^p] = 0$. Hence we have

$$d\left(\int_0^1 \mathrm{tr}(\Theta_i{}^p g_{ij}' g_{ij}^{-1})dt\right)$$

$$= \int_0^1 \mathrm{tr}([\Theta_i^p, \theta_i]g_{ij}' g_{ij}^{-1} + \Theta_i{}^p d(g_{ij}' g_{ij}^{-1}) -$$

$$- \Theta_i{}^p g_{ij}' \omega_{ij} g_{ij}^{-1})dt.$$

Then, since $\omega_{ij} = \theta_j - g_{ij}^{-1}\theta_i g_{ij}$, this right hand side is equal to

$$\int_0^1 \mathrm{tr}([\Theta_i{}^p, \theta_i]g_{ij}' g_{ij}^{-1} - ((\Theta_j{}^p)' \wedge \theta_j + (\Theta_j{}^p)' g_{ij}^{-1}\theta_i g_{ij}) \, dt$$

$$= \int_0^1 \mathrm{tr}\left((\Theta_i{}^p)' \wedge \theta_i - (\Theta_j{}^p)' \wedge \theta_j\right) dt.$$

Hence we have (22).   ■

Corollary. Let $\{\Psi_{p,i}\}$ be the 0-cochain determined by the Corollary of Lemma 6. Then the 0-cochain of $(2p+1)$-forms $\{\phi_{p,i}(\{\omega\})\}$ defined by

$$\phi_{p,i}(\{\omega\}) = \int_0^1 \mathrm{tr}(\Theta_i{}^p \wedge \theta_i') dt - d\Psi_{p,i},$$

gives a global closed $(2p+1)$-form $\phi_p(\{\omega\})$ on M.

Proof. We need only to show $d\left(\int_0^1 \mathrm{tr}(\Theta_i{}^p \wedge \theta_i') dt\right) = 0$.   Since $\int_0^1 \mathrm{tr}\left((\Theta_i{}^p)' \wedge \Theta_i\right) dt = 0$, we have $\int_0^1 \mathrm{tr}\left((\Theta_i{}^p)' \wedge d\theta_i\right) dt = -\int_0^1 \mathrm{tr}\left((\Theta_i{}^p)' \wedge \theta_i{}^2\right) dt$. Hence we get

$$d\left(\int_0^1 \mathrm{tr}(\Theta_i{}^p \wedge \theta_i') dt\right)$$

$$= -\int_0^1 \mathrm{tr}\left([\theta_i, \Theta_i{}^p] \wedge \theta_i' - (\Theta_i{}^p)' \wedge \theta_i^2\right) dt$$

$$= \int_0^1 \mathrm{tr}\left(\Theta_i{}^p \wedge (\theta_i' \wedge \theta_i + \theta_i \wedge \theta_i') - \Theta_i{}^p \wedge (\theta_i{}^2)'\right) dt = 0.$$

9.   **Theorem 2.** Let $\{\omega_{ij}\}$ be a representing cocycle of an element of $H^1(M, \mathcal{M}^1{}_{\Omega\mathfrak{g}})_0$, and $\mathrm{Ch}^p\left(\langle\omega\rangle^\natural\right)$ the p-th Chern character of $\langle\omega\rangle^\natural \in H^p(M \times S^1, \mathcal{M}^1)$ ([5], [6]). Then we have

$$(23) \qquad \langle\phi_{p,i}(\{\omega\})\rangle = -\left(2\pi\sqrt{-1}\right)^{p+1} p! \int_{S^1} \mathrm{Ch}^{p+1}\left(\langle\omega\rangle^\natural\right) dt.$$

Proof. For an $\Omega\mathfrak{g}$-valued differential form $\zeta = \sum \zeta_{i_1,\dots,i_p} dx_{i_1} \wedge \dots \wedge dx_{i_p}$, we set

$$\zeta^b = \sum \zeta_{i_1,\dots,i_p}{}^b dx_{i_1} \wedge \dots \wedge dx_{i_p}.$$

By using smooth partition of unity, for a connection form $\{\theta_i\}$ of $\{\omega_{ij}\}$, we can construct a connection form $\{\theta_i{}^\natural\}$ of $\{\omega_{ij}{}^\natural\}$ such that

$$\theta_i{}^\natural = \theta_i{}^b + f_i \, dt.$$

The curvature form $\{\Theta_i{}^\natural\}$ of $\{\theta_i{}^\natural\}$ takes the form

$$\Theta_i{}^\natural = \Theta_i{}^b + \left(df_i + [\theta_i{}^b, f_i] - \frac{\partial}{\partial t}\theta_i{}^b\right) dt.$$

By the Bianchi identity, we have $\mathrm{tr}\left(({\Theta_i}^b)^P \wedge [{\theta_i}^b, f_i]\right) = \mathrm{tr}\left(d({\Theta_i}^P)^b f_i\right)$. Hence we have

$$\int_0^1 \mathrm{tr}\left(({\Theta_i}^\natural)^{p+1}\right) dt$$

$$= -(p+1)\int_0^1 \mathrm{tr}({\Theta_i}^P \wedge {\theta_i}')dt + (p+1)\int_0^1 \mathrm{tr}(({\Theta_i}^b)^P \wedge (df_i +$$

$$+ [{\theta_i}^b, f_i]))dt$$

$$= -(p+1)\int_0^1 \mathrm{tr}({\Theta_i}^P \wedge {\theta_i}')dt + (p+1)d\left(\int_0^1 \mathrm{tr}({\Theta_i}^P {f_i}^\#)dt\right).$$

Here $f^\#$ is the LG-valued function defined by

(24) $$(f^\#(x))(t) = f(x,t).$$

Since f_i satisfies $({g_{ij}}^b)^{-1}\partial/\partial t({g_{ij}}^b) = f_j - ({g_{ij}}^b)^{-1}f_i {g_{ij}}^b$, we get

$$\int_0^1 \mathrm{tr}({\Theta_j}^P {f_j}^\#)dt - \int_0^1 \mathrm{tr}({\Theta_i}^b {f_i}^\#)dt = \int_0^1 \mathrm{tr}({\Theta_i}^P {g_{ij}}' {g_{ij}}^{-1})dt.$$

Hence we obtain

$$\int_0^1 \mathrm{tr}({\Theta_j}^P {f_j}^\#)dt - \Psi_{p,j} = \int_0^1 \mathrm{tr}({\Theta_i}^P {f_i}^\#)dt - \Psi_{p,i},$$

on $U_i \cap U_j$. Therefore we have (23). ∎

Definition 6. *The de Rham class of $\phi_p(\{\omega\})$ is called the p-th string class of $\langle\omega\rangle$ and denoted $\widetilde{c}^p(\langle\omega\rangle)$. If ξ is an ΩG-bundle over M, then we denote $\widetilde{c}^p(\rho^*(\xi))$ by $\widetilde{c}^p(\xi$ and call it the p-th string class of ξ.*

String classes of the elements of $H^1(M, \mathcal{M}^1{}_{L\mathfrak{g}})$ and LG-bundles are similarly defined. By using the notation $\widetilde{c}^p(\langle\omega\rangle)$, (23) is rewritten as

(23)' $$\widetilde{c}^p(\langle\omega\rangle) = -\left(2\pi\sqrt{-1}\right)^{p+1} p! \int_{S^1} \mathrm{Ch}^{p+1}\left(\langle\omega\rangle^\natural\right) dt.$$

By the definition of $\widetilde{c}^1(\langle\omega\rangle)$ and Proposition 2 we have

Theorem 3. *The image of $\langle\omega\rangle$ by the coboundary map δ : $H^1(M, \mathcal{M}^1{}_{\Omega\mathfrak{g}}) \to H^2(M, \Phi^1) = H^3(M, \mathbb{C})$ is $\widetilde{c}^1(\langle\omega\rangle)$.*

Corollary 1. *$\langle\omega\rangle$ is in the image of $j^* : H^1(M, \mathcal{M}^1{}_{\widetilde{\Omega g}}) \to H^1(M, \mathcal{M}^1{}_{\Omega g})$ if and only if $\widetilde{c}^1(\langle\omega\rangle) = 0$.*

Corollary 2. *If the structure group of an ΩG-bundle ξ can be lifted up $\widetilde{\Omega G}$, then $\widetilde{c}^1(\xi) = 0$.*

§4. Lifting of G-bundles on loop spaces

10. We denote by LM and ΩM the free and the based loop spaces over M, smooth Hilbert manifold modelled V. We assume that LM consists of Sobolev 1-loops (cf. [7]). Then if we set $\tilde{H}^1(S^1) = \{\gamma \in H^1(S^1), \gamma(0) = \gamma(1)\}$ and $\tilde{H}^1(S^1)_0 = \{\gamma \in H^1(S^1), \gamma(0) = \gamma(1) = 0\}$, LM and ΩM are Hilbert manifolds modeled by $\tilde{H}^1(S^1) \otimes V$ and $\tilde{H}^1(S^1)_0 \otimes V$. The connected component of the unit loop of ΩM is denoted by ΩM_e.

If g is a smooth G-valued function on M, we define a smooth LG-valued function g^L on LM by

$$(25) \qquad (g^L(\gamma))(t) = g(\gamma(t)).$$

We also define ΩG-valued functions $g^\Omega = g^\Omega_R$ and g^Ω_L on M by

$$(25)' \qquad g^\Omega(\gamma) = g(\gamma(0))^{-1} g^L(\gamma), \qquad g^\Omega_L(\gamma) = g^L(\gamma) g(\gamma(0))^{-1}.$$

Since $g(\gamma(0))$ is a constant, we have

Lemma 8. *If $\theta = g^{-1} dg = h^{-1} dh$, then we have*

$$(g^L)^{-1} d(g^L) = (h^L)^{-1} d(h^L),$$
$$(g^\Omega)^{-1} d(g^\Omega) = (g^L)^{-1} d(g^L).$$

Corollary. *If θ is an integrable form on M, then we can define an $\Omega\mathfrak{g}$-valued integrable form θ^L on ΩM_e by*

$$(26) \qquad \theta^\cdot|\Omega U = (g^L)^{-1} g(g^L), \qquad \theta = g^{-1} dg \text{ on } U.$$

Proof. We need only to show that there exists an open covering $\{U\}$ of M such that θ is integrated on U and $\{\Omega U\}$ covers ΩM_e. If γ belongs to ΩM_e, then γ is homotopic to 0. Hence it has a neighborhood $U(\gamma)$ such that θ is integrated on $U(\gamma)$. Since $\gamma \in \Omega U(\gamma)$, we have the Corollary. ∎

By this Corollary we get a map $\tilde{L} : H^0(M, \mathcal{M}^1) \to H^0(\Omega M_e, \mathcal{M}^1{}_{\Omega\mathfrak{g}})$. We can also define the maps $L : H^1(M, G_d) \to H^1(\Omega M_e, \Omega G_d)$ and $\tilde{L} : H^1(M, \mathcal{M}^1) \to H^1(\Omega M_e, \mathcal{M}^1\Omega\mathfrak{g})$. Then we have

Lemma 9. (i) $\widetilde{L}(H^1(M, \mathcal{M}^1))$ is contained in $H^1(\Omega M_e, \mathcal{M}^1{}_{\Omega\mathfrak{g}})_0$.

((ii) We can define L to be the map $L : H^1(M, G_d) \to H^1(\Omega M, LG_d)$.

(iii) The following diagrams are commutative:

$$
\begin{array}{ccccc}
H^0(M, \mathcal{M}^1) & \xrightarrow{\ \widetilde{L}\ } & H^0(\Omega M_e, \mathcal{M}^1{}_{\Omega\mathfrak{g}}) & \xrightarrow{\ i^*\ } & H^0(\Omega M_e, \mathcal{M}^1{}_{L\mathfrak{g}}) \\
\rho\uparrow & & & & \rho\uparrow \\
H^0(M, G_d) & & \xrightarrow{\quad\quad L\quad\quad} & & H^0(\Omega M_e, LG_d),
\end{array}
$$

$$
\begin{array}{ccccc}
H^1(M, \mathcal{M}^1) & \xrightarrow{\ \widetilde{L}\ } & H^1(\Omega M_e, \mathcal{M}^1{}_{\Omega\mathfrak{g}}) & \xrightarrow{\ i^*\ } & H^1(\Omega M_e, \mathcal{M}^1{}_{L\mathfrak{g}}) \\
\rho\uparrow & & & & \rho\uparrow \\
H^1(M, G_d) & & \xrightarrow{\quad\quad L\quad\quad} & & H^1(\Omega M_e, LG_d).
\end{array}
$$

Proof. (i) and (iii) follow from the definitions. Since a complex vector bundle over S^1 is always trivial, we have (ii). ∎

Lemma 10. Let $ev : \Omega M_e \times S^1 \to M$ be the evaluation map given by

$$ev(\gamma, t) = \gamma(t)$$

([9]). Then we have

(27)
$$\left(g^L\right)^b = ev^*(g), \qquad \left(\theta^L\right)^\natural = ev^*(\theta).$$

Proof. Since $\left(g^L\right)^b(\gamma, t) = (g^L(\gamma))(t) = g(\gamma(t))$, we have the first equality. Then we obtain the second equality by the definitions of θ^L and $\natural$. ∎

11. Since $\zeta(x) \in \Lambda^p V^* \otimes \mathfrak{g}$ if ζ is a $\mathfrak{g}$-valued p-form, if we define ζ^L by

$$(\zeta^L(\gamma))(t) = \zeta(\gamma(t)),$$

$\zeta^L(\gamma)$ belongs to $\mathrm{Map}(S^1, \Lambda^p V^* \otimes \mathfrak{g})$. Since $\mathrm{Map}(S^1, \Lambda^p V^* \otimes \mathfrak{g})$ is contained in $\Lambda^p(\mathrm{Map}(S^1, V^*)) \otimes \mathfrak{g}$, we may regard ζ^L as a p-form on LM. Since we get

$$f^L(\gamma + s\eta)(t) = f(\gamma(t)) + s\langle df(\gamma(t)), \eta(t)\rangle + o(s),$$

we have

(28)
$$d(f^L) = (df)^L.$$

If $\zeta = \sum \zeta_{i_1, \ldots, i_p}\, dx_{i_1} \wedge \ldots \wedge dx_{i_p}$, we may write

$$\zeta^L = \sum \zeta_{i_1, \ldots, i_p}{}^L\, dx_{i_1}{}^L \wedge \ldots \wedge dx_{i_p}{}^L, \qquad dx^L = d(x^L).$$

Hence we obtain

Lemma 11. *Let $\{\theta_i\}$ be a connection of $\langle\{\omega_{ij}\}\rangle \in H^1(M, \mathcal{M}^1)$. Then $\{\theta_i{}^L\}$ becomes a connections form of $\{\omega_{ij}{}^L\}$ and its curvature is $\{\Theta_i{}^L\}$, where $\{\Theta_i\}$ is the curvature of $\{\theta_i\}$.*

Theorem 4. *Let $\langle\omega\rangle$ be an element of $H^1(M, \mathcal{M}^1)$. Then we have*

$$(29) \qquad \widetilde{c}^p\left(\langle\omega\rangle^L\right) = -\left(2\pi\sqrt{-1}\right)^{p+1} p!\,\tau^{-1}\left(\mathrm{Ch}^{p+1}\left(\langle\omega\rangle\right)\right).$$

Here $\tau^{-1} : H^{q+1}(M, \mathbb{C}) \to H^q(\Omega M_e, \mathbb{C})$ is the inverse of the transgression map.

Proof. Since the diagram

$$
\begin{array}{ccc}
H^{q+1}(M, \mathbb{C}) & \xrightarrow{\;ev^*\;} & H^{q+1}(\Omega M_e \times S^1, \mathbb{C}) \\
& \searrow{\scriptstyle\tau^{-1}} & \downarrow{\scriptstyle\int_{S^1}} \\
& & H^q(\Omega M_e, \mathbb{C})
\end{array}
\qquad
\left(\int_{S^1}\phi\right)(\gamma) = \int_{\gamma\times S^1}\phi,
$$

is commutative ([9], [10]), we have by (23) and (27)

$$\widetilde{c}^p\left(\langle\omega\rangle^L\right) = -\left(2\pi\sqrt{-1}\right)^{p+1}\int_{S^1}\mathrm{Ch}^{p+1}\left(\left(\langle\omega\rangle^L\right)^\natural\right)$$

$$= -\left(2\pi\sqrt{-1}\right)^{p+1}\int_{S^1}\mathrm{Ch}^{p+1}\left(ev^*\left(\langle\omega\rangle\right)\right)$$

$$= -\left(2\pi\sqrt{-1}\right)^{p+1} p!\,\tau^{-1}\left(\mathrm{Ch}^{p+1}\left(\langle\omega\rangle\right)\right). \qquad \blacksquare$$

Corollary. *Let $c_p(\langle\omega\rangle)$ be the p-th Chern class of $\langle\omega\rangle$ (cf. [5], [6]). Then $\langle\omega\rangle^L$ is in the j^*-image if and only if*

$$c_1{}^2(\langle\omega\rangle) = 2c_2(\langle\omega\rangle).$$

Especially, ξ^L has an $\widetilde{\Omega}\mathfrak{g}$-valued connection if and only if $c_1{}^2(\xi) = 2c_s(\xi)$.

12. The map L is also defined for real vector bundles (In this case, L is only defined as the map from $H^1(M, GL(\mathbb{R})_d) \to H^1(\Omega M_e, LGL(\mathbb{R})_d)$). We denote by TM the tangent bundle of M. Then we have

$$(30) \qquad (TM)^L = T(\Omega M_e).$$

Therefore, denoting by $T^{\mathbb{C}}M$ the complexification of TM, we get

$$(30)' \qquad (T^{\mathbb{C}}M)^L = T^{\mathbb{C}}(\Omega M_e).$$

By $(30)'$ and the Corollary of Theorem 4 we obtain (cf. [11], [15], [22], [26], [27])

Theorem 5. ΩM_e *has an* $\widetilde{\Omega}\mathfrak{g}$-*valued connection if and only if* $p_1(M) = 0$. *Here* $p_1(M)$ *means the first (rational) Pontrjagin class of* M.

Note. The condition that ΩM_e has an $\widetilde{\Omega}\mathfrak{g}$-valued connection is weaker than the condition $T^{\mathbb{C}}(\Omega M_e)$ comes from an $\widetilde{\Omega}G$-bundle. Since we are working in de Rham cohomology, torsion parts of cohomology classes are ignored.

Theorem 4 shows that the $c_p(\langle w \rangle)$ are recovered from $\langle w \rangle^L$ if $p \geq 2$. We can also recover $c_1(\langle w \rangle)$ from $\langle w \rangle^L$ as follows: We denote by ΩG_0 the connected component of the identity of ΩG. The sheaf of germs of smooth ΩG_0-valued functions is denoted by $\Omega G_{0,d}$ and set $\rho(\Omega G_{0,d}) = \mathcal{M}^1{}_{\Omega \mathfrak{g}_0}$. Then if we set

$$(31) \qquad \qquad \widetilde{r}(\theta) = \frac{1}{2\pi\sqrt{-1}} \int \mathrm{tr}(\theta)dt,$$

we have the following commutative diagram with exact lines:

$$
\begin{array}{ccccccccc}
0 & \longrightarrow & M^1_{\Omega\mathfrak{g}_0} & \xrightarrow{\ i\ } & \mathcal{M}^1{}_{\Omega\mathfrak{g}} & \xrightarrow{\ \widetilde{r}\ } & \mathbb{C}_t & \longrightarrow & 0 \\
& & \rho\uparrow & & \rho\uparrow & & \uparrow & & \\
0 & \longrightarrow & \Omega G_{0,d} & \xrightarrow{\ i\ } & \Omega G_d & \xrightarrow{\ r\ } & \mathbb{Z}_t & \longrightarrow & 0.
\end{array}
$$

By this diagram, we obtain the following commutative diagram of cohomology sets:

$$
\begin{array}{ccccc}
H^1(M, \mathcal{M}^1{}_{\Omega\mathfrak{g}_0}) & \xrightarrow{\ i^*\ } & H^1(M, \mathcal{M}^1{}_{\Omega\mathfrak{g}}) & \xrightarrow{\ \widetilde{r}^*\ } & H^1(M, \mathbb{C}) \\
\uparrow & & \uparrow & & \uparrow \\
H^1(M, \Omega G_{0,d}) & \xrightarrow{\ i^*\ } & H^1(M, \Omega G_d) & \xrightarrow{\ \widetilde{r}^*\ } & H^1(M, \mathbb{Z}).
\end{array}
$$

Therefore we can define characteristic classes $\widetilde{r}^*(\langle w \rangle) \in H^1(M, \mathbb{C})$ and $R^*(\xi) \in H^1(M, \mathbb{Z})$ for $\langle w \rangle \in H^1(M, \mathcal{M}^1{}_{\Omega\mathfrak{g}})$ and $\xi \in H^1(M, \Omega G_d)$.

Theorem 4'. (i) $\widetilde{r}^*(\langle w \rangle)$ *is equal to* 0 *if and only if* $\langle w \rangle$ *is in the* i^*-*image*.
   (ii) $\widetilde{r}^*(\rho(\xi))$ *is an integral class*.
   (iii) *Let* $\tau^{-1} : H^2(M, \mathbb{C}) \to H^1(\Omega M_e, \mathbb{C})$ *be the inverse of the transgression. Then we have*

$$(29)' \qquad\qquad \widetilde{r}^*\left(\langle w \rangle^L\right) = 2\pi\sqrt{-1}\tau^{-1}(c_1(\langle w \rangle)).$$

Proof. (i) and (ii) follow from the definition. If $\{\omega_{ij}\}$ is a representing cocycle of $\langle \omega \rangle \in H^1(M, \mathcal{M}^1)$, $\{\operatorname{tr} \omega_{ij}\}$ represents $c_1(\langle \omega \rangle)$ in $H^1(M, \Phi^1)$. Hence we have $(29)'$ by (30) and the definition of τ^{-1}.

Theorems 4 and 4' show that $L : H^1(M, G_d) \rightarrow H^1(\Omega M_e, \Omega G_d)$ is injective if M is torsionfree. At this stage we do not know whether L is injective or not in general.

§5. The relation between β-classes and string classes and the Bott map in non-abelian de Rham theory

13. For an integrable form θ on M, we have defined its Chern–Simons type characteristic classes $\beta^p(\theta) \in H^{2p-1}(M, \mathbb{C})$ as the de Rham class of $(p-1)! / \left(2\pi\sqrt{-1} \right)^p (2p-1)! \operatorname{tr}(\theta^{2p-1})$ (cf. [7]). We know

$$(32) \qquad\qquad \beta^p(\theta) = g^*(e_p) \qquad \text{if } \theta = g^{-1} dg,$$

where e_p is the $(2p-1)$-th generator of $H^*(G, \mathbb{Z})$. By (32), $\beta^p(\theta)$ is equal to 0 if $g = e^f$ on M. Hence β-classes are defined as characteristic classes of the elements of $H^0(M, \mathcal{M}^1)/d^e(H^0(M, \mathfrak{g}_d))$ (cf. [7]).

If θ belongs to $H^0(M, \mathcal{M}^1)_0$, then its Gysin-image $\widetilde{\gamma}(\theta) \in H^1(M \times S^1, \mathcal{M}^1)$ is defined. By the definition of $\widetilde{\gamma}$, $\widetilde{\gamma}(\theta)|M \times (S^1 - \{0\})$ is trivial and it has a connection form $\{\zeta_1\}$ such that

$$\zeta_i(x, t) = d\left(h_i(x)^t \right) h_i(x)^{-t}, \qquad \text{on } U_i \times (S^1 - \{0\}) = U_i \times (0, 1).$$

Here $\theta = h_i^{-1} dh_i$ holds on U_i. Since $\|\theta_i\|$ is bounded on $U_i \times (0, 1)$, ζ_i defines a current on $U_i \times S^1$. Moreover, $\{\operatorname{tr}\left((d\zeta_i + \zeta_i \wedge \zeta_i)^p \right)\}$ defines a current on

$M \times S^1$, It is computed as follows:

$$[\operatorname{tr}(d\zeta_i + \zeta_i \wedge \zeta_i)^p](\Psi)$$

$$= \lim_{\varepsilon \to 0} \int_M \int_\varepsilon^{1-\varepsilon} (d(\operatorname{tr}(\zeta_i \wedge (d\zeta_i)^{p-1} + p\zeta_i{}^3 \wedge (d\zeta_i)^{p-2} + \cdots$$

$$+ p\zeta^{2p-1}) \wedge \Psi) + \operatorname{tr}(\zeta_i{}^{2p}) \wedge \Psi)$$

$$= \lim_{\varepsilon \to 0} \int_M \int_\varepsilon^{1-\varepsilon} d(\operatorname{tr}(\zeta_i{}^{2p-1}) \wedge \Psi$$

$$= \lim_{\varepsilon \to 0} \int_M (\operatorname{tr}(\zeta_i{}^{2p-1}(x, 1-\varepsilon)) \wedge \Psi(x, 1-\varepsilon) - \operatorname{tr}(\zeta_i{}^{2p-1}(x,\varepsilon)) \wedge \Psi(x,\varepsilon))$$

$$= \int_M (\operatorname{tr}\theta_i{}^{2p-1}) \wedge \Psi(x, 0).$$

As a collection of currents $\{\zeta_i\}$ gives a connection of $\tilde{\gamma}(\theta)$. Therefore the de Rham class of $\{\operatorname{tr}((d\zeta_i + \zeta_i \wedge \zeta_i)^p)\}$ (as a current) is $(2\pi\sqrt{-1})^p \, p! \operatorname{Ch}^p(\tilde{\gamma}(\theta))$. On the other hand, by the residue exact sequence ([4]), we have the following exact sequence

$$H^{2p-1}(M, \mathbb{C}) \xrightarrow{\delta_M} H^{2p}(M \times S^1, \mathbb{C}) \xrightarrow{i^*} H^{2p}(M \times (S^1 - \{0\}), \mathbb{C}).$$

By Künneth' formula, denoting e^p the generator of $H^p(S^1, \mathbb{C})$, we get

$$H^{2p})M \times S^1, \mathbb{C}) = H^{2p-1}(M, \mathbb{C}) \otimes e^1 \oplus H^{2p}(M, \mathbb{C}) \otimes e^0.$$

Therefore we obtain by the definitions of δ_M and i (cf. [4])

(33) $\delta_M : H^{2p-1}(M, \mathbb{C}) \cong H^{2p-1}(M, \mathbb{C}) \otimes e^1 \quad (\subset H^{2p}(M \times S^1, \mathbb{C})),$
  $i : H^{2p}(M, \mathbb{C}) \otimes e^0 \cong H^{2p}(M \times (S^1 - \{0\}), \mathbb{C}).$

Since $\delta_M(\langle\zeta\rangle)$ is represented by the current $T_{M,\zeta}$, $T_{M,\zeta}(\Psi) = \int_M \zeta(x) \wedge \Psi(x, 0)$, we get

(34) $$\operatorname{Ch}^p(\tilde{\gamma}(\theta)) = \frac{p!(p-1)!}{(2p-1)!} \delta_M(\beta^p(\theta)).$$

Theorem 6. *We have*

(35) $$\tilde{c}^p(\tilde{B}_0([\theta])) = -(2\pi\sqrt{-1})^{p+1} \frac{(2p+1)!}{(p+1)!} \beta^{p+1}(\theta).$$

Proof. Since $\widetilde{\gamma}(\theta) = \widetilde{B}_0([\theta])^{\natural}$ and δ_M is an isomorphism by (33), we have Theorem 6 by (34) and (23)′. ∎

Corollary. *If ξ is a loop group bundle and g is a G-valued function on M such that $\xi = B_0(g)$, then*

(36)
$$\widetilde{c}^p(\xi) = -\left(2\pi\sqrt{-1}\right)^{p+1} \frac{(2p+1)!}{(p+1)!} g^*(e_{p+1}).$$

Note. In [7], we defined maps

$$X : H^1(M, G_d) \to H^0(\Omega M_e, G_d)/\exp(H^0(\Omega M_e, \mathfrak{g}_d)),$$
$$\widetilde{X} : H^1(M, \mathcal{M}^1) \to H^0(\Omega M_e, \mathcal{M}^1)/d^e(H^0(\Omega M_e, \mathfrak{g}_d)).$$

$X(\xi)$ and $\widetilde{X}(\xi)$ are represented by representative functions of ΩM_e and $\widetilde{\Omega M_e}$, the universal covering space of ΩM_e, respectively ([7]). On the other hand, $(B_0)^{-1}(\xi^L) = g$ is given by

$$g = h_i(\gamma^2)(h_i(\gamma))^{-1}, \qquad \text{on } U_i,$$
$$\widetilde{g}_{ij}{}^L = h_i h_j{}^{-1}, \quad \widetilde{g}_{ij}{}^L(\gamma)(t) = g_{ij}(\gamma(t \bmod 1)).$$

Therefore we obtain

(37)
$$X(\xi) = (B_0)^{-1}(\xi^L), \qquad \widetilde{X}(\langle\omega\rangle) = (\widetilde{B}_0)^{-1}\left(\langle\omega\rangle^L\right).$$

14. We set $H_+ = \left\{\sum_{n>0} c_n e^{2n\pi\sqrt{-1}t}\right\}$ and $H_- = \left\{\sum_{n<0} c_n e^{2n\pi\sqrt{-1}t}\right\}$. Then we have $\widetilde{H}^1(S^1)_0 = H_+ \otimes H_-$. The algebra of all bounded linear operators on H_+ is denoted by $B(H_+)$. The ideal of all compact operators of $B(H_+)$ is denoted by $C(H_+)$. We also use the following notations (cf. [12]):

$$\mathrm{Cal} = B(H_+)/C(H_+),$$
$$F = F(H_+)/C(H_+), \qquad F(H_+) \text{ is the set of Fredholm operators.}$$

The connected component of the identity of F is denoted by F_0. By the imbedding of ΩG in GL_{res} the restricted general linear group on $\widetilde{H}^1(S^1)_0$ (cf. [23]), the $(1,1)$-component of $g \in \Omega G$ represents an element of F_0, that is, (1.1)-component of g has an inverse by a compact perturbation, if and only if $g \in \Omega G_0$. We also set

$$K = \{T \in GL(H_+)|\, T = I + C,\ C \text{ is a compact operator}\}.$$

K contains $GL(\infty) = \bigcup GL_n$ as a dense subgroup and the following sequence is exact:

$$0 \longrightarrow K \longrightarrow GL(H_+) \longrightarrow F_0 \longrightarrow 0.$$

Since $H^1(M, GL(H_+)) = \{0\}$ by a theorem of Kuiper ([17]), $\delta :$ $H^1(M, F_{0,d}) \to H^2(M, K_d)$ is injective by this sequence (for the definitions of δ and $H^2(M, K_d)$ cf. [5], [6]). Here $F_{0,d}$ and K_d are the sheaves of germs of smooth K and F_0 valued functions.

Definition 7. Let $q : H^1(M, G_{0,d}) \to H^1(M, F_{0,d})$ be the map induced by the projection to the $(1,1)$-component of th elements of ΩG_0 and $\delta_L :$ $H^1(M, F_{0,d}) \to H^2(M, K_d)_L$ the coboundary map in the left handed non-abelian cohomology sets (cf. [6]). Then we set

$$(38) \qquad\qquad B^1{}_L = \delta_L q.$$

$B^1{}_R$ is similarly defined.

We lift $B^1{}_R$ to be a map between non-abelian de Rham sets as follows: Let $\mathcal{M}^1{}_K$, $\mathcal{M}^1{}_{\mathfrak{gl}(H_+)}$ and $\mathcal{M}^1{}_{\mathrm{Cal}}$ be the image sheaves of K_d, $GL(H_+)_d$ and $F_{0,d}$ by $\rho : \rho(g) = g^{-1}dg$, and its induced map $\bar{\rho}$. We also set $\mathcal{M}^1{}_{K,L} = \rho_L(K_d)$, where $\rho_L(g) = (dg)g^{-1}$, and the induced map of q by $\tilde{q}$. Then we have the following commutative diagram:

$$
\begin{array}{ccccc}
H^1(M, \Omega G_{0,d}) & \xrightarrow{q} & H^1(M, F_{0,d}) & \xrightarrow{\delta} & H^2(M, K_d)_L \\
\rho \uparrow & & \bar{\rho} \uparrow & & \rho_L \uparrow \\
H^1(M, \mathcal{M}^1{}_{\Omega\mathfrak{g}}) & \xrightarrow{\tilde{q}} & H^1(M, \mathcal{M}^1{}_{\mathrm{Cal}}) & \xrightarrow{\delta_L} & H^2(M, \mathcal{M}^1{}_{K,L}).
\end{array}
$$

Hence if we define $\widetilde{B}^1{}_L : H^1(M, \mathcal{M}^1{}_{\Omega\mathfrak{g}}) \to H^2(M, \mathcal{M}^1{}_{K,L})$ by

$$(38)' \qquad\qquad \widetilde{B}^1{}_L = \rho_L \tilde{q},$$

we have the following commutative diagram:

$$
\begin{array}{ccc}
H^1(M, \mathcal{M}^1{}_{\Omega\mathfrak{g}}) & \xrightarrow{\widetilde{B}^1{}_L} & H^2(M, \mathcal{M}^1{}_{K,L}) \\
\rho \uparrow & & \rho_L \uparrow \\
H^1(M, \Omega G_{0,d}) & \xrightarrow{B^1{}_L} & H^2(M, K_d)_L.
\end{array}
$$

Definition 8. *We define the maps*

$$B_L : H^0(M, G_d)/\exp(H^0(M, \mathfrak{g}_d)) \to H^2(M, K_d)_L$$
$$\widetilde{B}_L : H^0(M, \mathcal{M}^1)_0/d^e(H^0(M, \mathfrak{g}_d)) \to H^2(M, \mathcal{M}^1{}_{K,L})$$

by

$$(39) \qquad\qquad B_L = B^1{}_L B_0, \qquad \widetilde{B}_L = \widetilde{B}^1{}_L \widetilde{B}_0.$$

Here B_0 means B_{0R}. B_R and $\widetilde{B}_R$ are similarly defined.

It seems that we may replace K by GL_∞ $(= GL(\infty))$ and $\mathcal{M}^1{}_K$ by $\mathcal{M}^1{}_\infty$ $(= \mathcal{M}^1{}_{\mathfrak{gl}(\infty)})$. If this is true, $\widetilde{B}_L$ maps (a subset of) the first non-abelian de Rham set of M into the (stable) third non-abelian de Rham set of M. Hence we may say $\widetilde{B}_L$ to be *the (left) Bott map* in non-abelian de Rham theory. In [6], to get a good de Rham correspondence in the third non-abelian de Rham theory, pairing of the elements of the right handed and left handed third non-abelian de Rham sets was considered. We expect that the meaning of this pairing will be clarified via $\widetilde{B}$ and the definition of $H^0(M, \mathcal{M}^1)_0$.

15. By using the Grassmannian model of ΩG ([23]) we define the maps

$$gr : H^0(M, \Omega G_d) \to H^1(M, G_{\infty,d}),$$
$$\Omega : H^1(M, G_d) \to H^0(M, G_{\infty,d})/H^0(M, G_{\infty,0,d}).$$

These maps are lifted as the maps

$$\widetilde{gr} : H^0(M, \mathcal{M}^1{}_{\Omega\mathfrak{g}}) \to H^1(M, \mathcal{M}^1{}_\infty),$$
$$\widetilde{\Omega} : H^1(M, \mathcal{M}^1) \to H^0(M, \mathcal{M}^1{}_{\Omega\mathfrak{g}_\infty})/H^0(M, \rho(\Omega G_{\infty,0,d})).$$

By using gr and Ω we define $\omega^b : H^1(M, \Omega G_d) \to H^1(\Omega M_e, G_{\infty,d})$ by

$$(40) \qquad\qquad \omega^b(\xi) = gr\left((B_0{}^{-1}(\xi))^L\right).$$

By   Theorem   1'   we   can   define   the   lift $\widetilde{\omega}^b$: $H^1(M, \mathcal{M}^1{}_{\Omega\mathfrak{g}})^\natural \to H^1(\Omega M_e, \mathcal{M}^1_\infty)$ of ω^b by

$$(40)' \qquad \widetilde{\omega}^b(\langle\omega\rangle) = \widetilde{gr}\left((\widetilde{B}_0{}^{-1}(\langle\omega\rangle))^L\right), \qquad H^1(M, \mathcal{M}^1{}_{\Omega\mathfrak{g}})^\natural = \operatorname{Im}\widetilde{B}_0.$$

Then by (2) and (35) we have

$$(41) \qquad \widetilde{c}^p(\langle\omega\rangle) = -\left(2\pi\sqrt{-1}\right)^{p+1} \frac{(2p+1)!}{p!(p+1)!} \tau^{-1}(\mathrm{Ch}^{p+1}(\widetilde{\omega}^b(\langle\omega\rangle))).$$

In conclusion, our results are summarized as *the trinity of β-classes (Chern–Simons classes), string classes and transgressed Chern classes* on the one hand, and the following two types of trinities of non-abelian de Rham sets (with characteristic classes) on the other hand.

(I). (a). *The first non-abelian de Rham set over M.*

(b). *The second non-abelian de Rham set over $M \times S^1$.*

(c). *The stable second non-abelian de Rham set over ΩM_e.*

From (a), (b) is mapped by the Gysin map and (c) is mapped by the inverse of transgression. Their composition is the evaluation map.

(II). (a). *The second non-abelian de Rham set over M.*

(b). *The stable first non-abelian de Rham set over $M \times S^1$.*

(c). *The first non-abelian de Rham set over ΩM_e.*

Note. In both cases trinities are not in the strict sense. In fact, we do not know whether the Gysin map, etc., are bijective or not.

(I) and (II) are visualized as the commutativity of the following diagrams:

$$
\begin{array}{ccc}
H^0(M, \mathcal{M}^1)_0/d^e(H^0(M, \mathfrak{g}_d)) & \xrightarrow{\ \widetilde{\gamma}\ } & H^1(M \times S^1, \mathcal{M}^1) \\
\downarrow & \searrow^{\widetilde{B}_0} & \uparrow \\
H^0(M, \mathcal{M}^1_\infty)/d^e(H^0(M, \mathfrak{g}_{\infty,d})) & & H^1(M, \mathcal{M}^1_{\Omega\mathfrak{g}})^\natural \\
\widetilde{L}\downarrow & & \uparrow\widetilde{\omega}^b \\
H^0(M_e, \mathcal{M}^1_{\Omega\mathfrak{g}})/d^e(\Omega M_e, \rho(\Omega G_{0,d})) & \xrightarrow{\ \widetilde{gr}\ } & H^1(\Omega M_e, \mathcal{M}^1_\infty),
\end{array}
$$

$$
\begin{array}{ccc}
H^1(M, \mathcal{M}^1) & \longrightarrow & H^0(M \times S^1, \mathcal{M}^1_\infty)/ \\
 & & d^e(H^0(M \times S^1, \mathfrak{g}_{\infty,d})) \\
\downarrow & \searrow & \uparrow \\
H^0(\Omega M_e, \mathcal{M}^1)/d^e(H^0(\Omega M_e, \mathfrak{g}_d)) & & H^0(M, \mathcal{M}^1_{\Omega\mathfrak{g},\infty})/ \\
 & & H^0(M, \rho(\Omega G_{\infty,0,d})) \\
\widetilde{B}_0\updownarrow & & \downarrow \\
H^1(\Omega M_e, \mathcal{M}^1_{\Omega\mathfrak{g}})^\natural & \longrightarrow & H^0(\Omega M_e, \mathcal{M}^1_\infty)/ \\
 & & d^e(H^0(\Omega M_e, \mathfrak{g}_{\infty,d})).
\end{array}
$$

A. ASADA

Note. The Bott map relates the third non-abelian de Rham set and the first non-abelian de Rham set. Hence we may regard the third non-abelian de Rham theory to be a gauge theory on ΩM_e (Loop gauge theory) or on $M \times S^1$ (cf. [8]).

References

[1]. ABE, K., On tortal torsion and a generic property of closed regular curves in Riemannian manifolds, Preprint.

[2]. ALVAREZ, O., KILLINGBACK, T. P., MANGANO, M. and WINDEY, P., String theory and loop space index theorems, *Commun. Math. Phys.,* bf 111 (1987), 1–10.

[3]. ANDERSSON, S. I., Pseudodifferential operators and characteristic classes for non-abelian cohomology, *Lect. Notes in Math.,* **1045** (1984), 1–10.

[4]. ASADA, A., Currents and residue exact sequences, *Journ. Fac. Sci. Shinshu univ.,* **3** (1968), 85–151.

[5]. ASADA, A., Non abelian de Rham theories, Coll. Math. Soc. János Bolyai 46, Topics in Differential Geometry, 83–115, North-Holland, 1988.

[6]. ASADA, A., Non abelian de Rham theory, Proc. Prospects of Math. Sci., World Sci., 1988. 13–40.

[7]. ASADA, A., Integrable forms on iterrated loop spaces and higher dimensional non abelian de Rham theory, *Lect. Notes in Math.,* **1910** (1989), 27–51.

[8]. AWADA, M. A., The exact equivalence of Chern–Simons theory with fermionic string theory, *Phys. Lett.,* **B221** (1989), 21–26.

[9]. BONOLA, L., COTTA–RAMUSIANO, P., RINALDI, M. and STASHEFF, J., The evaluation map in field theory, sigma-models and strings, I, II, *Commun. Math. Phys.,* **112** (1987), 237–282, **114** (1988), 381–437.

[10]. BOTT, R., The space of loops on a Lie group, *Michigan Math. Journ.,* **5** (1958), 35–61.

[11]. COQUEREAUX, R. and PILCH, K., String structures on loop bundles, *Commun. Math. Phys.,* **120** (1989), 353–378.

[12]. DOUGLAS. R. G., Banach Algebra Techniques in Operator Theory, Academic Press, 1972.

[13]. FLORATOS, E. G., ILIOPOULOS, J. and TIKTOPOULOS, G., A note on $SU(\infty)$ classical Yang–Mills theories, *Phys. Lett.,* **B217** (1989), 285–288.

[14]. GYSIN, W., Zur Homologie Theorie des Abbildungen und Faserungen von Mannigfaltigkeiten, *Commet. Math. Helv.,* **14** (1941), 61–121.

[15]. KILLINGBACK, T. P., World-sheet anomalies and loop geometry, *Nucl. Phys.*, **B288** (1987), 578–588.

[16]. KILLINGBACK, T. P., Quantum Chern–Simons theory, *Phys. Lett.*, **B219** (1989), 448–456.

[17]. KUIPER, N. H., The homotopy type of the unitary group of Hilbert space, *Topology* **3** (1965), 19–30.

[18]. LASHOF, R., Classification of fibre bundles by the loop space of the base, *Ann. Math.*, **64** (1956), 436–446.

[19]. MENSKY, M. B., Group of paths, Observations, Fields and Particles, Moscow, 1983 (in Russian).

[20]. MICKELSSON, J. and RAJEEV, S. G., Current algebras in $D + 1$-dimensions and determinant bundles over infinite-dimensional Grassmannians *Commun. Math. Phys.*, **116** (1988), 365–400.

[20]'. TANAKA, M. and FUJI, K., Universal Schwinger cocycles of current algebras in $(D + 1)$-dimensions, Preprint.

[21]. MILNOR, J., Construction of universal bundles, I. *Ann. Math.*, **63** (1956), 272–284.

[22]. PILCH, K. and WARNER, N. P., String structure and the index of the Dirac–Ramond operator on orbifolds, *Commun. Math. Phys.*, **115** (1988), 191–212.

[23]. PRESSLEY, A. and SEGAL, G., Loop Groups, Oxford, 1986.

[24]. RAJEEV, S. G., An exactly integrable algebraic model for $(3 + 1)$-dimensional Yang–Mills theory, *Phys. Lett.*, **209** (1988), 53–58.

[24]'. TANAKA, M. and FUJII, K., Note on algebraic analogue of Yang–Mills–Higgs theory, Preprint.

[25]. TAUBES, C. H., S^1 actions and elliptic genera, *Commun. Math. Phys.* **122** (1989), 455–526.

[26]. VAFA, C., Modular invariance and discrete torsion on orbifolds, *Nucl. Phys.*, **B273** (1986), 592–606.

[27]. WITTEN, E., Global anomalies in string theory, *Anomalies, Geometry and Topology*, 61–99, World Sci., 1985.

[28]. WITTEN, E., Quantum field theory and the Jones polynomial, *Commun. Math. Phys.*, **121** (1989), 351–399.

Added in proof. Prof. Michor kindly remarked:

(i) By his result, we need not the *Convention*.

(ii) Lemma 3 occurs in Varadarajan's book Lie groups, Lie algebras and their Representations.

Akira Asada

Department of Mathematics
Faculty of Science
Sinsyu University
Matumoto
Nagano Pref.
Japan

COLLOQUIA MATHEMATICA SOCIETATIS JÁNOS BOLYAI
56. DIFFERENTIAL GEOMETRY, EGER (HUNGARY), 1989

Eisenhart's Problem on the Cotangent Bundle[*]

Gh. ATANASIU

In order to obtain a unified field theory, A. Einstein [4] started from a nonsymmetric fundamental tensor. If a field is described by a (complex) tensor g_{ij} with Hermitian symmetry $g_{ij} = \overline{g}_{ji}$, we have a real symmetric tensor $g_{\underline{ij}}$ and a real alternate one $g_{\overline{ij}}$ from the splitting $g_{ij} = g_{\underline{ij}} + \sqrt{-1}g_{\overline{ij}}$.

On the other hand, L. P. Eisenhart [5] treated a generalized Riemannian space, that is, a space M associated with a nonsymmetric tensor $g_{ij}(x)$, and tried to solve the problem of finding the set of all linear connections compatible with such a space: $g_{ij|k} = 0$. In natural case this problem has been solved by R. Miron and Gh. Atanasiu [7] and for generalized Finsler spaces by Gh. Atanasiu, R. Miron and M. Hashiguchi [2] in their joint paper.

In this paper Eisenhart's problem is solved on the cotangent bundle in a particular case. The fundamental notions and notations concerning the cotangent bundle are all of R. Miron [6].

---

[*] This paper is in final form and no version of it will be submitted for publication elsewhere.

1. Eisenhart's problem on T^*M

Let $(x, p) = (x^i, p_i)$ $(i = 1, \ldots, n)$ be a point of the cotangent bundle (T^*M, π^*, M) of the C^∞-manifold $M, V = \operatorname{Ker} \pi^*$ the vertical distribution and $N = (N_{ij}(x, p))$ a nonlinear connection: $T_{(x,p)} T^*M = N_{(x,p)} \oplus V_{(x,p)}$.

We denote by $F^*\Gamma(\check{N}) = (F_j{}^i{}_k(x, p), C_i{}^{jk}(x, p))$ a d-connection (distinguished-connection, see [6]) on the cotangent bundle of M, having a fixed nonlinear connection $\overset{\circ}{N}$.

We investigate the d-connections for which we have

(1) $$g^{ij}{}_{|k} = 0, \qquad g^{ij}|^k = 0,$$

where

(2) $$g^{ij}(x, p) = g^{\underline{ij}}(x, p) + g^{\overline{ij}}(x, p)$$

is a d-tensor field of type (0,2) for which the symmetric part $g^{\underline{ij}}(x, p)$, (i.e. $g^{\underline{ij}} = g^{\underline{ji}}$) and respectively, the skew-symmetric part $g^{\overline{ij}}(x, p)$ (i.e. $g^{\overline{ij}} = -g^{\overline{ji}}$) are nondegenerate:

(3) $$\det \|g^{\underline{ij}}(x, p)\| . \|g^{\overline{ij}}(x, p)\| \neq 0.$$

We remember that

$$g^{ij}{}_{|k} = \delta_k g^{ij} + F_r{}^i{}_k g^{rj} + F_r{}^j{}_k g^{ir},$$
$$g^{ij}|^k = \dot{\partial}^k g^{ij} + C_r{}^{ik} g^{rj} + C_r{}^{jk} g^{ir},$$

where $\delta_i, \dot{\partial}^i$ is the local basis adapted to N and V

(4) $$\delta_i = \partial_i + N_{ji}(x, p)\dot{\partial}^j \qquad \left(\partial_i = \frac{\partial}{\partial x^i}, \quad \dot{\partial}^i = \frac{\partial}{\partial p_i} \right).$$

We denote

$$\|g^{\underline{ij}}(x, p)\|^{-1} = \|g_{\underline{ij}}(x, p)\|, \qquad \|g^{\overline{ij}}(x, p)\|^{-1} = \|g_{\overline{ij}}(x, p)\|,$$

and for a d-connection, which solves the Eisenhart problem on T^*M, we have from (1) the following equations:

(5) $$g^{\underline{ij}}{}_{|k} = 0, \qquad g^{\underline{ij}}|^k = 0, \qquad g^{\overline{ij}}{}_{|k} = 0, \qquad g^{\overline{ij}}|^k = 0,$$

which is equivalent to

(6) $\qquad g_{ij|\underline{k}} = 0, \qquad g_{ij}|^{\underline{k}} = 0, \qquad g_{\overline{ij}|\underline{k}} = 0, \qquad g_{\overline{ij}}|^{\underline{k}} = 0.$

We investigate the set of all d-connections $F^{*}\overset{\circ}{\Gamma}(N)$ for which we have (6) in the form:

(7) $\qquad F_{j}{}^{i}{}_{k} = \overset{\circ}{F}_{j}{}^{i}{}_{k} + B_{j}{}^{i}{}_{k}, \qquad C_{i}{}^{jk} = \overset{\circ}{C}_{i}{}^{jk} + D_{i}{}^{jk}$

where $F^{*}\overset{\circ}{\Gamma}(N) = (\overset{\circ}{F}_{j}{}^{i}{}_{k}(x,p), \overset{\circ}{C}_{i}{}^{jk}(x,p))$ is a fixed d-connection and $B_{j}{}^{i}{}_{k}(x,p)$, $D_{i}{}^{jk}(x,p)$ are arbitrary d-tensor fields of type (1,2) and (2,1) respectively.

We obtain for B and D the equations:

(8) $\qquad B_{i}{}^{r}{}_{k}g_{\underline{rj}} + B_{j}{}^{r}{}_{k}g_{\underline{ir}} = g_{\underset{ij}{\circ}|k}, \qquad D_{i}{}^{rk}g_{\underline{rj}} + D_{j}{}^{rk}g_{\underline{ir}} = g_{\underline{ij}}|^{\overset{\circ}{k}},$

(9) $\qquad B_{i}{}^{r}{}_{k}g_{\overline{rj}} + B_{j}{}^{r}{}_{k}g_{\overline{ir}} = g_{\underset{\overline{ij}}{\circ}|k}, \qquad D_{i}{}^{rk}g_{\overline{rj}} + D_{j}{}^{rk}g_{\overline{ir}} = g_{\overline{ij}}|^{\overset{\circ}{k}}.$

We do not know the general solution of the equation system (8) and (9). This is an open problem.

We give a solution for these equations in the following special case.

2. Solution of Eisenhart's problem on the cotangent bundle in the natural case

We shall here pay attention to a remarkable class.

Definition 1. An asymmetric metric $g^{ij}(x,p) = g^{\underline{ij}}(x,p) + g^{\overline{ij}}(x,p)$ is called *natural* if there is exist a function $\mu(x,p)$ such that

(10) $\qquad\qquad\qquad g_{\underline{ir}}g_{\underline{js}}g^{\overline{rs}} = \mu g_{\overline{ij}}.$

Examples.

1. In a Hamiltonian space $H^n = (M, H)$ (see [6]), where $H : T^* M \to R$ is the fundamental function of H^n, let $f_j^i(x, p)$ be a d-tensor field of type (1,1) which gives an almost complex d-structure: $f^2 = -\delta$. If we put

   (11) $$g_{\overline{i}j} = f_i^r g_{rj},$$

   where $\|g_{ij}(x, p)\| = \|g^{ij}(x, p)\|^{-1}$, and $g^{ij}(x, p) = \frac{1}{2} \dot{\partial}^i \dot{\partial}^j H$ is the fundamental metric of a Hamiltonian space, then $g_{\overline{i}\overline{j}}(x, p)$ is alternating, and $g_{ij} = g_{ij} + g_{\overline{i}\overline{j}}$ is an asymmetric metric on the cotangent bundle. In this case $\mu = -1$.

2. In a Hamiltonian space $H^n = (M, H)$, let $q_j^i(x, p)$ be a d-tensor field of type (1,1), which gives an almost product d-structure: $q^2 = -\delta$. If we put

   (12) $$g_{\overline{i}j} = q_i^r g_{rj}$$

where $\|g_{ij}(x, p)\| = \|g^{ij}(x, p)\|^{-1}$, and $g^{ij}(x, p) = \frac{1}{2} \dot{\partial}^i \dot{\partial}^j H$ is the fundamental metric of a Hamiltonian space, then $g_{\overline{i}\overline{j}}(x, p)$ is alternate, and $g_{ij} = g_{ij} + g_{\overline{i}\overline{j}}$ is an asymmetric metric on the cotangent bundle. In this case $\mu = +1$. This structure has been studied in [3].

Theorem 1. *If there exists a d-connection compatible with a natural asymmetric metric, then the function μ is constant.*

Definition 2. A natural asymmetric metric $g^{ij}(x, p) = g^{ij}(x, p) + g^{\overline{i}\overline{j}}(x, p)$ is called *elliptic* if $\mu = -c^2$ and *hyperbolic* if $\mu = c^2$, where c is a positive constant.

The elliptic case is reduced to Example 1 by putting $f_j^i = cg^{ir} g_{\overline{r}j}$ and the hyperbolic case is reduced to Example 2 by putting $q_j^i = cg^{ir} g_{\overline{r}j}$. Also the converse of Theorem 1 holds as follows:

Theorem 2. *If a natural asymmetric metric $g^{ij}(x, p) = g^{ij}(x, p) + g^{\overline{i}\overline{j}}(x, p)$ is elliptic or hyperbolic, then there exist d-connections $F^* \widetilde{\Gamma} = (\widetilde{N}_{ij}, \widetilde{F}_j{}^i{}_k, \widetilde{C}_i{}^{jk})$ compatible with $g^{ij}(x, p)$. Let $F^* \overset{\circ}{\Gamma} = (\overset{\circ}{N}_{ij}, \overset{\circ}{F}_j{}^i{}_k, \overset{\circ}{C}_i{}^{jk})$ be a given d-connection. Then $\widetilde{N}_{ij} = \overset{\circ}{N}_{ij}$, and in the elliptic case*

$$\widetilde{F}_j{}^i{}_k = \overset{\circ}{F}_j{}^i{}_k + \frac{1}{4} \left\{ g^{ir} g_{\underset{rj}{\circ}\,|\,k} + g^{\overline{i}r} g_{\underset{\overline{r}j}{\circ}\,|\,k} + f_j^r f^i_{\underset{r}{\circ}\,|\,k} \right\},$$

$$\widetilde{C}_i{}^{jk} = \overset{\circ}{C}_i{}^{jk} - \frac{1}{4} \left\{ g_{ir} g^{rj}{}_{\overset{\circ}{|}}{}^k + g_{\overline{i}r} g^{\overline{r}j}{}_{\overset{\circ}{|}}{}^k + f_r^j f_i^r{}_{\overset{\circ}{|}}{}^k \right\}.$$

In the hyperbolic case in these espressions $f_j^r f^i \underset{r\,|\,k}{\overset{\circ}{}}$, $f_r^j f_i^r \, |^k$ are replaced

by $-q_j^r q^i \underset{r\,|\,k}{\overset{\circ}{}}$, $-q_r^j q_i^r \, |^k$.

Now, for an asymmetric metric $g^{ij}(x,p)$ we can determine all solutions of Eisenhart's problem on T^*M. We put

(13) $\qquad \Lambda_{ij}^{kh} = 1/2(\delta_i^k \delta_j^h - g_{\underline{ij}} g^{\underline{kh}}), \qquad \Phi_{ij}^{kh} = 1/2(\delta_i^k \delta_j^h - g_{\overline{ij}} g^{\overline{kh}}).$

It is noted that an asymmetric metric g^{ij} is natural if and only if

(14) $\qquad\qquad\qquad\qquad \Lambda_{is}^{rk} \Phi_{rj}^{hs} = \Phi_{is}^{rk} \Lambda_{rj}^{hs}.$

Theorem 3. *The set of all d-connections $F^*\Gamma(\widetilde{N})$ compatible with an asymmetric metric $g^{ij}(x,p) = g^{\underline{ij}}(x,p) + g^{\overline{ij}}(x,p)$ is given by*

$$N_{ij} = \widetilde{N}_{ij}, \qquad F_j{}^i{}_k = \widetilde{F}_j{}^i{}_k + \Lambda^{qi}{}_{jp} \Phi^{rp}{}_{qs} Y^s{}_{rk},$$
$$C_i{}^{jk} = \widetilde{C}_i{}^{jk} + \Lambda^{jp}{}_{qi} \Phi^{qs}{}_{rp} Z_s{}^{rk}$$

where $F^*\widetilde{\Gamma}$ are the d-connections given in Theorem 2, Y_{jk}^i and Z_i^{jk} are arbitrary d-tensor fields.

There are some important particular cases. For example, if the symmetric part $g^{\underline{ij}}(x,p)$ of $g^{ij}(x,p)$ is given by $g^{\underline{ij}}(x,p) = \frac{1}{2}\dot\partial^i \dot\partial^j H$, where $H : T^*M \to R$ is the fundamental function of a Hamiltonian space $H^n = (M,H)$, we obtain a remarkable d-connection $F^*\overset{*}{\Gamma}$ compatible with the natural asymmetric metric $g^{ij}(x,p)$. This d-connection $F^*\overset{*}{\Gamma}$ can be called *canonical* for $g^{ij}(x,p)$ and is given by

Theorem 4. *The canonical d-connection $F^*\overset{*}{\Gamma}$ compatible with a natural asymmetric metric $g^{ij}(x,p)$ is given in the elliptic case by*

$$\overset{*}{N}_{ij} = \overset{m}{N}_{ij}, \qquad \overset{*}{F}_k{}^i{}_k = \overset{m}{F}_j{}^i{}_k + \frac{1}{4}\left\{ g^{i\overline{r}} g_{\underset{\overline{rj}\,|\,k}{}}{}^m + f_j^r f^i{}_{\underset{r\,|\,k}{}}{}^m \right\}$$

$$\overset{*}{C}_i{}^{jk} = \overset{m}{C}_i{}^{jk} - \frac{1}{4}\left\{ g^{i\overline{r}} g^{\overline{rj}}{}^m{}_{|}{}^k + f_r^j f j_i^r{}^m{}_{|}{}^k \right\}.$$

In the hyperbolic case in the formula for $\overset{*}{F}$, $f_j^r f^i \underset{r\,|\,k}{\overset{m}{}}$ is replaced by

$-q_j^r q^i \underset{r\,|\,k}{\overset{m}{}}$ and similarly in the formula for $\overset{*}{C}$, $f_r^j f_i^r{}^m{}_{|}{}^k$ is replaced by

$-q_r^j q_i^r \mid {}^k$. Here $M^*\Gamma = (\overset{m}{N}{}_{ij}, \overset{m}{F}{}_j{}^i{}_k, \overset{m}{C}{}_i{}^{jk})$ is the Miron metrical d-connection [3], [6]:

$$\overset{m}{N}{}_{ij} = -1/2 g_{jh}\{1/4 g_{ik}\dot{\partial}^k(H, \dot{\partial}^h H) + \dot{\partial}^h \partial_i H\}$$

$$\overset{m}{F}{}_j{}^i{}_k = 1/2 g^{ih}\{\delta\, g_{hj} + \delta_j g_{hk} - \delta_h g_{jk}\}$$

$$\overset{m}{C}{}_j{}^i{}_k = -1/2 g_{ih}\{\dot{\partial}g^{hj} + \dot{\partial}^j g^{hk} - \dot{\partial}^k g^{jk}\}$$

where $(f, g) = \partial_i f \dot{\partial}^i g - \dot{\partial}^i f \partial_i g$ is the Poisson bracket.

Remarks

1. This method can be applied also in the asymmetric case

    (a) $\det \|g^{ij}_{\underline{}}(x, p)\| \neq 0$,

    (b) $\operatorname{rank} g^{\overline{ij}}(x, p) = \dim M - k = 2p$.

2. This method is also true for the case

    (a) $\gamma^{ij}(x, p) = g^{ij}_{\underline{}}(x, p) + G^{ij}_{\underline{}}(x, p)$,

    (b) $\det \|g^{ij}_{\underline{}}(x, p)\| \cdot \|G^{ij}_{\underline{}}(x, p)\| \neq 0$,

    (c) $G^{ij}_{\underline{}}(x, p) = G^{ji}_{\underline{}}(x, p)$

analogously to the Finsler case studied by A. K. Arangizim and G. S. Asanov in [1].

References

[1] A. K. ARINGAZIM and G. S. ASANOV, Problems of Finsler theory of gauge fields and gravitation. *Rep. of Math. Physics* **2** (1988), 183–241.

[2] GH. ATANASIU, R. MIRON and M. HASHIGUCHI, Super-generalized Finsler spaces. *Rep. Fac. Sci. Kagoshima Univ.* **18** (1985), 1–16.

[3] GH. ATANASIU and F. C. KLEPP, Metrical almost product structures on the cotangent bundles. Colloquium on Differential Geometry, Eger, Hungary, August 20–26, 1989.

[4] A. EINSTEIN, A generalization of the relavistic theory of gravitation. *Ann. of Math.* **46** (1945), 578–584.

[5] L. P. EISENHART, Generalized Riemann spaces I, II. *Proc. Nat. Acad. Sci. USA*, **37** (1951) 311–315, **38** (1952), 505–508.

[6] R. MIRON, The geometry of Hamiltonian spaces. Proc. of the fifth Nat. Sem. on Finsler and Lagrange Spaces, Univ. Brasov, (1988) 249–278.

[7] R. Miron and Gh. Atanasiu, Existence at arbitrariété des connexions compatibles à une structure Riemann généralisée du type presque k-horsyplectique metrique. *Kodai Math. J.* **6** (1983), 228–237.

Gheorghe Atanasiu

Department of Mathematics
University of Brasov
2200 Brasov
Romania

COLLOQUIA MATHEMATICA SOCIETATIS JÁNOS BOLYAI
56. DIFFERENTIAL GEOMETRY, EGER (HUNGARY), 1989

Metrical Almost Product Structures on the Cotangent Bundle*

Gh. ATANASIU and F. C. KLEPP

Metrical structures on the cotangent bundle have been studied by R. Miron in connection with the geometry of Hamilton spaces [8], [9], and the almost product structures on the cotangent bundle have been studied by the authors in their paper [4].

Here, we introduce the notion of metrical almost product structure on the cotangent bundle and we investigate this structure starting from the similar notion on the base manifold. The Finsler case of these problems has been considered in [3], [4], [5], [6].

The terminology and notions of R. Miron [8], [9] are used with small modifications of the notations from [2].

---

* This paper is in final form and no version of it will be submitted for publication elsewhere.

§1. Distinguished geometrical objects on the cotangent bundle. Generalized Hamilton spaces. Hamilton spaces.

Let M be a real differentiable manifold of the class C^∞, with $\dim M = n$ and (T^*M, π^*, M) its cotangent bundle. Since a point of T^*M is a covector (x, p) in the point x of the base manifold M, a coordinate system $x = (x^i)$ on M induces a canonic coordinate system $(x, p) = (x^i, p_i)$ in T^*M by means of the Liouville 1-form $\widetilde{p} = p_i dx^i$. A local coordinate transformation on T^*M is given by

(1.1)
$$\begin{cases} \overline{x}^i = \overline{x}^i(x^1, x^2, \ldots, x^n), \quad \det \left\| \dfrac{\partial \overline{x}^i}{\partial x^j} \right\| \neq 0, \\[3mm] \overline{p}^i = \left(\dfrac{\partial x^i}{\partial \overline{x}^j} \right) p_j. \end{cases}$$

Let $\operatorname{Ker} \pi^* = V : (x, p) \in T^*M \to V_{(x,p)} \subset T_{(x,p)} T^*M$ be the vertical distribution on T^*M. Denoting by $\partial_i = \partial/\partial x^i$; $\dot{\partial}^i = \partial/\partial p_i$ the local natural basis of the module of the vector fields $\mathfrak{X}(T^*M)$, we observe that $(\dot{\partial}^i)$ is a local basis of V. A nonlinear connection N on T^*M is a distribution of the class C^∞ given by $(x, p) \in T^*M \to N_{(x,p)} \subset T_{(x,p)} T^*M$ such that $T_{(x,p)} T^*M = N_{(x,p)} \oplus V_{(x,p)}$. N will be called a horizontal distribution on T^*M and is characterized by $\delta_i = \partial_i + N_{ji}(x, p)\dot{\partial}^j$. The functions $N_{ij}(x, p)$ are the coefficients of the non-linear connection N. Then $(\delta_i, \dot{\partial}^i)$ is a local basis of the $\mathcal{F}(T^*M)$ module of the vector fields $\mathfrak{X}(T^*M)$ adapted to the supplementary distributions N and V. Its dual basis is $(dx^i, \delta p_i)$, where $\delta p_i = dp_i - N_{ij}(x, p)dx^j$.

A vector field $X \in \mathfrak{X}(T^*M)$ is uniquely written in the form $X = X^h + X^v$, $X^h \in N$, $X^v \in V$ and an 1-form ω is given by $\omega = \omega^h + \omega^v$; $\omega^h(X^v) = 0$; $\omega^v(X^h) = 0$.

A tensor field $t \in \mathfrak{T}^r_s(T^*M)$ is called distinguished on M (shortly a d-field) if $t(\omega_1, \ldots, \omega_r, X_1, \ldots, X_s) = 0$ for $\omega_i = \omega^h_i$ or $\omega_i = \omega^v_i$ ($i = \overline{1, r}$) and $X_j = X^h_j$ or $X_j = X^v_j$ ($j = \overline{1, s}$). For example X^h, X^v are d-vector fields and ω^h, ω^v are d-fields of 1-forms. The components of a tensor on T^*M determine d-tensor fields on M. Reciprocally a d-tensor field on M, i.e. a tensor field dependent on x and p, which relative to (1.1) has the classical transformation law, can be interpreted as a tensor field on T^*M. For example, if $R_{ijk} = -\{\delta_k N_{ij} - \delta_j N_{ik}\}$, then $R^1 = R_{ijk} dx^i \otimes dx^j \otimes dx^k$ is a tensor field of type $(0, 3)$ on T^*M.

We denote:

$$(1.2) \qquad \theta = \delta p_i \wedge dx^i.$$

A linear connection D on T^*M is called *distinguished* (shortly a *d*-connection), if $(D_X Y^h)^v = 0$; $(D_X Y^v)^h = 0$ and $D_X \theta = 0 \ \forall X, Y \in \mathfrak{X}(T^*M)$. We put: $D_X^h = D_{X^h}$; $D_X^v = D_{X^v}$, D^h and D^v being the h- and v-covariant derivatives in the algebra of d-tensor fields $T_d(T^*M) \subset T(T^*M)$.

A d-connection D on T^*M generates a triad $F^*\Gamma = (N_{jk}, F_j{}^i{}_k, C_i{}^{jk})$, which is called a d-connection on M, where [5]

$$D_{\delta_k} \delta_j = F_j{}^i{}_k \delta_i; \qquad D_{\dot{\partial}^k} \delta_i = C_i{}^{jk} \delta_j;$$
$$D_{\delta_k} \dot{\partial}^i = -F_j{}^i{}_k \dot{\partial}^j; \qquad D_{\dot{\partial}^k} \dot{\partial}^j = -C_i{}^{jk} \dot{\partial}^i.$$

The transformation formulas of N_{jk}, $F_j{}^i{}_k$, $C_i{}^{jk}$ relative to (1.1) are the following

$$(1.3) \qquad \overline{N}_{jk} = \left(\frac{\partial x^s}{\partial \overline{x}^j}\right)\left(\frac{\partial x^t}{\partial \overline{x}^k}\right) N_{st} + p_r\left(\frac{\partial^2 x^r}{\partial \overline{x}^j \partial \overline{x}^k}\right),$$

$$(1.4) \qquad \overline{F}_j{}^i{}_k = \left(\frac{\partial \overline{x}^i}{\partial x^r}\right)\left(\frac{\partial x^s}{\partial \overline{x}^j}\right)\left(\frac{\partial x^t}{\partial \overline{x}^k}\right) F_s{}^r{}_t + \left(\frac{\partial \overline{x}^i}{\partial x^r}\right)\left(\frac{\partial^2 x^r}{\partial \overline{x}^j \partial \overline{x}^k}\right),$$

$$(1.5) \qquad \overline{C}_i{}^{jk} = \left(\frac{\partial x^r}{\partial \overline{x}^i}\right)\left(\frac{\partial \overline{x}^j}{\partial x^s}\right)\left(\frac{\partial \overline{x}^k}{\partial x^t}\right) C_r{}^{st}.$$

Given a d-connection, the h- and v-covariant derivatives for a d-tensor field, e.g. K_j^i are

$$(1.6)$$
$$K^i_{j|k} = \delta_k K^i_j + K^r_j F_r{}^i{}_k - K^i_r F_j{}^r{}_k; \qquad K_i^j|^k = \dot{\partial}^k K_i^j + K_i^r C_r{}^{jk} - K_r^j C_i{}^{rk}.$$

The Ricci formulas are

$$(1.7) \quad \begin{cases} K^i_{j|k|\ell} - K^i_{j|\ell|k} = K^r_j R_r{}^i{}_{k\ell} - K^i_r R_j{}^r{}_{k\ell} - K^i_{j|r} T_k{}^r{}_\ell - K^i_j|^r R_{rk\ell}, \\ K^i_{j|k}|^\ell - K^i_j|^\ell{}_{|k} = K^r_j P_r{}^i{}_k{}^\ell - K^i_r P_j{}^r{}_k{}^\ell - K^i_{j|r} C_k{}^{r\ell} - K^i_j|^r P_{rk}\ell, \\ K^i_j|^k|^\ell - K^i_j|^\ell|^k = K^r_j S_r{}^{ik\ell} - K^i_r S_j{}^{rk\ell} - K^i_j|^k S_r{}^{k\ell}. \end{cases}$$

They introduce the *torsion* d-tensor fields of $F^*\Gamma$

$$(1.8) \qquad T_j{}^i{}_k = F_j{}^i{}_k - F_k{}^i{}_j; \qquad P_{jk}{}^i = -(\dot{\partial}^i N_{jk} - F_j{}^i{}_k);$$
$$S_i{}^{jk} = -(C_i{}^{jk} - C_i{}^{kj});$$

$C_i{}^{jk}$, R_{ijk}, and the *curvature* d-tensor fields of $F^*\Gamma$

(1.9)
$$\begin{cases} R_j{}^i{}_{k\ell} = \delta_\ell F_j{}^i{}_k - \delta_k F_j{}^i{}_\ell + F_j{}^s{}_k F_s{}^i{}_\ell - F_j{}^s{}_\ell F_s{}^i{}_k + C_j{}^{is} R_{sk\ell}, \\ P_j{}^i{}_k{}^\ell = \dot\partial^\ell F_j{}^i{}_k - C_j{}^{i\ell}{}_{|k} + C_j{}^{is} P_{sk}{}^\ell, \\ S_i{}^{jk\ell} = \dot\partial^\ell C_i{}^{jk} - \dot\partial^k C_i{}^{j\ell} + C_i{}^{sk} C_s{}^{j\ell} - C_i{}^{s\ell} C_s{}^{jk}. \end{cases}$$

We denote by $F^*\Gamma(N)$ a d-connection having N as the non-linear connection, and with $\overset{\circ}{{}_{|}}$, $\overset{\circ}{|}$, the h- and v-covariant derivatives with respect to a fixed d-connection $F^* \overset{\circ}{\Gamma}(N) = (\overset{\circ}{F}, \overset{\circ}{C})$.

If $F^*\Gamma = (N, F, C)$ and $F^*\overline\Gamma = (\overline N, \overline F, \overline C)$ are two d-connections on M, then a unique triad of d-tensor fields (A, B, D) is determined such that

(1.10)
$$\overline N_{ij} = N_{ij} - A_{ij}; \qquad \overline F_j{}^i{}_k = F_j{}^i{}_k + C_j{}^{ir} A_{rk} + B_j{}^i{}_k;$$
$$\overline C_i{}^{jk} = C_i{}^{jk} + D_i{}^{jk}$$

Reciprocally, if the d-connection $F^*\Gamma$ and the triad (A, B, D) of d-tensor fields are given, then $F^*\overline\Gamma$ given by (1.10) is a d-connection. The map $F^*\Gamma \to F^*\overline\Gamma$ defined by (1.10) is called a transformation on d-connections [7].

Definition 1.1. ([8]). *A generalized Hamilton space is a pair* $M^{*n} = (M, g^{ij}(x, p))$ *in which* $g^{ij}(x, p)$ *is a symmetric and nondegenerate d-tensor field;* $g^{ij}(x, p)$ *is called the metrical d-structure or the fundamental tensor of the space* M^{*n}.

Let us consider the d-tensor field $g_{ij}(x, p)$ uniquely determined by $\|g_{ij}(x, p)\| = \|g^{ij}(x, p)\|^{-1}$ and the d-tensor field

(1.11)
$$G^h = g_{ij}(x, p)dx^i \otimes dx^j;$$

G^h is globally defined on T^*M, symmetric and of rank n.

Theorem 1.1. ([8], [9]).

(i) There exists a unique d-connection $\overset{m}{D}$ on T^*M with the properties
$$\overset{m}{D}{}_{Xh} G^h = 0; \quad \overset{m}{D}{}_{Xv} G^h = 0.$$

(ii) In the basis $(\delta_i, \dot\partial^i)$ where $\overset{\circ}{N}$ is a fixed non-linear connection, $\overset{m}{D}{}^h$ and $\overset{m}{D}{}^v$ have the coefficients

(1.12)
$$\begin{cases} \overset{m}{F}_j{}^i{}_k = \frac{1}{2} g^{ih}(\overset{\circ}{\delta}_j g_{hk} + \overset{\circ}{\delta}_k g_{jh} - \overset{\circ}{\delta}_h g_{jk}), \\ \overset{m}{C}_i{}^{jk} = -\frac{1}{2} g_{ih}(\dot\partial^j g^{hk} + \dot\partial^k g^{jh} - \dot\partial^h g^{jk}). \end{cases}$$

We denote $M^*\Gamma(\overset{o}{N}) = (\overset{m}{F}{}_j{}^i{}_k, \overset{m}{C}{}_i{}^{jk})$ and we shall call it the Miron metrical d-connection [1], [2].

Theorem 1.2. ([8]). *There exists a unique d-connection $\overset{M}{D}$ on T^*M with the properties: $\overset{M}{D}_{Xh}G^h = 0;\ \overset{M}{D}_{Xv}G^h = 0$, whose d-torsion tensor fields $T_j{}^i{}_k,\ S_i{}^{jk}$ are given a priori. This is*

(1.13)
$$\begin{cases} \overset{M}{F}{}_j{}^i{}_k = \overset{m}{F}{}_j{}^i{}_k + \frac{1}{2}g^{ir}(g_{rh}T_j{}^h{}_k - g_{jh}T_r{}^h{}_k + g_{hk}T_j{}^h{}_r), \\[2mm] \overset{M}{C}{}_i{}^{jk} = \overset{m}{C}{}_i{}^{jk} - \frac{1}{2}g_{ir}(g^{rh}S_h{}^{jk} - g^{jh}S_h{}^{rk} + g^{hk}S_h{}^{jr}). \end{cases}$$

We shall call (1.13) the generalized Miron metrical d-connection and we shall denote it by $GM^*\Gamma(\overset{o}{N}) = (\overset{M}{F}{}_j{}^i{}_k, \overset{M}{C}{}_i{}^{jk})$ [2].

Definition 1.2. ([8], [9]). *A Hamilton space is a pair $H^n = (M, H)$, where $H : T^*M \to \mathbb{R}$ is a C^∞-function on $T^*M \setminus \{0\}$, continuous on the null section of the cotangent bundle (T^*M, π^*, M), with the property that the Hessian*

(1.14)
$$g^{ij}(x, p) = \dot\partial^i \dot\partial^j H / 2, \qquad \forall\, (x, p) \in T^*M$$

is nondegenerate, that is, $\operatorname{rank} \| g^{ij}(x, p) \| = n$.

Theorem 1.3. ([8], [9]).

(i) *There exists a non-linear connection $\overset{M}{N}$ on T^*M, globally defined on T^*M, determined only by the fundamental function $H(x, p)$ of a Hamilton space H^n.*

(ii) *The coefficients $\overset{M}{N}_{ij}$ on $\overset{M}{N}$ are given by the formula*

(1.15)
$$\overset{M}{N}_{ij} = -\frac{1}{2}g_{jh}\left\{ \frac{1}{4}g_{ik}\dot\partial^k(H, \dot\partial^h H) + \dot\partial^h \partial_i H \right\},$$

where $(H, \dot\partial^h H) = \partial_i H \dot\partial^i \dot\partial^h H - \dot\partial^i H \partial_i \dot\partial^h H$ is a Poisson bracket.

§2. Metrical almost product d-structures on M.

A pair of d-tensor fields $\left(g^{ij}(x,p), q_j^i(x,p)\right)$ on M, where $g^{ij}(x,p)$ is a metrical d-structure and $q_j^i(x,p)$ is an almost product d-structure, so that

$$(2.1) \qquad\qquad g^{rs} q_r^i q_s^j = g^{ij},$$

will be called a *metrical almost product d-structure on M*.

We consider a non-linear connection $\overset{\circ}{N}$ on T^*M, the d-tensor field G^h given by (1.11), and

$$(2.2) \qquad\qquad \overset{*}{Q} = q_j^i(x,p)\, \overset{\circ}{\delta}_i \otimes dx^j,$$

where $\overset{*}{Q}$ is of type (1.1), globally defined on T^*M and of rank n.

Theorem 2.1.

(i) There exists a d-connection $\widehat{D}$ on T^*M with the properties

$$(2.3) \qquad \widehat{D}_X^h G^h = 0, \quad \widehat{D}_X^v G^h = 0, \quad \widehat{D}_X^h \overset{*}{Q} = 0, \quad \widehat{D}_X^v \overset{*}{Q} = 0.$$

(ii) In the basis $(\overset{\circ}{\delta}_i, \dot{\partial}^i)$, $\widehat{D}^h$ and $\widehat{D}^v$ have the coefficients given by

$$(2.4) \qquad \begin{cases} \widehat{F}_j{}^i{}_k = \dfrac{1}{2} g^{ih}(\overset{\circ}{\delta}_j g_{hk} + \overset{\circ}{\delta}_k g_{jh} - \overset{\circ}{\delta}_h g_{jk}) + \dfrac{1}{2} q_r^i q_j^r{}_{\mid}{}^m{}_k, \\[2mm] \widehat{C}_i{}^{jk} = -\dfrac{1}{2} g_{ih}(\dot{\partial}^j g^{hk} + \dot{\partial}^k g^{jh} - \dot{\partial}^h g^{jk}) - \dfrac{1}{2} q_i^r q_r^j{}_{\mid}{}^{m}{}^k. \end{cases}$$

(iii) $\widehat{D}$ given by (2.4) depends on $\overset{\circ}{N}$, G^h and $\overset{*}{Q}$ only.

The proof is immediate.

We shall call $\widehat{D}$ given by (2.4) an *N-canonical metrical almost product d-connection*.

Now, we denote with $\overset{*}{g}{}^{ij}(x,p)$ the d-tensor field given by

$$(2.5) \qquad\qquad \overset{*}{g}{}^{ij} = q_r^i g^{rj},$$

which determines a new metrical d-structure on M, called induced d-structure, and we consider the d-tensor field

(2.6) $$\overset{*}{G}{}^h = \overset{*}{g}_{ij}(x,p)\,dx^i \otimes dx^j,$$

where $\left\| \overset{*}{g}_{ij}(x,p) \right\| = \left\| \overset{*}{g}{}^{ij}(x,p) \right\|^{-1}$. $\overset{*}{G}{}^h$ is of type $(0,2)$, symmetric, globally defined on T^*M and of rank n.

The Obata operators of $g_{ij}(x,p)$, $q^i_j(x,p)$ and $\overset{*}{g}_{ij}(x,p)$ are given by

(2.7) $$\underset{1}{\Lambda}{}^{kh}_{ij} = \frac{1}{2}(\delta^k_i \delta^h_j - g_{ij}g^{kh}); \qquad \underset{2}{\Lambda}{}^{kh}_{ij} = \frac{1}{2}(\delta^k_i \delta^h_j + g_{ij}g^{kh});$$

(2.8) $$\underset{1}{O}{}^{kh}_{ij} = \frac{1}{2}(\delta^k_i \delta^h_j + q^k_i q^h_j); \qquad \underset{2}{O}{}^{kh}_{ij} = \frac{1}{2}(\delta^k_i \delta^h_j - q^k_i q^h_j);$$

(2.9) $$\underset{1}{\Omega}{}^{kh}_{ij} = \frac{1}{2}(\delta^k_i \delta^h_j - \overset{*}{g}_{ij}\,\overset{*}{g}{}^{kh}); \qquad \underset{2}{\Omega}{}^{kh}_{ij} = \frac{1}{2}(\delta^k_i \delta^h_j + \overset{*}{g}_{ij}\,\overset{*}{g}{}^{kh}).$$

Proposition 2.1. *The following twelve commutativities hold*

$$\underset{\alpha}{\Lambda}\,\underset{\beta}{O} = \underset{\beta}{O}\,\underset{\alpha}{\Lambda}; \quad \underset{\alpha}{O}\,\underset{\beta}{\Omega} = \underset{\beta}{\Omega}\,\underset{\alpha}{O}; \quad \underset{\alpha}{\Omega}\,\underset{\beta}{\Lambda} = \underset{\beta}{\Lambda}\,\underset{\alpha}{\Omega} \qquad (\alpha,\beta = 1,2).$$

By direct calculation we prove the

Theorem 2.2. *Let $\overset{\circ}{D}$ be a fixed d-connection on T^*M with the coefficients $F^x\Gamma(\overset{\circ}{N}) = (\overset{\circ}{F}, \overset{\circ}{C})$. Then the d-connection $\widetilde{D}$ with the coefficients*

(2.10) $$\begin{cases} \widetilde{F}_j{}^i{}_k = \overset{\circ}{F}_j{}^i{}_k + \frac{1}{4}\left\{ q^i_r q^r_{j|k} + g^{ir} g_{\ \ rj|k}^{\ \circ} + \overset{*}{g}{}^{ir}\,\overset{*}{g}_{\ \ rj|k}^{\ \circ} \right\}, \\[2mm] \widetilde{C}_i{}^{jk} = \overset{\circ}{C}_i{}^{jk} - \frac{1}{4}\left\{ q^r_i q^j_r{}^{\overset{\circ}{|}k} + g_{ir}g^{rj}{}^{\overset{\circ}{|}k} + \overset{*}{g}_{ir}\,\overset{*}{g}{}^{rj}{}^{\overset{\circ}{|}k} \right\} \end{cases}$$

has the properties

(2.11) $$\widetilde{D}^h_X G^h = 0; \quad \widetilde{D}^v_X G^h = 0; \quad \widetilde{D}^h_X \overset{*}{Q} = 0; \quad \widetilde{D}^v_X \overset{*}{Q} = 0.$$

Remark 2.1. If we put $F^*\Gamma(\overset{\circ}{N}) = M^*\Gamma(\overset{\circ}{N})$ given by (1.12) in (2.10), we obtain the $\overset{\circ}{N}$-canonical metrical almost product d-connection (2.4).

Remark 2.2. We can put $F^*\Gamma(\overset{\circ}{N}) = GM^*\Gamma(\overset{\circ}{N})$ given by (1.13) in (2.10) and we obtain another metrical almost product d-connection.

If we use (1.1), we have

Theorem 2.3. *The set of all metrical almost product d-connections on*

$$T^*M : \quad D_X^h\, G^h = 0; \quad D_X^v\, G^h = 0; \quad D_X^h\, \overset{*}{Q} = 0; \quad D_X^v\, \overset{*}{Q} = 0,$$

is given by

$$(2.12) \qquad \begin{cases} N_{ij} = \overset{\circ}{N}_{ij} - X_{ij}, \\[4pt] F_j{}^i{}_k = \widetilde{F}_j{}^i{}_k + \widetilde{C}_j{}^{ir} X_{rk} + \underset{1}{\Lambda}{}^{mi}_{jt} \underset{1}{O}{}^{st}_{mr} Y_s{}^r{}_k, \\[4pt] C_i{}^{jk} = \widetilde{C}_i{}^{jk} + \underset{1}{\Lambda}{}^{tj}_{im} \underset{1}{O}{}^{rm}_{ts} Z_r{}^{sk}, \end{cases}$$

where $F^*\widetilde{\Gamma}(\overset{\circ}{N}) = (\widetilde{F}, \widetilde{C})$ *is given by (2.10) and* $X_{ij}, Y_j{}^i{}_k, Z_i{}^{jk}$ *are arbitrary d-tensor fields.*

Remark 2.3. *If* $g^{ij}(x,p) = \dot{\partial}^i \dot{\partial}^j H/2$ *is the fundamental tensor of the Hamilton space* H^n, *we can take* $\overset{\circ}{N} = \overset{M}{N}$ *everywhere in this paragraph, where* $\overset{M}{N}$ *is the Miron non-linear connection (1.15)* [1], [2].

Theorem 2.4. *If* H^n *is of local Minkowski type and has a d-tensor field* $q_j^i(x,p) = q_j^i(p)$ *such that* $q_k^i q_j^k = \delta_j^i$ *and the relation (2.1) is satisfied, then we have:*

1) *The* $\overset{M}{N}$*-canonical d-connection (2.4) has the form*

$$(2.13) \qquad \overset{c}{F}_j{}^i{}_k = 0; \qquad \overset{c}{C}_i{}^{jk} = -\frac{1}{2} g_{ih} \dot{\partial}^j g^{hk} - \frac{1}{2} q_i^r q_r^j \,\overset{m}{|}{}^k.$$

2) *The torsion tensor fields of (2.13) are given by*

$$(2.14) \qquad \begin{cases} \overset{c}{T}_j{}^i{}_k = 0; \quad \overset{c}{R}_{ijk} = 0; \quad \overset{c}{P}_{jk}{}^i = 0; \\[6pt] \overset{c}{C}_i{}^{jk}; \quad \overset{c}{S}_i{}^{jk} = \frac{1}{2} q_i^r (q_r^j \,\overset{m}{|}{}^k - q_r^k \,\overset{m}{|}{}^j). \end{cases}$$

3) *The cuvature tensor fields of (2.13) are given by*

$$(2.15) \quad \overset{c}{R}_j{}^i{}_{k\ell} = 0; \quad \overset{c}{P}_j{}^i{}_k{}^\ell = 0; \quad \overset{c}{S}_i{}^{jk\ell} = \underset{(k,l)}{\mathcal{Q}} \left\{ \dot{\partial}^\ell \overset{c}{C}_i{}^{jk} + \overset{c}{C}_s{}^{jk} \overset{c}{C}_i{}^{s\ell} \right\}$$

where $\mathcal{Q}$ *is the alternate summation.*

4) *The Bianchi identities of* (2.13) *are*

(2.16)
$$
\begin{cases}
\underset{hrs}{\mathcal{S}} \left\{ \overset{c}{S}_k{}^{hr}|^s + \overset{c}{S}_k{}^{h\ell}\, \overset{c}{S}_\ell{}^{rs} + \overset{c}{S}_k{}^{hrs} \right\} = 0 \\[2mm]
\underset{hrs}{\mathcal{S}} \left\{ \overset{c}{S}_j{}^{ihr}|^s + \overset{c}{S}_j{}^{ih\ell}\, \overset{c}{S}_\ell{}^{rs} \right\} = 0
\end{cases}
$$

where $\mathcal{S}$ *is the cyclic summation.*

The proof is elementary if we take into account that in this case we have $H(x,p) = H(p)$ and hence $\overset{M}{N}_{ij} = 0$.

Theorem 2.5.

(i) *A metrical almost product d-connection D has the properties*

$$
D_X^h\, \overset{*}{G}{}^h = 0; \quad D_X^v\, \overset{*}{G}{}^h = 0; \quad \forall X, Y \in \mathfrak{X}(T^*M).
$$

(ii) *The Obata operators* (2.7), (2.8) *and* (2.9) *are covariant constant with respect to D.*

(iii) *The d-tensor fields* $\underset{2}{\Lambda}{}^{hi}_{jr} R_h{}^r{}_{k\ell}$; $\underset{2}{O}{}^{hi}_{jr} R_h{}^r{}_{k\ell}$; $\underset{2}{\Omega}{}^{hi}_{jr} R_h{}^r{}_{k\ell}$ *(analogous in* $P_j{}^i{}_k{}^\ell$; $S_i{}^{jk\ell}$) *and their h- and v-covariant derivatives of every order vanish for any metrical almost product d-connection D.*

Proof. (i) and (ii) are evident. For (iii) we apply the Ricci formulas (1.7) to g_{ij}, q_j^i and $\overset{*}{g}_{ij}$. Taking into account (ii) we get the statement. ∎

We attract attention to the transformation $F^*\Gamma(\overset{\circ}{N}) \to F^*\overline{\Gamma}(\overset{\circ}{N})$ of metrical almost product d-connections. Owing to Theorem 2.3, they are given by

(2.17)
$$
\begin{cases}
\overline{N}_{ij} = N_{ij}; \qquad \overline{F}_j{}^i{}_k = F_j{}^i{}_k + \left(\underset{1}{\Lambda}\; \underset{1}{O} \right)^{si}_{jr} Y_s{}^r{}_k; \\[3mm]
\overline{C}_i{}^{jk} = C_i{}^{jk} + \left(\underset{1}{\Lambda}\; \underset{1}{O} \right)^{jr}_{si} Z_r{}^{sk}
\end{cases}
$$

where $Y_s{}^r{}_k$, $Z_i{}^{jk}$ are arbitrarily given d-tensor fields. Hence we have the following

Theorem 2.6. *The set of all transformations (2.17) and the mapping product form an Abelian group G_{map}, which is isomorphic to the additive group of the pairs of d-tensor fields $(\Lambda \underset{1}{O} Y, \Lambda \underset{1}{O} Z)$.*

The group G_{map} is a subgroup of the group G_{ap} (see [4], Theorem 2.4). The invariants $\overset{a}{T}(q)$, $\overset{a}{R}(q)$, $\overset{a}{C}(q)$, $\overset{a}{P}(q)$, $\overset{a}{S}(q)$, $(a = 1, 2)$ of the group G_{ap} (see [4], Theorem 2.5) are invariants also for G_{map}.

§3. Metrical almost product structures on T^*M.

The existence of the non-linear connection $\overset{\circ}{N}$ on T^*M, particularly of $\overset{\circ}{N} = \overset{M}{N}$ given by (1.15), permits us to consider the d-tensor fields

(3.1) $$G^v = g^{ij})x, p) \overset{\circ}{\delta} p_i \otimes \overset{\circ}{\delta} p_j$$

(3.2) $${}^*Q^* = q^j_i(x, p)\dot{\partial}^i \otimes \overset{\circ}{\delta} p_j$$

(3.3) $$\overset{*}{G}{}^v = \overset{*}{g}{}^{ij}(x, p) \overset{\circ}{\delta} p_i \otimes \overset{\circ}{\delta} p_j$$

which are globally defined on T^*M, of rank n, and of type $(0, 2)$, $(1, 1)$ and $(0, 2)$ respectively.

Then

$$G = G^h + G^v \qquad \text{and} \qquad \widetilde{G} = G^h - G^v$$

are Riemann structures on T^*M;

$$Q^I = \overset{*}{Q} + {}^*Q^* \qquad \text{and} \qquad Q^{II} = \overset{*}{Q} - {}^*Q^*$$

are almost product structures on T^*M, and

$$\overset{*}{G} = \overset{*}{G}{}^h + \overset{*}{G}{}^v \qquad \text{and} \qquad \overset{*}{\widetilde{G}} = \overset{*}{G}{}^h - \overset{*}{G}{}^v$$

are Riemann structures on T^*M.

Theorem 3.1. *The pairs of d-tensor fields (G, Q^I), $(\widetilde{G}, Q^I)$, (G, Q^{II}), $(\widetilde{G}, Q^{II})$ are metrical almost product structures on T^*M with the induced metrical d-structures $\overset{*}{G}$, $\overset{*}{\widetilde{G}}$, $\overset{*}{\widetilde{G}}$ and $\overset{*}{G}$, respectively.*

Definition 3.1. *A metrical almost product structure on T^*M is integrable if the almost product structure of the pair is integrable.*

Theorem 3.2.

(i) *The structure (G, Q^I) (resp. $(\widetilde{G}, Q^I)$) is integrable if and only if the following invariant d-tensor fields of the group G_{map} vanish:*

$$(3.4) \qquad \overset{1}{T}(q) = 0; \quad \overset{1}{R}(q) = 0; \quad \overset{1}{C}(q) = 0; \quad \overset{1}{P}(q) = 0; \quad \overset{1}{S}(q) = 0.$$

(ii) *The structure (G, Q^{II}) (resp. $(\widetilde{G}, Q^{II})$) is integrable if and only if the following invariant d-tensor fields of the group G_{map} vanish:*

$$(3.5) \qquad \overset{1}{T}(q) = 0; \quad \overset{2}{R}(q) = 0; \quad \overset{2}{C}(q) = 0; \quad \overset{2}{P}(q) = 0; \quad \overset{1}{S}(q) = 0.$$

References

[1] ATANASIU, GH., The Kawaguchi method of determination of the metrical connections in a Cartan space, *Memor. Secţ. St., seria IV, t.8, nr.1*, (1985), Ed. Acad. RSR, Bucureşti, (1987), 43–55.

[2] ATANASIU, GH. and HASHIGUCHI, M., Semi-symmetric Miron connections in dual Finsler Spaces, *Rep. Fac. Sci. Kagoshima Univ. (Math., Chem.)*, **20** (1987), 43–49.

[3] ATANASIU, GH. and KLEPP, F. C., Almost product Finsler Structures and Connections, *Studia Sci. Math. Hungarica*, **18** (1983), 43–56.

[4] ATANASIU, GH. and KLEPP, F. C., Almost product structures on cotangent bundle, to appear in Memor. Secţ. St. Ed. Acad. RSR, Bucureşti.

[5] KLEPP, F. C., Almost product Finsler structures, *Anal. Univ. "Al. Cuza" Iaşi*, **28** (1982), 59–67.

[6] KLEPP, F. C., Metrical almost product Finsler Structures, *Anal. Univ. "Al. I. Cuza" Iaşi*, **29** f.2. (1983), 21–25.

[7] MATSUMOTO, M., Foundation of Finsler Geometry and Special Finsler Spaces, Kaiseisha Press, Otsu, Japan, 1986.

[8] MIRON, R., Hamilton Geometry, *Sem. Mecanică*, **3**, Univ. Timişoara. 1987.

[9] MIRON, R., Sur la géométrie des espaces d'Hamilton, *C. R. Acad. Sci. Paris*, *t.306*, Série I, (1988), 195–198.

Gheorge Atanasiu

Department of Mathematics
University of Braşov
2200 Braşov
Romania

Francisco C. Klepp

Department of Mathematics
Polytechnical Institute Timişoara
1900 Timişoara
Romania

COLLOQUIA MATHEMATICA SOCIETATIS JÁNOS BOLYAI
56. DIFFERENTIAL GEOMETRY, EGER (HUNGARY), 1989

Pseudoconvexity in Lorentzian Geometry

J. K. BEEM*

1. Introduction

An n dimensional spacetime (M, g) of signature $(+, -, \ldots, -)$ is said to be *causally pseudoconvex* if given any compact set K in M, there is always a larger compact set K' such that all causal geodesic segments joining points of K must lie in K'. If no inextendible causal geodesic has either end imprisoned in a compact set, the spacetime satisfies the *causal disprisonment condition*. The concepts of pseudoconvexity and disprisonment have been used in the study of pseudodifferential equations [7]. These two conditions yield sufficiency conditions for global solvability in the distributional sense. Pseudoconvexity and disprisonment for Lorentzian manifolds have been investigated by Beem and Parker in [2], [3], and [4]. Geodesic completeness is not a stable property for general spacetimes [6], [8], but Beem and Ehrlich [1] have shown that causal geodesic completeness is stable for spacetimes which satisfy causal pseudoconvexity and causal disprisonment. In the present paper, we discuss some results that have been obtained using pseudoconvexity in the study of Lorentzian geometry.

---

* Supported in part by NSF grant DMS–8803511

2. Pseudoconvexity

Let Lor(M) denote the collection of all Lorentzian metrics on M and assume for each fixed nonnegative integer r this space of metrics is given the Whitney fine C^r topology, compare [3]. The following result by Beem and Parker [3] yields a stability theorem of the two conditions causal pseudoconvexity and causal disprisonment taken together.

Theorem 1. *If (M, g) is a spacetime which is both causally pseudoconvex and causally disprisoning, then there is a fine C^1 neighborhood $U(g)$ of g in Lor(M) such that all metrics g' in $U(g)$ are also both causally pseudoconvex and causally disprisoning.*

Recall that given local coordinates, the Christoffel symbols only depend on the components of the metric tensor and their first derivatives. Thus, metrics which are close in the fine C^1 topology will have geodesic systems which are close. It is easy to demonstrate the failure of C^r stability of causal disprisonment in the absence of causal pseudoconvexity for all nonnegative integers r.

Example 2. *Let $M' = R^1 \times S^1 = \{(t, \theta) | t \in R^1 \text{ and } 0 \le \theta \le 1\}$ with the usual identification of $(t, 0)$ with $(t, 1)$. Let $M = M' - \{(0, 0)\}$ and set $g = -dt\, d\theta + f(t)\, d\theta^2$ where $f(0) = 0$ and $f(t) = \exp(-t^{-2})$ for $t \ne 0$. All causal geodesicd of the original M', except the special null geodesic lying on the circle $t = 0$, have both ends leaving each fixed compact subset of M' and never returning. The deletion of one point of the circle $t = 0$, yields a spacetime (M, g) which is causally disprisoning. However, each fine C^r neighborhood $U(g)$ of g in Lor(M) has some metric h such that (M, h) contains a closed, and hence imprisoned, causal geodesic.*

The following result of Beem and Ehrlich [1] insures the stability of causal geodesic completeness.

Theorem 3. *Let (M, g) be a spacetime which is both causally pseudoconvex and causally disprisoning. If all causal geodesics of (M, g) are complete, then there is a fine C^1 neighborhood $U(g)$ of g in Lor(M) such that each metric g' in $U(g)$ is causally geodesically complete.*

Examples of causally geodesically complete spacetimes which fail to have fine C^r neighborhoods in $\text{Lor}(M)$ such that all metrics in the neighborhood are causally geodesically complete have been given by Williams [8], compare [1], [6]. In particular, Williams showed the torus $M = S^1 \times S^1 = \{(t, \theta) | 0 \le t \le 1, 0 \le \theta \le 1\}$ with the flat metric $g = dt \, d\theta$ has metrics arbitrarily close in each fine C^r topology which are incomplete.

One may also apply pseudoconvexity and disprisoning to the collection of all geodesics on a manifold M with affine connection ∇. One says (M, ∇) is *geodesically pseudoconvex* if given any compact set K there is a compact set K' such that all geodesic segments with both endpoints in K lie in K'. Furthermore, (M, ∇) is *geodesically disprisoning* if no inextendible geodesic has either end imprisoned in a compact set. Alternatively, one may define disprisonment by requiring that each and of each geodesic fail to have compact closure. The next result of Beem and Parker [5] is a generalization of the classical Hadamard–Cartan theorem.

Theorem 4. *Let (M, g) be both geodesically pseudoconvex and geodesically disprisoning. If (M, ∇) has no conjugate points, then the exponential map at each point is a diffeomorphism from it domain onto M and M is diffeomorphic to $\mathbb{R}^n$. Furthermore, (M, ∇) is geodesically connected in the sense that any two distinct points of M may be joined by a geodesic of (M, ∇).*

3. Causality

Causality has played an important role in relativity and it is natural to try and compare pseudoconvexity to causality conditions. In some sense, the causal pseudoconvexity condition rules out "holes" in the spacetime. Thus, it is somewhat close to the condition of global hyperbolicity. In fact, it is a strictly weaker condition. If (M, g) is a globally hyperbolic spacetime, then it is both causally pseudoconvex and causally disprisoning. However, unlike global hyperbolicity the condition of causal pseudoconvexity is not

Example 5. Let $M = \{(t,x) \mid t \in R^1$ and $-\pi/2 < x < \pi/2\}$ with $g = (\sec^2 x)(-dt^2 + dx^2)$ and $h = -dt^2 + dx^2$. Then (M, g) is two dimensional Anti–De Sitter spacetime and (M, h) is an open strip of two dimensional Minkowski spacetime. Let K be the compact set consisting of the two points $(0, 0)$ and $(\pi, 0)$. For (M, g), the timelike geodesics starting at the point $(0, 0)$ reconverge and pass though the point $(\pi, 0)$. The geodesic segments of (M, g) with endpoints $(0, 0)$ and $(\pi, 0)$ form a set that fails to have compact closure in M and thus (M, g) is not causally pseudoconvex. On the other hand, the spacetime (M, h) is causally pseudoconvex and is conformal to (M, g). Note also that (M, g) is an example of a spacetime that is causally pseudoconvex and yet fails to be globally hyperbolic.

With the signature convention $(+, -, \dots, -)$, we have the following result [2, p. 11] for spacetimes which have nonpositive sectional curvature on all timelike two dimensional linear subspaces of each tangent space.

Proposition 6. Let (M, g) be a distinguish and causally pseudoconvex spacetime. If (M, g) has everywhere nonpositive timelike sectional curvatures, then (M, g) is causally simple.

References

[1] J. K. BEEM and P. E. EHRLICH, Geodesic completeness and stability, *Math. Proc. Camb. Phil. Soc.* **102** (1987), 319–328.

[2] J. K. BEEM and P. E. PARKER, Klein–Gordon solvability and the geometry of geodesics, *Pac. J. Math.* **107** (1983), 1–14.

[3] J. K. BEEM and P. E. PARKER, Whitney stability of solvability, *Pac. J. Math.* **116** (1985), 11–23.

[4] J. K. BEEM and P. E. PARKER, Pseudoconvexity and general relativity, *J. of Geo. and Phys.* **4** (1987), 71–80.

[5] J. K. BEEM and P. E. PARKER, Pseudoconvexity and geodesic connectedness, *Ann. Mat. Pura Appl.* **155** (1989), 137–142.

[6] C. T. J. DODSON, *Categories, Bundles and Spacetime Topology*, second edition, Kluwer Academic Publ., Dordrecht, 1988.

[7] J. J. DUISTERMAAT and L. HORMANDER, Fourier integral operators II, *Acta*

[8] P. M. WILLIAMS, Instability of geodesic completeness and incompleteness, preprint, Dept. of Math., Univ. of Lancaster.

John K. Beem

Mathematics Department
University of Missouri
Columbia, MO. 65211
USA

COLLOQUIA MATHEMATICA SOCIETATIS JÁNOS BOLYAI

56. DIFFERENTIAL GEOMETRY, EGER (HUNGARY), 1989

The Bochner Curvature Tensor on a Hyperbolic Kähler Manifold*

C. L. BEJAN

Introduction

The conformal curvature tensor on a Riemannian manifold has a corresponding tensor on a Kähler manifold whose formula was established with respect to complex local coordinates by S. Bochner in [5] and then it was given with respect to real local coordinates by S. Tachibana in [11].

A. Avez proved in [1] that the Pontrjagin classes of a conformally flat Riemannian manifold are trivial. By using the same ideas as those contained in [8], C. L. Bejan showed in [2] that the Chern algebra of a Bochner flat Kähler manifold of constant Ricci scalar curvature is generated only by the first Chern form and by the fundamental 2-form of the manifold.

In the beginning of this note, we give the formula and we find some identities of the Bochner tensor on a hyperbolic Kähler manifold (also called para-Kähler manifold: see P. K. Rasevskii [10] and P. Libermann [7]). By analogy with [2] and using [3], we express here the Pontrjagin classes of a Bochner flat hyperbolic Kähler manifold of a constant Ricci scalar curvature

---

* This paper is in final form and no version of it will be submitted for publication elsewhere.

making use only of the fundamental 2-form φ of the manifold and an exact 2-form ψ. This is a generalization of a result obtained by C. L. Bejan in [3] regarding the Pontrjagin algebra of a hyperbolic Kähler manifold of constant F-sectional curvature.

§1. The Bochner tensor of a hyperbolic Kähler manifold

Everywhere in this paper, M is an n-dimensional hyperbolic Kähler manifold, i.e. M is endowed with a semi-Riemannian structure g and a product structure F such that $g(FX, Y) + g(X, FY) = 0$, $\forall X, Y \in \mathfrak{X}(M)$ and the Levi–Civita connection ∇ of g is an F-connection (that is $\nabla F = 0$), [7], [10]. Therefore $n = 2m$.

Let $R_{ij}{}^h{}_k$, R_{ij} and r be the curvature tensor, the Ricci curvature tensor and the scalar curvature, respectively. We take the usual notations for raising and lowering indices.

Like in the Kählerian case, we get:

$$R_{ij}{}^h{}_k F^i_l = -R_{li}{}^h{}_k F^i_j; \quad R_{ij}{}^h{}_k F^k_l = R_{ij}{}^k{}_l F^h_k; \quad F^a_i R_{aj} = -R_{ia} F^a_j;$$
$$F^a_i R^k_a = R^s_i F^k_s; \quad \nabla_a R_{ij}{}^a{}_k = \nabla_i R_{jk} - \nabla_j R_{ik} \text{ and}$$
$$(1) \qquad\qquad \nabla_i r = 2\nabla_a R^a_i.$$

We denote $S_{ij} = F^a_i R_{aj}$. Then we have $S_{ij} = -S_{ji}$ and from the Bianchi identity $\sum_{\text{cicl}(ijk)} \nabla_i R_{jklh} = 0$, we get:

$$(2) \qquad\qquad \sum_{\text{cicl}(ijk)} \nabla_i S_{jk} = 0 \text{ and therefore } S_{ij} \text{ is a closed 2-from.}$$

We denote $U_{ij}{}^h{}_k = R_{ij}{}^h{}_k - \frac{r}{n} W_{ij}{}^h{}_k$, where $W_{ij}{}^h{}_k = \frac{1}{n+2}(g_{ik}\delta^h_j - g_{jk}\delta^h_i + F_{jk}F^h_i - F_{ik}F^h_j - 2F_{ij}F^h_k)$.

Remark. [6] M is of constant F-sectional curvature if and only if $U = 0$.

As in the Kählerian case we have:

Proposition 1. *For $n > 2$, if $U = 0$ then M is Einstein (with respect to the semi-Riemannian structure g).* ∎

An example of an n-dimensional hyperbolic Kähler manifold of constant F-sectional curvature is given in [6] for any $n > 2$.

In dimension two another example is the pseudosphere S_1^2 endowed with the hyperbolic Kähler structure emphasized in [4] which has constant sectional curvature (hence it is of constant F-sectional curvature) and it is Einstein.

We introduce now the Bochner tensor on M by

(3) $B_{ijk}{}^h = R_{ijk}{}^h - T_{ijk}{}^h$, where

$$T_{ijk}{}^h = \frac{1}{n+4}\{R_{ik}\delta_j^h - R_{jk}\delta_i^h + g_{ik}R_j^h - g_{jk}R_i^h + S_{ki}F_j^h - S_{kj}F_i^h + {}$$
$$+ 2S_{ji}F_k^h - F_{ik}S_j^h + F_{jk}S_i^h - 2F_{ij}S_k^h - rW_{ijk}{}^h\}.$$

Remark that in an adapted coordinate system we have

$$B_{\alpha\beta\gamma}^\nu = B_{\overline{\alpha}\overline{\beta}\overline{\gamma}}^{\overline{\nu}} = 0, \text{ where}$$

the indices α, β, γ, ν run over the range $\overline{1, m}$ and the indices $\overline{\alpha}$, $\overline{\beta}$, $\overline{\gamma}$, $\overline{\nu}$ over the range $\overline{m+1, n}$ (corresponding to the two complementary distributions of M, given by the eigenvalues ± 1 of F).

By a straightforward calculation, we get the following identities satisfied by B as well as by T:

$$B_{ijk}{}^h = -B_{jik}{}^h; \quad B_{ijkh} = -B_{ijhk}; \quad \sum_{\text{cicl}(ijk)} B_{ijk}{}^h = 0;$$
$$B_{ijk}{}^j = 0; \quad B_{ijk}{}^k = 0; \quad B_{ijk}{}^h F_h^l = B_{ijh}{}^l F_k^h;$$
$$B_{ijk}{}^h F_j^i = -B_{lik}{}^h F_j^i; \quad B_{ijk}{}^h F_h^k = 0; \quad B_{ijk}{}^h F_h^i = 0.$$

As in the Kählerian case, we get:

Proposition 2. $B = U$ if and only if $R_{ij} = \frac{r}{n}g_{ij}$ (i.e. for $n > 2$, M is Einstein). ∎

As a consequence, we have:

Proposition 3. If M is of constant F-sectional curvature, then $B = 0$ and for $n > 2$ it follows that r is constant. ∎

Proposition 4. *Let M be Bochner flat. If M is Einstein, then M is of constant F-sectional curvature and for $n > 2$ the converse is also true.* ■

Like in the Kählerian case, we get the following formula:

$$\nabla_i B_{jh\,k}^{\ \ \ i} = \frac{n}{n+4}\left\{(\nabla_h R_{jk} - \nabla_j R_{hk}) + \frac{1}{2}W_{hj\,k}^{\ \ \ i}\nabla_i r\right\}.$$

§2. Calculus of the Pontrjagin classes

We deal here with the Pontrjagin algebra of M.

Let us recall from [3] that if ρ and Ω are the matrices associated to the tensor field R and FR, respectively, in the given local chart on M, i.e.

$$\rho = (\rho_k^i), \text{ where } \rho_k^i = \frac{1}{2}R_{j_1 j_2\,k}^{\ \ \ \ i}dx^{j_1} \wedge dx^{j_2} \text{ and } \Omega = (F_j^i \rho_k^j),$$

then the Pontrjagin class of order $2s$ given by $\Phi_s = [\text{trace } \rho^s]$ coincides with the so-called F-characteristic class defined by $[\text{trace } \Omega^s]$.

We denote by φ the fundamental 2-form of M, i.e. $\varphi = \frac{1}{2}F_{j_1 j_2}\,dx^{j_1} \wedge dx^{j_2}$. We remark that the class of the two-form $\psi = \frac{1}{2}S_{j_1 j_2}\,dx^{j_1} \wedge dx^{j_2}$ is up to a constant factor the Pontrjagin class of order two (see [3]) and hence $[\psi] = 0$.

Let us denote $\omega = \psi - \frac{r}{n+2}\varphi$.

From now on, we suppose $B = 0$.

By using (4) we get

(5) $\nabla_h R_j^u - \nabla_j R_h^u = A_{hj}^u$ and therefore

(6) $\nabla_b R_h^u = A_{hj}^k F_{kb}F^{ju}$, where $A_{hj}^u = -\frac{1}{2}W_{hj\,k}^{\ \ \ i}g^{ku}\nabla_i r.$

From the Ricci identity we get:

$$R_h^w R_{ab\,w}^{\ \ \ u} - R_w^u R_{ab\,h}^{\ \ \ w} =$$

$$= \frac{1}{2(n+2)}\{(\delta_b^u \delta_h^i - F_{hb}F^{iu} + g_{hb}g^{iu} + F_b^u F_h^i)\nabla_a\nabla_i r -$$

$$- (\delta_a^u \delta_h^i - F_{ha}F^{iu} + g_{ha}g^{iu} + F_a^u F_h^i)\nabla_b\nabla_i r\}.$$

By using the same technique as in [2] we get:

$$(7) \qquad \Omega_i^k = \frac{-2}{n+4}(\varphi R_i^k + \omega \delta_i^k) +$$

$$+ \frac{1}{n+4}(R_{j_1 i} F_{j_2}^k + g_{j_1 i} S_{j_2}^k + S_{i j_1} \delta_{j_2}^k - F_{j_1 i} R_{j_2}^k) dx^{j_1} \wedge dx^{j_2} -$$

$$- \frac{r}{(n+2)(n+4)}(g_{j_1 i} F_{j_2}^k + \delta_{j_1}^k F_{j_2 i}) dx^{j_1} \wedge dx^{j_2}$$

$$(8) \ \Omega_k^l \wedge \Omega_i^k = \Omega_k^l \wedge \left\{ \frac{-2}{n+4}(\varphi R_i^k + \omega \delta_i^k) + \right.$$

$$\left. + \frac{1}{n+4}(g_{j_3 i} S_{j_4}^k - F_{j_3 i} R_{j_4}^k) dx^{j_3} \wedge dx^{j_4} \right\}.$$

Now, by induction it follows:

Proposition 5. *If M is a Bochner flat hyperbolic Kähler manifold, then $\Omega^{s+1} = \Omega \wedge T^s$, where*

$$T_i^k = \frac{1}{n+4}\left\{ -2(\varphi R_i^k + \omega \delta_i^k) + (g_{j_3 i} S_{j_4}^k - F_{j_3 i} R_{j_4}^k) dx^{j_3} \wedge dx^{j_4} \right\}.$$

∎

We denote by P the matrix $P = (R_i^k)$.

Remark. If r is constant, by using (1), (4), and (2) we get $\nabla_i R_{jk} = 0$ and then $R_i^j \Omega_j^k = R_j^k \Omega_i^j$ (i.e. P and Ω commute) and therefore

$$\Omega^{s+1} = \left(\frac{-2}{n+4} \right)^s (\varphi P + \omega I)^s \wedge \Omega, \quad \forall s \geq 2 \text{ and}$$

$$P^s = m_s P + p_s I, \quad \forall s \geq 2, \text{ where}$$

$$\begin{cases} m_s = \beta^{s+1} + C_{s-2}^1 \beta^{s-3} \alpha + C_{s-3}^2 \beta^{s-5} \alpha^2 + \dots \\ p_s = \alpha(\beta^{s-2} + C_{s-3}^1 \beta^{s-4} \alpha + C_{s-4}^2 \beta^{s-6} \alpha^2 + \dots) \end{cases}$$

with

$$\begin{cases} \alpha = \frac{1}{n}\left(\text{trace } p^2 - \frac{r^2}{n+2} \right), \\ \beta = \frac{r}{n+2}. \end{cases}$$

Theorem. *If M is a Bochner flat hyperbolic Kähler manifold of constant Ricci scalar curvature, then*

$$\text{trace}\,\Omega^{s+1} = 2\left(\frac{-2}{n+4}\right)^s \{\omega^s \wedge \psi + s\varphi \wedge \omega^{s-1} \wedge (\alpha\varphi + \beta\psi)+$$

$$+ \sum_{i=2}^s C_s^i \varphi^i \wedge \omega^{s-i} \wedge [m_i(\alpha\varphi + \beta\psi) + p_s\psi]\}, \quad \forall\, s \geq 2.$$

■

Therefore the Pontrjagin classes $[\text{trace}\,\Omega^{s+1}]$, $s \geq 2$, can be expressed only with the forms φ and ψ.

By virtue of Proposition 3, in the particular case when M is an n-dimensional hyperbolic Kähler manifold of constant F-sectional curvature, C. L. Bejan has already proved in [3] that the Pontrjagin classes are all trivial.

References

1. A. AVEZ, Characteristic classes and Weyl tensor, Applications to general relativity, *Proc. Nat. Acad. Sci. U.S.A.* **66** (1970), 265–268.

2. C. L. BEJAN, On the Chern forms of a Bochner flat Kählerian manifold, *Rend. di Mat. Roma, VII*, vol. 4, f. 4. (1984), 593–599.

3. C. L. BEJAN, F-characteristic classes (to appear in *Tensor* vol. 50)

4. C. L. BEJAN, CR-Submanifolds of hyperbolical almost Hermitian manifolds (to appear in *Demonstratio Mathematica*, vol. 23, (1990))

5. S. BOCHNER, Curvature and Betti numbers, II, *Ann. of Math.* **50** (1949).

6. P. GADEA and A. MONTESINOS, Spaces of Constant para-holomorphic sectional curvature, *Pacific J. of Math.* vol. 136 no. 1 (1989), 85–101.

7. P. LIBERMANN, Sur le probléme d'équivalence de certaines structures infinitesimales, *Ann. di Mat.* **36** (1954), 27–120.

8. A. NEAGU and V. OPROIU, Chern forms and H-projective curvature of complex manifolds, Analele Univ. Iaşi, Tom XXIV, s. 1a, f. 1 (1978), 39–46.

9. P. PRVANOVIĆ, Holomorphically projective transformations in a locally product space, *Math. Balkanica* **1** (1971), 195–213.

10. P. K. RASEVSKII, The scalar field in a stratified space, *Trudy Sem. Vekt. Tenz. Anal.* **6** (1948), 225–248.

11. S. TACHIBANA, On the Bochner curvature tensor, *Nat. Sci. Rep. Ochanomizu Univ.* **18** (1967), 15–19.

Cornelia–Livia Bejan

Seminarul Matematic
Univ. "Al. I. Cuza"
6600 Iaşi, Romania

COLLOQUIA MATHEMATICA SOCIETATIS JÁNOS BOLYAI
56. DIFFERENTIAL GEOMETRY, EGER (HUNGARY), 1989

Generalized Gauge Theories[*]

A. BEJANCU

§0. Introduction

The purpose of our talk is to give a generalization of the classical gauge theory. More precisely, we construct a gauge theory with respect to some physical fields $Q^A(x^1, \ldots, x^m; y^1, \ldots, y^r)$, where (x^i) are the coordinates on a manifold and (y^α) are some *perturbation parameters*. That is, we consider physical fields defined on the total space of a vector bundle.

The background of our investigations is the tensor calculus induced by a horizontal distribution on a vector bundle. This is written in the first paragraph of the paper. Then we study the global gauge invariance of Lagrangians on a vector bundle. In §3 we prove that a local gauge invariant Lagrangian is obtained from a global gauge invariant Lagrangian by replacing $\frac{\delta Q^A}{\delta x^i}$ (see (1.6)) and $\frac{\partial Q^A}{\partial y^\alpha}$ by the horizontal and vertical gauge covariant derivatives $\overset{(h)}{D}_i Q^A$ and $\overset{(v)}{D}_\alpha Q^A$ given by (3.9) and (3.10) respectively. In §4 we construct strength fields (see (4.1)–(4.3), (4.12)–(4.14)) and Lagrangians for gauge fiedls (see (4.9)–(4.11), (4.16)–(4.18)). This enables us to propose the full Lagrangian (4.19) for our gauge theory. By using the

---

[*] This paper is in final form and no version of it will be submitted for publication elsewhere.

gauge H-connection GHC given by (1.22)–(1.25) we obtain in §5 the equations of motion (5.6), (5.14), (5.16) and the conservation laws (5.30). In the particular case of a trivial vector bundle (which seems to the author more appropriate to physics) we get the equations of motion (5.33), (5.34), (5.35) and the nice conservation laws (5.36). Finally, in the last paragraph we get four types of Bianchi identities (6.11)–(6.14).

The author believes that some physical theories should be reconsidered with respect to a gauge theory concerned with physical fields and Lagrangians which depend on some other coordinates besides the space-time coordinates. In fact we have already an example which supports the above assertion. Namely, the theory of *gravity* was reconsidered and the new theory of *supergravity* is nothing but a theory which is dealing with two groups of coordinates: the Bose coordinates and the Fermi coordinates.

§1. A tensorial calculus induced by a horizontal distribution on a vector bundle

Let $\xi = (E, \pi, M)$ be a vector bundle with M as a base space, E as the total space and $\pi : E \to M$ as the projection mapping. Suppose m is the dimension of the differentiable manifold M, and fibers of ξ are of dimension r. Then the coordinates on E are (x^i, y^α), $i = 1, \ldots, m$; $\alpha = 1, \ldots, r$, where (x^i) are the local coordinates on M. The general transformations of coordinates on E are the following

(1.1)
$$\begin{cases} \widetilde{x}^i = \widetilde{x}^i(x^1, \ldots x^m), \\ \widetilde{y}^\alpha = A^\alpha_\beta(x)y^\beta, \end{cases}$$

where $A^\alpha_\beta(x)$ are real differentiable functions defined locally on M.

Throughout the paper we shall use the following range for indices: $i, j, k, \ldots = 1, \ldots, m$; $\alpha, \beta, \gamma, \ldots = 1, \ldots, r$; $A, B, C, \ldots = 1, \ldots, p$; $a, b, c, \ldots = 1, \ldots, n$.

From (1.1) it follows

(1.2)
$$\frac{\partial}{\partial x^i} = B^j_i(x)\frac{\partial}{\partial \widetilde{x}^j} + \frac{\partial A^\alpha_\beta(x)}{\partial x^i}y^\beta\frac{\partial}{\partial \widetilde{y}^\alpha},$$

(1.3)
$$\frac{\partial}{\partial y^\alpha} = A^\beta_\alpha(x)\frac{\partial}{\partial \widetilde{y}^\beta},$$

where we put $B_i^j(x) = \frac{\partial \tilde{x}^j}{\partial x^i}$. In order to remove the non-homogeneity of (1.2) we consider the vertical bundle VE and we suppose that there exists a *horizontal distribution* HE on E, that is, we have

$$(1.4) \qquad\qquad TE = VE \oplus HE.$$

It is easy to see that HE is locally given by differentiable functions $N_i^\alpha(x,y)$ which satisfy

$$(1.5) \qquad N_i^\alpha(x,y)A_\alpha^\beta(x) = \tilde{N}_j^\beta(\tilde{x},\tilde{y})B_i^j(x) + \frac{\partial A_\alpha^\beta(x)}{\partial x^i}y^\alpha.$$

Then $\{\frac{\delta}{\delta x^1}, \ldots, \frac{\delta}{\delta x^m}\}$ defined by

$$(1.6) \qquad\qquad \frac{\delta}{\delta x^i} = \frac{\partial}{\partial x^i} - N_i^\alpha(x,y)\frac{\partial}{\partial y^\alpha},$$

is a local field of frames on HE, i.e. we have

$$(1.7) \qquad\qquad \frac{\delta}{\delta x^i} = B_i^j(x)\frac{\delta}{\delta \tilde{x}^j}.$$

A differentiable section of HE (resp. VE) is called a *horizontal* (resp. *vertical*) *vector field* on E. In a similar way, a differentiable section of the dual vector bundle HE^* (resp. VE^*) is called a *horizontal* (resp. *vertical*) *differential 1-form* on E.

Next, we denote by $\mathfrak{F}(E)$ the algebra of the differentiable functions on E and by $\Gamma(D)$ the $\mathfrak{F}(E)$-module of the differentiable sections of a vector bundle D over E. Then an $\mathfrak{F}(E)$-valued multilinear mapping T defined on

$$\underbrace{\Gamma(HE^*) \times \ldots \times \Gamma(HE^*)}_{l \text{ times}} \times \underbrace{\Gamma(VE^*) \times \ldots \times \Gamma(VE^*)}_{q \text{ times}} \times$$

$$\times \underbrace{\Gamma(HE) \times \ldots \times \Gamma(HE)}_{s \text{ times}} \times \underbrace{\Gamma(VE) \times \ldots \times \Gamma(VE)}_{t \text{ times}}$$

is called an *H-tensor field* of type $\begin{pmatrix} l & q \\ s & t \end{pmatrix}$. By using another approach, Miron [5] introduced such geometrical objects in order to develop a Finsler geometry on a vector bundle.

We give now the local expression of the above concepts. First, any horizontal vector field X and any vertical vector field Y is locally written as

$$(1.8) \qquad X = X^i(x,y)\frac{\delta}{\delta x^i} \qquad \text{and by}$$

$$(1.9) \qquad Y = Y^\alpha(x,y)\frac{\partial}{\partial y^\alpha} \qquad \text{respectively.}$$

On the other hand, any horizontal differential 1-form ω and any vertical differential 1-form Ω is given locally by

$$(1.10) \qquad \omega = \omega_i(x,y)\,dx^i, \qquad \text{and by}$$

$$(1.11) \qquad \Omega = \Omega_\alpha(x,y)\delta y^\alpha \qquad \text{respectively},$$

where we put

$$(1.12) \qquad \delta y^\alpha = dy^\alpha + N_i^\alpha\,dx^i.$$

Then, an H-tensor field T of type $\begin{pmatrix} l & q \\ s & t \end{pmatrix}$ is locally expressed as follows

$$(1.13) \quad T = T_{j_1\cdots j_s\,\beta_1\cdots\beta_t}^{i_1\cdots i_l\,\alpha_1\cdots\alpha_2}(x,y)\frac{\delta}{\delta x^{i_1}}\otimes\cdots\otimes\frac{\delta}{\delta x^{i_e}}\otimes\frac{\partial}{\partial y^{\alpha_1}}\otimes\cdots\otimes\frac{\partial}{\partial y^{\alpha_q}}$$

$$\otimes dx^{j_1}\otimes\cdots\otimes dx^{j_s}\otimes\delta y^{\beta_1}\otimes\cdots\otimes\delta y^{\beta_t}.$$

In order to get covariant derivatives of H-tensor fields we consider the linear connections $\overset{(h)}{\nabla}$ and $\overset{(v)}{\nabla}$ on the vector bundles HE and VE respectively. Then we call the triple $HC = \left(HE, \overset{(h)}{\nabla}, \overset{(v)}{\nabla}\right)$ an H-connection on E. Locally we put

$$(1.14) \quad \overset{(h)}{\nabla}_{\frac{\delta}{\delta x^j}}\frac{\delta}{\delta x^i} = F_i{}^k{}_j(x,y)\frac{\delta}{\delta x^k}; \quad \overset{(h)}{\nabla}_{\frac{\partial}{\partial y^\alpha}}\frac{\delta}{\delta x^i} = D_i{}^k{}_\alpha(x,y)\frac{\delta}{\delta x^k},$$

$$(1.15) \quad \overset{(v)}{\nabla}_{\frac{\delta}{\delta x^j}}\frac{\partial}{\partial y^\alpha} = L_\alpha{}^\gamma{}_j(x,y)\frac{\partial}{\partial y^\gamma}; \quad \overset{(v)}{\nabla}_{\frac{\partial}{\partial y^\beta}}\frac{\partial}{\partial y^\alpha} = C_\alpha{}^\gamma{}_\beta(x,y)\frac{\partial}{\partial y^\gamma}.$$

It is easy to check that an H-connection is locally given by differentiable functions $\left(N_i^\alpha, F_i{}^k{}_j, D_i{}^k{}_\alpha, L_\alpha{}^\gamma{}_j, C_\alpha{}^\gamma{}_\beta\right)$ which satisfy (1.5) and

$$(1.16) \qquad F_i{}^t{}_j(x,y)B_t^s(x) = \widetilde{F}_h{}^s{}_k(\tilde{x},\tilde{y})B_i^h(x)B_j^k(x) + B_i{}^s{}_j(x),$$

$$(1.17) \qquad D_i{}^k{}_\alpha(x,y)B_k^h(x) = \widetilde{D}_j{}^h{}_\gamma(\tilde{x},\tilde{y})B_i^j(x)A_\alpha^\gamma(x),$$

$$(1.18) \qquad L_\alpha{}^\gamma{}_j(x,y)A_\gamma^\beta(x) = \widetilde{L}_\mu{}^\beta{}_k(\tilde{x},\tilde{y})A_\alpha^\mu(x)B_j^k(x) + \frac{\partial A_\alpha^\beta(x)}{\partial x^j},$$

$$(1.19) \qquad C_\alpha{}^\gamma{}_\beta(x,y)A_\gamma^\varepsilon(x) = \widetilde{C}_\nu{}^\varepsilon{}_\mu(\tilde{x},\tilde{y})A_\alpha^\nu(x)A_\beta^\mu(x),$$

with respect to the transformations (1.1), where $B_i{}^s{}_j(x) = \frac{\partial^2 \widetilde{x}^s}{\partial x^i \, \partial x^j}$.

Next, we consider two semi–Riemannian metrics η and g on HE and VE respectively, locally given by

$$(1.20) \qquad \eta_{ij}(x,y) = \eta\left(\frac{\delta}{\delta x^i}, \frac{\delta}{\delta x^j}\right), \qquad \text{and by}$$

$$(1.21) \qquad g_{\alpha\beta}(x,y) = g\left(\frac{\partial}{\partial y^\alpha}, \frac{\partial}{\partial y^\beta}\right)$$

respectively.
Then we define

$$(1.22) \qquad F_i{}^k{}_j = \frac{1}{2}\eta^{kh}\left\{\frac{\delta\eta_{hi}}{\delta x^j} + \frac{\delta\eta_{hj}}{\delta x^i} - \frac{\delta\eta_{ij}}{\delta x^h}\right\},$$

$$(1.23) \qquad D_i{}^k{}_\alpha = 0,$$

$$(1.24) \qquad L_\alpha{}^\gamma{}_i = \frac{\partial N_i^\gamma}{\partial y^\alpha},$$

$$(1.25) \qquad C_\alpha{}^\gamma{}_\beta = \frac{1}{2}g^{\gamma\mu}\left\{\frac{\partial g_{\mu\alpha}}{\partial y^\beta} + \frac{\partial g_{\mu\beta}}{\partial y^\alpha} - \frac{\partial g_{\alpha\beta}}{\partial y^\mu}\right\},$$

where η^{kh} and $g^{\gamma\mu}$ are the entries of the inverse matrix of $[\eta_{ij}]$ and of $[g_{\alpha\beta}]$ respectively. Thus we have an H-connection

$$GHC = \left(N_i^\alpha, F_i{}^k{}_j, D_i{}^k{}_\alpha, L_\alpha{}^\gamma{}_i, C_\alpha{}^\gamma{}_\beta\right)$$

given by an arbitrary horizontal distribution $HE = (N_i^\alpha(x,y))$ and by (1.22)–(1.25). In view of its importance in the gauge theory we develop in the next sections, we call this GHC the *gauge H-connection* on E.

We are now concerned with covariant derivatives induced by an arbitrary H-connection. First, for any horizontal vector field given locally by (1.8) we define its *horizontal* and its *vertical H-covariant derivative* by

$$(1.26) \qquad X^i{}_{|j} = \frac{\delta X^i}{\delta x^j} + X^k F_k{}^i{}_j, \qquad \text{and by}$$

$$(1.27) \qquad X^i{}_{\|\alpha} = \frac{\partial X^i}{\partial y^\alpha} + X^k D_k{}^i{}_\alpha \qquad \text{respectively .}$$

Similarly, for any vertical vector field given by (1.9) we define its horizontal and vertical covariant derivative by

(1.28) $$Y^\alpha{}_{|i} = \frac{\delta Y^\alpha}{\delta x^i} + Y^\gamma L_\gamma{}^\alpha{}_i, \quad \text{and}$$

(1.29) $$Y^\alpha{}_{\|\beta} = \frac{\partial Y^\alpha}{\partial y^\beta} + Y^\gamma C_\gamma{}^\alpha{}_\beta$$

respectively.

The horizontal and vertical H-covariant derivatives of the horizontal differential 1-form w given by (1.10) and of the vertical differential 1-form Ω given by (1.11) are defined by

(1.30) $$w_{i|j} = \frac{\delta w_i}{\delta x^j} - w_k F_i{}^k{}_j,$$

(1.31) $$w_{i\|\alpha} = \frac{\partial w_i}{\partial y^\alpha} - w_k D_i{}^k{}_\alpha, \quad \text{and}$$

(1.32) $$\Omega_{\alpha|i} = \frac{\delta \Omega_\alpha}{\delta x^i} - \Omega_\gamma L_\alpha{}^\gamma{}_i,$$

(1.33) $$\Omega_{\alpha\|\beta} = \frac{\partial \Omega_\alpha}{\partial y^\beta} - \Omega_\gamma C_\alpha{}^\gamma{}_\beta \quad \text{respectively .}$$

We may easily form similar formulas for arbitrary H-tensor fields, however, we present here only the *horizontal* and the *vertical H-covariant derivatives* of an H-tensor field T with components $T^{i\alpha}_{j\beta}$ as follows

(1.34) $$T^{i\alpha}_{j\beta|k} = \frac{\delta T^{i\alpha}_{j\beta}}{\delta x^k} + T^{h\alpha}_{j\beta} F_h{}^i{}_k + T^{i\gamma}_{j\beta} L_\gamma{}^\alpha{}_k -$$
$$- T^{i\alpha}_{h\beta} F_j{}^h{}_k - T^{i\alpha}_{j\gamma} L_\beta{}^\gamma{}_k, \quad \text{and}$$

(1.35) $$T^{i\alpha}_{j\beta\|\gamma} = \frac{\partial T^{i\alpha}_{j\beta}}{\partial y^\gamma} + T^{k\alpha}_{j\beta} D_k{}^i{}_\gamma + T^{i\mu}_{j\beta} C_\mu{}^\alpha{}_\gamma -$$
$$- T^{i\alpha}_{k\beta} D_j{}^k{}_\gamma - T^{i\alpha}_{j\mu} C_\beta{}^\mu{}_\gamma.$$

An H-connection on E is said to be (h)-metrical (resp. (v)-metrical) if we have

(1.36) $$\eta_{ij|k} = 0 \quad (\text{resp. } g_{\alpha\beta\|\gamma} = 0).$$

It is easy to check that the gauge H-connection GHC is both (h)-metrical and (v)-metrical.

We say that the curvatures of the linear connections $\overset{(h)}{\nabla}$ and $\overset{(v)}{\nabla}$ are the curvatures of the H-connection $HC = \left(HE, \overset{(h)}{\nabla}, \overset{(v)}{\nabla} \right)$. It is now our goal to define torsions of HC. To this end we denote by h and v the projection morphism of TE to HE and to VE respectively. Then for any $X \in \Gamma(TE)$ we define the differential operators:

$$(1.37) \quad \overset{(h)}{\nabla}{}'_X : \Gamma(TE) \to \Gamma(TE); \quad \overset{(h)}{\nabla}{}'_X Y = \overset{(h)}{\nabla}{}_X hY,$$

$$(1.38) \quad \overset{(v)}{\nabla}{}'_X : \Gamma(TE) \to \Gamma(TE); \quad \overset{(v)}{\nabla}{}'_X Y = \overset{(v)}{\nabla}{}_X vY, \quad \forall Y \in \Gamma(TE).$$

It is easy to check that (1.37) and (1.38) define two *Otsuki connections* on TE (see Abe [1]). Then according to Nemoto [6] we get the torsion tensor fields

$$(1.39) \quad \overset{(h)}{T}{}'(X,Y) = \overset{(h)}{\nabla}{}'_X Y - \overset{(h)}{\nabla}{}'_Y X - h([X,Y]),$$

$$(1.40) \quad \overset{(v)}{T}{}'(X,Y) = \overset{(v)}{\nabla}{}'_X Y - \overset{(v)}{\nabla}{}'_Y X - v([X,Y]), \quad \forall X, Y \in \Gamma(TE).$$

We say that the torsion tensor fields $\overset{(h)}{T}{}'$ and $\overset{(v)}{T}{}'$ of $\overset{(h)}{\nabla}{}'$ and $\overset{(v)}{\nabla}{}'$ respectively give the torsions of HC.

In order to get the local components of the torsions of HC we first derive

$$(1.41) \qquad \left[\frac{\delta}{\delta x^i}, \frac{\delta}{\delta x^j} \right] = R^\alpha{}_{ij} \frac{\partial}{\partial y^\alpha}, \quad \text{and}$$

$$(1.42) \qquad \left[\frac{\delta}{\delta x^i}, \frac{\partial}{\partial y^\alpha} \right] = \frac{\partial N_i^\gamma}{\partial y^\alpha} \frac{\partial}{\partial y^\gamma},$$

where we put

$$(1.43) \qquad R^\alpha{}_{ij} = \frac{\delta N_i^\alpha}{\delta x^j} - \frac{\delta N_j^\alpha}{\delta x^i}.$$

Then by direct calculations from (1.14), (1.15) and (1.39)–(1.43) we obtain

$$(1.44) \qquad \overset{(h)}{T}{}' \left(\frac{\delta}{\delta x^j}, \frac{\delta}{\delta x^i} \right) = T_i{}^k{}_j \frac{\delta}{\delta x^k},$$

$$(1.45) \qquad \overset{(h)}{T}{}' \left(\frac{\partial}{\partial y^\alpha}, \frac{\delta}{\delta x^i} \right) = D_i{}^k{}_\alpha \frac{\delta}{\delta x^k},$$

(1.46)
$$\overset{(h)}{T}{}'\left(\frac{\partial}{\partial y^{\beta}},\frac{\partial}{\partial y^{\alpha}}\right)=0,$$

(1.47)
$$\overset{(v)}{T}{}'\left(\frac{\delta}{\delta x^{j}},\frac{\delta}{\delta x^{i}}\right)=R^{\gamma}{}_{ij}\frac{\partial}{\partial y^{\gamma}},$$

(1.48)
$$\overset{(v)}{T}{}'\left(\frac{\partial}{\partial y^{\alpha}},\frac{\delta}{\delta x^{i}}\right)=P^{\gamma}{}_{i\alpha}\frac{\partial}{\partial y^{\gamma}},$$

(1.49)
$$\overset{(v)}{T}{}'\left(\frac{\partial}{\partial y^{\beta}},\frac{\partial}{\partial y^{\alpha}}\right)=S^{\gamma}{}_{\alpha\beta}\frac{\partial}{\partial y^{\gamma}},$$

where we put

(1.50)
$$T_i{}^k{}_j = F_i{}^k{}_j - F_j{}^k{}_i,$$

(1.51)
$$P^{\gamma}{}_{i\alpha} = \frac{\partial N_i^{\gamma}}{\partial y^{\alpha}} - L_{\alpha}{}^{\gamma}{}_i,$$

(1.52)
$$S^{\gamma}{}_{\alpha\beta} = C_{\alpha}{}^{\gamma}{}_{\beta} - C_{\beta}{}^{\gamma}{}_{\alpha}.$$

Therefore, an H-connection has five torsions $T_i{}^k{}_j$, $D_i{}^k{}_{\alpha}$, $R^{\gamma}{}_{ij}$, $P^{\gamma}{}_{i\alpha}$ and $S^{\gamma}{}_{\alpha\beta}$ which are H-tensor fields on E. It is important to note that in case of a gauge H-connection all these torsions vanish except $R^{\alpha}{}_{ij}$.

§2. Global gauge invariance of Lagrangians on a vector bundle

Let $\xi = (E, \pi, M)$ be a vector bundle and $Q^A(x)$, $A = 1,\ldots,p$ be some physical fields on the base manifold M. As it is well-known (see Chaichian–Nelipa [4]), the simplest Lagrangian on M is of the following form

(2.1)
$$L_0(x) = L\left(Q^A(x), \frac{\partial Q^A}{\partial x^i}\right),$$

where L is a differentiable function on a domain of $\mathbb{R}^{p(1+m)}$. Next we consider some scalar fields $Q^A(x,y)$ on E. Then by replacing $Q^A(x)$ and $\frac{\partial Q^A}{\partial x^i}$ from (2.1) by $Q^A(x,y)$ and $\frac{\delta Q^A}{\delta x^i}$ respectively and by using (1.7) we obtain the following Lagrangian on the total space E:

(2.2)
$$L_0'(x,y) = L\left(Q^A(x,y), \frac{\delta Q^A}{\delta x^i}\right).$$

Thus we have a general procedure to get Lagrangians on the total space of a vector bundle from Lagrangians on the base space, provided there exists on E a horizontal distribution HE. It must be noted that the Lagrangian (2.2) contains both kinds of partial derivatives $\frac{\partial Q^A}{\partial x^i}$ and $\frac{\partial Q^A}{\partial y^\alpha}$ but incorporated in $\frac{\delta Q^A}{\delta x^i}$. However, according to the homogeneous transformations (1.3) we may consider Lagrangians on E of the following general form

$$(2.3) \qquad \mathcal{L}_0(x,y) = \mathcal{L}\left(\eta_{ij}(x,y),\ g_{\alpha\beta}(x,y),\ Q^A(x,y),\ \frac{\delta Q^A}{\delta x^i}(x,y), \right.$$
$$\left. \frac{\partial Q^A}{\partial y^\alpha}(x,y) \right),$$

where $\eta_{ij}(x,y)$ and $g_{\alpha\beta}(x,y)$ are the components of the semi–Riemannian metrics η and g on HE and VE respectively and $\mathcal{L}$ is a differentiable function on a domain of R^t, $t = m^2 + r^2 + p(1 + m + r)$.

Since η and g are H-tensor fields of type $\begin{pmatrix} 0 & 0 \\ 2 & 0 \end{pmatrix}$ and $\begin{pmatrix} 0 & 0 \\ 0 & 2 \end{pmatrix}$ respectively, we have

$$(2.4) \qquad\qquad \eta_{ij}(x,y) = \tilde{\eta}_{kh}(\tilde{x},\tilde{y}) B_i^k(x) B_j^h(x),$$
$$(2.5) \qquad\qquad g_{\alpha\beta}(x,y) = \tilde{g}_{\gamma\mu}(\tilde{x},\tilde{y}) A_\alpha^\gamma(x) A_\beta^\mu(x),$$

with respect to the transformations (1.1). Next, we define locally

$$(2.6) \quad H(x,y) = (|\det(\eta_{ij}(x,y))|)^{\frac{1}{2}}; \quad V(x,y) = (|\det(g_{\alpha\beta}(x,y))|)^{\frac{1}{2}},$$
$$(2.7) \qquad B(x) = \det(B_i^j(x)); \quad A(x) = \det(A_\alpha^\beta(x)).$$

Then by using (2.4)–(2.7) we infer

$$(2.8) \qquad\qquad H(x,y) = \tilde{H}(\tilde{x},\tilde{y})|B(x)|, \qquad \text{and}$$
$$(2.9) \qquad\qquad V(x,y) = \tilde{V}(\tilde{x},\tilde{y})|A(x)|.$$

Further, we define locally

$$(2.10) \qquad\qquad \mathcal{L}_0^*(x,y) = \mathcal{L}_0(x,y) H(x,y) V(x,y),$$

and by using (2.8)–(2.10) obtain

$$(2.11) \qquad\qquad \mathcal{L}_0^*(x,y) = \mathcal{L}_0^*(\tilde{x},\tilde{y}) B(x) A(x),$$

provided E is an orientable manifold. Thus $\mathcal{L}_0^*(x, y)$ is a Lagrangian density on E and we may claim that the functional

$$(2.12) \qquad I(\Omega) = \int_\Omega \mathcal{L}_0^*(x, y) \, dx^1 \wedge \ldots \wedge dx^m \wedge dy^1 \wedge \ldots \wedge dy^r,$$

where Ω is a compact domain of E, is independent of coordinates on E.

The variational principle

$$(2.13) \qquad \delta(I(\Omega)) = 0$$

implies the Euler–Lagrange equations for $Q^A(x, y)$:

$$(2.14) \qquad \frac{\partial \mathcal{L}_0^*}{\partial Q^A} - \frac{\partial}{\partial x^i}\left(\frac{\partial \mathcal{L}_0^*}{\left(\dfrac{\partial Q^A}{\partial x^i}\right)}\right) - \frac{\partial}{\partial y^\alpha}\left(\frac{\partial \mathcal{L}_0^*}{\partial\left(\dfrac{\partial Q^A}{\partial y^\alpha}\right)}\right) = 0.$$

As in (2.14), for a better printing we sometimes omit (x, y) from some symbols $S(x, y)$.

We next put

$$(2.15) \qquad \overset{(h)}{Q}{}_A{}^i = \frac{\partial \mathcal{L}}{\partial\left(\dfrac{\delta Q^A}{\delta x^i}\right)}, \qquad \text{and}$$

$$(2.16) \qquad \overset{(v)}{Q}{}_A{}^\alpha = \frac{\partial \mathcal{L}}{\partial\left(\dfrac{\partial Q^A}{\partial y^\alpha}\right)}\Bigg|\frac{\delta Q^A}{\delta x^i} = \text{const}$$

Then by using (1.3) and (1.7) we obtain

$$(2.17) \qquad \overset{\widetilde{(h)}}{Q}{}_A{}^j = \overset{(h)}{Q}{}_A{}^i B_i^j(x), \qquad \text{and}$$

$$(2.18) \qquad \overset{\widetilde{(v)}}{Q}{}_A{}^\beta = \overset{(v)}{Q}{}_A{}^\alpha A_\alpha^\beta(x),$$

with respect to (1.1). Hence

$$(2.19) \qquad \overset{(h)}{Q}{}_A = \overset{(h)}{Q}{}_A{}^i \frac{\delta}{\delta x^i}, \qquad \text{and}$$

$$(2.20) \qquad \overset{(v)}{Q}{}_A = \overset{(v)}{Q}{}_A{}^\alpha \frac{\partial}{\partial y^\alpha},$$

are p horizontal and vertical vector fields on E respectively.

Finally, by using (2.15), (2.16) and the covariant derivatives (1.26) and (1.29) with respect to the gauge H-connection GHC it is easy to see that (2.14) becomes

$$(2.21) \qquad \frac{\partial \mathcal{L}}{\partial Q^A} - \overset{(h)}{Q}{}_{A}{}^{i}{}_{|i} - \overset{(v)}{Q}{}_{A}{}^{\alpha}{}_{\|\alpha} = E_A,$$

where we put

$$(2.22) \qquad E_A = \left\{ \frac{1}{HV} \frac{\delta(HV)}{\delta x^i} - (L_\alpha{}^\alpha{}_i + F_i{}^j{}_j) \right\} \overset{(h)}{Q}{}_{A}{}^{i} +$$

$$+ \left\{ \frac{1}{HV} \frac{\partial(HV)}{\partial y^\alpha} - C_\alpha{}^\gamma{}_\gamma \right\} \overset{(v)}{Q}{}_{A}{}^{\alpha}.$$

Suppose now $E = M \times R^r$, that is, ξ is a trivial vector bundle over M. In this case we have the transformations of coordinates

$$(2.23) \qquad \begin{cases} \tilde{x}^i = \tilde{x}^i(x^1, \ldots, x^m) \\ \tilde{y}^\alpha = A^\alpha_\beta y^\beta \end{cases}$$

where A^α_β are real constants such that $\det(A^\alpha_\beta) \neq 0$. We next consider teh semi–Riemannian metrics $g = (g_{\alpha\beta}(x))$ and $\eta = (\eta_{ij}(x))$ on the vector bundle ξ and on M respectively. Then we define

$$(2.24) \qquad L_\beta{}^\alpha{}_i(x) = \frac{1}{2} g^{\alpha\mu}(x) \frac{\partial g_{\mu\beta}(x)}{\partial x^i}, \qquad \text{and}$$

$$(2.25) \qquad \overset{\circ}{N}{}_{i}^{\alpha}(x, y) = L_\beta{}^\alpha{}_i(x) y^\beta.$$

It is easy to see that the $\overset{\circ}{N}{}_{i}^{\alpha}(x, y)$ satisfy (1.5) with respect to (2.23). Hence in this case we have a particular horizontal distribution $\overset{\circ}{H}E$ given locally by (2.25). We may consider $\eta = (\eta_{ij}(x))$ as a particular semi–Riemannian metric on HE and the $F_i{}^k{}_j$ from (1.22) become just the coefficients of the Levi–Civita connection on M. Finally, we note that the $C_\alpha{}^\gamma{}_\beta$ from (1.25) are vanishing identically on E. Therefore, we get a gauge H-connection $\overset{\circ}{GHC} = (\overset{\circ}{N}{}_{i}^{\alpha}, F_i{}^k{}_j, D_i{}^j{}_\alpha, L_\alpha{}^\beta{}_i, C_\alpha{}^\gamma{}_\beta)$ given by (2.24), (2.25), (1.23),

$$(2.26) \qquad F_i{}^k{}_j = \frac{1}{2} \eta^{kh} \left\{ \frac{\partial \eta_{hi}}{\partial x^j} + \frac{\partial \eta_{hj}}{\partial x^i} - \frac{\partial \eta_{ij}}{\partial x^h} \right\}$$

and

(2.27) $$C_\alpha{}^\gamma{}_\beta = 0.$$

Further, by direct calculations using (2.6), (2.24) and (2.26) we obtain

(2.28) $$\frac{1}{V}\frac{\partial V}{\partial x^i} = L_\alpha{}^\alpha{}_i, \qquad \text{and}$$

(2.29) $$\frac{1}{H}\frac{\partial H}{\partial x^i} = F_i{}^j{}_j.$$

Taking into account that in this case H and V do not depend on (y^α) and using (2.27)–(2.29) in (2.22), we obtain the vanishing of E_A. Hence in case of the trivial vector bundle, the Euler–Lagrange equations become

(2.30) $$\frac{\partial \mathcal{L}}{\partial Q^A} - \overset{(h)}{Q}{}_A{}^i{}_{|i} - \overset{(v)}{Q}{}_A{}^\alpha{}_{\|\alpha} = 0,$$

where H-covariant derivatives are considered with respect to $\overset{\circ}{G}HC$.

We now come back to an arbitrary vector bundle $\xi = (E, \pi, M)$. Suppose G is an n-dimensional Lie group with a p-dimensional representation. Denote by (ε^a) the parameters of G and by $\{X_a\}$ a basis of the Lie algebra of G. Then we say that we have a *global gauge action* of G on the physical fields $Q^A(x, y)$ given by the infinitesimal transformations

(2.31) $$Q'^A(x, y) = Q^A(x, y) + \delta(Q^A(x, y)),$$

where we put

(2.32) $$\delta(Q^A(x, y)) = \varepsilon^a [X_a]^A_B Q^B(x, y).$$

From (2.31), by using (2.32) we obtain

(2.33) $$\frac{\delta Q'^A}{\delta x^i} = \frac{\delta Q^A}{\delta x^i} + \delta\left(\frac{\delta Q^A}{\delta x^i}\right), \qquad \text{and}$$

(2.34) $$\frac{\partial Q'^A}{\partial y^\alpha} = \frac{\partial Q^A}{\partial y^\alpha} + \delta\left(\frac{\partial Q^A}{\partial y^\alpha}\right),$$

where we put

(2.35) $$\delta\left(\frac{\delta Q^A}{\delta x^i}\right) = \varepsilon^a [X_a]^A_B \frac{\delta Q^B}{\delta x^i}, \qquad \text{and}$$

(2.36) $$\delta\left(\frac{\partial Q^A}{\partial y^\alpha}\right) = \varepsilon^a [X_a]^A_B \frac{\partial Q^B}{\partial y^\alpha}.$$

Next we locally define

$$(2.37) \qquad \overset{(h)}{J}{}_a{}^i = - \overset{(h)}{Q}{}_A{}^i [X_a]_B^A Q^B, \qquad \text{and}$$

$$(2.38) \qquad \overset{(v)}{J}{}_a{}^\alpha = - \overset{(v)}{Q}{}_A{}^\alpha [X_a]_B^A Q^B,$$

and note that

$$(2.39) \qquad \overset{(h)}{J}{}_a = \overset{(h)}{J}{}_a{}^i \frac{\delta}{\delta x^i}, \qquad \text{and}$$

$$(2.40) \qquad \overset{(v)}{J}{}_a = \overset{(v)}{J}{}_a{}^\alpha \frac{\partial}{\partial y^\alpha},$$

are n horizontal and vertical vector field on E respectively. We call $\overset{(h)}{J}{}_a$ and $\overset{(v)}{J}{}_a$ the *horizontal* and the *vertical currents* on E respectively.

Finally, we suppose that the Lagrangian $\mathcal{L}_0(x, y)$ from (2.3) is invariant with respect to the infinitesimal transformations (2.31), (2.33) and (2.34), that is

$$(2.41) \qquad \left\{ \frac{\partial \mathcal{L}}{\partial Q^A} Q^B + \overset{(h)}{Q}{}_A{}^i \frac{\delta Q^B}{\delta x^i} + \overset{(v)}{Q}{}_A{}^\alpha \frac{\partial Q^B}{\partial y^\alpha} \right\} [X_a]_B^A = 0.$$

Then by using (2.21), (2.37), (2.38) and (2.41) we obtain the following *conservation laws*

$$(2.42) \qquad \overset{(h)}{J}{}_a{}^i{}_{|i} + \overset{(v)}{J}{}_a{}^\alpha{}_{\|\alpha} = E_A [X_a]_B^A Q^B.$$

In case of the trivial vector bundle the conservation laws become

$$(2.43) \qquad \overset{(h)}{J}{}_a{}^i{}_{|i} + \overset{(v)}{J}{}_a{}^\alpha{}_{\|\alpha} = 0.$$

§3. Local gauge invariance of Lagrangians on a vector bundle

Suppose now that G acts on the physical fields $Q^A(x^i, y^\alpha)$ locally, that is, the constants ε^a from the previous paragraph are replaced by differentiable functions $\varepsilon^a(x^i, y^\alpha)$. Hence the local action of G is given by

$$(3.1) \qquad Q'^A(x, y) = Q^A(x, y) + \overset{*}{\delta}(Q^A(x, y)),$$

where we put

$$(3.2) \qquad \overset{*}{\delta}(Q^A(x, y)) = \varepsilon^a(x, y)[X_a]^A_B Q^B(x, y).$$

Then we obtain

$$(3.3) \qquad \frac{\delta Q'^A}{\delta x^i} = \frac{\delta Q^A}{\delta x^i} + \varepsilon^a [X_a]^A_B \frac{\delta Q^B}{\delta x^i} + \frac{\delta \varepsilon^a}{\delta x^i} [X_a]^A_B Q^B,$$

$$(3.4) \qquad \frac{\partial Q'^A}{\partial y^\alpha} = \frac{\partial Q^A}{\partial y^\alpha} + \varepsilon^a [X_a]^A_B \frac{\partial Q^B}{\partial y^\alpha} + \frac{\partial \varepsilon^a}{\partial y^\alpha} [X_a]^A_B Q^B.$$

Thus in general the global gauge invariant Lagrangian (2.3) is not invariant with respect to (3.3) and (3.4).

In order to get a local gauge invariant Lagrangian from $\mathcal{L}_0(x, y)$ we introduce the *horizontal* and the *vertical gauge fields* $H^a_i(x, y)$ and $V^a_\alpha(x, y)$ respectively. More precisely, we have the horizontal and vertical differential 1-forms:

$$(3.5) \qquad H^a = H^a_i(x, y)\, dx^i, \qquad \text{and}$$
$$(3.6) \qquad V^a = V^a_\alpha(x, y)\delta y^\alpha \qquad \text{respectively.}$$

Next, we assume that the local action of G on the gauge fields is given by

$$(3.7) \qquad \overset{*}{\delta}(H^a_i(x, y)) = \varepsilon^b C_b{}^a{}_c H^c_i(x, y) + \frac{\delta \varepsilon^a}{\delta x^i}(x, y),$$

and

$$(3.8) \qquad \overset{*}{\delta}(V^a_\alpha(x, y)) = \varepsilon^b C_b{}^a{}_c V^c_\alpha(x, y) + \frac{\partial \varepsilon^a}{\partial y^\alpha}(x, y) \qquad \text{respectively}$$

where $C_b{}^a{}_c$ are the structure constants of the Lie algebra of G with respect to the basis $\{X_a\}$. On the other hand, we define the *horizontal* and the *vertical gauge covariant derivatives* of the physical fields by

(3.9) $$\overset{(h)}{D}{}_i Q^A(x,y) = \frac{\delta Q^A}{\delta x^i}(x,y) - H_i^a(x,y)[X_a]_B^A Q^B(x,y),$$

and by

(3.10)
$$\overset{(v)}{D}{}_\alpha Q^A(x,y) = \frac{\partial Q^A}{\partial y^\alpha}(x,y) - V_\alpha^a(x,y)[X_a]_B^A Q^B(x,y) \quad \text{respectively .}$$

It is important to note that $\overset{(h)}{D}{}_i Q^A$ and $\overset{(v)}{D}{}_\alpha Q^A$ are the local components of the horizontal and of the vertical differential 1-forms

(3.11) $$\overset{(h)}{D} Q^A = \left(\overset{(h)}{D}{}_i Q^A\right) dx^i, \qquad \text{and}$$

(3.12) $$\overset{(v)}{D} Q^A = \left(\overset{(v)}{D}{}_\alpha Q^A\right) \delta y^\alpha,$$

respectively. Moreover, by (3.3) and (3.4) we have

(3.13) $$\overset{*}{\delta}\left(\frac{\delta Q^A}{\delta x^i}\right) = \varepsilon^a(x,y)[X_a]_B^A \frac{\delta Q^B}{\delta x^i} + \frac{\delta \varepsilon^a}{\delta x^i}[X_a]_B^A Q^B,$$

and

(3.14) $$\overset{*}{\delta}\left(\frac{\partial Q^A}{\partial y^\alpha}\right) = \varepsilon^a(x,y)[X_a]_B^A \frac{\partial Q^B}{\partial y^\alpha} + \frac{\partial \varepsilon^a}{\partial y^\alpha}[X_a]_B^A Q^B \qquad \text{respectively.}$$

Then by using (3.2), (3.7)–(3.10), (3.13) and (3.14) we obtain the homogeneous transformations

(3.15) $$\overset{*}{\delta}\left(\overset{(h)}{D}{}_i Q^A\right) = \varepsilon^a(x,y)[X_a]_B^A \overset{(h)}{D}{}_i Q^B,$$

(3.16) $$\overset{*}{\delta}\left(\overset{(v)}{D}{}_\alpha Q^A\right) = \varepsilon^a(x,y)[X_a]_B^A \overset{(v)}{D}{}_\alpha Q^B.$$

Finally, we consider the Lagrangian

(3.17)
$$\mathcal{L}_0'(x,y) = \mathcal{L}\left(\eta_{ij}(x,y), g_{\alpha\beta}(x,y), Q^A(x,y), \overset{(h)}{D}{}_i Q^A(x,y), \overset{(v)}{D}{}_\alpha Q^A(x,y)\right)$$

where $\mathcal{L}$ is the function in (2.6) and we claim that it is local gauge invariant with respect to the action of G. In fact, by using (3.2), (3.15) and (3.16) we obtain

$$(3.19) \qquad \overset{*}{\delta} \mathcal{L}'_0(x,y) = \left\{ \frac{\partial \mathcal{L}}{\partial Q^A} Q^B + \frac{\partial \mathcal{L}}{\partial \left(\overset{(h)}{D}_i Q^A \right)} \overset{(h)}{D}_i Q^B + \right.$$

$$\left. + \frac{\partial \mathcal{L}}{\partial \left(\overset{(v)}{D}_\alpha Q^A \right)} \overset{(v)}{D}_\alpha Q^B \right\} [X_a]^A_B \varepsilon^a.$$

Hence, taking account of (2.40) in (3.19) we infer $\overset{*}{\delta} \mathcal{L}'_0(x,y) = 0$, that is, $\mathcal{L}'_0(x,y)$ is a local gauge invariant Lagrangian.

Therefore, the local gauge invariant Lagrangian $\mathcal{L}'_0(x,y)$ is obtained from the global gauge invariant Lagrangian by a simple replacing of the derivatives $\frac{\delta Q^A}{\delta x^i}$ and $\frac{\partial Q^A}{\partial y^\alpha}$ by $\overset{(h)}{D}_i Q^A$ and by $\overset{(v)}{D}_\alpha Q^A$ respectively.

§4. Strength fields and Lagrangians for gauge fields

First, by using the structure constants $C_b{}^a{}_c$ of the Lie group G, the gauge fields $H^a_i(x,y)$ and $V^a_\alpha(x,y)$, and the horizontal distribution $HE = (N^\alpha_i(x,y))$, we locally define the differentiable functions

$$(4.1) \qquad R^a{}_{ij} = \frac{\delta H^a_i}{\delta x^j} - \frac{\delta H^a_j}{\delta x^i} - C_b{}^a{}_c H^c_i H^b_j + R^\gamma{}_{ij} V^a_\gamma,$$

$$(4.2) \qquad P^a{}_{i\alpha} = \frac{\partial H^a_i}{\partial y^\alpha} - \frac{\delta V^a_\alpha}{\delta x^i} - C_b{}^a{}_c H^c_i V^b_\alpha + \frac{\partial N^\gamma_i}{\partial y^\alpha} V^a_\gamma,$$

$$(4.3) \qquad S^a{}_{\alpha\beta} = \frac{\partial V^a_\alpha}{\partial y^\beta} - \frac{\partial V^a_\beta}{\partial y^\alpha} - C_b{}^a{}_c V^c_\alpha V^b_\beta,$$

where $R^\gamma{}_{ij}$ are the H-tensor fields given by (1.43).

It is noteworthy that $R^a{}_{ij}$, $P^a{}_{i\alpha}$ and $S^a{}_{\alpha\beta}$ are the components of some H-tensor fields of type $\begin{pmatrix} 0 & 0 \\ 2 & 0 \end{pmatrix}$, $\begin{pmatrix} 0 & 0 \\ 1 & 1 \end{pmatrix}$ and $\begin{pmatrix} 0 & 0 \\ 0 & 2 \end{pmatrix}$ respectively, for each $a \in \{1, \ldots, n\}$. Moreover, by direct calculations from (3.7) and (3.8) we obtain

(4.4) $$\overset{*}{\delta}(R^a{}_{ij}) = \varepsilon^b C_b{}^a{}_c R^c{}_{ij},$$

(4.5) $$\overset{*}{\delta}(P^a{}_{i\alpha}) = \varepsilon^b C_b{}^a{}_c P^c{}_{i\alpha},$$

(4.6) $$\overset{*}{\delta}(S^a{}_{\alpha\beta}) = \varepsilon^b C_b{}^a{}_c S^c{}_{\alpha\beta},$$

with respect to the local gauge action of G. We call $R^a{}_{ij}$, $P^a{}_{i\alpha}$, and $S^a{}_{\alpha\beta}$ the *horizontal*, the *mixed* and the *vertical strength fields* corresponding to the gauge theory we are dealing with.

Next, in order to construct Lagrangians for gauge fields we consider two semi–Riemannian metrics $\eta = (\eta_{ij}(x,y))$ and $g = (g_{\alpha\beta}(x,y))$ on HE and VE respectively and the Killing form K of G given by

(4.7) $$K_{ab} = C_a{}^d{}_e C_b{}^e{}_d.$$

Moreover, we suppose that G is a compact semisimple Lie group. Then K is nondegenerate on the Lie algebra of G and the structure constants of G are totally anti–symmetric, that is, we have

(4.8) $$C_b{}^a{}_c = -C_c{}^a{}_b = -C_a{}^b{}_c.$$

Finally, taking into account the tensorial character of strength fields we define the following Lagrangians:

(4.9) $$L_H(x,y) = -\frac{1}{4} K_{ab} \eta^{ij}(x,y) \eta^{kh}(x,y) R^a{}_{ik}(x,y) R^b{}_{jh}(x,y),$$

(4.10) $$L_{HV}(x,y) = -\frac{1}{2} K_{ab} \eta^{ij}(x,y) g^{\alpha\beta}(x,y) P^a{}_{i\alpha}(x,y) P^b{}_{j\beta}(x,y),$$

(4.11) $$L_V(x,y) = -\frac{1}{4} K_{ab} g^{\alpha\beta}(x,y) g^{\gamma\mu}(x,y) S^a{}_{\alpha\gamma}(x,y) S^b{}_{\beta\mu}(x,y).$$

Clearly, L_H, L_{HV} and L_V are differentiable scalar fields on E. Moreover, it is easy to check that they are local gauge invariant with respect to the action of G.

For the particular case of the 1-dimensional Lie group $U(1)$ of phase transformations we denote by $H_i(x,y)$ and $V_\alpha(x,y)$ the gauge fields on E. Then the strength fields are given by

(4.12) $$R_{ij} = \frac{\delta H_i}{\delta x^j} - \frac{\delta H_j}{\delta x^i} + R^\gamma{}_{ij} V_\gamma,$$

$$(4.13) \qquad P_{i\alpha} = \frac{\partial H_i}{\partial y^\alpha} - \frac{\delta V_\alpha}{\delta x^i} + \frac{\partial N_i^\gamma}{\partial y^\alpha} V_\gamma,$$

$$(4.14) \qquad S_{\alpha\beta} = \frac{\partial V_\alpha}{\partial y^\beta} - \frac{\partial V_\beta}{\partial y^\alpha}.$$

In case the H_i are functions of (x^i) alone and the V_α are vanishing identically, (4.12) becomes

$$(4.15) \qquad R_{ij} = \frac{\partial H_i}{\partial x^j} - \frac{\partial H_j}{\partial x^i},$$

which is nothing but the *electromagnetic tensor field*.

The local gauge invariant Lagrangians for gauge fields with respect to $U(1)$ are given by

$$(4.16) \qquad L_H(x, y) = -\frac{1}{4}\eta^{ij}(x, y)\eta^{kh}(x, y)R_{ik}(x, y)R_{jh}(x, y),$$

$$(4.17) \qquad L_{HV}(x, y) = -\frac{1}{2}\eta^{ij}(x, y)g^{\alpha\beta}(x, y)P_{i\alpha}(x, y)P_{j\beta}(x, y),$$

$$(4.18) \qquad L_V(x, y) = -\frac{1}{4}g^{\alpha\beta}(x, y)g^{\gamma\mu}(x, y)S_{\alpha\gamma}(x, y)S_{\beta\mu}(x, y).$$

As full Lagrangian for our gauge theory on the total space of an arbitrary vector bundle we propose

$$(4.19) \qquad \mathcal{L}(x, y) = \mathcal{L}'_0(x, y) + L_H(x, y) + L_{HV}(x, y) + L_V(x, y),$$

where $\mathcal{L}'_0(x, y)$ is the local gauge invariant Lagrangian (3.17).

§5. Equations of motion and conservation laws

In this section we consider the full Lagrangian (4.19) and define the Lagrangian density

$$(5.1) \qquad \mathcal{L}^*(x, y) = \mathcal{L}(x, y)H(x, y)V(x, y),$$

where H and V are differentiable functions given by (2.6). Then, as in §2, we consider the variational principle

$$(5.2) \qquad \delta\left(\int_\Omega \mathcal{L}^*(x, y)\, dx^1 \wedge \ldots \wedge dx^m \wedge dy^1 \wedge \ldots \wedge dy^r\right) = 0.$$

The same principle was considered by Asanov [2], p. 244, but with respect to some other Lagrangians.

Thus we have the following Euler–Lagrange equations

$$
\text{(5.3)} \qquad \frac{\partial \mathcal{L}^*}{\partial Q^A} - \frac{\partial}{\partial x^i}\left(\frac{\partial \mathcal{L}^*}{\partial\left(\frac{\partial Q^A}{\partial x^i}\right)}\right) - \frac{\partial}{\partial y^\alpha}\left(\frac{\partial \mathcal{L}^*}{\partial\left(\frac{\partial Q^A}{\partial y^\alpha}\right)}\right) = 0,
$$

$$
\text{(5.4)} \qquad \frac{\partial \mathcal{L}^*}{\partial H_i^a} - \frac{\partial}{\partial x^j}\left(\frac{\partial \mathcal{L}^*}{\partial\left(\frac{\partial H_i^a}{\partial x^j}\right)}\right) - \frac{\partial}{\partial y^\alpha}\left(\frac{\partial \mathcal{L}^*}{\partial\left(\frac{\partial H_i^a}{\partial y^\alpha}\right)}\right) = 0,
$$

$$
\text{(5.5)} \qquad \frac{\partial \mathcal{L}^*}{\partial V_\alpha^a} - \frac{\partial}{\partial x^i}\left(\frac{\partial \mathcal{L}^*}{\partial\left(\frac{\partial V_\alpha^a}{\partial x^i}\right)}\right) - \frac{\partial}{\partial y^\beta}\left(\frac{\partial \mathcal{L}^*}{\partial\left(\frac{\partial V_\alpha^a}{\partial y^\beta}\right)}\right) = 0.
$$

According to the theory we developped in §2, the equations (5.3) become

$$
\text{(5.6)} \qquad \frac{\partial \mathcal{L}}{\partial Q^A} - \overset{(h)}{Q}{}_A{}^i{}_{|i} - \overset{(v)}{Q}{}_A{}^\alpha{}_{\|\alpha} = E_A,
$$

where the covariant derivatives are considered with respect to the gauge H-connection GHC given by (1.22)–(1.25).

We are now interested in getting formulas similar to (5.6) in case of (5.4) and (5.5). To this end we first infer

$$
\text{(5.7)} \qquad \frac{\partial \mathcal{L}^*}{\partial\left(\frac{\partial H_i^a}{\partial x^j}\right)} = \frac{\partial \mathcal{L}}{\partial\left(\frac{\delta H_i^a}{\delta x^j}\right)} HV, \qquad \text{and}
$$

$$
\text{(5.8)} \qquad \frac{\partial \mathcal{L}^*}{\partial\left(\frac{\partial H_i^a}{\partial y^\alpha}\right)} = \left\{ \frac{\partial \mathcal{L}}{\partial\left(\frac{\partial H_i^a}{\partial y^\alpha}\right)}\bigg|_{\frac{\delta H_i^a}{\delta x^k}} = \text{const.} \right.
$$

$$
\left. - N_j^\alpha \frac{\partial \mathcal{L}}{\partial\left(\frac{\delta H_i^a}{\delta x^j}\right)} \right\} HV.
$$

Then it is easy to check that

(5.9)
$$\overset{(h)}{H}_a{}^{ij} = \frac{\partial \mathcal{L}}{\partial \left(\frac{\delta H_i^a}{\delta x^j} \right)}, \qquad \text{and}$$

(5.10)
$$\overset{(v)}{H}_a{}^{i\alpha} = \frac{\partial \mathcal{L}}{\partial \left(\frac{\partial H_i^a}{\partial y^\alpha} \right)} \Big|_{\frac{\delta H_i^a}{\delta x^j} = \text{const.}}$$

are the components of some H-tensor fields on type $\begin{pmatrix} 2 & 0 \\ 0 & 0 \end{pmatrix}$ and $\begin{pmatrix} 1 & 1 \\ 0 & 0 \end{pmatrix}$ respectively for each $a \in \{1, \ldots, n\}$. Moreover, $\overset{(h)}{H}_a{}^{ij}$ is an anti–symmetric H-tensor field, since we have

(5.11)
$$\overset{(h)}{H}_a{}^{ij} = 2\frac{\partial \mathcal{L}}{\partial R^a{}_{ij}}.$$

Further, taking into account that $T_i{}^k{}_j = 0$ with respect to GHC, and by using (1.23), (1.34) and (1.35) we obtain

(5.12)
$$\overset{(h)}{H}_a{}^{ij}{}_{|j} = \frac{\delta \overset{(h)}{H}_a{}^{ij}}{\delta x^j} + \overset{(h)}{H}_a{}^{ik} F_k{}^j{}_j,$$

(5.13)
$$\overset{(v)}{H}_a{}^{i\alpha}{}_{\|\alpha} = \frac{\partial \overset{(a)}{H}_a{}^{i\alpha}}{\partial y^\alpha} + \overset{(v)}{H}_a{}^{i\gamma} C_\gamma{}^\alpha{}_\alpha.$$

Finally, by using (5.7), (5.8), (5.9), (5.10), (5.12) and (5.13) we see that (5.4) becomes

(5.14)
$$\frac{\partial \mathcal{L}}{\partial H_i^a} - \overset{(h)}{H}_a{}^{ij}{}_{|j} - \overset{(v)}{H}_a{}^{i\alpha}{}_{\|\alpha} = \overset{(h)}{E}_a{}^i,$$

where we put

(5.15)
$$\overset{(h)}{E}_a{}^i = \left\{ \frac{1}{HV} \frac{\delta(HV)}{\delta x^k} - (L_\alpha{}^\alpha{}_k + F_k{}^j{}_j) \right\} \overset{(h)}{H}_a{}^{ik} +$$
$$+ \left\{ \frac{1}{HV} \frac{\partial(HV)}{\partial y^\gamma} - C_\gamma{}^\alpha{}_\alpha \right\} \overset{(v)}{H}_a{}^{i\gamma}.$$

In a similar way, (5.5) becomes

$$
(5.16) \qquad \frac{\partial \mathcal{L}}{\partial V_\alpha^a} + \overset{(h)}{V}{}_a{}^{\gamma i} L_\gamma{}^\alpha{}_i - \overset{(h)}{V}{}_a{}^{\alpha i}{}_{|i} - \overset{(v)}{V}{}_a{}^{\alpha \gamma}{}_{\|\gamma} = \overset{(v)}{E}{}_a{}^\alpha,
$$

where we put

$$
(5.17) \qquad \overset{(h)}{V}{}_a{}^{\alpha i} = \frac{\partial \mathcal{L}}{\partial \left(\dfrac{\delta V_\alpha^a}{\delta x^i} \right)},
$$

$$
(5.18) \qquad \overset{(v)}{V}{}_a{}^{\alpha \gamma} = \frac{\partial \mathcal{L}}{\partial \left(\dfrac{\partial V_\alpha^a}{\partial y^\gamma} \right)} \Bigg|_{\frac{\delta V_\alpha^a}{\delta x^k} = \text{const.}} \qquad \text{and}
$$

$$
(5.19) \qquad \overset{(v)}{E}{}_a{}^\alpha = \left\{ \frac{1}{HV} \frac{\delta (HV)}{\delta x^i} - (L_\gamma{}^\gamma{}_i + F_i{}^j{}_j) \right\} \overset{(h)}{V}{}_a{}^{\alpha i} +
$$

$$
+ \left\{ \frac{1}{HV} \frac{\partial (HV)}{\partial y^\gamma} - C_\gamma{}^\beta{}_\beta \right\} \overset{(v)}{V}{}_a{}^{\alpha \gamma}.
$$

In deriving (5.16) we make use of

$$
(5.20) \qquad \overset{(v)}{V}{}_a{}^{\alpha \gamma} = 2 \frac{\partial \mathcal{L}}{\partial S^a{}_{\alpha \gamma}},
$$

which implies that $\overset{(v)}{V}{}_a{}^{\alpha \gamma}$ are anti–symmetric H-tensor fields of type $\begin{pmatrix} 0 & 2 \\ 0 & 0 \end{pmatrix}$ for each $a \in \{1, \ldots, n\}$.

Next, since $\mathcal{L}(x, y)$ is local gauge invariant with respect to the action of G we have

$$
(5.21) \qquad \frac{\partial \mathcal{L}}{\partial Q^A} \overset{*}{\delta}(Q^A) + \overset{(h)}{Q}{}_A{}^i \overset{*}{\delta} \left(\frac{\delta Q^A}{\delta x^i} \right) + \overset{(v)}{Q}{}_A{}^\alpha \overset{*}{\delta} \left(\frac{\partial Q^A}{\partial y^\alpha} \right) +
$$

$$
+ \frac{\partial \mathcal{L}}{\partial H_i^a} \overset{*}{\delta}(H_i^a) + \overset{(h)}{H}{}_a{}^{ij} \overset{*}{\delta} \left(\frac{\delta H_i^a}{\delta x^j} \right) + \overset{(v)}{H}{}_a{}^{i\alpha} \overset{*}{\delta} \left(\frac{\partial H_i^a}{\partial y^\alpha} \right) +
$$

$$
+ \frac{\partial \mathcal{L}}{\partial V_\alpha^a} \overset{*}{\delta}(V_\alpha^a) + \overset{(h)}{V}{}_a{}^{\alpha i} \overset{*}{\delta} \left(\frac{\delta V_\alpha^a}{\delta x^i} \right) + \overset{(v)}{V}{}_a{}^{\alpha \beta} \overset{*}{\delta} \left(\frac{\partial V_\alpha^a}{\partial y^\beta} \right) = 0.
$$

Taking into account that $\overset{*}{\delta}$ commutes with both $\frac{\delta}{\delta x^i}$ and $\frac{\partial}{\partial y^\alpha}$, and by using (5.6), (5.14) and (5.16) in (5.21), we infer

$$(5.22) \quad \frac{\delta}{\delta x^i}\left\{ \overset{(h)}{Q}{}_A{}^i\,\overset{*}{\delta}(Q^A) + \overset{(h)}{H}{}_a{}^{ji}\,\overset{*}{\delta}(H_j^a) + \overset{(h)}{V}{}_a{}^{\alpha i}\,\overset{*}{\delta}(V_\alpha^a) \right\} +$$

$$\frac{\partial}{\partial y^\alpha}\left\{ \overset{(v)}{Q}{}_A{}^\alpha\,\overset{*}{\delta}(Q^A) + \overset{(v)}{H}{}_a{}^{i\alpha}\,\overset{*}{\delta}(H_i^a) + \overset{(v)}{V}{}_a{}^{\gamma\alpha}\,\overset{*}{\delta}(V_\gamma^a) \right\} +$$

$$+ \left\{ \overset{(h)}{Q}{}_A{}^k\,\overset{*}{\delta}(Q^A) + \overset{(h)}{H}{}_a{}^{jk}\,\overset{*}{\delta}(H_j^a) + \overset{(h)}{V}{}_a{}^{\alpha k}\,\overset{*}{\delta}(V_\alpha^a) \right\} F_k{}^i{}_i +$$

$$+ \left\{ \overset{(v)}{Q}{}_A{}^\beta\,\overset{*}{\delta}(Q^A) + \overset{(v)}{H}{}_a{}^{i\beta}\,\overset{*}{\delta}(H_i^a) + \overset{(v)}{V}{}_a{}^{\gamma\beta}\,\overset{*}{\delta}(V_\gamma^a) \right\} C_\beta{}^\alpha{}_\alpha +$$

$$+ \left\{ E_A\,\overset{*}{\delta}(Q^A) + \overset{(h)}{E}{}_a{}^i\,\overset{*}{\delta}(H_i^a) + \overset{(v)}{E}{}_a{}^\alpha\,\overset{*}{\delta}(V_\alpha^a) \right\} = 0.$$

We now replace $\overset{*}{\delta}(Q^A)$, $\overset{*}{\delta}(H_i^a)$ and $\overset{*}{\delta}(V_\alpha^a)$ from (3.2), (3.7) and (3.8) respectively in (5.22), and so taking into account the arbitrariness of $\varepsilon^a(x,y)$ we obtain

$$(5.23) \quad \frac{\delta}{\delta x^i}\left\{ \overset{(h)}{Q}{}_A{}^i[X_a]_B^A Q^B + \overset{(h)}{H}{}_b{}^{ji}C_a{}^b{}_c H_j^c + \overset{(h)}{V}{}_b{}^{\alpha i}C_a{}^b{}_c V_\alpha^c \right\} +$$

$$+\frac{\partial}{\partial y^\alpha}\left\{ \overset{(v)}{Q}{}_A{}^\alpha[X_a]_B^A Q^B + \overset{(v)}{H}{}_b{}^{i\alpha}C_a{}^b{}_c H_i^c + \overset{(v)}{V}{}_b{}^{\gamma\alpha}C_a{}^b{}_c V_\gamma^c \right\} +$$

$$+ \left\{ \overset{(h)}{Q}{}_A{}^k[X_a]_B^A Q^B + \overset{(h)}{H}{}_b{}^{jk}C_a{}^b{}_c H_j^c + \overset{(h)}{V}{}_b{}^{\alpha k}C_a{}^b{}_c V_\alpha^c \right\} F_k{}^i{}_i +$$

$$+ \left\{ \overset{(v)}{Q}{}_A{}^\beta[X_a]_B^A Q^B + \overset{(v)}{H}{}_b{}^{i\beta}C_a{}^b{}_c H_i^c + \overset{(v)}{V}{}_b{}^{\gamma\beta}C_a{}^b{}_c V_\gamma^c \right\} C_\beta{}^\alpha{}_\alpha +$$

$$+ \left\{ E_A[X_a]_B^A Q^B + \overset{(h)}{E}{}_b{}^i C_a{}^b{}_c H_i^c + \overset{(v)}{E}{}_b{}^\alpha C_a{}^b{}_c V_\alpha^c \right\} = 0,$$

$$(5.24) \quad \frac{\partial \mathcal{L}}{\partial H_i^a} + \overset{(h)}{Q}{}_A{}^i[X_a]_B^A Q^B + \overset{(h)}{H}{}_b{}^{ji}C_a{}^b{}_c H_j^c + \overset{(h)}{V}{}_b{}^{\alpha i}C_a{}^b{}_c V_\alpha^c = 0,$$

$$(5.25) \quad \frac{\partial \mathcal{L}}{\partial V_\alpha^a} + \overset{(h)}{V}{}_a{}^{\gamma i}L_\gamma{}^\alpha{}_i + \frac{1}{2}\overset{(h)}{H}{}_a{}^{ji}R^\alpha{}_{ij} + \overset{(v)}{Q}{}_A{}^\alpha[X_a]_B^A Q^B +$$

$$+ \overset{(v)}{H}{}_b{}^{i\alpha}C_a{}^b{}_c H_i^c + \overset{(v)}{V}{}_b{}^{\gamma\alpha}C_a{}^b{}_c V_\gamma^c = 0.$$

We next write

$$(5.26) \qquad \overset{(h)}{J}{}_a{}^i = \frac{\partial \mathcal{L}}{\partial H_i^a}, \qquad \text{and}$$

$$(5.27) \qquad \overset{(v)}{J}{}_a{}^\alpha = \frac{\partial \mathcal{L}}{\partial V_\alpha^a} + \overset{(h)}{V}{}_a{}^{\gamma i} L_\gamma{}^\alpha{}_i + \frac{1}{2}\overset{(h)}{H}{}_a{}^{ji} R^\alpha{}_{ij}.$$

Then by using (5.24)–(5.27) we obtain

$$(5.28) \qquad \overset{(h)}{J}{}_a{}^i = -\overset{(h)}{Q}{}_A{}^i [X_a]_B^A Q^B - \overset{(h)}{H}{}_b{}^{ji} C_a{}^b{}_c H_j^c - \overset{(h)}{V}{}_b{}^{\alpha i} C_a{}^b{}_c V_\alpha^c,$$

$$(5.29) \qquad \overset{(v)}{J}{}_a{}^\alpha = -\overset{(v)}{Q}{}_A{}^\alpha [X_a]_B^A Q^B - \overset{(v)}{H}{}_b{}^{i\alpha} C_a{}^b{}_c H_i^c - \overset{(v)}{V}{}_b{}^{\gamma\alpha} C_a{}^b{}_c V_\gamma^c.$$

Finally, taking into account (1.26), (1.29), (5.28) and (5.29) we see that (5.23) becomes

$$(5.30) \qquad \overset{(h)}{J}{}_a{}^i{}_{|i} + \overset{(v)}{J}{}_a{}^\alpha{}_{\|\alpha} = E_A[X_a]_B^A Q^B + \overset{(h)}{E}{}_b{}^i C_a{}^b{}_c H_i^c + \overset{(v)}{E}{}_b{}^\alpha C_a{}^b{}_c V_\alpha^c,$$

which are *conservation laws* we look for. We call

$$(5.31) \qquad \overset{(h)}{J}{}_a = \overset{(h)}{J}{}_a{}^i \frac{\delta}{\delta x^i}, \qquad \text{and}$$

$$(5.32) \qquad \overset{(v)}{J}{}_a = \overset{(v)}{J}{}_a{}^\alpha \frac{\partial}{\partial y^\alpha},$$

the *horizontal* and *vertical currents* respectively corresponding to the full Lagrangian $\mathcal{L}(x, y)$.

If, in particular, we consider a trivial vector bundle $E = M \times R^r$, then by using (5.15), (5.19) and by the study performed in §2 we obtain $\overset{(h)}{E}{}_a{}^i = 0$, $\overset{(v)}{E}{}_a{}^\alpha = 0$ and $E_A = 0$. Hence in this case the equations of motion become

$$(5.33) \qquad \frac{\partial \mathcal{L}}{\partial Q^A} - \overset{(h)}{Q}{}_A{}^i{}_{|i} - \overset{(v)}{Q}{}_A{}^\alpha{}_{\|\alpha} = 0,$$

$$(5.34) \qquad \frac{\partial \mathcal{L}}{\partial H_i^a} - \overset{(h)}{H}{}_a{}^{ij}{}_{|j} - \overset{(v)}{H}{}_a{}^{i\alpha}{}_{\|\alpha} = 0,$$

and

$$(5.35) \qquad \frac{\partial \mathcal{L}}{\partial V_\alpha^a} + \overset{(h)}{V}{}_a{}^{\gamma i} L_\gamma{}^\alpha{}_i - \overset{(h)}{V}{}_a{}^{\alpha i}{}_{|i} - \overset{(v)}{V}{}_a{}^{\alpha\gamma}{}_{\|\gamma} = 0 \qquad \text{respectively.}$$

Moreover, in this case the conservation laws (5.30) become

$$(5.36) \qquad \overset{(h)}{J}{}_a{}^i{}_{|i} + \overset{(v)}{J}{}_a{}^\alpha{}_{\|\alpha} = 0.$$

§6 Bianchi identities

In this section we obtain Bianchi identities for covariant derivatives of strength fields $R^a{}_{ij}$, $P^a{}_{i\alpha}$, $S^a{}_{\alpha\beta}$ with respect to an arbitrary H-connection $HC = (N^\alpha_i, F^k_{ij}, D^j_{i\alpha}, L_\alpha{}^\gamma{}_i, C_\alpha{}^\gamma{}_\beta)$ on E. To this end we first define the following *gauge covariant derivatives* of strength fields

$$(6.1) \qquad R^a{}_{ij|k} = \frac{\delta R^a{}_{ij}}{\delta x^k} + C_b{}^a{}_c R^b{}_{ij} H^c_k - R^a{}_{hj} F_i{}^h{}_k - R^a{}_{ih} F_j{}^h{}_k,$$

$$(6.2) \qquad R^a{}_{ij\|\alpha} = \frac{\partial R^a{}_{ij}}{\partial y^\alpha} + C_b{}^a{}_c R^b{}_{ij} V^c_\alpha - R^a{}_{hj} D_i{}^h{}_\alpha - R^a{}_{ih} D_j{}^h{}_\alpha,$$

$$(6.3) \qquad P^a{}_{i\alpha|j} = \frac{\delta P^a{}_{i\alpha}}{\delta x^j} + C_b{}^a{}_c P^b{}_{i\alpha} H^c_j - P^a{}_{h\alpha} F_i{}^h{}_j - P^a{}_{i\gamma} L_\alpha{}^\gamma{}_j,$$

$$(6.4) \qquad P^a{}_{i\alpha\|\beta} = \frac{\partial P^a{}_{i\alpha}}{\partial y^\beta} + C_b{}^a{}_c P^b{}_{i\alpha} V^c_\beta - P^a{}_{h\alpha} D_i{}^h{}_\beta - P^a{}_{i\gamma} C_\alpha{}^\gamma{}_\beta,$$

$$(6.5) \qquad S^a{}_{\alpha\beta|i} = \frac{\delta S^a{}_{\alpha\beta}}{\delta x^i} + C_b{}^a{}_c S^b{}_{\alpha\beta} H^c_i - S^a{}_{\gamma\beta} L_\alpha{}^\gamma{}_i - S^a{}_{\alpha\gamma} L_\beta{}^\gamma{}_i,$$

$$(6.6) \qquad S^a{}_{\alpha\beta\|\gamma} = \frac{\partial S^a{}_{\alpha\beta}}{\partial y^\gamma} + C_b{}^a{}_c S^b{}_{\alpha\beta} V^c_\gamma - S^a{}_{\mu\beta} C_\alpha{}^\mu{}_\gamma - S^a{}_{\alpha\mu} C_\beta{}^\mu{}_\gamma.$$

It is easy to verify that $R^a{}_{ij|k}$, $R^a{}_{ij\|\alpha}$, $P^a{}_{i\alpha|j}$, $P^a{}_{i\alpha\|\beta}$, $S^a{}_{\alpha\beta|i}$ and $S^a{}_{\alpha\beta\|\gamma}$ are the components of some H-tensor fields on E of type $\begin{pmatrix} 0 & 0 \\ 3 & 0 \end{pmatrix}$, $\begin{pmatrix} 0 & 0 \\ 2 & 1 \end{pmatrix}$, $\begin{pmatrix} 0 & 0 \\ 2 & 1 \end{pmatrix}$, $\begin{pmatrix} 0 & 0 \\ 1 & 2 \end{pmatrix}$, $\begin{pmatrix} 0 & 0 \\ 1 & 2 \end{pmatrix}$ and $\begin{pmatrix} 0 & 0 \\ 0 & 3 \end{pmatrix}$ respectively for each $a \in \{1, \ldots, n\}$. Moreover, the transformations of the gauge covariant derivatives with respect to the local gauge action of G are homogeneous and they are given by the adjoint representation of G.

Next, by using (1.41) and (1.42) it is easy to see that the Jacobi identities

$$(6.7) \qquad \sum_{(i,j,k)}^{\text{cyclic}} \left\{ \left[\left[\frac{\delta}{\delta x^i}, \frac{\delta}{\delta x^j} \right], \frac{\delta}{\delta x^k} \right] \right\} = 0, \qquad \text{and}$$

$$(6.8) \qquad \left[\left[\frac{\delta}{\delta x^i}, \frac{\delta}{\delta x^j} \right], \frac{\partial}{\partial y^\alpha} \right] + \left[\left[\frac{\delta}{\delta x^j}, \frac{\partial}{\partial y^\alpha} \right], \frac{\delta}{\delta x^i} \right] +$$

$$+ \left[\left[\frac{\partial}{\partial y^\alpha}, \frac{\delta}{\delta x^i} \right], \frac{\delta}{\delta x^j} \right] = 0,$$

are equivalent with

(6.9)
$$\sum_{(i,j,k)}^{\text{cyclic}} \left\{ \frac{\delta R^a{}_{ij}}{\delta x^k} + R^\gamma{}_{ij} \frac{\partial N^\alpha_k}{\partial y^\gamma} \right\} = 0,$$

and

(6.10)
$$\frac{\partial R^\gamma_{ij}}{\partial y^\alpha} = \frac{\delta}{\delta x^j} \left(\frac{\partial N^\gamma_i}{\partial y^\alpha} \right) - \frac{\delta}{\delta x^i} \left(\frac{\partial N^\gamma_j}{\partial y^\alpha} \right) +$$
$$+ \frac{\partial N^\mu_i}{\partial y^\alpha} \frac{\partial N^\gamma_j}{\partial y^\mu} - \frac{\partial N^\mu_j}{\partial y^\alpha} \frac{\partial N^\gamma_i}{\partial y^\mu} \qquad \text{respectively.}$$

Finally, by direct calculations from (1.43), (1.50)–(1.52), (4.1)–(4.3), (6.1)–(6.6), (6.9) and (6.10) we obtain the following types of Bianchi identities:

(6.11)
$$\sum_{(i,j,k)}^{\text{cyclic}} \left\{ R^a{}_{ij|k} + R^a{}_{ih} T_j{}^h{}_k + P^a{}_{i\alpha} R^\alpha{}_{jk} \right\} = 0,$$

(6.12)
$$\sum_{(\alpha,\beta,\gamma)}^{\text{cyclic}} \left\{ S^a{}_{\alpha\beta\|\gamma} + S^a{}_{\alpha\mu} S^\mu{}_{\beta\gamma} \right\} = 0,$$

(6.13)
$$P^a{}_{i\alpha\|\beta} - P^a{}_{i\beta\|\alpha} + S^a{}_{\alpha\beta|i} + S^a{}_{\beta\gamma} P^\gamma{}_{i\alpha} -$$
$$- S^a{}_{\alpha\gamma} P^\gamma{}_{i\beta} + P^a{}_{h\alpha} D_i{}^h{}_\beta - P^a{}_{h\beta} D_i{}^h{}_\alpha + P^a_{i\gamma} S^\gamma{}_{\alpha\beta} = 0,$$

(6.14)
$$P^a{}_{i\alpha|j} - P^a{}_{j\alpha|i} - R^a{}_{ij\|\alpha} + P^a{}_{j\gamma} P^\gamma{}_{i\alpha} - P^a{}_{i\gamma} P^\gamma{}_{j\alpha} -$$
$$- S^a{}_{\alpha\gamma} R^\gamma{}_{ij} + P^a{}_{h\alpha} T_i{}^h{}_j + R^a{}_{jh} D_i{}^h{}_\alpha - R^a{}_{ih} D_j{}^h{}_\alpha = 0.$$

In the particular case of a gauge H-connection GHC given by (1.22)–(1.25) we get the Bianchi identities

(6.15)
$$\sum_{(i,j,k)}^{\text{cyclic}} \left\{ R^a{}_{ij|k} + P^a{}_{i\alpha} R^\alpha{}_{jk} \right\} = 0,$$

(6.16)
$$\sum_{(\alpha,\beta,\gamma)}^{\text{cyclic}} \left\{ S^a{}_{\alpha\beta\|\gamma} \right\} = 0,$$

(6.17)
$$P^a{}_{i\alpha\|\beta} - P^a{}_{i\beta\|\alpha} + S^a{}_{\alpha\beta|i} = 0,$$

(6.18)
$$P^a{}_{i\alpha|j} - P^a{}_{j\alpha|i} - R^a{}_{ij\|\alpha} - S^a{}_{\alpha\gamma} R^\gamma{}_{ij} = 0.$$

In deriving (6.15)–(6.18) we use the vanishing of the torsions of GHC except $R^\alpha{}_{ij}$.

References

1. N. ABE, General Connections on Vector Bundles, *Kodai Math. J.* **8** (1985), 322–329.

2. G. S. ASANOV, Finsler Geometry, Relativity and Gauge Theories, D. Reidel Publishing Company, Dordrecht, 1985.

3. A. BEJANCU, Foundations of Direction-Dependent Gauge Theories, Preprint No. 13, Mechanics Sem., Timişoara, (1988) 60.

4. M. CHAICHIAN and N. F. NELIPA, Introduction to Gauge Field Theories, Springer Verlag, Berlin, 1984.

5. R. MIRON, Vector Bundles Finsler Geometry, *Proc. of the Nat. Sem. on Finsler Spaces*, Braşov, (1982), 147–188.

6. H. NEMOTO, On Differential Geometry of General Connections, *TRU Mathematics* **21** (1985), 67–94.

Aurel Bejancu

Department of Mathematics,
Polytechnic Institute of Iaşi
C. P. 17, Iaşi 1,
Romania

COLLOQUIA MATHEMATICA SOCIETATIS JÁNOS BOLYAI
56. DIFFERENTIAL GEOMETRY, EGER (HUNGARY), 1989

Geodesic Balls and Chern Numbers of Kähler Manifolds

N. BLAŽIĆ

1. Introduction

This is an exposition, in summary, of work which will appear in detail elsewhere (see [2]).

Let (M, g) be a p-dimensional Riemannian manifold and let $V_m(r)$ denote the volume of a small geodesic ball of radius r centered in $m \in M$. Then we can ask the following question:

What can be said about the geometry and global structure of (M, g) if all the volumes $V_m(r)$ are known?

For example, a 2-dimensional creature can use the volumes of small geodesic balls to recognize if it is living in flat space or in the standard sphere S^2. In this sense Gray and Vanhecke (see [5]) were concerned with the following conjecture:

Conjecture I. *Suppose*

$$(*) \qquad\qquad V_m(r) = \Omega_p r^p$$

for all $m \in M$ and all sufficiently small r. Then M is flat.

Here Ω_p is the volume of the unit ball in $\mathbb{R}^p$. In general case Conjecture I is not proved yet, but under some additional assumptions it can be shown to hold. In the first part of the paper we will consider similar questions when the condition (∗) is replaced by the condition (∗∗)

$$(\ast\ast) \qquad\qquad V_m(r) \geq \Omega_p r^p.$$

In the second part of the paper we introduce for a manifold M the notion of a geodesically-Einstein metric as a generalization of an Einstein metric. The main result is a topological obstruction for a compact complex surface to admit a geodesically-Einstein Kähler metric and, as a consequence, that $M = \mathbb{C}P^2 \# n$ does not admit a geodesically-Einstein Kähler metric for $n > 1$. We use the standard power series expansion for volumes of small geodesic balls to obtain our results.

I wish to thank O. Kowalsi and I. Vaisman for useful discussions and suggestions.

2. Preliminaries

In this paper we use the notations given in [5] and [4]. Let M be an n-dimensional analytic Riemannian manifold. Let $r_0 > 0$ be so small that the exponential map $\exp_m$ is a diffeomorphism on a ball of radius r_0 in the tangent space M_m. We put

$$V_m(r_0) = \text{ volume of } \{\exp_m(x) | x \in M_m, \|x\| \leq r_0\}.$$

In [5] it is shown (Theorem 3.3) that for $V_m(r)$ the following power series expansion holds

$$(2.1) \quad V_m(r) = \Omega_n r^n (1 - Ar^2 + Br^4 + O(r^6)) \qquad \text{where}$$

$$A = \frac{\tau}{6(n+2)},$$

$$B = \frac{1}{360(n+2)(n+4)}(-3\|R\|^2 + S\|\rho\|^2 + 5\tau^2 - 18\Delta\tau).$$

(Here Ω_n is the volume of the unit ball in $\mathbb{R}^n$.)

Suppose that M is a Kähler manifold of complex dimension n. Let $\theta^1, \ldots, \theta^n$ be a local field of unitary coframes. Then the Kähler metric may be written as $g = \Sigma(\theta^\alpha \otimes \bar{\theta}^\alpha + \bar{\theta}^\alpha \otimes \theta^\alpha)$ and the fundamental 2-form

$\phi(X, Y) = g(X, JY)$ is given by $\phi = \sqrt{-1} \Sigma \theta^\alpha \wedge \bar\theta^\alpha$. From now on we use the ranges $\alpha, \beta, \gamma, \delta, \ldots = 1, \ldots, n$. The form ϕ is closed. The fundamental class w of M is the de Rham cohomology class determined by ϕ. The curvature tensor R of M is the tensor field with local components $R_{\alpha\bar\beta\gamma\bar\delta}$. Then the $(1,1)$-forms Ω_β^α, defined by $\Omega_\beta^\alpha = \Sigma R_{\beta\gamma\bar\delta}^\alpha \theta^\gamma \wedge \bar\theta^\delta$, are closed. The Ricci tensor ρ and the scalar curvature τ are given by $\rho_{\alpha\bar\beta} = \Sigma R_{\alpha\bar\gamma\gamma\bar\beta}$ and $\tau = 2\Sigma \rho_{\alpha\bar\alpha}$. We denote by $\|R\|$ and $\|\rho\|$ the length of the curvature tensor and the Ricci tensor respectively, so that

$$\|R\|^2 = 4\Sigma R_{\alpha\bar\beta\gamma\bar\delta} R_{\beta\bar\alpha\delta\bar\gamma} \quad \text{and} \quad \|\rho\|^2 = 2\Sigma \rho_{\alpha\bar\beta} \rho_{\beta\bar\alpha}.$$

We define a closed $2k$-form γ_k by

$$\gamma_k = \frac{(-1)^k}{\left(2\pi\sqrt{-1}\right)^k k!} \Sigma \delta_{\beta_1 \ldots \beta_k}^{\alpha_1 \ldots \alpha_k} \Omega_{\alpha_1}^{\beta_1} \wedge \ldots \wedge \Omega_{\alpha_k}^{\beta_k}.$$

It is well-known that the k-th Chern class c_k is determined by the form γ_k. In particular, the first two Chern forms are given by

$$2\pi\gamma_1 = \sqrt{-1} \Sigma \Omega_\alpha^\alpha$$

and

$$-8\pi^2 \gamma_2 = \Sigma (\Omega_\alpha^\alpha \wedge \Omega_\beta^\beta - \Omega_\beta^\alpha \wedge \Omega_\alpha^\beta)$$

respectively.

Then we have

$$\gamma_1 \wedge \phi^{n-1} = \frac{\tau}{n\pi} \phi^n,$$

$$\gamma_1^2 \wedge \phi^{n-2} = \frac{1}{4n(n-1)\pi^2} (\tau^2 - 2\|\rho\|^2) \phi^n \quad \text{and}$$

$$\gamma_2 \wedge \phi^{n-2} = \frac{1}{8n(n-1)\pi^2} (\tau^2 - 4\|\rho\|^2 + \|R\|^2) \phi^n.$$

For a compact M, the generalized Chern numbers $w^{n-1} c_1(M)$, $w^{n-2} c_1^2(M)$ and $w^{n-2} c_2(M)$ are defined by $\int_M \gamma_1 \wedge \phi^{n-1}$, $\int_M \gamma_1^2 \wedge \phi^{n-2}$, and $\int_M \gamma_2 \wedge \phi^{n-2}$, respectively.

3. Characterization of complex space forms

Theorem 3.1. *Let (M, g, J) be a compact Kähler manifold of complex dimension n. Suppose that the generalized Chern numbers $\omega^{n-1} c_1$ and $\omega^{n-2} c_1^2$ are nonnegative. Then, if M satisfies the condition $(**)$, it is flat.*

Corollary 3.2. *Let M be a Kähler manifold as in Theorem 3.1. If the first Chern class $c_1(M)$ vanishes and if it satisfies the condition $(**)$, then M is flat.*

Let $M(\mu)$ be a Kähler manifold with complex dimension n and constant holomorphic sectional curvature $\mu \neq 0$. In [5] the following conjecture was stated:

Conjecture IV. *Let M be a Kähler manifold with complex dimension n and suppose that for all $m \in M$ and all sufficiently small $r > 0$, $V_m(r)$ is the same as that of an n-dimensional Kähler manifold with constant holomorphic sectional curvature μ. Then M has constant holomrphic sectional curvature μ.*

The following theorem proves one particular case of the Conjecture IV.

Theorem 3.3. *Let M be a compact Kähler manifold with complex dimension n, and suppose that for all $m \in M$ and all sufficiently small $r > 0$, $V_m(r)$ is the same as that of an n-dimensional compact Kähler manifold $M(\mu)$ with constant holomorphic sectional curvature μ. Let ω and ω_μ denote the fundamental classes of M and $M(\mu)$ respectively. If the following conditions*

(i) $\omega^{n-1} c_1(M) = \omega_\mu^{n-1} c_1(M(\mu))$,

(ii) $\omega^{n-2} c_1^2(M) \geq \omega_\mu^{n-2} c_1^2(M)$

are satisfied, then M has constant holomorphic sectional curvature μ.

Proof. Let τ_μ, $\|\rho_\mu\|^2$ and $\|R_\mu\|^2$ denote the appropriate functions for $M(\mu)$. Since $V_m(r) = V(r, \mu)$ we have

(4.3) $\tau = \tau_\mu$ and

(4.4) $3\left(\|R_\mu\|^2 - \|R\|^2\right) = 8\left(\|\rho_\mu\|^2 - \|\rho\|^2\right) \leq 0.$

The hypotheses (i) and (ii) imply that

$$(4.5) \qquad \int_M \tau \phi^n = \int_{M(\mu)} \tau_\mu \phi_\mu^n \quad \text{and}$$

$$(4.6) \qquad \int_M \left(\tau^2 - 2\|\rho\|^2 \right) \phi^n \geq \int_{M(\mu)} \left(\tau_\mu^2 - 2\|\rho_\mu\|^2 \right) \phi_\mu^n.$$

For $\mu = 0$, from (4.3), (4.6) and (4.4) it follows that $\tau = \|\rho\| = \|R\| = 0$ on M. So, in this case M is flat as we want to show. For $\mu \neq 0$ formulas (4.3) and (4.5) imply that

$$\int_M \phi^n = \int_{M(\mu)} \phi_\mu^n.$$

Then, using (4.4) and (4.6), we obtain

$$\int_M \|\rho\|^2 \phi^n \leq \int_{M(\mu)} \|\rho(\mu)\|^2 \phi^n.$$

This inequality and (4.4) give

$$\int_M \left(\|R\|^2 - \frac{4}{n+1}\|\rho\|^2 \right) \phi^n = \frac{4}{3}\left(\frac{3}{n+1} - 2 \right) \int_M \left(\|\rho_\mu\|^2 - \|\rho\|^2 \right) \phi^n \leq 0.$$

So $\|R\|^2 = \frac{4}{n+1}\|\rho\|^2$ on M and the required result follows from the lemma obtained in [4, p. 460]. ∎

Corollary 3.4. *Let $(M(\mu), g_\mu, J_\mu)$ be a compact n-dimensional Kähler manifold with constant holomorphic sectional curvature μ, fundamental 2-class ω_μ and almost complex structure J_μ. Suppose that $(M(\mu), g, J)$ is a Kähler manifold with fundamental 2-class ω and almost complex structure J. If*

(i) $V_m(r) \geq V(r, \mu)$ *for all $m \in M(\mu)$ and all sufficiently small $r > 0$,*

(ii) $\omega = \omega_\mu$,

(iii) $J = J_\mu$,

then (M, g) has constant holomorphic sectional curvature μ.

4. Geodesically-Einstein Kähler manifolds

Definition 4.1. Let M and M_ε be Riemannian manifolds of the same dimension p. We say that M is geodesically-Einstein with respect to the Einstein manifold M_ε if there exists a map $f : M \to M_\varepsilon$ such that

$$(4.1) \qquad\qquad V_m(r) = V_{f(m)}(r)$$

for all $m \in M$ and for all sufficiently small $r > 0$.

It is not easy to find nontrivial examples of geodesically-Einstein manifolds. But, anyway, it is interesting to consider the class of manifolds which satisfy the condition $V_m(r) = V_{f(m)}(r) + O(r^{p+6})$ instead of (4.1). Then, there are examples for Riemannian, non-Einstein, manifolds (M, g) with $V_m(r) = \Omega_p r^p (1 + O(r^6))$ obtained in [5] and their generalization is given in [7]. Kowalski introduced in [7] the notion of additive volume invariants of (M, g) and then constructed manifolds with $V_m(r) = \Omega_p r^p (1 + O(r^{16}))$.

It is to expect that geodesically-Einstein manifolds have some similar properties as Einstein manifolds. So, in this section we establish an inequality between Chern classes of geodesically-Einstein Kähler manifolds. Also geodesically-Einstein Kähler surfaces are considered.

Proposition 4.2. Let m and M_ε be compact, n-dimensional, $n \geq 2$, Kähler manifolds as it was supposed in Definition 4.1. If M is geodesically-Einstein with respect to M_ε, then

$$(4.2) \qquad\qquad \int_M \left\{ \gamma_2 - \frac{n}{2(n+1)} \gamma_1^2 \right\} \wedge \phi^{n-2} \geq 0.$$

For $n \geq 3$ the equality holds if and only if M is a complex space form. For $n = 2$, if M_ε is a homogeneous manifold, the equality holds if and only if M_ε is a complex space form.

Remark 4.3. The proof of this result utilizes only the first three nontrivial terms in the power series expansion of $V_m(r)$.

Remark 4.4. To prove Proposition 4.2. the map f in Proposition 4.2. can be arbitrary. Even, it is not necessary to be continuous.

Example 4.5. Here we will give an example of a non-Einstein Kähler manifold M for which

(4.3) $$V_m(r) = V(r, M_3) + O(r^{4p+6})$$

holds for all $m \in M$ and all small enough $r > 0$. Here M_3 is a complex space form of complex dimension $2p$, $p \geq 2$, and $V(r, M_3)$ is the volume of a geodesic ball of radius r in M_3. So let M_1 and M_2 be complex space forms of complex dimension p, with scalar curvatures equal to τ_1 and τ_2 respectively. Let M_3 have scalar curvature $\tau_1 + \tau_2$. Suppose that $\tau_2 = a\tau_1$ where $(1-p)(1+4p)a^2 - 2(1+p)(1-4p)a = (p-1)(1+4p)$. Then for $M = M_1 \times M_2$ we have (4.3). Since $\tau_1 \neq \tau_2$, $M_1 \times M_2$ is not an Einstein manifold. Due to last remark inequality (4.2) holds for $M = M_1 \times M_2$.

We consider now the consequences of this theorem for a compact Kähler surface M which satisfies (4.1). Let χ, σ and a denote its Euler characteristic, Hirzebruch signature and arithmetic genus, respectively. Then from the Gauss–Bonnet–Chern theorem, the Hirzebruch signature theorem and the Riemann–Roch–Hirzebruch theorem (see [1], [3], [6] and [8]), we have

$$\chi(M) = \int_M c_2,$$

$$\sigma(M) = \frac{1}{3} \int_M (c_1^2 - 2c_2),$$

$$a(M) = \frac{1}{12} \int_M (c_1^2 + c_2).$$

Since

$$\chi(M) - 3a(M) = a(M) - \sigma(M) = (1/4) \int_M (3c_2 - c_1^2) \geq 0,$$

we have the following corollary.

Corollary 4.6. *Let M be a compact Kähler surface satisfying the hypotheses of the Proposition 4.2. Then*

(i) $$\chi(M) \geq 3a(M) \quad \text{and}$$
(ii) $$a(M) \geq \sigma(M).$$

The equality holds in (i) or (ii) if and only if M_ε has constant holomorphic sectional curvature on $f(M) \subset M_\varepsilon$.

Remark 4.7. This corollary is a generalization of the Theorem 10.4 in [5].

Theorem 4.8. Let M be a complex surface. Then any surface $\overline{M}$ obtained from M by blowing up k points of M admits no geodesically-Einstein Kähler metric whenever either

$$k < \sigma - a \text{ or } k < (1/4)(3\sigma - \chi).$$

Now we can apply Corollary 4.6 on $M = \mathbb{C}P^2 \# n = \mathbb{C}P^2 \# \ldots \# \mathbb{C}P^2$.

Corollary 4.9. The manifold $M = \mathbb{C}P^2 \# n$ does not admit a geodesically-Einstein Kähler metric for $n \geq 1$.

References

[1] M. F. ATIYAH and I. M. SINGER, The index of elliptic operators III, *Ann. of Math.* **87** (1968), 546–604.

[2] N. BLAŽIĆ, The volumes of small geodesic balls and generalized Chern numbers of Kähler manifolds, *Nagoya Math. J.* **116** (1989), 181–189.

[3] B.–Y. CHEN, Some topological obstructions to Bochner–Kähler metrics and their applications, *J. Differential Geometry* **13** (1978), 547–558.

[4] B.–Y. CHEN and K. OGIUE, Some characterizations of complex space forms in terms of Chern classes, *Quart J. Math.* (Oxford) (3) 26 (1975), 459–464.

[5] A. GRAY and L. VANHECKE, Riemannian geometry as determined by the volumes of small geodesic balls, *Acta Math.* **142** (1979), 157–198.

[6] F. HIRZEBRUCH, Topological methods in algebraic geometry, Springer-Verlag, Berlin, 1966.

[7] O. KOWALSKI, Additive volume invariants of Riemannian manifolds, *Acta Math.* **145** (1980), 205–225.

[8] R. S. PALAIS, Seminar on Atiyah–Singer index theorem, Annals of Math. Studies, No. 57, Princeton University Press, Princeton, 1965.

Novica Blažić

Faculty of Mathematics
University of Belgrade
Studentski trg 16, P.P. 550,
11000 Beograd
Yugoslavia

COLLOQUIA MATHEMATICA SOCIETATIS JÁNOS BOLYAI
56. DIFFERENTIAL GEOMETRY, EGER (HUNGARY), 1989

Curvature Invariants and Applications[*]

N. BOKAN

1. Introduction

This paper is a survey about curvature invariants in the sense of representation theory and some applications. The study of these invariants is
interesting because they are used in several theories in differential geometry
such as the theory of the volumes of geodesic spheres and tubes, spectrum
theory and some topological and algebraic studies (e.g. the determination
of the Euler-Poincaré characteristic, the arithmetic genus, the Hirzebruch
signature, group representation theory etc.). Let us also mention that it
is possible to consider the index theorem of Atiyah-Bott-Patodi from the
curvature invariant point of view. (see e.g. [G 1], [KU]).

Since a curvature operator can be associated to an arbitrary connection
it is clear that it is possible to study curvature invariants for a Levi–
Civita connection, a metric connection, a formally holomorphic connection,
a symmetric connection etc. Moreover, it is possible to consider the action
of various groups: $GL(n, \mathbb{R})$, $SL(n, \mathbb{R})$, $U(n)$, $SO(n)$, $O(n)$, etc. In order to
understand all of this better we should explain deeper some facts from the

---

[*] This paper is in final form and no version of it will be submitted for publication
elsewhere.

theory of invariants of the classical groups (see e.g. [WE 2]), but because of the time restriction it is impossible to do this and hence our idea is the following: to explain roughly the notion of curvature invariants and mainly to say something about applications in topology. We will mention also some known results and some new results obtained in collaboration with N. Blažić. We refer to [B–G–M], [C], [G 2], [G 3], [S], [PA 1], [PA 2], [B–G], [B–G–O], [CH–V 2] etc. for applications of curvature invariants in spectral geometry.

It is a pleasure to acknowledge interesting discussions about the subject with N. Blažić and S. Nölker. Many thanks also to L. Vanhecke for careful reading of the manuscript and useful suggestions.

2. Curvature invariants

Let us start with the explanation of the notion "invariants".

Let F denote a field of characteristic zero, V a vector space over F and $\mathcal{G}$ a group acting linearly on V. An element $v \in V$ is called $\mathcal{G}$-invariant if

$$gv = v, \text{ for all } g \in \mathcal{G}.$$

It is convenient to consider also the induced $\mathcal{G}$-action on V^*, the dual space of V. Its $\mathcal{G}$-invariants are obviously the $\mathcal{G}$-equivariant functionals on V.

We know that the Riemannian curvature tensor R, associated to a Levi–Civita connection ∇, has the following properties

(2.1) $R(X, Y) = -R(Y, X),$

(2.2) $R(X, Y)Z + R(Z, X)Y + R(Y, Z)X = 0$

                              (the first Bianchi identity),

(2.3) $R(X, Y, Z, V) = R(Z, V, X, Y),$

where $X, Y, Z, \ldots$, are smooth vector fields defined on a Riemannian manifold $(\mathbb{M}^n, g)$. Let us denote by ρ and τ respectively the Ricci tensor and the scalar curvature of R. By definition a scalar valued curvature invariant, under the action of the orthogonal group $O(n)$, is a polynomial in the components of the curvature tensor and its covariant derivatives which does not depend on the choice of an orthonormal basis of $\mathbb{M}_m$. More general is a p-form valued curvature invariant (see [KU]). It is clear that a basis of the vector space of these invariants depends on the symmetry properties of R and the action of the group. Such a scalar valued invariant is said to have order k if it involves a total of k derivatives of the metric tensor g. Each component of the curvature tensor contains two derivatives. Using Weyl's theory of invariants, a basis for the invariants of low order has been computed in [B–G–M]. More precisely, Weyl's theorem implies that the invariant polynomials are contractions in the components of the curvature tensor and its covariant derivatives.

Let $I(k, n)$ denote the space of invariants of order $2k$ for manifolds of dimension n. The spaces $I(1, n)$ and $I(2, n)$ are well known (see for example [B–G–M, pp 76 and 79]). We have $\dim I(1, n) = 1$ for $n \geq 2$ and $\dim I(2, n) = 4$ for $n \geq 4$. Using the notations

$$\tau = \sum R_{ijij}, \quad \|\rho\|^2 = \sum \rho_{ij}^2, \quad \|R\|^2 = \sum R_{ijkl}^2, \quad \Delta\tau = \sum \nabla_{ii}^2 \tau,$$

$\{\tau\}$ is a basis for $I(1, n)$ and $\{\tau^2, \|\rho\|^2, \|R\|^2, \Delta\tau\}$ is a basis for $I(2, n)$. Δ denotes the Laplacian. The dimension of the space $I(3, n)$, invariants of order 6, is 17 provided $n \geq 6$. Gray and Vanhecke [GR–VA] gave a basis for $I(3, n)$. $\{\tau^2, \|\rho\|^2, \|R\|^2\}$ is a basis for the space of quadratic invariants, excluding covariant derivatives of R.

If we do not use the action of $O(n)$ but work with $U(n)$ we consider a Hermitian metric g on V and one can prove that $\{\tau^2, \|\rho\|^2, \|R\|^2\}$ is a basis for the space of quadratic invariants for Kähler manifolds too (see [B–G–M]). Tricerri and Vanhecke [TR–VA] have found a basis for the space of quadratic curvature invariants of a Riemannian curvature tensor on an almost Hermitian manifold under the action of the group $U(n)$.

Of course, if we do not consider Riemannian curvature tensors but curvature tensors corresponding to a symmetric connection, then they satisfy only the conditions (2.1) and (2.2). Since they have less symmetries than in the Riemannian case, the corresponding vector spaces of curvature invariants are not the same. Namely, the dimension of its vector space is larger than in the case of Riemannian curvature tensors. We have studied in [BO 3] (see also [BO 4]) quadratic invariants for these curvature tensors under the action of the group $SO(n)$.

Quadratic invariants for a curvature tensor corresponding to a symmetric connection on a Hermitian manifold, under the action of $U(n)$, have been studied in [NIK] and [MA], and on a normal almost contact manifold, under the action of $U(n) \times 1$, in [MA].

3. Curvature invariants and the representation theory

For a representation α of any group $\mathcal{G}$ on the vector spave V, we have an induced representation on the tensor space $\bigotimes^r V^*$, where V^* is the dual space of a space V. This induced representation $\tilde{\alpha}$ is given by

$$\tilde{\alpha}(a)T(x_1,\ldots,x_r) = T\left(\alpha(a^{-1})x_1,\ldots,\alpha(a^{-1})x_r\right), \quad a \in \mathcal{G}.$$

In order to prove the irreducibility of this induced representation (i.e. there exist no non trivial invariant subspaces) we can use the following characterization in terms of quadratic invariants:

An invariant subspace of $\bigotimes^r V^*$ is irreducible for the action of a subgroup $\mathcal{G}$ of $O(2n)$ if and only if the space of its quadratic invariants is 1-dimensional (see [B–G–M], [GR–HE]).

One can prove (see [B–G–M]) that all the quadratic invariants of a subspace of $\bigotimes^r V^*$ are restrictions of quadratic invariants of $\bigotimes^r V^*$. These last ones may be written as follows:

$$P(T) = \sum T(e_{i_1},\ldots,e_{i_r})T(e_{j_1},\ldots,e_{j_r})p(e_{i_1},\ldots,e_{i_r},e_{j_1},\ldots,e_{j_r})$$

where $T \in$ that subspace of $\bigotimes^r V^*$ and $p \in \bigotimes^{2r} V^*$ is an invariant. The invariant tensors of $\bigotimes^{2r} V^*$, under the action of $O(n)$, $U(n)$, $Sp(n)$ have been found by Iwahori [IW].

When $r \geq 4$, an interesting thing to study is the problem of irreducibility of the induced representation $\tilde{\alpha}$ on a vector space of curvature tensors. Then quadratic invariants are exactly curvature invariants. For more details see [BO 5].

4. Topology and curvature invariants

We define a closed $2k$-form γ_k by

$$\gamma_k = \frac{(-1)^k}{\left(2\pi\sqrt{-1}\right)^k k!} \sum \delta^{\alpha_1\ldots\alpha_k}_{\beta_1\ldots\beta_k} \Omega^{\beta_1}_{\alpha_1} \wedge \ldots \wedge \Omega^{\beta_k}_{\alpha_k}.$$

Ω is the curvature form of an m-dimensional Kähler manifold which can be expressed by a matrix-valued 2-form $(\Omega^{\alpha}_{\beta})$ where is given by

$$\Omega^{\alpha}_{\beta} = \sum R^{\alpha}_{\gamma\delta\bar{\beta}}\theta^{\gamma} \wedge \theta^{\delta}.$$

It is well known that the k-th Chern class c_k is determined by the form γ_k. In particular, the first two Chern forms are given by

$$2\pi\gamma_1 = \sqrt{-1}\sum_\alpha \Omega_\alpha^\alpha$$

and

$$-8\pi^2\gamma_2 = \sum\left(\Omega_\alpha^\alpha \wedge \Omega_\beta^\beta - \Omega_\beta^\alpha \wedge \Omega_\alpha^\beta\right)$$

respectively.

Then we have

(4.1) $\gamma_1 \wedge \Phi^{m-1} = \dfrac{\tau}{m\pi}\Phi^m,$

(4.2) $\gamma_1^2 \wedge \Phi^{m-2} = \dfrac{1}{4m(m-1)\pi^2}\left(\tau^2 - 2\|\rho\|^2\right)\Phi^m,$

and

(4.3) $\gamma_2 \wedge \Phi^{m-2} = \dfrac{1}{8m(m-1)\pi^2}\left(\tau^2 - 4\|\rho\|^2 + \|R\|^2\right)\Phi^m.$

The generalized Chern numbers (for a compact $\mathbb{M}$) $\omega^{m-1}c_1(\mathbb{M})$, $\omega^{m-2}c_1^2(\mathbb{M})$ and $\omega^{m-2}c_2(\mathbb{M})$ are defined by $\int_{\mathbb{M}}\gamma_1\wedge\Phi^{m-1}$, $\int_{\mathbb{M}}\gamma_1^2\wedge\Phi^{m-2}$ and $\int_{\mathbb{M}}\gamma_2\wedge\Phi^{m-2}$ respectively.

Let $\aleph(\mathbb{M})$, $\tau(\mathbb{M})$, $a(\mathbb{M})$ denote the Euler characteristic, the Hirzebruch signature and the arithmetic genus of a compact Kähler surface $\mathbb{M}^2$ respectively. Then. from the Gauss–Bonnet–Chern theorem, the Hirzebruch signature theorem and the Riemann–Roch–Hirzebruch theorem we have

(4.4) $\aleph(\mathbb{M}) = \displaystyle\int_{\mathbb{M}} c_2,$

(4.5) $\tau(\mathbb{M}) = \dfrac{1}{3}\displaystyle\int_{\mathbb{M}} (c_1^2 - 2c_2),$

(4.6) $a(\mathbb{M}) = \dfrac{1}{12}\displaystyle\int_{\mathbb{M}} (c_1^2 + c_2).$

(See [A–S], [HI], [P]).

The characterization of complex space forms in terms of the Bochner curvature tensor and Chern classes was considered by Blažić [BL 1], Chen [CH 2], Chen and Ogiue [CH–O 1], [CH–O 2]. The Euler–Poincaré characteristic, the Hirzebruch signature and the arithmetic genus for Kähler surfaces were investigated in terms of the Bochner curvature tensor by Chen in

[CH 1], [CH 3]. We consider also the conditions for a Kähler manifold to be biholomorphically covered by a flat manifold or to be a complex space form in terms of Chern classes and the complex conharmonic curvature tensor (see [PR]). Here are some theorems from papers mentioned above.

Theorem 4.1. [BL 1] *Let $\mathbb{M}$ be a compact Kähler manifold of complex dimension m. Then $\mathbb{M}$ is covered biholomorphically by a complex space $\mathbb{C}^m$ if and only if $\mathbb{M}$ is a locally symmetric space whose first Chern class $c_1(\mathbb{M})$ vanishes.*

Theorem 4.2. [CH 2] *An algebraic hypersurface of $\mathbb{CP}_5$ admits a BK-metric (i.e. a Bochner-flat metric) if and only if it is linear.*

Theorem 4.3. [CH 2] *A compact complex manifold of complex dimension m is covered biholomorphically by $\mathbb{C}^m$ if and only if its first Chern class vanishes and it admits a BK-metric.*

Theorem 4.4. [CH–O 1] *A complete intersection manifold with vanishing Bochner curvature tensor is a linear subspace.*

Theorem 4.5. [CH 1] *If a complex surface $\mathbb{M}$ admits a BK-metric, then the following inequalities are satisfied:*

(i) $\qquad\qquad\qquad\qquad \tau \geq 0,$

(ii) $\qquad\qquad\qquad\qquad \aleph \leq 4a,$

(iii) $\qquad\qquad\qquad\qquad \aleph \leq 3\tau,$

(iv) $\qquad\qquad\qquad\qquad a \leq \tau,$

with equality sign in (i) and (ii) if and only if $\mathbb{M}$ admits a Kähler metric and is locally a product of two complex curves with constant Gauss curvature K and $-K$ respectively.

Theorem 4.6. [BO 1] *Let $\mathbb{M}$ be a compact Kähler manifold of complex dimension $m > 1$. If the complex conharmonic curvature tensor H is zero, then $\omega^{m-2} c_2(\mathbb{M}) \leq 0$ with equality sign if and only if $\mathbb{M}$ is flat.*

Theorem 4.7. [BO 1] *Let $\mathbb{M}$ be a compact Kähler surface. If*

(i) $\quad \tau(\mathbb{M}) = 0,$

and

(ii) $\quad$ *the complex conharmonic curvature tensor is zero,*

then $\mathbb{M}$ is flat.

5. The volumes of geodesic spheres

There are many results about volumes of tubes and geodesic spheres obtained in terms of curvature invariants (see for example [GR 1], [GR 2], [KO], [CH–V 1], [GR–VA] etc.). But, several conjectures about these volumes are still open. Using some weaker assumptions than in these conjectures and adding some topological conditions it is possible to obtain some geometrical properties. In order to be more explicit we start with some results about geodesic spheres.

Let $G_m(r)$ be the geodesic sphere with the center m and sufficiently small radius r. Further, let $S_m(r)$ denotes the $(n-1)$-dimensional volume of $G_m(r)$. Then we have

$$S_m(r) = r^{n-1} \int_{S^{n-1}(1)} \theta_m(\exp_m(ru))\, du$$

where $S^{n-1}(1)$ denotes the unit sphere in $T_m M$. The volume $V_m(r)$ of the corresponding geodesic ball is given by

$$V_m(r) = \int_0^r S_m(t)\, dt.$$

Explicit expressions for θ_m are well known when (M, g) is a two-point homogeneous space. In this case one can write down explicit expressions for $S_m(r)$ too. We have (see, for example, [CH–V 1], [GR–VA])

$$S_m(r) = c_{n-1} r^{n-1}$$

for $\mathbb{E}_n$ and for the other two-point homogeneous spaces one has

$$S_m(r) = c_{n-1} \left(\frac{\sin \sqrt{\alpha} r}{\sqrt{\alpha}}\right)^p \left(\frac{2}{\sqrt{\alpha}} \sin \frac{\sqrt{\alpha}}{2} r\right)^q$$

with $p+q = n-1$ for the case of positive maximal sectional curvature. Here $p = n - 1, 1, 3$ or 7. c_{n-1} denotes the volume of $S^{n-1}(1)$, i.e.

$$c_{n-1} = \frac{n \pi^{\frac{n}{2}}}{\left(\frac{n}{2}\right)!}, \qquad \left(\frac{n}{2}\right)! = \Gamma\left(\frac{n}{2} + 1\right).$$

For a general Riemannian manifold we do not have an explicit expression for θ_m but we may write down some terms of the power series expansion for $S_m(r)$ and $V_m(r)$ in function of r. Then we obtain

$$(5.1) \qquad V_m(r) = \frac{(\pi r^2)^{\frac{n}{2}}}{\left(\frac{n}{2}\right)!} \left\{ 1 - \frac{\tau(m)}{6(n+2)} r^2 + \right.$$

$$\frac{1}{360(n+2)(n+4)} (-3\|R\|^2 + 8\|\rho\|^2 + 5\tau^2 - 18\Delta\tau)(m) r^4 + O(r^6) \Big\},$$

where τ denotes the scalar curvature and Δ is the Laplacian. We refer to [GR–VA] for the next term in this expansion.

It is worthwhile to note here that for an analytic $\mathbb{M}$, $V_m(r)$ is completely determined by the invariants of the curvature tensor. In (5.1) we have written down the terms involving linear and quadratic invariants.

Now we recall one of the volume conjectures as stated in [GR–VA].

The volume conjecture I. Let $(\mathbb{M}, g)$ be a Riemannian manifold such that for all $m \in \mathbb{M}$ and all sufficiently small $r > 0$ we have

$$V_m(r) = \omega r^n,$$

where $\omega = \frac{\pi^{\frac{n}{2}}}{\left(\frac{n}{2}\right)!}$. Then $(\mathbb{M}, g)$ is locally flat.

Here we compare $V_m(r)$ with the volume of a ball of radius r in Euclidean space. One may formulate similar conjectures when we take another two-point homogeneous space as model space (see [GR–VA]). Although these conjectures have been proved in several special cases, the general case still remains open. We mention some of these special cases.

Theorem 5.1. [GR–VA] *The volume conjecture I is true in any of the following cases:*

(a) $\dim \mathbb{M} \le 3$;

(b) $\mathbb{M}$ *is Einstein, or more generally if $\mathbb{M}$ has nonnegative or nonpositive Ricci curvature;*

(c) $\mathbb{M}$ *is conformally flat;*

(d) $\mathbb{M}$ *is a compact oriented four-dimensional manifold whose Euler–Poincaré characteristic and signature satisfy $\aleph(\mathbb{M}) \ge (-3/2)|\tau(\mathbb{M})|$;*

(e) $\mathbb{M}$ is a product of surfaces;

(f) $\mathbb{M}$ is a 4- or 5-dimensional manifold with parallel Ricci tensor;

(g) $\mathbb{M}$ is the product of symmetric spaces of classical type.

Theorem 5.2. [DI] *Let the model space $\mathbb{K}$ be a locally product of two Kähler manifolds $\mathbb{K}_1$ and $\mathbb{K}_2$ with constant holomorphic sectional curvatures H and $-H$ respectively. Let $\mathbb{M}_1 \times \mathbb{M}_2$ be the local product of the Kähler manifolds $\mathbb{M}_1$, $\mathbb{M}_2$ and let for all $p \in \mathbb{M}$ and all sufficiently small $r > 0$, the function $V_p(r)$ on $\mathbb{M}$ be equal to the corresponding function on the model space. If, in additional the relation*

$$||R||^2 = 4\frac{n(n+1) + m(m+1)}{n(n+1)^2 + m(m+1)^2}||\rho||^2,$$

holds, then $\mathbb{M}_1$ and $\mathbb{M}_2$ are also spaces of constant holomorphic sectional curvature. Here, R and ρ are the curvature and the Ricci tensor on $\mathbb{M}$ respectively and $n = \dim \mathbb{K}_1$, $m = \dim \mathbb{K}_2$.

Since the volume of geodesic balls and characteristic classes can be expressed in terms of invariants of the curvature tensor, we may expect some interdependence. These problems were investigated by Blažić [BL 2]. We mention some of the main results.

Theorem 5.3. *Let $(\mathbb{M}, g, J)$ be a compact Kähler manifold of complex dimension n. Suppose that generalized Chern numbers $\omega^{n-1}c_1$ and $\omega^{n-2}c_1^2$ are nonnegative. Then, if $\mathbb{M}$ satisfies one of the conditions*

(i) $\qquad\qquad V_m(r) \geq \Omega_{2n}r^{2n},$

(ii) $\qquad\qquad 2nV_m(r) \leq rS_m(r),$

then $\mathbb{M}$ is biholomorphically covered by $\mathbb{C}^n$.

(Here Ω_n is the volume of the unit ball in $\mathbb{E}^n$, that is

$$\Omega_n = \frac{\pi^{\frac{n}{2}}}{(\frac{n}{2})!}.$$

Blažić [BL 2] has also proved a particular case of the conjecture (IV) (stated in [GR–VA] by supposing

$$\omega^{n-1}c_1(\mathbb{M}) = \omega_\mu^{n-1}c_1(\mathbb{M}(\mu)),$$
$$\omega^{n-2}c_1^2(\mathbb{M}) \geq \omega_\mu^{n-2}c_1^2(\mathbb{M}(\mu)).$$

(Here $\mathbb{M}(\mu)$ is an n-dimensional compact Kähler manifold with constant holomorphic sectional curvature μ).

6. Metric connections and curvature invariants

A metric connection D on $\mathbb{M}$ is a linear connection which satisfies

$$Xg(Y,Z) = g(D_X Y, Z) + g(Y, D_X Z)$$

for every $X, Y, Z \in \mathfrak{X}(\mathbb{M})$.

The circature tensor S corresponding to the metric connection D has the following properties

(6.1) $$S(X,Y) = -S(Y,X),$$

(6.2) $$S(X,Y,Z,V) = -S(X,Y,V,Z),$$

where $S(X,Y,Z,V) = g(S(X,Y)Z,V)$.

Some special kinds of metric connections are very useful. For example, using the characteristic connection (see [G–B–N–V]) Miquel studied in [MI] the almost Hermitian geometry in relation with the volume $V_m D(r)$ of a small geodesic ball for this connection D.

The characteristic connection on an almost Hermitian manifold is the unique metric connection D satisfying

$$DJ = 0,$$

$$T(X,Y) + T(JX, JY) = 0,$$

where T is the torsion tensor of D.

A connection on a complex vector bundle over an almost complex manifold is called formally holomorphic provided

(i) D is complex, that is
   $D_U(J) = 0$, for $U \in \mathfrak{X}_\mathbb{C}(\mathbb{M})$;

(ii) D preserves the fiber metric h, that is
   $D_U(h) = 0$, i.e. $Uh(X,Y) = h(D_U X, Y) + h(X, D_U Y)$;
   for $U \in \mathfrak{X}_\mathbb{C}(\mathbb{M})$ and $X, Y \in \mathfrak{S}(\mathbb{E})$,

(iii) the curvature S of D satisfies
$$S_{JUJT} = S_{UT}, \text{ for } U, T \in \mathfrak{X}_{\mathbb{C}}(\mathbb{M}).$$

We mention here only some results related to topology and geometry for some manifolds where curvature invariants of a formally holomorphic connections are used intensively.

Theorem 6.1. [G–B–N–V] *Let* $\mathbb{M}$ *be a compact almost complex manifold of real dimension* $2n$, *and let* $\mathbb{E}$ *be a complex vector bundle over* $\mathbb{M}$. *Let* D *be a complex connection which is compatible with some metric on* $\mathbb{E}$. *Assume that* D *is formally holomorphic and the holomorphic bisectional curvature of* D *is nonnegative. Then the Chern numbers* $c_1^n(\mathbb{E})[\mathbb{M}]$, $c_1^{n-2} c_2(\mathbb{E})[\mathbb{M}]$ *and* $c_1^{n-4} c_2^2(\mathbb{E})[\mathbb{M}]$ *are nonnegative.*

Theorem 6.2. [BL 3] *Let* $(\mathbb{M}, g)$ *be a compact,* $2n$-*dimensional almost Kähler manifold. If a complex vector bundle* $\mathbb{E}$ *over* $(\mathbb{M}, g)$ *admits a metric* h *and a formally holomorphic connection* D *so that* $(\mathbb{E}, h, D, \mathbb{M}, g)$ *is an Einstein–Hermitian vector bundle, then the inequality*

$$\int_{\mathbb{M}} \left\{ (\lambda - 1)c_1(\mathbb{E})^2 - 2\lambda c_2(\mathbb{E}) \right\} \wedge \phi^{n-2} \leq 0$$

holds. The equality holds if and only if $(\mathbb{E}, h)$ *is projectively flat.*

Chern classes $c_1(\mathbb{E})$ and $c_2(\mathbb{E})$ are given by $[\gamma_1]$ and $[\gamma_2]$ respectively, where

$$\gamma_1^2 \wedge \phi^{n-2} = \frac{1}{4n(n-1)\pi^2}(\tau^{*2} - 2\|\widehat{\rho}\|^2)\phi^n,$$

$$\gamma_2 \wedge \phi^{n-2} = \frac{1}{8n(n-1)\pi^2}(\tau^{*2} - 2\|\widehat{\rho}\|^2 - 2\|\rho^*\|^2 + \|S\|^2)\phi^n,$$

S is the curvature tensor for D, $\widehat{\rho}$ and ρ^* are Ricci curvature tensors for S, i.e.

$$\widehat{\rho}(S)(x, y) = \sum S_{x\overline{y}\alpha\overline{\alpha}}, \qquad x, y \in \mathbb{M}_m \otimes \mathbb{C},$$

$$\rho^*(S)(t, u) = \sum S_{p\overline{p}t\overline{u}}, \qquad t, u \in \mathbb{E}_m,$$

and τ^* is the scalar curvature.

These results are a generalization of the corresponding results for the tangent bundle of an n-dimensional Kähler manifold [CH–O 2] and for an Einstein–Hermitian vector bundle $(\mathbb{E}, h)$ over a Kähler base space $(\mathbb{M}, g)$ [LÜ].

7. Symmetric connections and curvature invariants

In the study of projective geometry of an arbitrary manifold we deal with symmetric connections. In this section we consider some applications using curvature invariants in projective geometry of Riemannian and Hermitian manifolds and its relation with the topology of these manifolds.

Let us mention that a symmetric connection is a connection ∇ satisfying

$$\nabla_X Y - \nabla_Y X - [X, Y] = 0,$$

for $X, Y \in \mathfrak{X}(\mathbb{M})$.

The curvature tensor R, corresponding to such a connection ∇ has the following properties

(7.1) $R(X, Y) = -R(Y, X),$

(7.2) $\sigma R(X, Y)Z = 0$ (the first Bianchi identity)

where σ denotes the cyclic sum with respect to X, Y, Z.

We gave in [BO 3] the complete decomposition of the vector space $\mathfrak{R}(V)$ consisting of curvature tensors having the same symmetries as the curvature tensor associated with a symmetric connection on a Riemannian manifold. V is endowed with a positive definite metric g. We solved the problem under the action of $SO(n)$. The irreducibility of the components of the decomposition is based on a detailed treatment of the quadratic invariants of the curvature operator and some results from representation theory. Using equalities for some curvature invariants we have found some characterizations of Ricci flat and of projectively flat manifolds. We reformulated some of Ishihara's theorems [IS] about projective transformations in terms of our decomposition.

Now it is naturally to put the question: what can we say about eventual relations between the topology and the projective geometry in terms of quadratic invariants? We give some examples using the first Chern characteristic class and the Ricci curvature for Kähler manifolds and Hermitian surfaces.

First of all let us recall the notion of a projective transformation and a group of projective transformations.

A map $f : (\widetilde{M}, \widetilde{\nabla}) \to (M, \nabla)$ of manifolds with symmetric connections $\widetilde{\nabla}$ and ∇ is called projective if for each geodesic γ of $\widetilde{\nabla}$, $f \circ \gamma$ is a reparametrization of a geodesic of ∇, that is, there must exists a strictly increasing C^∞ function h on some open interval such that $f \circ \gamma \circ h$ is a ∇-geodesic. Linear connections $\widetilde{\nabla}$ and ∇ on M are said to be projectively equivalent if the identity map of M is projective. A projective transformation of (M, ∇) is a diffeomorphism which is projective (see [PO], [WE 1]). The transformation s is projective on M, if the pull back $s^*\nabla$ of the connection is projectively related to ∇, i.e. if there exists a global 1-form $\pi = \pi(s)$ on M such that

(7.3) $$s^*\nabla_X Y = \nabla_X Y + \pi(X)Y + \pi(Y)X,$$

for arbitrary smooth vector fields $X, Y \in \mathfrak{X}(M)$. By virtue of (7.3), if s and t are two projective transformations, we find

$$\pi(st) = \pi(s) + \widehat{s}\pi(t),$$

where $\widehat{s}$ is the cotangent map i.e. $[\widehat{s}\pi]_{s(p)} = \widehat{s}[\pi]_p$.

Let G be a group of projective trasnformations of the affine connection ∇. If there exists a certain projectively related affine connection for which G is a group of affine transformations, then the group G is said to be essentially affine with respect to the connection ∇. If a transformation s of M preserves geodesics and the affine character of the parameter on each geodesic, then s is called an affine transformation of the connection ∇ or simply of the manifold M, and we say that s leaves the connection ∇ invariant.

Ishihara studied in [IS] the group of projective transformations, the group of affine transformations and the group of isometries of M. Blažić and the author have studied these groups in connection with characteristic classes in [BL–BO].

Since on a Hermitian manifold of real dimension 4 there exists a symmetric connection D compatible with an almost complex structure J (i.e. $DJ = 0$) (see [GR–HE] and [BO 2]) it is possible, using invariance theory, to find formulas for the square of the first Chern form, for the second Chern form and for the first Pontryagin form too. We have [BL–BO]

(7.4) $$\gamma_1^2 = \frac{1}{4\pi^2} \left[(\tau^2 - \tau^{*2}) - \sum (\rho_{qp}\rho_{\overline{pq}} + \rho_{pq}\rho_{qp}) \right] dV,$$

(7.5) $$\gamma_2 = \frac{1}{8\pi^2} \left[-\sum R_{pqrs} R_{pqsr} - 2\sum \widehat{\rho}_{pq}\widehat{\rho}_{qp} - \right.$$

$$\sum \left(\rho_{qp}\rho_{\overline{pq}} + \rho_{pq}\rho_{qp} \right) + \left(\tau^2 - \tau^{*2} \right) \Big] dV,$$

$$p_1 = \gamma_1^2 - 2\gamma_2 = -\frac{1}{8\pi^2} \left[\sum R_{pqrs}R_{pqsr} + 2\sum \hat{\rho}_{pq}\hat{\rho}_{qp} \right] dV.$$

Here we have used the following notations: R, ρ τ are the curvature tensor, the Ricci tensor and the scalar curcature respectively, associated to the connection D, and $\hat{\rho}$ and τ^* are given by

$$\hat{\rho}(R)(X,Y) = \hat{\rho}(X,Y) = \sum_i R(e_i, X, e_i, Y),$$

$$\tau^*(R) = \tau^* = \sum_i R(e_i, e_j, Je_j, e_i) = \sum_i \rho(e_j, Je_j),$$

where $(e_1, \ldots, e_n, Je_1, \ldots, Je_n)$ is an arbitrary adapted orthonormal basis of $T_p(M)$.

Using the formulas (7.4) and (7.5) we can study a dependence of the projective geometry of a 4-dimensional Hermitian manifold M on its topology. Our main results are given in the following theorems.

Theorem 7.1. *If a 4-dimensional Hermitian manifold M endowed with the symmetric connection D is compact and if*

(i) $\rho(X,Y) = -\rho(Y,X),$

(ii) $\rho(X,Y) = \rho(JX, JY), \quad \tau^* = 0,$

(iii) $c_1^2(M) \leq 0,$

then the group of projective diffeomorphisms of the connection D coincides with the group of affine diffeomorphisms of the connection D.

Theorem 7.2. *If a compact and irreducible 4-dimensional Hermitian manifold M, endowed with the symmetric connection D, satisfies*

(i) $\rho(X,Y) = -\rho(Y,X),$

(ii) $\rho(X,Y) = \rho(JX, JY), \quad \tau^* = 0,$

(iii) $c_1^2(M) \leq 0,$

then the group of projective diffeomorphisms of the connection D coincides with the group of isometries of M.

For a Kähler manifold of real dimension $2n$ we can use the formulas (4.1), (4.2) and (4.3) for the Chern forms study some dependence of the group of projective diffeomorphisms of the Levi Civita connection ∇ on the Chern characteristic classes.

Our main results are the following.

Theorem 7.3. *If a Kähler manifold* $\mathbb{M}$ *is compact and*

(i) $\tau = \text{const},$

(ii) $c_1(\mathbb{M}) = 0,$

then the group of all projective diffeomorphisms of the Levi Civita connection is coincides with the group of all affine diffeomorphisms of the same connection.

Theorem 7.4. *If a Kähler manifold* $\mathbb{M}$ *is compact and irreducible Riemannian manifold and if moreover*

(i) $\tau = \text{const},$

(ii) $c_1(\mathbb{M}) = 0,$

then the group of all projective diffeomorphisms of the Levi Civita connection coincides with the group of all isometries of $\mathbb{M}$.

References

[A–S] ATIYAH M. F. and SINGER I. M., The index of elliptic operators, III, *Ann. of Math.* **87** (1968), 546–604.

[B–G–M] BERGER M., GAUDUCHON P. and MAZET E., Le spectre d'une variete riemannienne, *Lecture Notes in Math.* **194**, Springer-Verlag, Berlin and New York, 1971.

[BL 1] BLAŽIĆ N., Chern classes and locally symmetric Hermitian manifolds, *Colloquia Mathematica Societatis J. Bolyai, 46. Topics in Diff. Geom., Debrecen,* (1984), 203–212.

[BL 2] BLAŽIĆ N., The volumes of small geodesic balls and generalized Chern numbers of Kähler manifolds, *Nagoya J. of Math.* **116** (1989), 181–189.

[BL 3] BLAŽIĆ N., Chern classes of complex vector bundles over almost complex manifolds, *Boll. U.M.I.,* **7**, 3–B (1989), 939–951.

[BL–BO]   BLAŽIĆ N. and BOKAN N., The Chern characteristic classes and a group of projective transformations, preprint 1989.

[BO 1]   BOKAN N., Complex conformal connection of Kähler manifold, its small geodesic balls and generalized Chern numbers, *J. Ramanujan Math. Soc.* **4(1)** (1989), 93–108.

[BO 2]   BOKAN N., Curvature tensors on Hermitian manifolds, *Colloquia Mathematica Societatis J. Bolyai, 46. Topics in Diff. Geom., Debrecen (Hungary),* (1984), 213–239.

[BO 3]   BOKAN N., On the complete decomposition of curvature tensors of Riemannian manifolds with symmetric connection, Rendiconti del Circolo Matematico di Palermo (3) **39** (1990) (to appear).

[BO 4]   BOKAN N., Quadratic invariants and locally affine symmetric manifolds, *Proc. Int. Conf. Diff. Geom. Appl., Dubrovnik (Yugoslavia), June 26 – July 3, 1988,* 69–82.

[BO 5]   BOKAN N., The decomposition theory and its applications, The mathematical heritage of C.F. Gauss, *World Scientific Publishing Company, Singapore, 1990,* to appear.

[B–G]   BRANSON T. P. and GILKEY P. B., The asymptotics of the Laplacian on a manifold with boundary, preprint 1989.

[B–G–O]   BRANSON T. P., GILKEY P. B. and ORSTED B., Leading terms in the heat invariants, preprint 1989.

[C]   CHAVEL I., Eigenvalues in Riemannian geometry, Academic Press, New York, 1984.

[CH 1]   CHEN B. Y., Surfaces admettant une métrique kählerienne de Bochner, C.R. Acad. Sci. Paris **282** (1976), 643–644.

[CH 2]   CHEN B. Y., Chern Classes and Kaehler metric, *Bull. Inst. Math. Acad. Sinica* **6** (1978), 259–270.

[CH 3]   CHEN B. Y., Some topological obstructions to Bochner–Kähler metrics and their applications, *J. Differential Geometry* **13** (1978), 547–558.

[CH–O 1]   CHEN B. Y. and OGIUE K., A characterization of complex space forms in terms of Bochner curvature tensor and Chern classes and its application, *Indiana Univ. Math. J.* **24**, No 11 (1975), 1087–1091.

[CH–O 2]   CHEN B. Y. and OGIUE K., Some characterization of complex space forms in terms of Chern classes, *Quart. J. Math. Oxford (3)* **26** (1975), 459–464.

[CH–V 1]   CHEN B. Y. and VANHECKE L., Differential geometry of geodesic spheres, *J. Reine Angew. Math.* **325** (1981), 28–67.

[CH–V 2]   CHEN B. Y. and VANHECKE L., The spectrum of the Laplacian of Kähler manifolds, *Proc. Amer. Math. Soc.* **79**, No 1 (1980), 82–86.

[DI] DIMITRIĆ I., A remark on the volume of geodesic balls in the product of Kähler manifolds, *Colloquia Mathematica Societatis J. Bolyai, 46. Topics in Diff. Geom., Debrecen (Hungary),* (1984), 299–304.

[G 1] GILKEY P. B., Invariance theory, the heat equation, and the Atiyah–Singer index theorem, Publish or Perish, Wilmington, 1984.

[G 2] GILKEY P. B., Spectral geometry and the Kähler condition for complex manifolds, *Inven. Math.* **26** (1974), 231–258.

[G 3] GILKEY P. B., The spectral geometry of a Riemannian manifold, *J. Differential Geometry* **10**, No 4 (1975) 601–618.

[GR 1] GRAY A., The volume of a small geodesic ball in a Riemannian manifold, *Michigan Math. J.* **20** (1973), 329–344.

[GR 2] GRAY A., Tubes, Eddison – Wesley Publishing Company, 1988.

[G–B–N–V] GRAY A., BAROS M., NAVEIRA A. M. and VANHECKE L., The Chern numbers of holomorphic vector bundles and formally holomorphic connections of complex vector bundles over almost complex manifolds, *J. Reine Angew. Math.* **314** (1980), 84–98.

[GR–HE] GRAY A. and HERVELLA L. M., The sixteen classes of almost Hermitian manifolds, *Ann. Mat. Pura Appl.* **123** (1980), 35–58.

[GR–VA] GRAY A. and VANHECKE L., Riemannian geometry as determined by the volumes of small geodesis balls, *Acta Math.* **142** (1979), 157–198.

[HI] HIRZEBRUCH F., Topological methods in algebraic geometry, Springer-Verlag, Berlin, 1966.

[IS] ISHIHARA S., Groups of projective transformations and groups of conformal transformations, *J. Math. Soc. Japan* **9**, No 2 (1957), 195–227.

[IW] IWAHORI N., Some remarks on tensor invariants of $O(n)$, $U(n)$, $Sp(n)$, *J. Math. Soc. Japan* **10** (1958), 145–160.

[KO] KOWALSKI O., Additive volume invariants of Riemannian manifolds, *Acta Math.* **145** (1980), 205–225.

[KU] KULKARNI R. S., Index theorems of Atiyah–Bott–Patodi and curvature inavriants, Les Presses de l'Universite de Montreal, Montreal, 1975.

[LÜ] LÜBKE M., Chernklassen von Hermite–Einstein–Vector–Bündeln, *Math. Ann.* **260** (1982), 133–141.

[MA] MATZEU P., La decomposizione completa di $\mathcal{R}(V)$, preprint 1989.

[MI] MIQUEL V., Volumes of certain small geodesis balls and almost Hermitian geometry, *Geometriae Dedicata* **15** (1984), 261–267.

[NIK] NIKČEVIĆ S., On the decomposition of curvature tensor fields on Hermitian manifolds, *Coll. Diff. Geom., August 20–26, 1989, Eger (Hungary),* Abstracts, p. 39 and preprint 1989.

[P] PALAIS R. S., Seminar on Atiyah – Singer index theorem, *Ann. of Math. Stud.* **57**, Princeton University Press, Princeton, 1965.

[PA 1] PATODI V. K., Curvature and the eigenforms of the Laplace operator, *J. Differential Geometry* **5** (1971), 233–249.

[PA 2] PATODI V. K., Curvature and the fundamental solution of the heat equation, *J. Indian Math. Soc.* **34** (1970) 269–285.

[PO] POOR W. A., Differential geometric structures, McGraw–Hill Book Company, New York, 1981.

[PR] PRVANOVIĆ M., A generalization of the Bochner and contact Bochner curvature tensor, *Review of Research Fac. of Science, Univ. Novi Sad,* **12** (1982), 349–367.

[S] SAKAI T., On eigenvalues of Laplacian and curvature of Riemannian manifold, *Tôhoku Math. J.* **23** (1971), 589–603.

[TR–VA] TRICERRI F. and VANHECKE L., Curvature tensors on almost Hermitian manifolds, *Trans. Amer. Math. Soc.* **267**, No 2 (1981), 365–398.

[WE 1] WEYL H., Zur Infinitesimalgeometrie, Einordnung der projektiven und der konformen Auffassung, *Göttinger Nach.* (1921), 99–112.

[WE 2] WEYL H., Classical groups, their invariants and representations, *Princeton Univ. Press*, Princeton, N.J., 1946.

Neda Bokan

Faculty of Mathematics
University of Belgrade
Studentski trg 16, PP 550
11000 Belgrade, Yugoslavia

Adapted Basis in D Recurrent Lagrange Space[*]

I. ČOMIĆ

1. Adapted basis

Let E be a $2n$ dimensional differentiable manifold and any $u(x,y) \in E$ in some coordinate system has coordinates $(x^1, \ldots, x^n, y^1, \ldots, y^n)$. We shall consider the coordinate transformation of the form

$$(1.1) \qquad x^{i'} = x^{i'}(x^1, \ldots, x^n), y^{i'} = y^i \frac{\partial x^{i'}}{\partial x^i}, \qquad \text{rank} \left[\frac{\partial x^{i'}}{\partial x^i} \right] = n.$$

The tangent space of E, $T(E)$ is spanned at each point u by vectors $\partial_i = \frac{\partial}{\partial x^i}$ and $\dot{\partial}_i = \frac{\partial}{\partial y^i}$. Any vector $X \in T(E)$ may be written in this basis in the form

$$(1.2) \qquad X = \tilde{X}^i \partial_i + \tilde{\tilde{X}}^i \dot{\partial}_i.$$

The elements of the basis $\{\partial_i, \dot{\partial}_i\}$ with respect to the coordinate transformation (1.1) are not transforming as vectors, so using the nonlinear connection $N_j^i(x,y)$ and the tensor field $M_j^i(x,y)$, the new basis $\{\delta_i, \dot{\delta}_i\}$ is introduced. We shall suppose that the tensor

$$(1.3) \qquad I_j^i(x,y) = \delta_j^i - M_j^i(x,y)$$

---

[*] This paper is in final form and no version of it will be submitted for publication elsewhere.

has the property $\det I_j^i \neq 0$ and $J_j^i(x,y)$ is defined by

(1.4) $$J_j^i I_h^j = \delta_h^i, \qquad I_j^i J_h^j = \delta_h^i.$$

The new basis $\{\delta_i, \dot{\delta}_i\}$ in $T(E)$ is introduced by

(1.5)   (a) $\delta_i = I_i^j(\partial_j - N_j^h \dot{\partial}_h) \Leftrightarrow \partial_j = J_j^h \delta_h + N_j^k J_k^h \dot{\delta}_h$
         (b) $\dot{\delta}_i = I_i^j \dot{\partial}_j \Leftrightarrow \dot{\partial}_j = J_j^h \dot{\delta}_h.$

From (1.2) it follows that X, in the basis $\{\delta_i, \dot{\delta}_i\}$, has the form

(1.6) $$X = X^i \delta_i + \overline{X}^i \dot{\delta}_i,$$

where

(1.7) $$X^i = \widetilde{X}^h J_h^i, \qquad \overline{X}^i = (\widetilde{\widetilde{X}}^k + \widetilde{X}^j N_j^k) J_k^i.$$

The natural basis in $T^*(E)$ is $\{dx^i, dy^i\}$ and any 1-form $w \in T^*(E)$ in this basis has the form

(1.8) $$w = \widetilde{w}_i \, dx^i + \widetilde{\widetilde{w}}_i \, dy^i.$$

The elements of the basis $\{dx^i, dy^i\}$ are not transforming as co-vectors with respect to (1.1) so that we introduce the new basis $\{\widehat{D}x^i, \widehat{D}y^i\}$ by

(1.9)   (a) $\widehat{D}x^i = J_j^i \, dx^j \Leftrightarrow dx^i = I_j^i \widehat{D}x^j$
         (b) $\widehat{D}y^i = J_j^i(dy^j + N_k^j \, dx^k) \Leftrightarrow dy^i = I_j^i \widehat{D}y^j - N_k^i I_j^k \widehat{D}x^j.$

With respect to the basis $\{\widehat{D}x^i, \widehat{D}y^i\}$ the 1-form w given by (1.8) has the form

(1.10) $$w = w_i \widehat{D}x^i + \overline{w}_i \widehat{D}y^i,$$

where

(1.11)   (a) $w_i = (\widetilde{w}_h - \widetilde{\widetilde{w}}_k N_h^k) I_i^h$ (b) $\overline{w}_i = \widetilde{\widetilde{w}}_j I_i^j.$

The vector spaces spanned by δ_i and $\dot{\delta}_i$ $(i = 1, \ldots, n)$ we denote by $T_H(E)$ and $T_V(E)$ respectively, where

$$T(E) = T_H(E) \oplus T_V(E).$$

The basis vectors $\widehat{D}x^i$ and $\widehat{D}y^i$ $(i = 1,\ldots,n)$ span $T_H^*(E)$ and $T_V^*(E)$ respectively where

$$T^*(E) = T_H^*(E) \oplus T_V^*(E).$$

Lemma 1.1. *The bases* $\{\delta_i, \dot{\delta}_i\}$ *and* $\{\widehat{D}x^i, \widehat{D}y^i\}$ *defined by* (1.5) *and* (1.9) *respectively are dual to each other, i.e.*

$$< \widehat{D}x^i, \delta_j >= \delta_j^i, \qquad < \widehat{D}x^i, \dot{\delta}_j >= 0,$$
$$< \widehat{D}y^i, \delta_j = 0 >, \qquad < \widehat{D}y^i, \dot{\delta}_j >= \delta_j^i.$$

Proof. The proof is obtained by direct calculation using (1.5) and (1.9). ∎

Lemma 1.2. *The elements of the adapted basis* $\{\delta_i, \dot{\delta}_i\}$, $\{\widehat{D}x^i, \widehat{D}y^i\}$, *the coordinates of the vector* X *and the 1-form* w *determined by* (1.6) *and* (1.10) *are transforming as tensors with respect to the coordinate transformation* (1.1).

Proof. The proof is obtained by direct calculation using the transformation law of nonlinear connection

$$(1.12) \qquad N_i^j(x,y) = N_{i'}^{j'}(x',y')\frac{\partial x^{i'}}{\partial x^i}\frac{\partial x^j}{\partial x^{j'}} - \frac{\partial^2 x^j}{\partial x^{i'}\partial x^{k'}}\frac{\partial x^{i'}}{\partial x^i}y^{k'}.$$

∎

2. The metric tensor

Let us suppose that in $T^*(E) \otimes T^*(E)$ a tensor g is given by

$$(2.1) \qquad g = g_{ij}\widehat{D}x^i \otimes \widehat{D}x^j + \overline{g}_{ij}\widehat{D}y^i \otimes \widehat{D}y^j,$$

where $[g_{ij}]$ and $[\overline{g}_{ij}]$ are symmetric, positive definite matrices. If $[g^{ij}]$ and $[\overline{g}^{ij}]$ are inverse matrices of $[g_{ij}]$ and $[\overline{g}_{ij}]$ respectively, then we raise and lower indices in the following way:

$$(2.2) \qquad X_j = g_{ij}X^i \qquad \overline{X}_j = \overline{g}_{ij}\overline{X}^i$$
$$w^i = g^{ij}w_j \qquad \overline{w}^i = \overline{g}^{ij}\overline{w}_j.$$

The scalar product of the two vectors

(2.3) $\qquad\qquad$ (a) $X = X^i \delta_i + \overline{X}^i \dot{\delta}_i$

$\qquad\qquad\qquad\qquad$ (b) $Y = Y^i \delta_i + \overline{Y}^i \dot{\delta}_i$

in the notation (X, Y) is determined by

$$(X, Y) = g_{ij} X^i Y^j + \overline{g}_{ij} \overline{X}^i \overline{Y}^j.$$

The length of the vector X, $|X|$ is defined by $|X|^2 = (X, X)$ and if θ is the angle between X and Y then

$$\cos\theta = \frac{(X, Y)}{|X| |Y|}.$$

If we define that the vector X and Y are orthogonal with respect to g if $\cos\theta = 0$, then any vector from $T_H(E)$ is orthogonal to any vector from $T_V(E)$ with respect to g.

3. Connection coefficients

We shall define the linear connection ∇ in the following way:

(3.1) $\qquad\qquad \nabla_{\delta_j} \delta_i = F_{ij}^k \delta_k, \qquad \nabla_{\delta_j} \dot{\delta}_i = \overline{F}_{ij}^k \dot{\delta}_k,$

$\qquad\qquad\qquad \nabla_{\dot{\delta}_j} \delta_i = C_{ij}^k \delta_k, \qquad \nabla_{\dot{\delta}_j} \dot{\delta}_i = \overline{C}_{ij}^k \dot{\delta}_k.$

If X and Y are vector fields given by (2.3), then using the usual rule for the connection ∇ we obtain

(3.2) $\quad \nabla_X Y = \left(Y_{|j}^k X^j + Y^k |_j \overline{X}^j \right) \delta_k + \left(\overline{Y}_{|j}^k X^j + \overline{Y}^k |_j \overline{X}^j \right) \dot{\delta}_k,$

where

(3.3) $\quad$ (a) $Y^k |_{|j} = \delta_j Y^k + F_{ij}^k Y^i,$ $\qquad$ (b) $\overline{Y}_{|j}^k = \delta_j \overline{Y}^k + \overline{F}_{ij}^k \overline{Y}^i,$

$\qquad\quad$ (c) $Y^k |_j = \dot{\delta}_j Y^k + C_{ij}^k Y^i,$ $\qquad$ (d) $\overline{Y}^k |_j = \dot{\delta}_j \overline{Y}^k + \overline{C}_{ij}^k \overline{Y}^i.$

Lemma 3.1. $Y^k |_{|j}$, $Y^k |_j$, $\overline{Y}_{|j}^k$, $\overline{Y}^k |_j$ defined by (3.3) are transforming as tensor with respect to (1.1) if C_{ij}^k, $\overline{C}_{ij}^k$ are transforming as tensors and if F_{ij}^k, $\overline{F}_{ij}^k$ are transforming in the following way:

(3.4) $\qquad\qquad F_{ij}^k \dfrac{\partial x^j}{\partial x^{j'}} \dfrac{\partial x^{k'}}{\partial x^k} = F_{i'j'}^{k'} \dfrac{\partial x^{i'}}{\partial x^i} + I_j^n \dfrac{\partial^2 x^{k'}}{\partial x^i \partial x^n} \dfrac{\partial x^j}{\partial x^{j'}}.$

The formula obtained by substitution of F by $\overline{F}$ on both sides of (3.4) is also correct.

Proof. The proof is obtained by (1.1), (3.3) and Lemma 1.2.　■

All the tensors and connection coefficients mentioned in Lemma 3.1. are also functions of (x, y).

From (3.4) we can find that F_{ij}^k is transforming as connection coefficient with respect to (1.1) when $I_j^n = \delta_j^n$ i.e. when $M_j^n = 0$. In this case from (1.3)–(1.5) and (1.8) follows

(3.5)　　　(a) $I_j^i = J_j^i = \delta_j^i$　　(b) $\delta_i = \partial_i - N_i^j \dot{\delta}_j$　　(c) $\dot{\delta}_i = \dot{\partial}_i$

　　　　　　(d) $\widehat{D}x^i = dx^i$　　(e) $\widehat{D}y^i = dy^i + N_j^i \, dx^j$.

The linear connection ∇ defined on $T(E)$ by (3.1) introduces in the usual way a connection ∇^* (denoted also by ∇) in $T^*(E)$, where

(3.6)　　　$\nabla_{\delta_j} \widehat{D}x^k = -F_{ij}^k \widehat{D}x^i, \qquad \nabla_{\delta_j} \widehat{D}y^k = -\overline{F}_{ij}^k \widehat{D}y^i,$

　　　　　　$\nabla_{\dot{\delta}_j} \widehat{D}x^k = -C_{ij}^k \widehat{D}x^i, \qquad \nabla_{\dot{\delta}_j} \widehat{D}y^k = -\overline{C}_{ij}^k \widehat{D}y^i.$

The connection ∇ defined in $T^*(E)$ by (3.6) is consistent with the raising and lowering if indices given by (2.2).
If X is given by (1.6) and g by (2.1) then we have
(3.7)
$$\nabla_X g = \left(g_{ij|k} X^k + g_{ij}|_k \overline{X}^k\right) \widehat{D}x^i \otimes \widehat{D}x^j + \left(\overline{g}_{ij|k} X^k + \overline{g}_{ij}|_k \overline{X}^k\right) \widehat{D}y^i \otimes \widehat{D}y^j,$$

where

(3.8)　　　　　(a) $g_{ij|k} = \delta_k g_{ij} - F_{ik}^h g_{hj} - F_{jk}^h g_{ih}$

　　　　　　　　(b) $g_{ij}|_k = \dot{\delta}_k g_{ij} - C_{ik}^h g_{hj} - C_{jk}^h g_{ih}$.

If in (3.8) we substitute g for $\overline{g}$, F for $\overline{F}$, C for $\overline{C}$, the obtained formulae are also valid.

4. D symbolism

If X is the tangent vector of the curve $x^i = x^i(t)$, $y^i = y^i(t)$ i.e.

(4.1) $$X = dx^i \partial_i + dy^i \partial_i,$$

then using (1.5) and (1.9) we can prove that X determined by (4.1) may be written as the form

(4.2) $$X = \widehat{D}x^i \delta_i + \widehat{D}y^i \dot{\delta}_i,$$

Definition 4.1. For any tensor field T defined on E we define

$$DT = \nabla_X T$$

where X is given by (4.2). If in (3.2) we substitute X given by (4.2) then we can define DY, DY^i and $D\overline{Y}^i$ by the following relations

(4.3) $DY = \nabla_X Y = \left(DY^i\right)\delta_i + \left(D\overline{Y}^i\right)\dot{\delta}_i,$

(4.4) $DY^i = Y^i_{|k}\widehat{D}x^k + Y^i|_k\widehat{D}y^k, \qquad D\overline{Y}^i = \overline{Y}^i_{|k}\widehat{D}x^k + \overline{Y}^i|_k\widehat{D}y^k.$

In a similar way using (3.7) and (4.2) we define Dg, Dg_{ij} and $D\overline{g}_{ij}$ by

(4.5) $Dg = (Dg_{ij})\widehat{D}x^i \otimes \widehat{D}x^j + (D\overline{g}_{ij})\widehat{D}y^i \otimes \widehat{D}y^j,$

(4.6) $Dg_{ij} = g_{ij|k}\widehat{D}x^k + g_{ij}|_k\widehat{D}y^k, \qquad D\overline{g}_{ij} = \overline{g}_{ij|k}\widehat{D}x^k + \overline{g}_{ij}|_k\widehat{D}y^k.$

5. Recurrent Lagrange spaces

Up till now in $T(E)$ and $T^*(E)$ a special adapted basis was introduced, with respect to this basis the connection ∇ is defined and a metric tensor g is given. We can determine the connection coefficients in a different manner.

Definition 5.1. We shall denote by $E(N, M, \nabla, g)$ a differentiable manifold E supplied with adapted bases $\{\delta_i, \dot{\delta}_i\}$ and $\{\widehat{D}x^i, \widehat{D}y^i\}$, a linear connection ∇ and the metric tensor g determined above.

Let $K(x,y)$ and $\overline{K}(x,y)$ be the following 1-forms respectively:

(5.1)
$$K = K(x,y) = \lambda_k(x,y)\widehat{D}x^k + \mu_k(x,y)\widehat{D}y^k,$$
$$\overline{K} = \overline{K}(x,y) = \overline{\lambda}_k(x,y)\widehat{D}x^k + \overline{\mu}_k(x,y)\widehat{D}y^k.$$

Definition 5.2. *The space* $E(N, M, \nabla, g)$ *in which the connection coefficients are determined under conditions*

(5.2)
$$Dg_{ij} = Kg_{ij}, \qquad D\overline{g}_{ij} = \overline{K}\overline{g}_{ij},$$

(K and $\overline{K}$ are given by (5.1) which are equivalent to the following conditions

(5.3)
$$g_{ij|k} = \lambda_k g_{ij}, \qquad g_{ij}|_k = \mu_k g_{ij},$$
$$\overline{g}_{ij|k} = \overline{\lambda}_k \overline{g}_{ij}, \qquad \overline{g}_{ij}|_k = \overline{\mu}_k \overline{g}_{ij}.)$$

we shall call the recurrent Lagrange space.

From (3.8) and (5.3) we obtain the following

Theorem 5.1. *In the recurrent Lagrange space the connection coefficients have the form:*
(5.4)
$$2F_{ijk} = (\delta_i g_{jk} + \delta_k g_{ij} - \delta_j g_{ik}) - (\lambda_i g_{jk} + \lambda_k g_{ij} + \lambda_j g_{ik}) + \widetilde{F}_{ijk} - \widetilde{F}_{jki} + \widetilde{F}_{kij},$$

(5.5)
$$2C_{ijk} = (\dot{\delta}_i g_{jk} + \dot{\delta}_k g_{ij} - \dot{\delta}_j g_{ik}) - (\mu_i g_{jk} + \mu_k g_{ij} - \mu_j g_{ik}) + \widetilde{C}_{ijk} - \widetilde{C}_{jki} + \widetilde{C}_{kij}$$

where

(5.6)
$$F_{ijk} = g_{jh}F^h_{ik}, \qquad C_{ijk} = g_{jh}C^h_{ik},$$

(5.7)
$$\widetilde{F}^j_{ik} = F^j_{ik} - F^j_{ki}, \qquad \widetilde{C}^j_{ik} = C^j_{ik} - C^j_{ki}.$$

The formulae (5.4)–(5.7) are valid if we substitute F, g, λ, $\widetilde{F}$, C, μ, $\widetilde{C}$ *respectively by* $\overline{F}$, $\overline{g}$, $\overline{\lambda}$, $\overline{\widetilde{F}}$, $\overline{C}$, $\overline{\mu}$, $\overline{\widetilde{C}}$.

Definition 5.3. *The recurrent Lagrange space in which the connection coefficients are determined by (5.4)–(5.7) we shall denote by*

$E(N, M, \nabla, g, \lambda, \mu, \overline{\lambda}, \overline{\mu}, \widetilde{F}, \overline{\widetilde{F}}, \widetilde{C}, \overline{\widetilde{C}})$. M, g, λ, μ, $\overline{\lambda}$, $\overline{\mu}$, $\widetilde{C}$, $\overline{\widetilde{C}}$ are tensors with the properties described above and N is a nonlinear connection.

From (3.4) and (5.7) we can see that $\widetilde{F}^j_{ik}$ and $\overline{\widetilde{F}}^j_{ik}$ are not transforming as tensors under a coordinate transformation (1.1) because

(5.8) $\quad \widetilde{F}^{k'}_{i'j'} = \widetilde{F}^k_{ij} \dfrac{\partial x^i}{\partial x^{i'}} \dfrac{\partial x^j}{\partial x^{j'}} \dfrac{\partial x^{k'}}{\partial x^k} - \left(I^n_j \dfrac{\partial^2 x^{k'}}{\partial x^i \partial x^n} - I^n_i \dfrac{\partial^2 x^{k'}}{\partial x^j \partial x^n} \right) \dfrac{\partial x^j}{\partial x^{j'}} \dfrac{\partial x^i}{\partial x^{i'}}.$

The formula obtained from (5.8) by substitution of $\widetilde{F}$ with $\overline{\widetilde{F}}$ is also valid. From (5.8) it is obvious that under the condition

(5.9) $\qquad\qquad\qquad I^n_j = \delta^n_j \qquad$ i.e. $\qquad M^n_j = 0$

$\widetilde{F}^k_{ij}$ and $\overline{\widetilde{F}}^k_{ij}$ are transforming as tensors.

If we take

(5.10) $\qquad\qquad \lambda_i = 0, \qquad \mu_i = 0, \qquad \overline{\lambda}_i = 0, \qquad \overline{\mu}_i = 0,$

then we obtain a metric connection, because in this case (5.3) reduces to the form

(5.11) $\qquad\quad g_{ij|k} = 0, \qquad \overline{g}_{ij|k} = 0, \qquad g_{ij}|_k = 0, \qquad \overline{g}_{ij}|_k = 0.$

If we suppose that F_{ijk}, C_{ijk}, $\overline{F}_{ijk}$ and $\overline{C}_{ijk}$ are symmetric in the first and the third indices, then from (5.7) follows

(5.12) $\qquad \widetilde{F}_{ijk} = 0, \qquad \overline{\widetilde{F}}_{ijk} = 0, \qquad \widetilde{C}_{ijk} = 0, \qquad \overline{\widetilde{C}}_{ijk} = 0.$

If the conditions (5.9), (5.10) and (5.12) are satisfied, then the space
$E(N, M, g, \nabla, \lambda, \mu, \overline{\lambda}, \overline{\mu}, \widetilde{F}, \overline{\widetilde{F}}, \widetilde{C}, \overline{\widetilde{C}})$ reduces to the space
$E(N, 0, g, \nabla, 0, 0, 0, 0, 0, 0, 0, 0)$ in which the connection coefficients are determined by

(5.13) $\qquad\qquad\qquad 2F_{ijk} = \delta_i g_{jk} + \delta_k g_{ij} - \delta_j g_{ik}$

(5.14) $\qquad\qquad\qquad 2C_{ijk} = \dot{\partial}_i g_{jk} + \dot{\partial}_k g_{ij} - \dot{\partial}_j g_{ik}.$

If F, g and C are substituted respectively by $\overline{F}$, $\overline{g}$ and $\overline{C}$ in (5.13) and (5.14), the obtained formulae are also valid.

6. Recurrent Finsler spaces as subclasses of recurrent Lagrange spaces

In Finsler space all the tensors and vector spaces are functions of (x, y), but they are homogeneous of degree zero in y. This restriction is not necessary in Lagrange spaces. In Finsler space there exists a metric function $L(x, y)$ homogeneous of degree one in y. The Finsler vector and tensor fields can be interpreted as horizontal or vertical fields in Lagrange space. We shall choose the horizontal case. We shall determine the horizontal metric tensor by

$$(6.1) \qquad g_{ij}(x, y) = 2^{-1} \dot{\partial}_i \dot{\partial}_j L^2(x, y)$$

and the unit vector l by

$$(6.2) \qquad l^i = L^{-1}(x, y) y^i, \qquad l_i = \dot{\partial}_i L.$$

Using the homogeneity condition, we obtain

$$(6.3) \qquad g_{ij} l^i l^j = g_{00} = 1,$$

where "0" means the contraction by l. g_{00} is homogeneous of degree two in l (l is given by (6.2)).

In this way a Finsler space may be considered as a $2n - 1$ dimensional space in which all the tensor fields T are functions of (x, l), and they are homogeneous of degree zero in l, except the nonlinear connection $N(x, l)$ and the metric function $L(x, l)$ which are homogeneous of degree one in l. In the Finsler space we have:

$$(6.4) \qquad \delta_i = I_i^j(x, l) \left(\partial_j - N_j^k(x, l) \, {}'\dot{\partial}_k \right) \qquad {}'\delta_i = I_i^j(x, l) L' \dot{\partial}_j,$$

where

$$(6.5) \qquad \partial_j = \frac{\partial}{\partial x^j} \qquad {}'\dot{\partial}_j = \frac{\partial}{\partial l^j}$$

and

$$(6.6) \qquad \widehat{D}x^i = J_j^i(x, l) \, dx^j, \qquad \widehat{D}l^i = L^{-1} J_j^i(x, l) \left(dl^j + N_h^j(x, l) \, dx^h \right).$$

From (6.4) and (6.6) follows

(6.7) (a) $\delta_j = J_j^h \delta_h + L^{-1} N_j^k J_k^{h'} \dot{\delta}_h$ (b) $'\dot{\delta}_j = L^{-1} J_j^{h'} \dot{\delta}_h$

(6.8)   (a) $dx^i = I_j^i \widehat{D} x^j$ (b) $dl^i = L I_j^i \widehat{D} l^j - N_k^i I_j^k \widehat{D} x^j$.

The vectors $\{\delta_i, '\dot{\delta}_i\}$ and $\{\widehat{D} x^i, \widehat{D} l^i\}$ are dual to each other i.e.

$$< \widehat{D} x^i, \delta_j >= \delta_j^i, \qquad < \widehat{D} x^i, '\dot{\delta}_j >= 0,$$
$$< \widehat{D} l^i, \delta_j >= 0, \qquad < \widehat{D} l^i, '\dot{\delta}_j >= \delta_j^i,$$

but they do not form the basis of $T(E)$ and $T^*(E)$ respectively, because they are not linearly independent. As we are dealing only with the horizontal vector fields, we define the connection ∇ by

(6.9) $\nabla_{\delta_j} \delta_i = F_{ij}^k \delta_k \qquad \nabla_{'\dot{\delta}_j} \delta_i = C_{ij}^k \delta_k.$

δ_i, $'\dot{\delta}_i$, $\widehat{D} x^i$, $\widehat{D} l^i$, F_{ij}^k, C_{ij}^k are homogeneous of degree zero in l.

We restrict our attention to the case $\overline{Y}^i = 0$, $\overline{g}_{ij} = 0$ i.e. when

$$Y = Y^i \delta_i, \qquad g = g_{ij} \widehat{D} x^i \otimes \widehat{D} x^j, \qquad l = l^i \delta_i.$$

From the condition (4.4) and (4.6) follows that in this case

$$D \overline{Y}^i = 0, \quad \overline{Y}_{|k}^i = 0, \quad \overline{Y}^i|_k = 0, \quad D \overline{g}_{ij} = 0, \quad \overline{g}_{ij|k} = 0, \quad \overline{g}_{ij}|_k = 0.$$

If Y is horizontal vector field, $Y = Y^i \delta_i$ and X is given by

(6.10) $X = dx^i \partial_i + dl^{i'} \dot{\partial}_i = \widehat{D} x^i \delta_i + \widehat{D} l^{i'} \dot{\delta}_i,$

then

(6.11) $\nabla_X Y = (DY^k) \delta_k \qquad DY^k = Y_{|j}^k \widehat{D} x^j + Y^k|_j \widehat{D} l^j,$

$Y_{|j}^k$ and $Y^k|_j$ are given by

(6.12) $Y_{|j}^k = \delta_j Y^k + F_{ij}^k Y^i \qquad Y^k|_j = '\dot{\delta}_j Y^k + C_{ij}^k Y^i.$

For the horizontal metric tensor we have

(6.13) $Dg = (Dg_{ij}) \widehat{D} x^i \otimes \widehat{D} x^j, \qquad Dg_{ij} = g_{ij|k} \widehat{D} x^k + g_{ij}|_k \widehat{D} l^k,$

where

(6.14) $$g_{ij|k} = \delta_k g_{ij} - F_{ik}^h g_{hj} - F_{jk}^h g_{ih},$$

(6.15) $$g_{ij}|_k = {}'\dot{\delta}_k g_{ij} - C_{ik}^h g_{hj} - C_{jk}^h g_{ih}.$$

Substituting (6.12) into (6.11) and using (6.4) and (6.5), we obtain

(6.16) $$DY^i = dY^i + F_{jk}^i Y^j \widehat{D}x^k + C_{jk}^i \widehat{D}l^k,$$

where

$$dY^i = \partial_j Y^i \, dx^j + {}'\partial_j Y^i \, dl^j.$$

As (6.16) is correct for any horizontal vector field Y, so it is valid also for the vector field l. So we have

(6.17) $$Dl^i = dl^i + F_{ok}^i \widehat{D}x^k + C_{ok}^i \widehat{D}l^k.$$

Substituting dl^i from (6.8b) into (6.17) we get

(6.18) $$Dl^i = \left(F_{oj}^i - N_k^i I_j^k\right) \widehat{D}x^j + \left(LI_j^i + C_{oj}^i\right) \widehat{D}l^j.$$

Here it is supposed that $N_k^i = N_k^i(x, l)$ is the nonlinear connection coefficient and $I_j^k = \delta_j^k - M_j^k(x, l)$ is a tensor.

From (1.12) and (3.4) we obtain that $F_{oj}^i - N_k^i I_j^k$ with respect to (1.1) is transforming as tensor, but this expression is homogeneous of degree one in l.

Theorem 6.1. *If*

$$C_{oj}^i = LM_j^i, \qquad F_{oj}^i = N_k^i I_j^k = N_k^i(\delta_j^k - M_j^k)$$

then $Dl^i = L\widehat{D}l^i$, *and* $\widehat{D}l^j$ *and* $\widehat{D}x^j$ *are connected by the relation*

(6.19) $$\left(F_{oj}^i - N_k^i I_j^k\right) \widehat{D}x^j + \left(C_{oj}^i - LM_j^i\right)\widehat{D}l^j = 0.$$

Proof. Substituting (1.3) into (6.18), we get

(6.20) $$Dl^i = L\widehat{D}l^i + \left(F_{oj}^i - N_k^i I_j^k\right) \widehat{D}x^j + \left(C_{oj}^i - LM_j^i\right) \widehat{D}l^j.$$

From (6.20) it follows the statement. ■

Definition 6.1. *The recurrent Finsler space denoted by*
$F(N, M, D, \lambda, \mu, \widetilde{F}, \widetilde{C})$ *is obtained if the connection coefficients are determined from the relation*

(6.21) $$Dg_{ij} = \left(\lambda_k \widehat{D}x_k + \mu_k \widehat{D}l^k \right) g_{ij} = g_{ij|k} \widehat{D}x^k + g_{ij}|_k \widehat{D}l^k.$$

For the operator D the product rule is valid, so from (6.3) we obtain

(6.22) $$(Dg_{ij}) l^i l^j + 2g_{ij} l^i Dl^j = 0.$$

Substituting (6.20), (6.21) into (6.22) and using (6.3), we obtain
(6.23)
$$\left[\lambda_k g_{oo} + 2l_i \left(F^i_{ok} - N^i_j I^h_k \right) \right] \widehat{D}x^k + \left[\mu_k g_{oo} + 2l_i \left(L\delta^i_k + C^i_{ok} - LM^i_k \right) \right] \widehat{D}l^k = 0$$

This is the crucial relation which gives the connection between $\widehat{D}x^k$ and $\widehat{D}l^k$ in the recurrent Finsler space. If the Finsler space is not recurrent, i.e. $\lambda_k = 0$ and $\mu_k = 0$ then (6.23) reduces to

(6.24) $$l_i \left(F^i_{oj} - N^i_k I^k_j \right) \widehat{D}x^j + l_i \left(L\delta^i_j + C^i_{oj} - LM^i_j \right) \widehat{D}l^j = 0.$$

If $L\widehat{D}l^j = Dl^j$, then using (6.19) from (6.24) we obtain the well-known relation

(6.25) $$l_j Dl^j = 0.$$

If we multiply (6.23) with θ_{ij} ($\theta_{ij} = \theta_{ij}(x, l)$), $\theta_{ij} = \theta_{ji}$, tensor homogeneous of degree -2 with respect to l, and add to the left hand side of the relation

$$\left(\lambda_k \widehat{D}x^k + \mu_k \widehat{D}l^k \right) g_{ij} = g_{ij|k} \widehat{D}x^k + g_{ij}|_k \widehat{D}l^k,$$

then equating the coefficients beside $\widehat{D}x^k$ and $\widehat{D}l^k$ we obtain

(6.26) $$g_{ij|k} = \lambda_k g_{ij} + \theta_{ij} \left[\lambda_k g_{oo} + 2l_h \left(F^h_{ok} - N^h_m I^m_k \right) \right],$$
(6.27) $$g_{ij}|_k = \mu_k g_{ij} + \theta_{ij} \left[\mu_k g_{oo} + 2l_h \left(L\delta^h_k + C^h_{ok} - LM^h_k \right) \right].$$

Theorem 6.2. *In the recurrent Finsler space the connection coefficients are determined by*
(6.28)

$$F_{ijk} = -2^{-1} \left(I^h_k N^{m'}_h \partial_m g_{ij} + I^h_i N^{m'}_h \partial_m g_{jk} - I^h_j N^{m'}_h \partial_m g_{ki} \right)$$
$$- 2^{-1} (\lambda_k g_{ij} + \lambda_i g_{jk} - \lambda_j g_{ki}) + \Gamma_{ijk}$$
$$- (1 + \theta_{oo})^{-1} [\theta_{ij} (\gamma^o_{ok} - N^o_m I^m_k) + \theta_{jk} (\Gamma^o_{oi} - N^o_m I^m_i) -$$
$$- \theta_{ki} (\Gamma^o_{ok} - N^o_m I^m_k)]$$
$$+ 2^{-1} (\widetilde{F}_{ijk} - \widetilde{F}_{jki} + \widetilde{F}_{kij})$$

(6.29)

$$C_{ijk} = 2^{-1} L \left(I_k^{h'} \dot{\partial}_h g_{ij} + I_i^{h'} \dot{\partial}_h g_{jk} - I_j^{h'} \dot{\partial}_h g_{ki} \right)$$
$$- 2^{-1} \left(\mu_k g_{ij} + \mu_i g_{jk} - \mu_j g_{ki} \right)$$
$$- (1 + \theta_{oo})^{-1} L \left[\theta_{ij} (l_k - M_k^o) + \theta_{jk} (l_i - M_i^o) - \theta_{ki} (l_j - M_j^o) \right]$$
$$+ 2^{-1} \left(\widetilde{C}_{ijk} - \widetilde{C}_{jki} + \widetilde{C}_{kij} \right),$$

where

(6.30) $\qquad \Gamma_{ijk} = 2^{-1} \left[I_k^h \dot{\partial}_h g_{ij} + I_i^h \dot{\partial}_h g_{jk} - I_j^h \dot{\partial}_h g_{ki} \right], \qquad '\dot{\partial}_m = \partial / \partial l^m.$

Proof. Substituting (6.14) into (6.26) we obtain the connection coefficients F_{ik}^j in the form

(6.31)

$$F_{ijk} = -2^{-1} \left(I_k^h N_h^{m'} \dot{\partial}_m g_{ij} + I_i^h N_h^{m'} \dot{\partial}_m g_{jk} - I_j^h N_h^{m'} \dot{\partial}_m g_{ki} \right)$$
$$- 2^{-1} \left[\lambda_k (g_{ij} + \theta_{ij} g_{oo}) + \lambda_i (g_{jk} + \theta_{jk} g_{oo}) - \lambda_j (g_{ki} + \theta_{ki} g_{oo}) \right]$$
$$- \left[\theta_{ij} (F_{ok}^o - N_m^o I_k^m) + \theta_{jk} (F_{oi}^o - N_m^o I_i^m) - \theta_{ki} (F_{oj}^o - N_m^o I_j^m) \right]$$
$$+ 2^{-1} \left(\widetilde{F}_{ijk} - \widetilde{F}_{jki} + \widetilde{F}_{kij} \right) + \Gamma_{ijk},$$

Using the homogeneity condition for the metric tensor g determined by (6.1), from (6.30) and (6.31) the expression for F_{0jk} is obtained and from that it follows

(6.32) $\qquad F_{ook} = -2^{-1} \lambda_k (1 + \theta_{oo}) g_{oo} - \theta_{oo} (F_{ok}^o - N_m^o I_k^m) + \Gamma_{ook}.$

From (6.32) after some calculation we get

(6.33) $\qquad (F_{ok}^o - N_m^o I_k^m) = (1 + \theta_{oo})^{-1} (\Gamma_{ok}^o - N_m^o I_k^m) - 2^{-1} \lambda_k g_{oo},$

where by the assumption $\theta_{oo} \neq -1$, and substituting (6.33) into (6.31) we obtain (6.28).
Substituting (6.15) into (6.27), we obtain C_{ijk} in the form

(6.34)

$$C_{ijk} = 2^{-1} L \left(I_k^{h'} \dot{\partial}_h g_{ij} + I_i^{h'} \dot{\partial}_h g_{jk} - I_j^{h'} \dot{\partial}_h g_{ki} \right) -$$
$$- 2^{-1} \left[\mu_k (g_{ij} + \theta_{ij} g_{oo}) + \mu_i (g_{jk} + \theta_{jk} g_{oo}) - \mu_j (g_{ki} + \theta_{ki} g_{oo}) \right] -$$
$$- \left[\theta_{ij} (L l_k + C_{ok}^o - L M_k^o) + \theta_{jk} (L l_i + C_{oi}^o - L M_i^o) - \right.$$
$$\left. - \theta_{ki} (L l_j + C_{oj}^o - L M_j^o) \right] +$$
$$+ 2^{-1} \left(\widetilde{C}_{ijk} - \widetilde{C}_{jki} + \widetilde{C}_{kij} \right).$$

Using the homogeneity condition of the metric tensor from (6.34), we have

$$C_{ook} = 2^{-1}\mu_k(1+\theta_{oo})g_{oo} - \theta_{oo}(Ll_k + C^o_{ok} - LM^o_k)$$

and from this relation we obtain

(6.35) $\quad Ll_k + C^o_{ok} - LM^o_k = -2^{-1}\mu_k g_{oo} + (1+\theta_{oo})^{-1}L(l_k - M^o_k).$

Substituting (6.35) into (6.34), we get (6.29).
In the special case when $M^i_j = 0$ the vectors $\{\delta_i, \dot\delta_i\}$ of $T(E)$ and $\{Dx^i, Dy^i\}$ of $T^*(E)$ have the form

$$I^i_j = J^i_j = \delta^i_j, \qquad \delta_i = \partial_i - N^j_i \dot\partial_j, \qquad \dot\delta_i = L\dot\partial_i,$$

$$Dx^i = dx^i, \qquad Dl^i = L^{-1}(dl^i + N^i_j dx^j).$$

In this case the connection coefficients in the recurrent Finsler space are determined by

(6.36)
$$
\begin{aligned}
F_{ijk} =\;& \gamma_{ijk} - 2^{-1}\left(N^m_k \partial_m g_{ij} + N^m_i \partial_m g_{jk} - N^m_j \partial_m g_{ki}\right) \\
& - 2^{-1}\left(\lambda_k g_{ij} + \lambda_i g_{jk} - \lambda_j g_{ki}\right) \\
& - (1+\theta_{oo})^{-1}\left[\theta_{ij}\left(\gamma^o_{ok} - N^o_k\right) + \theta_{jk}\left(\gamma^o_{oi} - N^o_i\right) - \theta_{ki}\left(\gamma^o_{ok} - N^o_k\right)\right] \\
& + 2^{-1}(\tilde F_{ijk} - \tilde F_{jki} + \tilde F_{kij})
\end{aligned}
$$

where γ_{ijk} is the Christoffel symbol and

(6.37)
$$
\begin{aligned}
C_{ijk} =\;& 2^{-1}L\left(\partial_k g_{ij} + \partial_i g_{jk} - \partial_j g_{ki}\right) \\
& - 2^{-1}\left(\mu_k g_{ij} + \mu_i g_{jk} - \mu_j g_{ki}\right) \\
& - (1+\theta_{oo})^{-1}L\left(l_k \theta_{ij} + l_i \theta_{jk} - l_j \theta_{ki}\right) \\
& + 2^{-1}(\tilde C_{ijk} - \tilde C_{jki} + \tilde C_{kij}).
\end{aligned}
$$

In this case by the coordinate transformation (1.1) F_{ijk} is transforming as connection coefficient and C_{ijk} as tensor. For $\lambda_k = 0$ $\mu_k = 0$ we obtain the generalized metric connection in the Finsler space. For $\lambda = 0$ and $\mu = 0$ we have $Dg_{ij} = 0$ but $g_{ij|k} \neq 0$ and $g_{ij}|_k \neq 0$. For $\lambda = 0$ and $M = 0$ using (6.33), (6.33) and (6.35), we obtain the new form of (6.26) and (6.27):

(6.26a) $\qquad\qquad g_{ij|k} = 2\theta_{ij}(1+\theta_{oo})^{-1}(\gamma^o_{ok} - N^o_k),$

(6.27) $$g_{ij}|_k = 2\theta_{ij}(1 + \theta_{oo})^{-1} Ll_k.$$

In this case from (6.26a), (6.27a) and $Dg_{ij} = 0$ we get

(6.38) $$2\theta_{ij}(1 + \theta_{oo})^{-1} \left[\left(\gamma_{ok}^o - N_k^o\right)\widehat{D}x^k + Ll_k\widehat{D}l^k\right] = 0$$

For $\theta_{ij} \neq 0$ (6.38) reduces to

(6.39) $$\left(\gamma_{ok}^o - N_k^o\right)\widehat{D}x^k + Ll_k\widehat{D}l^k = 0,$$

which is a special case of (6.24). For $F_{ok}^j = N_k^j$ (6.39) is reduced to the well-known relation $Ll_k Dl^k = l_k Dl^k = 0$ (where $Dl^k = dl^k + F_{om}^k \, dx^m$) in the Finsler space. ∎

For $\widetilde{F}_{ijk} = 0$, $\widetilde{C}_{ijk} = 0$, $\theta_{ij} = 0$, $\lambda_k = 0$, $\mu_k = 0$, $N_k^m = g^{mj} F_{ojk}$, (6.36) and (6.37) are reduced to the Cartan connection coefficients:

$$F_{ijk} = \gamma_{ijk} - 2^{-1}\left(F_{ok}^m \dot{\partial}_m g_{ij} + F_{oi}^m \dot{\partial}_m g_{jk} - F_{oj}^m \dot{\partial}_m g_{ki}\right)$$

$$C_{ijk} = 2^{-1} L\left(\dot{\partial}_k g_{ij} + \dot{\partial}_i g_{jk} - \dot{\partial}_j g_{ki}\right).$$

(6.36) and (6.37) correspond respectively to (1.17) and (1.18) from [3] if we put $M = 0$ in (1.18). For $\theta = 0$, $F_{0k}^j = N_k^j$, (6.36) and (6.37) are reduced to the connection coefficients which are very similar to those obtained by A. Moór in [6]. The difference is coming from another assumption concerning the symmetry of indices.

References

[1] G. S. ASANOV, Finsler Geometry, Relativity and Gauge Theories, D. Reidel Publ. Comp. (1987).

[2] I. ČOMIĆ, Generalization of d-connection, *Tensor N. S.*, **48** (1989), 1–10.

[3] I. ČOMIĆ, Affine connections in the Finsler space (to be published in Univ. u Novom Sadu Zb. rad. Prirod.-mat. fak. Ser. Mat. 19, 1 (1989).

[4] M. MASTSUMOTO, Foundations of Finsler Geometry and Special Finsler Spaces, Kaiseisha Press, Japan, (1986).

[5] R. MIRON, Vector Bundles, Finsler Geometry, The Proceedings of the National Sem. on Finsler spaces (1983), 147–189.

[6] A. MOÓR, Über eine übertragungstheorie der Linienelementräume mit recurrent Grundtensor, *Tensor N. S.*, **29** (1975), 47–64.

Irena Čomić

Faculty of Technical Sciences
21000 Novi Sad
Yugoslavia

Examples of Spaces of Connections and Universal Connections[*]

L. A. CORDERO, C. T. J. DODSON and P. E. PARKER

This is a preliminary report on our recent work regarding *spaces* of principal connections, by which we mean sets or systems of connections with some additional structure, and their relation to the universal connections of García [3]. We recall the

Theorem of García. *There exists a natural bijection between principal connections on the principal G-bundle P and the sections of $JP/G \to B$. Moreover, every principal connection on P is the pullback of a universal connection ω_Λ on $JP/G \times P \twoheadrightarrow JP/G$.*

Also relevant is the [4]

Theorem of Kobayashi. *There exists a bijection between classes of principal $\mathbb{T}^r$-bundles over a closed manifold M and $H^2(M, \mathbb{Z}^r)$.*

Throughout this paper, $\mathbb{T}^r$ is the r-torus $S^1 \times \cdots \times S^1$ with r factors.

Our work seeks to combine these as possible and display the universal formulae *explicitly*. As a consequence, we also provide an explicit illustration of the Chern–Weil theory. This note closely follows Parker's talk at Eger and describes four examples. Complete proofs and additional details will appear elsewhere [1, 2].

[*] Partially supported by NATO Research Grant 0741/88.

1. Heisenberg bundles

Let H_0^3 denote $\mathbb{R}^3$ considered as a 3-dimensional Abelian Lie group, and for each integer $n \neq 0$ let H_n^3 denote the 3-dimensional Lie group given by matrix multiplication of the matrices

$$
\begin{bmatrix}
1 & x^1 & -\dfrac{x^3}{n} \\[2mm]
0 & 1 & x^2 \\[1mm]
0 & 0 & 1
\end{bmatrix},
$$

so that (x^1, x^2, x^3) are global coordinates.

Theorem 1.1. *Up to isomorphism, all principal S^1 bundles over $\mathbb{T}^2$ appear as $H_n^3/\mathbb{Z}^3 \xrightarrow{\pi} \mathbb{R}^2/\mathbb{Z}^2$, where $\pi : (x^1, x^2, x^3) \mapsto (x^1, x^2)$.* ∎

We also shall denote the total space of this bundle by K_n.

The left invariant vectorfields are given by ∂_1, ∂_3, and $\partial_2 - nx^1\partial_3$ in obvious notation, and we denote the dual 1-forms by ω^1, ω^3, ω^2, respectively. The canonical connection on K_n is given by $\omega = \omega^3$ with curvature $\Omega = d\omega^3 = n\omega^1 \wedge \omega^2$. Let $\tilde{\pi} : H_n^3 \to \mathbb{R}^2$ be the projection covering π, and observe that a section of $J(\tilde{\pi})$ may be represented by (x, s, σ). We may then write the horizontal lift map in the form

$$
\begin{bmatrix}
1 & 0 \\
\sigma_1 & \sigma_2
\end{bmatrix}.
$$

Passing to the quotient, we see that σ_1 and σ_2 are periodic so that s is doubly periodic.

Theorem 1.2. *The space of principal connections on the principal circle bundle $K_n \to \mathbb{T}^2$ is given by the $\mathbb{R}^2/\mathbb{Z}^2$-valued functions σ on $\mathbb{T}^2$ such that σ_2 is nonvanishing.* ∎

Theorem 1.3. *The universal connection for K_n is*

$$
\omega_\Lambda = \omega^3 = \sigma_1\omega^1 + (1 - \sigma_2)\omega^2
$$

with universal curvature

$$
\Omega_\Lambda = n\omega^1 \wedge \omega^2 - d(\sigma_1\omega^1 + \sigma_2\omega^2).
$$
∎

Note in particular the coboundary term in the universal curvature. This gives an explicit illustration of Chern–Weil theory.

Similarly, one may analyze principal circle bundles over surfaces of genus two or more.

2. Hopf connections

For these examples, we recall the Hopf fibration $S^3 \xrightarrow{\pi} S^2$. We shall regard $S^3 \cong SO_4/SO_3$ and $S^2 \cong SO_3/SO_2$. Recall that $SO_4 \cong SO_3 \times S^3$ with corresponding Lie algebra decomposition $so_4 \cong so_3 \oplus so_3 = \langle \omega^1, \omega^2, \omega^3 \rangle \oplus \langle \omega^4, \omega^5, \omega^6 \rangle$ for left-invariant 1-forms on SO_3 and S^3, respectively. We let $\tilde{\pi} : SO_4 \to SO_3$ be the projection covering π. For $S^2 \cong SO_3/SO_2$, we write $so_3 = \langle \alpha^1, \alpha^2, \alpha^3 \rangle$ with $\tilde{\pi}^* \alpha^i = \omega^{i+3}$ and with $\tilde{\pi}^* \langle \alpha^1 + \alpha^2 + \alpha^3 \rangle = \langle \omega^4 + \omega^5 + \omega^6 \rangle$. If we further take $\pi^* \frac{1}{3}(\alpha^1 + \alpha^2 + \alpha^3) = n(\omega^4 + \omega^5 + \omega^6)$ for $n \neq 0$, then $n = 1$ gives the Hopf bundle and the other n give the principal circle bundles over S^2 with total space the lens space $S^3/\mathbb{Z}_n$. For $n = 0$, we take the trivial circle bundle over S^2. Up to isomorphism, this accounts for all principal circle bundles over S^2.

The Hopf connection on $S^3 \to S^2$ is given by $\omega_H = \omega^4 + \omega^5 + \omega^6$ and has curvature $\Omega_H = d\omega_H$. Then the connection $\omega = n\omega_H$ on bundle n has curvature $n\Omega_H$, corresponding to Euler class n of bundle n.

Theorem 2.1. *The space of connections on the Hopf bundle is given by the space of functions σ on S^2 having values in the set of 3×3 matrices of rank 3 with row-sums 1.* ∎

Theorem 2.2. *The universal connection for the bundle $\tilde{\pi} : SO_4 \to SO_3$ is*

$$\tilde{\omega}_\Lambda = \tilde{\omega}_H + (1 - \sigma) \cdot \begin{bmatrix} \omega^4 \\ \omega^5 \\ \omega^6 \end{bmatrix},$$

where the dot denotes matrix multiplication followed by summation of the components of the resulting column vector. The universal curvature again has the form of $\tilde{\Omega}_H$ plus a coboundary term. To pass to the quotient bundle π with Euler class n, use $n\omega_H$ and $n\Omega_H$. ∎

The Heisenberg bundles are the model for certain principal bundles over compact nilmanifolds, and the Hopf bundles for certain principal bundles over spaces homeomorphic to a symmetric space. For convenience, we shall refer to the latter as *symmanifolds*, by analogy with nilmanifolds and solvmanifolds, since the Lie algebras involved are usually called *symmetric*.

3. Compact nilmanifolds

These are manifolds of the form G/Γ where G is a simply connected nilpotent Lie group and Γ is a lattice in G. We construct a tower of principal torus-bundles with principal connections. Let $\{G_i\}$ denote the ascending central series of G and let $\Gamma_i = \Gamma \cap G_i$.

$$
\begin{array}{ccccc}
\mathbb{T}^u & \hookrightarrow & G_m/\Gamma_m & = & G/\Gamma \\
& & \downarrow & & \\
& & \vdots & & \\
& & \downarrow & & \\
\mathbb{T}^t & \hookrightarrow & G_4/\Gamma_4 & & \\
& & \downarrow & & \\
\mathbb{T}^s & \hookrightarrow & G_3/\Gamma_3 & & \\
& & \downarrow & & \\
\mathbb{T}^r & \hookrightarrow & G_2/\Gamma_2 & & \\
& & \downarrow & & \\
& & G_1/\Gamma_1 & \cong & \mathbb{T}^p
\end{array}
$$

Theorem 3.1. *Up to isomorphism, any tower of principal torus-bundles with principal connections comes from a nilmanifold in this way.* ∎

The principal connection on each principal bundle $\mathbb{T}^k \hookrightarrow G_i/\Gamma_i \twoheadrightarrow G_{i-1}/\Gamma_{i-1}$ is obtained from the left-invariant 1-forms by the method used for the Heisenberg bundles.

One may now proceed to compute the universal connections and universal curvatures.

4. Symmanifolds

Let G be a Lie group with Lie algebra $\mathfrak{G}$ and H a closed subgroup with Lie subalgebra $\mathfrak{H}$. If we write $\mathfrak{G} = \mathfrak{H} \oplus \mathfrak{M}$, then $[\mathfrak{H}, \mathfrak{H}] \subseteq \mathfrak{H}$ is equivalent to H being a subgroup. Additional requirements that may be imposed are $[\mathfrak{H}, \mathfrak{M}] \subseteq \mathfrak{M}$, which means that $\mathfrak{G}$ or G/H is reductive, and $[\mathfrak{M}, \mathfrak{M}] \subseteq \mathfrak{H}$, which means that $\mathfrak{G}$ or G/H is symmetric. We are concerned with symmetric, reductive G/H.

Observe that in any case there is the principal H-bundle $H \hookrightarrow G \twoheadrightarrow G/H$. Let $\{\alpha^i\}$ be a basis of left-invariant 1-forms for H as a subgroup of G and let $\{e_i\}$ be the corresponding (part of a) dual frame.

Theorem 3.1. *If G/H is symmetric and reductive, then $\omega = \sum \alpha^i \otimes e_i$ is a principal connection on $H \hookrightarrow G \twoheadrightarrow G/H$.* ∎

This case covers all the preceding ones. Note that we do not need or use any of the usual symmetric space structures on G/H.

Again, one may now proceed to compute the universal connections and curvatures.

References

[1] L.A. CORDERO, C.T.J. DODSON and P.E. PARKER, Connections on principal circle bundles over the torus and the sphere, *preprint Depto. Xeom. Top.*, Univ. Santiago 1989.

[2] L.A. CORDERO, C.T.J. DODSON and P.E. PARKER, Connections on principal bundles over symmetric compacta, in preparation.

[3] P.L. GARCÍA, Connections and 1-jet fiber bundles, *Rend. Sem. Mat. Univ. Padova* **47** (1972), 227–242.

[4] S. KOBAYASHI, Principal fibre bundles with the 1-dimensional toroidal group, *Tôhoku Math. J.* **8** (1956), 29–45.

L.A. Cordero

Depto. Xeom. Top.
Universidade de Santiago
Santiago de Compostela
Spain

C.T.J. Dodson

Math. Dept.
University of Toronto
Toronto M5S 1A1
Canada

P.E. Parker

Math. Dept.
Wichita State University
Wichita KS 67208
USA

COLLOQUIA MATHEMATICA SOCIETATIS JÁNOS BOLYAI
56. DIFFERENTIAL GEOMETRY, EGER (HUNGARY), 1989

Supergravity, Supersymmetry: a Geometric Unitary Spinor Theory[*]

A. CRUMEYROLLE

0. Introduction

In field theory, since fifteen years or more an avalanche of works concerns the infinitesimal invariance called "supersymmetry". Supersymmetry introduces graded Lie algebras or superalgebras and these subjects are very popular now in the scientific world. It was convenient to define these superalgebras by means of non-commutative analysis, and this method tends to define new objects called "supermanifolds". Some successes in this direction spread the belief that supersymmetry necessitates anticommutative analysis. We intend to develop here a different idea.

Considering spinors as elements of minimal left ideals in a Clifford algebra, some geometric properties in pseudo-riemannian context give naturally supersymmetry (Rarita–Schwinger theory and generalizations called supergravitations).

We develop in this paper a unitary gravitational theory including some spinor fields, superalgebras come from an enlarged triality principle that

---

[*] This paper is in final form and no version of it will be submitted for publication elsewhere.

we introduced several years ago in our works [2,a]. This construction is "off-shell".

Supersymmetries appear naturally for tensor-spinor fields, connected with Bianchi identities and Einstein–Hilbert equations.

If one compares our methods with the usual approach (the reader can refer to Y. Choquet–Bruhat, for a short but significative account [1]) they bring three main pecularities:

a) the definition of spinor bundles uses the Clifford algebra bundle (r-isotropic method) of the manifold,

b) graded anticommutative analysis and supermanifolds are useless,

c) connections in the spinor fields are naturally associated with pseudo–euclidean connections, the torsion tensor is completely antisymmetric.

— However we have to make this last c) choice which imposes a supplementary but acceptable condition (cf. (23) or (23 bis) below) in order to preserve supersymmetry in the non torsion free case.

Our methods extend naturally to any signature and dimension, and furnish a coherent and general method in the construction of N-supergravitation theories ($N = 2^k$, $k = 0, 1, 2, \ldots$); we get ordinary N-supergravitation and Kaluza–Klein N-supergravitation. This approach is geometric, natural, and always the same in any dimension, except in some details: we note in particular that lagrangians are constructed without artificial addition of terms and have always the same type. We intended to develop a super Kaluza–Klein theory and a super Yang–Mills theory, in a next paper, using exactly the same methods.

The Appendix gives some results about the internal coherence of the theory (a well posed physical system taking account of the normalization conditions) and some details in dimension 4 and (3, 1) signature.

We suppose that the reader is well acquainted with general relativity, spinor structures in the modern context and differential geometry.

1. Preliminaries

$E_{3,1}$ is the standard Minkowski space, with pseudo-metric g, $g_{\alpha\beta}$ components, and a Clifford algebra $C_{3,1}$ admitting the following table:

$$e_\alpha e_\beta + e_\beta e_\alpha = 2g_{\alpha\beta}.$$

If (e_0, e_1, e_2, e_3) is a normed frame then $(e_0)^2 = -1$ $(e_1)^2 = (e_2)^2 = (e_3)^2 = 1$. The $(+++-)$ -signature permits to define G-invariant Majorana spinors
(G is the usual Clifford group) [2,b].

$V_{3,1}$ is the usual curved space-time, with local frames (e_α), and pseudo-
metric g, adopting a classical abuse of notations.

Let D be a pseudo-euclidean connection $(Dg = 0)$ and ∇ the Riemann–
Christoffel torsion free connection ($R - C$-connection).

In this paper Latin indices correspond to natural frames in the tangent
space of $V_{3,1}$.

$\Gamma^\alpha_{\beta\gamma}$ such that $D_\beta e_\alpha = \Gamma^\lambda_{\alpha\beta} e_\lambda$ are the components of D and we recall
some classical formulas:

(1)
$$\begin{cases} \Gamma^i_{jk} = \{^i_{jk}\} + B^i_{jk} \\ B^i_{jk} = g^{ir}(g_{\ell j} S^\ell_{rk} + g_{\ell k} S^\ell_{rj}) + S^i_{jk}, \end{cases}$$

where $\{^i_{jk}\}$ are Christoffel symbols, and S^i_{jk} components of the torsion
tensor.

$$B_{ijk} = S_{jik} + S_{kij} + S_{ijk}.$$

Here we *suppose that* $S_{ijk} = g_{\ell i} S^\ell_{jk}$ *is completely antisymmetric*; this particular case intervenes in the Einstein–Dirac theory, and leads to $B^i_{jk} = S^i_{jk}$
and $S^k_{ki} = S_i = 0$.

$\widehat{R}$ and R designate the curvature tensors of D and ∇ respectively, and
it is easy to obtain:

(2) $$\widehat{R}_{\nu\sigma,\rho\mu} = R_{\nu\sigma,\rho\mu} - \nabla_\mu B_{\nu\sigma\rho} + \nabla_\rho B_{\nu\sigma\mu} - B^\chi_{\sigma\rho} B_{\nu\chi\mu} + B^\chi_{\sigma\mu} B_{\nu\chi\rho}$$

(3) $$\widehat{R}_{\nu\sigma,\rho\mu} = -\widehat{R}_{\nu\sigma,\mu\rho}, \qquad \widehat{R}_{\nu\sigma\rho\mu} \neq -\widehat{R}_{\sigma\nu\rho\mu}, \qquad \text{generally} .$$

However, if the S_{ijk} components are totally antisymmetric:

(4) $$\widehat{R}_{\nu\sigma,\rho\mu} = -\widehat{R}_{\sigma\nu,\rho\mu} \qquad \text{and} \qquad \widehat{R}^\nu_{\nu\rho\mu} = 0.$$

If η is the volume 4-form, with the same hypothesis:

(5)
$$\eta^{\nu\sigma\rho\mu}\,\widehat{R}_{\nu\sigma,\rho\mu} = \eta^{\rho\mu\nu\sigma}\,\widehat{R}_{\rho\mu,\nu\sigma}\,.$$

Finally, if $\widehat{\mathcal{R}}_{\alpha\beta}$ and $\mathcal{R}_{\alpha\beta}$ are the Ricci tensors components:

(6)
$$\widehat{\mathcal{R}}_{\alpha\beta} = \mathcal{R}_{\alpha\beta} + \nabla_\lambda B^\lambda_{\alpha\beta} - B^\sigma_{\alpha\lambda} B^\lambda_{\sigma\beta}\,.$$

Now, we suppose that $V_{3,1}$ is orientable and possesses a spinor structure. But we need the existence of a global sesquilinear hermitian form $\mathcal{H}$, also we suppose that $V_{3,1}$ owns a time orientation and a spatial orientation, this hypothesis permitting to define $\mathcal{H}$; $\mathcal{H}$ is left invariant for the natural action of the special spinorial group, and also for the G_0-action (G_0 is the reduced Clifford group) [2,b].

$\mathcal{H}(\varphi, \psi)$ has purely imaginary values if φ and ψ are Majorana spinors; $\mathcal{H}$ is neutral [2,b].

We introduce also vector-spinor fields

$$\psi = \psi_\lambda \otimes e^\lambda,$$

where (ψ_λ) are cross-sections in the spinor bundle and there exists a natural extension of $\mathcal{H}$ to these fields:

$$(\psi; \varphi) = \mathcal{H}(\psi^\lambda, \varphi^\mu)\, g(e_\lambda, e_\mu) = \mathcal{H}(\psi^\lambda, \varphi^\mu) g_{\lambda\mu}$$
$$= (\psi^\lambda; \varphi^\mu) g_{\lambda\mu}$$

(7)   with $(\psi^\lambda; \varphi^\mu) = (\psi^\mu; \varphi^\lambda).$

The connection D is lifted to a connection over the spinor bundle. If ω^α_β is the connection form of D, the connection form of D in spinor fields is given by:

$$u = \frac{1}{4}\omega^\alpha_\beta\, e_\alpha e^\beta \qquad \text{(natural lifting)},$$

and if (γ^α) define the components of the fundamental tensor-spinor associated to e^α, then $D\gamma^\alpha = 0$. [2,b].

With classical notations, the volume 4-form η admits the components:

$$\eta^{\lambda\mu\nu\rho} = -\frac{1}{\sqrt{|g|}}\varepsilon^{\lambda\mu\nu\rho}, \qquad \eta_{\lambda\mu\nu\rho} = \sqrt{|g|}\,\varepsilon_{\lambda\mu\nu\rho}$$

where $|g| = |\det((g_{\alpha\beta}))|$; $D\eta = 0$ is a classical result.

If e_α, $\alpha = 0,1,2,3$ constitute an orthonormed local frame in the tangent bundle of $V_{3,1}$, we define $e^N = e^0 e^1 e^2 e^3$, $(e^N)^2 = -1$, e^N is a global density field (weight 1) and we have the important formula:

(8) $$(e_\mu e_\alpha e_\lambda - e_\lambda e_\alpha e_\mu) = 2e^N \eta_{\mu\alpha\lambda\beta} e^\beta \qquad [1] \ .$$

Indeed, if two indices are equal, this is evident; and if the indices are distinct two by two:

$$e_\mu e_\alpha e_\lambda = e^N \eta_{\mu\alpha\lambda\beta} e^\beta$$

$$\eta^{\mu\alpha\lambda\sigma} e_\mu e_\alpha e_\lambda = 3! \delta_\beta^\sigma e^N e^\beta = 3! e^N e^\sigma,$$

one obtains the result putting $\sigma = 0, 1, 2$, or 3.

2. Rarita–Schwinger theory in curved space-time $V_{3,1}$

$\psi = \psi_\lambda \otimes e^\lambda$ is a vector-spinor field, $\lambda = 0,1,2,3$, namely a spin $3/2$ field, which has to satisfy:

(9) $$A^\lambda \equiv \eta^{\lambda\mu\nu\rho} e_\mu D_\nu \psi_\rho = 0$$

(10) $$\chi \equiv e^\lambda \psi_\lambda = 0.$$

If ψ is a G-Majorana spinor, also A^λ, and $i(A^\lambda; \psi_\lambda)$ is real. Using (8), we can write by (9):

(9 bis) $$(e^\lambda e^\nu e^\rho - e^\rho e^\nu e^\lambda) D_\nu \psi_\rho = 0,$$

where (e^α) is an orthonormed frame.

It is easy to prove that (9) and (10) are equivalent to:

(10) and $e^\alpha D_\alpha \psi_\rho = 0$ (9 ter)

(9 bis and (10) give also $D_\rho \psi^\rho = 0$).

We introduced over $V_{3,1}$ a spinor structure, lifting the tangent bundle frame with structural group the connected component of the Lorentz group, to a spinor frame bundle. Here we are using our results about spinor structures: there exist in the complexified tangent bundle an isotropic 2-field

f, f is locally defined by a product $y_1^* y_2^*$, if (y_1, y_2, y_1^*, y_2^*) constitute a Witt decomposition of the tangent space, $g(y_\alpha, y_{\beta^*}) = h_{\alpha\beta^*}$, $h_{\alpha\beta} = h_{\alpha^* \beta^*} = 0$, and we put $h = \det((h_{\alpha\beta^*}))$, $\sqrt{|h|}$ defines a density field with weight $1/2$. A local field of spinors is written $x \to (uf)_x$, where u_x is a cliffordian element.

$x \to (uf, vf)_x$ is also a density field with weight $1/2$. One knows that $\sqrt{|g|}$ admits weight 1.

The previous considerations explain the choice of the coefficients in the next lagrangian:

(11) $$\mathcal{L} = \int \left[\sqrt{|g|}\, \widehat{\mathcal{R}} + ia\sqrt{|h|}(A^\lambda \, ; \, e^N \psi_\lambda) \right] d^4 x,$$

$d^4 x = dx^0 \wedge dx^1 \wedge dx^2 \wedge dx^3$, $\widehat{\mathcal{R}} = g^{\alpha\beta}\widehat{\mathcal{R}}_{\alpha\beta}$, a is a real constant; one considers the integral over some compact domain in $V_{3,1}$.

We obtain the equations of de Rarita–Schwinger theory by varying independently $S_{\alpha\beta\gamma}$, ψ_λ, $g^{\alpha\beta}$. (However we have to vary $g^{\alpha\beta}$ according to an infinitesimal isometry because we have to preserve the spinor structure).

a) We vary the $S_{\alpha\beta\gamma}$, but we take into account the antisymmetry.

Coming from $\widehat{\mathcal{R}}$ we obtain $\nabla_\lambda(S^\lambda_{\alpha\beta} g^{\alpha\beta})$, a divergence giving 0, according to Stokes' theorem.

Varying $(-B^\sigma_{\alpha\lambda} B^\lambda_{\sigma\beta})$, we get $-2(\delta(S_{\alpha\beta\gamma}))S^{\alpha\beta\gamma}$.

From $(A^\lambda; e^N \psi_\lambda)$, using orthonormed frames, taking account of (8) and (10), appears:
$-4(\psi^\mu; e^\lambda \psi^\nu)\delta S_{\mu\lambda\nu} + (e^\nu e^\mu e^\lambda \psi_\rho; \psi^\rho)\delta S_{\mu\lambda\nu}$, and according to (7)

(12) $$S^{\lambda\mu\nu} = \frac{ai}{2}(e^\lambda e^\mu e^\nu \psi_\rho \, ; \, \psi^\rho).$$

b) We vary the ψ_λ:

$$A^\lambda = D_\nu(\eta^{\lambda\mu\nu\rho} e_\mu \psi_\rho) = D_\nu(\sigma^{\lambda\nu}), \qquad \text{car} \qquad D_\nu \eta = 0.$$

It is well known that $D\mathcal{H} = 0$, so we obtain:

$$D_\nu(\delta\sigma^{\lambda\nu} \, ; \, e^N \psi_\lambda) - (\delta\sigma^{\lambda\nu} \, ; \, D_\nu e^N \psi_\lambda) + (D_\nu \sigma^{\lambda\mu} \, ; \, e^N \delta\psi_\lambda),$$

where the first term gives 0 after integration.

We note also $D_\nu(e^N \psi_\lambda) = e^N D_\nu \psi_\lambda$, and nullifying the arbitrary variation we obtain:

$$(-e^N \delta\psi_\rho \; ; \; A^\rho) + (A^\rho \; ; \; e^N \delta\psi_\rho) = 0, \qquad \text{namely}$$

$$(A^\rho \; ; \; e^N \delta\psi_\rho) = 0,$$

the spinors being Majorana-spinors.

Finally $A^\rho = 0$, as we need.

c) We vary the $(g^{\alpha\beta})$ now, according to infinitesimal isometry. In $(A_\mu \; ; \; e^N \psi_\lambda)$ nothing changes because the scalar product is invariant according to the isometries.

A classical checking gives:

$$(13) \qquad \begin{cases} \widehat{\mathcal{R}}_{(\lambda\mu)} - \dfrac{1}{2}(\mathcal{R} - \widehat{S})g_{\lambda\mu} = T_{(\lambda\mu)} \\[2mm] \text{or}: \widehat{\mathcal{R}}_{(\lambda\mu)} - \dfrac{1}{2}\widehat{\mathcal{R}}g_{\lambda\mu} = T_{(\lambda\mu)} \end{cases}$$

where : $\widehat{\mathcal{R}}_{(\lambda\mu)} = \frac{1}{2}(\widehat{\mathcal{R}}_{\lambda\mu} + \widehat{\mathcal{R}}_{\mu\lambda})$

$$\widehat{S} = S_{\alpha\beta\gamma} S^{\alpha\beta\gamma}, \qquad \widehat{\mathcal{R}} = \mathcal{R} = \widehat{\mathcal{R}}_{\lambda\mu} g^{\lambda\mu}.$$

$$(14) \qquad T_{(\lambda\mu)} = \frac{-ai}{\sqrt{|h|}} \left[(A_\lambda \; ; \; e^N \psi_\mu) - \frac{1}{4}(A^\sigma \; , \; e^N \psi_\sigma) g_{\lambda\mu} \right].$$

But taking account of $A_\lambda = 0$, (13) reduces to the Einstein–Hilbert equation:

$$(15) \qquad \widehat{\mathcal{R}}_{(\lambda\mu)} - \frac{1}{2}(\mathcal{R} - \widehat{S})g_{\lambda\mu} = 0 \text{ or } \widehat{\mathcal{R}}_{(\lambda\mu)} - \frac{1}{2}\widehat{\mathcal{R}}g_{\lambda\mu} = 0.$$

3. Supersymmetry

a) We check first $D_\lambda A^\lambda$ in the symmetric case: $D = \nabla$ and $\widehat{R} = R$.

$$\nabla_\lambda A^\lambda = \frac{1}{2}\eta^{\lambda\mu\nu\rho} e_\mu (\nabla_\lambda \nabla_\nu - \nabla_\nu \nabla_\lambda)\psi_\rho$$

$$= \frac{1}{8}\eta^{\lambda\mu\nu\rho} e_\mu R^\alpha_{\beta\lambda\nu} e_\alpha e^\beta \psi_\rho - \frac{1}{2}\eta^{\lambda\mu\nu\rho} e_\mu R^\alpha_{\rho\lambda\nu} \psi_\alpha .$$

(For this checking one chooses natural frames, and one applies the curvature definition to $\psi = \psi_\rho \otimes e^\rho$).
But the Bianchi identities give 0 for the last term. Now, we are using orthonormal frames, and (8) gives:

$$\nabla_\lambda A^\lambda = -\frac{e^N}{16} R^\alpha_{\beta\lambda\nu} (e^\rho e^\lambda e^\nu - e^\nu e^\lambda e^\rho) e_\alpha e^\beta \psi_\rho$$

$$= -\frac{e^N}{8} e^\rho R_{\beta\alpha,\lambda\nu} e^\lambda e^\alpha e^\nu e^\beta \psi_\rho + \frac{e^N}{4} R_{\beta\alpha,\lambda\nu} e^\lambda e^\alpha e^\beta \psi^\nu +$$

$$+ \frac{e^N}{4} e^\rho R_{\beta\alpha,\lambda\nu} e^\lambda e^\beta g^{\alpha\nu} \psi_\rho .$$

We can prove, using again Bianchi identities, that:

(16) $R_{\beta\alpha\lambda\nu} e^\lambda e^\alpha e^\nu = 0.$

and we obtain:

$$\nabla_\lambda A^\lambda = \frac{e^N}{4} e^\rho R_{\beta\lambda} e^\lambda e^\beta \psi_\rho - \frac{e^N}{4} R_{\alpha\beta\lambda\nu} e^\lambda e^\alpha e^\beta \psi^\nu$$

and using (16) again:

(17) $\nabla_\lambda A^\lambda = -\frac{e^N}{2} (R^\alpha_\beta - \frac{1}{2} g^\alpha_\beta R) e^\beta \psi_\alpha .$

$\nabla_\lambda A^\lambda$ is null, if we take account of (15).
The condition (10) is useless in the symmetric case.

b) *D is non torsion free.*

Of course $S_{\alpha\beta\gamma}$ is antisymmetric totally. Curvature definition gives, like in a) above:

$$D_\lambda A^\lambda = \frac{1}{8}\eta^{\lambda\mu\nu\rho} e_\mu \widehat{R}^\alpha_{\beta\lambda\nu} e_\alpha e^\beta \psi_\rho - \frac{1}{2}\eta^{\lambda\mu\nu\rho} e_\mu \widehat{R}^\alpha_{\rho\lambda\nu} \psi_\alpha .$$

Using (8) and routine checking in a Clifford algebra, we obtain, with orthonormal frames now:

$$\frac{1}{8}\eta^{\lambda\mu\nu\rho}\,e_\mu\widehat{R}^\alpha_{\beta,\lambda\nu}\,e_\alpha e^\beta\psi_\rho = -\frac{e^N}{16}(e^\lambda e^\nu e^\rho - e^\rho e^\nu e^\lambda)e^\alpha e^\beta\widehat{R}_{\alpha\beta,\lambda\nu}\,\psi_\rho =$$

$$= -e^N\widehat{R}_{\alpha\beta}\,e^\beta\psi^\alpha - \frac{e^N}{2}\widehat{R}_{\alpha\beta,\lambda\nu}\,e^\nu e^\beta e^\lambda\psi^\alpha -$$

$$- \frac{e^N}{4}\widehat{R}_{\alpha\beta,\lambda\nu}\,e^\nu e^\alpha e^\beta\psi^\lambda,$$

meanwhile:

$$\frac{1}{2}\eta^{\lambda\mu\nu\rho}\,e_\mu\widehat{R}^\alpha_{\rho\lambda\nu}\psi_\alpha = -\frac{e^N}{2}\widehat{R}_{\alpha\beta,\lambda\nu}\,e^\nu e^\beta e^\lambda\psi_\alpha.$$

Thus we obtain:

(18)
$$D_\lambda A^\lambda = -e^N\widehat{R}_{\alpha\beta}\,e^\beta\psi^\alpha - \frac{e^N}{4}\widehat{R}_{\alpha\beta,\lambda\nu}\,e^\nu e^\alpha e^\beta\psi^\lambda.$$

The Bianchi identities give:

$$\sum_{\substack{\text{c.p}\\ \lambda,\mu,\nu}}\widehat{R}^\alpha_{\lambda\mu\nu} + 2D_\lambda S^\alpha_{\underline{\mu\nu}} = 0.$$

where μ and ν are not submitted to any derivation.
But $D_\beta(S_{\alpha\underline{\lambda\nu}}) = D_\beta(S_{\alpha\lambda\nu}) = ia(e_\alpha e_\lambda e_\nu D_\beta\psi_\theta : \psi^\theta)$,
as you can see, with $D\gamma_\alpha = 0$, $D\mathcal{H} = 0$, $Dg = 0$, and Majorana property of spinors.
In particular, we have

(19)
$$D_\sigma S^\sigma_{\alpha\beta} = 0 \qquad \text{equivalent to} \qquad \mathcal{R}_{[\alpha\beta]} = 0,$$

because $D_\sigma S^\sigma_{\alpha\beta} = 0$ is equivalent to:
$(e^\sigma e_\alpha e_\beta D_\sigma\psi_\theta ; \psi^\theta) = 0$, and it taking account of (9 ter) above we get:
$e^\sigma D_\sigma\phi_\theta = 0$.
From : $\widehat{R}_{\alpha\beta\lambda\nu} + \widehat{R}_{\alpha\nu\beta\lambda} + \widehat{R}_{\alpha\lambda\nu\beta} = -2\sum_{\substack{\text{c.p}\\ \beta,\lambda,\nu}}D_\beta S_{\alpha\lambda\nu}$, it results, as we can

check that

$$e^\lambda e^\beta e^\nu (D_\beta S_{\alpha\lambda\nu}) = \frac{1}{2}(e^\lambda e^\beta e^\nu - e^\nu e^\beta e^\lambda)D_\beta S_{\alpha\lambda\nu} - 2D_\lambda S^\lambda_{\alpha\beta}e^\beta,$$

and the first term in the right gives, modulo a scalar factor:

$$e^N e_\sigma(e^N e_\alpha e_\lambda e_\nu(e^\nu e^\lambda e^\beta - e^\beta e^\lambda e^\nu)D_\beta\psi_\theta : \psi^\theta)$$

and using (9 ter) again we obtain easily:

$$e^\lambda e^\beta e^\nu (D_\beta S_{\alpha\lambda\nu}) = -2D_\lambda S^\alpha_{\alpha\beta} e^\beta = 0$$
$$\widehat{R}_{\alpha\beta\lambda\nu} e^\alpha e^\beta e^\nu = 4D_\lambda S^\lambda_{\alpha\beta} e^\beta = 0.$$

According to (2) above:

$$\widehat{R}_{\alpha\beta,\lambda\nu} - \widehat{R}_{\lambda\nu,\alpha\beta} = \nabla_\lambda S_{\alpha\beta\nu} - \nabla_\nu S_{\alpha\beta\lambda} - \nabla_\alpha S_{\lambda\nu\beta} + \nabla_\beta S_{\lambda\nu\alpha}$$
$$e^\nu e^\alpha e^\beta \widehat{R}_{\alpha\beta,\lambda\nu} = - \widehat{R}_{\lambda\alpha\nu\beta} e^\nu e^\alpha e^\beta - 2R_{\lambda\beta} e^\beta +$$
$$+ e^\nu e^\alpha e^\beta (\nabla_\beta S_{\lambda\nu\alpha} - \nabla_\alpha S_{\lambda\nu\beta} - \nabla_\nu S_{\alpha\beta\lambda}) +$$
$$+ e^\nu e^\alpha e^\beta \nabla_\lambda S_{\alpha\beta\nu},$$

and some checking in the Clifford algebra gives:

$$e^\nu e^\alpha e^\beta \widehat{R}_{\alpha\beta,\lambda\nu} = 2D_\sigma(S^\sigma_{\lambda\beta}) e^\beta - 2R_{\lambda\beta} e^\beta + e^\nu e^\alpha e^\beta (\nabla_\lambda S_{\alpha\beta\nu}).$$

From (18) and (19) we obtain finally:

$$D_\lambda A^\lambda = -\frac{e^N}{2} \widehat{R}_{\lambda\beta} e^\beta \psi^\lambda - \frac{e^N}{4} e^\nu e^\alpha e^\beta (\nabla_\lambda S_{\alpha\beta\nu})$$

(20)
$$D_\lambda A^\lambda = -\frac{e^N}{2} (\widehat{R}_{(\alpha\beta)} - \frac{1}{2} \widehat{R} g_{\alpha\beta}) e^\beta \psi^\alpha$$

taking account of (10) and of a supplementary normalisation condition:

$$e^\nu e^\alpha e^\beta \nabla_\lambda (S_{\alpha\beta\nu}) \cdot \psi^\lambda = 0.$$

If ν, α, β, σ are distinct two to two:

(21)
$$e^\sigma \nabla_\lambda (X_\sigma) \cdot \psi^\lambda = 0$$

where (X_σ) are the components of the covector X corresponding to the three-form of torsion with $S_{\alpha\beta\nu}$ components.

Finally, according to (15), $D_\lambda A^\lambda = 0$ if (21) is postulated.

Remark. It is immediate to deduct from (19) that $\nabla_\sigma S^\sigma_{\alpha\beta} = 0$ and $\nabla_\lambda X_\sigma - \nabla_\sigma X_\lambda = 0$, also (21) becomes, taking account of (9 ter):

(9 bis)
$$e^\sigma \nabla_\sigma (X_\lambda \psi^\lambda) = 0 :$$

it is a Dirac equation for $\varphi = X_\lambda \psi^\lambda$.

Let us now consider again:

$$\mathcal{L} = \int \left[\sqrt{|g|}\widehat{\mathcal{R}} + ia\sqrt{|h|}(A^\lambda : e^N \psi_\lambda) \right] d^4x.$$

Varying the ψ_ρ, we obtained, in order to nullify, modulo a divergence:

$$(-e^N \delta\psi_\rho ; A^\rho) + (A^\rho ; e^N \delta\psi_\rho).$$

Let us put:

(22) $$\delta\psi_\rho = D_\rho\varepsilon$$

where ε is a Majorana spinor field, and we obtain:

$$(A^\rho ; e^N D_\rho\varepsilon) - (e^N D_\rho\varepsilon ; A^\rho),$$

but:

$$D_\rho(A^\rho ; e^N \varepsilon) = (D_\rho A^\rho ; e^N \varepsilon) + (A^\rho ; e^N D_\rho\varepsilon)$$
$$D_\rho(e^N \varepsilon ; A^\rho) = (e^N D_\rho\varepsilon ; A^\rho) + (e^N \varepsilon ; D_\rho A^\rho),$$

also, modulo a divergence we come back to:

$$(e^N \varepsilon ; D_\rho A^\rho) - (D_\rho A^\rho ; e^N \varepsilon).$$

As we have $D_\rho A^\rho = -\frac{e^N}{2}\widehat{S}_{(\lambda\mu)} e^\mu \psi^\lambda, \widehat{S}_{(\lambda\mu)} = \widehat{\mathcal{R}}_{\lambda\mu} - \frac{1}{2}\widehat{\mathcal{R}}g_{\lambda\mu},$

$$(e^N)^2 = -1,$$

we get back:

$$ai\sqrt{|h|}\widehat{S}_{(\lambda\mu)} \left((e^\mu \psi_\lambda ; \varepsilon) - (\varepsilon, e^\mu \psi^\lambda) \right)$$

and the variation of the $g^{\lambda\mu}$ yields:

$$\sqrt{|g|}\widehat{S}_{(\lambda\mu)} \delta g^{\lambda\mu} , (A^\lambda = 0).$$

If we choose:
$\delta g^{\lambda\mu} = \frac{ia}{2\sqrt{|h|}}\left((\varepsilon , e^\mu \psi^\lambda) - (e^\mu \psi^\lambda ; \varepsilon) \right)+$ symmetric terms in λ and μ namely:

(23) $$\delta g^{\lambda\mu} = \frac{a}{2\sqrt{|h|}} \operatorname{Im}(\varepsilon ; e^\mu \psi^\lambda + e^\lambda \psi^\mu)$$

we have $\delta \mathcal{L} = 0$.

Physicists speak about "infinitisemal supersymmetry".

   We have obtained this *supersymmetry is a spinor background, without graded analysis.*

   Only in the asymmetric case, do we have to suppose the new condition (21).

Remark. Physicists often write indices a, b, c, ... relative to orthonormed frames and indices λ, μ, ν, ... relative to any sort of frames. They put:

$$e_a = e_a^\lambda e_\lambda , \; e_\lambda = e_\lambda^b e_b,$$

and call the (e_a^λ) a tetrad field, varying the (e_a^λ) arbitrarily .

Taking:

(24) $$\delta(e_a^\lambda) = \frac{1}{\sqrt{|h|}} \, \mathrm{Im}(\varepsilon \, ; \, e^\lambda \psi_a), \qquad a = 2,$$

the conditions (22) and (24) are the "physicist supersymmetries" in the usual form.

   Note also that supersymmetries exist in the pseudo-riemannian case, with torsion-free connection, and condition (10) and (21) are useless in this particular case.

   The following paragraph is devoted to show that supersymmetries belong to a graded Lie algebra or super-Lie algebra.   .

A geometric frame for supersymmetries: the enlarged triality principle and the interaction algebra

We suppose that $\mathcal{H}$ is G_0-invariant, if G_0 is the subgroup of the Clifford group G, with elements having spinor norm 1. According to our papers [2,b] the sesquilinear forms $\mathcal{H}$ constitute a complex vector space with dimension one, and can be identified for Majorana spinors with bilinear forms $\mathcal{B}$, modulo $a \pm 1$, $\pm i$ factor; thus we can replace $i\mathcal{H}$, by $\mathcal{B}$, $\mathcal{B}$ is antisymmetric [2,b], and we extend $\mathcal{B}$ to the tensor-spinors. We recall that if spinors are defined by uf (f is an r-isotropic vector, $n = 2r$):

$$\mathcal{B}(uf, vf)f = \beta(uf)vf$$

(β: principal anti-involution).

We intend to work with the Poincaré group (subgroup of a conformal group), also according our technics [2,b] we add to the Minkowski space, a 2-dimensional hyperbolic space $E_{1,1}$, generated by e_4 and e_5 with $(e_4)^2 = 1$, $(e_5)^2 = -1$, $e_4 e_5 + e_5 e_4 = 0$, in a natural extension of the minkowskian Clifford algebra.

We take Witt's standard frame:

$$x_0 = \frac{e_4 + e_5}{2}, \quad x_1 = \frac{e_3 + e_0}{2}, \quad x_2 = \frac{ie_1 + e_2}{2},$$

$$y_0 = \frac{e_4 - e_5}{2}; \quad y_1 = \frac{e_3 - e_0}{2}, \quad y_2 = \frac{-ie_1 + e_2}{2}$$

and $\widehat{f} = y_0 y_1 y_2$ determines the standard spinors. Naturally, over $V_{3,1}$, we have to add a trivial vector bundle ξ_2, with rank 2, with standard fiber $E_{1,1}$; if ξ is the tangent bundle to $V_{3,1}$.

$$\xi_1 = \xi \oplus \xi_2$$

is the "increased tangent bundle", provided with pseudo-euclidean and spinor structures. With the previous hypothesis there exists a spinor structure over ξ_1, we denote by $\widehat{s}$ the field of spinors over ξ_1, and the roof $\frown$ is used for the enlarged structure.

A particular frame for minkowskian Majorana spinors, where $f = y_1 y_2$ is:

$$s_1 = f - ix_2 f$$
$$s_2 = x_1 x_2 f - ix_1 f$$
$$s_3 = if - x_2 f$$
$$s_4 = ix_1 x_2 f - x_1 f$$

the conjugation $\mathcal{C}$ is defined by $\mathcal{C}(uf) = \overline{u}\gamma f$, with $\gamma = e_1$, they exist globally over $V_{3,1}$, with our hypothesis [2,b].

It is the same with the enlarged structure, $\widehat{f} = y_0 y_1 y_2 = y_0 f$ and a frame for Majorana spinors is now:

$$\widehat{s}_1 = \widehat{f} - ix_2 \widehat{f}$$
$$\widehat{s}_2 = x_1 x_2 \widehat{f} - ix_1 \widehat{f} \qquad \text{—etc}$$

and $x_0\widehat{s}_1,\ x_0\widehat{s}_2,\ x_0\widehat{s}_3,\ x_0\widehat{s}_4$.

The Lie algebra of the Poincaré group corresponds to the space generated by $(e_\alpha e_\beta)\ \alpha, \beta = 0,1,2,3$ and $(x_0 e_\alpha)$, $\alpha = 0,1,2,3$, belonging to the Lie algebra $0(4,2)$.

By means of the "enlarged triality principle [2,a] we can construct a conformo-sympletic minkowskian Lie algebra (C.S.M.L.A). It is a graded Lie algebra (or super Lie algebra), with [,] brackets and { , } anti-brackets. We note by $\widehat{o}$ the product coming from the enlarged triality principle.

We recall here the multiplication table of the (C.S.M.L.A), where we write only non null products:

(25)

$\widehat{o}$	$x_0\widehat{f}$	$x_1\widehat{f}$	$x_2\widehat{f}$	$x_0 x_1 x_2\widehat{f}$
$\widehat{f}$	$2y_1y_2$	$-2y_0y_2$	$2y_0y_1$	$-\frac{J_0+J_1+J_2+3i\hat{e}_N}{2}$
$x_0 x_1\widehat{f}$	$2x_0 y_2$	$2x_1 y_2$	$\frac{J_0-J_1+J_2+3i\hat{e}_N}{2}$	$2x_0 x_1$
$x_0 x_2\widehat{f}$	$-2x_0 y_1$	$\frac{J_0-J_1+J_2-3i\hat{e}_N}{2}$	$-2x_2 y_1$	$2x_0 x_2$
$x_1 x_2\widehat{f}$	$\frac{J_0-J_1-J_2+3i\hat{e}_N}{2}$	$-2y_0 x_1$	$-2y_0 x_2$	$2x_1 x_2$

$$\text{Where}:\ J_0 = x_0 y_0 - y_0 x_0$$
$$J_1 = x_1 y_1 - y_1 x_1$$
$$J_2 = x_2 y_2 - y_2 x_2$$
$$\widehat{e}_N = e_4 e_5 e_N$$

$\{u\widehat{f},\ v\widehat{f}\} = -(u\widehat{f}\,\widehat{o}\,v\widehat{f})$
$\{u\widehat{f}\,v\widehat{f}\} = 0$, if $u\widehat{f}$ and $v\widehat{f}$ have the same parity.

(26)
$$\begin{cases} [x_0 u\widehat{f},\ a] = -a x_0 u\widehat{f},\ a \in \mathcal{L}(0'(1,3)) \\ [u\widehat{f},\ a] = -a u\widehat{f} \\ [x_0 u\widehat{f},\ x_0 e_k] = [u\widehat{f},\ y_0 e_k] = 0 \\ [x_0 u\widehat{f},\ y_0 e_k] = e_k u\widehat{f},\ [u\widehat{f},\ x_0 e_k] = e_k x_0 u\widehat{f} \\ [x_0 u\widehat{f},\ J_0] = -x_0 u\widehat{f},\ [u\widehat{f},\ J_0] = u\widehat{f}. \end{cases}$$

In general $[u\widehat{f},\ x] = -x u\widehat{f}$, $x \in \mathcal{L}(0'\ (4,2)) \oplus \widehat{e}_N$ and $[x,\ y]$ is defined in $\mathcal{L}(0'\ (4,2)) \oplus \widehat{e}_N$, identified, by means of orthonormed frames with exterior algebra.

$\mathcal{L}(0'\,(1,3))$ and $\mathcal{L}(0'\,(4,2))$ are respectively the complexified algebras of $0(1,3)$ and of $0(4,2)$.

The $\widehat{e}_N$-action is evident.

With Majorana spinors we obtain thus a real super-algebra with 24 dimensions (8 spinor dimensions, 15 conformal Lie algebra dimensions and $\widehat{e}_N$).

The hatching boxes give after the identifications:

$$x_0 u \widehat{f} \quad \text{and} \quad -uf$$
$$x_0 e_\alpha \quad \text{and} \quad (-i)e_\alpha$$

$$(-i \text{ replaces } x_0 \text{ or } y_0)$$

the following new table:

$\widehat{o}$	f	$x_1 x_2 f$
$x_1 f$	$-2iy_2$	$-2ix_1$
$x_2 f$	$2iy_1$	$-2ix_2$

(27)

and for Majorana spinors:

$\widehat{o}$	s_1	s_2	s_3	s_4
s_1	$4y_1$	$-2e_2$	0	$2e_1$
s_2	$-2e_2$	$-4x_1$	$-2e_1$	0
s_3	0	$-2e_1$	$4y_1$	$-2e_2$
s_4	$2e_1$	0	$-2e_2$	$-4x_1$

(28)

Meanwhile from $[x_0 u \widehat{f}, x_0 e_\alpha] = 0$ we obtain

$$[uf,\ e_\alpha] = 0 \quad \text{and also, analogously:}$$
$$[uf,\ a] = -auf,\ a \in \mathcal{L}(0(3,1)).$$

With the usual table of the Lie Poincaré algebra, we get a "super-algebra of Poincaré". This superalgebra, in finite dimension contains any terms coming from some of them after derivation according any parameter; this remark will be useful below.

We note that it is possible to obtain the table (27) by means of the enlarged triality in the Minkowski case, directly, with the condition:

$$i\mathcal{B}(xuf, vf) = g(uf \mathbin{\dot{\circ}} vf, x)$$

$\forall x \in E'$, E' complexified of $E_{3,1}$, $uf \in S^+$, $vf \in S^-$, with the Majorana spinors:

$$\mathcal{H}(xuf, vf) = g(uf \mathbin{\dot{\circ}} vf, x) \quad [2,b].$$

The definition above extend naturally to cross-sections in the bundles.

After these recalls (for the reader not aware of our previous papers), we arrive to the geometric interpretation of supersymmetry.

From (24) we deduce:

$$i\sqrt{|h|}\delta e_\lambda^a = \mathcal{H}(\varepsilon \,;\, e^a \psi_\lambda) = \mathcal{H}(e^a \varepsilon \,;\, \psi_\lambda)$$
$$\sqrt{|h|}\delta e_\lambda^a = \mathcal{B}(e^a \varepsilon \,;\, \psi_\lambda)$$
$$= \mathcal{B}(e^a \varepsilon^+ \,;\, \psi_\lambda^-) + \mathcal{B}(e^a \varepsilon^- \,;\, \psi_\lambda^+)$$
$$= \mathcal{B}(e^a \varepsilon^+ \,;\, \psi_\lambda^-) - \mathcal{B}(e^a \psi_\lambda^+, \varepsilon^-)$$
$$= (\varepsilon^+ \mathbin{\dot{\circ}} \psi_\lambda^-)^a - (\psi_\lambda^+ \mathbin{\dot{\circ}} \varepsilon^-)^a$$

$\sqrt{|h|}\delta e_\lambda^a$ is the ath covariant component of:

$$(\varepsilon^+ \mathbin{\dot{\circ}} \psi_\lambda^-) - (\varepsilon^- \mathbin{\dot{\circ}} \psi_\lambda^+).$$

Moreover, we observe that if $\delta\psi_\lambda = D_\lambda\varepsilon$, D_λ is obtained writing:

$$\frac{\partial\varepsilon}{\partial\lambda} + \widehat{\Gamma}_{\beta\lambda}^\alpha e_\alpha e^\beta \varepsilon = \left(\frac{\partial}{\partial\lambda} + \mu_\lambda\right)\varepsilon$$

$\frac{\partial}{\partial\lambda}$ being a pfaffian derivative, μ_λ is a form with values in $\mathcal{L}(0(3,1))$.

Both elements act in the Lie superalgebra. Finally we can see that supersymmetry works in an orthosympletic Lie algebra, constructed over cross-sections in the Clifford bundle of ξ_1, and this result confirms our conjectures in [2,a].

Note that if the existence of supersymmetry depends on some equations of the fields, the superalgebra is constructed without the help of these equations (off-shell construction).

Generalization

Let us consider a pseudo-riemannian manifold V, with arbitrary even dimension, $n = 2r$, and signature (p, q).

We suppose that V satisfies the same hypothesis given above for $V_{3,1}$. However the existence of G-invariant Majorana spinors needs a condition, also we suppose that we are working now with general spinors.

We have to change condition (9) above into (9 bis), and we define:

$$(30) \qquad B^\lambda \equiv (e^\lambda e^\nu e^\rho - e^\rho e^\nu e^\lambda) D_\nu \psi_\rho, \qquad \text{and suppose} \quad B^\lambda = 0.$$

The new lagrangian is:

$$(31) \qquad \mathcal{L} = \int \sqrt{|g|} \widehat{\mathcal{R}} + \sqrt{|h|} a(B^\lambda \, ; \, \psi_\lambda) + \sqrt{|h|} \overline{a}\overline{(B^\lambda \, ; \, \psi_\lambda)} d^n x.$$

Varying the $S_{\alpha\beta\gamma}$, we obtain:

$$(32) \qquad\qquad S^{\lambda\mu\nu} = \frac{a}{4}(e^\lambda e^\mu e^\nu \psi_\rho \, ; \, \psi^\rho) + \text{c.c.}$$

Also, varying the ψ_λ, $(\delta\psi_\lambda$ is not a Majorana spinor), we obtain again $B^\rho = 0$, and by means of the $\delta g^{\alpha\beta}$, we get (15) again.

What about supersymmetry?

$$D_\lambda B^\lambda = (e^\lambda e^\nu e^\rho - e^\rho e^\nu e^\lambda) D_\lambda D_\nu \psi_\rho$$

and we obtain again the same results.

However, in the (21) condition, the X_σ are the components of an $(n-3)$-form, and σ is a compound-index

$$\sigma = (\alpha_1, \alpha_2, \ldots, \alpha_{n-3}).$$

We have to put (22) again, and (23) needs straightforward modification taking account of the lagrangian (31).

The enlarged triality principle is constructed with detailed modifications. The reader has to use our remarks in [2,a], orthosymplestic superalgebra is the tool you need again.

In the odd dimension, we have to postulate the existence of a rank one bundle (in the complexified tangent bundle to V). We suppose here that the rank one bundle is a real bundle, imitating the usual Kaluza–Klein theory.

In this context we can develop an equivalent theory for the odd dimension.

Physicists often suppose that V is a bundle over $V_{3,1}$, and they often consider a trivial bundle: in local language the two cases are identical.

If $n = 2r$, or $n = 2r + 1$, spinor fields are of dimension $2^r = 2^{2+k}$, $k = r - 2$.

If V is a trivial bundle over $V_{3,1}$, with rank $2k$ (or $2k + 1$), the cross-sections in the spinor bundle over V are direct sums of 2^k cross-section in the spinor tangent bundle of $V_{3,1}$. If we put $2^k = N$, the theory is called N-supergravitation. However $n = 2r$ and $n = 2r + 1$ give the same N, also we have to distinguish:

$$\textit{Ordinary } N - \textit{super-gravitation,} \quad n = 2r, \quad k = r - 2$$
$$N = 2^k, \quad k = 0, 1, 2, \cdots$$

$$\textit{Kaluza–Klein } N - \textit{supergravitation :} \quad n = 2r + 1, \quad k = r - 2$$
$$N = 2^k, \quad k = 0, 1, 2, \cdots$$

both generalize the Rarita–Schwinger theory, (with torsion or torsion free). Thus if:

$$N = 1, \quad k = 0, \quad n = 4 \text{ we have} \quad \text{:. a vector-spinor } \psi_\mu, \quad \mu = 0, 1, 2, 3$$
$$\text{(gravitino)}$$
$$\text{.a pseudo-metric field } g_{\mu\nu}$$
$$\text{(graviton)}$$

. a vector field (X_σ) associated
with the torsion (completely antisymmetric).

The field equations are given above.

$N = 1, k = 0, n = 5,$ we have:

. a vector-spinor ψ_μ , $\mu = 0, 1, 2, 3,$ spin 3/2

. a spinor-field ψ_a , $a = 4$ (spin 1/2)

. a pseudo-metric field $g_{\mu\nu}$, $\mu, \nu = 0, 1, 2, 3.$

. an electro-magnetic potential $(g_{\mu a})$, $a = 4$

. a vector-field X_σ

. a two-form: $X_{\alpha\beta a}$, $\alpha, \beta = 0, 1, 2, 3,$ $a = 4,$

$N = 8, k = 3, n = 10:$ we obtain:

. 8 spin 3/2 fields: ψ_μ^A , $\mu = 0, 1, 2, 3.$

. 48 spin 1/2 fields: ψ_a^A , $a = 4, 5, 6, 7, 8, 9.$

. 1 gravitation $g_{\lambda\nu}$, $\lambda, \nu = 0, 1, 2, 3.$

. 6 vector-fields: $g_{\mu a}.$

. 21 scalar-fields: $g_{ab}.$

. a vector-field X_σ, (or three-form $X_{\alpha\beta\gamma}$, $\alpha, \beta, \gamma = 0, 1, 2, 3$).

. 6 two forms $A_{\alpha\beta a}.$

. 15 vector-fields $A_{\alpha ab}.$

$N = 8, k = 3, n = 11,$ gives:

. 8 spin 3/2 fields $\psi_\mu^A.$

. 56 spin 1/2 fields $\psi_a^A.$

. 1 graviton $g_{\mu\nu}.$

. 7 vector-fields $g_{\mu a}.$

. 28 scalar-fields $g_{ab}.$

. a vector-field X_σ (or a three-form $X_{\alpha\beta\gamma}$).

. 7 two-forms $A_{\alpha\beta a}.$

. 21 vector-fields $A_{\alpha ab}.$

Field equations come from the first case $(N = 1)$ by natural generalizations. There are no "zoo-formulas".

There is no limit for dimension, in the purely mathematic context. Some authors give such a limit, but arguing with physical reasons.

Appendix

The system (9)–(10) with (3–1) signature is well posed:

$$(9) \ A^\lambda \equiv \eta^{\lambda\mu\nu\rho} e_\mu D_\nu \psi_\rho = 0$$
$$(10) \ \chi \equiv e^\lambda \psi_\lambda = 0.$$

We can replace (9) by:

(9.ter) $$e^\alpha D_\alpha \psi_\rho = 0.$$

Our method is inspired by [1].
First, if $V_{3,1} = E_{3,1}$, (9) (10) are replaced by:

$$\chi = 0, \quad f_\rho \equiv e^\alpha \partial_\alpha \psi_\rho = 0.$$

The Cauchy data for $t = 0$ are $\psi_\lambda(0, \cdots)$ and they satisfy:

$$A^0 = \varepsilon^{ijk}_{123} e_i (\partial_j \psi_k) = 0 \ (i, j, k, = 1, 2, 3) \text{ and } \chi = 0, \text{ if } t = 0.$$

But (9 bis) gives:
$$e^\alpha e^\lambda (\partial_\alpha \chi) - e^\alpha e^\lambda f_\alpha = 0.$$

If $f_\alpha = 0$, $A^\lambda = 0$ implies $e^\alpha e^\lambda \partial_\lambda \partial_\alpha (\chi) = 0$, $g^{\lambda\alpha} \partial_{\lambda\alpha}(\chi) = 0$.
For $t = 0$, $\chi = 0$ and then $\partial_i x = 0$; thus $f_\alpha = 0$, $A^\lambda = 0$ implies $\partial_0 \chi = 0$ for $t = 0$.
Equation $g^{\lambda\alpha} \partial_{\lambda\alpha}(\chi) = 0$ admits the solution $\chi = 0$. $\chi = 0$ is the unique solution satisfying the Cauchy data and $\chi = 0$, $A^0 = 0$ are preserved by this solution.
Now you suppose that $V_{3,1}$ is a curved space, but with symmetric connection. We can consider $\chi = 0$ and $f_\rho \equiv e^\alpha \nabla_\alpha \psi_\rho = 0$ which implies:

$$e^\nu e^\lambda \nabla_\lambda \nabla_\nu (\chi) = 0.$$

But $\nabla_\lambda \nabla_\nu \psi_\rho - \nabla_\nu \nabla_\lambda \psi_\rho \equiv \frac{1}{4} R^\alpha_{\beta\lambda\nu} e_\alpha e^\beta \psi_\rho - R^\alpha_{\rho\lambda\nu} \psi_\alpha$, leads to:

$$\nabla_\lambda \nabla_\nu \chi - \nabla_\nu \nabla_\lambda \chi \equiv \frac{1}{4} R_{\alpha\beta,\lambda\nu} e^\lambda e^\beta \chi,$$

after left multiplication by e^ρ and some checking in the Clifford algebra.

Thus if (ψ_λ) satisfies (9 ter) and (10), $\nabla_\lambda \nabla_\nu(\chi) = \nabla_\nu \nabla_\lambda(\chi)$ and $g^{\lambda\nu} \nabla_\lambda \nabla_\nu(\chi) = 0$. For $t = 0$, $\chi = 0$ and $\partial_i \chi = 0$, also $f_\rho = 0$ and $e^\nu e^\lambda \nabla_\nu(\chi) = 0$ implies that $\partial_0 \chi = 0$ for $t = 0$.

Equation $g^{\lambda\nu} \nabla_\lambda \nabla_\nu(\chi) = 0$ admits the solution $\chi = 0$, and it is the solution satisfying Cauchy data, the constraint $\chi = 0$ (for $t = 0$) and $A^0 = 0$ are preserved.

Finally, we consider the general case, with torsion, $S_{\alpha\beta\gamma}$ totally anti-symmetric

$$\chi = 0, \quad f_\rho \equiv e^\alpha D_\alpha \psi_\rho = 0 \quad \text{implies} \quad e^\nu e^\lambda D_\lambda D_\nu(\chi) = 0$$

$$D_\lambda D_\nu(\chi) - D_\nu D_\lambda(\chi) \equiv \frac{1}{4} \widehat{R}_{\alpha\beta,\lambda\nu} \, e^\rho e^\alpha e^\beta \psi_\rho - \widehat{R}_{\alpha\rho\lambda\nu} \, e^\rho \psi^\alpha$$

$$\equiv \frac{1}{4} R_{\alpha\beta,\lambda\nu} \, e^\alpha e^\beta \chi$$

(we are using here (4)).

Thus we obtain the same situation again. However, we have to choose a solution of the Einstein–Hilbert equation (15) and we have to satisfy condition (21) or (21 bis).

Equations (15) concern the manifold $V_{3,1}$; what about (21)?

(21) appears for the supersymmetry condition only, in the general case, not in the symmetric case.

It is easy, using the Majorana frame s_1, s_2, s_3, s_4, and putting: $\psi_\rho = A_\rho^\alpha S_\alpha$, to obtain the detailed expression of (10):

(33)
$$A_0^1 = A_3^1 - A_2^2 + A_1^4$$
$$A_0^2 = -A_2^1 - A_3^2 - A_1^3$$
$$A_0^3 = -A_1^2 + A_3^3 - A_2^4$$
$$A_0^4 = A_1^1 - A_2^3 - A_3^4$$

and (9 ter) gives a differential system between the A_ρ^α, $\alpha = 1, 2, 3, 4$, $\rho = 1, 2, 3$, where we can suppose A_ρ^α real. This system is:

(34) $\qquad \partial_i(A_\rho^\alpha) a_\alpha^{i\theta} + A_\rho^\alpha B_\alpha^\theta = 0, \quad \theta = 1, 2, 3, 4, \quad \rho = 1, 2, 3,$

and for $\rho = 0$, we have to take into account (33).

We define: $e^i s_\alpha = a_\alpha^{i\theta} s_\theta$, where the $a_\alpha^{i\theta}$ are constant.
$\frac{1}{4} e^\alpha \widehat{\Gamma}_{\mu\alpha}^\lambda e_\lambda e^\mu s_\theta = B_\alpha^\theta s_\theta$, B_α^θ are scalar, depending on the torsion coefficients.

With (12) we check $S^{\lambda\mu\nu}$. Modulo a constant factor, we consider then $(e^\rho e^N \psi_\sigma \; ; \; \psi^\sigma)$,

$$e^N s_1 = s_3, \quad e^N s_2 = s_4, \quad e^N s_3 = -s_1, \quad e^N s_4 = -s_2,$$

(because $e^N f = i f$).

With Majorana spinors and $\mathcal{H} = i\mathcal{B}$, we obtain the associated vector X to the torsion tensor. Modulo a constant factor:

$$X = X^0 e_0 + X^1 e_1 + X^2 e_2 + X^3 e_3$$
$$X^0 = \sum_\alpha A_\sigma^\alpha A^{\alpha\sigma}, \quad X^1 = 2(A_\sigma^1 A^{4\sigma} - A_\sigma^2 A^{3\sigma})$$
$$X^2 = -2(A_\sigma^1 A^{2\sigma} + A_\sigma^3 A^{4\sigma}), \quad X^3 = A_\sigma^1 A^{1\sigma} + A_\sigma^3 A^{3\sigma} - A_\sigma^2 A^{2\sigma} - A_\sigma^4 A^{4\sigma}.$$

Thus X is generally non null.

But the A_ρ^α satisfy also (21):

$$e_\sigma(\nabla_i X^\sigma) \cdot \psi^i = 0, \quad \psi^i = A^{\alpha i} s_\alpha$$

$e_\sigma[(\partial_i X^\sigma) + \Gamma_{\alpha i}^\sigma X^\alpha] \cdot A^{\lambda i} s_\lambda = 0$, giving 4 differential equations.

After use of (33), we can consider the field equations: $e^\alpha D_\alpha \psi_\rho = 0$ and $e^\sigma \nabla_\sigma (X_\lambda \psi^\lambda) = 0$ between 12 coefficients A_ρ^α, $\rho \neq 0$; in the symmetric case the last conditions are evanescent.

References

[1] CHOQUET–Y. BRUHAT, Causality of classical supergravity. *Proceed. of A. Eddington cent. Symp.*, **vol.2** World SC. Publ. Singapour, 1986.

[2] A. CRUMEYROLLE,

    a) Construction d'algèbres de Lie graduées orthosympleticques et confor-mosymplectiques minkowskiennes. Lect. Notes in Math. n° 1165. Springer, p.52–83.

    b) Algèbres de Clifford, Structures Spinorielles, Orthogonales et Symplec-tiques. Preprint — Toulouse, 1987.

    c) Orthogonal and symplectic Clifford algebras, spinor structures (Kluwer A. P. Dordrecht 1990) (Translation of b) above).

Albert Crumeyrolle

Université Paul Sabatier
Mathématiques
118, route de Narbonne
31062 Toulouse Cedex
France

COLLOQUIA MATHEMATICA SOCIETATIS JÁNOS BOLYAI

56. DIFFERENTIAL GEOMETRY, EGER (HUNGARY), 1989

Connections Deduced from Mechanical System of Second Order*

A. DEKRÉT

Let L be a real function on the tangent bundle $T^r M$ of order r, that is, the space of all r-jets from $\mathbb{R}$ into a smooth manifold M with source $0 \in \mathbb{R}$. The main geometrical idea of the Lagrangian formalism in classical mechanics on M consists in a construction of the mechanical system S_L of L on $T^{2r-1} M$ the integral curves of which satisfy the Euler differential equation of order $2r$, see [5, 3] if $r = 1$ and [8] if r is arbitrary. In the case of the regular Lagrangian L of order r, Rodrigues and Leon, [8], have constructed a connection Γ_L on $T^{2r-1} M \to T^{2r-2} M$, see also [1, 2, 5] if $r = 1$. In this paper we show that there is a 3-parameter or 4-parameter or 3-parameter family of all natural operators of first order from the space of all differential equations of order four into the space of connections on $T^3 M \to T^2 M$ or on $T^3 M \to TM$ or on $T^3 M \to M$, respectively. We collect properties of these connections in the case of the mechanical system S_L of a given regular Lagrangian L of second order. Our considerations are in the category C^∞.

---

* This paper is in final form and no version of it will be submitted for publication elsewhere.

Canonical objects on $T^2 M$

Throughout this paper (π) or $(V\pi)$ or $C^\infty \pi$ will be the abreviated notation for a fibred manifold $\pi : Y \to M$ or for the space of all vertical vectors on Y or for the set of smooth sections of π, respectively, p_N will denote the canonical tangent projection $TN \to N$ and Tf will be the tangent prolongation of a map f.

There are the canonical vector fields $C_1, \ldots, C_r$ and the canonical vector bundle morphisms $J_1, \ldots, J_r$ on $T^r M$, [8]. We introduce another constructions of these objects in the case of $T^2 M$.

First we recall that on a vector bundle $\pi : E \to M$ with a chart (x^i, y^α) there are the Liouville field $V = y^\alpha \frac{\partial}{\partial y^\alpha}$ induced by the homotheties on E and the canonical identification $VE = E \times_M E$ which determines for any $h \in E$ an embedding $^h\pi : E_{\pi h} \to VE$, $(e) \mapsto (h, e)$, $^h\pi(x^i, y^\alpha) = (x^i, h^\alpha, 0, y^\alpha)$.

A chart (x^i) on M induces the chart $(x^i, x_{10}^i, x_{01}^i, x_{11}^i)$ on TTM. According to two vector bundle structures (p_{TM}) and (T_{p_M}) there are two Liouville fields $V_1 = x_{01}^i \frac{\partial}{\partial x_{01}^i} + x_{11}^i \frac{\partial}{\partial x_{11}^i}$ and $V_2 = x_{10}^i \frac{\partial}{\partial x_{10}^i} + x_{11}^i \frac{\partial}{\partial x_{11}^i}$ on TTM.

Recall the canonical involution $i_2(x^i, x_{10}^i, x_{01}^i, x_{11}^i) = (x^i, x_{01}^i, x_{10}^i, x_{11}^i)$ on TTM, [3]. Let $X \in T_h(TTM)$. Put $\varphi_2(X) =^h p_{TM} T_{p_{TM}}(X)$, $\varphi_1(X) = {}^h T_{p_M} \circ i_2 \circ TT_{p_M}(X)$. We get two vector bundle morphisms

$$\varphi_2 = dx^i \otimes \frac{\partial}{\partial x_{01}^i} + dx_{10}^i \otimes \frac{\partial}{\partial x_{11}^i}$$

$$\varphi_1 = dx^i \otimes \frac{\partial}{\partial x_{10}^i} + dx_{01}^i \otimes \frac{\partial}{\partial x_{11}^i} \quad \text{on } TTM.$$

It is clear that $T^2 M$ is a submanifold of TTM such that $h \in T^2 M$ if and only if $i_2(h) = h$, $x_{10}^i = x_{01}^i$. It is not hard to check that the restriction of $V_1 + V_2$ or $\varphi_1 + \varphi_2$ on $T^2 M$ is a vector field or a vector bundle morphism on $T^2 M$, respectively. In the induced chart (x_0^i, x_1^i, x_2^i), where $x_1^i = x_{10}^i = x_{01}^i$,

$x_2^i = x_{11}^i$ we get

$$C_1 := (V_1 + V_2)|_{T^2 M} = x_1^i \frac{\partial}{\partial x_1^i} + 2x_2^i \frac{\partial}{\partial x_2^i},$$

$$J_1 := (\varphi_1 + \varphi_2)|_{T^2 M} = dx^i \otimes \frac{\partial}{\partial x_1^i} + 2dx_1^i \otimes \frac{\partial}{\partial x_2^i}.$$

Denote $J_2 : \frac{1}{2}J_1^2 = dx_0^i \otimes \frac{\partial}{\partial x_2^i}$, $C_2 = \frac{1}{2}J_1(C_1) = x_1^i \frac{\partial}{\partial x_2^i}$.

Remark 1. It arises the problem of the unicity of the fields C_1, C_2 and of the vector bundle morphisms J_1, J_2 on $T^2 M$. In order to answer this question it is suitable to use the natural bundle theory, [6, 7, 9]. Recall that if F, G are two natural functor in the category M_m of m-dimensional manifolds and local diffeomorphisms then a natural transformation from F into G is a family of mappings ϕ such that for any local diffeomorphism f from M into N it holds $Gf \circ \phi_M = \phi_N \circ Ff$. A canonical vector field on $T^2 M$ can be treated as a natural transformation C from T^2 into TT^2 such that $p_{T^2 M} \circ C_M = id_{T^2 M}$, $M \in M_m$. Since $T^2 M \to M$ or $TT^2 M \to M$ is a fibre bundle associated to the principal fibre bundle $H^2 M$ or $H^3 M$ with the standard fibre $(T^2 \mathbb{R}^m)_0$ or $(TT^2 \mathbb{R}^m)_0$, respectively, where $H^r M$ denotes the space of all r-frames on M with the structure group L_m^r, then the natural transformations C are in bijection with the L_m^3-equivariant maps φ from $(T^2 \mathbb{R}^m)_0$ into $(TT^2 \mathbb{R}^m)_0$ such that $p_{T^2 \mathbb{R}^m} \varphi = id_{T^2 \mathbb{R}_0^m}$, [7, 9]. Using the methods, developed by a number of authors, see for example [6], it is possible to prove that all natural transformation C are of the form $aC_1 + bC_2$, $a, b \in \mathbb{R}$. Quite analogously, it can be shown that all natural transformations from $TT^2 M$ into $TT^2 M$ over $id_{T^2 M}$ form five parameter family $a_1 C_1 + a_2 C_2 + b_1 J_1 + b_2 J_2 + b_3 id_{TT^2 M}$ and thus all canonical vector bundle morphisms on $T^2 M$ are of the form $b_1 J_1 + b_2 J_2 + b_3 id_{TT^2 M}$, $b_i \in \mathbb{R}$.

In the same way as above we can construct the canonical Liouville field C_1 and the canonical vector bundle morphism J_1 on $T^r M$. In the induced chart $(x_0^i, x_1^i, \ldots, x_r^i) = j_0^r \gamma$, $x_k^i = \frac{d^k \gamma^i(0)}{dt^k}$, we have

$$C_1 = \sum_{k=1}^{r} kx_k^1 \frac{\partial}{\partial x_k^i}, \qquad J_1 = \sum_{k=1}^{r} k\, dx_{k-1}^i \otimes \frac{\partial}{\partial x_k^i}$$

Put

$$J_k := \frac{1}{k!} J_1^k = \sum_{u=k}^{r} \binom{u}{k} dx_{u-k}^j \otimes \frac{\partial}{\partial x_u^i}$$

$$C_k := \frac{1}{k!} J_1^{k-1}(C_1) = \sum_{u=k}^{r} \binom{u}{k} x_{n-k+1}^j \frac{\partial}{\partial x_u^j}, \quad k = 2, r,$$

compare with [8].

Let us recall that the differential equation of order $r+1$ is a vector field S on $T^r M$ such that $J_1(S) = C_1$, [8]. Its coordinate expression is

$$S = \sum_{j=0}^{r-1} x_{j+1}^i \frac{\partial}{\partial x_j^i} + b^i \frac{\partial}{\partial x_r^i}.$$

The connection determined by the Lie derivation $L_S J_1$ of the canonical morphism J_1 according to S has been studied by a number of authors, see for example [1, 2, 5], if $r = 1$ and [8] in a general case. It arises a question whether there are another connections induced by S. We will solve this problem if $r = 3$ because we are interested in mechanics of second order.

At first we recall needed facts about connections on a fibre manifold $\pi : Y \to M$. A 1-form ω on Y is said to be π-semibasic if $\omega(X) = 0$ for any $X \in V\pi$. A connection Γ on Y can be given by its horizontal form h_Γ that is a π-semibasic vector value 1-form on Y such that $T\pi h_\Gamma(X) = T\pi(X)$. In a chart (x^i, y^α) on Y, $h_\Gamma = dx^i \otimes \frac{\partial}{\partial x^i} + \Gamma_i^\alpha dx^i \otimes \frac{\partial}{\partial y^\alpha}$. Then $v_\Gamma := id_{TY} - h_\Gamma = (dy^\alpha - \Gamma_i^\alpha dx^i) \otimes \frac{\partial}{\partial y^\alpha}$ is called the vertical form of Γ and $H^\Gamma := \text{Im} \, h_\Gamma$ denotes the vector bundle of all Γ-horizontal vectors on Y. It is known that if φ is a π-semibasic and $V\pi$-value 1-form on Y then $h_\Gamma + \varphi$ is the horizontal form of the other connection which will be denoted by $\Gamma + \varphi$.

Further we will use the well known canonical injection $j_k : T^k M \to TT^{k-1} M$, $j_0^k \gamma \mapsto j_0^1 (t \mapsto j_{s=0}^{k-1} \gamma(t+s))$, $j_k(x_s^i) = (x_0^i, \ldots, x_{k-1}^i, x_1^i, \ldots, x_k^i)$.

There is bijection between the set of all differential equations $S : T^3 M \to TT^3 M$ and the set of all sections of (π_3^4), $S \mapsto \overline{S}$, $S = j_4 \circ \overline{S}$, where $\pi_k^r : T^r M \to T^k M$ is the canonical projection of r-jets into their k-subjects.

Our geometrical constructions of connections on T^3M are based on some mappings that we shall introduce. Let J_1 be the canonical vector bundle morphism on T^4M, $h \in T^4M$, $\pi_3^4 h = u$. Let $J_1^h : T_u T^3M \to V_h \pi_0^4$ be a vector morphism such that $J_1^h(Y) = J_1(Z)$, $Z \in T_h T^4M$, $T\pi_3^4(Z) = Y$. Then we define a vector bundle morphism $\overline{J}_1 : V\pi_0^4 \to TT^3M$ such that for $W \in V_h \pi_0^4$, $\overline{J}_1(W) = Y \in T_u T^3M$, where $J_1^h(Y) = W$.

Let $S = \sum_{p=1}^{3} x_p^i \frac{\partial}{\partial x_{p-1}^i} + b^i \frac{\partial}{\partial x_3^i}$ be a differential equation of order four. Then $\overline{S}(x_p^i) = (x_p^i, b^i)$, $p = 0, 1, 2, 3$. Restricting $T\overline{S}$ on $(V\pi_0^3)$ we define the vector bundle morphism $\tau_S : (V\pi_0^3) \to TT^3M$ putting

$$\tau_S := \overline{J}_1 \cdot (T\overline{S}|_{V\pi_0^3}) = dx_1^i \otimes \frac{\partial}{\partial x_0^i} + \frac{1}{2} dx_2^i \otimes \frac{\partial}{\partial x_1^i} + \frac{1}{3} dx_3^i \otimes \frac{\partial}{\partial x_2^i} +$$
$$+ \frac{1}{4}(b_{j_1}^i \, dx_1^j + b_{j_2}^i \, dx_2^j + b_{j_3}^i \, dx_3^j) \otimes \frac{\partial}{\partial x_3^i},$$

where $b_{j_s}^i := \frac{\partial b^i}{\partial x_s^j}$.

According to three canonical projections π_i^3, $i = 0, 1, 2$, there are three types of connections on T^3M. In order to find them we use the theory of natural operators. First we will deal with connections on (π_2^3). Let $J(\pi_2^3)$ denote the first jet prolongation of (π_2^3). It is known that a connection $^2\Gamma$ on (π_2^3) can be viewed as a global cross-section of $J(\pi_2^3) \overset{\beta}{\to} T^3M$. We find a family of operators $\phi : C^\infty(\pi_3^4) \to C^\infty(\beta)$ such that for any $s, s_1, s_2 \in C^\infty(\pi_3^4)$ it holds

1. $\phi_N(T^4 f \cdot s \cdot (T^3 f)^{-1}) = J\pi_2^3(f) \circ \phi_M(s) \circ (T^3 f)^{-1}$ for every local diffeomorphism from M into N, where T^r or $J\pi_2^3$ denotes the natural functor $M \mapsto T^r M$ or $M \mapsto J(T^3 M \to T^2 M)$, respectively,

2. $j_h^1 s_1 = j_h^1 s_2$ implies $\phi s_1(h) = \phi s_2(h)$

3. $\phi_U(s|_U) = (\phi_{M^s})|_U$ for every open subset $U \subset M$, and smoothly parametrized families of sections are transformed into smoothly parametrized ones.

The images under these operators ϕ are called the connections 1-dependent on S. These operators are in bijection with the L_m^5-equivariant maps $\varphi : J(T^4 R^m \to T^3 R^m)_0 \to (J\pi_2^3 R^m)_0$ over $id_{T^3 R_0^m}$. We find the expression of φ.

Let $g = j_0^5 f = (f_j^i, \ldots, f_{jkstu}^i) \in L_m^5$, $h = j_u^1 \sigma = (x_s^i, b^i, b_{k_0}^i, b_{k_s}^i) \in J(T^4 R^m \to T^3 R^m)_0$, $q = j_v^1 \varepsilon = (x_s^i, \Gamma_{k_0}^i, \Gamma_{k_1}^i, \Gamma_{k_2}^i) \in (J\pi_2^3 R^m)_0$, $s = 1, 2, 3$, and $\tilde{g} = j_0^5 \tilde{f}$ be the inverse of g. Then formula of the action of the group L_m^5 on $J(T^4 R^m \rightarrow T_2^3 R^m)_0$ or on $(J\pi_2^3 R^m)_0$ is given as follows

(1) $(g, h) \mapsto j_x^1(u \mapsto T^4 f \circ \sigma \circ T^3 \tilde{f}(u)) = \bar{h}$ or

(2) $(g, q) \mapsto j_v^1(z \mapsto T^3 f \circ \varepsilon \circ T^2 \tilde{f}(z)) = \bar{q}$, respectively .

By a tedious calculation we get the coordinate forms $\bar{x}_{k_s}^i = F_{k_s}^i(h, g)$, $\bar{b}^i = B^i(h, g)$, $\bar{b}_{k_0}^i = B_{k_0}^i(h, g)$, $\bar{b}_{k_s}^i = B_{k_s}^i(h, g)$, $\bar{\Gamma}_{k_j}^i = H_{k_j}^i(h, q)$, $s = 1, 2, 3$, $j = 0, 1, 2$, of the formulas (1) and (2). For the sake of their length we do not introduce their exact expressions.

We need to find all functions $\Gamma_{t_q}^i = \varphi_{t_q}^i(b_k, b_{j_s}^k)$, $q = 0, 1, 2$, $s = 0, 1, 2, 3$, satisfying the L_m^5-equivariance condition $\bar{\Gamma}_{t_q}^i = \varphi_{t_q}^i(\bar{b}^k, \bar{b}_{j_s}^k)$. We proceed by [6].

Equivariance with respect to the subgroup of all homotheties or to the subgroup $\mathrm{Ker}\, \pi_4^5$ implies that $\varphi_{t_q}^i$ are independent on the variables b^k or $b_{j_0}^k$, respectively.

With respect to $\mathrm{Ker}\, \pi_3^5$ the equation of equivariance gives

$$\varphi_{t_0}^i = \frac{1}{4} b_{t_1}^i + c_{t_0}^i(b_{j_2}^u, b_{j_3}^u)$$

and the independence of the functions $\varphi_{t_1}^i$ and $\varphi_{t_2}^i$ on $b_{j_1}^i$.

According to the $\mathrm{Ker}\, \pi_2^5$-equivariance we deduce that the functions $\varphi_{t_2}^i$ do not depend on $b_{s_2}^k$. Then the $\mathrm{Ker}\, \pi_1^5$-equivariance gives $\varphi_{t_2}^i = \frac{3}{4} b_{t_3}^i + c_t^i$. Finally the full L_m^5-equivariance yields $c_t^i = c_1 \delta_t^i$.

Then the condition of the $\mathrm{Ker}\, \pi_2^5$-equivariance implies

$$\varphi_{t_1}^i = \frac{1}{2} b_{t_2}^i + c_t^i(b_{j_3}^k), \quad \varphi_{t_0}^i = \frac{1}{4} b_{t_1}^i - \frac{1}{6} c_1 b_{t_2}^i + h_t^i(b_{j_3}^i)$$

With respect to $\mathrm{Ker}\, \pi_1^5$-equivariance $\varphi_{t_1}^i = \frac{1}{2} b_{t_2}^i - \frac{1}{2} c_1 b_{t_3}^i + c_t^i$. Finally the full L_m^5-equivariance gives $c_s^i = c_2 \delta_s^i$. Then according to $\mathrm{Ker}\, \pi_1^5$-equivariance $\varphi_{t_0}^i = \frac{1}{4} b_{t_1}^i - \frac{1}{6} b_{t_2}^i - \frac{1}{4} c_2 b_{t_3}^i + c_t^i$. At last L_m^5-equivariance implies $c_t^i = c_3 \delta_t^i$. We conclude that all connections on (π_2^3) which are 1-dependent on S form a 3-parameter family $^2\gamma$ given by the equation $dx_3^i = \varphi_{t_0}^i \, dx_0^t + \varphi_{t_1}^i \, dx_1^t + \varphi_{t_2}^i \, dx_2^t$.

In order to construct these connections calculate the Lie derivations $L_S J_1$, $L_S J_2$, $L_S J_3$ of the canonical vector bundle morphism J_1, J_2, J_3 with respect to S. Then

$$\frac{1}{4}(Id_{TT^3 M} + L_S J_1) = \left(dx^i_s - \frac{1}{4}(b^i_{x^i_1} \, dx^i_0 + 2b^i_{j_2} \, dx^j_1 + 3b^i_{j_3} \, dx^j_2)\right) \otimes \frac{\partial}{\partial x^i_3}$$

is the vertical form of the connection $^2\Gamma_3$ on (π^3_2) which coincides with one in [8].

The 1-forms

$$J_1 + L_S J_2 = \left(6dx^i_2 - b^i_{j_2} \, dx^j_0 - 3b^i_{j_3} \, dx^j_1\right) \otimes \frac{\partial}{\partial x^i_3}$$

$$J_2 + L_S J_3 = (4dx^i_1 - b^i_{j_3} \, dx^j_0) \otimes \frac{\partial}{\partial x^i_3}$$

are π^3_2-semibasic and $V\pi^3_2$-valued. Now it is not hard to control

Proposition 1. *Let S be a differential equation of order four. Then every connection $^2\Gamma$ on (π^3_2) which is 1-dependent on S is of the form*

$$^2\Gamma = {}^2\Gamma_3 + c_1(J_1 + L_S J_2) + c_2(J_2 + L_S J_3) + c_3 J_3,$$

where c_1, c_2, c_3 are real constants.

Remark 2. It is interesting that the horizontal form of connection $^2\Gamma_3$ can be expressed in the form $\tau_S \circ J_1$, where the vector bundle morphism τ_S was introduced above.

Calculating $\tau^2_S \circ J_2$ or $\tau^3_S \circ J_3$, we immediately check that

$$2\tau^2_S \cdot J_2 = dx^i_0 \otimes \frac{\partial}{\partial x^i_0} + dx^i_1 \otimes \frac{\partial}{\partial x^i_1} + \left(\frac{1}{6}b^i_{j_2} \, dx^j_0 + \frac{1}{2}b^i_{j_3} \, dx^j_1\right) \otimes \frac{\partial}{\partial x^i_2} +$$
$$+ \left[\left(\frac{1}{4}b^i_{j_1} + \frac{1}{8}b^i_{s_3} b^s_{j_2}\right) dx^j_0 + \left(\frac{1}{2}b^i_{j_2} + \frac{3}{8}b^i_{s_3} b^s_{j_3}\right) dx^j_1\right] \otimes \frac{\partial}{\partial x^i_3}$$

or

$$6\tau^3_S \cdot J_3 = dx^i_0 \otimes \frac{\partial}{\partial x^i_0} + dx^s_0 \otimes \left[\frac{1}{4}b^i_{s_3} \frac{\partial}{\partial x^i_1} + \left(\frac{1}{6}b^i_{s_2} + \frac{1}{8}b^i_{k_3} b^k_{s_3}\right) \frac{\partial}{\partial x^i_2} +\right.$$
$$+ \left.\left(\frac{1}{4}b^i_{s_1} + \frac{1}{8}b^i_{k_2} b^k_{s_3} + \frac{1}{8}b^i_{k_3} b^k_{s_2} + \frac{3}{32}b^i_{k_3} b^k_{j_3} b^j_{s_3}\right) \frac{\partial}{\partial x^i_3}\right]$$

is the horizontal form of a connection $^1\Gamma_S$ or $^0\Gamma_S$ on (π^3_1) or (π^3_0), respectively.

By the quite analogous procedure as above we obtain

Proposition 2. *Let S be a differential equation of order four. Then a connection Γ on (π_1^3) belongs to the family $^1\gamma$ of all connections 1-dependent on S if and only if there is real numbers c_1, c_2, c_3, c_4 such that*

$$\Gamma = {}^1\Gamma_S + c_1\tau_S(J_2 + L_S J_3) + c_2\left(\frac{1}{4}J_2 - \frac{3}{4}L_S J_3\right) + c_3(J_2 + L_S J_3) + c_4 J_3.$$

Proposition 3. *Let S be a differential equation of order four. Then every connection on (π_0^3) which is 1-dependent on S is of the form*

$$\Gamma = {}^0\Gamma_S + c_1\tau_S^2 \cdot J_3 + c_2\tau_S \cdot J_3 + c_3 J_3, \qquad c_i \in R.$$

The family of these connections is denoted by $^0\gamma$.

Connections induced by mechanical systems of second order

In the sequel, we will give a brief survey about the needed notions of classical mechanics on M. We restrict to the geometry of a regular Lagrangian L of second order in the autonomous case. We refer to [8] for another informations on higher mechanics.

The space $\pi_{k-1}^k : T^k M \to T^{k-1}$ is an affine fibred manifold. A function L on $T^k M$ is called the regular Lagrangian if the Hesian of the restriction $L(\pi_{k-1}^k)_h$ on the fibre over $h \in T^{k-1} M$ is regular for any h, i.e. if $\det L_{i_k j_k} \neq 0$.

Every J_S on $T^k M$ determines a derivation of first order $d_{J_s} = [i_{J_s}, d] = i_{J_s} d - di_{J_s}$, where d denotes the standard exterior differential and

$$i_{J_s}\omega(X_1,\ldots,X_p) = \sum_{j=1}^{p} \omega(X_1,\ldots,J_s(X_j),\ldots X_p).$$

Let $f : T_1^k M \to R$ be a smooth function. Then $df : TT^k M \to R$ is a function on $TT^k M$. Let $j_{k+1} : T^{k+1} M \to TT^k M$ be the canonical embedding. Then $d_T f = i_{k+1}^*(df) = \sum_{p=0}^{k} f_{i_p} x_{p+1}^i$ is a function on $T^{k+1} M$. It is clear that d_T is a derivation operator which can be extended to a derivation (due to Tulczyjew) of order 0 and commutative with d on the algebra λ that is the quotient set of $\bigcup_k \lambda(T^k M)$ by the equivalence relation according to which two forms $\alpha \in (T^k M)$, $\beta \in (T^j M)$, $k \geq j$, are equivalent if $\alpha = (\pi_j^k)^* \beta$.

Let L be a given regular Lagrangian of order k. Then the mechanical system of L is a vector fields S_L on $T^{2k-1} M$ such that $i_{S_L} d\omega_L = -dE$, where

$$\omega_L = \sum_{j=1}^{k} (-1)^j d_T^{j-1} d_{J_j} L, \qquad E = \sum_{j=1}^{k} (-1)^j d_T^{j-1} C_j(L) + L.$$

The vector field S_L is a semispray and its integral curves satisfy the Euler differential equation.

It is known [2, 5] that if $k = 1$, then the connection Γ_{S_L} given by $L_{S_L} J_1$ is Lagrangian, i.e. $\Gamma_{S_L} = \text{Orth}\,\Gamma_{S_L}$, where $\text{Orth}\,\Gamma_{S_L}$ denotes the connection on TM which is $d\omega_L$-orthogonal to Γ_{S_L}. We focus on some analogous properties of all connections determined by S_L, if $k = 2$. In this case

$$\omega_L = \left(\sum_{p=0}^{2} L_{i_2 k_p} x_{p+1}^k - L_{i_1} \right) dx_0^i - L_{i_2} dx_1^i,$$

$$S_L = \sum_{p=0}^{2} x_{p+1}^i \frac{\partial}{\partial x_p^i} + b^i \frac{\partial}{\partial x_3^i}, \qquad \text{where}$$

$$b^i = \tilde{L}^{ik} \left(\sum_{p=0}^{2} L_{k_1 j_p} x_{p+1}^j - \sum_{p,p=0}^{2} L_{k_2 j_p u_s} x_{p+1}^j x_{s+1}^u - \right.$$
$$\left. - L_{k_2 j_0} x_2^j - L_{k_2 j_1} x_3^j - L_{k_0} \right),$$

where $L_{i_2 k_2} \tilde{L}^{ks} = \delta_i^s$.

By Propositions 1, 2, 3 there are three families of connections 1-dependent on S_L. For sake of simplicity, in order to investigate $d\omega_L$-properties of these connections we use the Legendre transformation of the Langrangian L.

Every π_1^3-semibasic form $\omega = a_i dx_0^i + b_i dx_1^i$ on $T^3 M$ determines a morphism $\varphi_\omega : T^3 M \to T^* TM$ over id_{TM} such that $\varphi_\omega(h)$ is a form on

TM such that $\varphi_\omega(h)(Y) = \omega(X)$, $X \in T_h T^3 M$, $T\pi_1^3(X) = Y$. If $(x_0^i, x_1^i, z_i^0, z_i^1)$ is a chart on T^*TM then the expression of φ_ω is

$$z_i^0 = a_i, \qquad z_i^1 = b_i.$$

There are three canonical submersions on T^*TM:

$$\pi : T^*TM \to TM, \qquad (x_0^i, x_1^i, z_i^0, z_1^0) \mapsto (x_0^i, x_1^i)$$

$$p_M \cdot \pi : T^*TM \to M, \qquad (x_0^i, x_1^i, z_i^0, z_i^1) \mapsto (x_0^i)$$

$$\xi : T^*TM \to (VTM)^*, \qquad \xi(z)(Y) = z(Y),$$

$$Y \in VTM, \qquad (x_0^i, x_1^i, z_i^0, z_i^1) \mapsto (x_0^i, x_1^i, z_i^1).$$

Let $\lambda = z_i^0 \, dx_i^0 + z_i^1 \, dx_i^1$ be the Liouville 1-form on T^*TM, that is $\lambda(X) = z(T\pi X)$, $x \in T_z(T^*TM)$.

According to the submersions ξ, π, $p_M \circ \pi$ there are three types of connections on T^*TM as follows

a) The coordinate expression of a connection $^2\overline{\Gamma}$ on (ξ) is $dz_i^0 = \Gamma_{ij}^0 \, dx_0^j + \Gamma_{ij}^1 \, dx_1^j + \Gamma_i^j \, dz_j^1$. Then it is easy to check that by the equation $dx_1^i = -\Gamma_j^i \, dx_0^j$, $dz_i^1 = \Gamma_{ji}^1 \, dx_0^j$, $dz_i^0 = \Gamma_{ji}^0 \, dx_0^j$ is given the connection $\mathrm{Orth}_{d\lambda}(^2\overline{\Gamma})$ which is $d\lambda$-orthogonal to $^2\overline{\Gamma}$, i.e. $d\lambda(X,Y) = 0$ for any $^2\overline{\Gamma}$-horizontal tangent vector X and any $\mathrm{Orth}_{d\lambda}(^2\overline{\Gamma})$-horizontal vector Y. Therefore $\mathrm{Orth}_{d\lambda}(^2\overline{\Gamma}) \subset^2 \overline{\Gamma}$, (it means that every $\mathrm{Orth}_{d\lambda}(^2\overline{\Gamma})$-horizontal vector is $^2\overline{\Gamma}$-horizontal as well). if and only if

$$(3) \qquad \Gamma_{ij}^0 - \Gamma_{is}^1 \Gamma_j^s = \Gamma_{ji}^0 - \Gamma_{js}^1 \Gamma_i^s.$$

b) If $dz_i^1 = {}^0F_{ij} \, dx_0^j + {}^1F_{ij} \, dx_1^j$, $dz_i^0 = {}^0G_{ij} \, dx_0^j + {}^1G_{ij} \, dx_1^j$ are the equations of a connection $^1\overline{\Gamma}$ on (π) then $dz_i^1 = {}^1G_{ji} \, dx_0^j + {}^1F_{ji} \, dx_1^j$, $dz_i^0 = {}^0G_{ji} \, dx_0^j + {}^0F_{ji} \, dx_1^j$ is the expression for $\mathrm{Orth}_{d\lambda}(^1\overline{\Gamma})$. Consequently the equation

$$(4) \qquad {}^0F_{ij} = {}^1G_{ji}, \qquad {}^1F_{ij} = {}^1F_{ji}, \qquad {}^0G_{ij} = {}^0G_{ji}$$

is a sufficient and necessary condition for $^1\overline{\Gamma}$ to be Lagrangian, that is, $^1\overline{\Gamma} = \mathrm{Orth}_{d\lambda}(^1\overline{\Gamma})$.

c) Finally, for the sake of completion we introduce, (see the a,-case), that if $dx_1^i = H_j^i \, dx_0^j$, $dz_i^0 = H_{ij}^0 \, dx_0^j$, $dz_i^1 = H_{ij}^1 \, dx_0^j$ is the expression of a connection $^0\overline{\Gamma}$ on $(p_M \cdot \pi)$ then $dz_j^0 = H_{ij}^0 \, dx_0^i + H_{ij}^1 \, dx_1^i - H_j^1 \, dz_i^1$ is the equation of $\mathrm{Orth}_{d\lambda}(^0\overline{\Gamma})$ and $^0\overline{\Gamma} \subset \mathrm{Orth}_{d\lambda}(^0\overline{\Gamma})$ if and only if $H_{ij}^0 + H_{is}^1 H_j^s = H_{ji}^0 + H_{js}^1 H_i^s$.

In the case of the regular Lagrangian L of second order the form ω_L on $T^3 M$ is π_1^3-semibasic and the T-prolongation of the Legendre transformation φ_{ω_L} is of the form

$$z_i^1 = -L_{i_2}, \quad z_i^0 = L_{i_2 k_p} x_{p+1}^k - L_{x_1^i}, \quad dz_i^1 = -L_{i_2 k_p} \, dx_p^k$$

$$dz_i^0 = \left(L_{i_2 k_p j_0} x_{p+1}^k - L_{i_1 j_0} \right) dx_0^j + \left(L_{i_2 k_p j_1} x_{p+1}^k + L_{i_2 j_0} - L_{i_1 j_1} \right) dx_1^j +$$
$$+ \left(L_{i_2 k_p j_2} x_{p+1}^k + L_{i_2 j_1} - L_{i_1 j_2} \right) dx_2^j + L_{i_2 j_2} \, dx_3^j,$$

where $p = 0, 1, 2$. It gives

Lemma. *If L is a regular Lagrangian then the Legendre transformation is a local symplectic isomorphism of the symplectic spaces $(T^3 M, d\omega_{L_3})$ and $(T * TM, d\lambda)$ according to all couples (π_2^3, ξ), (π_1^3, π), $(\pi_0^3, p_M \cdot \pi)$ of the canonical submersions.*

Let $^2\overline{\Gamma} \in {}^2\gamma$ be a connection on (π_2^3) 1-dependent on the mechanical system S_L. Denote $^2\overline{\Gamma}$ the image of $^2\Gamma$ under the Legendre transformation. Let Γ_{ij}^0, Γ_{ij}^1, Γ_i^q be the Christoffels of $^2\overline{\Gamma}$. It is possible to show that (3) is not satisfied, in general. If we calculate the Christoffels H_j^i, H_{ij}^0, H_{ij}^1 of the connection $^0\overline{\Gamma} = \varphi_{\omega_L}(^0\Gamma)$, where $^0\Gamma \in {}^0\gamma$ is a connection on (π_0^3) 1-dependent on S_L, we get that $\Gamma_{ij}^1 \neq H_{ji}^1$. Because of it $\mathrm{Orth}\,^2\overline{\Gamma} \notin \varphi_{\omega_L}(^0\gamma)$. Recall that the difference of two connections on $T^*TM \to M$ is a $V(p_M \circ \pi)$-value $p_M \circ \pi$-semibasic 1-form. Since $-\Gamma_j^i = \frac{1}{4} b_{j_3}^i + c_1 \delta_j^i$, $H_j^i = \frac{1}{4} b_{j_3}^i + \overline{c}_1 \delta_j^i$ the difference $\mathrm{Orth}_{d\lambda}\,{}^2\Gamma - \varphi_{\omega_L}(^0\Gamma)$ is a $V\pi$-value $p_M \circ \pi$-semibasic iff $c_1 = \overline{c}_1$. We can formulate

Proposition 4. *Let $^0\gamma$ be the family of all connections on (π_0^3) or (π_2^3) which are 1-dependent on S_L. Let $^2\Gamma \in {}^2\gamma$. Then, in general, $\mathrm{Orth}_{d\omega_L}(^2\Gamma) \notin {}^0\gamma$. If $^2\gamma$ is determined by parameters c_1, c_2, c_3 then for any $^0\Gamma \in {}^0\gamma$ with parameters $\overline{c}_1 = c_1, \overline{c}_2, \overline{c}_3$ the difference $\mathrm{Orth}_{d\omega_L}(^2\Gamma) - {}^0\Gamma$ is a π_0^3-semibasic vector 1-form with values in $(V\pi_1^3)$. In general, $\mathrm{Orth}_{d\omega_L}\,{}^2\Gamma \not\subset {}^2\Gamma$.*

Remark 3. In the case of a regular Lagrangian of first order, the constructions of the system $^0\gamma$ and $^2\gamma$ give the same 1-parameter family of connections Γ on TM. As a consequence of Proposition 4 it holds that Γ is a Lagrangian.

Quite analogously as above, using (4) it can be deduced

Proposition 5. *A connection $^1\Gamma \in {}^1\gamma$ is not Lagrangian, in general. The difference $^1\Gamma - \mathrm{Orth}_{d\omega}(^1\Gamma)$ is $(V\pi_1^3)$-value π_1^3-semibasic form such that its restriction on VTM is a $(V\pi_2^3)$-value form.*

References

[1] M. CRAMPIN, Alternative Lagrangians in particle dynamics, Proc. Conf. *Diff. Geom. and Its Appl. Brno 1986* published by the J. E. Purkyně University, Brno (1987) 1–12.

[2] A. DEKRÉT, Mechanical structures and connections, Proc. Conf. Diff. Geom., Dubrovnik (1988) to appear.

[3] C. GODBILLON, Géometrie différentielle et mécanique analytic, Paris (1969).

[4] J. KLEIN, Espaces variationnels et mécanique, *Ann. Inst. Fourier*, **12** (1969) 1–124.

[5] J. KLEIN, Almost symplectic structures in dynamics, Proc. Conf. Diff. Geom. and Its Appl. Brno 1986 published by the J. E. Purkyně University, Brno (1987) 79–90.

[6] I. KOLÁŘ, Some natural operators in differential geometry, Proc. Conf. Diff. Geom. and Its Appl. Brno 1986 published by the J. E. Purkyně University, Brno (1987) 80–110.

[7] D. KRUPKA, Elementary theory of differential invariants, *Arch. Math.*, **4**, SCRIPTA Fac. Sci. Nat. UJEP Brunensis, XIV, (1978) 207–214.

[8] M. DE LEÓN and P. R. RODRIGUES, Generalized Classical Mechanics and Field Theory, *North-Holland Mathematics Studies*, Amsterdam, **112** (1985) 290 p.

[9] R. S. PALAIS and C. L. TERNG, Natural bundles have finite order, *Topology*, **16** (1977) 271–277.

Anton Dekrét

Department of Mathematics
and Physics
VŠLD, Maxova 24
960 53 Zvolen
Czechoslovakia

On some Construction of Harmonic Maps between Pseudo-Riemannian Manifolds

L. DI TERLIZZI and J. J. KONDERAK

§ **1.** In the present paper we shall use the Hopf construction to give some new examples of harmonic maps between pseudo-Riemannian manifolds. It was first observed by R. T. Smith [S] that the Hopf construction leads to harmonic maps.

Let (M, g), (N, h) be pseudo-Riemannian manifolds and let $f : M \to N$ be a smooth map. Then df is a section of the bundle

$$T^* M \otimes f^{-1}(TN).$$

This bundle is equipped with the linear connection ∇ induced by the Levi–Civita connections on M and N. The section (∇df) of

$$T^* M \otimes f^{-1}(TN)$$

is called *the second fundamental form of* f. Then there is the section

$$\tau(f) := Trace_g \nabla df$$

of the pull-back bundle $f^{-1}(TN)$; this section $\tau(f)$ is called the *tension field of* f. It is said that f is harmonic iff

$$\tau(f)(x) = 0$$

for all points $x \in M$. For a review of general properties of harmonic maps and techniques which are used in that theory look to [EL1], [EL2].

§ **2.** We denote by $\mathbf{R}^{p,q}$ the $(p+q)$-dimensional vector space with the scalar product $< , >$ such that

$$< (x_1, \ldots, x_p, x_{p+1} \ldots, x_{p+q}), (y_1, \ldots, y_p, y_{p+1} \ldots, y_{p+q}) >=$$

$$= -\sum_{\alpha=1}^{p} x_\alpha y_\alpha + \sum_{\alpha=p+1}^{p+q} x_\alpha y_\alpha.$$

Let $\nu_1, \nu_2 \in \{-1, 1\}$ then we define:

$$\Sigma(\nu_1) : = \{x \in \mathbf{R}^{p,q} \,|\, < x, x >= \nu_1\}$$
$$\Xi(\nu_2) : = \{x \in \mathbf{R}^{s,t} \,|\, < x, x >= \nu_2\}$$

where p, q, s, t are positive integers.

Then the space $\Sigma(\nu_1)$ is a pseudo-Riemannian manifold of signature

$$\begin{cases} (p-1, q) & \text{if} \quad \nu_1 = -1; \\ (p, q-1) & \text{if} \quad \nu_1 = 1 \end{cases}$$

and has constant sectional curvature equal to ν_1. The space $\Sigma(1)$ is called the pseudosphere and $\Sigma(-1)$ is called the pseudo-Riemannian hyperbolic space. They will be denoted by $\mathbf{S}^{p,q-1}$ and $\mathbf{H}^{p-1,q}$ respectively. We have the following diffeomorphisms of manifolds

$$\mathbf{S}^{p,q-1} \cong \mathbf{R}^p \times \mathbf{S}^{q-1}$$
$$\mathbf{H}^{p-1,q} \cong \mathbf{S}^{p-1} \times \mathbf{R}^q.$$

Hence $\mathbf{S}^{p,q-1}$ is connected iff $q \geq 2$ and $\mathbf{H}^{p-1,q}$ is connected iff $q \geq 2$. Otherwise those manifolds have two components.

The properties of $\Xi(\nu_2)$ may be described in the same way as of $\Sigma(\nu_1)$. Namely we have that

$$\Xi(1) = \mathbf{S}^{s,t-1}$$
$$\Xi(-1) = \mathbf{H}^{s-1,t}$$

and they are regular submanifolds of $\mathbf{R}^{s,t}$ etc.

We shall construct here harmonic maps between the spaces of the above type (cf. §[6]).

§ **3.** Let

$$w = (w^1, \ldots, w^{s+t})$$

denote a function

$$w : \mathbf{R}^{p,q} \to \mathbf{R}^{s,t}$$

such that w^α is a k-homogenous polynomial for all $\alpha = 1, \ldots, (s+t)$. Then we have the following

Proposition 3.1. *If all w^α are harmonic functions $(\alpha = 1, \ldots, s+t)$ and $w(\Sigma(\nu_1)) \subset \Xi(\nu_2)$, then*

$$w|_{\Sigma(\nu_1)} : \Sigma(\nu_1) \to \Xi(\nu_2)$$

is a harmonic map.

For a proof of this Proposition look [K2] (Corollary I.3.7.) and [B].

Remark 3.2. In Proposition 3.1. the harmonic functions on $\mathbf{R}^{p,q}$ are functions f such that

$$\triangle^{p,q}(f) = -\sum_{a=1}^{p} \frac{\partial^2 f}{\partial x_a^2} + \sum_{a=p+1}^{m} \frac{\partial^2 f}{\partial x_a^2} = 0.$$

In this formula $\triangle^{p,q}$ denotes the Laplacian on $\mathbf{R}^{p,q}$ induced by the indefinite metric. This Laplacian may be obtained as a tension field associated with the indefinite metric of $\mathbf{R}^{p,q}$.

§ **4.** We consider an algebra A over $\mathbb{R}$ of finite dimension with a multiplicative unit 1 and equipped with a bilinear symmetric nondegenerated product $< , >$ such that

$$< xy, xy > = < x, x >< y, y >$$

for all $x, y \in A$. Such an algebra we call *a seminormed algebra* (cf. [HL]). The algebra A may be nonassociative.

For a given seminormed algebra A we have an orthogonal decomposition

$$A = (\mathbb{R} \cdot 1) \oplus (\mathbb{R} \cdot 1)^\perp.$$

Hence for each $x \in A$ we have a decomposition $x = x_1 + x_2 \in (\mathbb{R}\cdot 1) \oplus (\mathbb{R} \cdot 1)^{\perp}$ and we denote $\operatorname{Re} x = x_1$ and $\operatorname{Im} x = x_2$. There is naturally defined conjugation in A, namely

$$\overline{x_1 + x_2} = x_1 - x_2.$$

If we demand that A has the scalar product $< , >$ positively defined then by the classical Hurwitz theorem we get that A is isometrically isomorphic to $\mathbf{R}$, $\mathbf{C}$, $\mathbf{H}$ or $\mathbf{O}$.

In the case when $< , >$ is nonpositively defined the classification of seminormed algebras may be done using Clifford algebras (cf. [K1]). By $\mathbf{R}_{p,q}$ we shall denote the universal Clifford algebra associated with $\mathbf{R}_{p,q}$ (cf. [P]). The algebra $\mathbf{R}_{p,q}$ may be defined as an associative algebra over $\mathbb{R}$ with unit which has multiplicative generators $e_1, \ldots, e_{p+q}$ such that

$$e_1^2 = \ldots = e_p^2 = -1,$$
$$e_{p+1}^2 = \ldots = e_{p+q}^2 = 1$$

and

$$e_\alpha e_\beta = -e_\beta e_\alpha$$

for $\beta \neq \alpha$. A linear basis of $\mathbf{R}_{p,q}$ consists of elements

$$\{e_{i_1} \cdot \ldots \cdot e_{i_r} \mid \ 1 \le i_1 < \ldots < i_r \le p+q, \qquad 0 \le r \le p+q\}.$$

Let us notice that the inclusion map

$$\mathbf{u} : \mathbf{R}^{p,q} \to \mathbf{R}_{p,q}$$

such that

$$\mathbf{u}(x_1, \ldots, x_{p+q}) = \sum_{\alpha=1}^{p+q} x_\alpha e_\alpha$$

has the property that

$$(\mathbf{u}(x_1, \ldots, x_{p+q}))^2 = < \mathbf{u}(x_1, \ldots, x_{p+q}), \ \mathbf{u}(x_1, \ldots, x_{p+q}) > \cdot \mathbf{1}$$

where $\mathbf{1}$ is the unit of $\mathbf{R}_{p,q}$. In the Clifford algebra there is defined a linear operation of conjugation in the following way:

$$\overline{e_{i_1} \cdot \ldots \cdot e_{i_r}} = (-e_{i_r}) \cdot \ldots \cdot (-e_{i_1}).$$

Having conjugation we may try to define an inner product in the algebra $\mathbf{R}_{p,q}$ in the following way: $g : \mathbf{R}_{p,q} \times \mathbf{R}_{p,q} \to \mathbf{R}_{p,q}$,

$$g(x,y) := \frac{1}{2}(x\bar{y} + y\bar{x}).$$

This formula gives a real scalar product if and only if $p + q \le 2$.

If A is a seminormed algebra with the scalar product $<\,,>$ nonpositively defined then A is isometrically isometric to one of the following algebras:

$$\mathbf{R}_{0,1}, \; \mathbf{R}_{0,2} \quad \text{or} \quad \mathbf{R}_{0,2} \oplus \mathbf{R}_{0,2}\varepsilon_1.$$

(cf. [K1]). By

$$\mathbf{R}_{0,2} \oplus \mathbf{R}_{0,2}\varepsilon_1$$

we denote a seminormed algebra obtained from $\mathbf{R}_{0,2}$ via the Cayley–Dickson process. The product, conjugation and scalar product in $\mathbf{R}_{0,2} \oplus \mathbf{R}_{0,2}\varepsilon_1$ are defined as follows:

$$(a + b\varepsilon_1) \cdot (c + d\varepsilon_1) = (ac + \bar{d}b) + (da + b\bar{c})\varepsilon_1$$

$$\overline{(a + b\varepsilon_1)} = (\bar{a} - b\varepsilon_1)$$

$$< (a + b\varepsilon_1), (c + d\varepsilon_1) > = \frac{1}{2}\left((a + b\varepsilon_1)\overline{(c + d\varepsilon_1)} + (c + d\varepsilon_1)\overline{(a + b\varepsilon_1)}\right).$$

§ 5. Let A be a seminormed algebra which is isometric to $\mathbf{R}^{p,q}$ then the function

$$w : A \times A \to \mathbf{R} \times A$$

defined as

$$w(x,y) = (< x, x > - < y, y >, 2x\bar{y})$$

consists of $p + q + 1$ harmonic polynomials (cf. [K2]).

We equip the space $\mathbf{R} \times A$ with the nondegenerated scalar product such that

$$\ll (t, x), (\tau, y) \gg = t\tau + < x, y > .$$

Then with respect to this new scalar product in $\mathbf{R} \times A$ we have that

$$w(\Sigma(\nu)) \subset \Xi(1)$$

for all $\nu = \pm 1$. It is because

$$\ll w(x,y), w(x,y) \gg = \ll (< x, x > - < y, y >, 2x\bar{y}),$$
$$(< x, x > - < y, y >, 2x\bar{y}) \gg$$
$$= (< x, x > - < y, y >)^2 + 4 < x\bar{y}, x\bar{y} >$$
$$= (< x, x > - < y, y >)^2 + 4 < x, x >< y, y >$$
$$= 1.$$

Hence by Proposition 3.1. we get that

$$w|_{\Sigma(\nu)} : \Sigma(\nu) \to \Xi(1)$$

is a harmonic map.

§ **6.** In the last section of this paper we shall give some examples. In [K2] we used the seminormed algebras $\mathbf{R}_{0,1}$, $\mathbf{R}_{0,2}$ to obtain harmonic maps. Now we shall exploit the algebra $\mathbf{R}_{0,2} \oplus \mathbf{R}_{0,2} \varepsilon_1$. Let us observe that as a vector space this algebra is isometric to $\mathbf{R}^{4,4}$. Let $x, y, z \in \mathbf{R}_{0,2} \oplus \mathbf{R}_{0,2} \varepsilon_1$.

Example 6.1. We have the following harmonic maps:

$$f_1 : \mathbf{S}^{8,7} \to \mathbf{S}^{4,4}$$
$$f_2 : \mathbf{H}^{7,8} \to \mathbf{S}^{4,4}$$

where

$$f_1(x,y) = f_2(x,y) = (< x, x > - < y, y >, 2x\bar{y}).$$

Example 6.2. There is the following pseudo-Riemannian generalization of the Veronese map:

$$f_3 : \mathbf{S}^{12,11} \to \mathbf{S}^{12,13}$$
$$f_4 : \mathbf{H}^{11,12} \to \mathbf{S}^{12,13}$$

where

$$f_3(x,y,z) = f_4(x,y,z)$$
$$= \sqrt{3} \left(x\bar{y}, x\bar{z}, y\bar{z}, \frac{1}{2}(x\bar{x} - y\bar{y}), \frac{1}{2\sqrt{3}}(x\bar{x} + y\bar{y} - 2z\bar{z}) \right).$$

f_3, f_4 are harmonic because they are restrictions of functions which are harmonic 2-homogeneous polynomials.

Example 6.3. Let

$$f_5 : \mathbf{S}^{8,7} \to \mathbf{S}^{12,11}$$
$$f_6 : \mathbf{H}^{7,8} \to \mathbf{S}^{12,11}$$

where

$$f_5(x,y) = f_6(x,y) = (x^2, y^2, \sqrt{2}x\overline{y}).$$

This map is also a 2-homogenous harmonic polynomial which preserves pseudo-Riemannian space forms, respectively, hence f_5, f_6 are harmonic maps.

References

[B] P. BAIRD, Harmonic maps with symmetries, harmonic morphisms, and deformation of metrics, *Research Notes in Math.* No. **87**, Pitman (1983).

[EL1] J. EELLS and L. LEMAIRE, Selected topics in harmonic maps, CBMS Regional Conf. Series, No. **50**, (1983).

[EL2] J. EELLS and L. LEMAIRE, Another report on harmonic maps, *Bull. London Math. Soc.* No. **86**, Vol. 20, Part 5 (1988), p. 385–524.

[HL] R. HARVEY and H. B. LAWSON, Calibrated geometries, *Acta Math.*, **148** (1982), 47–157.

[K1] J. J. KONDERAK, Hurwitz theorem for seminormed algebras, to appear in Atti del. Sem. Mat. e Fisico, Univ. di Modena (1991).

[K2] J. J. KONDERAK, Constructions of harmonic maps between pseudo-Riemannian spheres and hyperbolic spaces, *Proc. Amer. Math. Soc.*, Vol. 109, No. **2**, June 1990.

[P] I. R. PORTEUS, Topological Geometry, Cambridge Univ. Press, Cambridge–London–N.York (1981).

[R] A. RATTO, Equvariant harmonic maps between manifolds with metrics of (p,q) signature, preprint (1988).

[S] R. T. SMITH, Harmonic maps of spheres, Thesis, Warwick University, (1972).

Luigia Di Terlizzi

Dipartimento di Matematica,
Università degli Studi di Bari,
Via G. Fortunato,
70125 Bari, ITALY

Jerzy J. Konderak

Instituto di Matematica,
Facoltà di Scienze
Università di Salerno,
84100 Salerno, ITALY

COLLOQUIA MATHEMATICA SOCIETATIS JÁNOS BOLYAI
56. DIFFERENTIAL GEOMETRY, EGER (HUNGARY), 1989

Metric Fibrations of Lobachevsky–Bolyai Space[*]

V. A. EFREMOVIČ and E. M. GORELIK

1. What was the first (historically) topological fibration? Hopf constructed a fibration of 3-sphere. And, about fifty years before Clifford had constructed the fibration of the elliptic space into so-called Cliffordian parallels, and this was practically just the same construction.

The concept of topological fibration is a generalization of those two constructions. But being topological one it can't keep their metric properties: isometry and parallelness of different fibers. So, V. A. Efremovič has defined metric fibration of a metric space.

2. Definition 1. *Let M be a metric space, $A, B \subset M$. We call A and B parallel $(A \| B)$ if*

$$\qquad 1) \qquad \forall\, a \in A \qquad \exists!\, a^* \in B : aa^* = d(A, B)$$
$$\text{and} \quad 2) \qquad \forall\, b \in B \qquad \exists!\, b^* \in A : bb^* = d(A, B),$$

where $d(A, B)$ is the distance between A and B (see [1]).

(It means that for each point $a \in A$ there exists the unique nearest point $a^* \in B$ and the distance aa^* is constant; and the same holds for each point $b \in B$).

---

[*] This paper is in final form and no version of it will be submitted for publication elsewhere.

Examples:

1) Parallel straight lines in Euclidean geometry are parallel under this definition.

2) Parallel planes in Euclidean space are parallel, too.

3) Two concentric circles are parallel, but a circle A and the one-point set B of the centre of A are not parallel, because for $b \in B$ the nearest $b^* \in A$ is not unique.

4) An axis of a cylinder and a screw line on this cylinder are parallel.

5) Different fibers of Hopf's fibration are parallel with the only exception: two fibers which are the farthest ones from each other are not parallel.

6) Just the same situation holds for Cliffordian parallels in elliptic space.

7) Parallel straight lines in Lobachevsky–Bolyai plane H^2 of course are not parallel under our definition, but two concentric horocycles (i.e. horocycles with the same center at the absolute of H^2) are parallel (Figure 1).

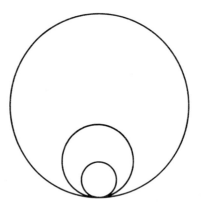

Figure 1.

8) In a similar way, two concentric horospheres in H^3 are parallel.

3. Definition 2. *Let M be an arbitrary metric space which is a union of its subsets, called fibers $M = \bigcup_{F \in B} F$, so that two conditions are satisfied:*

1) *Any two fibers are isometric as subsets of M : $F' \equiv F''$.*

2) *Two fibers, which are sufficiently near, are parallel:*

$$\exists \, c > 0 : d(F', F'') < c \Rightarrow F' \| F''.$$

Then we call this union a *metric fibration of the total space M into the fibers F with the base $\mathcal{B}$* [1]. *(Here base is the set of all the fibers).*

Examples:

1) The fibration of Euclidean space E^n into parallel k-planes $(k = 0, 1, \ldots, n)$, here 0-plane is a point.

2) The fibration of H^2 into the family of concentric horocycles and the fibration of H^n into the family of concentric horospheres of codimension 1.

3) The fibration of a cylinder in E^3 into screw lines.

4) The classical construction of Hopf and Clifford.

Remark. In the examples 1), 2) the constant $c > 0$ from the definition 2 can be choosen arbitrary large, in the examples 3), 4) it is not the case.

4. The classification of metric fibrations of E^n is described in the following theorem.

Theorem 1. *The only metric fibrations of E^n with connected fibers are the unions of parallel k-planes $(k = 0, 1, \ldots, n)$ (see [2]).*

Hyperbolic geometry is much more interesting from this point of view. Even for the plane H^2 the fibration into horocycles is not only one: There are two other fibrations. Let us describe them.

We consider an axis $\ell \subset H^2$ which divides H^2 into two halfplanes: left and right. O_1, O_2 — the ends of ℓ — are points of the absolute. For each point $x \in \ell$ we construct two figures:

1) a half-horocycle in the left half-plane with the center O_1, containing x;

2) the analogous curve in the right half-plane with the center O_2.

The union of these half horocycles is called a *broken horocycle* Γ_x (Figure 2).

Theorem 2. $H^2 = \bigcup\limits_{x \in \ell} \Gamma_x$ *is a metric fibration (see [3]).*

A broken horocycle is not an analytic curve, but it is a C_1-curve.

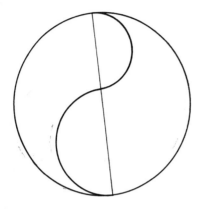

Figure 2.   Figure 3.

For construction the third fibration of H^2 let us fix an axis ℓ and an acute angle α. Let $\mathcal{N}$ be a family of all the straight lines intersecting the axis ℓ in the angle α. Orthogonal trajectory to the family $\mathcal{N}$ is called "the fifth line" (Figure 3). If we denote by D_x the fifth line containing $x \in \ell$, then the following theorem holds.

Theorem 3. *The union $H^2 = \bigcup_{x \in \ell} D_x$ is a metric fibration (see [3]).*

This fibration was constructed by A. G. Vainstein.

V. A. Efremovič called D_x the fifth line because of its importance in hyperbolic geometry (the straight line, the circle, the horocycle and the equidistant are the first four lines).

Opposite to the broken horocycle the fifth line is an analytic curve. In Cartesian coordinates of the first genus ([6]) its equation is:

$$x = \int_0^y \sqrt{\tanh^2 t + \tan^2 \alpha}\, dt.$$

The natural equation of the fifth line is:

$$k = \frac{s}{\sqrt{s^2 + \tan^2 \alpha}},$$

here k is the curvature, s is the natural parameter (the length).

We have a classification theorem

Theorem 4. *There are exactly three non-trivial metric fibrations (that means different from that of one-pointed fibers and of the whole plane) of H^2 with connected fibers as follows:*

1) *the fibration into horocycles,*

2) *the fibration into broken horocycles,*

3) *the fibration into the fifth lines (see [3]).*

The fibers of all these fibrations are topologically equivalent: they are homeomorphic to the straight line. But there are two different uniform classes: 1) horocycle, 2) broken horocycle and the fifth line.

And if we consider a curve completed by its absolute points then we obtain two different topological classes: horocycle becomes a circle, and the two other become segments.

5. In the Lobachevsky–Bolyai space H^3 the fibration into horospheres is not the only fibration. We shall construct another one.

Let us consider a horisontal plane in H^3 and a horocycle Γ with the center O in it is called "the spine". For each axis ℓ of Γ we construct the vertical plane containing ℓ. In that plane there are two horocycles with axis ℓ that intersect Γ: one — with the center O, and the other with the opposite center. If we take the first one then the union of all those vertical horocycles will be the horosphere with the center O (Figure 4). But in the second case the union of all those horocycles is another surface (Figure 5). It is called horotorus Ω_Γ with the spine Γ.

Theorem 5. *The union $H^3 = \bigcup_\Gamma \Omega_\Gamma$ — for all horizontal horocycles Γ with the fixed center O — is a metric fibration of H^3 into horotori. [4]*

6. The horotorus is called so because it is a limit of tori with a fixed point in the shortest parallel and both radii increasing infinitely. In Cartesian coordinates the equation of horotorus is:

$$\exp x = \frac{\cosh y}{\cosh z}.$$

It is asymmetric in y, z — due to asymmetry of the spine and the meridian. The equation of horosphere is symmetric:

$$\exp x = \cosh y \cdot \cosh z.$$

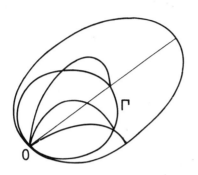

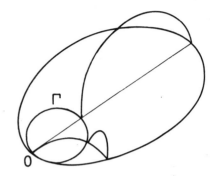

Figure 4.   **Figure 5.**

Horotorus is a Riemannian manifold with the first quadratic form: $ds^2 = d\eta^2 + (1 + \eta^2)^2 \, d\xi^2$, where ξ is the natural coordinate on the spine and η is that one on the meridian.

The group $\mathrm{Iso}(\Omega)$ of isometries of horotorus is isomorphic to $\mathbb{Z}_2 \times \mathrm{Iso}(E^1)$.

For $n \geq 3$ we can construct a metric fibration of H^n into $n-1$-dimensional horotori with k-dimensional horosphere as a spine $(k = 1, \ldots, n-2)$ and $(n-k-1)$-dimensional horosphere as a meridian (we consider horocycle as an 1-horosphere).

7.   Now, let us study the 4-dimensional space $H^2 \times H^2$ — the Euclidean product of two hyperbolic planes. (We call the direct product $Z = X \times Y$ "Euclidean product", if for arbitrary $x_1, x_2 \in X$, $y_1, y_2 \in Y$ the distance between two points of the product $Z : (x_1, y_1)$ and (x_2, y_2) — is equal to $\sqrt{(x_1 x_2)^2 + (y_1 y_2)^2}$, where $x_1 x_2$ and $y_1 y_2$ are distances in factors.) In the first factor H^2 we have horocyclic coordinates x, y with the metric $ds^2 = dy^2 + e^{2y} \, dx^2$; in the second factor these are u, v.

So we have coordinates x, y, u, v for $H^2 \times H^2$ with the metric $ds^2 = e^{2y} \, dx^2 + e^{2v} \, du^2 + dy^2 + dv^2$.

In these coordinates the equation $y = v =: z$ defines a 3-dimensional manifold with the first quadratic form:

$$ds^2 = 2 \, dz^2 + e^{2z} \, (dx^2 + du^2).$$

This is a Lobachevsky–Bolyai space H^3 of curvature $-1/2$.

In a similar way the equation $y - v = c$ defines an embedding of H^3 (with curvature $-1/2$) into $H^2 \times H^2$ we denote it H^3_c for each constant c.

Theorem 6. $H^2 \times H^2 = \bigcup_c H^3_c$ is a metric fibration of $H^2 \times H^2$ (see [4]).

In the same coordinates the equation $y + v = c$, where c is a constant, defines a 3-dimensional manifold with the first quadratic form: $ds^2 = 2\ dz^2 + e^{2z}\ dx^2 + e^{2c}e^{-2z}\ du^2$. It is a *Sol*-manifold from among the 8 homogeneous 3-dimensional manifolds in Thurston classification (see [5]).

Theorem 7. $H^2 \times H^2 = \bigcup_c Sol_c$, where Sol_c is a manifold defined by the equation $y + v = c$, is a metric fibration of $H^2 \times H^2$.

8. The system:
$$\begin{cases} y = a \\ v = b \end{cases}$$
where a, b are constants — defines a 2-dimensional surface $M_{a,b}$ in the above space $H^2 \times H^2$.

Definition. $M_{a,b}$ is called a homogeneous horotorus.

$M_{a,b}$ is a product of two orthogonal horocycles in $H^2 \times H^2$.

Theorem 8. The union $H^2 \times H^2 = \bigcup_{a,b} M_{a,b}$ is a metric fibration of $H^2 \times H^2$ into 2-dimensional fibers: homogeneous horotori (see [4]).

The proof is based on the following. First of all $M_{a,b}$ and $M_{c,d}$ are, of course, isometric. And they are parallel too, because for each point $p = (x, a, u, b)$ the nearest point in $M_{c,d}$ is $p^* = (x, c, u, d)$. That is why the metric is $ds^2 = e^{2y}\ dx^2 + dy^2 + e^{2v}\ du^2 + dv^2$, so the distance between p and arbitrary point $q \in M_{c,d}$ is $\sqrt{(a - c)^2 + (b - d)^2}$ and the equality takes place if and only if $q = p^* = (x, c, u, d)$.

We can also construct H^3_c and Sol_c in such a way.

$$\bigcup_{a-b=c} M_{a,b} = H^3_c \qquad\qquad \bigcup_{a+b=c} M_{a,b} = Sol_c$$

The first of these unions is a metric fibration of H^3 and the second one of *Sol* is that into the same fibers. Hence, H^3 and *Sol* (as subsets of

$H^2 \times H^2$) consist of the same fibers: homogeneous horotori. (In H^3 $M_{a,b}$ is an horosphere).

9. There is an important difference between those two fibrations: for H^3 mapping of one fiber F' to another F'' $x \rightarrow x^*$, where $x^* \in F''$ is the nearest point to $x \in F'$ — satisfies the following condition: "equal distances transform to equal distances" — it means that

$$\forall\, x, y, z, t \in F' : xy = zt \Rightarrow x^* y^* = z^* t^*.$$

And for *Sol* it is not true. So, each isometry of a fiber in H^3 can be extended to the whole H^3 but in *Sol* it is not so.

This is the cause of the difference between their groups of isometries from the point of view of the metric fibrations.

References

[1] V. A. EFREMOVIČ, A. G. VAINSTEIN, E. M. GORELIK and E. A. LOGINOV, Metric fibrations of Riemannian manifolds, Colloquia Math. Soc. J. Bolyai, Budapest, 1979, 195–198.

[2] E. M. GORELIK, Metric fibrations of Euclidean space, Proc. of Yaroslavl University, Yaroslavl, 1980, (Russian).

[3] A. G. VAINSTEIN, E. M. GORELIK and V. A. EFREMOVIČ, On metric fibrations, Proc. of Moscow University Seminar of Vector and Tensor Analysis, Moscow, 1981, 55–59 (Russian).

[4] E. M. GORELIK, Metric fibrations of Lobachevsky space, Dokl. Akad. Nauk SSSR, 301:6 (1988), (Russian).

[5] P. SCOTT, The Geometries of 3-manifolds, *Bulletin of London Math. Soc.* **15** (1983), 401–487.

[6] V. F. KAGAN, The elements of geometry, Gostehizdat, Moscow–Leningrad, 1949.

V. A. Efremovič

E. M. Gorelik

ul. Taskentnaja 4–2–7
109444 Moscow
USSR

COLLOQUIA MATHEMATICA SOCIETATIS JÁNOS BOLYAI

56. DIFFERENTIAL GEOMETRY, EGER (HUNGARY), 1989

Reducibility of G-invariant Linear Connections in Principal G-bundles*

M. FERRARIS, M. FRANCAVIGLIA and L. GATTO

1. Introduction

The present paper is devoted to investigate a nice geometrical problem which arises in view of possible applications to unified theories of gravitation and other physical interactions.

As is well known, essentially two main possibilities were investigated in the early time to generalize the geometric framework of Einstein's theory of gravitation, which was based on a pseudo-Riemannian metric g on a four dimensional space-time M [4]. The first direction (taken by Kaluza [21] and Klein [22] in the hope to unify gravity and electromagnetism) was to allow "more dimensions", i.e., to assume that Physics should be described by a "unified field" taken to be a metric g in a space P of dimension larger than four, suitably projecting onto space-time (M, g); in their work, Kaluza and Klein did in fact assume what in modern geometrical language amounts to say that P is a $U(1)$-bundle over M. The second possible direction (undertaken by Einstein [5] and Eddington [3]) was to allow "more degrees of freedom" but still in four dimensions, i.e., to remain in space-time M but

---

* This paper is in final form and no version of it will be submitted for publication elsewhere.

to choose a more general geometrical object (basically, a linear connection in M) to represent the "unified field". Neither one of these two approaches lead in fact to a satisfactory unified theory of gravity and electromagnetism, so that in early 50's both were almost abandoned by Physicists (see, e.g., [24],[27]).

The interest on principal bundles in Physics was later revived by the rise of so-called "gauge theories", whereby gauge fields (i.e., Yang–Mills fields) are represented by principal connections on a G-bundle P over space-time M, G being a physically meaningful Lie group (e.g., $SU(2)$ or $SU(3)$). This is in fact, especially in its "supersymmetric version" and in connection with so-called "string theory" one of the leading themes in the search of a unifying theory of all physical fields (see, e.g.,[2]).

An interesting result concerning the unification program was recently found by one of us and J. Kijowski ([9]), with the determination of a four-dimensional unified theory of gravitation and electromagnetism, based on a first-order Lagrangian depending on a linear connection with torsion, in exactly the way which was unfruitfully pursued by Einstein, Eddington and Schrödinger [26]. It is possible to show that field equations so obtained are dynamically equivalent, by means of a suitable Legendre transformation, to Einstein equations coupled with Maxwell equations. However, from the very structure of this approach, it seems rather clear that other fundamental interactions cannot be unified to gravity in this four-dimensional framework.

This raises a challenging problem: is it possible to combine the Kaluza–Klein framework and the Ferraris–Kijowski ideas to reproduce, after some suitable Legendre transformation, reasonable physical field equations? More precisely, the hope is to formulate a Lagrangian field theory based on a G-invariant linear connection on a principal G-bundle (where the choice of G will be dictated by physical requirements, so to allow standard Yang–Mills fields) in such a way that the field equations of the given linear connection will suitably generate the unifying scheme searched for. Of course this possibility heavily depends on the choice of the Lagrangian, and for the moment we cannot predict whether any class of such suitable Lagrangians exists. In any case, before discussing "dynamics" one has to check that "kinematics" is reasonable, i.e., one has to investigate under which conditions does a G-invariant linear connection on a principal G-bundle contain (in an intrinsic way) both a principal connection and a linear connection in the base manifold. This has led us to introduce and investigate the notion of *totally vertical (linear) connection* and of *(dimensionally) reducible connection*, which shall hereafter discussed, and to show that, essentially, reducibility from a linear connection in a G-bundle P onto a principal connection is possible (under mild regularity conditions) whenever G is a reducible sub-group of some linear group. This formed the core of the Doctoral Dissertation [18]. A discussion of the case $G = U(1)$ (which is particularly relevant for electromagnetism and classical Kaluza–Klein theory) was already given in [12] and [15]. The general case was discussed in [17], in a coordinate language. This paper will instead be devoted to an intrinsic coordinate-free discussion (although some useful coordinate expressions will be given in the appendixes).

2. Splitting of linear connections

To set-up a purely geometric framework for our discussion we shall first recall some basic facts about "operatorial" decompositions of a linear connection ∇ on a (finite dimensional) manifold P endowed with two complementary smooth distributions.

Let us recall that a (smooth) d-dimensional *distribution* $\mathcal{D}$ on P is a (smooth) map $p \longmapsto \mathcal{D}_p$ which associates a d-dimensional subspace $\mathcal{D}_p \subseteq T_pP$ to each point $p \in P$ (with $1 \le d \le \dim(P)$). A vectorfield $X \in \mathfrak{X}(P)$ is $\mathcal{D}$-compatible if $X(p) \in \mathcal{D}_p$ for all $p \in P$. The subset $\mathfrak{X}_d(P) \subseteq \mathfrak{X}(P)$ of all $\mathcal{D}$-compatible vectorfields is a $\mathcal{F}(P)$-submodule of $\mathfrak{X}(P)$ but in general it is not a Lie-subalgebra, unless $\mathcal{D}$ is *integrable*. A distribution $\mathcal{D}$ defines a vector subbundle $(\mathcal{D}, P, \tau_{P/\mathcal{D}})$ of the tangent bundle (TP, P, τ_P), whose sections coincide with all $\mathcal{D}$-compatible vectorfields on P. It is thence defined a canonical embedding $\iota_\mathcal{D} : \mathfrak{X}_\mathcal{D}(P) \to \mathfrak{X}(P)$, which in turn allows to define *adapted charts* of $\mathcal{D}$ and *adapted bases* of $\mathfrak{X}_\mathcal{D}(P)$.

Let us then consider two *complementary distributions* $\mathcal{H}_1$ and $\mathcal{H}_2$, i.e., such that:

$$(\mathcal{H}_1)_p \oplus (\mathcal{H}_2)_p = T_p(P) \quad , \quad \forall p \in P;$$
$$\dim[(\mathcal{H}_1)_p] = \dim(P) - \dim[(\mathcal{H}_2)_p] = d \quad , \quad \forall p \in P.$$

Given any linear connection ∇ on P, we define as follows its "operatorial" decomposition. Let η_1 and η_2 be the projections of $\mathfrak{X}(P)$ onto $\mathfrak{X}_1 \equiv \mathfrak{X}_{\mathcal{H}_1}(P)$

and $\mathfrak{X}_2 \equiv \mathfrak{X}_{\varkappa_2}(P)$ respectively. Following [7]–[8] we can thence introduce eight kinds of "adapted" operators

$$_k\nabla^{ij} : \mathfrak{X}_i \times \mathfrak{X}_j \longrightarrow \mathfrak{X}_k \qquad (i,j,k=1,2)$$

by setting:

(2.1) $$_k\nabla^{ij}(X,Y) = \eta_k(\nabla_X Y)$$

for all $X \in \mathfrak{X}_i(P)$ and $Y \in \mathfrak{X}_j(P)$. With this definition one has:

(2.2) $$\nabla_X Y = \sum_{k=1}^{2}\left[\sum_{i,j=1}^{2} {_k\nabla^{ij}}(\eta_i(X).\eta_j(Y))\right].$$

It is not difficult to show that $_1\nabla^{11}$ and $_2\nabla^{22}$ behave again as linear connections, while the remaining six "mixed" operators have a tensorial behaviour (see [7]–[8] for further details). This remark will be useful in the sequel.

3. A decomposition theorem for G-invariant linear connections in principal G-bundles

The case which is of interest for us is the following. Let P be the total space of a principal G-bundle over a manifold M, with projection $\pi : P \longrightarrow M$. We set $m = \dim(M)$, $r = \dim(G)$ (so that $\dim(P) = m + r$). As is well known, there is a canonical G-invariant distribution

$$V : p \in P \longmapsto V_p(P) \equiv Ker(T_p\pi) \ , \qquad \forall p \in P \ ,$$

where $V_p(P)$ is called the *vertical space* at $p \in P$. The corresponding compatible vectorfields $\mathfrak{X}_V(P)$ are called the *vertical vectorfields* of P. A *principal connection* on P is a G-invariant distribution complementary to the (canonical) vertical distribution V; we often denote it by $\mathcal{H}_\omega$ and call it the *horizontal distribution* of ω $(T_p P = V_p(P) \oplus (\mathcal{H}_\omega)_p)$. Here ω is the *connection 1- form*, i.e., a 1-form over P with values in the Lie algebra $\mathcal{G}$ of the group G and satisfying the following property

(3.1) $$(R_g)^* \omega = Ad_{g^{-1}} \circ \omega \ , \qquad \forall g \in G \ ,$$

where R_g is the right-multiplication defined on P by the natural action of G,* denotes the pullback and $Ad_{g^{-1}}$ is the adjoint representation of G into its Lie algebra $\mathcal{G}$. In fact, given any such ω the distribution $\mathcal{H}_\omega$ associates to each point $p \in P$ the kernel of ω_p itself. Whenever the context is clear we shall identify $\mathcal{H}_\omega$ with ω.

In particular, for any *linear* connection ∇ in (P, M, G) we shall be able to apply the "operatorial" decomposition discussed above by projecting along the two complementary distributions V and $\mathcal{H}_\omega$, ω being any (arbitrary) connection 1-form. We recall the following definition: *A linear connection ∇ on a principal G-bundle (P, M, G) is G-invariant if and only if the following holds*

$$(3.2) \qquad (R_g)_*(\nabla_X Y) = \nabla_{(R_g)_* X}[(R_g)_* Y] \quad ,$$

for all $g \in G$, $X \in \mathfrak{X}(P)$ and $Y \in \mathfrak{X}(P)$. (Here $(R_g)_*$ denotes the push-forward of R_g). A useful criterion for testing G-invariance of ∇ by looking at its local components with respect to a (local) vectorfield basis will be discussed in the Appendix A.

Let us also recall that linear connections on a manifold P can be thought of as sections of the (affine) bundle $\mathcal{C}_L(P)$ of all linear connections, which is isomorphic to the quotient bundle $J^1 L(P)/Gl(m + r)$ (see [14]). Here $L(P)$ is the principal bundle of linear frames of P, J^1 denotes the first jet-prolongation over P and $Gl(m + r)$ is the linear group (i.e., the structure group of $L(P)$, being $m + r = \dim(P)$). In particular, the natural action of G onto P induces a natural action $\mathcal{R}$ of G onto $\mathcal{C}_L(P)$:

$$\mathcal{R} : \mathcal{C}_L(P) \times G \longrightarrow \mathcal{C}_L(P)$$
$$(\nabla, g) \longmapsto \mathcal{R}_g(\nabla)$$

defined by:

$$(3.3) \qquad [\mathcal{R}_g(\nabla)]_X Y = (R_g)^* \{\nabla_{(R_g)_* X}[(R_g)_* Y]\} \quad .$$

With this definition, we see immediately that: ∇ *is G-invariant iff $\mathcal{R}_g(\nabla)$ coincides with ∇ for all $g \in G$.* Moreover, it will be meaningful to consider the quotient bundle $\mathcal{C}_L(P)/G$, whose sections will therefore represent all G-invariant linear connections on P.

Let us now prove the following:

Theorem 3.1. *The quotient bundle $\mathcal{C}_L(P)/G$ is a fiber bundle over M which admits infinitely many fibered-product decompositions.*

(3.4)

$$\mathcal{C}_L(P)/G \simeq \mathcal{C}_L(M) \times (T \otimes T^* \otimes A^*) \times (T \otimes A^* \otimes T^*) \times (T \otimes A^* \otimes A^*)$$
$$\times (A \otimes T^* \otimes T^*) \times (A \otimes T^* \otimes A^*) \times (A \otimes A^* \otimes T^*) \times (A \otimes A^* \otimes A^*)$$

(where: $\times \equiv \underset{M}{\times}$, $\otimes \equiv \underset{M}{\otimes}$, $A \equiv (VP)/G$ *denotes the adjoint vector bundle of P, $A^* \equiv (V^*P)/G$ is the dual of A, $T \equiv TM$ and $T^* \equiv T^*M$ are the tangent bundle and the cotangent bundle of M respectively).*

Proof.   The strategy of our proof will be the following. We first show that local (smooth) decompositions of this kind always exist at least on trivializing open subsets $U \subseteq M$. We shall then use any arbitrary *principal* connection ω to make global the local construction.

Consider then an open subset $U \subseteq M$, admitting a local section $\sigma : U \to P$. (For our purposes it will not be restrictive to assume that U is the domain of a local trivialization of P). Let $X \in \mathfrak{X}(M)$ be a vectorfield (possibly defined only in U). There exists a G-invariant lift $\widetilde{X}_\sigma \in \mathfrak{X}(\pi^{-1}(U))$ such that:

(3.5) $T_p\pi \circ \widetilde{X}_\sigma(p) = X(\pi(p))$, $\forall p \in \pi^{-1}(U) \subseteq P.$

In fact, let $x \in U$ be a point of $U \subseteq M$. Since G acts on $\pi^{-1}(U) \equiv U \times G$, any point $p \in \pi^{-1}(x)$ in the fiber over x has the form $p = \sigma(x) \cdot g$ for some $g \in G$. We define thence $\widetilde{X}_\sigma(p)$ by setting:

(3.6) $\widetilde{X}_\sigma(p) \equiv \widetilde{X}_\sigma(\sigma(x)g) \equiv T_pR_g \circ T_x\sigma \circ X(x).$

With this definition we see immediately that G-invariance of $\widetilde{X}_\sigma$ is satisfied, as well as smoothness (which follows from the smoothness of R_g and σ). To see that $\widetilde{X}_\sigma$ is a local vector field in P, i.e., a section of $T\pi^{-1}(U)$, it is enough to remark that $\pi \circ R_g = \pi$ and that the following diagram is commutative

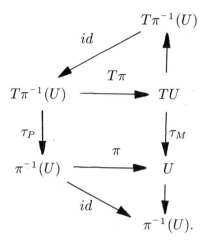

Consider now a further vectorfield $Y \in \mathfrak{X}(M)$ and let $\widetilde{Y}_\sigma$ be its (G-invariant) lift as above. We have thus by definition:

$$T_p\pi \circ \widetilde{X}_\sigma(p) = X(\pi(p)),$$
$$T_p\pi \circ \widetilde{Y}_\sigma(p) = Y(\pi(p)).$$

Therefore, the following definition is meaningful:

(3.7) $\qquad (\widetilde{\nabla}(\sigma)_X Y)(x) = T_{\sigma(x)}\pi \circ (\nabla_{\widetilde{X}_\sigma} \widetilde{Y}_\sigma)(\sigma(x))$

(recall that $\nabla_{\widetilde{X}_\sigma} \widetilde{Y}_\sigma$ is a G-invariant vectorfield in $\pi^{-1}(U)$, since $\nabla, \widetilde{X}_\sigma$ and $\widetilde{Y}_\sigma$ are all G-invariant). We have hence defined a (smooth) operator $\widetilde{\nabla}(\sigma)$ over $U \subseteq M$, which depends on the choice of σ. It is easy to see that $\widetilde{\nabla}(\sigma)$ is in fact a linear connection in $U \subseteq M$, so that $\nabla \longmapsto \widetilde{\nabla}(\sigma)$ defines a projection from $\mathcal{C}_L(\pi^{-1}(U))/G$ onto $\mathcal{C}_L(U)$. Smoothness of this projection is obvious (e.g., it can be easily checked by reverting to local coordinates in the trivialization $\pi^{-1}(U) \simeq U \times G$ and expressing $\widetilde{\nabla}$ in the natural induced coordinates of $J^1L(\pi^{-1}(U)) \simeq J^1L(P)$). In a completely analogous way (which we leave as an exercise to the reader) one can define (σ-dependent) smooth projections from $\mathcal{C}_L(\pi^{-1}(U))/G$ onto (the restrictions to U of) the other factors in the right hand side of equation (3.4). Roughly speaking, in local fibered coordinates $z^A = (x^i, y^\lambda)$ this amounts to split the set of $(m+r)^3$ connection components Γ^A_{BC} ($A, B, C = 1, \ldots, m + r = \dim(P)$) into the m components $\widetilde{\Gamma}^i_{jk}$ of a local linear connection in M and a number of "mixed" terms. Counting the dimensions of $\mathcal{C}_L(P)/G$ and of the product on the right hand side of eqn. (3.4) one immediately checks that the set of all these local projections establishes in fact a *local* bundle isomorphism, *which heavily depends on the choice of σ.*

Let us now remark that the mapping

$$p \longmapsto \mathrm{im}(T_{\pi(p)}\,\sigma)$$

defines, for each local section σ, a local principal connection in $\pi^{-1}(U)$ (since the subspaces are complementary to each $V_p P$ and define a G-invariant distribution). Accordingly, the local procedure we described above extends immediately to the whole of P whenever we fix, instead of a local section σ, a globally defined mapping $\Sigma : TM \to TP$ satisfying the following conditions:

(i) $\dim(\Sigma_p(T_{\pi(p)}\,M)) = m = \dim(M)\,, \quad \forall p \in P \;\;;$

(ii) $p \longmapsto \mathrm{im}(\Sigma_p) \qquad$ is a G-invariant distribution.

As is well known, this is equivalent to assigning a principal connection ω_Σ in P, so that we end up with global projections which all together define, for any given principal connection ω_Σ, a global (non-canonical) bundle isomorphism of the required type. This ends our proof. ∎

At this point it is natural to put forward the following question: *Under which conditions does a G-invariant linear connection ∇ in P decompose canonically according to (3.4)?*

In accordance with our discussion above, the question amounts in turn to establish conditions under which a G-invariant linear connection ∇ in P generates "canonically" a principal connection ω in P. In the rest of the present paper we shall discuss a set of sufficient conditions which allow to select in $\mathcal{C}_L(P)/G$ a family of linear connections which, following a terminology introduced in [20] for G-invariant metrics, will be called *dimensionally reducible connections.*

4. Totally vertical linear connections

Let us then consider a (G-invariant) linear connection ∇ in P and let $X, Y \in \mathfrak{X}_V(P)$ be two vertical vectorfields. In general, the covariant derivative $\nabla_X Y$ will no longer be vertical, so that the following definition makes sense:

Definition. *A linear connection ∇ in a principal G-bundle (P, M, G) is totally vertical iff for all pairs (X, Y) of vertical vectorfields the covariant derivative $\nabla_X Y$ is also a vertical vectorfield.*

Accordingly, a totally vertical linear connection ∇ is characterized by the following property:

$$(4.1) \qquad \langle Z, \nabla_X Y \rangle = 0 \quad, \forall X, Y \in \mathfrak{X}_V(P), \forall Z \in \Omega^1_{\text{hor}}(P)$$

where $\Omega^1_{\text{hor}}(P) \subset \Omega^1(P)$ is the set of all horizontal 1-forms of P. A further way of characterizing totally vertical connections is given by the following remark. Let $p \longmapsto \mathcal{H}(p)$ be any complementary distribution for the vertical distribution $p \longmapsto V_p(P)$ (e.g., the horizontal distribution associated to any principal connection ω of P). According to (2.1) a tensorial operator

$$_1\nabla^{22} : \mathfrak{X}_V(P) \times \mathfrak{X}_V(P) \longrightarrow \mathfrak{X}_{\text{hor}}(P, \mathcal{H})$$

is defined between pairs of sections of $V(P)$ and vectorfields in P which are horizontal with respect to $\mathcal{H}$ (we have taken $\mathcal{H} = \mathcal{H}_1$ and $V = \mathcal{H}_2$). Hence we have: *A (G-invariant) linear connection ∇ is totally vertical if and only if for all principal connections $\mathcal{H}_\omega$ in P the tensorial operator $_1\nabla^{22}$ associated to the splitting $TP = V(P) \oplus \mathcal{H}_\omega$ is identically vanishing.*

Do totally vertical linear connections exist globally? This is certainly the case whenever P is paracompact and G is a compact Lie group. In fact, provided G is compact, one can easily construct a totally vertical local connection in a closed subset $\pi^{-1}(C) = C \times G$ over a closed subset $C \subseteq U \subseteq M$, where U is any trivializing open domain in M. Paracompactness of P thence ensures that such local objects can be glued together over a trivializing atlas to cover the whole of P with a linear connection extending all the local ones. This connection is necessarily a totally vertical connection, since the remark above about the tensoriality of each $_1\nabla^{22}$ ensures that a totally vertical connection is fully characterized by its restrictions.

As an example, let γ be a Riemannian metric on M, $G = S^1$, ϕ be the standard metric of S^1 and ω be a principal connection in a S^1-

bundle (P, M, S^1) with projection $\pi : P \longrightarrow M$. A Kaluza–Klein metric on (P, M, S^1) is any metric of the form:

$$(4.2) \qquad\qquad g = \pi^*(\gamma) + \phi \circ (\omega \times \omega)$$

i.e., any S^1-invariant metric in P. Then it is easy to check that the Levi–Civita connection of each Kaluza–Klein metric is totally vertical.

5. Dimensional Reducibility in $P(M, Gl(n, \mathbb{R}))$.

Let us now fix our attention to the case in which $G = Gl(n, \mathbb{R})$, where n has no a priori relation whatsoever with the dimension m of the basis M. Let us consider a $Gl(n, \mathbb{R})$-bundle P and a principal connection ω. Let us denote by

$$H_\omega : \mathfrak{X}(P) \longrightarrow \mathfrak{X}_{\mathrm{hor}}(P, \omega)$$

and

$$V_\omega : \mathfrak{X}(P) \longrightarrow \mathfrak{X}_V(P)$$

the projections defined by ω, from vectorfields in P to ω-horizontal and vertical vectorfields respectively. We have of course $H_\omega \oplus V_\omega = 1$.

As in Section 2 we define the operator $_2\nabla^{21}$ (assuming $\mathcal{H}_1 \equiv \mathcal{H}_\omega$ and $\mathcal{H}_2 \equiv V$) and we extend it to a mapping

$$_2\nabla^{21} : \mathfrak{X}(P) \times \mathfrak{X}(P) \longrightarrow \mathfrak{X}_V(P)$$

by setting:

$$(5.1) \qquad\qquad _2\nabla^{21} \equiv V_\omega \circ \nabla \circ (V_\omega \times H_\omega).$$

The following proposition shall be proved directly, although it is an immediate consequence of the results recalled in Section 2. Giving a direct proof is in fact instructive for our present purposes.

Proposition 5.1. *The application* $_2\nabla^{21} : \mathfrak{X}(P) \times \mathfrak{X}(P) \longrightarrow \mathfrak{X}_V(P)$ *is* $\mathcal{F}(P)$-*bilinear.*

Proof: (i) Let $f \in \mathcal{F}(P)$. We have:

$$\begin{aligned}
_2\nabla^{21}(f \cdot X, Y) &= V_\omega[\nabla_{V_\omega(f \cdot X)}(H_\omega(Y))] = V_\omega[\nabla_{f \cdot V_\omega(X)}(H_\omega(Y))] \\
&= V_\omega[f \cdot \nabla_{V_\omega(X)}(H_\omega(Y))] = f \cdot V_\omega[\nabla_{V_\omega(X)}(H_\omega(Y))] \\
&= f \cdot {_2\nabla^{21}}(X, Y)
\end{aligned}$$

by $\mathcal{F}(P)$-linearity of V_ω and $\mathcal{F}(P)$-linearity of $Z \longmapsto \nabla_Z$.

(ii) Analogously we have:

$$
\begin{aligned}
{}_2\nabla^{21}(X, f \cdot Y) &= V_\omega[\nabla_{V_\omega(X)}(H_\omega(f \cdot Y))] = V_\omega[\nabla_{V_\omega(X)}(f \cdot H_\omega(Y))] \\
&= V_\omega[f \cdot \nabla_{V_\omega(X)}(H_\omega(Y)) + (V_\omega(X)(f)) \cdot H_\omega(Y)] \\
&= f \cdot V_\omega[\nabla_{V_\omega(X)}(H_\omega(Y))] + (V_\omega(X)(f)) \cdot V_\omega[H_\omega(Y)] \\
&= f \cdot {}_2\nabla^{21}(X, Y)
\end{aligned}
$$

because of $\mathcal{F}(P)$-linearity of both V_ω and H_ω, and being $V_\omega(H_\omega(Y)) \equiv 0$. ∎

The above proposition allows us to consider ${}_2\nabla^{21}$ as operating "point-wise" on tangent vectors, as ${}_2\nabla^{21}(X, Y)$ will just depend on the values of X and Y at $p \in P$. Let us then fix a vectorfield $Y \in \mathfrak{X}(P)$ and let us consider the $\mathcal{F}(P)$-linear application from $\mathfrak{X}_V(P)$ to $\mathfrak{X}_V(P)$ induced by ${}_2\nabla^{21}$, i.e.:

$$
X \in \mathfrak{X}_V(P) \longmapsto {}_2\nabla^{21}(X, Y) \equiv V_\omega(\nabla_X H_\omega(Y)).
$$

Being $H_\omega(Y) \equiv Y - V_\omega(Y)$ we have thus:

$$(5.2) \qquad {}_2\nabla^{21}(X, Y) = V_\omega(\nabla_X Y) - V_\omega(\nabla_X V_\omega(Y)).$$

If ∇ is a totally vertical linear connection eqn. (5.2) simplifies to:

$$(5.3) \qquad {}_2\nabla^{21}(X, Y) = V_\omega(\nabla_X Y) - \nabla_X(V_\omega Y).$$

Notice that the right hand side of (5.3) is linear non-homogeneous in the projection operator V_ω. Accordingly, for all points $p \in P$ it is meaningful to define a linear application

$$
\Lambda_\omega : T_p P \longrightarrow \operatorname{End}(\mathfrak{gl})
$$

where $\mathfrak{gl} \equiv \mathfrak{gl}(n, \mathbb{R})$ is the Lie algebra of $Gl(n, \mathbb{R})$, by setting:

$$(5.4) \qquad \Lambda_\omega(w) = {}_2\nabla^{21}(\cdot, H_\omega(w)) \qquad w \in T_p P$$

and using the (natural) isomorphism between $V_p(P)$ and $gl(n, \mathbb{R})$ (in fact, $\Lambda_\omega(w)$ as defined by (5.4) maps $V_p(P)$ into itself). We should have indicated also the dependence on the point $p \in P$, but for simplicity of notation this will be hereafter omitted without any danger of confusion. This linear mapping $\Lambda_\omega : T_p P \to \mathrm{End}(\mathfrak{gl})$ can be of course interpreted as a point of the tensor product $\mathrm{End}(\mathfrak{gl}) \otimes T_p^* P$. We notice, however, that $\Lambda_\omega(w)$ does not depend on the whole of $w \in T_p P$, but only on its ω-horizontal component $H_\omega(w)$. Accordingly this induces the possibility of interpreting Λ_ω as a "horizontal" 1-form with values in $\mathrm{End}(\mathfrak{gl})$, i.e., a point in the tensor product $\mathrm{End}(\mathfrak{gl}) \otimes \mathrm{Hor}_p(P)$ (where $\mathrm{Hor}_p(P)$ is the set of horizontal 1-forms at $p \in P$). In local coordinates $z^A = (x^i, y^\lambda)$ $(i = 1, \ldots, m \equiv \dim(M)$, $\lambda = 1, \ldots, n^2 \equiv \dim(Gl(n, \mathbb{R}))$ an element θ of $\mathrm{Hor}_p(P)$ has the form $\theta = \theta_i(x^k, y^\mu) dx^i)$. From an enumerative viewpoint, thus, Λ_ω is represented by $m \cdot n^4$ independent components (at each point $p \in P$), being $\dim[\mathrm{End}(\mathfrak{gl})] = n^4$.

Let us now suppose that the principal connection ω varies while the totally vertical $Gl(n, \mathbb{R})$-invariant linear connection ∇ remains fixed. We have thus a family $\omega \longmapsto \Lambda_\omega$ of applications from (each) $T_p P$ into $\mathrm{End}(\mathfrak{gl})$. We aim at choosing ω (or, equivalently, the projection operator V_ω) in such a way that ∇ satisfy a further "verticality condition" which shall be hereafter specified. We first remark that the Lie-algebra $\mathfrak{gl} \equiv \mathfrak{gl}(n, \mathbb{R})$ is isomorphic to $\left(\mathbb{R}^{n^2}\right) \otimes \left(\mathbb{R}^{n^2}\right)^*$ and there exist canonical isomorphisms:

$$\mathrm{End}(\mathfrak{gl}) \simeq \mathfrak{gl} \otimes \mathfrak{gl}^* \simeq \left(\mathbb{R}^{n^2}\right) \otimes \left(\mathbb{R}^{n^2}\right)^* \otimes \left(\mathbb{R}^{n^2}\right)^* \otimes \left(\mathbb{R}^{n^2}\right).$$

We define now a projection operator

$$\Pi : \mathrm{End}(\mathfrak{gl}) \longrightarrow \mathfrak{gl}$$

as the linear map which is uniquely induced by the following multilinear map:

(5.5) $$\Pi : (u, \alpha, \beta, v) \longmapsto \alpha(v) \cdot (u \otimes \beta).$$

It is easily seen that Π is the only (linear) operator which associates the matrix product $A \circ B$ to the point $A \otimes B$ of $\mathrm{End}(\mathfrak{gl})$ with $\mathfrak{gl} \otimes \mathfrak{gl}^*$ and we consider $n \times n$ matrices as elements of $\mathfrak{gl}$ itself.

We can finally give the following definitions:

Definitions 5.2. *A totally vertical $Gl(n, \mathbb{R})$-invariant linear connection ∇ on $(P, M, Gl(n, \mathbb{R}))$ is dimensionally reducible if there exists a principal*

connection ω *such that (for all points* $p \in P$) *the composition* $\Pi \circ \Lambda_\omega$:
$T_p P \to \mathfrak{gl}(n, \mathbb{R})$ *vanishes identically.*

Remark 5.3. Let us consider the following system:

$$(5.6) \qquad \Pi \circ \Lambda_\omega(w) = \Pi \circ {}_2\nabla^{21}(., H_\omega(w)) = \mathbf{0} \in \mathfrak{gl}, \qquad \forall w \in T_p P.$$

From (5.3) and our remarks on Λ_ω as an element of $\text{End}(\mathfrak{gl}) \otimes \text{Hor}_p(P)$ it follows that ${}_2\nabla^{21} = 0$ can be considered as a (inhomogeneous) linear system of dimension $m \cdot n^4$ in the unknown projection operator V_ω, which has instead $m \cdot n^2$ components. This system could of course be incompatible. Applying Π to ${}_2\nabla^{21} = 0$ reduces the dimension from $m \cdot n^4$ to $m \cdot n^2$, so that (5.6) is in fact a system of $m \cdot n^2$ equations in $m \cdot n^2$ unknowns. Hence, if eqn. (5.6) admits a unique solution ω, then the system has to be regular. This justifies the following:

Definition 5.4. *A totally vertical connection in* $(P, M, Gl(n, \mathbb{R}))$ *is regular in an open subset* $U \subseteq M$ *if for all points* $p \in U \subseteq M$ *the affine mapping* $\mathbf{M} : \omega \mapsto \Pi \circ \Lambda_\omega$ *is invertible.*

The definition is meaningful. In fact both ω and $\Pi \circ \Lambda_\omega$ have the same number of independent components ($m \cdot n^2$) at each point $p \in U$, since both can be considered as $\mathfrak{gl}$-valued horizontal 1-forms in P. Hence, the regularity of ∇ can be stated by requiring the condition

$$(5.7) \qquad\qquad\qquad \det(\mathbf{M}') \neq 0$$

where $\mathbf{M}'$ is the linear part of the affine mapping $\omega \mapsto \Pi \circ \Lambda_\omega$ above. Such a regularity condition will be satisfied by all connections ∇ belonging to an open dense subset of $\mathcal{C}_L(P)/G$, provided this space is endowed with a "reasonable" topology (this is true, e.g., in the compact-open topology). The local coordinate expression of this regularity condition will be discussed in the Appendix B.

We can thence conclude this Section by stating the following result, which has been implicitly proved in the above discussion.

Proposition 5.5. *If a totally vertical connection* ∇ *in* $(P, M, Gl(n, \mathbb{R}))$ *is regular in the sense of definition 5.4, then it is dimensionally reducible.*

6. Dimensionally reducible linear connections

Let now (P, M, G) be a principal G-bundle having as structure group a Lie group G of matrices, which we assume to be embedded into a suitable linear group $Gl(n, \mathbb{R})$ for some n (having no a priori relations with the dimensions of P and M). We first recall the following:

Definition 6.1. *Let H be a Lie group and $G \subseteq H$ a Lie subgroup of H. We say that G is reductive into H if there exists a direct sum vector-space decomposition*

$$(6.1) \qquad\qquad\qquad \mathcal{H} = \mathcal{G} \oplus \mathcal{M}$$

of the Lie algebra $\mathcal{H}$ of H, where $\mathcal{G}$ is the Lie algebra of $G \subseteq H$ and $\mathcal{M}$ is a vector subspace of $\mathcal{H}$ invariant under the adjoint representation of G into $\mathcal{G}$ (i.e., $Ad_g(\mathcal{M}) \subseteq \mathcal{M} \; \forall g \in G$).

The scope of this Section is to extend the notion of "dimensional reducibility" discussed in Section 5 to the case of totally vertical linear connections in bundles (P, M, G) having as structure group a reducible subgroup G of some linear group $Gl(n, \mathbb{R})$. To this aim let us first recall the following result, whose proof can be found in [23].

Proposition 6.2. *Let $G \subseteq H$ be a reductive subgroup of H and let (P, M, G) be a principal G-subbundle of a principal H-bundle (Q, M, H). Let $\eta_1 : \mathcal{H} \to \mathcal{G}$ and $\eta_2 : \mathcal{H} \to \mathcal{M}$ be the projections of the direct sum splitting $\mathcal{H} = \mathcal{G} \oplus \mathcal{M}$. If ω is a principal connection in (P, M, H) then its projection $\eta_1(\omega_{|P})$ is a principal connection in (P, M, G).*

We can now proceed as follows. Given a fiber bundle (P, M, G) with G reductive in some $Gl(n, \mathbb{R})$, we shall define a new principal fiber bundle $(Q, M, Gl(n, \mathbb{R}))$ as the associate bundle to P via the natural action of G onto $Gl(n, \mathbb{R})$ induced by the group inclusion. This defines an embedding $i : P \to Q$ which is of course a bundle morphism (in fact a principal bundle reduction) which preserves the transition functions. We shall then assign a principal connection ω in the "larger" bundle Q to generate a $Gl(n, \mathbb{R})$-invariant splitting of the tangent spaces $T_q Q$ into their ω-horizontal and vertical parts.

Considering a connection ω which reduces to P will in turn generate a G-invariant splitting of the tangent spaces T_pP into their vertical and ω-horizontal parts. In this framework one can still define linear operators Λ_ω (at each point $p \in P$) as we did in the previous Section 5, by setting again

$$\Lambda_\omega : T_pP \longrightarrow \text{End}(\mathfrak{gl}),$$

$$\Lambda_\omega = {}_2\nabla^{21}(., H_\omega(w)), \qquad w \in T_pP \simeq T_{i(p)}Q.$$

This allows us to give the following definition:

Definition 6.3. *A totally vertical linear connection ∇ in a G-bundle (P, M, G) is dimensionally reducible (with respect to $Gl(n, \mathbb{R})$) if there exist:*

(i) a linear group $Gl(n, \mathbb{R})$ such that G is one of its reducible subgroups;

(ii) a principal connection ω in $(Q, M, Gl(n, \mathbb{R}))$ defined as above, which reduces to P and such that $\Pi \circ \Lambda_\omega$ vanishes at all points $p \in P$.

With formally the same reasoning as in the previous Section 5, we easily conclude that (for any given bundle Q reducing to P) if a connection ω satisfying (ii) above exists, it is unique, provided the given linear connection ∇ in P satisfies a regularity condition, which also in this case amounts to require the regularity of the linear mapping $\omega \mapsto \Pi \circ \Lambda_\omega$ which is here considered as a mapping from $\mathcal{G}$-valued horizontal 1-forms into the same space, by means of the projection from $\mathfrak{gl}$ onto its subalgebra $\mathcal{G}$. With the obvious modification definition 5.4 carries through, to give following:

Definition 6.4. *A linear connection ∇ in (P, M, G) is regular (with respect to $Gl(n, \mathbb{R})$) if:*

(i) there exists a $Gl(n, \mathbb{R})$ as in definition (6.3);

(ii) the appropriate generalization of condition (5.7) holds for the mapping $\omega \rightarrow \Pi \circ \Lambda_\omega$.

As in section 5, the analogous of proposition 5.5 holds true:

Proposition 6.5. *If ∇ is a totally vertical linear connection regular in the sense of def. 6.4 then it is dimensionally reducible in the sense of def. 6.3.*

At this point, being G a reductive subgroup of the Lie group $Gl(n, \mathbb{R})$, using the above proposition 6.5 and applying the result of proposition 6.2 we have immediately the following:

Theorem 6.6. *Let (P, M, G) be a principal bundle, having as structure group a reductive subgroup G of some linear group $Gl(n, \mathbb{R})$. Let ∇ be a dimensionally reducible linear connection in the sense of definition 6.3 (e.g., a totally vertical and regular G-invariant connection). Let $\omega = \omega(\nabla)$ be the unique principal connection induced by ∇ in the bundle $(Q, M, Gl(n, \mathbb{R}))$ which reduces to (P, M, G). If $\eta_1 : \mathfrak{gl}(n, \mathbb{R}) \to \mathcal{G}$ is the projection from the Lie algebra of $Gl(n, \mathbb{R})$ onto the Lie algebra $\mathcal{G}$ of G, the projection $\widetilde{\omega} \equiv \eta_1(\omega_{|P})$ is a principal connection in (P, M, G).*

Let us remark here once again that the reduction method we rely on and the ensuing "projected connection" $\widetilde{\omega}$ do in principle depend on the choice of the embedding group $Gl(n, \mathbb{R})$ and of the extending bundle $(Q, M, Gl(n, \mathbb{R}))$. An important question to be asked is therefore the following: *to what extent is the reduced connection $\widetilde{\omega}$ uniquely defined?* For the moment we are not able to answer this problem, which is probably hard to be solved completely. We conjecture that $\widetilde{\omega}$ is uniquely defined in each equivalence class of group representations $G \to Gl(n, \mathbb{R})$ and that natural embeddings $Gl(n, \mathbb{R}) \to Gl(n', \mathbb{R})$ $(n < n')$ will not affect the result, but we have no clear ideas how $\widetilde{\omega}$ will change if inequivalent group representations are chosen.

As far as the physical applications are concerned we have however to stress that usually Physics dictates a canonical choice of both the embedding group $Gl(n, \mathbb{R})$ (e.g., this holds for all $G = SU(r)$), so that in these cases a kind of "*canonical projected connection*" will be at hands.

Appendix A. Local representations

We first discuss a useful local criterion to check whether a given linear connection ∇ in a principal G-bundle (P, M, G) is G-invariant. We shall make extensive use of the notation and techniques introduced by us in [16].

Let $U \subseteq M$ be a trivializing open subset and $V \subseteq \pi^{-1}(U) \subseteq P$ be a local chart. Coordinates in V will be denoted by $z^A = (x^i, y^\lambda)$, with $A = 1, \ldots, \dim(P) = m + r$, $i = 1, \ldots, m = \dim(M)$ and $\lambda = 1, \ldots, r = \dim(G)$. Let $\partial_A \equiv (\partial_i, \partial_\lambda)$ the corresponding natural basis in $\mathfrak{X}(V)$. The fields ∂_i are G-invariant $((R_g)_* \partial_i = \partial_i)$ under the right action of G onto P, while the remaining ones ∂_λ are not. A right-invariant *local* basis (r_λ) for $\mathfrak{X}_V(\pi^{-1}(U))$ can be constructed as the image of a basis (e_λ) of the Lie algebra $\mathcal{G} \simeq V_p P$

under the left action on $\pi^{-1}(U)$ induced by the right action of G onto P and by the explicit choice of a trivialization (see [26] for details). Hence $(r_A) \equiv (\partial_i, r_\lambda)$ is a right invariant basis in $V \subseteq \pi^{-1}(U)$. Let (ρ^A) be the dual basis of (local) 1-forms over $V \subseteq P$. The connections symbols of ∇ will be the (anholonomic) components:

(A.1) $$\Gamma^A_{BC} \equiv \langle \rho^A | \nabla_{r_C} r_B \rangle$$

whit respect to the right-invariant bases (r_A) and (ρ^A).

We have (see [16] and [18]):

Proposition A.1. *A linear connection ∇ in (P, M, G) is G-invariant if and only if its connection symbols Γ^A_{BC} with respect to the (local) right invariant bases (r_A) of each trivialization are independent on group variables, i.e.:*

(A.2) $$\partial_\mu(\Gamma^A_{BC}) = 0 \quad , \quad \mu = 1, \ldots, r = \dim(G).$$

Let now ω be a principal connection in (P, M, G). The corresponding expression in local right invariant bases has the simple form:

(A.3) $$\omega_{|V} = [A^\mu_i(x^j)\rho^i + \rho^\mu] \otimes \underline{e}_\mu,$$

where $(\underline{e}_\mu)$ is a basis of $\mathcal{G}$ and $A^\mu_i(x^j)$ are the so-called *gauge potentials*. These are local functions

$$A^\mu_i : U \subseteq M \longrightarrow \mathbb{R}$$

satisfying the following transformation rules
(A.4)
$$A^{\mu'}_{i'}(x') = \{[Ad_{m_{(\alpha\beta)}(x)}]^{\mu'}_\mu A^\mu_i(x)\}_{\big|_{x=x(x')}} \chi^i_{i'}(x') + \{[(m_{(\alpha\beta)})^* \theta_L]^{\mu'}_{i'}(x)\}_{\big|_{x=x(x')}}$$

where: Ad_g is the adjoint representation of G into $\text{End}(\mathcal{G})$; $m_{(\alpha\beta)} : U_{(\alpha)} \cap U_{(\beta)} \to G$ are local transition functions in two different trivializations; $\chi^i_{i'}$ is the Jacobian of a local change of coordinates in $U_{(\alpha)} \cap U_{(\beta)} \subseteq M$; $\theta_L : \mathcal{G} \to \mathcal{G} \otimes T_* G$ is the *left invariant canonical 1-form on G* defined by:

(A.5) $$\theta_L(g) = T_g \circ L_{g^{-1}}.$$

Using (A.3) one can easily find "adapted" bases for ω. The most natural one is the following:

(A.6) $$\mathbf{R}_i = r_i - A^\mu_i(x)r_\mu \quad , \quad \mathbf{R}_\mu = r_\mu,$$

with dual co-basis:

$$(A.6')\qquad\qquad \widehat{\mathbf{R}}^i = r^i, \qquad \widehat{\mathbf{R}}^\mu = \rho^\mu + A_i^\mu(x)\rho^i.$$

Denoting by $\widetilde{\Gamma}^A_{BC}$ the connection symbols with respect to the newly defined bases $(\mathbf{R}_A)$ and $(\widehat{\mathbf{R}}^A)$ (i.e., $\widetilde{\Gamma}^A_{BC} \equiv \langle \widehat{\mathbf{R}}^A | \nabla_{\mathbf{R}_C}\mathbf{R}_B \rangle$) one finds the following relations (see [16] and [17] for details):

$$(A.7)\qquad \widetilde{\Gamma}^k_{ij} = \Gamma^k_{ij} - \Gamma^k_{i\sigma} A^\sigma_j - \Gamma^k_{\rho j} A^\rho_i + \Gamma^k_{\rho\sigma} A^\rho_i A^\sigma_j,$$

$$(A.8)\qquad \widetilde{\Gamma}^k_{i\mu} = \Gamma^k_{i\mu} - \Gamma^k_{\rho\mu} A^\rho_i,$$

$$(A.9)\qquad \widetilde{\Gamma}^k_{\beta j} = \Gamma^k_{\beta j} - \Gamma^k_{\beta\sigma} A^\sigma_j,$$

$$(A.10)\qquad \widetilde{\Gamma}^k_{\beta\mu} = \Gamma^k_{\beta\mu},$$

$$(A.11)\qquad \widetilde{\Gamma}^\alpha_{ij} = A^\alpha_k(\Gamma^k_{ij} - \Gamma^k_{i\sigma} A^\sigma_j - \Gamma^k_{\rho j} A^\rho_i + \Gamma^k_{\rho\sigma} A^\rho_i A^\sigma_j)+$$
$$+ \Gamma^\alpha_{ij} - \Gamma^\alpha_{i\sigma} A^\sigma_j - \Gamma^\alpha_{\rho j} A^\rho_i + \Gamma^\alpha_{\rho\sigma} A^\rho_i A^\sigma_j - \partial_j A^\alpha_i,$$

$$(A.12)\qquad \widetilde{\Gamma}^\alpha_{i\mu} = \Gamma^\alpha_{i\mu} + A^\alpha_k\Gamma^k_{i\mu} - \Gamma^\alpha_{\sigma\mu} A^\sigma_i - A^\alpha_k\Gamma^k_{\sigma\mu} A^\sigma_i,$$

$$(A.13)\qquad \widetilde{\Gamma}^\alpha_{\beta j} = \Gamma^\alpha_{\beta j} + A^\alpha_k\Gamma^k_{\beta j} - \Gamma^\alpha_{\beta\sigma} A^\sigma_j - A^\alpha_k\Gamma^k_{\beta\sigma} A^\sigma_j,$$

$$(A.14)\qquad \widetilde{\Gamma}^\alpha_{\beta\mu} = \Gamma^\alpha_{\beta\mu} + A^\alpha_k\Gamma^k_{\beta\mu}.$$

The eight objects locally defined by the right hand sides of (A.7)–(A.14) are nothing but the local representations of the eight operators $_i\nabla^{jk}$ $(i,j,k = 1,2)$ defined in Section 2. As an example, for $_1\nabla^{22}$ we have:

$$_1\nabla^{22}(X,Y) = H_\omega[\nabla_{V_\omega(X)}(V_\omega(Y))] = \langle \widehat{\mathbf{R}}^i | \nabla_{X^\mu\mathbf{R}_\mu}(Y^\nu\mathbf{R}_\nu)\rangle \mathbf{R}_i$$
$$= (\widetilde{\Gamma}^i_{\mu\nu} X^\mu Y^\nu)\mathbf{R}_i$$

(and so on). Therefore: *A linear connection* ∇ *is totally vertical if and only if the following holds*

$$(A.15)\qquad\qquad\qquad \widetilde{\Gamma}^k_{\beta\mu} = 0.$$

(This follows from $_1\nabla^{22}(X,Y) = (\widetilde{\Gamma}^k_{\mu\nu} X^\mu Y^\nu)\mathbf{R}_k = 0$, being X, Y arbitrary and $(\mathbf{R}_k)$ linearly independent). Moreover, using (A.10) and the tensoriality of $_1\nabla^{22}$ (which is manifest when checking the transformation rules for the components $\Gamma^k_{\mu\nu}$) one has also the following: *A linear connection* ∇ *is totally vertical if and only if in one (and hence all) trivializations it is*

$$(A.16)\qquad\qquad\qquad \Gamma^k_{\beta\mu} = 0.$$

Appendix B. Local characterization of dimensionally reducible connections

Let $(P, M, Gl(n, \mathbb{R}))$ be a principal $Gl(n, \mathbb{R})$-bundle. In a trivialization $\pi^{-1}(U)$ we shall use local coordinates $z^A = (x^i, y^\lambda)$, with $A = 1, \ldots, m + n^2 = \dim(P)$, $i = 1, \ldots, m = \dim(M)$ and $\lambda = 1, \ldots, n^2 = \dim(Gl(n, \mathbb{R}))$. Since $Gl(n, \mathbb{R})$ will be considered as a group of matrices, the Greek index λ will be replaced by a couple of Latin indices $\binom{a}{b}$ with $a, b = 1, \ldots, n$ (taken in the first half of the Latin alphabet). To avoid any possible confusion, upper and lower indices of P will be separated by vertical slashes (e.g., if $\lambda = \binom{a}{b}$, Γ^λ_{ij} will be denoted by $\Gamma|^a_b|_{ij}$).

Let ∇ be a linear connection in P. Setting as usual $\Gamma^A_{BC} \equiv \langle \rho^A | \nabla_{r_C} r_B \rangle$ and adopting the aforementioned notation we end up with eight families of connection symbols, i.e.:

(A.7) $\qquad \Gamma^k_{ij}, \Gamma^k_i|^e_f|, \Gamma^k|^c_d|_j, \Gamma^k|^c_d|^e_f|, \Gamma|^a_b|_{ij}, \Gamma|^a_b|_i|^e_f|, \Gamma|^a_b|^c_d|_j, \Gamma|^a_b|^c_d|^e_f|.$

If ∇ is G-invariant the above functions will depend only on the coordinates (x^i) and will be independent on $y^\lambda \equiv y|^a_b|$. The appropriate forms of the adapted basis $\mathbf{R}_A$ and cobasis $\widehat{\mathbf{R}}^A$ are:

$$\mathbf{R}_i = \frac{\partial}{\partial x^i} - A|^a_b|_i(x)\, y|^b_c| \frac{\partial}{\partial y|^a_c|} \quad , \quad \mathbf{R}|^b_a| = y|^b_c| \frac{\partial}{\partial y|^a_c|} \quad ,$$

$$\widehat{\mathbf{R}}^i = dx^i \quad , \quad \widehat{\mathbf{R}}|^a_b| = dy|^a_c|(y^{-1})|^c_b| + A|^a_b|_i(x) dx^i.$$

Let us now assume ∇ is totally vertical, i.e.: $\Gamma^k|^c_d|^e_f| \equiv 0$, in all trivializations. In this hypothesis one has:

(B.1) $\qquad \widetilde{\Gamma}|^a_b|^c_d|_j\, dx^j = \left(\Gamma|^a_b|^c_d|_j + A|^a_b|_k \Gamma^k|^c_d|_j - \Gamma|^a_b|^c_d|^f_e| A|^f_e|_j \right) dx^j.$

This is in fact the appropriate local representation of eqn. (5.6) of Section 5, because of the very definition of $_2\nabla^{21}$ (recall that "2" refers to "vertical" indices while "1" refers to "horizontal" ones). Therefore $\|\widetilde{\Gamma}|^a_b|^c_d|_j\, dx^j\|$ is a matrix of $\mathfrak{gl}(n, \mathbb{R})$-valued 1-forms which represents in fact the operator Λ_ω defined by (5.4). Applying the projection operator $\Pi : \mathrm{End}(\mathfrak{gl}) \to \mathfrak{gl}$ amounts, in our notation, to take the contraction over the two Latin indices b and c. Therefore, the linear non-homogeneous system of equations $\Pi \circ \Lambda_\omega = 0$ is in fact expressed as follows:

(B.2) $\qquad \Gamma|^a_b|^b_d|_j + A|^f_e|_k \left(\delta^a_f \Gamma^k|^e_d|_j - \delta^k_i \Gamma|^a_b|^b_d|^e_f| \right) = 0$

which is in fact a system of mn^2 equations in the mn^2 unknowns $A|_e^f|_k$ $(e, f = 1, \ldots, n$ and $j = 1, \ldots, m)$. Let us set:

(B.3)
$$\mathbf{M}^k|_f^e|_d^a|_j = \delta_f^a \Gamma^k|_d^e|_j - \delta_i^k \Gamma|_b^a|_d^b|_f^e|$$

and consider $\mathbf{M}$ as a $(mn^2 \times mn^2)$ square matrix. The regularity condition (5.7) amounts then to require that the determinant of this matrix be different from zero. Under this hypothesis the linear system

(B.4)
$$\Gamma|_b^a|_d^b|_j + A|_e^f|_k \mathbf{M}^k|_f^e|_d^a|_j = 0$$

admits a unique solution $A|_e^f|_k$, which represents the gauge potentials of the unique ω induced by ∇. As thoroughly discussed in [16] and [17], in fact, the following properties hold:

(a) The condition $\det(\mathbf{M}) \neq 0$ is intrinsic. In fact from (A.7)–(A.14) it follows that the condition is independent from the trivialization.

(b) If $A|_e^f|_k$ and $A|_{e'}^{f'}|_{k'}$ are solutions of (B.4) in two different trivializations, then they are related by the appropriate form of the gauge transformation (A.4).

In the more general case of a reductive subgroup $G \subseteq Gl(n, \mathbb{R})$ the local expressions of the system $\Pi \circ \Lambda_\omega = 0$ and the "reduced" connection $\widetilde{\omega}$ are formally the same as (B.4) above, although now the mn^2 "components" will satisfy further constraints expressing (in the local coordinates $y|_b^a|$ of $Gl(n, \mathbb{R})$) the embedding of G into $Gl(n, \mathbb{R})$ (one can use, in a sense, the functions $y|_b^a|$ as redundant parameters in place of the true group coordinates. See [17] for details).

References

[1] BRILL, D., Kaluza–Klein Theories, in *"Proceedings of the 11th International Conference on General Relativity and Gravitation"*, M.A.H. MacCallum ed.; Cambridge University Press (Cambridge, UK, 1987), 320–325.

[2] DUFF, M. J., Supergravity, Kaluza–Klein and Supersrtings, in *"Proceedings of the 11th International Conference on General Relativity and Gravitation"*, M.A.H. MacCallum ed.;Cambridge University Press (Cambridge, UK, 1987), 18–60.

[3] EDDINGTON, A. S., The Mathematical Theory of Relativity, *Cambridge University Press*, 2nd. ed. (Cambridge, UK, 1924).

[4] EINSTEIN, A., Die Grundlagen der allgemeinen Relatitätstheorie; *Ann. Phys., Lpz.*, **49**, 769 (1916).

[5] EINSTEIN, A., Zur affinen Feldtheorie, *Sitzungsber. Preuss. Akad. Wiss.* (Berlin), 137–140 (1923).

[6] EYRAUD, H., Les équations de la dynamique de l'éther, Blanchard (Paris, 1926).

[7] FAVA, F., Distribuzioni e strutture indotte, *Atti della Academia delle Scienze di Torino* **103**, 93-120 (1969).

[8] FAVA, F., Derivazioni di algebre graduate e connessioni non lineari di ordine qualsiasi, in *"Conferenze del Seminario di Matematica dell'Università di Bari"*, Laterza (1979).

[9] FERRARIS, M. and KIJOWSKI, J., General Relativity is a Gauge Type Theory, *Letters in Mathematical Physics* **5**, 127–135 (1981).

[10] FERRARIS, M. and KIJOWSKI, J., Unified Geometric Theory of Electromagnetic and Gravitational Interactions, *Journal of General Relativity and Gravitation* **14** (1), 37–47 (1982).

[11] FERRARIS, M., Affine Unified Theories of Gravitation and Electromagnetism, in *"Proceedings of Journées Relativistes 1983"*, Torino, May 5–8, 1983, S. Benenti, M. Ferraris & M. Francaviglia eds.; Pitagora (Bologna, 1985), 125–153.

[12] FERRARIS, M., FRANCAVIGLIA, M. and GATTO, L., Towards an Affine Approach to Kaluza–Klein Theory, *Rend. Sem. Mat. Univ. Modena XXXVII*, 131–145 (1989).

[13] FRANCAVIGLIA M., Elements of Differential and Riemannian Geometry, Monographs and Textbooks in Physical Sciences, Lecture Notes **4**, Bibliopolis (Napoli, 1988).

[14] GARCIA P. L., Connections and 1-jet Fiber Bundles, *Rend. Sem. Mat. Univ. Padova* **47**, 227–242 (1972).

[15] GATTO, L., FERRARIS, M. and FRANCAVIGLIA, M., A Natural Decomposition of S^1-Invariant Linear Connections on S^1-Bundles, *Rend. Ist. Lomb. Sc. Lett.* A 122, 65–77 (1988).

[16] GATTO, L., FERRARIS, M. and FRANCAVIGLIA, M., Remarks on Right-Invariant Vectorfield Bases in Principal Bundles and their Applications to Decomposition Theorems, *Rend. Sem. Mat. Univ. Pol. Torino* **46** (3), 309–322 (1988).

[17] GATTO, L., FERRARIS, M. and FRANCAVIGLIA, M., Principal Connections from G-Invariant Linear Connections, Boll UMI (7) 4–B, 905–926 (1990).

[18] GATTO, L., Connessioni lineari G-invarianti su un fibrato principale: decomposizioni naturali e possibili applicazioni in fisica, Doctoral Dissertation, Univ. of Torino (1987, unpublished).

[19] GLASHOW S. L., Towards a Unified Theory: Threads in a Tapestry, *Rev Mod. Phys.* **52**, 539 (1980).

[20] JADCZYK, A., Symmetry of Einstein–Yang–Mills Sysytems and Dimensional Reduction, *Journ. Geom. and Phys.* **1**(2), 97–126 (1984).

[21] KALUZA, T., *Sitzungsber. Preuss. Akad. Wiss.,* Berlin, Math. Phys., k.1, 1966 (1921)).

[22] KLEIN, O., Z. Phys. **27**, 895 (1925).

[23] KOBAYASHI, S. and NOMIZU, K., Foundations of Differential Geometry, Vol. I, Interscience Publishers (New York, 1963).

[24] LICHNEROWICZ, A., Théories relativistes de la gravitation et de l'electro-magnétisme, Masson & Cie Editeurs (Paris, 1955).

[25] SALAM, A., Gauge Unification of Fundamental Forces, *Rev. Mod. Phys.* **52**, 525–538 (1980).

[26] SCHRÖDINGER, E., The Relation between Metric and Affinity, *Proc. R. Irish. Acad.* **51 A**, 147–150 (1947).

[27] TONNELAT, M. A., Les théories unitaires de l'électromagnétisme et de la gravitation, Gauthier-Villars (Paris, 1965).

[28] WEINBERG, S., Conceptual Foundations of the Unified Theory of Weak and Electromagnetic Interactions, *Rev. Mod. Phys.* **52**, 512–524 (1980).

[29] WEYL, H., Raum-Zeit-Materie, 1st, 2nd and 3rd eds., Springer-Verlag (Berlin, 1918,1919,1921); Space-Time-Matter, Dover (New York, 1950).

Marco Ferraris

Dipartimento di Matematica,
Università di Cagliari,
Via Ospedale 72,
09134 CAGLIARI,
Italy

Mauro Francaviglia

Istituto di Fisica Matematica
"J.–L. Lagrange".
Via Carlo Alberto 10,
10123 TORINO,
Italy

Letterio Gatto

Istituto di Fisica Matematica
"J.–L. Lagrange".
Via Carlo Alberto 10,
10123 TORINO,
Italy

Some Properties of Surfaces with Zero Normal Torsion in Euclidean Space[*]

V. T. FOMENKO

In this paper some new properties of a surface in Euclidean space are given which are connected with the normal torsion of the surface at the point in a given direction. The tensor of Darboux's type is considered and the surfaces of Darboux's type having zero tensor of Darboux's type are determined. The full classification of two-dimensional surfaces of Darboux's type is given.

1. Let F^ℓ be a 1-dimensional surface in m-dimensional Euclidean space E^m given in the neighbourhood of every point $x \in F^\ell$ by the equation

$$\bar{r} = \bar{r}(u^1, u^2, \ldots, u^\ell)$$

where $\bar{r} \in C^3$. Let $g_{ij} = (\bar{r}_i, \bar{r}_j)$ be a metric tensor of F^ℓ where $\bar{r}_i = \partial_i \bar{r}$ and $\partial_i = \partial/\partial u_i$. We denote Christoffel's symbols and the covariant derivatives respectively by Γ_{ij}^k, ∇_i. Below we suppose that the indexes $i, j, k, \ldots$ run from 1 up to ℓ. Let $\{\bar{n}_\sigma\}_{\ell+1}^m$ be $(m - \ell)$ vector fields of unit normals of the neighbourhood of the point orthogonal to each other at each point. We denote the second quadratic forms of F^ℓ with respect to the normal $\bar{n}_\sigma$ by $b_{\sigma ij} = -(\bar{n}_{\sigma i}, \bar{r}_j)$ where $\bar{n}_{\sigma i} = \partial_i \bar{n}_\sigma$, and the metric tensor of the fibre

---

[*] This paper is in final form and no version of it will be submitted for publication elsewhere.

$N_x^{m-\ell}$ of the normal bundle NF^ℓ over the surface F^ℓ at the point x by $g_{\sigma\tau}(x) = (\bar{n}_\sigma(x), \bar{n}_\tau(x))$, and the coefficients of the normal connection by $\Gamma_{i\sigma}^\tau$, and the covariant derivatives of van der Warden–Bortolotti's connection along the surface F^ℓ by $\overline{\nabla}_i$.

It is known that the vector fields $\bar{r}_i$, $\bar{n}_\sigma$ and the tensor fields g_{ij}, $b_{\sigma ij}$ on the surface F^ℓ in E^m are connected by the following Gauss–Weingarten–Codazzi's equations:

$$\overline{\nabla}_i \bar{r}_j = b_{ij}^\sigma \bar{n}_\sigma, \qquad \overline{\nabla}_i \bar{n}_\sigma = -g^{kj} b_{\sigma ik} \bar{r}_j,$$

$$\overline{\nabla}_i b_{jk}^\sigma = \overline{\nabla}_j b_{ik}^\sigma$$

where $\overline{\nabla}_i \bar{r}_j = \partial_i \bar{r}_j - \Gamma_{ij}^k \bar{r}_k$,

$\overline{\nabla}_i \bar{n}_\sigma = \partial_i \bar{n}_\sigma - \Gamma_{i\sigma}^\tau \bar{n}_\tau$,

$$\overline{\nabla}_i b_{jk}^\sigma = \partial_i b_{jk}^\sigma - \Gamma_{ij}^m b_{mk}^\sigma - \Gamma_{ik}^m b_{jm}^\sigma + \Gamma_{i\tau}^\sigma b_{jk}^\tau,$$

$$b_{ij}^\sigma = g^{\sigma\tau} b_{\tau ij}, \qquad g^{\sigma\tau} g_{\lambda\tau} = \delta_\lambda^\tau,$$

the tensor δ_λ^τ is the Kronecker symbol.

2. Let TF^ℓ be a tangent bundle over the surface F^ℓ and T_x^ℓ be a plane tangent to the surface F^ℓ at the point x. Let $\bar{t}$ be the unit vector at the point x from T_x^ℓ, $\bar{t} = t^i \bar{r}_i$. We shall write $(x, \bar{t}) \in TF^\ell$ supposing that $x \in F^\ell$, $\bar{t} \in T_x^\ell$. The vector $\bar{t}$ and the normal space $N_x^{m-\ell}$ of F^ℓ at x determine the $(m-\ell+1)$-dimensional vector subspace $E(x, \bar{t})$ of E^m. The intersection of F^ℓ and $E(x, \bar{t})$ gives a curve $\gamma_n(x, \bar{t})$ in the neighbourhood of x which is called the normal section of F^ℓ at x in the direction $\bar{t}$: $\gamma_n(x, \bar{t}) = F^\ell \cap E(x, \bar{t})$.

In general, the normal section $\gamma_n(x, \bar{t})$ is a twisted curve in $E(x, \bar{t})$ having the first, second and etc. curvatures which will be denoted by $k_1, k_2, \ldots$ respectively. The curvatures $k_1 = k_n(x, \bar{t})$ and $k_2 = \kappa_n(x, \bar{t})$ are called the normal curvature and the normal torsion of the surface at the point x in the direction $\bar{t}$ respectively. The vector $\bar{k}_1 = k_1 \bar{n}$ where $\bar{n} = \bar{n}(x, \bar{t})$ is the first unit normal vector of $\gamma_n(x, \bar{t})$ at the point x is called the vector of the normal curvature of F^ℓ at x in the direction $\bar{t}$. Let $u^i = u^i(s)$ be the equations of the section $\gamma_n(x, \bar{t})$ where s is a natural parameter and $u = (u^1, u^2, \ldots, u^\ell)$ and $\bar{\tau} = \{\tau^1, \tau^2, \ldots, \tau^\ell\}$ and $\tau^i = du^i(s)/ds$, and $u(0) = x$ and $\tau^i(0) = t^i$. We shall say that the vector $\bar{a}$ is transferred parallelly from the point x in the direction $\bar{t}$ in NF^ℓ if the vector field $\bar{a}(u(s), \tau(s))$ formed as a result of the transfer has the following property: $(\bar{a}', (x, \bar{t}))^{\perp} = 0$ where $(\ldots)^{\perp}$ is an operation of taking the normal component of the vector and the mark "$'$" denotes the derivative along the curve $\gamma_n(x, \bar{t})$ across the parameter s. We shall obtain

Theorem 1. *The normalized vector $\bar{n}(x, \bar{t}) = \bar{k}_1/k_1$ of the normal curvature of the surface F^{ℓ} at the point x in the direction $\bar{t}$ is transferred parallelly from the point x in the direction $\bar{t}$ in NF^{ℓ} if and only if $\kappa_n(x, \bar{t}) = 0$.*

3. The surface F^{ℓ} in E^m is called a surface of Darboux's type if there is a vector field $\bar{\lambda}$ such that

$$g_{\sigma\tau}(x) \underset{\lambda}{\Theta}_{ijk}^{\sigma}(x) \underset{\lambda}{\Theta}_{tmn}^{\tau}(x) \equiv 0 \qquad \forall\, x \in F^{\ell}$$

where $\qquad \underset{\lambda}{\Theta}_{ijk}^{\sigma}(x) = \overline{\nabla}_i b_{jk}^{\sigma}(x) - 3\lambda_{(i}(x) b_{jk)}^{\sigma}(x), \qquad \lambda_i = (\bar{\lambda}, \bar{r}_i).$

The tensor $\underset{\lambda}{\Theta}_{ijk,tmn} = g_{\sigma\tau} \underset{\lambda}{\Theta}_{ijk}^{\sigma} \underset{\lambda}{\Theta}_{tmn}^{\tau}$ is called the tensor of Darboux's type with respect to the vector field $\bar{\lambda}$. If λ_i is a gradient tensor, then the surface of Darboux's type is called a gradient surface and it is called a non-gradient surface in the opposite case.

Below we shall prove

Theorem 2. *On the surfaces of Darboux's type and only on them the following properties hold:*
   1) $\kappa_n(x, \bar{t}) \equiv 0 \qquad \forall\, (x, \bar{t}) \in TF^{\ell}$;
   2) for any geodesic line $\mathcal{L}_g : u^i = u^i(s), s \in [s_0, s_1]$ on the surface there is a constant C such that

$$k(u(s)) = C \exp \int_{s_0}^{s} \lambda_i(u(s))\, du^i(s)$$

where k is the first curvature of $\mathcal{L}_g$ as a line in E^m.

Theorem 3. *Let F^2 be a two-dimensional surface in E^m. Then F^2 is a surface of Darboux's type if and only if F^2 is one of the following surfaces:*
   1) a surface of Darboux's type which lies in a 3-subspace E^3 of E^m (it is the gradient surface of Darboux's type with $\lambda_i = \partial_i \ln \sqrt[4]{K}$ where K is

the Gaussian curvature of F^2);

    2) a cylindrical or cornical surface which lies in a 3-subspace E^3 of E^m (it is a gradient surface of Darboux's type with $\lambda_i \equiv 0$);

    3) a surface which consists of all the tangent straight lines of the curve in a 3-subspace E^3 of E^m (it is a non-gradient surface of Darboux's type);

    4) a piece of Clifford's torus in E^4 of E^m which is an open portion of the product surface of two planar circles (it is a gradient surface of Darboux's type with $\lambda_i \equiv 0$);

    5) an open portion of Veronese's surface in E^5 of E^m (it is a gradient surface of Darboux's type with $\lambda_i \equiv 0$).

Theorem 4. Let F^ℓ be a surface in E^m with the parallel second fundamental form in NF^ℓ. Then F^ℓ is a surface of Darboux's type in E^m.

4. Proof of Theorem 1.

Let $\{\bar{t}_1, \bar{t}_2, \ldots, \bar{t}_\ell, \bar{n}_{\ell+1}, \ldots, \bar{n}_m\}$ be a frame along the curve $\gamma_n(x, \bar{t})$ where $\{\bar{t}_1, \bar{t}_2, \ldots, \bar{t}_\ell\}$ is Schmidt's frame in a tangent plane with respect to the tangent vector field $\bar{\tau}$ along the curve $\gamma_n(x, \bar{t})$ and $\{\bar{n}_{\ell+1}, \bar{n}_{\ell+2}, \ldots, \bar{n}_m\}$ is an orthogonalized frame in the normal plane with $\bar{n}_{\ell+1} = \bar{n}(x, \bar{t})$ at the point x. If $\overline{R}(s) \equiv \bar{r}(u(s))$ is the equation of the curve $\gamma_n(x, \bar{t})$ with the natural parameter s, then we have

$$\overline{R}' = \tau,$$
$$\overline{R}'' = k_g \bar{t}_2 + k_n \bar{n}_{\ell+1},$$
$$\overline{R}''' = k_g \bar{t}_2' + k_g' \bar{t}_2 + k_n' \bar{n}_{\ell+1} + k_n \bar{n}_{\ell+1}'.$$

Then we believe that

$$\bar{n}_{\ell+1}' = A_1 \bar{t}_1 + \ldots + A_\ell \bar{t}_\ell + B_{\ell+2} \bar{n}_{\ell+2} + \ldots + B_m \bar{n}_m$$

at the point x. Since we have $k_g = 0$, $\bar{n}_{\ell+1} = \bar{n}(x, \bar{t})$, $\overline{R}'' \in E(x, \bar{t})$ at the point x, we obtain at the point x

$$\overline{R}' \wedge \overline{R}'' \wedge \overline{R}''' = (\bar{t} \wedge \bar{n}_{\ell+1} \wedge \bar{n}_{\ell+2}) k_n(x, \bar{t}) B_{\ell+2} + \ldots +$$
$$+ (\bar{t} \wedge \bar{n}_{\ell+1} \wedge \bar{n}_m) k_n(x, \bar{t}) B_m.$$

Since the second curvature $|\kappa|$ of the curve $\gamma_n(x, \bar{t})$ at the point x is given by the formula

$$|\kappa| = |\overline{R}' \wedge \overline{R}'' \wedge \overline{R}'''| / k_n^2(x, \bar{t}),$$

we find that $|\kappa|^2 = \left(\sum_{\ell+2}^{m} B_6^2 \right) \Big/ k_n^2(x,\bar{t}).$

Hence we obtain $\kappa_n(x,\bar{t}) = 0$ if and only if $B_\sigma = 0$, $\sigma = \ell+2,\ldots,m$ at the point x or $\left(\bar{n}'(x,\bar{t}) \right)^\perp = 0$. This proves Theorem 1. ■

5. Let F^ℓ be a surface in $E^{\ell+p_1}$ of $E^{\ell+p_2}$ where $p_1 < p_2$. We may easily obtain the following:

if F^ℓ is a surface of Darboux's type in $E^{\ell+p_1}$, then F^ℓ is a surface of Darboux's type in $E^{\ell+p_2}$. If F^ℓ lies in $E^{\ell+p_1}$ of $E^{\ell+p_2}$ and F^ℓ is a surface of Darboux's type in $E^{\ell+p_2}$ then F^ℓ is a surface of Darboux's type in $E^{\ell+p_1}$. Indeed we choose a local field of the orthogonalized frame $\{\bar{n}_\sigma\}_{\ell+1}^{\ell+p_2}$ of F^ℓ along $\gamma_n(x,\bar{t})$ in $E^{\ell+p_2}$ such that the vectors $\bar{n}_{\ell+p_1+1},\ldots,\bar{n}_{\ell+p_2}$ are normal to $E^{\ell+p_1}$ and constant in $E^{\ell+p_2}$. Then we have

$$b_{ij}^\sigma \equiv 0, \qquad \sigma = \ell+p_1+1,\ldots,\ell+p_2;$$
$$\Gamma_{\sigma i}^\tau \equiv 0, \qquad \tau = \ell+p_1+1,\ldots,\ell+p_2; \qquad \sigma = \ell+1,\ldots,\ell+p_2.$$

Thus, it follows that $\underset{\lambda}{\Theta}{}_{ijk}^\sigma = \nabla_i b_{jk}^\sigma - 3\lambda_{(i} b_{jk)}^\sigma \equiv 0$,
$\sigma = \ell+p_1+1,\ldots,\ell+p_2$ and

$$\underset{\lambda}{\Theta}{}^s{}_{ijk} = \bar{\nabla}_i b_{jk}^\sigma + \sum_{\tau=\ell+1}^{\ell+p_2} \Gamma_{\tau i}^\sigma b_{jk}^\tau - 3\lambda_{(i} b_{jk)}^\sigma =$$

$$= \bar{\nabla}_i b_{jk}^\sigma + \sum_{\tau=\ell+1}^{\ell+p_1} \Gamma_{\tau i}^\sigma b_{jk}^\tau - 3\lambda_{(i} b_{jk)}^\sigma \equiv \underset{\lambda}{\widetilde{\Theta}}{}_{ijk}^\sigma$$

where $\underset{\lambda}{\widetilde{\Theta}}{}_{ijk}^\sigma$ is the tensor of Darboux's type with respect to the normal $\bar{n}_\sigma$ and calculated with respect to the space $E^{\ell+p_1}$.

Hence the tensor of Darboux's type $\underset{\lambda}{\Theta}{}_{ijk,\ell mn}$ of the surface F^ℓ in $E^{\ell+p_2}$ is the tensor of Darboux's type $\underset{\lambda}{\widetilde{\Theta}}{}_{ijk,\ell mn}$ of the surface F^ℓ in $E^{\ell+p_1}$.

Lemma. *Let F^ℓ be a surface in E^m. Then the formula*

$$\underset{\lambda}{\Theta}{}^2(x,\bar{t}) = k_n^2(x,\bar{t})\kappa_n^2(x,\bar{t}) + \left(\dot{k}_n(x,\bar{t}) - (\bar{\lambda}(x),\bar{t})k_n(x,\bar{t}) \right)^2$$

takes the place where the mark "·" denotes the derivative along the geodesic line going from the point x in the direction $\bar{t}$.

Proof. Put

$$\underset{\lambda}{\Theta}{}^{\sigma} = \nabla k^{\sigma} - 3(\bar{\lambda}, \bar{t})k^{\sigma}$$

where $\nabla k^{\sigma} = \nabla_i b^{\sigma}_{jk} t^i t^j t^k$, $\bar{\lambda} = \bar{\lambda}(x)$. Then we have

$$k_n^4 \kappa_n^2 = \sum_{\sigma < \tau}\left(k^{\tau} \underset{\lambda}{\Theta}{}^{\sigma} - k^{\sigma} \underset{\lambda}{\Theta}{}^{\tau} \right)^2.$$

Since $k_n \dot{k}_n = \sum_{\sigma} k^{\sigma} \underset{\lambda}{\Theta}{}^{\sigma} + 3\sum_{\sigma}(\bar{\lambda}, \bar{t})k^{\sigma} k^{\sigma}$, the following relation is true

$$\sum_{\sigma} k^{\sigma} \underset{\lambda}{\Theta}{}^{\sigma} = k_n \dot{k}_n - 3(\bar{\lambda}, \bar{t})k_n^2.$$

From the previous relation we find that

$$k_n^4 \kappa_n^2 + \left(k_n \dot{k}_n - 3(\bar{\lambda}, \bar{t})k_n^2 \right)^2 =$$

$$= \sum_{\sigma=\ell+1}^{m} k^{\ell+1} k^{\ell+1} \underset{\lambda}{\Theta}{}^{\sigma} \underset{\lambda}{\Theta}{}^{\sigma} + \sum_{\sigma=\ell+1}^{m} k^{\ell+2} k^{\ell+2} \underset{\lambda}{\Theta}{}^{\sigma} \underset{\lambda}{\Theta}{}^{\sigma} +$$

$$+ \ldots + \sum_{\sigma=\ell+1}^{m} k^m k^m \underset{\lambda}{\Theta}{}^{\sigma} \underset{\lambda}{\Theta}{}^{\sigma}.$$

The latter relation can be reduced to the form

$$k_n^4 \kappa_n^2 + \left(k_n \dot{k}_n - 3(\bar{\lambda}, \bar{t})k_n^2 \right)^2 = \sum_{\sigma=\ell+1}^{m} \left(\underset{\lambda}{\Theta}{}^{\sigma} \right)^2 \cdot \sum_{\tau=\ell+1}^{m} (k^{\tau})^2$$

which proves the Lemma. ∎

6. Proof of Theorem 2.

It follows from the Lemma that on the surfaces of Darboux's type and only on them we have

1) $\kappa_n(x, \bar{t}) \equiv 0 \qquad \forall\, (x, \bar{t}) \in TF^{\ell},$

2) $\dot{k}_n(x, \bar{t}) - 3(\bar{\lambda}, \bar{t})k_n(x, \bar{t}) \equiv 0 \qquad \forall\, (x, \bar{t}) \in TF^{\ell}.$

Since along any geodesic line $\mathcal{L}_g$: $u^i = u^i(s)$, $s \in [s_0, s_1]$, the relation $k(u(s)) = k_n(u, \overline{\tau})$ is true where $\overline{\tau} = \left\{ \frac{du^1}{ds}, \ldots, \frac{du^\ell}{ds} \right\}$ and integrating along $\mathcal{L}_g$, we find

$$k(u(s)) = C \exp \int_{s_0}^{s} \lambda_i(u(s)) \, du^i(s), \qquad C = \text{const}.$$

This proves Theorem 2. ∎

7. Proof of Theorem 3.

Since on the surface F^2 with $\underset{\lambda}{\Theta}_{ijk,\ell mn} \equiv 0$ we have $\kappa_n(x, \overline{t}) \equiv 0$, then the surface of Darboux's type lines in E^3 or it is Clifford's torus in E^4 or it is Veronese's surface in E^5 (see [1], [2]).

It is easy to show that the Clifford torus in E^4 is a gradient surface of Darboux's type. Indeed the formula $\overline{\nabla}_k b_{ij}^\sigma \equiv 0$ corresponds to the Clifford torus. Taking $\lambda_i \equiv 0$ we find that the Clifford torus in E^4 is a gradient surface of Darboux's type.

We shall show now that the second fundamental form of the Veronese surface is parallel in the normal bundle. The fundamental tensors of the Veronese surface of curvature $1/3$ lying on the hypersphere $S_0^4(1)$ in E^5 are given by the formulas:

$$\|g_{ij}\| = \begin{pmatrix} E & 0 \\ 0 & E \end{pmatrix}; \qquad \|b_{ij}^3\| = \begin{pmatrix} \frac{E}{\sqrt{3}} & 0 \\ 0 & -\frac{E}{\sqrt{3}} \end{pmatrix};$$

$$\|b_{ij}^4\| = \begin{pmatrix} 0 & \frac{E}{\sqrt{3}} \\ \frac{E}{\sqrt{3}} & 0 \end{pmatrix}; \qquad \|b_{ij}^5\| = \begin{pmatrix} E & 0 \\ 0 & E \end{pmatrix};$$

$$\|\Gamma_{4i}^3\| = \|\partial_2 \ln E, -\partial_1 \ln E\|; \qquad \|\Gamma_{5i}^3\| =$$

$$= \|\Gamma_{5i}^4\| = \|0, 0\|; \qquad E = \frac{12}{(1 + u^2 + v^2)^2}; \qquad (u, v) \in \mathbb{R}^2$$

$$\text{and} \quad (u, v) \equiv \left(-\frac{u}{u^2 + v^2}, -\frac{v}{u^2 + v^2} \right).$$

It results from here that $\overline{\nabla}_i b_{jk}^\sigma \equiv 0$.

Taking $\lambda_i \equiv 0$, we find that the Veronese surface in E^5 is a gradient surface of Darboux's type.

Now we shall consider the case $F^2 \subset E^3$. We denote b_{ij}^3 by b_{ij}. It is known that there are curvature lines on the surface F^2 in E^3. Without loss of generality we may suppose that it is a coordinate net. Then $g_{12} = 0$, $b_{12} = 0$ and the condition $\Theta_{\substack{ijk,\ell mn \\ \lambda}} = 0$ gives the equations

$$\overline{\nabla}_1 b_{22} = \lambda_1 b_{22}; \qquad \overline{\nabla}_1 b_{11} = 3\lambda_1 b_{11};$$
$$\overline{\nabla}_1 b_{12} = \lambda_2 b_{11}; \qquad \overline{\nabla}_2 b_{22} = 3\lambda_2 b_{22}$$

where we must find the vector $\overline{\lambda} = \{\lambda_1, \lambda_2\}$. This system is transformed to the form

$$\partial_1 \left(\frac{b_{22}}{g_{22}}\right) = \lambda_1 \left(\frac{b_{22}}{g_{22}}\right); \qquad \partial_1 \left(\frac{b_{11}}{g_{11}}\right) = 3\lambda_1 \left(\frac{b_{11}}{g_{11}}\right);$$
$$\partial_2 \left(\frac{b_{22}}{g_{22}}\right) = 3\lambda_2 \left(\frac{b_{22}}{g_{22}}\right); \qquad \partial_2 \left(\frac{b_{11}}{g_{11}}\right) = \lambda_2 \left(\frac{b_{11}}{g_{11}}\right).$$

If the main curvatures of F^2 are denoted by k_1, k_2 then we may write the latter system in the following form

$$\partial_1 k_2 = \lambda_1 k_2, \qquad \partial_1 k_1 = 3\lambda_1 k_1,$$
$$\partial_2 k_2 = 3\lambda_2 k_2, \qquad \partial_2 k_1 = \lambda_2 k_1.$$

Let $K \neq 0$, then we find that

$$\partial_1 \ln \left(\frac{|k_2|}{\sqrt[3]{|k_1|}}\right) = 0, \qquad \partial_2 \ln \left(\frac{|k_1|}{\sqrt[3]{|k_2|}}\right) = 0.$$

It means that along the curvature lines the relations

$$k_2^3 = C_1(v) k_1; \qquad k_1^3 = C_2(u) k_2$$

are true where $C_1(v)$, $C_2(u)$ are some functions of their parameters.

These relations characterize Darboux's surface in E^3 and only them.

Let $K \equiv 0$, then we may consider that $k_1 \equiv 0$, $k_2 \neq 0$. (It is clear that the case $k_1 \equiv 0$, $k_2 \equiv 0$ results in the plane in E^3).

a) The case of the cylindrical surface.
Let F^2 be a cylindrical surface in E^3 given by the equation $\overline{r} = \overline{p}(u) + v\overline{a}$ where $\overline{a} = \overline{\text{const}}$, $|\overline{a}| = 1$, and u is the natural parameter and $\overline{a} \perp \overline{p}'$. Since $g_{11} = 1$, $g_{12} = 0$, $g_{22} = 1$; $b_{11} = k(u)$, $b_{12} = 0$, $b_{22} = 0$ where $k(u)$

is the curvature of the curve with the equation $\bar{\rho} = \bar{\rho}(u)$, we find that $\lambda_1 = \partial_1 \ln k(u)$ and $\lambda_2 = 0$. Hence the surface F^2 is a gradient surface of Darboux's type.

b) The case of a conical surface.

Let F^2 be a conical surface in E^3 given by the equation $\bar{r} = v\bar{\rho}(u)$ where $|\bar{\rho}| = 1$ and u is a natural parameter. Since $g_{11} = v^2$, $g_{12} = 0$, $g_{22} = 1$, $b_{11} = \frac{v^2 k(u)\varphi(u)}{\sqrt{1+v^2}}$, $b_{12} = b_{22} = 0$ where $k(u)$ is the curvature of the curve with the equation $\bar{\rho} = \bar{\rho}(u)$ and

$$\varphi(u) = (\bar{\rho}(u), \bar{\rho}'(u), \bar{\rho}''(u)) / |\bar{\rho}''(u)|$$

we find that
$$\lambda_2 = -\partial_1 \ln \sqrt{1 + v^2};$$
$$\lambda_1 = \partial_2 \ln |\psi(u)|^{1/3}.$$

Since $\partial_1 \lambda_2 = \partial_2 \lambda_1$, the surface F^2 is a gradient surface of Darboux's type.

c) The case of the surface formed by the tangent straight lines of the curve.

Let F^2 be given by the equation $\bar{r} = \bar{\rho}(u) + v\bar{\rho}'(u)$ where u is a natural parameter. In this case we have $g_{11} = 1 + k^2 v^2$, $g_{12} = 1$, $g_{22} = 1$, $b_{11} = k\kappa v$, $b_{12} = 0$, $b_{22} = 0$ where k and κ are the curvature and the torsion of the curve $\mathcal{L} : \bar{\rho} = \bar{\rho}(u)$ respectively. The condition $\underset{\lambda}{\Theta}_{ijk,\ell mn} = 0$ means that

$\nabla_1 b_{11} = 3\lambda_1 b_{11}$; $\nabla_2 b_{11} = \lambda_2 b_{11}$; $\nabla_2 b_{12} = 0$, $\nabla_1 b_{12} = \lambda_2 b_{11}$, $\nabla_2 b_{22} = 0$, $\nabla_1 b_{22} = 0$. It follows from here that

$$\lambda_2 = -1/v; \qquad \lambda_1 = \frac{1}{3}\left(\ln\left|\frac{\kappa}{k}\right|\right)' - \frac{2}{3}\cdot\frac{1}{v}.$$

Since $\partial_1 \lambda_2 \neq \partial_2 \lambda_1$, the surface F^2 is a non-gradient surface of Darboux's type.

8. Proof of Theorem 4.

If the second fundamental form is parallel on the surface F^ℓ in E^m, then $\nabla_i b^\sigma_{jk} \equiv 0$. Then, for $\lambda_i \equiv 0$ F^ℓ is the gradient surface of Darboux's type.

Consequently, the theorem is proved. ∎

References

[1] V. T. FOMENKO, Some properties of two-dimensional surfaces with zero normal torsion in E^4. Math. Sbornik, vol. 106 (148) N4 (Russian); English transl., Math. USSR Sbornik vol. 35 (1979) N2, 251–256.

[2] BANG–YEN–CHEN and SHI–JIE LI, Classification of surfaces with pointwise planar normal sections and its application to Fomenko's conjecture, *Journal of Geometry,* vol. 26 (1986), 21–34.

V. T. Fomenko

ul. Vichnevaja 56. kv. 42
347937 Taganrog
USSR

COLLOQUIA MATHEMATICA SOCIETATIS JÁNOS BOLYAI

56. DIFFERENTIAL GEOMETRY, EGER (HUNGARY), 1989

Projective Structures in Sinyukov Manifolds*

S. FORMELLA

1. Introduction

Let (M, g) be a connected n-dimensional $(n \geq 3)$ Riemannian manifold of class C^∞ with not necessarily definite metric g. Let $\mathfrak{F}(M)$ be the algebra of functions on M and $\mathfrak{X}(M)$ the $\mathfrak{F}$-module of vector fields on M. If on a manifold the condition

(1)
$$(\nabla_X S)(Y, Z) = \sigma(X)g(Y, Z) + \nu(Y)g(X, Z) + \nu(Z)g(X, Y), \quad X, Y, Z \in \mathfrak{X}(M)$$

holds, where $S(X, Y)$ is the Ricci tensor of (M, g) and $\sigma(X)$, $\nu(X)$ are certain 1-forms, then the manifold is said to be either an L_n space ([9], [10]) or a generalized Einstein manifold ([3]). In this paper such manifolds will be called Sinyukov manifolds. Manifolds of this kind were first considered by N. S. Sinyukov in [9] and E. N. Sinyukova in the paper [10]. For an arbitrary Sinyukov manifold (M, g) there always exists a Riemannian manifold which is geodesically equivalent to it. A Sinyukov manifold is a manifold with harmonic conformal curvature tensor [3].

In this paper we shall study some properties of Sinyukov manifolds.

---

* This paper is in final form and no version of it will be submitted for publication elsewhere.

2. Preliminaries

Let (M, g) and $(M, \bar{g})$ be two n-dimensional Riemannian or pseudo-Riemannian manifolds. A mapping $\gamma : (M, g) \to (M, \bar{g})$ is called geodesic, and the metrics g and $\bar{g}$ are called geodesically corresponding if γ preserves the geodesics. The following theorem is well-known. A mapping $\gamma : (M, g) \to (M, \bar{g})$ is geodesic if and only if one of the following conditions is satisfied

$$(2) \qquad \overline{\nabla}_X Y = \nabla_X Y + \psi(X)Y + \psi(Y)X,$$

$$(3) \quad (\nabla_X \bar{g})(Y, Z) = 2\psi(X)\bar{g}(Y, Z) + \psi(Y)\bar{g}(X, Z) + \psi(Z)\bar{g}(X, Y),$$
$$X, Y, Z \in \mathfrak{X}(M),$$

where ∇ and $\overline{\nabla}$ are the Levi–Civita connections on (M, g) and $(M, \bar{g})$ respectively and ψ is locally a gradient,

$$(4) \qquad (\nabla_X a)(Y, Z) = \lambda(Y)g(X, Z) + \lambda(Z)g(X, Y),$$

where "a" is a symmetric covariant tensor field of degree 2 and $\lambda(X)$ is locally a gradient ([9]).

In [9] N. S. Sinyukov has proved that if (M, g) admits a geodesic mapping onto $(M, \bar{g})$, then (M, a) admits a geodesic mapping onto $(M, \tilde{g} = \exp(2\psi)g)$ with the same 1-form ψ, where "a" satisfies (4).

In the local coordinate system (U, x^i) the tensors a, g and $\bar{g}$ are connected with each other in the following form

$$(5) \qquad a_{ij} = \exp(2\psi)\bar{g}^{\alpha\beta} g_{\alpha i} g_{\beta j},$$
$$(6) \qquad \lambda_i = -\exp(2\psi)\bar{g}^{\alpha\beta} g_{\alpha i} \psi_\beta,$$

where $\bar{g}^{ij}$ denotes the components $(\bar{g}_{ij})^{-1}$.

If $\psi_i \neq 0$ (it is equivalent to $\lambda_i \neq 0$), then the geodesic mapping is called non-trivial. In the sequel the geodesic mapping of (M, g) onto $(M, \bar{g})$ will be denoted by $\gamma : (M, g) \xrightarrow{\psi} (M, \bar{g})$. The curvature tensors and the Ricci tensors of (M, g) and $(M, \bar{g})$ are related by

$$(7) \qquad \overline{R}^l{}_{ijk} = R^l{}_{ijk} + \delta^l_k \psi_{ij} - \delta^l_j \psi_{ik},$$

(8)
$$(n-1)\psi_{ij} + S_{ij} = \overline{S}_{ij}, \qquad S_{ij} = R^t{}_{ijt},$$

(9)
$$\psi_{ij} = \psi_{i,j} - \psi_i\psi_j$$

where the comma in (9) denotes covariant differentiation with respect to the metric g. We recall that the projective curvature tensor W defined by

(10)
$$W^l{}_{ijk} = R^l{}_{ijk} - \frac{1}{n-1}[S_{ij}\delta^l_k - S_{ik}\delta^l_j]$$

is invariant under geodesic mappings, i.e. $W^l{}_{ijk} = \overline{W}^l{}_{ijk}$.

3. Some classifications of Sinyukov manifolds

In [3] it was shown that a manifold (M, g) is a Sinyukov manifold if and only if one of the following two conditions is satisfied

(i) $(M, \widetilde{g} = \exp(2\psi)g)$ is an Einstein manifold admitting a geodesic mapping.

(ii) $(M, \widetilde{g})$ is also a Sinyukov manifold.

We prove the following

Theorem 1. *Let (M, g) be a Sinyukov manifold. If $\gamma : (M, g) \xrightarrow{\psi} (M, \overline{g})$ is a geodesic mapping, then*

(a) *in the case (i) we have*

$$\Psi = \frac{-\tau}{n-2}a + Ag$$

(b) *in the case (ii) we obtain*

$$\Psi = \frac{-\tau}{n-2}a + Bg + \frac{\widetilde{\tau}}{n-2}\overline{g},$$

where Ψ is the tensor field with components ψ_{ij} (s. (9)), $\tau, \widetilde{\tau} = \text{const.} \neq 0$ and $A, B \in \mathfrak{F}(M)$.

Proof. In [3] the author of this paper has obtained the following theorem: a necessary and sufficient condition for a manifold (M, g) to be a Sinyukov manifold is that

(11)
$$S_{ij} = \tau a_{ij} + (\sigma + d)g_{ij}$$

holds where $\tau, d = \text{const.} \neq 0$, $\sigma \in \mathfrak{F}(U)$.

If the metric $\tilde{g}$ is also Sinyukov metric, then the equalities

(12) $$\overline{g}_{ij,k} = \widetilde{\lambda}_i \widetilde{g}_{jk} + \widetilde{\lambda}_j \widetilde{g}_{ik} \qquad \text{(s. (4))}$$

and

(13) $$\widetilde{S}_{ij} = \widetilde{\tau}\overline{g}_{ij} + (\widetilde{\sigma} + \widetilde{d})\widetilde{g}_{ij}$$

are fulfilled on some neighborhood U of x, where $\widetilde{\tau}, \widetilde{d} = \text{const.} \neq 0$, $\widetilde{\sigma} \in \mathfrak{F}(U)$, the semicolon denotes covariant differentiation with respect to the metric $\widetilde{g}$. Since $\widetilde{g} = \exp(2\psi)g$, we have the relation

(14) $$\widetilde{S}_{ij} = S_{ij} + (n-2)\psi_{ij} + [\triangle_2\psi + (n-2)\triangle_1\psi]g_{ij},$$

where

$$\triangle_2\psi = g^{\alpha\beta}\psi_{\alpha,\beta}, \qquad \triangle_1\psi = g^{\alpha\beta}\psi_\alpha\psi_\beta.$$

Now, from (11), (13) and (14) we obtain (b). If $(M,\widetilde{g})$ is an Einstein manifold, then (a) follows from (14) and (11). This completes the proof. ∎

Lemma ([3]). *If (M, g) is a Sinyukov manifold then the relation*

(15) $$\lambda_{i,j} = \frac{1}{n-2}\left[a_{it}S^t_j - \frac{r}{n(n-1)}a_{ij} - \frac{a}{n}S_{ij} + (n-2)\varphi g_{ij}\right]$$

holds on (U, g), where

$$\varphi = \frac{\lambda}{n} - \frac{1}{n(n-2)}a_{tu}S^{tu} + \frac{ra}{n(n-1)(n-2)}, \qquad \lambda = \lambda_{t,u}g^{tu},$$

$a = a_{tu}g^{tu}$ and r is the scalar curvature of (M, g).

4. Projective structures in Sinyukov manifolds

We consider a principal fiber bundle $P^2(M)$ of 2-frames over M with structure group $G^2(n)$ — the group of second order jets on $\mathbb{R}^n$ (s. [6]). A projective structure on a manifold M is a subbundle P of $P^2(m)$ with structure group $G(n) \subset G^2(n)$, where

$$G(n) = \left\{ \begin{pmatrix} A & 0 \\ u & c \end{pmatrix} \in SL(n+1; \mathbb{R}) \right\} \Big/ \, Center,$$

where $c \in \mathbb{R}$, $A \in GL(n; \mathbb{R})$ and u is an n-row vector. In real case the group G is the transformation group of n-dimensional real projective space $RP^n = PGL(n; \mathbb{R})/G(n)$, where $PGL(n; \mathbb{R}) = SL(n+1; \mathbb{R})/Center$.

Let $\mathfrak{U}$ be the set of all affine connections on M whose torsion is zero. We introduce a relation ρ in $\mathfrak{U}$ as follows: $\overline{\nabla}\rho\nabla$ iff there exists a 1-form ψ on M such that the condition (2) holds on M. It is obvious that ρ is an equivalent relation. As an immediate corollary to Proposition 12 of [6], we have

Proposition. *The class $\mathfrak{U}/\rho$ is in a one-to-one correspondence with the projective structures of M.*

Two affine connections without torsion are projectively related if they belong to the same projective structure. If two affine connections $\overline{\nabla}$ and ∇ without torsion belong to the same projective structure, then they have the same geodesics.

Let Γ^0 and ∇ be respectively a tensor field of type $(0, 2)$ and an affine connection on M. Denote by Γ^0_{ij} and Γ^k_{ij} respectively the components of Γ^0 and ∇ in the local coordinate system (U, x^i). If there exists a tensor field Γ^0 and an affine connection ∇ on M, then a projective connection ω can be constructed in the same way as in the paper [5]. The functions Γ^α_{ij} are called the components of the projective connection ω with respect to (U, x^i).

Let (Γ^0, ∇) and $(\overline{\Gamma}^0, \overline{\nabla})$ be two pairs of tensor fields of type $(0, 2)$ and affine connections on M. We assume moreover that there exists a 1-form ψ on M and the following relations are valid

$$\tag{16} \overline{\Gamma}^k_{ij} = \Gamma^k_{ij} + \delta^k_i \psi_j + \delta^k_j \psi_i,$$

$$\tag{17} \overline{\Gamma}^0_{ij} = \Gamma^0_{ij} - \psi_i \psi_j + \psi_{i,j},$$

where

$$\psi_{i,j} = \partial_j \psi_i - \Gamma^t_{ij} \psi_t \qquad \text{(s. (2))}.$$

When these conditions are satisfied, we say that (Γ^0, ∇) is projectively equivalent to (Γ^0, ∇). Denote by R^l_{ijk} the components of the curvature tensor field of the affine connection Γ^k_{ij} without torsion. Putting $S_{ij} = R^t_{ijt}$, we define a tensor field Γ^0 of type $(0, 2)$ by

$$(18) \qquad \Gamma^0_{ij} = \frac{n}{n^2 - 1} S_{ij} + \frac{1}{n^2 - 1} S_{ji}.$$

Using Γ^0 and Γ^k_{ij} we can define a normal projective connection ω on M ([5]). In this case we say that ω is the normal projective connection of the affine connection ∇. If two pairs (Γ^0, ∇) and $(\overline{\Gamma}^0, \overline{\nabla})$ define respectively the normal projective connections of the affine connections ∇ and $\overline{\nabla}$ satisfying (16), then these two pairs are projectively equivalent. For each projective structure P of a manifold M, there is a unique normal projective connection ([5], [6]).

Let (M, g) be a Riemannian or pseudo-Riemannian manifold covered by a system of coordinate neighborhoods (U, x^i). We consider an associated $(n + 1)$-dimensional manifold M' and define a set of functions on M' by

$$(19) \qquad \begin{aligned} \Pi^\alpha_{ij} &= \Gamma^\alpha_{ij}, \\ \Pi^\beta_{\alpha 0} &= \Pi^\beta_{0\alpha} = C\delta^\beta_\alpha, \end{aligned}$$

where $C = \text{const.} \neq 0$, $\alpha, \beta = 0, 1, 2, \ldots, n$; $i, j, k = 1, 2, \ldots, n$ and Γ^k_{ij} are the Christoffel symbols of (M, g) ([1]). The functions (19) are the components of an affine connection in the manifold M'. The components $\Pi^\delta_{\alpha\beta\gamma}$ of the curvature tensor field of the connection Π vanish identically save

$$(20) \qquad \Pi^0_{ijk} = \Gamma^0_{ik,j} - \Gamma^0_{ij,k}$$

and

$$(21) \qquad \Pi^l_{ijk} = R^l_{ijk} + C(\delta^l_j \Gamma^0_{ik} - \delta^l_k \Gamma^0_{ij}),$$

where R are the components of the curvature tensor of (M, g) and the comma denotes covariant differentiation with respect to g. If the projective connection (19) is a normal connection on M then $\Gamma^0_{ij} = \frac{1}{C(n-1)} S_{ij}$ (i.e. $\Pi_{ij} = \Pi^\alpha_{ij\alpha} = 0$). In this case the curvature tensor (21) is the Weyl projective curvature tensor (10), it depends only on the projective structure P. After straightforward calculations we have

Theorem 2. *The manifold* (M', Π) *is a Riemannian or pseudo-Riemannian manifold and* Π *is the Levi–Civita connection iff the tensor*

$$(22) \qquad C P_{ij} = \mu'_{,ij} - \mu \Gamma^0_{ij}, \qquad \mu' = \frac{\mu}{2C},$$

satisfies the condition

$$(23) \qquad P_{ij,k} = \mu'_{,i} \Gamma^0_{jk} + \mu'_{,j} \Gamma^0_{ik},$$

where $\mu \in \mathfrak{F}(U)$ *and the comma denotes covariant differentiation with respect to* g. *The metric of* (M', Π) *is of the form*

$$(24) \qquad \begin{aligned} p_{00} &= \exp(2Cx^0)\mu, \\ p_{0i} &= p_{i0} = \frac{1}{2C} \exp(2Cx^0)\mu_{,i}, \\ p_{ij} &= \exp(2Cx^0)P_{ij}. \end{aligned}$$

Theorem 3. *If a manifold* (M, g) *is a Sinyukov manifold, then the manifold* (M', Π) *is a pseudo-Riemannian manifold with metric tensor* $p_{\alpha\beta}$ *of the form* *(24) and*

$$(25) \qquad \begin{aligned} CP_{ij} &= \frac{1}{2} \frac{n}{n-2} [\tau a_i^t a_{tj} + h_1 a_{ij} + h_2 g_{ij}], \\ \Gamma^0_{ij} &= \tau' a_{ij} + (\sigma + d) g_{ij}, \end{aligned}$$

where $h_1, h_2, d, \tau' = \text{const.} \neq 0, \sigma \in \mathfrak{F}(U)$.

Proof. Every Sinyukov manifold satisfies the conditions (11), (15) and ([3])

$$\sigma = \frac{n}{(n-1)(n+2)} r + d_1,$$

$$\nu = \frac{n-2}{2n} \sigma + d_2,$$

$$\nu_i = \tau \lambda_i, \qquad a_{it} \lambda^t = \theta(\lambda)\lambda_i, \qquad \theta'(\lambda) = 2,$$

$$(n-2)\varphi = \frac{4(n-1)}{n-2} \tau \lambda^2 + d_3,$$

where r is the scalar curvature on (M, g), $d_1, d_3, d_2 = \text{const.}$, $\lambda \in \mathfrak{F}(U)$. Hence, in view of (4), we obtain

$$\frac{n}{2} \lambda_{,ij} = CP_{ij} + \lambda \Gamma^0_{ij},$$

$$CP_{ij,k} = \frac{\lambda_i}{2} \Gamma^0_{jk} + \frac{\lambda_j}{2} \Gamma^0_{ik},$$

where P and Γ^0 are of the form (25) and $\tau' = \frac{n\tau}{n-2}$. Using Theorem 2 we obtain our assertion. ∎

Now Theorem 1 and Theorem 3 imply

Theorem 4. *A manifold* (M, g) *is a Sinyukov manifold iff the normal projective connections* (Γ^0, ∇) *and* $(\overline{\Gamma}^0, \overline{\nabla})$, *where*

$$C(n-1) \cdot \Gamma^0_{ij} = \tau a_{ij} + (\sigma + d) g_{ij},$$
$$C(n-1) \cdot \overline{\Gamma}^0_{ij} = \sigma_1 g_{ij}$$
$$\text{or} \quad C(n-1) \cdot \overline{\Gamma}^0_{ij} = \overline{\tau g}_{ij} + \sigma_2 g_{ij},$$

$\sigma_1, \sigma_2 \in \mathfrak{F}(U)$, $\overline{g}$ *is the metric tensor and* $\overline{\nabla}$ *is the Levi–Civita connection with respect to* $\overline{g}$, *are projectively equivalent.*

References

[1] L. BERWALD, On the Projective Geometry of Paths, *Annals of Math.* **37** (1936), 879–898.

[2] S. FORMELLA, On Geodesic Mappings in Einstein Manifolds, Coll. Math. Soc. J. Bolyai 46, 1984, 483–492.

[3] S. FORMELLA, Generalized Einstein Manifolds, *Rend. Circ. Matem. Palermo* (to appear).

[4] S. FORMELLA, Projective Structures in Riemannian and Einstein Manifolds (to appear).

[5] S. ISHIHARA, Groups of Projective Transformations on a Projectively Connected Manifold, *Japan J. Math.* **25** (1955), 37–80.

[6] S. KOBAYASHI and T. NAGANO, On Projective Connections, *J. Math. Mech.* **13** (1964), 215–236.

[7] J. MIKEŠ, On Geodesic Mappings of Einstein Spaces, *Matem. Zamiet.* **28** (1980), 935–938.

[8] J. MIKEŠ, Geodesic Mappings of Special Riemannian Spaces, Coll. Mathem. Soc. J. Bolyai, 46, 1984, 793–813.

[9] N. S. SINYUKOV, Geodesic Mappings of Riemannian Spaces, Moscow, Nauka, 1979.

[10] E. N. SINYUKOVA, Geodesic Mappings of Spaces L_n, *Izv. Vyssh. Ucheb. Zav. Mat.* **3** (1982), 57–61.

[11] P. VENZI, Über konforme und geodätische Abbildungen, *Result. der Mathemat.* **5** (1982), 184–198.

Stanisław Formella

ul. Klemensiewicza 10/4.
PL – 70-028
Szczecin

COLLOQUIA MATHEMATICA SOCIETATIS JÁNOS BOLYAI

56. DIFFERENTIAL GEOMETRY, EGER (HUNGARY), 1989

Horizontal Lift of Linear Connections to Vector Bundles Associated with the Principal Bundle of Linear Frames

J. GANCARCEWICZ and N. RAHMANI

0. Preliminaries

Let $\pi : E \to M$ be a vector bundle with a standard fibre F — (where F is a real vector space of dimension N) — associated with the principal fibre bundle of linear frames LM. We denote by $\underline{E}$ the set of sections of class C^∞ of E and by $X(M)$ (resp. $X(E)$) the $C^\infty(M)$-module of all vectors fields of class C^∞ on M (resp. on E), where $C^\infty(M)$ is the set of real-valued functions on M of class C^∞.

A connection on the total space E may be viewed as a mapping

$$\widetilde{\nabla} : X(M) \times \underline{E} \to \underline{E}$$
$$(X, S) \to \widetilde{\nabla}_X S$$

satisfying the following conditions:

(i) $\widetilde{\nabla}_{fX} S = f \widetilde{\nabla}_X S,$ (ii) $\widetilde{\nabla}_{X+Y} S = \widetilde{\nabla}_X S + \widetilde{\nabla}_Y S,$

(iii) $\widetilde{\nabla}_X fS = f\widetilde{\nabla}_X S + X(f)S,$ (iv) $\widetilde{\nabla}_X (S_1 + S_2) = \widetilde{\nabla}_X S_1 + \widetilde{\nabla}_X S_2$

for all vector fields X, Y, functions f on M and sections S on E, for complementary results and information see [1], [3] volume I and [4].

Let (V, Φ) be a chart on M, where M is a C^∞ manifold of dimension n, so a trivialization of E over U is a diffeomorphism $\widetilde{\Phi}$ such that the following diagram

$$E|_U = \pi^{-1}(U) \quad \xrightarrow{\;\widetilde{\Phi}\;} \quad U \times F$$
$$\pi \searrow \qquad \swarrow P_1$$
$$U$$

commutes, that is $P_1 \circ \widetilde{\Phi} = \pi \Leftrightarrow P_1 = \pi \circ \widetilde{\Phi}^{-1}$.

Let $E_1, \ldots, E_N$ be a fixed basis of F; then we consider the sections $\rho_1, \ldots, \rho_N$ of $E|_U$ defined by

$$(1) \qquad\qquad \rho_a(x) = \widetilde{\Phi}^{-1}(x, E_a),$$

they are called the adapted sections to the trivialization of E over U.

Let $(U, x^1, \ldots, x^n)$ be the local coordinates on M. There is a uniquely determined Γ^b_{ia} on U such that

$$(2) \qquad\qquad \widetilde{\nabla}_{\partial_i} \rho_a = \Gamma^b_{ia} \rho_b.$$

Let $\gamma : (a, b) \to M$ be a curve of class C^∞ and let $J_\gamma(E)$ be the set of all sections of E defined along γ, that is an element of $J_\gamma(E)$ is a mapping $\rho : (a, b) \to E$ of class C^∞ such that $\pi \circ \rho = \gamma$. So a connection $\widetilde{\nabla}$ in E defines a mapping

$$\widetilde{\nabla}_{\dot{\gamma}} : J_\gamma(E) \to J_\gamma(E)$$

called the covariant differentiation along γ. If $S = S^a (\rho_a \circ \gamma)$ is an element of $J_\gamma(E)$ then for local coordinates $(U, x^1, \ldots, x^n)$ in M we have

$$(3) \qquad\qquad \widetilde{\nabla}_{\dot{\gamma}} S = \left\{ \frac{d}{dt} S^a + (\Gamma^a_{ib} \circ \gamma) \frac{d}{dt} \gamma^i S^b \right\} \rho_a$$

where $\gamma^i = x^i \circ \gamma$, $i = 1, \ldots, n$; $a = 1, \ldots, N$.

Proposition 0.1. *If $\gamma : (a, b) \to M$ is a curve, y is an element of $E_{\gamma(t_0)} = \pi^{-1}(\gamma(t_0))$ for some $t_0 \in (a, b)$, then there is one and only one section $S \in J_\gamma(E)$ such that*

$$\text{(i)} \quad S(t_0) = y, \qquad \text{(ii)} \quad \widetilde{\nabla}_{\dot{\gamma}} S = 0.$$

Proof. This is a classical result which could be found in [1].   ∎

Let y be a fixed element of E and $x = \pi(y)$. We denote by H_y the set of all velocity vectors $\dot{S}(0)$, where $S : (-\varepsilon, \varepsilon) \to E$ is a section along $\gamma = \pi \circ S$ satisfying the conditions (i) and (ii) of Proposition 0.1 with $t_0 = 0$.

Let $(U, x^1, \ldots, x^n)$ be the local coordinates in M and let $\widetilde{\Phi} : E|_U = \pi^{-1}(U) \leftrightarrow U \times F$ be a trivialization of E over U, then we can define the induced local coordinates $(\pi^{-1}(U), x^i, y^\alpha)$ on E by

$$x^i(y) = x^i(\pi(y)) \quad ; \quad y = y^\alpha(y)\rho_a$$

for all $y \in \pi^{-1}(U)$.

Let $\partial_1, \ldots, \partial_n, \delta_1, \ldots, \delta_N$ be the canonical frame associated to the induced chart $(\pi^{-1}(U), x^i, y^\alpha)$. If $X = \dot{\gamma}(0)$ is a velocity vector of γ and S is the unique section defined along γ satisfying the conditions (i) and (ii) of Proposition 0.1, then

$$\dot{S}(0) = X^i \partial_i - X^i \Gamma^b_{ia} y^a S_b. \tag{4}$$

If we denote for every $y \in E$, the vertical subspace of $T_y E$, by $V_y E = \mathrm{Ker}\, d_y \pi = T_y(E_{\pi(y)})$, we have the following decomposition of $T_y E$

$$T_y E = V_y E \oplus H_y, \tag{5}$$

in particular $\pi|_{H_y} : H_y \to T_{\pi(y)} M$ is an isomorphism so if X is a vector field on M, we can define the horizontal lift X^H of X to E by the formula

$$X^H y = \left(d_y \pi|_{H_y}\right)^{-1} \left(X_{\pi(y)}\right). \tag{6}$$

It is easy to verify the following proposition:

Proposition 0.2.

$$(X + Y)^H = X^H + Y^H; \qquad (fX)^H = f^V X^H; \qquad X^H(f^V) = (Xf)^V$$

for every $X, Y \in \mathfrak{X}(M)$ and $f \in C^\infty(M)$, where $f^V = f \circ \pi$.
So in local induced coordinates $(\pi^{-1}(U), x^i, y^\alpha)$,

$$X^H y = X^i(\pi(y))\partial_i - X^i(\pi(y))\Gamma^b_{ia}(\pi(y))y^a \delta_b. \tag{7}$$

Proof. This is an obvious result which could be found as the referee informs us in Poor's book [5] page 58 Proposition 2.31. ∎

1. Vertical lift of sections of E

Let $\pi : E \to M$ be a vector bundle with standard fibre F. A vector field $\tilde{X}$ on E is called projectable on M if there is a vector field X on M such that

$$(8) \qquad\qquad d\pi \circ \tilde{X} = X \circ \pi.$$

X is called the projection of $\tilde{X}$ and X is uniquely determined by $\tilde{X}$.

The set of all projectable vector fields on M is a Lie algebra and the projection mapping is a Lie homomorphism. We have the following proposition:

Proposition 1.1. *A vector field $\tilde{X}$ on E is projectable on a vector field X on M if and only if:*

$$(9) \qquad\qquad \left(\tilde{X}(f^V)\right) = (X(f))^V$$

for every $f \in C^\infty (M)$, where $f^V = f \circ \pi$.

Proof. See the notion of f-related vector fields in [4] volume I. ∎

A vector field $\tilde{X}$ is called vertical if for each point y of E, $\tilde{X}_y$ is a vertical vector, that is, $\tilde{X}_y$ belongs to $V_y E$. A vertical vector field on E is projectable on M and its projection is zero. We have the following obvious corollary:

Corollary 1.2. *Let $\tilde{X}$ be a vector field on E. $\tilde{X}$ is a vertical vector field on E if and only if:*

$$\tilde{X}\left(f^V\right) = 0 \text{ for every function } f \text{ on } M.$$

Since $E_{\pi(y)} = \pi^{-1}(\pi(y))$ is a vector space, for each point y of E there is a natural isomorphism, the canonical parallel translation,

$$\Psi_y : V_y E = T_y \left(E_{\pi(y)}\right) \to E_{\pi(y)}.$$

If $S : M \to E$ is a section of E (that is $\pi \circ S = Id$), then we can define a vector field on E, denoted by S^V and called vertical lift of S to E, by the formula

$$(10) \qquad S_y^V = \psi_y^{-1}\left(S_{\pi(y)}\right).$$

Using an induced chart on E, we have

$$(11) \qquad S^V = S^a \delta_a,$$

where $S = S^a \rho_a$ and $\rho_1, \ldots, \rho_N$ are the adapted sections. The definition by the formula (10) generalizes the definition of vertical lifts of vector field to the tangent bundle, defined by K. Yano and S. Ishihara [7] and vertical lifts of 1-forms to the cotangent bundle, defined by K. Yano and E. M. Patterson [8]. This definition generalizes also the definition of vertical lifts of tensors introduced by Jacek Gancarcewicz and Noureddine Rahmani [2] and [6]. The referee informs us in his report that a similar definition of vertical lifting of sections defined by the formula (10), could be found in Poor's book [5] page 38, 1.64 but we learned the existence of this result only after the completion of the present paper.

Proposition 1.3. *If S, S' are sections of E and f, g are functions on M then*

$$(S + S')^V = S^V + S'^V, \qquad (fS)^V = f^V S^V, \qquad S^V\left(f^V\right) = 0,$$

where $f^V = f \circ \pi$.

2. Horizontal lifting of linear connections on M to the total space E of the vector bundle $\pi : E \to M$ associated with principal bundle of linear frames

The main theorem of this paper is the following one:

Theorem 2.1. *Let Γ be a linear connection on M. If $\pi : E \to M$ is a vector bundle, then there is one and only one linear connection $\widetilde{\nabla}$ on the manifold E such that*

$$(i) \ \widetilde{\nabla}_{X^H} Y^H = (\nabla_X Y)^H, \qquad (ii) \ \widetilde{\nabla}_{X^H} S^V = (\nabla_X S)^V,$$
$$(iii) \ \widetilde{\nabla}_{S^V} X^H = 0, \qquad (iv) \ \widetilde{\nabla}_{S^V} S'^V = 0$$

for all vector fields X, Y on M and all sections S, S' of E, where ∇ is the covariant differentiation, with respect to vector fields on M, of sections of vector bundles associated to the principal fibre bundle of linear frames LM.

The article of R. Crittenden [1] is very helpful to understand the last sentence above. To prove this theorem we need the following lemma.

Lemma 2.2. Let Γ be a linear connection on M and $\widetilde{\nabla}$ be a linear connection on E. For a chart (U, x^i) on M we denote by

$$(12) \qquad \widetilde{\nabla}_{\partial_i}\partial_j = \widetilde{\Gamma}^k_{ij}\partial k + \widetilde{\Gamma}^a_{ij}\delta_a,$$

$$(13) \qquad \widetilde{\nabla}_{\partial_i}\delta_a = \widetilde{\Gamma}^j_{ia}\partial_j + \widetilde{\Gamma}^b_{ia}\delta_b,$$

$$(14) \qquad \widetilde{\nabla}_{\delta_a}\partial_i = \widetilde{\Gamma}^j_{ai}\partial_j + \widetilde{\Gamma}^b_{ai}\delta_b,$$

$$(15) \qquad \widetilde{\nabla}_{\delta_a}\delta_b = \widetilde{\Gamma}^i_{ab}\partial_i + \widetilde{\Gamma}^c_{ab}\delta_c$$

the Christoffel symbols of $\widetilde{\nabla}$ with respect to the induced chart on E. If the conditions of Theorem 2.1 are satisfied, then

$$(16) \qquad \widetilde{\Gamma}^k_{ij} = \Gamma^k_{ij}$$

$$(17) \qquad \widetilde{\Gamma}^a_{ij} = \left(\partial_i\Gamma^a_{jb} + \Gamma^a_{ic}\Gamma^c_{jb} - \Gamma^k_{ij}\Gamma^a_{kb}\right)y^b,$$

$$(18) \qquad \widetilde{\Gamma}^i_{ja} = 0,$$

$$(19) \qquad \widetilde{\Gamma}^b_{ia} = \Gamma^b_{ia},$$

$$(20) \qquad \widetilde{\Gamma}^j_{ia} = 0,$$

$$(21) \qquad \widetilde{\Gamma}^b_{ai} = \Gamma^b_{ia},$$

$$(22) \qquad \widetilde{\Gamma}^i_{ab} = 0,$$

$$(23) \qquad \widetilde{\Gamma}^a_{bc} = 0,$$

where Γ^i_{jk} are Christoffel symbols of the linear connection Γ and Γ^a_{ib}; $i = 1, \ldots, n$; $a, b = 1, \ldots, N$ are defined by the equation (2).

Proof. It can be seen that $\delta_\alpha = (\rho_\alpha)^V$, and we have from equation (7) that

$$(\partial_i)^H = \partial_i - \Gamma^b_{ic}y^c\delta_b \quad \text{so}$$

$$\widetilde{\nabla}_{\delta_a}(\partial_i)^H = \widetilde{\nabla}_{\delta_a}\partial_i = \delta_a\left(\Gamma^b_{ic}y^c\right)\delta_b - \Gamma^b_{ic}y^c\widetilde{\nabla}_{\delta_a}\delta_b.$$

From the conditions of Theorem 2.1 it follows that

$$\widetilde{\nabla}_{\delta_a}(\partial_i)^H = \widetilde{\nabla}_{\delta_a}\partial_i - \Gamma^b_{ia}\delta_b = 0,$$

so we obtain $\tilde{\Gamma}^j_{ai} = 0$ and $\tilde{\Gamma}^b_{ai} = \tilde{\Gamma}^b_{ia}$.

From the fact that $\tilde{\nabla}\delta_a \delta_b = \tilde{\Gamma}^l_{ab}\partial_l + \tilde{\Gamma}^c_{ab}\delta_c = 0$

it follows that $\tilde{\Gamma}^l_{ab} = 0$ and $\tilde{\Gamma}^c_{ab} = 0$ because $(\partial_1, \ldots, \partial_n, \delta_1, \ldots, \delta_N)$ is a

basis at every point $y \in \pi^{-1}(U)$. We also obtain $\tilde{\Gamma}^c_{ia} = \Gamma^c_{ia}$ and $\tilde{\Gamma}^l_{ia} = 0$

from $\tilde{\nabla}_{(\partial_i)^H} \delta_a = \tilde{\nabla}_{\partial_i}\delta_a - \Gamma^c_{ib}y^b \tilde{\nabla}_{\delta_c}\delta_a$

$$= \tilde{\Gamma}^l_{ia}\partial_l + \tilde{\Gamma}^c_{ia}\delta_c = (\nabla_{\partial_i}\rho_a)^V = \Gamma^c_{ia}\delta_c,$$

since $(\partial_1, \ldots, \partial_n, \delta_1, \ldots, \delta_N)$ is a basis at every $y \in \pi^{-1}(U)$. Finally we

consider

$$\tilde{\nabla}_{(\partial_i)^H}(\partial_j)^H = \tilde{\nabla}_{(\partial_i - \Gamma_i{}^a{}_b y\delta_a)}\partial_j - \Gamma^d_{jc}y^c\delta_d$$

$$= \tilde{\Gamma}^l_{ij}\partial_l + \tilde{\Gamma}^d_{ij}\delta_d - \Gamma^a_{jb}\Gamma^d_{ia}y^b\delta_d - \partial_i\Gamma^b_{jc}y^c\delta_b$$

$$= (\nabla_{\partial_i}\partial_j)^H = \Gamma^l_{ij}\partial_l - \Gamma^l_{ij}\Gamma^b_{lc}y^c\delta_b,$$

thus

$$\tilde{\Gamma}^l_{ij} = \Gamma^l_{ij},$$

$$\tilde{\Gamma}^d_{ij} = \left(\partial_i\Gamma^d_{jc} + \Gamma^a_{jc}\Gamma^d_{ia} - \Gamma^l_{ij}\Gamma^d_{lc}\right)y^c.$$

∎

Proof of Theorem 2.1. The uniqueness of a linear connection $\tilde{\nabla}$ on E satisfying conditions (i), (ii), (iii) and (iv) of Theorem follows, because according to Lemma 2.2 the Christoffel symbols of $\tilde{\nabla}$ are uniquely determined by the given linear connection on M. Thus we need to prove only the existence of $\tilde{\nabla}$.

Let (U, x^i) be a chart on M. We can define a linear connection $\tilde{\nabla}$ on $E|_U$ such that its Christoffel symbols with respect to the induced chart are given by formulas (16) to (23). This linear connection $\tilde{\nabla}$ on $E|_U$ satisfies these conditions

(24)

$$\tilde{\nabla}_{(\partial_i)^H}(\partial_j)^H = (\nabla_{\partial_i}\partial_j)^H,$$

$$\tilde{\nabla}_{(\partial_i)^H}(\rho_a)^V = (\nabla_{\partial_i}\rho_a)^V,$$

$$\tilde{\nabla}_{(\rho_a)^V}(\partial_i)^H = 0,$$

$$\tilde{\nabla}_{(\rho_a)^V}(\rho_\beta)^V = 0$$

for $i, j = 1, \ldots, n$ and $\alpha, \beta = 1, \ldots, N$. It is easy to prove that

(25)

$$\tilde{\nabla}_{X^H}Y^H = (\nabla_X Y)^H,$$

$$\tilde{\nabla}_{X^H}S^V = (\nabla_X S)^V,$$

$$\tilde{\nabla}_{S^V}X^H = \tilde{\nabla}_{S^V}S'^V = 0$$

for all vector fields X, Y on U and all sections S, S' of $E|_U$. For example we prove the first formula amongst (25). Let X and Y be vector fields on U. If we denote by

$$X = X^i \partial_i, \qquad Y = Y^j \partial_j$$

the coordinates of X and Y with respect to the chart (U, x_i), according to Proposition 0.2 we have

$$X^H = (X^i)^V (\partial_i)^H, \qquad Y^H = (Y^j)^V (\partial_j)^H$$

and hence, using the formulas (24), we obtain

$$
\begin{aligned}
\tilde{\nabla}_{X^H} Y^H &= (X^i)^V \left((\partial_i)^H \right) \left((Y^j)^V \right) (\partial_j)^H + (Y^j)^V \tilde{\nabla}_{(\partial_i)^H} (\partial_i)^H \\
&= (X^i)^V \left((\partial_i Y^j)^V \right) (\partial_j)^H + (Y^j)^V \left(\nabla_{\partial_i} \partial_j^H \right)^H \\
&= \left(X^i (\partial_i Y^j) \partial_j + Y^j \nabla_{\partial_i} \partial_j \right)^H \\
&= (\nabla_X Y)^H.
\end{aligned}
$$

The other formulas of (25) can be established in a similar way. If (U, x^i) and (U', x'^i) are two charts on M, then we can define two linear connections $\tilde{\nabla}$ and $\tilde{\nabla}'$ respectively on $E|_U$ and $E|_{U'}$. From (25) we have

$$
\begin{aligned}
\tilde{\nabla}_{X^H} Y^H &= (\nabla_X Y)^H = \tilde{\nabla}'_{X^H} Y^H, \\
\tilde{\nabla}_{X^H} S^V &= (\nabla_X S)^V = \tilde{\nabla}'_{X^H} S^V, \\
\tilde{\nabla}_{S^V} X^H &= \tilde{\nabla}'_{S^V} X^H = 0, \\
\tilde{\nabla}_{S^V} S'^V &= \tilde{\nabla}'_{S^V} S'^V = 0
\end{aligned}
$$

for all vector fields X, Y on $U \cap U'$ and all sections S, S' of $E|_{U \cap U'} = (E|_U) \cap (E|_{U'})$. Hence by Lemma 3.1 the linear connections $\tilde{\nabla}$ and $\tilde{\nabla}'$ coincide on $E|_{U \cap U'}$. Using an atlas on M, we can define linear connection $\tilde{\nabla}$ on E. This connection $\tilde{\nabla}$ satisfies the conditions of Theorem 2.1 and the proof is complete. ∎

The linear connection $\tilde{\nabla}$ on E satisfying conditions (i), (ii), (iii), (iv) of Theorem 2.1 is called horizontal lift of Γ from M to E. The following three corollaries are immediate consequences of Theorem 2.1.

Corollary 2.3. (K. Yano, S. Ishihara [7]). *If ∇ is a linear connection on M, then there is one and only one linear connection $\widetilde{\nabla}$ on TM such that*

$$\widetilde{\nabla}_{X^H} Y^H = (\nabla_X Y)^H, \quad \widetilde{\nabla}_{X^H} Y^V = (\nabla_X Y)^V, \quad \widetilde{\nabla}_{X^V} Y^H = \nabla_{X^V} Y^V = 0$$

for all vector fields X, Y on M, where X^H is the horizontal lift of X to TM with respect to ∇.

Corollary 2.4. (K. Yano, E. M. Patterson [8]). *If ∇ is a linear connection on M, then there is one and only one linear connection $\widetilde{\nabla}$ on T^*M such that*

$$\widetilde{\nabla}_{X^H} Y^H = (\nabla_X Y)^H; \quad \widetilde{\nabla}_{X^H} \phi^V = (\nabla_X \phi)^V; \quad \widetilde{\nabla}_{\phi^V} X^H = \widetilde{\nabla}_{\phi^V} \omega^V = 0$$

for all vectors fields X, Y on M and all 1-forms ϕ, ω on M where X^H is the horizontal lift of X to T^*M with respect to ∇.

Next we will study the torsion tensor and the curvature tensor of the horizontal lift of the linear connection ∇ to the vector bundle $E(M, \pi, F)$ associated with principal fibre bundle of linear frames LM.

Using the local coordinates, we can easily prove that

$$(26) \qquad \qquad \left[X^H, S^V\right] = (\nabla_X S)^V$$
$$(27) \qquad \qquad \left[S^V, S'^V\right] = 0$$

for every vector field X on M and sections S and S' of E.

We shall denote the difference $\left[X^H, Y^H\right] - [X, Y]^H$ by $(R(X, Y))^\square$ where $R(X, Y)$ is the curvature transformation of the linear connection ∇. We have then the following proposition:

Proposition 2.5. *Let E be a vector bundle associated to LM and ∇ be a linear connection on M. If $\widetilde{\nabla}$ is the horizontal lift of ∇ to E and $\widetilde{T}$ is the torsion tensor of $\widetilde{\nabla}$, then we have*

$$\widetilde{T}\left(X^H, Y^H\right) = (T(X, Y))^H - (R(X, Y))^\square,$$
$$\widetilde{T}\left(X^H, S^V\right) = \widetilde{T}\left(S^V, S'^V\right) = 0$$

for all vector fields X, Y on M and all sections S, S' of E, where T is the torsion tensor of ∇.

Proof. Using Theorem 2.1 and the formulas (26) and (27), we have

$$\tilde{T}\left(X^H, Y^H\right) = \tilde{\nabla}_{X^H} Y^H - \tilde{\nabla}_{Y^H} X^H - \left[X^H, Y^H\right]$$
$$= (\nabla_X Y)^H - (\nabla_Y X)^H - [X, Y]^H - \left(\left[X^H, Y^H\right] - [X, Y]^H\right)$$
$$= (T(X, Y))^H - (R(X, Y))^\square,$$
$$\tilde{T}\left(X^H, S^V\right) = \tilde{\nabla}_{X^H} S^V - \tilde{\nabla}_{S^V} X^H - \left[X^H, S^V\right] =$$
$$= (\nabla_X S)^V - (\nabla_X S)^V = 0,$$
$$T\left(S^V, S'^V\right) = \tilde{\nabla}_{S^V} S'^V - \tilde{\nabla}_{S'^V} S^V - \left[S^V, S'^V\right] = 0.$$

∎

To calculate the curvature tensor of $\tilde{\nabla}$, it is useful to remark that we can demonstrate, using local coordinates, that

(28) $$\tilde{\nabla}_{[R(X,Y)]^\square} S^V = 0 \quad \text{and}$$

(29) $$\tilde{\nabla}_{[R(X,Y)]^\square} Z^H = 0$$

where X, Y, Z are vector fields on M, S is a section of E and $R(X,Y)$ is a curvature transformation of the connection ∇ given on M. Then we have the following proposition:

Proposition 2.6. *If $\tilde{\nabla}$ is the horizontal lift of the linear connection given on M to a vector bundle E associated with LM and $\tilde{R}$ is the curvature tensor of $\tilde{\nabla}$, then*

$$\tilde{R}\left(X^H, Y^H\right) Z^H = (R(X, Y)Z)^H$$
$$\tilde{R}\left(X^H, Y^H\right) S^V = (R(X, Y)S)^V$$
$$\tilde{R}\left(X^H, S^V\right) = \tilde{R}\left(S^V, S'^V\right) = 0$$

for all vector fields X, Y, Z on M and all sections S, S' of E, where $R(X, Y)$ is the curvature transformation of ∇.

Proof. Using Theorem 2.1 and formulas (28) and (29), we have

$$\tilde{R}\left(X^H, Y^H\right) Z^H = \tilde{\nabla}_{X^H}\left(\tilde{\nabla}_{Y^H} Z^H\right) - \tilde{\nabla}_{Y^H}\left(\tilde{\nabla}_{X^H} Z^H\right) -$$
$$- \tilde{\nabla}_{[X^H, Y^H]} Z^H$$
$$= (\nabla_X \nabla_Y Z)^H - (\nabla_Y \nabla_X Z)^H - (\nabla_{[X,Y]} Z)^H - \tilde{\nabla}_{R(X,Y)^\square} Z^H$$
$$= (R(X, Y)Z)^H,$$

$$\widetilde{R}\left(X^H, Y^H\right) S^V = \widetilde{\nabla}_{X^H}\left(\widetilde{\nabla}_{Y^H} S^V\right) - \widetilde{\nabla}_{Y^H}\left(\widetilde{\nabla}_{X^H} S^V\right) - \nabla_{[X^H, Y^H]} S^V$$
$$= \left(\nabla_X \nabla_Y S\right)^V - \left(\nabla_Y \nabla_X S\right)^V - \left(\nabla_{[X,Y]} S\right)^V - \nabla_{R(X,Y)} \square S^V$$
$$= (R(X,Y)S)^V$$

Similarly we can prove

$$\widetilde{R}\left(X^H, S^V\right) X^H = 0 \quad , \quad \widetilde{R}\left(X^H, S^V\right) S'^V = 0,$$
$$\widetilde{R}\left(S^V, S'^V\right) X^H = 0 \quad , \quad \widetilde{R}\left(S^V, S'^V\right) S''^V = 0$$

for all vector fields X, Y, Z on M and all sections S, S', S'' of E. ∎

From Proposition 2.5, we have:

Proposition 2.7. *Let $\widetilde{\nabla}$ be the horizontal lift of the linear connection ∇ on M to a vector bundle E associated with LM. If the linear connection $\widetilde{\nabla}$ is without torsion, then $\widetilde{\nabla}$ is without torsion if and only if the curvature transformation $R(X,Y)$ is zero for all vector fields X, Y on M.*

From Proposition 2.6 we have:

Proposition 2.8. *Let $\widetilde{\nabla}$ be the horizontal lift of a linear connection ∇ on M to a vector bundle E associated with LM. Then the linear connection $\widetilde{\nabla}$ is flat (that is, $\widetilde{R} = 0$) if and only if the curvature transformation $R(X,Y)$ of ∇ is zero for all vector fields X, Y on M.*

Proposition 2.7 and 2.8 generalize similar propositions proved by K. Yano and S. Ishihara in the case of tangent bundles [7] and by K. Yano and E. M. Patterson in the case of cotangent bundles [8].

References

[1] R. CRITTENDEN, Covariant differentiation, *Quart. J. Math. Oxford (2)*, **13** (1962), 285–298.

[2] J. GANCARCEWICZ and N. RAHMANI, Relèvement horizontal de champs de tenseurs de type (1,1) de M à $T^*M \otimes TM$, (sous presse).

[3] S. KOBAYASHI and K. NOMIZU, Foundations of Differential Geometry I, New-York–London, (1963).

[4] PHAM–MAN–QUAN, Introduction à la géométrie des variétés différentiables, Dunod–Paris, (1969).

[5] W. A. POOR, Differential Geometric Structures, Mc Graw–Hill, (1981).

[6] N. RAHMANI, Relèvement horizontal de champs de tenseurs de type (1,1) sur M à $T_q^p(M)$, Proceedings of the conference on differential geometry and its applications, Nové Mesto na moravé, Czechoslovakia, September 5–9, (1983).

[7] K. YANO and S. ISHIHARA, Horizontal lifts of tensor fields and connections to tangent bundles, *Journ. Math. and Mech.* **16** (1967), 1015–1030.

[8] K. YANO and E. M. PATTERSON, Horizontal lift from a manifold to its cotangent bundle, *Journ. Math. Soc. Japan* **19** (1967), 185–198.

Noureddine Rahmani

Laboratoire de Mathématiques et
Informatique
4, rue des Frères Lumière
68093 Mulhouse Cédex
France

Jacek Gancarcewicz

Istitut Matematyki
Universytet Jagiellonski
UR. Reymonta P. V
Krakow, Poland

Functorality and Heat Equation Asymptotics[*]

P. B. GILKEY [**]

§0 Introduction

Let M be a bounded domain in $\mathbb{R}^2$ and let $\Delta = \delta d$ be the Laplacian on functions. In his famous paper "Can one hear the shape of a drum" [Ka], M. Kac asked the question of whether the spectrum determines the geometry of M up to isometry. One thinks of M as comprising the head of a drum; the spectrum of Δ with Dirichlet boundary conditions gives the frequencies of the drum. Hence the somewhat fanciful title of the article. More generally, one can ask the same question for an arbitrary compact Riemannian manifold M of arbitrary dimension m and the question is more tractable in this setting.

The asymptotics of the heat equation are a fundamental tool in spectral geometry. We suppose for the moment $dM = 0$ to simplify the discussion. Let $\{\lambda_\nu\}$ be the eigenvalues of Δ where each eigenvalue is repeated according to multiplicity. The fundamental solution of the heat equation, $e^{-t\Delta}$, is of

---

[*] This paper is in final form and no version of it will be submitted for publication elsewhere.

[**] Research partially supported by the NSF and NSA

trace class in L^2 and as $t \to 0^+$ there is an asymptotic series of the form:

$$Tr_{L^2}(e^{-t\Delta}) = \sum_{\nu} e^{-t\lambda_\nu} \simeq \sum_{n \geq 0} t^{n-m/2}\, a_n(\Delta).$$

The $a_n(\Delta)$ are locally computable i.e. there exist local invariants $a_n(x, \Delta)$ so

$$a_n(\Delta) = \int_M a_n(x, \Delta).$$

Although the eigenvalues of Δ reflect global information, the $a_n(x, \Delta)$ are determined by the local geometry of the manifold. They are complicated expressions in the curvature tensor and its covariant derivatives. Let R_{ijkl} be the components of the curvature tensor of the Levi–Civita connection. We use the sign convention $R_{1212} = -1$ on the unit sphere in $\mathbb{R}^3$. We shall sum over repeated indices with respect to a local orthonormal frame for the tangent space. Let

$$\rho_{ij} = -R_{ikjk} \qquad \text{and} \qquad \tau = \rho_{ii}$$

be the Ricci tensor and the scalar curvature. Then (see [Gi–4])

$$a_0(\Delta) = (4\pi)^{-m/2} \int_M 1$$

$$a_1(\Delta) = 6^{-1}(4\pi)^{-m/2} \int_M \tau$$

$$a_2(\Delta) = 360^{-1}(4\pi)^{-m/2} \int_M 5\tau^2 - 2|\rho|^2 + 2|R|^2$$

$$a_3(\Delta) = 45360^{-1}(4\pi)^{-m/2} \int_M -142|\nabla\tau|^2 - 26|\nabla\rho|^2 - 7|\nabla R|^2 +$$

$$+ 35\tau^3 - 42\tau|\rho|^2 + 42\tau|R|^2 - 36|\rho^3| - 20\rho_{ij}\rho_{kl}R_{ikjl} -$$

$$- 8\rho_{ij}R_{ikln}R_{jkln} - 24R_{ijkl}R_{ijnp}R_{klnp}.$$

The first term in the asymptotic expansion is the volume of M so one can "hear" both the volume and the dimension. If $m = 2$, the Euler Poincare characteristic $\chi(M)$ is a spectral invariant by the Gauss Bonnet theorem; consequently one can "hear" the underlying topological type of a Riemann surface. One can "hear" a number of other geometric properties. For example, in low dimensions, Berger has shown the unit sphere in $\mathbb{R}^{m+1}$ is characterized by its spectrum [Be]. Similarly, Patodi [Pa] has shown that by taking into account the spectrum of the Laplacian $\Delta_p = d\delta + \delta d$ on p forms, one can "hear" constant scalar curvature and constant sectional curvature. There are many other results in this direction. We sketch Berger's argument in dimension 2 and Patodi's argument in general to illustrate the flavor.

If P is a local invariant, let $P[M] = \int_M P$ be the corresponding integrated invariant. Suppose first $m = 2$. Then $1[M]$, $\tau[M]$, and $\tau^2[M]$ are spectral invariants since these are non zero multiples of $a_0(\Delta)$, $a_1(\Delta)$, and $a_2(\Delta)$. Since M has constant scalar curvature c if and only if

$$(c - \tau)^2[M] = c^2[M] - 2c\tau[M] + \tau^2[M] = 0,$$

this property is spectrally determined. The sphere S^2 and projective space RP^2 are the only Riemann surfaces of constant curvature $+1$ and they are distinguished by their volume; thus one can "hear" S^2 and RP^2. In higher dimension, there are three linearly independent quadratic integrated invariants $\tau^2[M]$, $|\rho|^2[M]$, and $|R|^2[M]$. Patodi showed

$$a_2(\Delta_p) = c_1(m, p)\tau^2[M] + c_2(m, p)|\rho|^2[M] + c_3(m, p)|R|^2[M]$$

and computed the coefficients $c_i(m, p)$. The matrix of coefficients for $p = 0, 1, 2$ in non singular so $\tau^2[M]$, $|\rho|^2[M]$, and $|R|^2[M]$ are determined by the spectrum of the Laplacians Δ_p for $p = 0, 1, 2$; his result now follows.

This gives the flavor of such results; one computes the asymptotics of the Laplacian for certain geometrically occurring Laplacians and then asks what geometrical data are determined. Most of the work in this direction has been for manifolds without boundary as it is only very recently that the appropriate formulas for manifolds with boundary have become available.

Unfortunately, the answer to Kac's question in complete generality is no. Milnor [Mi] showed there exist two 16 dimensional tori which are isospectral but not isometric. Milnor's isospectral tori have the same underlying topology. It was very surprising therefore when Vigneras [Vi] constructed examples of 3 dimensional manifolds with constant curvature -1 which are isospectral and which have different fundamental groups. Thus one can not "hear" the topology of a manifold. Vigneras also constructed isospectral but not isometric Riemann surfaces with constant curvature -1; examples are now known for any genus $g \geq 4$ [Br]. Ikeda [Ik] and Urakawa [Ur] provided similar examples of isospectral but not isometric manifolds with constant curvature $+1$ and 0. Later Sunada [Su] outlined a general procedure for finding such examples. We refer to the survey by R. Brooks [Br] for further details.

The examples cited above form discrete families. Gordon and Wilson [GiWi] have constructed continuous families of isospectral non isometric manifolds. Gordon [Go] also constructed manifolds which are isospectral on functions but not in 1 forms so one has to be a bit careful about what category one is working in. Recently Brooks and Gordon [BrGo] have constructed continuous families of isospectral non isometric manifolds all of which are contained in a conformal class.

All the known families can be compactified. This leads one to ask whether in general isospectral metrics form compact sets modulo gauge equivalence. This question has been answered in the affirmative by Osgood et al [OsPhSa] for compact Riemann surfaces and for planar domains. Brooks at al [BrPeYa, CaYa] have a similar compactness theorem for isospectral metrics within a fixed conformal class if the dimension is 3. These compactness theorems are an interesting mixture of global analysis and local analysis. The heat equation asymptotics again play a crucial role, both for beginning the arguments and also for providing the inductive Sobolev arguments. For these arguments one needs not only knowledge of the asymptotics a_0, a_1, and a_2 but also information concerning the *leading terms* in the asymptotic expansion. Let

$$\varepsilon_n = (4\pi)^{-m/2}(-1)^n/\{2^{n+1} \cdot 1 \cdot 3 \cdot \ldots \cdot (2n+1)\}.$$

Then modulo terms which involve lower order jets of the metric, if $n \geq 3$

$$a_n(\Delta) = \varepsilon_n \int_M (n^2 - n - 1)|\nabla^{n-2}\tau|^2 + 2|\nabla^{n-2}\rho|^2 + \ldots.$$

Since $\int_M |\nabla^{n-2}R|^2 = \int_M 4|\nabla^{n-2}\rho|^2 - |\nabla^{n-2}\tau|^2 + \ldots$, this controls the higher order jets of the curvature tensor.

In this paper, we will discuss a functorial approach to computing heat equation asymptotics for manifolds with boundary; we hope these results will lead eventually to an understanding of manifolds with boundary similar to that for manifolds without boundary. We refer to work by Kennedy et al [KeCrDo], and Moss et al [MoDo] for other recent approaches to this problem. We shall also determine the leading terms in the asymptotics since the same functorial methods apply to this as well. We shall work quite generally to obtain formulas valid for the Laplacian on p forms, the spin Laplacian, the Dolbeault Laplacian, etc. We shall wherever possible emphasize the functorial nature as opposed to purely computational details.

In §1, we state the main theorem of this paper concerning boundary asymptotics. In $2, we outline the functorial properties of the heat equation we will use to compute the boundary asymptotics. In $3, we compute the interior asymptotics and the asymptotics associated with Dirichlet boundary conditions. In §4, we study Neumann boundary conditions and mixed boundary conditions. In §5 we compute the leading terms in the asymptotics of the heat equation. We conclude with a fairly extensive bibliography as a guide to further results. We refer to joint work with Branson and Orsted ([BrGi] and [BrGiOr]) for further details concerning the methods we will use here.

§1 Local formulas

We consider a class of operators which includes all Laplacians which occur in differential geometry. Let M be a compact Riemannian manifold with boundary dM and let V be a smooth vector bundle over M. Let P be an arbitrary second order differential operator with leading symbol given by the metric tensor. In other words, assume P has the form

$$P = -\left(g^{ij}\frac{\partial^2}{\partial x_i \partial x_j} + P^k \frac{\partial}{\partial x_k} + Q\right).$$

It is convenient to work more tensorially. There exists a unique connection and endomorphism (∇^V, E) which are determined by P so

$$P\phi = -(g^{ij}\phi_{;ij} + E\phi).$$

Let $\Gamma_{ij}{}^k$ be the Christoffel symbols of the Levi–Civita connection. Let ω_l be the connection 1 form of ∇^V, and let Ω be the curvature of ∇^V. Then:

$$\Gamma_{ij}{}^k = \frac{1}{2}g^{kl}\left(g_{il/j} + g_{jl/i} - g_{ij/l}\right)$$

$$\omega_l = \frac{1}{2}\left(g_{il}P^i + g_{il}g^{jk}\Gamma_{jk}{}^i\right)$$

$$E = Q - g^{ij}\omega_{i/j} - g^{ij}\omega_i\omega_j + g^{jk}\omega_i\Gamma_{jk}{}^i$$

$$\Omega_{ij} = \omega_{j/i} - \omega_{i/j} + \omega_i\omega_j - \omega_j\omega_i.$$

This is a familiar construction in differential geometry; $P_0\phi = -g^{ij}\phi_{;ij}$ is the "rough" or Bochner Laplacian. If P is a "natural" Laplacian, ∇^V is the Levi–Civita connection and E is given by curvature. For example, if P is the Laplacian on 1 forms, the Weitzenboch formula shows E is minus the Ricci tensor. If P is the spin Laplacian, the Lichnerowicz formula shows E is minus one fourth the scalar curvature.

We must impose boundary conditions if $dM \neq 0$. Let $\phi \in C^\infty(V)$. Let

$$B_D(\phi) = \phi|_{dM} = 0$$

be Dirichlet boundary conditions. Let N be the inward unit normal and let S be an endomorphism of V. Neumann boundary conditions are

$$B_N(\phi) = (\phi_{;N} + S\phi)|_{dM} = 0.$$

We restrict to Neumann or Dirichlet boundary conditions for the moment; we will discuss mixed boundary conditions in §4.

We consider local formulas $a(x, P)$ which are polynomial in the jets of the total symbol of P with coefficients which are smooth functions of the metric tensor; $a(x, P)$ is invariant if the value is independent of the local frame for V and local coordinate system on M. Since we can always restrict to geodesic normal coordinates, the appropriate structure group is $O(m)$ since we will work invariantly on V. When considering boundary points $y \in dM$, the notion of invariance changes slightly. Let $y = (x_1, \ldots x_{m-1})$ be a system of local coordinates on dM; we choose the final coordinate x_m to be geodesic distance to the boundary so the curves $\gamma(t) = (y, t)$ are the geodesic normal rays from the boundary. By restricting to geodesic normal coordinates on the boundary, we can reduce the structure group to $O(m-1)$. This reduction of the structure group to $O(m)$ or $O(m-1)$ permits us to use the invariance theory of H. Weyl.

We invoke the calculus of pseudo-differential operators as discussed by Seely [Se] and Greiner [Gr] to study the short time behavior of the heat equation. Let $f \in C^\infty(M)$ be a scalar function and let $N^\nu(f)$ be the νth covariant normal derivative of f. We integrate with respect to the volume forms on M and dM defined by the Riemannian metric.

Lemma 1.1.

(a) $Tr_{L^2}(fe^{-tP}) \simeq \sum_n a_n(f, P)t^{n-m/2}$.

(b) *There exist invariant local formulas* $a_n(x, P)$ *and* $a_{n,\nu}(y, P)$ *so*

$$a_n(f, P) = \int_M f a_n(x, P) + \sum_\nu \int_{dM} N^\nu(f) a_{n,\nu}(y, P).$$

Remark. In general, $n = 0, \frac{1}{2}, 1, \ldots$ ranges over the non-negative half integers in (a); if $dM = 0$, however, the half integer coefficients vanish so $n = 0, 1, 2, \ldots$. We introduce the auxiliary function f to recapture divergence terms which would otherwise be lost.

Let $\{\phi_\nu, \lambda_\nu\}$ be a complete spectral resolution of P. The fundamental solution of the heat equation is

$$K(t, x, y, P) = \sum_\nu e^{-t\lambda_\nu} \phi_\nu(x) \otimes \phi_\nu(y).$$

Consequently

$$Tr_{L^2}(fe^{-tP}) = \int_M f(x) \sum_\nu e^{-t\lambda} |\phi_\nu(x)|^2.$$

In particular, if we take $f = 1$, then $Tr_{L^2}(e^{-tP}) = \sum_\nu e^{-t\lambda_\nu}$ so this is a spectral invariant of P. If $dM = 0$, we can always integrate by parts to eliminate derivatives of the auxiliary function f; this is not possible if $dM \neq 0$ so the normal derivatives of f enter naturally. In the interior of the manifold, the kernel $K(t, x, x, D)$ is independent of the boundary conditions modulo an error term which vanishes to infinite order in t; the fiber trace has an asymptotic expansion

$$Tr_V K(t, x, x, P) \simeq \sum_n a_n(x, P)t^{n-m/2}.$$

Near the boundary, the kernel becomes highly dependent on the boundary conditions. We choose local frames for V near the boundary which are parallel along the geodesic normal rays. Let $x_m \in [0, \varepsilon]$; we integrate the fiber trace along the geodesic rays to define the boundary contribution:

$$\int_0^\varepsilon f(x) Tr_V K(t, x, x, P) \, dx_m \simeq \sum_{n, \nu} t^{n - m/2} N^\nu(f) a_{n, \nu}(y, P).$$

The fiber trace of the heat kernel behaves like a distribution near the boundary and the behavior is detected through the use of the auxiliary function f.

The local nature of these invariants is crucial. The interior invariants $a_n(x, P)$ are built universally and polynomially from the metric tensor, its inverse, and the covariant derivatives of $\{R, \Omega, E\}$. When considering the boundary invariants $a_{n, \nu}(y, P)$ we must also introduce the second fundamental form L and when considering Neumann boundary conditions we must also consider the tensor S; we only differentiate $\{L, S\}$ tangentially and set $S = 0$ for Dirichlet boundary conditions. Let A be a monomial term of degree $(k_R, k_\Omega,, k_E, k_L, k_S)$ in (R, Ω, E, L, S) and let k_∇ explicit covariant derivatives appear. The degree of A in the jets of the total symbol of P is

$$2(k_R + k_\Omega + k_E) + k_L + k_S + k_\nabla.$$

By using dimensional analysis, it follows that the invariants of Lemma 1.1 are homogeneous of degree $2n$; more specifically each monomial of $a_n(x, P)$ is homogeneous of degree $2n$ and each monomial of $a_{n, \nu}(x, P)$ is homogeneous of degree $2n - \nu - 1$. Since n is a half integer, $2n$ can be odd. It would perhaps have been better to have chosen a different indexing convention, but we have followed the one most prevalent in the literature.

We use H. Weyl's theorem on the invariants of the structure group $O(m)$ and $O(m - 1)$ to express $a_n(x, P)$ and $a_{n, \nu}(y, P)$ tensorially. We begin our study by examining the invariants a_0 and $a_{.5}$; the situation is particularly simple for these invariants. We use the homogeneity to see there exist universal constants $\gamma_0(m)$ and $\gamma_1(m)$ so that

$$a_0(f, P) = \dim(V) \gamma_0(m) \int_M f, \qquad \text{and}$$

$$a_{.5}(f, P) = \dim(V) \gamma_1(m) \int_{dM} f.$$

We use the method of universal examples to determine $\gamma_0(m)$ and $\gamma_1(m)$. Let

$$M = S^1 \quad \text{and} \quad P = -\frac{\partial^2}{\partial \theta^2};$$

$\text{spec}(P) = \{n^2\}_{n \in Z}$ so

$$Tr(e^{-tP}) = \sum_{n \in Z} e^{-tn^2}.$$

Let $\phi(x) = e^{-x^2}$. We use Riemann sums to approximate

$$Tr(e^{-tP}) = 1/\sqrt{t} \sum_{n \in Z} \sqrt{t}\phi\left(n\sqrt{t}\right) \simeq$$

$$\simeq 1/\sqrt{t} \int_{-\infty}^{\infty} \phi(x)\, dx = \sqrt{\pi/t}.$$

Since all the derivatives of the symbol of P vanish and there are no boundary terms, this approximation is valid to infinite order in t. More generally, if

$$M = S^1 \times \ldots \times S^1 \quad \text{and} \quad P = -\sum_{\nu} \frac{\partial^2}{\partial \theta_\nu^2},$$

the same argument shows $Tr(e^{-tP}) \simeq (\pi/t)^{m/2}$. Since $\text{vol}(M) = (2\pi)^m$,

$$a_0(1, e^{-tP}) = (4\pi t)^{-m/2} \int_M 1.$$

This shows $\gamma_0(m) = (4\pi)^{-m/2}$.

We can use the same argument to study boundary value problems. Let

$$M = [0, \pi] \quad \text{and} \quad P = -\frac{\partial^2}{\partial \theta^2}.$$

With Neumann boundary conditions, $\{n^2, \cos(n\theta)\}_{n \geq 0}$ is a complete spectral resolution of P so

$$Tr(e^{-tP}) = \sum_{n \geq 0} e^{-tn^2} = \frac{1}{2}\left\{\sum_{n \in Z} e^{-tn^2}\right\} + \frac{1}{2} \simeq$$

$$\simeq \frac{1}{2}\sqrt{\pi/t} + \frac{1}{2}.$$

Since the boundary of M consists of 2 pieces, $a_{.5}(1, P) = \frac{1}{4} \text{vol}(dM)$. We take the product with a flat $m - 1$ dimensional torus to see

$$a_{.5}(1, P) = \frac{1}{4}(4\pi)^{-(m-1)/2} \text{vol}(dM)$$

so $\gamma_1(m) = 4^{-1}(4\pi)^{-(m-1)/2}$ for Neumann boundary conditions. For Dirichlet boundary conditions, the zero eigenvalue is missing; $\{n^2, \sin(n\theta)\}_{n>0}$ is a complete spectral resolution of P. This changes the signs;

$$Tr(e^{-tP}) = \sum_{n>0} e^{-tn^2} = \frac{1}{2}\left\{\sum_{n \in Z} e^{-tn^2}\right\} - \frac{1}{2}$$

$$\simeq \frac{1}{2}\sqrt{\pi/t} - \frac{1}{2}.$$

Consequently $\gamma_1(m) = -4^{-1}(4\pi)^{-(m-1)/2}$ with Dirichlet boundary conditions.

The higher order asymptotics are more complicated and encode geometric information. Let indices $\{i, j, \ldots\}$ range from 1 through m and index a local orthonormal frame $\{e_1, \ldots, e_m\}$ for the tangent bundle of M. Near the boundary of M, we shall assume $e^m = N$ is the inward unit geodesic normal. Let indices $\{a, b, \ldots\}$ range from 1 through $m - 1$ and index a local orthonormal frame $\{e_1, \ldots, e_{m-1}\}$ for the tangent bundle of the boundary. Let

$$L_{ab} = (\nabla_{e_a} e_b, N) = \Gamma_{ab}{}^n = -\frac{1}{2}(g^{nn} g^{aa} g^{bb})^{\frac{1}{2}} g_{ab/n}$$

be the second fundamental form. Let ";" be multiple covariant differentiation with respect to the Levi–Civita connection of M and let ":" be multiple covariant differentiation tangentially with respect to the Levi–Civita connection of the boundary. The tensors L and S are only defined on dM and can only be differentiated tangentially.

Let $\delta = +4^{-1}$ for Neumann and $\delta = -4^{-1}$ for Dirichlet boundary conditions.

Lemma 1.2. There exist universal constants $\{\alpha_i, \beta_j, b_i, c_i, d_i, e_i\}$ so

$$a_0(f, P) = (4\pi)^{-m/2} \int_M f$$

$$a_{.5}(f, P) = \delta(4\pi)^{-(m-1)/2} \int_{dM} f$$

$$a_1(f, P) = (4\pi)^{-m/2} 6^{-1} \left\{ \int_M \alpha_1 fE + \alpha_2 f\tau + \int_{dM} b_0 fL_{aa} + \right.$$

$$\left. + b_1 f_{;N} + b_2 fS \right\}$$

$$a_{1.5}(f, P) = \delta(4\pi)^{-(m-1)/2} 96^{-1} \left\{ \int_{dM} f(c_0 E + c_1 \tau + c_2 f R_{aNaN} + \right.$$

$$+ c_3 f L_{aa} L_{bb} +$$

$$+ c_4 f L_{ab} L_{ab} + c_7 S L_{aa} + c_8 S^2) + f_{;N}(c_5 L_{aa} + c_9 S) +$$

$$\left. + c_6 f_{;NN} \right\}$$

$$a_2(f, P) = (4\pi)^{-m/2} 360^{-1} \left\{ \int_M f(\alpha_3 E_{;kk} + \alpha_4 \tau E + \alpha_5 E^2 + \right.$$

$$+ \alpha_6 \tau_{;kk} + \alpha_7 \tau^2$$

$$+ \alpha_8 \rho^2 + \alpha_9 R^2 + \alpha_{10} \Omega^2) + \int_{dM} f(d_1 E_{;N} + d_2 \tau_{;N} + d_3 L_{aa:bb} +$$

$$+ d_4 L_{ab:ab} + d_5 E L_{aa} + d_6 \tau L_{aa} + d_7 R_{aNaN} L_{bb} +$$

$$+ d_8 R_{aNbN} L_{ab} +$$

$$+ d_9 R_{abcb} L_{ac} + d_{10} L_{aa} L_{bb} L_{cc} + d_{11} L_{ab} L_{ab} L_{cc} +$$

$$+ d_{12} L_{ab} L_{bc} L_{ac} +$$

$$+ d_{13} \Omega_{in:i} + d_{14} SE + d_{15} S\tau + d_{16} S R_{aNaN} +$$

$$+ d_{17} S L_{aa} L_{bb} +$$

$$+ d_{18} S L_{ab} L_{ab} + d_{19} S^2 L_{aa} + d_{20} S^3 + d_{21} S_{:aa}) +$$

$$+ f_{;N}(e_1 E + e_2 \tau +$$

$$+ e_3 R_{aNaN} + e_4 L_{aa} L_{bb} + e_5 L_{ab} L_{ab} + e_8 S L_{aa} +$$

$$+ e_9 S^2) +$$

$$\left. + f_{;NN}(e_6 L_{aa} + e_{10} S) + e_7 f_{;iiN} \right\}$$

Remark. The normalizing constants $(4\pi)^{-m/2}$ and $\delta(4\pi)^{-(m-1)/2}$ are motivated by the calculation of a_0 and $a_{.5}$ discussed above. The constants α_i determine the interior asymptotics and are independent of the boundary conditions chosen; the remaining constants depend on the boundary conditions chosen. We set $S = 0$ for Dirichlet boundary conditions as this tensor is not relevant. The normalizing constants of 6^{-1}, 96^{-1}, and 360^{-1} are included to simplify later formulas. The essential normalization is the factors of $(4\pi)^{-m/2}$ and $(4\pi)^{-(m-1)/2}$; these ensure the coefficients which appear are *independent* of m; this is proved by taking products with flat tori as was done in the evaluation of the constans $\gamma_0(m)$ and $\gamma_1(m)$. For the most part, we will be working with line bundles so we shall omit the fiber trace in our formulas; the dimension of the vector bundle plays no role in these formulas.

In §2 we will develop functorial properties of these invariants and in §3 and §4 we will evaluate the unknown coefficients and prove:

Theorem 1.3. *Interior Asymptotics*

$\alpha_1 = 6$	$\alpha_2 = 1$	$\alpha_3 = 60$	$\alpha_4 = 60$	$\alpha_5 = 180$
$\alpha_6 = 12$	$\alpha_7 = 5$	$\alpha_8 = -2$	$\alpha_9 = 2$	$\alpha_{10} = 30$

Theorem 1.4. *With Dirichlet boundary conditions:*

$b_0 = 2$	$b_1 = -3$	$b_2 = 0$	$c_0 = 96$	$c_1 = 16$
$c_2 = 8$	$c_3 = 7$	$c_4 = -10$	$c_5 = -30$	$c_6 = 24$
$c_7 = 0$	$c_8 = 0$	$c_9 = 0$	$d_1 = -120$	$d_2 = -18$
$d_3 = 24$	$d_4 = 0$	$d_5 = 120$	$d_6 = 20$	$d_7 = 4$
$d_8 = -12$	$d_9 = 4$	$d_{10} = 40/21$	$d_{11} = -88/7$	$d_{12} = 320/21$
$d_{13} = 0$	$d_{14} = 0$	$d_{15} = 0$	$d_{16} = 0$	$d_{17} = 0$
$d_{18} = 0$	$d_{19} = 0$	$d_{20} = 0$	$d_{21} = 0$	$e_1 = -180$
$e_2 = -30$	$e_3 = 0$	$e_4 = -180/7$	$e_5 = 60/7$	$e_6 = 24$
$e_7 = -30$	$e_8 = 0$	$e_9 = 0$	$e_{10} = 0$	

Theorem 1.5. *With Neumann boundary conditions:*

$b_0 = 2$	$b_1 = 3$	$b_2 = 12$	$c_0 = 96$	$c_1 = 16$
$c_2 = 8$	$c_3 = 13$	$c_4 = 2$	$c_5 = 6$	$c_6 = 24$
$c_7 = 96$	$c_8 = 192$	$c_9 = 96$	$d_1 = 240$	$d_2 = 42$
$d_3 = 24$	$d_4 = 0$	$d_5 = 120$	$d_6 = 20$	$d_7 = 4$
$d_8 = -12$	$d_9 = 4$	$d_{10} = 40/3$	$d_{11} = 8$	$d_{12} = 32/3$
$d_{13} = 0$	$d_{14} = 720$	$d_{15} = 120$	$d_{16} = 0$	$d_{17} = 144$
$d_{18} = 48$	$d_{19} = 480$	$d_{20} = 480$	$d_{21} = 120$	$e_1 = 180$
$e_2 = 30$	$e_3 = 0$	$e_4 = 12$	$e_5 = 12$	$e_6 = 24$
$e_7 = 30$	$e_8 = 72$	$e_9 = 240$	$e_{10} = 120$	

We shall concentrate for the most part on the terms a_1, $a_{1.5}$, and the interior contribution of a_2; the calculations for the boundary terms comprising a_2 are much more complicated but similar in nature and not particularly illuminating; we refer to [BrGi] for the full calculation of a_2.

§2 Functorial properties

We begin with some functorial properties of the heat invariants:

Lemma 2.1.

(a) Let $M = M_1 \times M_2$, $P = P_1 \otimes 1 + 1 \otimes P_2$, $dM_2 = 0$, and $f(x) = f(x_1)f(x_2)$.

$$a_n(x, P) = \sum_{n=p+q} a_p(x_1, P_1)a_q(x_2, P_2)$$

$$a_{n,\nu}(y, P) = \sum_{n=p+q} a_{p,\nu}(y_1, P_1)a_q(x_2, P_2).$$

(b) If $P(\varepsilon) = e^{-2\varepsilon f} P$ then $\frac{d}{d\varepsilon}\big|_{\varepsilon=0} a_n(1, P) = (m - 2n)a_n(f, P)$.

(c) If $P(\varepsilon) = P - \varepsilon F$, then $\frac{d}{d\varepsilon}\big|_{\varepsilon=0} a_n(1, P) = a_{n-1}(F, P)$.

(d) If $m = 2n + 2$, then $\frac{d}{d\varepsilon}\big|_{\varepsilon=0} a_n(e^{-2\varepsilon f} F, e^{-2\varepsilon f} P) = 0$.

(e) Let Δ_p be the Laplacian on p forms and assume $dM = 0$. Then

$$\sum_p (-1)^p a_n(1, \Delta_p) = \begin{cases} 0 & \text{if } 2n \neq m, \\ \chi(M) & \text{if } 2n = m. \end{cases}$$

Remark. We will generalize (e) to manifolds with boundary by taking suitable mixed boundary conditions; we postpone a discussion of these boundary conditions until §4.

Proof. The proof of (a)–(d) is a formal calculation; the necessary analytic details can be justified using the techniques of Gilkey–Smith [GiSm]. If P is as in (a), then:

$$Tr_{L^2}(fe^{-tP}) = Tr_{L^2}(f_1 e^{-tP_1}) Tr_{L^2}(f_2 e^{-tP_2});$$

we compare powers of t in the resulting asymptotic expansions to prove (a). To prove (b), consider the variation $P(\varepsilon) = e^{-2\varepsilon f} P$. We compute:

$$\frac{d}{d\varepsilon}\Big|_{\varepsilon=0} Tr_{L^2}(e^{-tP}) = -t\, Tr_{L^2}\left(\frac{d}{d\varepsilon}\Big|_{\varepsilon=0} (P)e^{-tP}\right)$$

$$= 2t\, Tr_{L^2}(fPe^{-tP})$$

$$= -2\frac{\partial}{\partial t} Tr_{L^2}(fe^{-tP});$$

(b) now follows by equating terms in the asymptotic series. Let $P(\varepsilon) = P - \varepsilon F$. We prove (c) by equating terms in the asymptotic series

$$\frac{d}{d\varepsilon}\Big|_{\varepsilon=0} Tr_{L^2}(e^{-tP}) = t\, Tr_{L^2}(Fe^{-tP}).$$

We use (b, c) to prove (d). Let $m = 2n + 2$. Let $P(\varepsilon, \delta) = e^{-2\varepsilon f}\{P - \delta F\}$;

$$0 = \frac{\partial^2}{\partial\delta\partial\varepsilon}\{a_{n+1}(1, P(\varepsilon, \delta))\}$$

$$= \frac{\partial^2}{\partial\varepsilon\partial\delta}\{a_{n+1}(1, P(\varepsilon, \delta))\}$$

$$= \frac{\partial}{\partial\varepsilon}\{a_n(e^{-2\varepsilon} F, e^{-2\varepsilon P})\}.$$

We use index theory to prove (e). Let

$$E(\lambda, p) = \{\omega \in C^\infty \Lambda^p(M) : \Delta_p \omega = \lambda\omega\}$$

be the eigenspaces. The 0 eigenspace has a cohomological interpretation;

$$E(0,p) \simeq H^p(M)$$

is isomorphic to the pth De Rham cohomology group of M. If $\lambda \neq 0$, exterior differentiation d preserves these eigenspaces and provides a long exact sequence:

$$0 \to E(\lambda, 0) \to E(\lambda, 1) \to \dots \to E(\lambda, m) \to 0.$$

Consequently:

$$\sum_p (-1)^p \dim E(\lambda, p) = \begin{cases} 0 & \text{if } \lambda \neq 0 \\ \chi(M) & \text{if } \lambda = 0 \end{cases}$$

We complete the proof by equating powers of t in the asymptotic expansion:

$$\sum_p (-1)^p Tr(e^{-t\Delta_p}) = \sum_{\lambda,p} (-1)^p \dim(E(\lambda, p))e^{-t\lambda}$$

$$= \chi(M)$$

$$\simeq \sum_n \left\{ \sum_p (-1)^p a_n(1, \Delta_p) \right\} t^{n-m/2}.$$

∎

Lemma 2.1(b) involves conformal variations:

$$g(\varepsilon) = e^{2\varepsilon f} g_0 \qquad \text{and} \qquad P(\varepsilon) = e^{-2\varepsilon f} P.$$

Let $\Delta = -g^{-1} \frac{\partial}{\partial x_i} g g^{ij} \frac{\partial}{\partial x_j}$ be the scalar Laplacian. If we vary the metric,

$$\frac{d}{d\varepsilon}\bigg|_{\varepsilon=0} \text{dvol}_k = kf \, \text{dvol}_k,$$

$$\frac{d}{d\varepsilon}\bigg|_{\varepsilon=0} \omega_m = \frac{1}{2}(2-m)f_{;N}$$

$$\frac{d}{d\varepsilon}\bigg|_{\varepsilon=0} \Delta = -2f\Delta - (m-2)g^{ij} f_{;i} \frac{\partial}{\partial x_j}, \text{ and}$$

$$\frac{d}{d\varepsilon}\bigg|_{\varepsilon=0} R_{ijkl} = \delta_{jl} f_{;ik} + \delta_{ik} f_{;jl} - \delta_{il} f_{;jk} - \delta_{jk} f_{;il}.$$

To make Neumann boundary conditions conformally invariant, we must also vary the auxiliary endomorphism S. The inward unit normal is $N(\varepsilon) = e^{-\varepsilon f}\frac{\partial}{\partial x_m}$. Therefore:

$$\nabla_N = \frac{\partial}{\partial x_m} + \omega_m$$

where $\omega_m = \frac{1}{2}(g_{im}P^m + g_{im}g^{jk}\Gamma_{jk}{}^m)$. We set

$$S(\varepsilon) = e^{-\varepsilon f}\{\omega_m(0) - \omega_m(\varepsilon)\};$$

this ensures B_N is unchanged during this variation. This leads to the following variational formulas; as the calculations are straightforward we shall omit the proof.

Lemma 2.2. Let $F(\varepsilon) = e^{-2\varepsilon f}F$ and $P(\varepsilon) = e^{-2\varepsilon f}F$.

Label	$\frac{d}{d\varepsilon}\big|_{\varepsilon=0}$	Answer
b_0	L_{aa}	$-fL_{aa} - (m-1)f_{;N}$
b_1	$F_{;N}$	$-3fF_{;N} - 2f_{;N}F$
b_2	$S(\varepsilon)$	$-fS + \frac{1}{2}(m-2)f_{;N}$
α_1, c_0	E	$-2fE + \frac{1}{2}(m-2)f_{;ii}$
α_2, c_1	τ	$-2f\tau - 2(m-1)f_{;ii}$
c_2	R_{aNaN}	$-2fR_{aNaN} + f_{;aa} + (m-1)f_{;NN}$
c_3	$L_{aa}L_{bb}$	$-2fL_{aa}L_{bb} - 2(m-1)f_{;N}L_{aa}$
c_4	$L_{ab}L_{ab}$	$-2fL_{ab}L_{ab} - 2f_{;N}L_{aa}$
c_5	$F_{;N}L_{aa}$	$-4fF_{;N}L_{aa} - 2f_{;N}FL_{aa} - (m-1)f_{;N}F_{;N}$
c_6	$F_{;NN}$	$-4fF_{;NN} - 5f_{;N}F_{;N} - 2f_{;NN}F + f_{;a}F_{;a}$
c_7	SL_{aa}	$-2fS + \frac{1}{2}(m-2)f_{;N}L_{aa} - (m-1)Sf_{;N}$
c_8	S^2	$-2fS^2 + (m-2)f_{;N}S$
c_9	$F_{;N}S$	$-4F_{;N}S + \frac{1}{2}(m-2)f_{;N}F_{;N} - 2f_{;N}FS$
α_3	$E_{;kk}$	$-4fE_{;kk} - 2f_{;kk}E + \frac{1}{2}(m-2)f_{;iijj} + (m-6)f_{;k}E_{;k}$
α_4	τE	$-4f\tau E + \frac{1}{2}(m-2)f_{;ii}\tau - 2(m-1)f_{;ii}E$
α_5	E^2	$-4fE^2 + (m-2)f_{;ii}E$
α_6	$\tau_{;kk}$	$-4f\tau_{;kk} - 2f_{;kk}\tau - 2(m-1)f_{;iijj} + (m-6)f_{;i}\tau_{;i}$
α_7	τ^2	$-4f\tau^2 - 4(m-1)f_{;ii}\tau$
α_8	ρ^2	$-4\rho^2 - 2f_{;ii}\tau - 2(m-2)f_{;ij}\rho_{ij}$
α_9	R^2	$-4R^2 - 8f_{;ij}\rho_{ij}$
α_{10}	Ω^2	$-4f\Omega^2$

§3 Interior asymptotics and Dirichlet boundary conditions

In this section, we will use Lemma 2.1 to complete the proof of Theorems 1.3 and 1.4; it is somewhat surprising that the simple relations of 2.1 suffice. We first compute the interior asymptotics. We suppress the normalizing constants $(4\pi)^{-m/2}$ for the moment and let $dM = 0$. We apply 2.1(c). Under the variation $P(\varepsilon) = P - \varepsilon F$, we must replace E by $E + \varepsilon F$. Consequently:

$$\frac{d}{d\varepsilon}\bigg|_{\varepsilon=0} a_1(1, P - \varepsilon F) = 6^{-1} \int_M \alpha_1 F = a_0(F, P) = \int_M F$$

$$\frac{d}{d\varepsilon}\bigg|_{\varepsilon=0} a_2(1, P - \varepsilon F) = 360^{-1} \int_M \alpha_4 F\tau + 2\alpha_5 FE =$$

$$= a_1(F, P) = 6^{-1} \int_M \alpha_1 FE + \alpha_2 F\tau.$$

Consequently $\alpha_1 = 6$, $\alpha_5 = 180$, and $\alpha_4 = 60\alpha_2$. We apply 2.1(d) with $m = 4$ and $n = 1$. By Lemma 2.2:

$$0 = \frac{d}{d\varepsilon}\bigg|_{\varepsilon=0} a_1(e^{-2\varepsilon f} F, e^{-2\varepsilon f} P) = 6^{-1} \int_M F(\alpha_1 - 6\alpha_2) f_{;ii}.$$

Consequently $\alpha_2 = 1$ and $\alpha_4 = 60$.

Let P be the Laplacian on a product manifold $M = M_1 \times M_2$. $E = \Omega = 0$. We use 2.1(a) and compute

$$a_2(1, P) = a_2(1, P_1) + a_2(1, P_2) + a_1(1, P_1)a_1(1, P_2)$$
$$\tau^2(M_1 \times M_2) = \tau^2(M_1) + \tau^2(M_2) + 2\tau(M_1)\tau(M_2)$$
$$\rho^2(M_1 \times M_2) = \rho^2(M_1) + \rho^2(M_2)$$
$$R^2(M_1 \times M_2) = R^2(M_1) + R^2(M_2).$$

The only cross term arises from τ^2. Consequently $180^{-1}\alpha_7 = (6^{-1}\alpha_2)^2$ so $\alpha_7 = 5$.

We use 2.1(d) with $m = 6$ and $n = 2$ to compute:

$$0 = 360 \frac{d}{d\varepsilon}\bigg|_{\varepsilon=0} a_2(e^{-2\varepsilon f} F, e^{-2\varepsilon f} P) =$$

$$= \int_M F\{(-2\alpha_3 - 10\alpha_4 + 4\alpha_5)f_{;kk} E + (2\alpha_3 - 10\alpha_6)f_{;iijj} +$$
$$+ (2\alpha_4 - 2\alpha_6 - 20\alpha_7 - 2\alpha_8)f_{;ii}\tau + (-8\alpha_8 - 8\alpha_9)f_{;ij}\rho_{ij}\}.$$

We set the coefficients to zero to get the relations:

$$0 = -2\alpha_3 - 600 + 720, \qquad 0 = 2\alpha_3 - 10\alpha_6,$$
$$0 = 120 - 2\alpha_6 - 100 - 2\alpha_8, \qquad 0 = -8\alpha_8 - 8\alpha_9.$$

This shows $\alpha_3 = 60$, $\alpha_6 = 12$, $\alpha_8 = -2$, and $\alpha_9 = 2$.

By 2.1(a), $0 = 2a_2(1, \Delta_0) - a_2(1, \Delta_1)$. Consequently

$$0 = \int_M Tr\{\alpha_4 \tau E(\Delta_1) + \alpha_5 E(\Delta_1)^2 + \alpha_{10} \Omega(\Delta_1)^2\}.$$

By the Weitzenboch formulas, $\Delta_1(\phi) = -\phi_{;kk} + \frac{1}{2}\tau\phi$. Consequently $E(\Delta_1) = -\frac{1}{2}\tau I$ so:

$$Tr(E(\Delta_1)\tau) = -\tau^2, \qquad Tr(E(\Delta_1)^2) = \frac{1}{2}\tau^2, \qquad \text{and}$$

$$Tr(\Omega^2) = Tr(\Omega_{ij}\Omega_{ij}) = R_{ijkl}R_{ijkl} = -\tau^2.$$

This shows $-\alpha_4 + \frac{1}{2}\alpha_5 - \alpha_{10} = 0$ so $\alpha_{10} = 30$; this proves Theorem 1.3. ∎

We begin the proof of Theorem 1.4 with:

Lemma 3.1. *For either Neumann or Dirichlet boundary conditions:*

(a) $c_0 = 96$ and $c_1 = 16$.

(b) $0 = -b_0(m - 1) - b_1(m - 2) + \frac{1}{2}b_2(m - 2) - (m - 4)$.

(c) $0 = 48(m - 2) - 32(m - 1) + c_2(m - 1) - c_6(m - 3)$.

(d) $0 = -48(m-2)+32(m-1)-c_2-2c_3(m-1)-2c_4-c_5(m-3)+\frac{1}{2}c_7(m-2)$.

(e) $0 = -c_7(m - 1) + c_8(m - 2) - c_9(m - 3)$.

(f) $0 = -4c_5 - 5c_6 + 3/2 \cdot c_9$.

Proof. By Lemma 2.1(a), $96^{-1}c_0 = 6^{-1}\alpha_1$ and $96^{-1}c_1 = 6^{-1}\alpha_2$. This proves (a). We use Lemma 2.1(b) with $n = 1$. We prove (b) by using

Lemma 2.2 to compute:

$$0 = 6(4\pi)^{m/2} \left\{ \frac{d}{d\varepsilon} \bigg|_{\varepsilon=0} a_1(F, e^{-2\varepsilon f} P) - (m-2)a_1(Ff, P) \right\}$$

$$= \int_M \frac{d}{d\varepsilon}\bigg|_{\varepsilon=0} (6E + \tau) + \int_{dM} \frac{d}{d\varepsilon}\bigg|_{\varepsilon=0} (b_0 L_{aa} + b_2 S) -$$
$$- (m-2)a_1(Ff, P)$$

$$= \int_M (3(m-2) - 2(m-1)) f_{;ii}$$

$$+ \int_{dM} \left(-b_0(m-1) - b_1(m-2) + \frac{1}{2}(m-2)b_2 \right) f_{;N}$$

$$= \int_{dM} \left(-b_0(m-1) - b_1(m-2) + \frac{1}{2}(m-2)b_2 - (m-4) \right) f_{;N}$$

Next we use 2.1(b) with $n = 1.5$; there are no interior terms. We prove (c), (d), and (e) by studying the coefficients of $f_{;NN}$, $f_N L_{aa}$, and $f_{;N} S$ and substituting $c_0 = 96$ and $c_1 = 16$. We omit details as the calculations are similar. We use Lemma 2.1(d) with $m = 5$ and $n = 1.5$ and study the coefficient of $F_{;N} f_{;N}$ to prove (e). ■

We use Lemma 3.1 to prove Theorem 1.4. We set the coefficients of all terms involving S to 0. Consequently $b_2 = 0$, and $c_i = 0$ for $i \geq 7$. By 3.1(a), $c_0 = 96$ and $c_1 = 16$. We use 3.1(b) to see $b_0 = 2$ and $b_1 = -3$. We use the remaining equations of Lemma 3.1 to see

$$c_2 = 8, \quad c_6 = 24 \quad c_5 = -30, \quad c_3 = 7, \quad \text{and} \quad c_4 = -10.$$

This determines a_0, $a_{.5}$, a_1, and $a_{1.5}$ for Dirichlet boundary conditions; the calculation of the boundary contribution in a_2 is exactly the same and is therefore omitted; we refer to [BrGi] for details.

§4 Neumann and mixed boundary conditions

Lemma 2.1 does not suffice to determine the boundary contribution for Neumann boundary conditions; we saw in our discussion of Dirichlet boundary conditions that we got a consistent set of equations by setting the coefficients of all the terms involving S to zero. We need some additional relationships particular to dimension $m = 1$. We follow the development of [Gi-1]. Let M be a compact 1-dimensional Riemannian manifold with boundary; M is either the circle or a closed interval. Let $b \in C^\infty(M)$ be a real function. We form

$$A = \frac{\partial}{\partial x} - b, \qquad A^* = -\frac{\partial}{\partial x} - b,$$

$$D_1 = A^* A = -\left\{ \frac{\partial^2}{\partial x^2} + (-b_x - b^2) \right\}, \qquad \text{and}$$

$$D_2 = A A^* = -\left\{ \frac{\partial^2}{\partial x^2} + (b_x - b^2) \right\}.$$

We take Dirichlet boundary conditions for D_1 and Neumann boundary conditions $A^* \phi|_{dM} = 0$ for D_2; consequently $S(D_2) = b$. Let f_x and f_{xx} be the first and second derivatives of f.

Lemma 4.1. $(2n - 1)\{a_n(f, D_1) - a_n(f, D_2)\} = a_{n-1}(f_{xx} + 2bf_x, D_1)$.

Proof. Let $\{\theta_\nu, \lambda_\nu\}$ be a spectral resolution for D_1. The heat kernel of D_1 is:

$$K(t, x, x, D_1) = \sum_\nu e^{-t\lambda_\nu} \theta_\nu(x)^2.$$

We differentiate with respect to t;

$$\frac{\partial}{\partial t} K(t, x, x, D_1) = -\sum_\nu \lambda_\nu e^{-t\lambda_\nu} \theta_\nu(x^2) = -\sum_\nu e^{-t\lambda_\nu} D_1\theta_\nu \cdot \theta_\nu.$$

As $\left\{ A\theta_\nu / \sqrt{\lambda_\nu}, \lambda_\nu \right\}_{\lambda_\nu \neq 0}$ is a spectral resolution of D_2 on $\ker(D_2)^\perp$,

$$\frac{\partial}{\partial t} K(t, x, x, D_2) = -\sum_{\lambda_\nu \neq 0} e^{-t\lambda_\nu} A\theta_\nu \cdot A\theta_\nu.$$

Since $A\theta_\nu = 0$ if $\lambda_\nu = 0$ we may sum over all ν. This shows

$$2\frac{\partial}{\partial t}\{K(t, x, x, D_1) - K(t, x, x, D_2)\} =$$

$$= -2\sum_\nu e^{-t\lambda_\nu}\{D_1\theta_\nu \cdot \theta_\nu - A\theta_\nu \cdot A\theta_\nu\} =$$

$$= 2\sum_\nu e^{-t\lambda_\nu}\{(\theta''_\nu - b'\theta_\nu\theta_\nu - b^2\theta_\nu\theta_\nu) +$$

$$+ (\theta'_\nu\theta'_\nu - 2b\theta'_\nu\theta_\nu + b^2\theta_\nu\theta_\nu)\} =$$

$$= \frac{\partial}{\partial x}\left(\frac{\partial}{\partial x} - 2b\right) K(t, x, x, D_1).$$

We integrate by parts. We have chosen Dirichlet boundary conditions for D_1 so $K(t, x, x, D_1)$ vanishes to second order on dM. Consequently the boundary terms vanish:

$$\int_M 2f \frac{\partial}{\partial t}\{K(t, x, x, D_1) - K(t, x, x, D_2)\}\, dx$$

$$= \int_M f \frac{\partial}{\partial x}\left(\frac{\partial}{\partial x} - 2b\right) K(t, x, x, D_1)\, dx$$

$$= \int_M (f_{xx} + 2bf_x) K(t, x, x, D_1)\, dx.$$

this shows

$$2\frac{\partial}{\partial t}\{Tr(fe^{-tD_1}) - Tr(fe^{-tD_2})\} = Tr((f_{xx} + 2bf_x)e^{-tD_1}).$$

We now equate coefficients in the asymptotic series. ∎

We use Lemma 4.1 to prove

Lemma 4.2. *For Neumann boundary conditions*

$b_1 = 3$	$b_2 = 12$			
$c_0 = 96$	$c_1 = 16$	$c_6 = 24$	$c_8 = 192$	$c_9 = 96$

Proof. Let $M = [0, 1]$ and let $m = 1$. By Lemma 3.1, $c_0 = 96$ and $c_1 = 16$. We omit factors of $(4\pi)^*$. By Lemma 2.1 and Theorems 1.3 and 1.4,

	Dirichlet	Neumann
$a_0(f, P)$	$\int_M f$	$\int_M f$
$4a_{.5}(f, P)$	$-\int_{dM} f$	$\int_{dM} f$
$6a_1(f, P)$	$\int_M 6fE + \int_{dM} -3f_x$	$\int_M 6fE + \int_{dM} b_2 fS + b_1 f_x$
$4 \cdot 96a_{1.5}(f, P)$	$-\int_{dM} 96fE + 24f_{xx}$	$\int_{dM} 96fE + c_8 fS^2$
		$+c_9 f_x S + c_6 f_{xx}$

We adopt the notation of Lemma 4.1; let b vanish to infinite order at $x = 1$. Then:

$$E(D_1) = -b_x - b^2, \quad E(D_2) = b_x - b^2, \quad S(D_1) = 0, \text{ and } S(D_2) = b.$$

If $n = 1$, then:

$$0 = \int_M 6f(E(D_1) - E(D_2)) - 6f_{xx} - 12bf_x$$

$$+ \int_{dM} -b_2 fS + (-3 - b_1)f_x$$

$$= \int_{dM} (12 - b_2)fb + (6 - 3 - b_1)f_x.$$

Thus $b_2 = 12$ and $b_1 = 3$. If $n = 1.5$, then:

$$0 = -4^{-1}96^{-1}2\left\{ \int_{dM} 96fE(D_1) + 24f_{xx} \right\}$$

$$-4^{-1}96^{-1}2\left\{ \int_{dM} 96fE(D_2) + c_8 fB^2 + c_9 f_x b + c_6 fxx \right\}$$

$$+4^{-1}\left\{ \int_{dM} f_{xx} + 2bf_x \right\}$$

$$=-192^{-1}\int_{dM} 96f(E(D_1) + E(D_2)) + c_8 fb^2 +$$

$$+ (24 + c_6 - 48)f_{xx} + (c_9 - 96)f_x b$$

$$=192^{-1}\int_{dM} (c_8 - 192)fb^2 + (c_6 - 24)f_{xx} + (c_9 - 96)f_x b.$$

Thus $c_6 = 24$, $c_8 = 192$, and $c_9 = 96$. ∎

We can now prove Theorem 1.5. By Lemma 4.2, $b_1 = 3$ and $b_2 = 12$. By Lemma 3.1(b)

$$0 = -b_0(m - 1) - 3(m - 2) + 6(m - 2) - (m - 4).$$

Thus $b_0 = 2$. By Lemma 4.2, $c_0 = 96$, $c_1 = 16$, $c_6 = 24$, $c_8 = 192$, and $c_9 = 96$. By Lemma 3.1(c, e),

$$0 = 48(m - 2) - 32(m - 1) + c_2(m - 1) - 24(m - 3)$$

$$0 = -c_7(m - 1) + 192(m - 2) - 96(m - 3).$$

Thus $c_2 = 8$ and $c_7 = 96$. Finally by Lemma 3.1(d, f)

$$0 = -4c_5 - 5 \cdot 24 + \frac{1}{2}96 \cdot 3$$

$$0 = -48(m - 2) + 32(m - 1) - 8 - 2c_3(m - 1) - 2c_4 - c_5(m - 3) +$$

$$+ 48(m - 2).$$

Thus $c_3 = 13$, $c_5 = 6$, and $c_4 = 2$. This completes the determination of the boundary contribution to a_n for Neumann boundary conditions for $n \leq 1.5$; the calculation of a_2 is the same. We omit the details and refer to [BrGi].

We conclude this section by discussing mixed boundary conditions. We combine the results of §1 into a single theorem although at the cost of notational complexity. We motivate mixed boundary conditions by discussing the De Rham complex; we refer to Gilkey [Gi–3] for details. We also refer to Luckock and Moss [LM] for another example where such mixed boundary conditions appear in the context of quantum gravity.

Let $y = (x_1, \ldots, x_{m-1})$ be a system of local coordinates on the boundary dM and let x_m be the geodesic distance to the boundary; $x = (y, x_m)$ is a system of coordinates near dM and $ds^2 = g_{ab} \, dx_a \, dx_b + dx_m^2$. Let

$$I = \{1 \leq a_1 < a_2 < \ldots < a_p \leq m-1\} \qquad \text{and} \qquad dy^I = dx_{a_1} \wedge \ldots \wedge dx_{a_p}.$$

Decompose the exterior algebra $\Lambda(T^* M) = \Lambda_N \oplus \Lambda_D$ where

$$\Lambda_n = \text{span}\left\{dy^I\right\} \qquad \text{and} \qquad \Lambda_D = \text{span}\left\{dx^m \wedge dy^I\right\};$$

Λ_n are the tangential differential forms and Λ_D those forms which vanish on the boundary. Absolute boundary conditions are defined by taking Neumann boundary conditions on Λ_N and Dirichlet boundary conditions on Λ_D. More precisely, if $\omega \in C^\infty (\Lambda(T^* M))$, decompose $\omega = \sum_I \{f_I \, dy^I + g_I \, dx_m \wedge dy^I\}$. Absolute boundary conditions $B_A \omega = 0$ are defined by

$$\left. \frac{\partial}{\partial x_m} f_I \right|_{dM} = g_I|_{dM} = 0.$$

B_A is independent of the coordinate system chosen for the boundary; relative boundary conditions B_R are defined by dualizing these boundary conditions using the Hodge $*$ operator; $B_R(\omega) = B_A(*\omega)$.

Exterior differentiation preserves absolute boundary conditions; interior differentiation preserves relative boundary conditions. Let $\Delta_{p,-} = (d\delta + \delta d)_p$ be the associated Laplacian with absolute or relative boundary conditions; $\Delta_{p,-}$ is a self adjoint elliptic operator. Let

$$E(\lambda, p, -) = \{\omega \in C^\infty (\Lambda^p(M)) : \Delta_p \omega = \lambda \omega \text{ and } B_- \omega = 0\}$$

be the associated eigenspaces. $E(0, p, A) \simeq H^p(M)$ and $E(0, p, R) \simeq H^p(M, dM)$. The Hodge operator

$$* : E(0, p, A) = H^p(M) \simeq E(0, m, -p, R) = H^{m-p}(M, dM)$$

defines Poincare duality between the absolute and relative cohomology groups. If $\lambda \neq 0$, exterior differentiation d preserves these eigenspaces and provides a long exact sequence

$$0 \to E(\lambda, 0-) \to E(\lambda, 1, -) \to \ldots \to E(\lambda, m, -) \to 0.$$

Consequently Lemma 2.1(e) holds if we adopt absolute boundary conditions; for relative boundary conditions we would have to replace $\chi(M)$ by $\chi(M, dM)$.

Neumann boundary conditions on Λ_N are not defined invariantly, but are expressed relative to a special frame. Extend L to the exterior algebra. We must take $S = -L$ since

$$\nabla_N(\omega_N) = \left(\frac{\partial}{\partial x_m} + L \right)(\omega_N).$$

We generalize this example. Let V be a vector bundle over M. We assume given a decomposition $V = V_N \oplus V_D$ near dM which is parallel with respect to the normal geodesic rays. We take Neumann boundary conditions on V_N modified by some endomorphism S of V_N and Dirichlet boundary conditions on V_D. Let Π_N and Π_D be the corresponding projection operators and let

$$\psi = \Pi_N - \Pi_D.$$

Such boundary conditions are elliptic and give rise to asymptotics of the heat equation. We combine the results of §1 into a single formula by introducing some additional terms which reflect the second fundamental form of the splitting $V = V_N \oplus V_D$ or equivalently the tangential covariant derivatives of the endomorphism ψ. We shall assume S is self adjoint and ∇^V is a Riemannian to simplify the invariance theory; this is the case for the natural examples appearing in differential geometry.

Theorem 4.3.
$$a_0(f, P) = (4\pi)^{-m/2} \int_M Tr_V(f)$$

$$a_{.5}(f, P) = 4^{-1}(4\pi)^{-(m-1)/2} \int_{dM} Tr_V(\psi f)$$

$$a_1(f, P) = (4\pi)^{-m/2} 6^{-1} \left\{ \int_M Tr_V(6fE + f\tau) + \right.$$

$$+ \int_{dM} Tr_V(2fL_{aa} + 3\psi f_{;N} + 12fS) \right\}$$

$$a_{1.5}(f, P) = 4^{-1}(4\pi)^{-(m-1)/2} 96^{-1} \left\{ \int_{dM} Tr_V \{ f(96\psi E + 16\psi \tau + \right.$$

$$+ 8f\psi R_{aNaN} + (13\Pi_N - 7\Pi_D)L_{aa}L_{bb} +$$

$$+ (2\Pi_N + 10\Pi_D)L_{ab}L_{ab} + 96SL_{aa} + 192S^2 - 12\psi_{:a}\psi_{:a}) +$$

$$+ f_{;N}((6\Pi_N + 30\Pi_D)L_{aa} + 96S) + 24\psi f_{;NN} \} \right\}$$

$$a_2(f, P) = (4\pi)^{-m/2} 360^{-1} \left\{ \int_M Tr_V \{ f(60E_{;kk} + 60\tau E + 180E^2 + \right.$$

$$+ 30\Omega^2 + 12\tau_{;kk} + 5\tau^2 - 2\rho^2 + 2R^2) \} +$$

$$+ \int_{dM} Tr_V \{ f(240\Pi_N - 120\Pi_D)E_{;N} +$$

$$+ (42\Pi_N - 18\Pi_D)\tau_{;N} + 24L_{aa:bb} + 0L_{ab:ab} + 120EL_{aa} +$$

$$+ 20\tau L_{aa} + 4R_{aNaN}L_{bb} - 12R_{aNaN}L_{ab} + 4R_{abcb}L_{ac} + 0\Omega_{in:i} +$$

$$+ \{ (280\Pi_N + 40\Pi_D)L_{aa}L_{bb}L_{cc} +$$

$$+ (168\Pi_N - 264\Pi_D)L_{ab}L_{ab}L_{cc} +$$

$$+ (224\Pi_N + 320\Pi_D)L_{ab}L_{bc}L_{ac} \}/21 + 720SE + 120S\tau +$$

$$+ 0SR_{aNaN} + 144SL_{aa}L_{bb} + 48SL_{ab}L_{ab} +$$

$$+ 480S^2 L_{aa} + 480S^3 + 120S_{:aa} + 60\psi\psi_{:a}\Omega_{aN} -$$

$$- 42\psi_{:a}\psi_{:a}L_{bb} + 6\psi_{:a}\psi_{:b}L_{ab} - 120\psi_{:a}\psi_{:a}S) + 0\Psi_{:a}\Omega_{aN} +$$

$$+ f_{;N}(180\psi E + 30\psi\tau + 0R_{aNaN} +$$

$$+ (84\Pi_N - 180\Pi_D)/7 \cdot L_{aa}L_{bb} +$$

$$+ (84\Pi_N + 60\Pi_D)/7 \cdot L_{ab}L_{ab} + 72SL_{aa} + 240S^2 -$$

$$- 18\psi_{:a}\psi_{:a}) + f_{;NN}(24L_{aa} + 120S) + 30\psi f_{;iiN} \} \right\}.$$

Proof. Much of this formula follows from the results of §1; we must introduce additionally $Tr(f\psi_{:a}\psi_{:a})$ in $a_{1.5}$ and the terms

$$Tr\{ f\psi\psi_{:a}\Omega_{aN}, \ f\psi_{:a}\psi_{:a}L_{bb}, \ f\psi_{:a}\psi_{:b}L_{ab}, \ f\psi_{:a}\psi_{:a}S, \ f\psi_{:a}\Omega_{aN}, \ f_{;N}\psi_{:a}\psi_{:a} \}$$

in a_2. We compute $a_{1.5}$ and refer to [BrGi] for the computation of a_2. Denote the coefficient of $f\psi_{:a}\psi_{:a}$ by β_1. We use Lemma 2.1 with $m = 2$ and $n = 1.5$

to evaluate β_1; we take absolute boundary conditions. Λ^0 and Λ^2 are trivial so attention is focussed on Λ^1. Let $\kappa = L_{11}$ be the geodetic curvature. We study the coefficient of $\kappa\tau$ and omit other terms. We compute:

T	$T(\theta_1)$	$T(\theta_2)$	T	$T(\theta_1)$	$T(\theta_2)$
∇_1	$\kappa\theta_2$	$-\kappa\theta_1$	E	$-\frac{1}{2}\tau\theta_1$	$-\frac{1}{2}\tau\theta_2$
∇_2	$\kappa\theta_1$	0	Ω_{12}	$-\frac{1}{2}\tau\theta_2$	$\frac{1}{2}\tau\theta_1$
S	$-\kappa\theta_1$	0	$S_{:1}$	$-\kappa^2\theta_2$	$-\kappa^2\theta_1$
ψ	θ_1	$-\theta_2$	$\psi_{:1}$	$2\kappa\theta_2$	$2\kappa\theta_1$

As $2n \neq m$,

$$0 = \int_{dM} Tr(96SL_{aa} + 192S^2 + \beta_1\psi_{:a}\psi_{:a})$$

$$= -\int_{dM} (-96 + 192 + 8\beta_1)\kappa^2.$$

This shows $\beta_1 = -12$. This formula for $a_{1.5}$ agrees with that of Luckock and Moss [LM, see equation C29]. ∎

§5 Leading terms in the heat equation

We begin by generalizing the variational formulas of Lemma 2.2; we omit the details as the calculations are entirely straightforward. We work modulo lower order terms and assume $dM = 0$.

Lemma 5.1. Let $F(\varepsilon) = e^{-2\varepsilon f} F$ and $P(\varepsilon) = e^{-2\varepsilon f} F$.

$\int_M \{\frac{d}{d\varepsilon}\big|_{\varepsilon=0} + (2n+2)f\}$	$\int_M F\Delta^n f$	
$F\Delta^{n-1}E$	$-\frac{1}{2}(m-2)$	
$F\Delta^{n-1}\tau$	$2(m-1)$	
$\int_M \{\frac{d}{d\varepsilon}\big|_{\varepsilon=0} + 2nf\}$	$\int_M \nabla^{n-2}\Delta f \cdot \nabla^{n-2}\tau$	$\int_M \nabla^{n-2}\Delta f \cdot \nabla^{n-2}E$
$\|\nabla^{n-2}\tau\|^2$	$4(m-1)$	0
$\|\nabla^{n-2}\rho\|^2$	m	0
$\nabla^{n-2}\tau \cdot \nabla^{n-2}E$	$-\frac{1}{2}(m-2)$	$2(m-1)$
$\nabla^{n-2}\Omega \cdot \nabla^{n-2}\Omega$	0	0
$\nabla^{n-2}E \cdot \nabla^{n-2}E$	0	$-(m-2)$

Let $\varepsilon_n = (-1)^n/\{2^{n+1} \cdot 1 \cdot 3 \cdot \ldots \cdot (2n+1)\}$. We work modulo lower order terms.

Theorem 5.2. Let $n \geq 3$ and let $dM = 0$.

(a) $a_n(f,P) = \varepsilon_n(4\pi)^{-m/2} \int_M f\Delta^{n-1} Tr\{-2n\tau I - 4(2n+1)E\} + \ldots$

(b) $a_n(1,P) = \varepsilon_n(4\pi)^{-m/2} \int_M Tr\{(n^2 - n - 1)|\nabla^{n-2}\tau|^2 I + 2|\nabla^{n-2}\rho|^2 I$
$\qquad + 4(2n+1)(n-1)\nabla^{n-2}\tau \cdot \nabla^{n-2}E + 2(2n+1)\nabla^{n-2}\Omega \cdot \nabla^{n-2}\Omega$
$\qquad + 4(2n+1)(2n-1)\nabla^{n-2}E \cdot \nabla^{n-2}E\} + \ldots$

Proof. We use invariance theory to see there exist universal constants $\delta_i(n)$ which are independent of the dimension m so:

$$a_n(f,P) = (4\pi)^{-m/2} \int_M f\Delta^{n-1} Tr(\delta_1(n)\tau \cdot I + \delta_2(n)E) + \ldots$$

where we work modulo terms which are quadratic in the jets of the total symbol of P and which consequently involve fewer covariant derivatives.

We adopt the notation of Lemma 4.1. We omit factors of $(4\pi)^{-1/2}$ and compute:

$$(2n-1)\{a_n(f, D_1) - a_n(f, D_2)\}$$

$$= (2n-1)\delta_1(n) \int_M f(\Delta^{n-1}E(D_1) - \Delta^{n-1}E(D_2)) + \ldots$$

$$= (2n-1)\delta_1(n) \int_M -2f\Delta^{n-1}b_x + \ldots$$

$$= a_{n-1}(f_{xx} + 2bf_x, D_1)$$

$$= \delta_1(n-1) \int_M -f_{xx}\Delta^{n-2}b_x + \ldots$$

$$= \delta_1(n-1) \int_M f\Delta^{n-1}b_x + \ldots$$

Consequently $-2(2n-1)\delta_1(n) = \delta_1(n-1)$. Since $\delta_1(1) = 1$,

$$\delta_1(n) = (-1)^{n-1}/\{2^{n-1} \cdot 1 \cdot 3 \cdot \ldots \cdot (2n-1)\} = -4(2n+1)\varepsilon_n.$$

To compute $\delta_2(n)$, we use Lemma 2.1(d) and Lemma 5.1. Let $m = 2n + 2$.

$$0 = \frac{1}{2}a_n(e^{-2\varepsilon f}F, e^{-2\varepsilon f}P)$$

$$= \int_M F\Delta^n f(-n\delta_1(n) + 2(2n+1)\delta_2(n)) + \ldots$$

This shows $\delta_2(n) = n\delta_1(n)/\{2(2n+1)\} = 2n\varepsilon_n$ and proves (a).

Let $n \geq 3$. Since $dM = 0$, we may integrate by parts; for example:

$$\int_M |\nabla^{n-2}R|^2 = \int_M 4|\nabla^{n-2}\rho|^2 - |\nabla^{n-2}\tau|^2 + \ldots$$

We omit terms with fewer than $n-1$ covariant derivatives of the curvature tensor. We use invariance theory, the Bianci identities, and integration by parts to see that modulo lower order terms,

$$a_n(1, P) = \varepsilon_n \int_M Tr\{\beta_1(n)|\nabla^{n-2}\tau|^2 I + \beta_2(n)|\nabla^{n-2}\rho|^2 I +$$

$$+ \beta_3(n)\nabla^{n-2}\tau \cdot \nabla^{n-2}E + \beta_4(n)\nabla^{n-2}\Omega \cdot \nabla^{n-2}\Omega +$$

$$+ \beta_5(n)\nabla^{n-2}E \cdot \nabla^{n-2}E\} + \ldots$$

Clearly $\int_M f\Delta^{n-1}\phi = \int_M \nabla^{n-1} f \cdot \nabla^{n-1}\phi$. We use Lemmas 2.1(b) and Lemma 5.1 and study the coefficients of $\nabla^{n-2} f_{;ii} \cdot \nabla^{n-2}\tau$ and $\nabla^{n-2} f_{;ii} \cdot \nabla^{n-2} E$ to see:

$$0 = 4(m-1)\beta_1(n) + m\beta_2(n) - \frac{1}{2}(m-2)\beta_3(n) + (m-2n)2n$$
$$0 = 2(m-1)\beta_3(n) - (m-2)\beta_5(n) + (m-2n)4(2n+1)$$

We solve these equations to compute $\beta_1(n)$, $\beta_2(n)$, $\beta_3(n)$, and $\beta_5(n)$.

We use Lemma 2.1(e) to calculate $\beta_4(n)$. On 1 forms,

$$E = -\frac{1}{2}\tau \qquad \text{and} \qquad \Omega_{ij}(dx^k) = R_{ijkl}\, dx^l.$$

We use the coefficient of $|\nabla^{n-2}\tau|^2$ to see $-\beta_3(n) - \beta_4(n) + \frac{1}{2}\beta_5(n) = 0$ and determine $\beta_5(n)$. ∎

References

[Be] M. BERGER, Eigenvalues of the Laplacian, *Proc. Symp. in Pure Math.* **V XVI** (1968), 121–126.

[BrGi] T. BRANSON and P. B. GILKEY, The asymptotics of the Laplacian on a Manifold with Boundary, *Comm. in PDE* **15** (1990), 245–272.

[BrGiOr] T. BRANSON, P. B. GILKEY, and B. OERSTED, Leading Terms in the Heat Invariants, *Proceedings AMS* **109** (1990), 437–450.

[BrGo] R. BROOKS and C. GORDON, Isospectral Families of Conformally Equivalent Riemannian Metrics, to appear *Bull. AMS.*

[Br] R. BROOKS, Constructing isospectral manifolds, *AM. Math. Mon* **95** (1988), 823–839.

[BrPeYa] R. BROOKS, P. PERRY and P. YANG, Isospectral sets of conformally equivalent metrics, *Duke Math. J.* **58** (1989), 131–150.

[CaYa] S.-Y. A. CHANG and P. C. YANG, Isospectral conformal metrics on 3 manifolds, J. Amer. Math. Soc. **3** (1990), 117–145; Compactness of isospectral conformal metrics on S^3, *Comment. Math. Helv.* **64** (1989), 363–374; The conformal deformation equation and isospectral sets of conformal metrics; *Contemp. Math.* **101** (1989), 165–178.

[Gi-1] P. B. GILKEY, Recursion relations and the asymptotic behavior of the eigenvalues of the Laplacian, *Compositio Math.* **38** (1979), 201–240.

[Gi-2] P. B. GILKEY, Invariance Theory, the Heat Equation, and the Atiyah–Singer Index Theorem, Publish or Perish, Wilmington, Delaware, 1984.

[Gi-3] P. B. GILKEY, The Boundary Integrand in the Formula for the Signature and Euler Characteristic of a Riemannian Manifold with Boundary, *Advances In Math.* **V 15 No 3** (1975), 334–360.

[Gi-4] P. B. GILKEY, The spectral Geometry of a Riemannian Manifold, *J. Diff. Geo.* **V10** (1975), 601–618.

[GiSm] P. B. GILKEY and L. SMITH, The eta invariant for a Class of Elliptic Boundary Value Problems, *Comm. on Pure and Appl. Math.* **V XXXVI** (1983), 85–131.

[Go] C. GORDON, Riemannian manifolds isospectral on functions but not on 1 forms, *J. Diff. Geo.* **24** (1986), 79–96.

[GoWi] C. GORDON and E. WILSON, Isospectral deformations of compact solv-manifolds, *J. Diff. Geo.* **19** (1984), 241–256.

[Gr] P. GREINER, An asymptotic expansion for the heat equation, *Arch. Rat. Mech. Anal.* **41** (1971), 163–218.

[Ik] A. IKEDA, On Spherical Space Forms which are Isospectral but not Isometric, *J. Math. Soc. Japan* **V35** (1983), 437–444.

[Ka] M. KAC, Can one hear the shape of a drum?, *American Math. Monthly*, **V783** (1966), 1–23.

[KeCrDo] G. KENNEDY, R. CRITCHLEY and J. S. DOWKER, Finite Temperature Field Theory with Boundaries: Stress Tensor and Surface Action Renormalization, *Annals of Physics* **125** (1980), 346–400.

[LuMo] H. LUCKOCK and I. MOSS, The Quantum Geometry of Random Surfaces and Spinning Membranes, to appear in Classical and Quantum Gravity.

[McSi] H. P. MCKEAN and I. M. SINGER, Curvature and the eigenvalues of the Laplacian, *J. Diff. Geo.* **1** (1967), 43–69.

[Me] J. MELMED, J. Phys. A21 (1989) L1131.

[Mi] J. MILNOR, Eigenvalues of the Laplace operator on certain manifolds, *Proc. Nat. Acad. Sci.* USA **51** (1964) 542.

[Mo] I. C. MOSS, Class and Quantum Gravity 6 (1989) 659.

[MoDo] I. C. MOSS and J. S. DOWKER, The correct B_4 Coefficient, *Phys. Letts.* **B 229** (1989), 261–263.

[OsPhSa] B. OSGOOD, R. PHILLIPS and P. SARNAK, Compact isospectral sets of surfaces, *J. Funct. Anal.* **80** (1988), 212–234.

[Pa] V. K. PATODI, Curvature and the fundamental solution of the heat operator, *J. Indian Math.* Soc. **34** (1970), 269–285.

[Pl] A. PLEIJEL, A Study of Certain Green's functions with Applications in the Theory of Vibrating Membranes, *Ark. Mat.* **2** (1954), 553–569.

[Se] R. SEELEY, The resolvent of an elliptic boundary value problem, *Amer. J. Math.* **91** (1960), 889–920.

[Sm] L. SMITH, The Asymptotics of the Heat Equation for a Boundary Value Problem, *Invent. Math.* **63** (1981), 467–493.

[StWa] K. STEWARTSON and R. T. WAECHETER, On hearing the shape of a drum, *Cambridge Philosophical Society* **69** (1971), 353–363.

[Su] T. SUNADA, Riemannian coverings and isospectral manifolds, *Ann. Math.* **121** (1985), 169–186.

[Ur] H. URAKAWA, Bounded domains which are isospectral but not congruent, *Ann. Scient. Ec. Norm. Sup* **15** (1982), 441–456.

[Vi] M. VIGNERAS, Variete Riemanniennes Isospectrales et non isometriques, *Ann. Math.* **112** (1980), 21–32.

Peter B. Gilkey

Mathematics Department
University of Oregon
Eugene, OR 97403
USA

COLLOQUIA MATHEMATICA SOCIETATIS JÁNOS BOLYAI

56. DIFFERENTIAL GEOMETRY, EGER (HUNGARY), 1989

Rank Problems for Webs $W(d,2,r)$ [1]

V. V. GOLDBERG

0. Introduction

A d-web $W(d,n,r)$ is given in an open domain D of a differentiable manifold X^{nr} of dimension nr by d foliations $X_\xi, \xi = 1, \ldots, d$, of codimension r in D if the tangent r-planes to the leaves (web surfaces) of X_ξ through a point in D are in general position.

If foliations X_ξ are determined by the completely integrable system $\underset{\xi}{\omega^i} = 0, i = 1, \ldots, r$, then the equation of the form

$$\sum_{\xi=1}^{d} \underset{\xi}{A_{i_1 \ldots i_q}} \underset{\xi}{\omega^{i_1}} \wedge \ldots \wedge \underset{\xi}{\omega^{i_q}} = 0$$

is said to be an *abelian q-equation* [Gr 77]. The maximum number of linearly independent abelian q-equations is called the *q-rank* of the web $W(d,n,r)$.

There are two fundamental problems in web geometry:

1. Finding an upper bound for the q-rank.

---

[1] This paper is in final form and no version of it will be submitted for publication elsewhere.

2. Determination of the maximum q-rank webs.

Both problems were solved for webs $W(d, 2, 1)$ and $W(d, 3, 1)$ (in these cases there is only one possible value for q: $q = 1$) as well as for d-webs of curves in a three-dimensional space during intensive development of web geometry in the 30's (see [BB 38] where one can find further references).

The first problem was solved for the webs $W(d, n, 1)$ by S. S. Chern [C 36] in 1936. In 1978 S. S. Chern and P. A. Griffiths solved the second problem for so-called normal webs $W(d, n, 1)$ (see [CG 78a], [CG 81]). D. Damiano in [D 83] solved both problems for d-webs of curves in an n-dimensional space taking into consideration abelian $(n - 1)$-equations. In particular, he proved that a so-called exceptional $(n + 3)$-web of curves is of maximum $(n - 1)$-rank and it is the unique non-linearizable web among quadrilateral d-webs, $d \geq n \geq 3$. (A quadrilateral d-web is defined as a d-web of curves for which every n-subweb generates a coordinate system in the ambient n-space.)

The first problem was solved by S. S. Chern and P. A. Griffiths for the webs $W(d, n, r)$, $r > 1$, and $q = r$ [CG 78b]. To our knowledge, none of two problems was studied anywhere in the case $1 \leq q \leq r$ except of the papers [D 83] and [CG 78b].

In the present paper we consider webs $W(d, 2, r)$. In Section 1 we introduce necessary notions and formulas. In particular, we

- find conditions for a web $W(d, 2, 2)$, $d \geq 3$, to be isoclinic or Grassmannizable;

- introduce almost Grassmannizable webs as webs $W(d, 2, r)$ whose all basis affinors are scalar and prove that such webs are isoclinic with one exception for $W(4, 2, 2)$ which can be both isoclinic and non-isoclinic;

- define almost algebraizable webs as isoclinic almost Grassmannizable webs satisfying (1.74), and find under what condition such webs are algebraizable;

- give the invariant geometric characterization of condition for an isoclinic almost Grassmannizable web to be almost algebraizable and find the analytic condition for a non-isoclinic almost Grassmannizable web to possess the same geometric property;

- give a new characterization of conditions for an isoclinic three-web $W(3, 2, r)$ to be hexagonal or algebraizable;

- prove that a non-isoclinic three-web $W(3, 2, 2)$ satisfying certain conditions can be uniquely extended to a non-isoclinic almost Grassmannizable web $AGW(4, 2, 2)$ and find the location of the fourth foliation X_4 of $AGW(4, 2, 2)$ with respect to the foliations X_1, X_2 and X_3, and

- give examples of isoclinic three-webs which cannot be extended to a non-isoclinic almost Grassmannizable web $AGW(4, 2, 2)$.

In Section 2 we prove the following statements:

- Webs $W(d, 2, r)$ of maximum r-rank are almost Grassmannizable.

- The maximum r-rank $\pi(d, 2, r)$ of an almost Grassmannizable web $AGW(d, 2, r)$ is equal to 0 if $r > 2$ and $d \leq r + 1$, and equal to $(d-1)(d-2)(d-3)/6$ if $r = 2$ (the last result can be found in [CG 78b]).

- A web $W(4, 2, 2)$ is of maximum 2-rank if and only if it is an almost algebraizable web or non-isoclinic almost Grassmannizable web for which any of four listed affine connections is equiaffine, and

- An almost Grassmannizable web $AGW(d, 2, 2), d > 4$, is of maximum 2-rank if and only if it is algebraizable.

The main results of the present paper were presented without proofs in author's paper [G 83].

1. Webs of codimension r on a $2r$-dimensional differentiable manifold

1.1. Basic notions and equations

In an open domain D of a differentiable manifold X^{2r} of dimension $2r$ a d-web $W(d, 2, r), d \geq 3$, *of codimension r* is given by d foliations $X_\xi, \xi = 1, \ldots, d$, of codimension r in D if the tangent r-planes to the leaves $V_\xi \subset X_\xi$ through a point in D are in general position.

Two webs $W(d, 2, r)$ and $\widetilde{W}(d, 2, r)$ are *equivalent* to each other if there exists a local diffeomorphism $\phi : D \to \widetilde{D}$ of their domains transferring the foliations of W into the foliations of $\widetilde{W}$.

The foliations X_ξ, $\xi = 1, \ldots, d$, of the web $W(d, 2, r)$ can be given by d completely integrable systems of Pfaffian equations

$$\underset{\xi}{\omega^i} = 0, \qquad \xi = 1, \ldots, d; \qquad i = 1, \ldots, r,$$

where the forms $\underset{1}{\omega^i}$ and $\underset{2}{\omega^i}$ are the basis forms of the manifold X^{2r} and

(1.1) $\qquad -\underset{3}{\omega^i} = \underset{1}{\omega^i} + \underset{2}{\omega^i}, \qquad \underset{\alpha}{\omega^i} = \underset{\alpha}{\lambda^i_j} \underset{1}{\omega^j} + \underset{2}{\omega^i}, \qquad \alpha = 4, \ldots, d,$

$$\det(\underset{\alpha}{\lambda^i_j}) \neq 0, \qquad \det(\delta^i_j - \underset{\alpha}{\lambda^i_j}) \neq 0, \qquad \det(\underset{\alpha}{\lambda^i_j} - \underset{\beta}{\lambda^i_j}) \neq 0,$$

(1.2) $\qquad\qquad\qquad\qquad\qquad \alpha \neq \beta, \qquad \alpha, \beta = 4, \ldots, d$

(see [G 77], [G 80]) where the quantities $\underset{\alpha}{\lambda^i_j}$, $i, j = 1, \ldots, r$, form an $(1, 1)$-tensor for any $\alpha = 4, \ldots, d$, and these $d - 3$ tensors $\underset{\alpha}{\lambda^i_j}$ are distinct. They are called the *basis affinors* of a web $W(d, 2, r)$ (see [G 77], [G 80]).

It is easy to show (see [G 77], [G 80] for $d = 4$) that for the web $W(d, 2, r)$ the following equations hold:

(1.3) $\qquad \begin{cases} d\underset{1}{\omega^i} = \underset{1}{\omega^j} \wedge \omega^i_j + a^i_{jk} \underset{1}{\omega^j} \wedge \underset{1}{\omega^k}, \\ d\underset{2}{\omega^i} = \underset{2}{\omega^j} \wedge \omega^i_j - a^i_{jk} \underset{2}{\omega^j} \wedge \underset{2}{\omega^k}, \end{cases}$

(1.4) $\qquad\qquad d\omega^i_j - \omega^k_j \wedge \omega^i_k = b^i_{jkl} \underset{1}{\omega^k} \wedge \underset{2}{\omega^l},$

(1.5) $\qquad\qquad \nabla a^i_{jk} = b^i_{[j|l|k]} \underset{1}{\omega^k} + b^i_{[jk]l} \underset{2}{\omega^l},$

(1.6) $\qquad\qquad a^i_{jk} = -a^i_{kj},$

(1.7) $\qquad\qquad b^i_{[jkl]} = 2a^m_{[jk} a^i_{|m|l]},$

(1.8) $\qquad\qquad \nabla \underset{\alpha}{\lambda^i_j} = \underset{\alpha}{\lambda^i_{jk}} \underset{1}{\omega^k} + \underset{\alpha}{\mu^i_{jk}} \underset{2}{\omega^k}, \qquad \alpha = 4, \ldots, d,$

(1.9) $\qquad \underset{\alpha}{\lambda^i_{[jk]}} - \underset{\alpha}{\mu^i_{[j|p|}} \underset{\alpha}{\lambda^p_{k]}} = \underset{\alpha}{\lambda^i_p} a^p_{jk} + \underset{\alpha}{\lambda^p_{[k}} \underset{\alpha}{\lambda^q_{j]}} a^i_{pq}, \qquad \alpha = 4, \ldots, d,$

where the quantities a^i_{jk} and b^i_{jkl} form the *torsion and curvature tensors* of the 3-subweb $[1, 2, 3]$ determined by the foliations X_1, X_2 and X_3, and ∇ is the symbol of the covariant differentiation in the connection γ_{123} induced by the subweb $[1,2,3]$ with connection forms ω^i_j, so that, for example,

$$\nabla a^i_{jk} = da^i_{jk} - a^i_{mk}\omega^m_j - a^i_{jm}\omega^m_k + a^m_{jk}\omega^i_m.$$

The tensors $\{a^i_{jk}\}$ and $\{b^i_{jkl}\}$ are said to be the *torsion and curvature tensors* of the web $W(d, 2, r)$.

In the case of webs $W(4, 2, r)$ the expressions of the connection forms, torsion and curvature tensors of the connection $\gamma_{\xi\eta\zeta}$ induced by the three-subweb $[\xi, \eta, \zeta], \xi, \eta, \zeta = 1, \ldots, d$, generated by the foliations X_ξ, X_η and X_ζ were found in [G 77], [G 80].

Later on we will need the expressions of the connection forms of the connections $\gamma_{12\alpha}$, $\gamma_{31\alpha}$, and $\gamma_{23\alpha}$, $\alpha = 4, \ldots, d$:

(1.10)
$$\begin{cases} \underset{12\alpha}{\omega}{}^i_j = \omega^i_j - \mu^i_{mk}\underset{\alpha}{\widetilde{\lambda}}{}^m_j\underset{2}{\omega}{}^k, \\[2mm] \underset{31\alpha}{\omega}{}^i_j = \omega^i_j + 2a^i_{jk}\underset{1}{\omega}{}^k - \mu^i_{mk}\underset{\alpha}{\widetilde{\lambda}}{}^m_j\underset{3}{\omega}{}^k, \\[2mm] \underset{23\alpha}{\omega}{}^i_j = \omega^i_j - \underset{\alpha}{\widetilde{\lambda}}{}^l_j\lambda^i_{lk}\underset{3}{\omega}{}^k + \underset{\alpha}{\widetilde{\lambda}}{}^l_j(\lambda^i_{lk} - \mu^i_{lk} - 2\underset{\alpha}{\lambda}{}^i_p a^p_{lk})\underset{2}{\omega}{}^k; \end{cases}$$

in (1.10) the matrices $(\underset{\alpha}{\widetilde{\lambda}}{}^i_j)$ and $(\underset{\alpha}{\widetilde{\widetilde{\lambda}}}{}^i_j)$ are the inverse matrices of $(\underset{\alpha}{\lambda}{}^i_j)$ and $(\delta^i_j - \underset{\alpha}{\lambda}{}^i_j)$.

1.2. Isoclinic, transversally geodesic, almost Grassmannizable and Grassmannizable webs

Let us consider an isoclinic r-surface V^r of the manifold X^{2r} which is determined by the following system of Pfaffian equations:

$$\lambda\underset{1}{\omega}{}^i + \underset{2}{\omega}{}^i = 0$$

where λ is a function of a point $p \in X^{2r}$. For a tangent vector dx to X^{2r} at the point p we have:

(1.11)
$$dx = \underset{1}{\omega}{}^i e^1_i + \underset{2}{\omega}{}^i e^2_i.$$

It follows from (1.11) that the vectors e_i^2, e_i^1, $e_i^3 = e_i^1 - e_i^2$, and $e_i^4 = e_i^1 - \lambda_\alpha^j e_j^2$, $\alpha = 4, \ldots, d$, are tangent to the leaves V_1, V_2, V_3, and V_α passing through the point p. On the surface V^r we have

$$(1.12) \qquad dx = \underset{1}{\omega}^i (e_i^1 - \lambda e_i^2).$$

A web $W(d, 2, r), d > 3$ whose basis affinors $\underset{\alpha}{\lambda}_j^i$ are scalar:

$$(1.13) \qquad \underset{\alpha}{\lambda}_j^i = \delta_j^i \underset{\alpha}{\lambda}$$

is said to be an *almost Grassmannizable* web. We will denote such webs by $AGW(d, 2, r)$.

The transversal vectors $\eta^\xi = \eta^i e_i^\xi$, $\xi = 1, \ldots, d$, are tangent to the leaves V_ξ at the point p. It is easy to see that for an $AGW(d, 2, r)$ the vectors η^ξ lie in a 2-plane.

The bivector $\eta^1 \wedge \eta^2$ determined by η^i is said to be a *transversal bivector* of $AGW(d, 2, r)$.

It follows from (1.12) that the tangent r-plane of V^r intersects the transversal bivector $\eta^1 \wedge \eta^2$ in the direction of the vector $\eta^i (e_i^1 - \lambda e_i^2)$. The cross ratio of this vector and the three vectors η^1, η^2, η^3 (η^α) is equal to λ (respectively λ/λ_α). This cross ratio does not depend on quantities η^i giving the direction of $\eta^1 \wedge \eta^2$. Because of this, the surfaces V^r defined by (1.11) are called *isoclinic surfaces* of a web $AGW(d, 2, r)$.

A web $AGW(d, 2, r)$ is said to be *isoclinic* if there exists a one-parameter family of isoclinic surfaces through any point $p \in D$.

Note that the definition of an isoclinic web $AGW(d, 2, r)$ is exactly the same as that for its 3-subweb $[1, 2, 3]$ (see [A 74]). Because of this, an analytical criterion for an $AGW(d, 2, r)$ be isoclinic is the same as that for its 3-subweb $[1, 2, 3]$: *a web* $W(3, 2, r)$, $r > 2$, *is isoclinic if and only if its torsion tensor* a_{jk}^i *has the following special structure:*

$$(1.14) \qquad a_{jk}^i = a_{[j} \delta_{k]}^i.$$

As we noted above, the same condition (1.14) is necessary and sufficient for a web $AGW(d, 2, r)$, $d > 3$, $r > 2$, to be isoclinic. Note that in [A 74] the restriction $r > 2$ was missed. This restriction was included in necessary and sufficient conditions for a web $W(n+1, n, r)$, $n \geq 2$, to be isoclinic in [G 77] and for $W(3, 2, r)$ in [AS 81].

For later use we will need necessary and sufficient conditions for a web $AGW(d, 2, 2)$ to be isoclinic. Thus we set $r = 2$. In this case the torsion tensor a^i_{jk} always has the form (1.14). Because of this, the structure equations (1.3) can be written in the form

(1.15)
$$\begin{cases} d\underset{1}{\omega}^i = \underset{1}{\omega}^j \wedge \underset{1}{\omega}^i_j + a_j \underset{1}{\omega}^j \wedge \underset{1}{\omega}^i, \\ d\underset{2}{\omega}^i = \underset{2}{\omega}^j \wedge \omega^i_j - a_j \underset{2}{\omega}^j \wedge \underset{2}{\omega}^i. \end{cases}$$

Exterior differentiation of (1.15) gives

(1.16)
$$\begin{cases} \Omega^i_j \wedge \underset{1}{\omega}^j - \nabla a_j \wedge \underset{1}{\omega}^j \wedge \underset{1}{\omega}^i = 0, \\ \Omega^i_j \wedge \underset{2}{\omega}^j + \nabla a_j \wedge \underset{2}{\omega}^j \wedge \underset{2}{\omega}^i = 0, \end{cases}$$

where
$$\Omega^i_j = d\omega^i_j - \omega^k_j \wedge \omega^i_k, \quad \nabla a_j = da_j - a_k \omega^k_j.$$

In the case $r = 2$ the general form of Ω^i_j and ∇a_j satisfying (1.16) is (1.4) and

(1.17)
$$\nabla a_i = p_{ij} \underset{1}{\omega}^j + q_{ij} \underset{2}{\omega}^j,$$

where by means of (1.5) and (1.14) we have

(1.18)
$$b^i_{[j|l|k]} = \delta^i_{[k} p_{j]l}, \quad b^i_{[jk]l} = \delta^i_{[k} q_{j]l}.$$

Note that in the case $r = 2$ the equations (1.16) do not imply a symmetry of p_{ij} and q_{ij}.

Proposition 1.1. *A web $AGW(d, 2, 2)$, $d > 3$, is isoclinic if and only if the quantities p_{ij} and q_{ij} in (1.17) are symmetric.*

Proof. The equations (1.11) and their differential consequences are satisfied on a surface V^r. Exterior differentiation of (1.11) by means of (1.11) and (1.15) leads to

(1.19)
$$[d\lambda + (\lambda - \lambda^2) a_j \underset{1}{\omega}^j] \wedge \underset{1}{\omega}^i = 0.$$

On V^r the differential $d\lambda$ is expressed in terms of the forms $\underset{1}{\omega}^i$ only. This and (1.19) give

(1.20)
$$d\lambda = (\lambda^2 - \lambda) a_j \underset{1}{\omega}^j.$$

Exterior differentiation of (1.20) by means of (1.11), (1.20), and (1.17) gives

(1.21) $$p_{[jk]} - \lambda q_{[jk]} = 0.$$

Since λ is arbitrary, it follows from (1.21) that

(1.22) $$p_{jk} = p_{kj}, \quad q_{jk} = q_{kj}.$$

Conversely, the exterior differentiation of (1.20) leads to the identity using (1.17) and (1.22). ∎

Note that Proposition 1.1 for a three-web $W(3,2,r)$ appeared in the book [AS 81].

Since, by (1.22), p_{jk} and q_{jk} are symmetric for an isoclinic web $AGW(d,2,2)$, for the webs $AGW(d,2,r)$, $r \geq 2$, we can use the same procedure which was used in [A 74] in the case $r > 2$ for the isoclinic webs $W(3,2,r)$.

Let

(1.23) $$p_{jk} = f_{jk} - h_{jk}, \; q_{jk} = g_{jk} - h_{jk},$$

where f_{jk}, g_{jk}, and h_{jk} are symmetric $(0,2)$-tensors.

Using (1.23), we can rewrite equations (1.17) and (1.18) in the form

(1.24) $$\nabla a_i = (f_{ij} - h_{ij})\underset{1}{\omega}^j + (g_{ij} - h_{ij})\underset{2}{\omega}^j,$$

(1.25) $$b^i_{[j|l||k]} = (f_{l[j} - h_{l[j})\delta^i_{k]}, \; b^i_{[jk]l} = (g_{l[j} - h_{l[j})\delta^i_{k]}.$$

For any tensor b^i_{jkl} the following identity

(1.26) $$b^i_{jkl} = b^i_{(jkl)} + \frac{1}{3}b^i_{[jk]l} + \frac{1}{3}b^i_{[jl]k} + b^i_{[lk]j} + \frac{4}{3}b^i_{[j|k|l]} + \frac{2}{3}b^i_{[k|l|j]}$$

can be checked by inspection. Using (1.25), we can write the identity (1.26) in the form

(1.27) $$b^i_{jkl} = a^i_{jkl} + f_{jk}\delta^i_l + g_{lj}\delta^i_k + h_{kl}\delta^i_j$$

where

(1.28) $$a^i_{jkl} = b^i_{jkl} - (f_{(jk} + g_{(jk} + h_{(jk})\delta^i_{l)}.$$

If we impose the following restriction for the tensor a^i_{jkl} :

(1.29)
$$a^i_{ikl} = 0,$$

it allows us to determine h_{kl}. In fact, it follows from (1.28) and (1.29) that

(1.30)
$$h_{kl} = \frac{3}{4} b^i_{(ikl)} - (f_{kl} + g_{kl}).$$

Note that (1.30) is equivalent to (1.29).

Exterior differentiation of (1.4), where b^i_{jkl} is substituted from (1.27), and the application of Cartan's lemma give the following equations

(1.31)
$$\begin{cases} \dot{\nabla} f_{ij} = \underset{1}{f_{ijk}}\underset{1}{\omega^k} + \underset{2}{f_{ijk}}\underset{2}{\omega^k}, \\[2mm] \dot{\nabla} g_{ij} = \underset{1}{g_{ijk}}\underset{1}{\omega^k} + \underset{2}{g_{ijk}}\underset{2}{\omega^k}, \\[2mm] \dot{\nabla} h_{ij} = \underset{1}{h_{ijk}}\underset{1}{\omega^k} + \underset{2}{h_{ijk}}\underset{2}{\omega^k}, \\[2mm] \dot{\nabla} a^i_{jkl} = \underset{1}{a^i_{jklm}}\underset{1}{\omega^m} + \underset{2}{a^i_{jklm}}\underset{2}{\omega^m}, \end{cases}$$

where we introduced the new differential operator (see [A 74])

(1.32)
$$\dot{\nabla} = \nabla + a_m (\underset{1}{\omega^m} - \underset{2}{\omega^m}),$$

$\underset{s}{f_{ijk}}, \underset{s}{g_{ijk}}, \underset{s}{h_{ijk}}, s = 1, 2$, are symmetric with respect to i, j, k, and

(1.33)
$$\underset{1}{a^i_{jl[km]}} = \underset{1}{g_{lj[k}}\delta^i_{m]}, \qquad \underset{2}{a^i_{jl[km]}} = \underset{2}{f_{jl[k}}\delta^i_{m]},$$

(1.34)
$$a_m a^m_{ijk} = \underset{2}{f_{ijk}} - \underset{1}{g_{ijk}} + \underset{2}{h_{ijk}} - 3a_{(i} h_{jk)}.$$

A web $AGW(d, 2, r)$ is said to be *transversally geodesic* if for any bivector $\eta^1 \wedge \eta^2$ there exists a two-dimensional surface V^2 tangent to $\eta^1 \wedge \eta^2$ at p and each bivector $\eta^1 \wedge \eta^2$ is tangent to one and only one V^2.

Note again that this definition of a transversally geodesic web $AGW(d, 2, r)$ is the same as that for its 3-subweb $[1, 2, 3]$.

It was proved in [A 74] for $W(3, 2, r)$ — and the same proof can be developed for $AGW(d, 2, r), d \geq 3$, — that an *isoclinic web* $AGW(d, 2, r)$, $d \geq 3, r \geq 2$, is transversally geodesic if and only if

(1.35)
$$a^i_{jkl} = 0.$$

A web $AGW(d, 2, r)$ which is isoclinic and transversally geodesic is a *Grassmannizable web* i.e. it is equivalent to a *Grassmann d-web* (see [A 80] or [G 82a]) which is formed by d foliations of Schubert varieties of codimension r on the Grassmannian $G(1, r + 1)$ in a projective space $P^N, N = \binom{r+2}{2} - 1$. Each Schubert variety of the foliation X_ξ is the image of a bundle of straight lines of a projective space P^{r+1} of dimension $r + 1$. Each foliation X_ξ is the image of such bundles with vertices on the hypersurface V_ξ of P^{r+1}. If d hypersurfaces V_ξ belong to an algebraic hypersurface V_d^r of degree d, then the corresponding Grassmann d-web is called *algebraic*. A web $W(d, 2, r)$ which is equivalent to algebraic d-web is said to be *algebraizable*.

It follows from this the following

Proposition 1.2. *A web $AGW(d, 2, 2)$ is Grassmannizable if and only if the Pfaffian derivatives p_{jk} and q_{jk} of its torsion tensor are symmetric and the quantities a^i_{jkl} vanish.*

It follows from (1.35), (1.31), and (1.33) that for a Grassmannizable web $W(d, 2, 2)$ we have

$$\underset{1}{g}_{lj[k} \delta^i_{m]} = 0, \qquad \underset{2}{f}_{jl[k} \delta^i_{m]} = 0$$

i.e., for $r \geq 2$,

(1.36) $\underset{1}{g}_{ijk} = 0, \qquad \underset{2}{f}_{ijk} = 0.$

In addition, in this case it follows from (1.34), (1.35), and (1.36) that

(1.37) $\underset{1}{h}_{ijk} = \underset{2}{h}_{ijk} + 3a_{(i}h_{jk)}.$

1.3. Almost Grassmannizable and almost algebraizable webs

In the next theorem we will prove that, with one exception, the webs $AGW(d, 2, r)$ are always isoclinic.

Theorem 1.3. *An almost Grassmannizable web $AGW(d,2,r)$ is isoclinic if $r > 2$, $d \geq 4$ or $r = 2$, $d > 4$. A web $AGW(4,2,2)$ can be isoclinic or non-isoclinic.*

Proof. Suppose that a web $W(d,2,r)$ is almost Grassmannizable. Then its basis affinors have the structure (1.13) where, by (1.2),

$$(1.38) \qquad\qquad \underset{\alpha}{\lambda} \neq 0, 1.$$

It follows from (1.13) and (1.8) that

$$(1.39) \qquad\qquad d\underset{\alpha}{\lambda} = \underset{\alpha 1}{\lambda}_i \underset{1}{\omega}^i + \underset{\alpha 2}{\lambda}_i \underset{2}{\omega}^i$$

Equations (1.13), (1.39), and (1.8) imply

$$(1.40) \qquad\qquad \underset{\alpha}{\lambda}^i_{jk} = \delta^i_j \underset{\alpha 1}{\lambda}_k, \qquad \underset{\alpha}{\mu}^i_{jk} = \delta^i_j \underset{\alpha 2}{\lambda}_k.$$

Substituting (1.40) and (1.13) into (1.9), we obtain (1.14) where

$$(1.41) \qquad\qquad a_i = \left(-\underset{\alpha 1}{\lambda}_i + \underset{\alpha}{\lambda}\underset{\alpha 2}{\lambda}_i \right) \Big/ \left(\underset{\alpha}{\lambda} - \underset{\alpha}{\lambda}^2 \right).$$

This proves Theorem 1.3 for $r > 2$. To prove it for $r = 2$, we note that equations (1.41) imply

$$(1.42) \qquad\qquad \underset{\alpha 1}{\lambda}_i = \underset{\alpha}{\lambda}\underset{\alpha}{\lambda}_i + (\underset{\alpha}{\lambda}^2 - \underset{\alpha}{\lambda})a_i,$$

where we denoted $\underset{\alpha 2}{\lambda}_i$ by $\underset{\alpha}{\lambda}_i$. Using equations (1.42), we can rewrite equations (1.39) in the form

$$(1.43) \qquad\qquad d\underset{\alpha}{\lambda} = \underset{\alpha}{\lambda}(\underset{\alpha}{b}_i - a_i)\underset{1}{\omega}^i + (\underset{\alpha}{b}_i - \underset{\alpha}{\lambda}a_i)\underset{2}{\omega}^i$$

where

$$(1.44) \qquad\qquad \underset{\alpha}{b}_i = \underset{\alpha}{\lambda}_i + \underset{\alpha}{\lambda}a_i.$$

Exterior differentiation of equation (1.43) gives

$$(1.45) \qquad \begin{aligned} &[\nabla \underset{\alpha}{b}_i + (\underset{\alpha}{b}_i a_j - p_{[ij]} - \underset{\alpha}{\lambda}p_{(ij)} + \underset{\alpha}{\lambda}q_{(ij)} - \underset{\alpha}{b}_i \underset{\alpha}{b}_j)\underset{1}{\omega}^j \\ &- \underset{\alpha}{\lambda}q_{[ij]}\underset{2}{\omega}^j] \wedge (\underset{\alpha}{\lambda}\underset{1}{\omega}^i + \underset{2}{\omega}^i) + (1 - \underset{\alpha}{\lambda})(p_{[ij]} + \underset{\alpha}{\lambda}q_{[ji]}\underset{1}{\omega}^j \wedge \underset{1}{\omega}^i = 0 \end{aligned}$$

where

$$\nabla_\alpha b_i = db_i - b_j \omega_i^j.$$

Equation (1.45) shows that $\nabla_\alpha b_i$ is expressed in terms of ω_1^k and ω_2^k :

(1.46) $$\nabla_\alpha b_i = s_{ik} \omega_1^k + b_{ik} \omega_2^k.$$

Substitution of (1.46) into (1.45) gives by means of linear independence of $\omega_1^i \wedge \omega_1^j$, $\omega_2^i \wedge \omega_2^j$ and $\omega_1^i \wedge \omega_2^j$ the following relations:

(1.47) $$s_{[ij]} = p_{[ij]} - b_{[i} a_{j]},$$

(1.48) $$b_{[ij]} = \lambda q_{[ij]},$$

(1.49) $$s_{ij} = \lambda b_{ji} + b_i(b_j - a_j) + \lambda(p_{ji} - q_{ji}).$$

Equations (1.47)–(1.49) imply by means of the inequalities (1.39) the relations

(1.50) $$p_{[ij]} - \lambda q_{[ij]} = 0.$$

The equations (1.50) are satisfied identically for $r > 2$ because in this case the quantities p_{ij} and q_{ij} are symmetric.

Suppose that $r = 2$. Then, if $d > 4$, the relations (1.50) imply (1.22), and almost Grassmannizable webs $AGW(d, 2, 2)$, $d > 4$, are isoclinic.

For the almost Grassmannizable webs $AGW(4, 2, 2)$ there are two possibilities:

(a) $p_{[ij]} = q_{[ij]} = 0$. In this case the webs $AGW(4, 2, 2)$ are isoclinic.

(b) $p_{[ij]} \neq 0$, $q_{[ij]} \neq 0$. In this case we have a non-isoclinic almost Grassmannizable web. For such a web we have (1.13), (1.17), and the equations

(1.51) $$\begin{cases} \nabla_4 b_i = [\lambda b_{ji} + b_i(b_j - a_j) + \lambda(p_{ij} - q_{ji})] \omega_1^j + b_{ij} \omega_2^j, \\ b_{[ij]} = \lambda q_{[ij]}, \quad p_{[ij]} = \lambda q_{[ij]}, \end{cases}$$

which follow from (1.46), (1.48)–(1.50).

Note that Theorem 1.3 extends for $n = 2$, $r \geq 2$ the content of Theorem 3 announced in [A 80] for $n \geq 2$, $r \geq 3$ and stated there in terms of almost Grassmann structures associated with a web. Note also that synthetical proofs of all theorems of [A 80] but Theorem 3, were published in [A 82] (see also [A 83a]): it was noted there that Theorem 3 can be proved analytically. This analytical proof was published in [A 81].

For the isoclinic almost Grassmannizable webs $AGW(d, 2, r)$, $d > 4$, the equations (1.22), (1.23), and (1.49) allow us to write the equations (1.46) in the form

$$(1.52) \qquad \nabla_{\alpha} b_i = b_{ij} (\lambda \omega^j_{\alpha 1} + \omega^j_{2}) + [b_i (b_j - a_j) + \lambda (f_{ij} - g_{ij})] \omega^j_{\alpha 1}$$

where $b_{ij} = b_{ji}$.

We introduce new quantities k_{ij}, $k_{ij} = k_{ji}$, such that

$$b_{ij} = g_{ij} - k_{ij}.$$

Using them, we can write the equations (1.52) in the form

$$(1.53) \qquad \nabla_{\alpha} b_i = [b_i (b_j - a_j) + \lambda (f_{ij} - k_{ij})] \omega^j_{1} + (g_{ij} - k_{ij}) \omega^k_{1}.$$

Exterior differentiation of (1.53) and the application of Cartan's lemma give

$$(1.54) \quad \dot{\nabla}_{\alpha} k_{ij} = [k_{ijk} (\lambda \omega^k_{\alpha 1} + \omega^k_{2}) + (b_t a^t_{ijk} + g_{ijk} - \lambda f_{ijk} + 3b_{(k} k_{ij)}) \omega^k_{1}.$$

Thus, in the case of an isoclinic $AGW(d, 2, r)$ we have (1.13), (1.38), (1.14), (1.15), (1.24), (1.25), (1.27), (1.29), (1.31), (1.33), (1.34), and (1.43). Equations (1.13) and (1.43) imply

$$(1.55) \qquad \begin{cases} \lambda^i_{\alpha j k} = \delta^i_j \lambda / (b_k - a_k), & \mu^i_{\alpha j k} = \delta^i_j (b_k - \lambda a_k), \\ \tilde{\lambda}^i_{\alpha j} = \delta^i_j / \lambda, & \tilde{\tilde{\lambda}}^i_{\alpha j} = \delta^i_j / (1 - \lambda), \qquad \alpha = 4, \dots, d. \end{cases}$$

By means of (1.13) and (1.55), the equations (1.10) can be written in the form

$$(1.56) \qquad \begin{cases} \omega^i_{12\alpha j} = \omega^i_j + \delta^i_j (a_k - b_k / \lambda) \omega^k_{2}, \\ \omega^i_{31\alpha j} = \omega^i_j + a_j \omega^i_1 - \delta^i_j a_k \omega^k_1 + (b_k - \lambda a_k)(\omega^k_1 + \omega^k_2)/(1 - \lambda), \\ \omega^i_{23\alpha j} = \omega^i_j + \delta^i_j \lambda (b_k - a_k)(\omega^k_1 + \omega^k_2)/(1 - \lambda) + \\ \qquad + [\delta^i_j b_k (1 - 1/\lambda) + 2\delta^i_{[j} a_{k]}] \omega^k_2. \end{cases}$$

Lemma 1.4. *If* $\underset{\xi\eta\zeta}{\omega}{}^i_j$ *are the connection forms of the affine connection* $\gamma_{\xi\eta\zeta}$ *induced by the 3-subweb* $[\xi,\eta,\zeta]$, *then*

(1.57) $$d\,\underset{\xi\eta\zeta}{\omega}{}^i_j - \underset{\xi\eta\zeta}{\omega}{}^k_j \wedge \underset{\xi\eta\zeta}{\omega}{}^i_k = (a^i_{jkl} + \delta^i_l \underset{\xi}{k}_{jk} + \delta^i_k \underset{\eta}{k}_{jl} + \delta^i_j \underset{\zeta}{k}_{kl})\underset{1}{\omega}{}^k \wedge \underset{2}{\omega}{}^l$$

where

$$\underset{1}{k}_{ij} = f_{ij}, \quad \underset{2}{k}_{ij} = g_{ij}, \quad \underset{3}{k}_{ij} = h_{ij}.$$

Proof. We shall distinguish two cases:

i) *At least two out of the three indices* ξ, η, ζ *are equal to* 1, 2, *or* 3. In this case we have the connection forms $\underset{123}{\omega}{}^i_j = \omega^i_j$, $\underset{12\alpha}{\omega}{}^i_j$, $\underset{31\alpha}{\omega}{}^i_j$, $\underset{23\alpha}{\omega}{}^i_j$, $\alpha = 4,\ldots,d$, and 20 forms obtained from these forms by permutations of lower indices. For the first four forms the equation (1.57) can be proved by the straightforward exterior differentiation of ω^i_j and (1.56).

To prove (1.57) for the 20 other forms, first we must find the expressions of these forms. In [AS 71b] the following relations were found:

(1.58) $$\begin{cases} \underset{\xi\eta\zeta}{\omega}{}^i_j = \underset{\eta\xi\zeta}{\omega}{}^i_j, \\[2mm] \underset{1\alpha2}{\omega}{}^i_j = \underset{21\alpha}{\omega}{}^i_j - 2\,\underset{21\alpha}{a}{}^i_{jk}\underset{1}{\omega}{}^k, \\[2mm] \underset{3\alpha1}{\omega}{}^i_j = \underset{31\alpha}{\omega}{}^i_j + 2\,\underset{31\alpha}{a}{}^i_{jk}(\underset{1}{\omega}{}^k + \underset{2}{\omega}{}^k). \end{cases}$$

In [G 80] (see also [G 82a]) we found the following expressions of $\underset{21\alpha}{a}{}^i_{jk}$ and $\underset{31\alpha}{a}{}^i_{jk}$:

(1.59) $$\begin{cases} \underset{21\alpha}{a}{}^i_{jk} = -\underset{12\alpha}{a}{}^i_{jk} = -(a^i_{jk} + \mu^i_{m[j}\underset{\alpha}{\widetilde{\lambda}}{}^{m]}_k), \\[2mm] \underset{31\alpha}{a}{}^i_{jk} = -a^i_{jk} + \mu^i_{m[j}\underset{\alpha}{\widetilde{\widetilde{\lambda}}}{}^{m]}_k. \end{cases}$$

The equations (1.59) and (1.58) allow us to rewrite (1.58) in the form:

(1.60) $$\begin{cases} \underset{\xi\eta\zeta}{\omega}{}^i_j = \underset{\eta\xi\zeta}{\omega}{}^i_j, \\[2mm] \underset{1\alpha2}{\omega}{}^i_j = \underset{12\alpha}{\omega}{}^i_j + 2\underset{\alpha}{b}_{j[}\delta^i_{k]}\underset{1}{\omega}{}^k, \\[2mm] \underset{3\alpha1}{\omega}{}^i_j = \underset{31\alpha}{\omega}{}^i_j - 2\delta^i_{[k}(\underset{\alpha}{b}^i_{j]} - a_{j]})(\underset{1}{\omega}{}^k + \underset{2}{\omega}{}^k)/(1 - \underset{\alpha}{\lambda}). \end{cases}$$

Taking exterior derivatives of (1.60) and using (1.59), we can get the other 20 relations (1.57).

ii) *All the indices ξ, η, ζ are greater than 3.* The foliations X_α, X_β, and X_γ of a 3-subweb $[\alpha, \beta, \gamma]$ are determined by the equations $\underset{a}{\omega^i} = 0$ where

$$\underset{a}{\omega^i} = -(\lambda \underset{a}{\underset{1}{\omega^i}} + \underset{2}{\omega^i}), \quad a = \alpha, \beta, \gamma.$$

Eliminating $\underset{1}{\omega^i}$ and $\underset{2}{\omega^i}$ from these three equations, we obtain

(1.61) $$(\underset{\beta}{\lambda} - \underset{\gamma}{\lambda})\underset{\alpha}{\omega^i} + (\underset{\gamma}{\lambda} - \underset{\alpha}{\lambda})\underset{\beta}{\omega^i} + (\underset{\alpha}{\lambda} - \underset{\beta}{\lambda})\underset{\gamma}{\omega^i} = 0.$$

The equation (1.61) can be written in the form

(1.62) $$\underset{\alpha}{\overline{\omega}^i} + \underset{\beta}{\overline{\omega}^i} + \underset{\gamma}{\overline{\omega}^i} = 0$$

if we use the substitution

(1.63) $$\underset{\alpha}{\overline{\omega}^i} = \underset{\beta\gamma}{A}\underset{\alpha}{\omega^i}, \quad \underset{\beta}{\overline{\omega}^i} = \underset{\gamma\alpha}{A}\underset{\beta}{\omega^i}, \quad \underset{\gamma}{\overline{\omega}^i} = \underset{\alpha\beta}{A}\underset{\gamma}{\omega^i}$$

where $\underset{\beta\gamma}{A} = \underset{\beta}{\lambda} - \underset{\gamma}{\lambda}$ etc. The equations (1.62) for the three-web $[\alpha, \beta, \gamma]$ are similar to the equations $\underset{1}{\omega^i} + \underset{2}{\omega^i} + \underset{3}{\omega^i} = 0$ for the three-web $[1, 2, 3]$.

Exterior differentiation of the first two equations of (1.63) gives

(1.64) $$\begin{cases} d\underset{\alpha}{\overline{\omega}^i} = \underset{\alpha}{\overline{\omega}^k} \wedge \underset{\alpha\beta\gamma}{\omega^i_k} + \underset{\alpha\beta\gamma}{a}\underset{j}{}\underset{\alpha}{\overline{\omega}^j} \wedge \underset{\alpha}{\overline{\omega}^i}, \\[2mm] d\underset{\beta}{\overline{\omega}^i} = \underset{\beta}{\overline{\omega}^k} \wedge \underset{\alpha\beta\gamma}{\omega^i_k} - \underset{\alpha\beta\gamma}{a}\underset{j}{}\underset{\beta}{\overline{\omega}^j} \wedge \underset{\beta}{\overline{\omega}^i}, \end{cases}$$

where

(1.65) $$\begin{aligned} \underset{\alpha\beta\gamma}{\omega^i_k} &= \omega^i_k + \delta^i_k a_j (2\underset{2}{\omega^j} + \underset{1}{\omega^j}) - \underset{\alpha\beta\gamma}{A}(\underset{\alpha\beta}{A}\delta^i_k b_j + \underset{\gamma\alpha}{A}\delta^i_j b_k + \underset{\beta\gamma}{A}\delta^i_k b_j)\underset{\alpha}{\overline{\omega}^j} \\ &\quad + \underset{\alpha\beta\gamma}{A}(\underset{\beta\gamma}{A}\delta^i_j b_k + \underset{\alpha\beta}{A}\delta^i_k b_j + \underset{\gamma\alpha}{A}\delta^i_k b_j)\underset{\beta}{\overline{\omega}^j}, \end{aligned}$$

(1.66) $$\underset{\alpha\beta\gamma}{a}\underset{j}{} = \underset{\alpha\beta\gamma}{A}(\underset{\alpha\beta\gamma}{A}b_j + \underset{\beta\gamma\alpha}{A}b_j + \underset{\gamma\alpha\beta}{A}b_j),$$

(1.67) $$\underset{\alpha\beta\gamma}{A} = [(\underset{\alpha}{\lambda} - \underset{\beta}{\lambda})(\underset{\beta}{\lambda} - \underset{\gamma}{\lambda})(\underset{\gamma}{\lambda} - \underset{\alpha}{\lambda})]^{-1}.$$

Comparison of (1.64) and (1.15) shows that the forms $\underset{\alpha\beta\gamma}{\omega^i_j}$ are the connection forms of the affine connection $\gamma_{\alpha\beta\gamma}$ induced by the 3-subweb $[\alpha, \beta, \gamma]$ and

(1.68) $$\underset{\alpha\beta\gamma}{a}\underset{jk}{^i} = \underset{\alpha\beta\gamma}{a}\underset{[j}{}\delta^i_{k]},$$

where the $\underset{\alpha\beta\gamma}{a}\underset{j}{}$ are determined by (1.66), is the torsion tensor of $[\alpha, \beta, \gamma]$.

Exterior differentiation of (1.65) leads now to (1.57). ∎

Corollary 1.5. *The curvature tensor of the three-subweb* $[\alpha, \beta, \gamma]$ *is*

(1.69)
$$\underset{\alpha\beta\gamma}{b}{}^i_{jkl} = \underset{\alpha\beta\gamma}{A} (a^i_{jkl} + \delta^i_l \underset{\alpha}{k}_{jk} + \delta^i_k \underset{\beta}{k}_{jl} + \delta^i_j \underset{\gamma}{k}_{kl}).$$

Proof. In fact, it follows from (1.63) that

(1.70)
$$\begin{cases} \underset{1}{\omega}^i = (-\underset{\alpha}{\overline{\omega}}^i / \underset{\beta\gamma}{A} + \underset{\beta}{\overline{\omega}}^i / \underset{\gamma\alpha}{A}) / \underset{\alpha\beta}{A}, \\ \underset{2}{\omega}^i = (\lambda \underset{\beta\alpha}{\overline{\omega}}^i / \underset{\beta\gamma}{A} - \lambda \underset{\alpha\beta}{\overline{\omega}}^i / \underset{\gamma\alpha}{A}) / \underset{\alpha\beta}{A}. \end{cases}$$

Substituting (1.70) into (1.57), we obtain

(1.71) $$d \underset{\alpha\beta\gamma}{\omega}{}^i_j - \underset{\alpha\beta\gamma}{\omega}{}^k_j \wedge \underset{\alpha\beta\gamma}{\omega}{}^i_k = \underset{\alpha\beta\gamma}{A} (a^i_{jkl} + \delta^i_l \underset{\alpha}{k}_{jk} + \delta^i_k \underset{\beta}{k}_{jl} + \delta^i_j \underset{\gamma}{k}_{kl}) \underset{\alpha}{\omega}^k \wedge \underset{\beta}{\omega}^l.$$

Comparison of (1.71) and (1.4) gives (1.69). ∎

Corollary 1.6. *The exterior differential of the contracted connection forms* $\underset{\xi\eta\zeta}{\omega}{}^i_i$ *is*

(1.72)
$$d \underset{\xi\eta\zeta}{\omega}{}^i_i = (\underset{\xi}{k}_{kl} + \underset{\eta}{k}_{kl} + r\underset{\zeta}{k}_{kl}) \underset{1}{\omega}^k \wedge \underset{2}{\omega}^l.$$

Proof. This result immediately follows from (1.57). ∎

Corollary 1.7. *The connection* $\gamma_{\xi\eta\zeta}$ *is equiaffine if and only if*

(1.73)
$$\underset{\xi}{k}_{ij} + \underset{\eta}{k}_{ij} + r\underset{\zeta}{k}_{ij} = 0.$$

Proof. In fact, the connection forms $\underset{\xi\eta\zeta}{\omega}{}^I_J$ of $\gamma_{\xi\eta\zeta}$ are

$$\left(\underset{\xi\eta\zeta}{\omega}{}^I_J \right) = \begin{pmatrix} \underset{\xi\eta\zeta}{\omega}{}^i_j & 0 \\ 0 & \underset{\xi\eta\zeta}{\omega}{}^i_j \end{pmatrix}, \qquad I, J = 1, \ldots, 2r; \quad i, j = 1, \ldots, r.$$

Because of this, we have $\underset{\xi\eta\zeta}{\omega}{}^I_I = 2 \underset{\xi\eta\zeta}{\omega}{}^i_i$. The connection $\gamma_{\xi\eta\zeta}$ is equiaffine if and only if $d \underset{\xi\eta\zeta}{\omega}{}^I_I = 0$ (see [G 66]). In our case $d \underset{\xi\eta\zeta}{\omega}{}^I_I = 2 \underset{\xi\eta\zeta}{\omega}{}^i_i$, and it follows from (1.72) that this expression vanishes if and only if we have (1.73). ∎

A web $W(d, 2, r)$ is said to be *almost algebraizable* if it is almost Grassmannizable and its tensors satisfy the following relation

$$(1.74) \qquad K_{ij} = \sum_{\xi=1}^{d} k_{ij} = 0.$$

We will denote such webs by $AAW(d, 2, r)$.

Note that for a three-web $W(3, 2, r)$ the condition (1.74) has the form $f_{ij} + g_{ij} + h_{ij} = 0$ and this is the necessary and sufficient condition for hexagonality (and algebraizability) for an isoclinic $W(3, 2, r)$ (see [AS 81]). Therefore, *an almost algebraizable web $W(3, 2, r)$ is algebraizable.*

In the case $d > 3$ the conditions (1.74) do not imply (1.35), i.e., an almost algebraizable web is not necessarily transversally geodesic and consequently it is not necessarily algebraizable.

We will find now under what condition an $AAW(d, 2, r)$ must be algebraizable.

Theorem 1.8. *An almost algebraizable web $AAW(d, 2, r)$, $d > 3$, is algebraizable if and only if it is transversally geodesic.*

Proof. The *necessity* of the theorem follows from the fact that an algebraizable web $W(d, 2, r)$ is Grassmannizable and consequently is transversally geodesic. To prove the *sufficiency*, we note that an almost algebraizable transversally geodesic web $W(d, 2, r)$ is a Grassmannizable web for which the conditions (1.74) hold. It was proved in [G 83] for $d = 4$ and in [W 82] (see also [W 84]) for $d > 4$ that the condition (1.74) is necessary and sufficient for a Grassmannizable web $W(d, 2, r)$ to be algebraizable. ∎

Note that in [W 82] and [W 84] the conditions (1.74) were treated as necessary and sufficient conditions for d local hypersurfaces V_ξ of a real projective space P^{r+1} of dimension $r + 1$ to belong to an algebraic hypersurface of degree d. The tensors $k_{ij} \atop \xi$ become the second fundamental tensors of V_ξ. For $d = 4$ these conditions were established earlier by the author [G 83]. Note that in 1983 in the paper [A 83b] M. A. Akivis found similar condition for d surfaces of codimension $s \geq 1$ in a real projective space, and that a comparable theorem for a complex projective space was proved by P. A. Griffiths [G 66] in 1976.

We will now prove the theorem giving some criteria for an isoclinic $AGW(d, 2, r)$ to be almost algebraizable.

Theorem 1.9. *An isoclinic almost Grassmannizable web $AGW(d,2,r)$ is almost algebraizable if and only if one of the following affine connections is equiaffine:*

(i) The middle connection of the canonical affine connections

$\gamma_{123}, \gamma_{234}, \ldots, \gamma_{d12}.$

(ii) The middle connection of all the $3!\binom{d}{3}$ affine connections $\Gamma_{\xi\eta\zeta}$.

(iii) For $r = d-2$ the middle connection of the affine connections $\Gamma_{\xi_0\eta_0\zeta}$ where ξ_0 and η_0 are fixed.

(iv) For $r = d-2$ the middle connection of the affine connections $\gamma_{\xi_0\eta\zeta}$ where ξ_0 is fixed and the pairs η, ζ are all neighboring pairs of the sequence $1, 2, \ldots, \xi_0 - 1, \xi_0 + 1, \ldots, d.$

Proof. All these statements are equivalent to the condition (1.74). In fact, the contracted connection forms in these cases respectively are

(i) $\quad \underset{1}{\Omega_I^I} = 2(\underset{123}{\omega_i^i} + \underset{234}{\omega_i^i} + \ldots + \underset{d12}{\omega_i^i})/d,$

(ii) $\quad \underset{2}{\Omega_I^I} = 2\sum_{\xi,\eta,\zeta} \underset{\xi\eta\zeta}{\Omega_i^i}/[d(d-1)(d-2)],$

(iii) $\quad \underset{3}{\Omega_I^I} = 2\sum_{\xi} \underset{\xi_0\eta_0\zeta}{\omega_i^i}/d,$

(iv) $\quad \underset{4}{\Omega_I^I} = 2(\underset{\xi_0 12}{\omega_i^i} + \underset{\xi_0 23}{\omega_i^i} + \ldots + \underset{\xi_0,\xi_0-1,\xi_0+1}{\omega_i^i} + \ldots + \underset{\xi_0 d1}{\omega_i^i})/(d-1).$

Using (1.72) and the condition $r = d-2$ for (iii) and (iv), we obtain

(1.75) $\qquad d\underset{s}{\Omega_I^I} = \underset{s}{a} K_{ij}\underset{1}{\omega^i} \wedge \underset{2}{\omega^j}, \qquad s = 1,2,3,4,$

where

$K_{ij} = \sum_{\xi} \underset{\xi}{k_{ij}} \quad \text{and} \quad \underset{1}{a} = 2(r+2)/d, \quad \underset{2}{a} = r+2, \quad \underset{3}{a} = 2r/(r+2), \quad \underset{4}{a} = 2.$

The relation (1.75) shows that each of the equations $d\underset{s}{\Omega_I^I} = 0$ is equivalent to (1.74). ∎

Combining the results of Theorems 1.8 and 1.9, we obtain the following condition of algebraizability of an isoclinic $AGW(d,2,r)$:

Corollary 1.10. *An isoclinic almost Grassmannizable web $AGW(d,2,r)$ is algebraizable if and only if it is transversally geodesic and one of the affine connections listed in Theorem 1.9 is equiaffine.* ∎

Using the definition of an almost Grassmannizable web and Corollary 1.10, we obtain

Corollary 1.11. *A web $W(d,2,r)$ is algebraizable if and only if it is transversally geodesic, isoclinic, has scalar basis affinors and one of the affine connections listed in Theorem 1.9 is equiaffine.* ∎

Note that if $d = 3$, then

$$d\Omega_{1}^{I} = d\Omega_{2}^{I} = \frac{2}{3}(f_{ij} + g_{ij} + h_{ij})\omega^{i} \wedge \omega^{j}.$$

As we mentioned before, the condition $f_{ij} + g_{ij} + h_{ij} = 0$ is a necessary and sufficient condition for hexagonality (and consequently algebraizability) of an isoclinic web $W(3,2,r)$.

Using Theorem 1.9 and the last equation, we obtain from this the following two propositions for a three-web $W(3,2,r)$:

Proposition 1.12. *An isoclinic web $W(3,2,r)$ is hexagonal if and only if the middle connection $\gamma = \frac{1}{3}(\gamma_{123} + \gamma_{231} + \gamma_{312})$ is equiaffine.* ∎

Proposition 1.13. *A web $W(3,2,r)$ is algebraizable if and only if it is isoclinic and the middle connection γ is equiaffine.* ∎

These two results are new for three-webs $W(3,2,r)$.

We will conclude this subsection by giving analytical consequences of the condition (1.74) for almost algebraizability and the conditions (1.74) and (1.35) for algebraizability of an isoclinic $AGW(d,2,r)$.

Differentiating (1.74) and using (1.31) and (1.54), we obtain by means of the linear independence of ω^{i}_{1} and ω^{i}_{2} that

$$(1.76) \qquad K_{1\,ijk} + b_{t}a^{t}_{\alpha\,ijk} + g_{1\,ijk} - \lambda_{\alpha} f_{2\,ijk} = 0, \qquad K_{2\,ijk} = 0,$$

where

$$(1.77) \quad \begin{cases} K_{1\,ijk} = f_{1\,ijk} + g_{1\,ijk} + h_{1\,ijk} + \sum_{\alpha} \lambda_{\alpha} k_{\alpha\,ijk} + 3\sum_{\alpha} b_{\alpha(i} k_{\alpha\,jk)}, \\[2mm] K_{2\,ijk} = f_{2\,ijk} + g_{2\,ijk} + h_{2\,ijk} + \sum_{\alpha} k_{\alpha\,ijk}. \end{cases}$$

If, in addition to (1.74), we have (1.35), this gives us (1.36), and the equations (1.76) and (1.77) become

$$(1.78) \qquad\qquad K_{1\,ijk} = 0, \qquad K_{2\,ijk} = 0,$$

$$(1.79) \quad \begin{cases} \underset{1}{K}_{ijk} = \underset{1}{f}_{ijk} + h_{ijk} + \sum_{\alpha} \underset{\alpha}{\lambda} \underset{\alpha}{k}_{ijk} + 3h_{(i}a_{jk)} + 3\sum_{\alpha} \underset{\alpha}{b}_{(i} \underset{\alpha}{k}_{jk)}, \\[2ex] \underset{2}{K}_{ijk} = \underset{2}{g}_{ijk} + h_{ijk} + \sum_{\alpha} \underset{\alpha}{k}_{ijk} \end{cases}$$

where $\underset{1}{f}_{ijk} = f_{ijk}$, $\underset{2}{g}_{ijk} = g_{ijk}$, $\underset{2}{h}_{ijk} = h_{ijk}$.

In Section 2 we will need the following identity which is obtained from (1.78) and (1.79):

$$(1.80) \quad f_{ijk} - \underset{\alpha_0}{\lambda} g_{ijk} + (1 - \underset{\alpha_0}{\lambda}) h_{ijk} + \sum_{\alpha} (\underset{\alpha}{\lambda} - \underset{\alpha_0}{\lambda}) \underset{\alpha}{k}_{ijk} = -3\sum_{\alpha} \underset{\alpha}{b}_{(i} \underset{\alpha}{k}_{jk)} - 3a_{(i}h_{jk)}$$

where α_0 is fixed.

1.4. Non-isoclinic almost Grassmannizable 4-webs

According to Theorem 1.3, among the webs $AGW(d, 2, r)$ only the webs $AGW(4, 2, 2)$ can be of both kinds: isoclinic and non-isoclinic. We considered the non-isoclinic case for any $r > 2$ in the subsection 3. Now we will study non-isoclinic webs $AGW(4, 2, 2)$.

Let $AGW(4, 2, 2)$ be a non-isoclinic almost Grassmannizable web. For such a web we have the equations (1.15), (1.4), (1.17), (1.18), (1.43) and (1.51). Denote $p_{[12]}$ by p and $q_{[12]}$ by q:

$$(1.81) \qquad\qquad p_{[12]} = p, \qquad q_{[12]} = q.$$

Since $X_4 \neq X_s$, $s = 1, 2, 3$, we have

$$(1.82) \qquad\qquad p \neq 0, \qquad q \neq 0, \qquad p \neq q.$$

The equation (1.51) shows that

$$(1.83) \qquad\qquad \lambda = p/q.$$

This means that the foliation X_4 of a non-isoclinic web $AGW(4, 2, 2)$ is uniquely determined by its 3-subweb $[1, 2, 3]$ provided that the inequalities (1.81) hold.

Let us find the location of the foliation X_4 with respect to the foliations X_1, X_2, and X_3.

Let V_1, V_2 and V_3 be leaves of the X_1, X_2, and X_3 passing through a point $p \in X_4$. The two-dimensional tangent subspaces $T_p(V_\alpha)$ generate in the tangent space $T_p(X^4)$ the cone C_p of second order of signature $(2,2)$ and the field of these cones defines a pseudo-conformal structure $CO(2,2)$ on X^4 (see [A 83b]). The subspaces $T_p(V_\alpha)$ are determined in $T_p(X^4)$ by the systems $\underset{1}{w^i} = 0$, $\underset{2}{w^i} = 0$, and $\underset{1}{w^i} + \underset{2}{w^i} = 0$. The equation of the cone C_p has the form

(1.84)
$$\underset{1}{w^1}\underset{2}{w^2} - \underset{1}{w^2}\underset{2}{w^1} = 0.$$

This is the isotropic cone of the structure $CO(2,2)$ (the light cone in Physics terminology). It follows from (1.84) that the cone C_p has two families of two-dimensional flat generators (α- and β-planes according to [P 68]):

(1.85)
$$s\underset{1}{w^1} + \underset{2}{w^1} = 0, \qquad s\underset{1}{w^2} + \underset{2}{w^2} = 0;$$

(1.86)
$$t\underset{1}{w^1} + \underset{1}{w^2} = 0, \qquad t\underset{2}{w^1} + \underset{2}{w^2} = 0.$$

For the three-web [1,2,3] these two 2-planes are isoclinic and transversally geodesic 2-planes respectively (see [A 74] and [AS 81].

The tensor C of conformal structure of any $CO(2,2)$ has 10 independent components A_s and B_s, $s = 0,1,2,3,4$. It is splitted into two subtensors C^+ and C^- determined by A_s and B_s respectively (see [A 83b] or [AHS 78]).

The relative conformal curvature of α- and β-planes are determined by the polynomials $C^+(s)$ and $C^-(t)$ each of degree four:

$$C^+(s) = A_4 s^4 - 4A_3 s^3 + 6A_2 s^2 - 4A_1 s + A_0,$$
$$C^-(t) = B_4 s^4 - 4B_3 s^3 + 6B_2 s^2 - 4B_1 s + B_0,$$

where s and t are parameters of α- and β-planes (see (1.85) and (1.86)). There exists four null α-planes and four null β-planes: for them $C^+(s) = 0$ and $C^-(t) = 0$ respectively (see [A 83b] or [AHS 78]).

For the structure $CO(2,2)$ associated with the three-subweb [1,2,3] the equation $C^+(s) = 0$ has the form (see [A 83b]):

(1.87)
$$-qs^3 + (q - p)s^2 + p = 0$$

where p and q are defined by (1.81) and (1.82).

It follows from (1.11) and (1.85) that null α-planes are determined by the vectors

$$(1.88) \qquad\qquad h_i = se_i^2 - e_i^1, \qquad i = 1, 2$$

where s is a root of (1.87). The roots of (1.87) are 0, ∞, 1, and $-p/q$. Therefore, according to (1.88) the first three null α-planes are tangent to the leaves V_1, V_2, and V_3 passing through $p \in X^4$ and the fourth root $-p/q$ determines the fourth invariant distribution. These four null α-planes are determined respectively by the vectors $e_i^1, e_i^2, e_i^2 - e_i^1$, and $-pe_i^1 + qe_i^2$.

On the other hand, it follows from (1.11), (1.1), and (1.85) that the tangent 2-plane to the leaf V_4 of X_4 passing through $p \in X^4$ is determined by the vecrors $pe_i^2 + qe_i^1$.

The 1st, 2nd, 4th null α-planes and the tangent 2-plane to the leaf $V_4 \in X_4$ intersect the only bivector $\eta^1 \wedge \eta^2$ (see subsection 1.2) of our $AGW(4, 2, 2)$ in the direction of the vectors:

$$(1.89) \quad E^1 = \eta^i e_i^1, \;\; E^2 = \eta^i e_i^2, \;\; F = \eta^i (qe_i^1 - pe_i^2), \;\; G = \eta^i (qe_i^1 + pe_i^2).$$

It is easy to see from (1.89) that $(E^1, E^2; F, G) = -1$.

Let us find out under what conditions a non-isoclinic three-web $W(3, 2, 2)$ can be extended to a non-isoclinic almost Grassmannizable four-web $AGW(4, 2, 2)$.

For this we need the prolongations of equations (1.17). Exterior differentiation of (1.17) by means of (1.15), (1.4), and (1.17) and application of Cartan's lemma lead to the following equations:

$$(1.90) \qquad \nabla p_{ij} = p_{ijk} \underset{1}{\omega^k} + p_{ijk} \underset{2}{\omega^k}, \qquad \nabla q_{ij} = q_{ijk} \underset{1}{\omega^k} + q_{ijk} \underset{2}{\omega^k}$$

where

$$(1.91) \qquad \begin{cases} p_{i[jk]} + p_{i[j} a_{k]} = 0, \\ \underset{1}{} \\ q_{i[jk]} - q_{i[j} a_{k]} = 0, \\ \underset{1}{} \\ p_{ijk} = q_{ikj} + a_l b_{ijk}^l = 0, \\ \underset{2}{} \quad \underset{1}{} \end{cases}$$

and

$$\nabla p_{ij} = dp_{ij} - p_{kj} \omega_i^k - p_{ik} \omega_j^k, \;\; \nabla q_{ij} = dq_{ij} - q_{kj} \omega_i^k - q_{ik} \omega_j^k.$$

It follows from (1.90) and (1.81) that

(1.92)
$$\begin{cases} dp = p\omega_i^i + \underset{1}{p_i}\underset{1}{\omega^i} + \underset{2}{p_i}\underset{2}{\omega^i}, \\ dq = q\omega_i^i + \underset{1}{q_i}\underset{1}{\omega^i} + \underset{2}{q_i}\underset{2}{\omega^i} \end{cases}$$

where

(1.93)
$$\underset{k}{p_i} = \underset{k}{p_{[12]i}}, \quad \underset{k}{q_i} = \underset{k}{q_{[12]i}}, \quad i, k = 1, 2.$$

Differentiating (1.83) by means of (1.92) and (1.43), we get

(1.94)
$$\underset{1}{p_i} - \lambda \underset{1}{q_i} = p(b_i - a_i), \qquad \underset{2}{p_i} - \lambda \underset{2}{q_i} = qb_i - pa_i.$$

If we eliminate b_i from (1.94), we will get the condition

(1.95)
$$q(q\underset{1}{p_i} - p\underset{1}{q_i}) - p(q\underset{2}{p_i} - p\underset{2}{q_i}) = pq(p - q)a_i.$$

In addition, such a web $W(3, 2, 2)$ has to satisfy the conditions $\lambda \neq 0$, ∞, 1, which, by virtue of (1.83), give (1.82). Note that for the web $AGW(4, 2, 2)$ we can find the quantities b_i from (1.94) and b_{ij} from (1.51).

We proved the following theorem:

Theorem 1.14. *A non-isoclinic three-web $W(3, 2, 2)$ given in X^4 and satisfying the condition (1.94) and the inequalities (1.82) can be uniquely extended to a non-isoclinic almost Grassmannizable four-web $AGW(4, 2, 2)$. The fourth foliation X_4 of the $AGW(4, 2, 2)$ is determined by the equation $p\underset{1}{\omega^i} + q\underset{2}{\omega^i} = 0$. The directions at which the 1st, 2nd, 4th null α-planes of the pseudo-conformal structure $CO(2, 2)$ associated with the $AGW(4, 2, 2)$ and the tangent 2-plane to the leaf V_4 of X_4 passing through $p \in X^4$ intersects all bivectors $\eta^1 \wedge \eta^2$ of the $AGW(4, 2, 2)$ form a harmonic quadruple. Such a web $AGW(4, 2, 2)$ is determined by the equations (1.15), (1.4), (1.17), (1.18), (1.43), (1.51), (1.90), (1.94), (1.95) and (1.84).* ∎

Note that if we depart from any 3-subweb $[\xi, \eta, \zeta]$ of a non-isoclinic $AGW(4, 2, 2)$, then its extension to the four-web is the given web $AGW(4, 2, 2)$.

To show this, we need to calculate for $[\xi, \eta, \zeta]$ the quantities similar to $\underset{1}{\omega^i}$, $\underset{2}{\omega^i}$, $\underset{4}{\omega^i}$, λ, ω_j^i, a_i, b_i, p_{ij}, q_{ij}, and b_{ij} and check that for them the equations (1.51) hold.

Let us consider, for example, the 3-subweb $[1, 2, 4]$. If for this subweb one denotes all the quantities mentioned above by the same letters with bar, then it is easy to see that

(1.96)
$$
\begin{cases}
\underset{1}{\bar{\omega}}{}^i = \lambda \underset{1}{\omega}{}^i, \quad \underset{2}{\bar{\omega}}{}^i = \underset{1}{\omega}{}^i, \quad \underset{4}{\bar{\omega}}{}^i = \underset{3}{\omega}{}^i, \quad \bar{\lambda} = 1/\lambda, \\[2mm]
\bar{\omega}_j^i = \omega_j^i - \delta_j^i (b_k - \lambda a_k) \underset{2}{\omega}{}^k / \lambda, \quad \bar{a}_i = b_i / \lambda, \quad \bar{b}_i = a_i / \lambda, \\[2mm]
\bar{p}_{ij} = (b_{ji} + p_{ij} - q_{ji})/\lambda, \quad \bar{q}_{ij} = b_{ij}/\lambda, \quad \bar{b}_{ij} = q_{ji}/\lambda.
\end{cases}
$$

Using (1.96), one can check that $\bar{b}_{[ij]} = \bar{p}_{[ij]} = \lambda \bar{q}_{[ij]}$ provided that (1.51) holds for the 3-subweb $[1, 2, 3]$.

A similar calculation can be done for any $[\xi, \eta, \eta]$.

We will consider now examples of isoclinic three-webs $W(3, 2, 2)$ and non-isoclinic three-webs $W(3, 2, 2)$ which can not be extended to a non-isoclinic $AGW(4, 2, 2)$ since they do not satisfy conditions of Theorem 1.14. Note that in [G 85], [G 86], [G 87] we constructed examples of webs $W(3, 2, 2)$ which can be extended to a non-isoclinic $AGW(4, 2, 2)$.

In each example we define a $W(3, 2, 2)$ by its closed form equations $z^i = f^i(x^j, y^k)$, calculate its torsion tensor a_{jk}^i and the connection forms ω_j^i by means of the following formulas (see equations (40) and (41) in [AS 81] or (11) and (12) in [AS 71a]):

(1.97)
$$
a_{jk}^i = \Gamma_{[jk]}^i, \quad \omega_j^i = \Gamma_{kj}^i \underset{1}{\omega}{}^k + \Gamma_{jk}^i \underset{2}{\omega}{}^k
$$

where

(1.98)
$$
\Gamma_{jk}^i = -\frac{\partial^2 f^i}{\partial x^l \partial y^m} \bar{g}_j^l \tilde{g}_k^m
$$

and $(\bar{g}_j^l)$ and $(\tilde{g}_k^m)$ are the inverse matrices of the matrices $(\bar{f}_j^i) = (\partial f^i / \partial x^j)$ and $(\tilde{f}_j^i) = (\partial f^i / \partial y^j)$.

Then, using the equations (1.14), (1.17), (1.81), (1.98) and (1.97), we find consecutively a_i, p_{ij}, q_{ij}, p and q.

Example. (see [B 35]). A web $W(3, 2, 2)$ is given by

(1.99)
$$
f^1 = x^1 + y^1, \qquad f^2 = (x^2 + y^2)(y^1 - x^1).
$$

For this web we have

(1.100) $a_1 = 2/(y^1 - x^1)$, $a_2 = 0$, $p_{12} = q_{12} = p_{21} = q_{21} = 0$.

Because of (1.100), the web (1.99) is isoclinic.

Example 1.16. Consider the web $W(3, 2, 2)$ given by

(1.101) $$f^1 = x^1 y^1 - x^2 y^2, \quad f^2 = x^1 y^2 + x^2 y^1.$$

Calculations give

$$a_1 = 2(x^1 y^1 + x^2 y^2)/(\Delta_1 \Delta_2), \quad p = 0,$$
(1.102) $$a_2 = 2(x^2 y^1 - x^1 y^2)/(\Delta_1 \Delta_2), \quad q = -4y^1 y^2/(\Delta_1^2 \Delta_2),$$

where $\Delta_1 = y_1^2 + y_2^2$, $\Delta_2 = x_1^2 + x_2^2$.

Example 1.17. A web $W(3, 2, 2)$ is given by

(1.103) $$f^1 = x^2 e^{x^1 y^1}, \quad f^2 = x^2 + y^2.$$

We have

(1.104) $$\begin{cases} a_1 = 0, \ a_2 = -(x^1 + 1)/(x^1 x^2), \\ p = -[2(x^1 x^2)^2 y^1 e^{x^1 y^1}]^{-1}, \quad q = 0. \end{cases}$$

Example 1.18. Suppose that a web $W(3, 2, 2)$ is given by

(1.105) $$f^1 = x^1 + y^1, \qquad f^2 = x^1 y^1 + x^2 y^2.$$

In this case

(1.106) $$a_1 = (x^1 - y^1)/(x^2 y^2), \quad a_2 = 0, \quad p = q = (y^1 - x^1)/[2(x^2 y^2)^2].$$

It follows from (1.102), (1.104), and (1.106) that the webs (1.101), (1.103), and (1.105) are non-isoclinic (the quantities p and q do not vanish simultaneously) but they cannot be extended to to a non-isoclinic $AGW(4, 2, 2)$ since the inequalities (1.81) do not hold for them.

Example 1.19. A web $W(3, 2, r)$ is given by

(1.107) $$f^1 = (x^1 + y^1)^3/6 + [(x^1)^2 + (y^1)^2 + 2x^2 y^2]/2, \quad f^2 = x^2 + y^2.$$

The calculations give

(1.108) $$a_1 = 0, \ a_2 = -(x^1 + y^1)(x^2 - y^2)/(\Delta_1 \Delta_2)$$

where

(1.109) $\qquad \Delta_1 = \dfrac{1}{2}(x^1 + y^1)^2 + x^1, \qquad \Delta_2 = \dfrac{1}{2}(x^1 + y^1)^2 + y^1$

and

(1.110)

$$\begin{cases} 2p = \{1 - (x^1 + y^1)[(x^1 + y^1)^3 + 3(x^1 + y^1)^2/2 + y^1]/(\Delta_1\Delta_2)\} \cdot \\ \qquad \cdot (x^2 - y^2)/(\Delta_1^2\Delta_2), \\ 2q = \{1 - (x^1 + y^1)[(x^1 + y^1)^3 + 3(x^1 + y^1)^2/2 + x^1]/(\Delta_1\Delta_2)\} \cdot \\ \qquad \cdot (x^2 - y^2)/(\Delta_1\Delta_2^2). \end{cases}$$

We can see from (1.109) and (1.110) that the three-web (1.107) is non-isoclinic. However, this web cannot be extended to a non-isoclinic $AGW(4,2,2)$ since it does not satisfy the conditions (1.95). To see this, we note that the equations (1.95) for the web (1.107) are equivalent to the equations

(1.111) $\qquad \dfrac{\partial \lambda}{\partial x^1} \Delta_2 = (x^1 + y^1)(\lambda - \lambda^2), \qquad \dfrac{\partial \lambda}{\partial y^1} \Delta_1 = (x^1 + y^1)(1 - \lambda)$

where $\lambda = p/q$ and p, q are determined by (1.110). A straightforward calculation shows that the conditions (1.95) are not satisfied for the web (1.107).

We will conclude this subsection by finding an analytic condition under which for a non-isoclinic web $AGW(4,2,2)$ all four affine connections mentioned in Theorem 1.9 are equiaffine.

Theorem 1.20. *For a non-isoclinic almost Grassmannizable web $AGW(4,2,2)$ the following statements are equivalent:*

(i) *Each of the four affine connections indicated in Theorem 1.9 is equiaffine.*

(ii) *The form $\underset{4}{w_k^k} + (\underset{4}{a_k} - \underset{2}{b_k}/\lambda)\underset{2}{w^k}$ is a total differential.*

(iii) *The curvature tensor $\underset{4}{b_{jkl}^i}$ of $AGW(4,2,2)$ satisfies the following equation:*

(1.112) $\qquad\qquad\qquad\qquad \underset{4}{b_{kij}^k} = \underset{4}{b_{ij}} - q_{ij}.$

Proof. Using (1.56) and (1.60), we find the expressions of the forms $\Omega^I_{\underset{\xi}{I}}, \xi = 1, 2, 3, 4$ (see the proof of Theorem 1.9):

(1.113)
$$
\begin{cases}
\Omega^I_{\underset{1}{I}} = 2\omega^i_i + 2(a_i - \underset{4}{b}/\underset{4}{\lambda})\omega^i_{\underset{2}{}} - d\ln(1 - \underset{4}{\lambda})^{3/2}, \\
\Omega^I_{\underset{2}{I}} = 2\omega^i_i + 2(a_i - \underset{4}{b}_i/\underset{4}{\lambda})\omega^i_{\underset{2}{}} - d\ln[\underset{4}{\lambda}/(1 - \underset{4}{\lambda})^5], \\
\Omega^I_{\underset{3}{I}} = 2\omega^i_i + 2(a_i - \underset{4}{b}_i/\underset{4}{\lambda})\omega^i_{\underset{2}{}}, \\
\Omega^I_{\underset{4}{I}} = 2\omega^i_i + 2(a_i - \underset{4}{b}_i/\underset{4}{\lambda})\omega^i_{\underset{2}{}} + 2d\ln[\underset{4}{\lambda}/(1 - \underset{4}{\lambda})],
\end{cases}
$$

where $\Omega^I_{\underset{3}{I}}$ and $\Omega^I_{\underset{4}{I}}$ are calculated correspondingly for $\xi_0 = 1$, $\eta_0 = 2$, and $\xi_0 = 1$.

It is easy to check that $d\Omega^I_{\underset{\xi}{I}} = 0$ if and only if the form $\omega^i_i + (a_i - \underset{4}{b}_i/\underset{4}{\lambda})\omega^i_{\underset{2}{}}$ is a total differential. Using (1.17) and (1.51), we obtain for this form:

(1.114) $\qquad d[\omega^i_i + (a_i - \underset{4}{b}_i/\underset{4}{\lambda})] = (b^i_{ijk} - \underset{4}{b}_{jk} + q_{jk})\omega^j_{\underset{1}{}} \wedge \omega^k_{\underset{2}{}}.$

The equation (1.114) shows that the form $\omega^i_i + (a_i - \underset{4}{b}_i/\underset{4}{\lambda})\omega^i_{\underset{2}{}}$ is a total differential if and only if the condition (1.112) holds.

2. Webs $W(d,2,r)$ of maximum r-rank

2.1. The r-rank of webs $W(d,2,r)$, $r > 2$

In this section we shall study the r-rank problems for webs $W(d,2,r)$. Let us first to define the r-rank of webs $W(d,n,r)$.

Suppose that the leaves of the ξth foliation of a web $W(d,n,r)$ are given as level sets of functions $u_\xi(x)$:

$$u^i_\xi(x) = \text{const.}, \qquad \xi = 1, \ldots, d.$$

The functions $u_\xi(x)$ are defined up to a local diffeomorphism in the space of $u_\xi(x)$.

An exterior r-equation of the form

(2.1) $\qquad \displaystyle\sum_{\xi=1}^{d} f_\xi(u^j_\xi) du^1_\xi \wedge \ldots \wedge du^r_\xi = 0, \qquad j = 1, \ldots, r,$

is said to be an *abelian r-equation*. The maximum number R_r of linearly independent abelian r-equations admitted by the $W(d, n, r)$ is called the *r-rank* of the web $W(d, n, r)$.

It follows from the definition that the coefficients f_ξ are constant on the leaves of the ξth foliation of $W(d, n, r)$.

If there exists an upper bound $\pi_r(d, n, r)$ of R_r, then $R_r \leq \pi_r(d, n, r)$.

These definitions are due to P. A. Griffiths who gave in [Gr 77] a more general definition of the q-rank for the webs $W(d, n, r)$ of codimension r, $1 \leq q \leq r$.

Lemma 2.1. *Suppose that the leaves of the ξth foliation of a web $W(d, n, r)$ are given by a completely integrable system $\underset{\xi}{\omega} = 0$, $i = 1, \ldots, r$; $\xi = 1, \ldots, d$.*

An r-equation of the form

$$(2.2) \qquad \sum_\xi F_\xi \underset{\xi}{\omega^1} \wedge \ldots \wedge \underset{\xi}{\omega^r} = 0$$

is an abelian equation if and only if the exterior differential of each its term vanishes.

Proof. First of all we have

$$(2.3) \qquad \underset{\xi}{\omega^i} = \underset{\xi}{A^i_j} d\underset{\xi}{u^j}, \qquad \det(\underset{\xi}{A^i_j}) \neq 0, \qquad \xi = 1, \ldots, d.$$

If equation (2.2) is an abelian r-equation, then after substitution (2.3) ξth term of the sum (2.2) has the form $f_\xi d\underset{\xi}{u^1} \wedge \ldots \wedge d\underset{\xi}{u^r}$, and according to the lemma condition, its exterior differential vanishes: $df_\xi \wedge d\underset{\xi}{u^1} \wedge \ldots \wedge d\underset{\xi}{u^r} = 0$. It follows from this that df_ξ is a linear combination of $d\underset{\xi}{u^i}$ only, f_ξ depends only on $\underset{\xi}{u^j}$ and (2.2) is an abelian r-equation in the sense of the definition of the r-rank given above.

Conversely, suppose we have an equation (2.1). It is obvious that the exterior differential of each of its terms vanishes. After we substitute $d\underset{\xi}{u^i}$ expressed in terms of $\underset{\xi}{\omega^i}$, each term of the equation (2.1) will turn into the corresponding term of (2.2), and its exterior differential vanishes since it is still the same term.

Note that there are two differences between (2.1) and (2.2): F_ξ does not only depend on the $\underset{\xi}{u^j}$ and, in general, the $d\underset{\xi}{\omega^i}$ do not vanish. Suppose again that the leaves of the foliations X_ξ of a web $W(d, 2, r)$ are given by the equations (1.1) and (1.2). Using these equations, we can write an abelian r-equation for the web $W(d, 2, r)$ in the form:

$$\alpha \underset{1}{\omega^1} \wedge \ldots \wedge \underset{1}{\omega^r} + \beta \underset{2}{\omega^1} \wedge \ldots \wedge \underset{2}{\omega^r} + \gamma(\underset{1}{\omega^1} + \underset{2}{\omega^1}) \wedge \ldots \wedge (\underset{1}{\omega^r} + \underset{2}{\omega^r})$$

(2.4)
$$+ \sum_{\alpha=4}^{d} \underset{\alpha}{\sigma}(\underset{\alpha}{\lambda_j^1} \underset{1}{\omega^j} + \underset{2}{\omega^1}) \wedge \ldots \wedge (\underset{\alpha}{\lambda_j^r} \underset{1}{\omega^j} + \underset{2}{\omega^r}) = 0,$$

where, according to Lemma 2.1, each term is a closed r-form.

Equating to zero coefficients of $\underset{1}{\omega^i} \wedge \underset{2}{\omega^2} \wedge \ldots \underset{2}{\omega^r}$, $i \neq 1$; $\underset{2}{\omega^1} \wedge \underset{1}{\omega^j} \wedge \ldots \wedge \underset{2}{\omega^r}$, $j \neq 2$; $\ldots$, $\underset{1}{\omega^1} \wedge \underset{2}{\omega^2} \wedge \ldots \wedge \underset{2}{\omega^r}$, $\underset{2}{\omega^1} \wedge \underset{1}{\omega^2} \wedge \ldots \wedge \underset{2}{\omega^r}$, $\ldots$ in (2.4) we obtain

(2.5)
$$\sum_{\alpha} \underset{\alpha}{\sigma} \underset{\alpha}{\lambda_j^i} = 0, \quad i \neq j, \quad \gamma = -\sum_{\alpha} \underset{\alpha}{\sigma} \underset{\alpha}{\lambda_i^i} \quad \text{(no summation in } i\text{),}$$

(2.6)
$$\sum_{\alpha} \underset{\alpha}{\sigma}(\underset{\alpha}{\lambda_i^i} - \underset{\alpha}{\lambda_j^j}) = 0, \quad i \neq j; \quad \text{(no summation in } i \text{ and } j\text{).}$$

We will state now the theorem recently proved by J. B. Little [L 86].

Theorem 2.2. *Every maximum r-rank web $W(d, n, r)$, $r \geq 2$, $d > r(n - 1) + 2$, is almost Grassmannizable.* ∎

Applying Theorem 2.2 to the webs $W(d, 2, r)$, i.e., to the case $n = 2$, we see that the webs $W(d, 2, r)$, $r \geq 2$, $d > r + 2$, are almost Grassmannizable. It follows from this result that in all these cases $d > 4$ since $r \geq 2$. So, to cover all possible cases of webs $W(d, 2, r)$ of maximum r-rank, we must consider separately the cases $d = 4$ and $d \leq r + 2$.

For webs $W(4, 2, r)$ we shall prove that such webs are almost Grassmannizable if they admit at least one abelian r-equation, i.e., they are of non-zero (not necessarily maximum) r-rank.

Proposition 2.3. *Webs $W(4, 2, r)$ admitting at least one abelian r-equation are almost Grassmannizable.*

Proof. Suppose that $d = 4$. If $\underset{4}{\sigma} = 0$, it follows from (2.4) that $\alpha = \beta = \gamma = 0$. In this case the $W(4, 2, r)$ does not admit abelian r-equations. Suppose that $\underset{4}{\sigma} \neq 0$. The equations (2.5) and (2.6) imply (1.13) where

$\lambda = \lambda_1^1 = \ldots = \lambda_r^r$. This means that webs $W(4, 2, r)$ admitting at least one
abelian r-equation are almost Grassmannizable. ▨

If $d > 4$, the equations (2.5) and (2.6) are satisfied identically for almost
Grassmannizable webs $AGW(d, 2, r)$ of non-zero r-rank.

From now on we will restrict ourselves to almost Grassmannizable webs
$AGW(d, 2, r)$ of maximum r-rank. In them the first equation (2.5) and the
equation (2.6) are satisfied identically. By Theorem 2.3 and Proposition
2.4, this assumption is natural for webs $W(d, 2, r)$, $r \geq 2$, $d > r + 2$, and
webs $W(4, 2, r)$ of maximum r-rank. It is still unknown whether the webs
$W(d, 2, r)$, $d = r + 2$, of maximum r-rank are almost Grassmannizable or
not.

Proposition 2.4. *The r-rank of a web $AGW(d, 2, r)$ is equal to zero if
$r > 2$ and $d \leq r + 1$.*

Proof.   For webs $AGW(d, 2, r)$ we have the equations (1.13) and the
equation

$$(2.7) \qquad\qquad \gamma = -\sum_{\alpha} \sigma_{\alpha} \lambda_{\alpha}$$

which follows from (2.5) and (1.13). In addition, comparison of coefficients
of $\underset{1}{\omega^1} \wedge \ldots \wedge \underset{1}{\omega^r}$ and $\underset{2}{\omega^1} \wedge \ldots \wedge \underset{2}{\omega^r}$ gives by means of (2.7) and (1.13) that

$$(2.8) \qquad \alpha = -\sum_{\alpha} \sigma_{\alpha}(\lambda_{\alpha} - \lambda^r), \qquad \beta = -\sum_{\alpha} \sigma_{\alpha}(\lambda_{\alpha} - 1).$$

Let $r > 2$. Comparing coefficients of $\underset{1}{\omega^1} \wedge \underset{1}{\omega^2} \wedge \underset{2}{\omega^3} \wedge \ldots \wedge \underset{2}{\omega^r}, \underset{1}{\omega^1} \wedge \underset{1}{\omega^2} \wedge$
$\underset{1}{\omega^3} \wedge \underset{2}{\omega^4} \wedge \ldots \wedge \underset{2}{\omega^r}, \ldots$, we obtain the system

$$(2.9) \qquad\qquad \sum_{\alpha} \sigma_{\alpha}(\lambda_{\alpha}^k - \lambda_{\alpha}) = 0, \qquad k = 2, \ldots, r - 1,$$

of $r - 2$ equations in $d - 3$ quantities $\underset{\alpha}{\sigma}$. It is easy to see that if $d \leq r + 1$, then
rank $(\lambda_{\alpha}^k - \lambda_{\alpha}) = d - 3$, and the system (2.9) has only the trivial solution. ■

By Proposition 2.4, almost Grassmannizable webs $AGW(d, 2, r)$, $r > 2$,
may have a non-zero r-rank if $r = 2$ or $r > 2$, $d > r + 1$. We will study here
only the case $r = 2$.

2.2. Almost Grassmannizable webs $AGW(d,2,2)$ of maximum 2-rank

For webs $AGW(d,2,2)$ of non-zero r-rank we have equations (1.13), (2.7), and (2.8). In addition, we note that for $r = 2$ the torsion tensor a^i_{jk} has always the form (1.14) and the structure equations (1.3) become (1.15). Further, as was proved in Section 1, equations (1.13) imply (1.43) and (1.52).

In order to find the maximum 2-rank webs $AGW(d,2,2)$, we note that the left member of the abelian 2-equation (2.4) is the sum of d summands: $\sum_{\xi=1}^{d} \Omega_\xi = 0$. In the case $r = 2$ we have (see (2.6), (2.7), and (2.8)):

$$
(2.10) \quad
\begin{cases}
\Omega_1 = \sum_\alpha \underset{\alpha}{\sigma}(\underset{\alpha}{\lambda} - \lambda^2)\underset{1}{\omega^1} \wedge \underset{1}{\omega^2}, \\[2mm]
\Omega_2 = \sum_\alpha \underset{\alpha}{\sigma}(\underset{\alpha}{\lambda} - 1)\underset{2}{\omega^1} \wedge \underset{2}{\omega^2}, \\[2mm]
\Omega_3 = \sum_\alpha \underset{\alpha}{\sigma}\underset{\alpha}{\lambda}(\underset{1}{\omega^1} + \underset{2}{\omega^1}) \wedge (\underset{1}{\omega^2} + \underset{2}{\omega^2}), \\[2mm]
\Omega_\alpha = \sum_\alpha \underset{\alpha}{\sigma}(\underset{\alpha}{\lambda}\underset{1}{\omega^1} + \underset{2}{\omega^1})(\underset{\alpha}{\lambda}\underset{1}{\omega^2} + \underset{2}{\omega^2}), \quad \alpha = 4, \ldots, d,
\end{cases}
$$

where, according to Lemma 2.1, Ω_ξ, $\xi = 1, \ldots, d$, are closed 2-forms.

Exterior differentiation of (2.10) gives the following independent cubic exterior equations:

$$
(2.11) \quad
\begin{cases}
\sum_\alpha [(\underset{\alpha}{\lambda} - \lambda^2)(d\underset{\alpha}{\sigma} - \underset{\alpha}{\sigma}\omega^i_i) - \underset{\alpha}{\sigma}(\underset{\alpha}{\lambda} - 1)(b_i - \underset{\alpha}{\lambda}a_i)\omega^i] \wedge \underset{1}{\omega^1} \wedge \underset{1}{\omega^2} = 0, \\[2mm]
\sum_\alpha [(\underset{\alpha}{\lambda} - 1)(d\underset{\alpha}{\sigma} - \underset{\alpha}{\sigma}\omega^i_i) + \underset{\alpha}{\sigma}\underset{\alpha}{\lambda}(b_i - a_i)\omega^i] \wedge \underset{2}{\omega^1} \wedge \underset{2}{\omega^2} = 0, \\[2mm]
d\underset{\alpha}{\sigma} - \underset{\alpha}{\sigma}(\omega^k_k + 2a_i\underset{2}{\omega^i} + b_i\omega^i) \wedge (\underset{\alpha}{\lambda}\underset{1}{\omega^1} + \underset{2}{\omega^1}) \wedge (\underset{\alpha}{\lambda}\underset{1}{\omega^2} + \underset{2}{\omega^2}) = 0.
\end{cases}
$$

It follows from (2.11) that

$$
(2.12) \quad d\underset{\alpha}{\sigma} - \underset{\alpha}{\sigma}\omega^k_k - \underset{\alpha}{\sigma}(b_i\omega^i + 2a_i\underset{2}{\omega^i}) = \underset{\alpha}{\sigma}_i(\underset{\alpha}{\lambda}\underset{1}{\omega^i} + \underset{2}{\omega^i}),
$$

$$
(2.13) \quad \sum_\alpha (\underset{\alpha}{\lambda} - \lambda^2)\underset{\alpha}{\sigma}_i = \sum_\alpha \underset{\alpha}{\sigma}[(2\underset{\alpha}{\lambda} - 1)b_i - \underset{\alpha}{\lambda}a_i].
$$

Note that two quantities $\underset{\alpha_0}{\sigma}_i$, α_0 fixed, out of the total number $2(d - 3)$ quantities $\underset{\alpha}{\sigma}_i$, can be determined from (2.13) because $\underset{\alpha}{\sigma} \neq 0$ and $\lambda \neq 0, 1$.

Exterior differentiation of (2.12) and the application of Cartan's lemma give

$$(2.14) \quad \nabla_{\alpha}\sigma_i - \sigma_i\omega_k^k + \sigma(3f_{ik} + k_{ik} - 2\sigma_{(k}b_{i)})\omega^k_1 - 3\sigma_i a_k \omega^k_2 = \sigma_{ik}(\lambda\omega^k_1 + \omega^k_2)$$

where $\nabla_{\alpha}\sigma_i = d\sigma_i - \sigma_j\omega_i^j$ and the σ_{ij} are symmetric with respect to i and j.

Differentiating (2.13) and using (2.14), we obtain, by means of the linear independence of the ω^k_1 and ω^k_2, the following equations:

(2.15)
$$\begin{cases} \sum_{\alpha}(\lambda^2 - \lambda^3)\sigma_{ik} = \sum_{\alpha}\{(3\lambda^2 - 2\lambda)(2\sigma_{(i}b_{k)} - \sigma k_{ik} + \sigma(6\lambda - 2)b_i b_k \\ \qquad\qquad - 2\lambda^2\sigma_{(i}a_{k)} - 4\sigma\lambda^2 b_{(i}a_{k)} + \sigma(\lambda - \lambda^2)f_{ik} + \sigma\lambda h_{ik}\}, \\ \sum_{\alpha}(\lambda - \lambda^2)\sigma_{ik} = \sum_{\alpha}\{(2\lambda - 1)[\sigma(g_{ik} - k_{ik}) + 2\sigma_{(i}b_{k)}] \\ \qquad\qquad + 2\sigma(b_i b_k - b_{(i}a_{k)}) + \lambda[\sigma(h_{ik} - g_{ik}) - 2\sigma_{(i}a_{k)}]\}. \end{cases}$$

Note that six quantities σ_{ik}, σ_{ik}, $\alpha_0 \neq \beta_0$, α_0, β_0 fixed, out of the total number $3(d - 3)$ quantities σ_{ik} can be determined from (2.15) since the corresponding determinant is equal to

$$\sigma_{\alpha_0\beta_0}\sigma_{\alpha_0}(\lambda_{\alpha_0} - \lambda^2_{\beta_0})(\lambda_{\beta_0} - \lambda^2_{\beta_0})(\lambda_{\beta_0} - \lambda_{\alpha_0})$$

and different from zero because of $\sigma \neq 0$, $\lambda \neq \lambda$ and $\lambda \neq 0,1$.

Using the same procedure, we can get $3 \cdot 4$ out of the total number $4(d - 3)$ quantities σ_{ijk}, $4 \cdot 5$ out of the total number $5(d - 3)$ quantities σ_{ijkl}, etc. and all $(d - 2)(d - 3)$ quantities $\sigma_{i_1 i_2 \ldots i_{d-3}}$.

If we substitute all the $\sigma_{i_1 \ldots i_{d-3}}$ obtained from the differential equations for $\sigma_{i_1 \ldots i_{d-4}}$, we will have in these equations only functions whose differentials are known. Exterior differentiation of these equations gives equations of the form

$$(2.16) \qquad\qquad A_{i_1 \ldots i_{d-4}jk}\omega^j_1 \wedge \omega^k_2 = 0$$

where $A_{i_1 \ldots i_{d-4}jk}$ are linear functions of σ, σ_i, σ_{jk}, $\ldots$, $\sigma_{i_1 \ldots i_{d-4}}$. Since all products $\omega^i_1 \wedge \omega^k_2$ are linearly independent, we have

$$(2.17) \qquad\qquad A_{i_1 \ldots i_{d-4}jk} = 0.$$

Differentiation of (2.17) gives by means of the linear independence of the ω^j_1 and ω^j_2 two equations similar to (2.17), i.e., linear equations in σ_α, $\sigma_{i_1}_\alpha, \ldots, \sigma_{i_1 \ldots i_{d-4}}_\alpha$. Differentiation of these two equations leads to four new equations linear in σ_α, $\sigma_{i_1}_\alpha, \ldots, \sigma_{i_1 \ldots i_{d-4}}_\alpha$.

The 2-rank R_2 of a web $AGW(d, 2, 2)$ is equal to the number of independent Pfaffian equations (2.12), (2.14), ... which is

$$(d-3) \sum_{k=0}^{d-4} (k+1) - \sum_{k=1}^{d-4} k(k+1) - \varepsilon$$

where the first term is the number of equations (2.12), (2.14), ..., the second term is the number of relations (2.13), (2.15), ... and the last term is the number of independent equations in the system consisting of (2.17) and its differential consequences where all the quantities $\sigma_{i_1}_\alpha$, $\sigma_{i_1 i_2}_\alpha, \ldots, \sigma_{i_1 \ldots i_{d-4}}_\alpha$ which can be found from (2.13), (2.15), ... are substituted.

This 2-rank has the maximum value $\pi_2(d, 2, 2)$ if and only if ε is minimal. It can be proved that the minimal value for ε is 0. This means that all coefficients of (2.17) vanish. In fact, we will see that the vanishing of the coefficient in $\sigma_{i_1 \ldots i_{d-4}}_\alpha$ implies (1.74) if $d = 4$ and (1.74), (1.35) if $d > 4$, and the vanishing of all the other coefficients of (2.17) is a differential consequence of (1.74). Therefore, the maximum 2-rank

$$\pi_2(d, 2, 2) = (d-3) \sum_{k=0}^{d-4} (k+1) - \sum_{k=1}^{d-4} k(k+1) = (d-1)(d-2)(d-3)/6.$$

This matches the result of S. S. Chern and P. A. Griffiths [CG 78b] in the case $n = r = 2$.

To clarify our considerations, we consider the cases $d = 4$ and $d > 4$ separately because, as we mentioned before, the conditions for webs $W(4, 2, 2)$ and $W(d, 2, 2)$, $d > 4$ to be of maximum 2-rank differ for these two cases.

2.3. Four-webs $W(4,2,2)$ of maximum 2-rank

Theorem 2.5. *An isoclinic web $W(4,2,2)$ is of maximum 2-rank if and only if it is almost algebraizable. A non-isoclinic web $W(4,2,2)$ is of maximum 2-rank if and only if it is almost Grassmannizable web and any of the four affine connections mentioned in Theorem 1.9 is equiaffine.*

Proof. For webs $W(4,2,2)$ of maximum 2-rank we have $\underset{4}{\sigma} \neq 0$ and we can express all $\underset{4}{\sigma}_i$ from (2.13). Substituting them into (2.12), we obtain

$$d\underset{4}{\sigma} - \underset{4}{\sigma}\omega_k^k - \underset{4}{\sigma}(\underset{4}{b}_i\omega^i + 2\underset{1}{a}_i\omega^i) = \underset{4}{\sigma}[(2\lambda-1)\underset{4}{b}_i - \lambda\underset{4}{a}_i](\lambda\underset{1}{\omega}^i + \underset{2}{\omega}^i)/(\underset{4}{\lambda} - \underset{4}{\lambda}^2)$$

or

(2.18)
$$d\ln\underset{4}{\sigma} = \omega_k^k - d\ln(\underset{4}{\lambda}-1) + (\underset{4}{a}_i - \underset{4}{b}_i/\lambda)\underset{2}{\omega}^i.$$

By Proposition 2.4, webs $W(4,2,2)$ of maximum 2-rank are almost Grassmannizable. According to Theorem 1.3, we must distinguish two cases:

a) *A web $AGW(4,2,2)$ is isoclinic.* In this case we have (1.24), (1.53), (1.31), and (1.54). Exterior differentiation of (2.18) gives

(2.19)
$$(b_{kij}^k - h_{ij} + \underset{4}{k}_{ij})\underset{1}{\omega}^i \wedge \underset{2}{\omega}^j = 0.$$

Since the $\underset{1}{\omega}^i \wedge \underset{2}{\omega}^j$ are linearly independent, it follows from (2.19) that

(2.20)
$$b_{kij}^k - h_{ij} + \underset{4}{k}_{ij} = 0.$$

But (1.27) and (1.29) show that

(2.21)
$$b_{kij}^k = f_{ij} + g_{ij} + 2h_{ij}.$$

It follows from (2.21) that condition (2.20) can be written in the form (1.74). Therefore, isoclinic webs $W(4,2,2)$ of maximum 2-rank are characterized by (1.13), (1.24), (1.52), and (1.74). Under these conditions the equation (2.18) is completely integrable and its solution depends on one constant. This means that the maximum 2-rank of isoclinic webs $W(4,2,2)$ is equal to one: $\pi_2(4,2,2) = 1$, and an isoclinic web $W(4,2,2)$ is of maximum 2-rank if and only if it is almost algebraizable.

b) *A web $W(4, 2, 2)$ is not isoclinic.* In this case we have (1.17) and (1.51). Exterior differentiation of (2.18) gives

$$(2.22) \qquad (b^k_{kij} - b_{ij} + q_{ij})\underset{1}{\omega^i} \wedge \underset{2}{\omega^j} = 0.$$

Equation (2.22) implies (1.112). Therefore, non-isoclinic webs $W(4, 2, 2)$ of maximum 2-rank are characterized by equations (1.13), (1.17), (1.51), and (1.112). Under these conditions, equation (2.18) is completely integrable and maximal 2-rank is again equal to one: $\pi_2(4, 2, 2) = 1$.

According to Theorem 1.15, a non-isoclinic web $W(4, 2, 2)$ is of maximum 2-rank if and only if any of four affine connections of Theorem 1.9 is equiaffine.

Note that in both cases which were considered in the proof of this theorem, we can write equation (2.12) in the following form:

$$(2.23) \qquad d\ln[\underset{4}{\sigma}(\underset{4}{\lambda} - 1)] = \underset{4}{\omega^k_k} + (a_i - \underset{4}{b_i}/\underset{4}{\lambda})\underset{2}{\omega^i}.$$

Under conditions (1.43), (1.24), (1.53), and (1.74) or (1.17), (1.51), and (1.112) the exterior differential of the right member of (2.23) vanishes. This means that under any of these conditions equation (2.23) is completely integrable.

The conditions mentioned above distinguish two classes of webs $W(4, 2, 2)$ of maximum 2-rank.

We are able to write now the only abelian 2-equation for both kinds of webs $W(4, 2, 2)$ of maximum 2-rank:

$$(\underset{4}{\lambda} - \underset{4}{\lambda^2})\sigma\underset{4}{\omega^1} \wedge \underset{1}{\omega^2} + (\underset{4}{\lambda} - 1)\sigma\underset{4}{\omega^1} \wedge \underset{2}{\omega^2} - \underset{4}{\lambda}\sigma(\underset{4}{\omega^1} + \underset{1}{\omega^1}) \wedge (\underset{2}{\omega^2} + \underset{1}{\omega^2})$$

$$(2.24) \qquad + \sigma(\underset{4}{\lambda}\underset{4}{\omega^1} + \underset{1}{\omega^1}) \wedge (\underset{4}{\lambda}\underset{1}{\omega^2} + \underset{2}{\omega^2}) = 0.$$

This equation holds for any $W(4, 2, 2)$ — it is an identity — but it is an abelian 2-equation if and only if the function $\underset{4}{\sigma}$ is a solution of (2.18) and a web $W(4, 2, 2)$ is almost algebraizable (satisfies (1.13), (1.43), (1.24), (1.53), (1.74)) or a web $W(4, 2, 2)$ satisfies conditions (1.13), (1.17), (1.51), (1.112). Under these conditions Lemma 2.1 can be applied.

This situation is similar to one for $W(3, 2, r)$: the equation $\underset{1}{\omega^i} + \underset{2}{\omega^i} + \underset{3}{\omega^i} = 0$ holds for any $W(3, 2, r)$. However, it will be an abelian 1-equation only for parallelizable webs — for them this equation can be reduced to the form $du^i_1 + du^i_2 + du^i_3 = 0$ (if $r = 1$, the equation $\underset{1}{\omega} + \underset{2}{\omega} + \underset{3}{\omega} = 0$ can be reduced to the form $du_1 + du_2 + du_3 = 0$ if and only if a web $W(3, 2, 1)$ is hexagonal).

2.4. Webs $W(d, 2, 2)$, $d > 4$, of maximum 2-rank

Theorem 2.6. *A web $W(d, 2, 2)$, $d > 4$, is of maximum 2-rank $\pi_2(d, 2, 2)$ if and only if it is algebraizable.*

Proof. First, note that by Theorem 2.3, webs $W(d, 2, 2)$, $d > 4$, of maximum 2-rank are almost Grassmannizable. We will consider webs $W(5, 2, 2)$ — the general case $d > 4$ is similar. For $W(5, 2, 2)$ we have equations (2.13) and (2.15) where $\alpha = 4, 5$. We can find $\underset{5}{\sigma_i}$ (or $\underset{4}{\sigma_i}$), $\underset{4}{\sigma_{ik}}$, and $\underset{5}{\sigma_{ik}}$ from these equations:

$$(2.25) \qquad \underset{\alpha}{\sigma_i} = \left\{ \sum_{\gamma=4}^{5} \left[(2\lambda - 1)\underset{\gamma}{b} - \lambda \underset{\gamma}{a_i} \right] \underset{\gamma}{\sigma} - (\lambda - \lambda^2)\underset{\beta}{\sigma_i} \right\} / (\lambda - \lambda^2),$$

$$
\begin{aligned}
\underset{\alpha}{\sigma_{ik}} = \Big\{ & f_{ik} \sum_{\gamma} \underset{\gamma}{\sigma}(\underset{\gamma}{\lambda} - \lambda^2) + g_{ik}\underset{\beta}{\lambda} \sum_{\gamma} \underset{\gamma}{\sigma}(1 - \underset{\gamma}{\lambda}) + h_{ik}(1 - \underset{\beta}{\lambda}) \sum_{\gamma} \underset{\gamma}{\sigma}\underset{\gamma}{\lambda} \\
& + (\sigma \underset{\beta\beta}{k_{ik}} - 2\sigma_i \underset{\beta\beta}{b_k})(\lambda - \underset{\beta}{\lambda^2}) + (3\underset{\alpha}{\lambda^2} - 2\underset{\alpha}{\lambda} - 2\underset{\alpha\beta}{\lambda\lambda} + \underset{\beta}{\lambda})(2\sigma_{(i}\underset{\alpha}{b_k)} - \sigma \underset{\alpha\alpha}{k_{ik}}) \\
& + \sigma \underset{\beta\beta\beta}{b_i b_k}(4\underset{\beta}{\lambda} - 2) + \sigma \underset{\alpha\alpha\alpha}{b_i b_k}(6\underset{\alpha}{\lambda} - 2\underset{\beta}{\lambda} - 2) + 2\sigma_{(i}\underset{\alpha\alpha}{a_k)}(\underset{\alpha}{\lambda} - \underset{\beta}{\lambda}) \\
(2.26) \quad & - 2\sigma\lambda \underset{\beta\beta\beta}{b_{(i}a_k)} + \sigma \underset{\alpha\alpha}{b_{(i}a_k)}(2\underset{\beta}{\lambda} - 4\underset{\alpha}{\lambda}) \Big\} / [(\lambda - \lambda^2)(\lambda - \lambda)]
\end{aligned}
$$

where $\beta \ne \alpha$. We have now only two independent equations (2.14) for $\underset{4}{\sigma_i}$ — two others are their consequences because of (2.25). Substituting $\underset{4}{\sigma_{ik}}$ from (2.26) into these equations and taking exterior derivatives of the equations obtained, we get equations of the form (2.16) and (2.17). The latter one is:

$$
\begin{aligned}
& \underset{4}{\sigma_t}(a_{ijk}^t + 3\delta_i^{(t}\delta_j^l\delta_k^{m)}K_{lm})(\underset{4}{\lambda} - \underset{4}{\lambda^2})(\underset{4}{\lambda} - \underset{5}{\lambda}) \\
& \quad + \sigma[\underset{4}{\phi_{ijk}} - \lambda\underset{5}{\phi_{ijk}} + 3(\underset{4}{\lambda} - \underset{5}{\lambda})(1 - 2\lambda)(\underset{4}{b_{(i}}\underset{4}{K_{jk)}} - \underset{4}{b_{(i}}\underset{4}{k_{jk)}}) \\
& \quad - 3(\underset{4}{\lambda} - \underset{4}{\lambda^2})(\underset{5}{b_{(i}}K_{jk)} - \underset{5}{b_{(i}}\underset{5}{k_{jk)}}) + 3\lambda(\underset{4}{\lambda} - \underset{4}{\lambda})(a_{(i}\underset{5}{K_{jk)}} - a_{(i}h_{jk)})] \\
(2.27) \quad & + \underset{5}{\sigma}(\underset{5}{\lambda} - \underset{5}{\lambda^2})[\theta_{ijk} - \lambda\underset{4}{\underset{2}{\theta_{ijk}}} - 3\underset{4}{b_{(i}}K_{jk)} + 3\underset{4}{b_{(i}}\underset{4}{k_{jk)}}] = 0
\end{aligned}
$$

where

(2.28)

$$
\begin{cases}
\underset{1}{\phi}_{ijk} = (\underset{4}{\lambda} - \underset{4}{\lambda^2})\underset{1}{f}_{ijk} + \underset{5}{\lambda}(1 - \underset{4}{\lambda})\underset{1}{g}_{ijk} + \underset{4}{\lambda}(1 - \underset{5}{\lambda})\underset{1}{h}_{ijk} \\
\qquad + (-3\underset{4}{\lambda^2} + 2\underset{4}{\lambda} + 2\underset{4}{\lambda}\underset{5}{\lambda} - \underset{5}{\lambda})(\underset{4}{\lambda}\underset{4}{k}_{ijk} + 3\underset{5}{b}_{(i}\underset{4}{k}_{jk)} + \underset{4}{b}_t\underset{4}{a}^t_{ijk} + \underset{1}{g}_{ijk} - \underset{4}{\lambda}\underset{2}{f}_{ijk}), \\[4pt]
\underset{2}{\phi}_{ijk} = \underset{5}{\lambda}^{-1}\Big[(3\underset{5}{\lambda} - 2\underset{4}{\lambda})(\underset{4}{\lambda} - \underset{4}{\lambda^2})\underset{2}{f}_{ijk} + \underset{5}{\lambda}(\underset{4}{\lambda} - \underset{4}{\lambda^2})\underset{2}{g}_{ijk} + \underset{4}{\lambda^2}(1 - \underset{5}{\lambda})\underset{2}{h}_{ijk} \\[4pt]
\qquad - \underset{4}{\lambda^2}(2\underset{4}{\lambda} - \underset{5}{\lambda} - 1)\underset{4}{k}_{ijk}\Big], \\[4pt]
\underset{1}{\theta}_{ijk} = \underset{1}{f}_{ijk} + \underset{1}{g}_{ijk} + \underset{1}{h}_{ijk} + \underset{5}{\lambda}\underset{5}{k}_{ijk} + 3\underset{5}{b}_{(i}\underset{5}{k}_{jk)} + \underset{5}{b}_t\underset{}{a}^t_{ijk} + \underset{1}{g}_{ijk} - \underset{5}{\lambda}\underset{2}{f}_{ijk}, \\[4pt]
\underset{2}{\theta}_{ijk} = \underset{2}{f}_{ijk} + \underset{2}{g}_{ijk} + \underset{2}{h}_{ijk} + \underset{5}{k}_{ijk}.
\end{cases}
$$

As we mentioned earlier, differentiation of (2.28) gives us two new equations linear in σ_i, $\underset{4}{\sigma}$, $\underset{5}{\sigma}$. Their differentiation gives four new equations, etc. For a web $AGW(5,2,2)$ of maximum 2-rank $\underset{4}{\sigma}$ and $\underset{5}{\sigma}$ are independent and our system is a system for $\underset{4}{\sigma}_i$ only. The 2-rank of $AGW(5,2,2)$ is equal to the number of independent equations from (2.12) and (2.14). By means of (2.25), the number of independent equations from (2.14) is $4 - \varepsilon$ where ε is the rank of the matrix of the coefficients of the system in $\underset{4}{\sigma}_i$ which we discussed above. It is obvious that $0 \le \varepsilon \le 2$.

We obtain the conditions for a maximum 2-rank web $AGW(5,2,2)$ requiring $\varepsilon = 0$ if possible. This means that the equation (2.27) must vanish. It follows from (2.27) that in this case we have first of all

(2.29)
$$
\underset{4}{\sigma}_t(a^t_{ijk} + 3\delta_i^{(t}\delta_j^k\delta_k^{m)}K_{lm}) = 0.
$$

Consider the eight equations (2.29) for all possible values $i,j,k = 1,2$. We obtain the following conditions:

(2.30)
$$
\begin{cases}
a^1_{111} + 3K_{11} = 0, & a^2_{111} = 0, \\
a^2_{222} + 3K_{22} = 0, & a^1_{222} = 0, \\
a^1_{112} + 2K_{12} = 0, & a^1_{211} + K_{11} = 0, \\
a^2_{212} + 2K_{12} = 0, & a^1_{122} + K_{22} = 0.
\end{cases}
$$

The relations (1.29) show that conditions (2.30) imply

(2.31)
$$
K_{ij} = 0, \qquad a^i_{jkl} = 0,
$$

i.e., conditions (1.74) and (1.35).

Let us show that under conditions (2.31) two other terms of (2.27) vanish. In fact, (2.31) implies (1.36), (1.78), (1.79), (1.80), and therefore

(2.32)
$$
\begin{cases}
\underset{1}{\phi_{ijk}} - \underset{5}{\lambda}\underset{2}{\phi_{ijk}} = 3\underset{5}{\lambda}(\underset{4}{\lambda} - \underset{5}{\lambda})a_{(i}h_{jk)} + 3(\underset{4}{\lambda} - \underset{5}{\lambda})(1 - 2\underset{5}{\lambda})b_{(i}\underset{4}{k}_{jk)} \\
\qquad\qquad - 3(\underset{4}{\lambda} - \underset{4}{\lambda^2})b_{(i}\underset{5}{k}_{jk)} \\
\underset{1}{\theta_{ijk}} - \underset{4}{\lambda}\underset{2}{\theta_{ijk}} = - 3b_{(i}\underset{4}{k}_{jk)}.
\end{cases}
$$

The equations (2.32) prove the vanishing two last terms of (2.27).

Thus, conditions (2.31) are necessary and sufficient for a web $W(5,2,2)$ to be of maximum 2-rank $\pi_2(5,2,2) = 4$. The conditions (2.31) together with Theorem 1.8 prove that a web $W(5,2,2)$ of maximum 2-rank is algebraizable.

We conclude the paper by the following remark:

Remark 2.7. The case $d = 4$ for isoclinic webs differs from the case $d > 4$ since the final Pfaffian equations which do not contain new functions (equations (2.18) for $d = 4$ and (2.14), (2.27) for $d = 5$) contain only the contracted connection form w_k^k for $d = 4$ and contain the forms w_j^i and w_k^k for $d > 4$. After exterior differentiation in the first case only the contracted curvature tensor b_{kij}^k appears whose expression does not contain the tensor a_{jkl}^i (because of (1.35)) while in the second case the general curvature tensor b_{jkl}^i appears whose expression contains the tensor a_{jkl}^i.

The webs $W(4,2,2)$ of maximum 2-rank are exceptional in the sense that they are not necessarily algebraizable while the webs $W(d,2,2)$, $d > 4$, of maximum 2-rank are algebraizable.

In the papers [G 85], [G 86], [G 87] we proved the existence of webs $W(4,2,2)$ of maximum 2-rank by giving examples of both kinds (isoclinic and non-isoclinic) of webs $W(4,2,2)$ of maximum 2-rank. Their existence disproves P. A. Griffiths' conjecture that webs $W(d,n,r)$ of maximum r-rank are algebraizable.

References

[A 74] Akivis, M. A.: On isoclinic three-webs and their interpretation in a ruled space of projective connection. (Russian) *Sibirsk. Mat. Zh.* **15**

(1974), no. 1, 3–15. English translation: *Siberian Math. J.* **15** (1974), no. 1, 1–9.

[A 80] AKIVIS, M. A.: Webs and almost Grassmann structures. (Russian) *Dokl. Akad. Nauk SSSR* **252** (1980), no. 2, 267–270. English translation: *Soviet Math. Dokl.* **21** (1980), no. 3, 707-709.

[A 81] AKIVIS, M. A.: A geometric condition of isoclinity of a multidimensional web. (Russian) Webs and Quasigroups, Kalinin. Gos. Univ., Kalinin, 1981, 3–7.

[A 82] AKIVIS, M. A.: Webs and almost Grassmann structures. (Russian) *Sibirsk. Mat. Zh.* **23** (1982), no. 6, 6–15. English translation: *Siberian Math. J.* **23** (1982), no. 6, 763–770.

[A 83a] AKIVIS, M. A.: Completely isotropic submanifolds of a four-dimensional pseudo-conformal structure. (Russian) *Izv. Vyssh. Uchebn. Zaved. Mat.* **1983**, no. 1(248), 3–11. English translation: *Soviet Math. (Iz. VUZ)* **27** (1983), no. 1, 1–11.

[A 83b] AKIVIS, M. A.: A local condition of algebraizability of a system of submanifolds of a real projective space. (Russian) *Dokl. Akad. Nauk SSSR* **272** (1983), no. 6, 1289–1291. English translation: *Soviet Math. Dokl.* **28** (1983), no. 2, 507–509.

[AS 71a] AKIVIS, M. A. and SHELEKHOV, A. M.: The computation of the curvature and torsion tensors of a multidimensional three-web and of the associator of the local quasigroup that is connected with it. (Russian) *Sibirsk. Mat. Zh.* **12** (1971), no. 5, 953–960. English translation: *Siberian Math. J.* **12** (1971), no. 5, 685–689.

[AS 71b] AKIVIS, M. A. and SHELEKHOV, A. M.: Local differentiable quasigroups and connections that are associated with a three-web of multidimensional surfaces. (Russian) *Sibirsk. Mat. Zh.* **12** (1971), no. 6, 1181–1191. English translation: *Siberian Math. J.* **12** (1971), no. 6, 845–892.

[AS 81] AKIVIS, M. A. and SHELEKHOV, A. M.: "Foundations of the theory of webs." (Russian) Kalinin. Gos. Univ., Kalinin, 1981, 88 pp.

[AHS 78] ATIYAH, M. F.; HITCHIN, N. I. and SINGER, I. M.: Self-duality in four-dimensional Riemannian geometry. *Proc. Roy. Soc. London Ser. A* **362** (1978), 425–461.

[BB 38] BLASCHKE, W. and BOL, G.: "Geometrie der Gewebe." Springer-Verlag, Berlin, 1938, viii+339 pp.

[B 35] BOL, G.: Über 3-Gewebe in vierdimensionalen Raum. *Math. Ann.* **110** (1935), 431–463.

[C 36] CHERN, S. S.: Abzählungen für Gewebe. *Abh. Math. Sem. Univ. Hamburg* **11** (1936), no. 1–2, 163–170.

[CG 78a] CHERN, S. S. and GRIFFITHS, P. A.: Abel's theorem and webs. *Jahresber. Deutsch. Math.-Verein.* **80** (1978), no. 1–2, 13–110.

[CG 78b] CHERN, S. S. and GRIFFITHS, P. A.: An inequality for the rank of a web and webs of maximum rank. *Ann. Scuola Norm. Sup. Pisa Cl. Sci.* (4) **5** (1978), no. 3, 539–557.

[CG 81] CHERN, S. S. and GRIFFITHS, P. A.: Corrections and addenda to our paper "Abel's theorem and webs". *Jahresber. Deutsch. Math.-Verein.* **83** (1981), 78–83.

[D 83] DAMIANO, D. B.: Webs and characteristic forms on Grassmann manifolds. *Amer. J. Math.* **105** (1983), 1325–1345.

[G 66] GOLDBERG, V. V.: On a normalisation of p-conjugate systems of an n-dimensional projective space. (Russian) *Trudy Geom. Sem.* **1** (1966), 89–109.

[G 77] GOLDBERG, V. V.: On the theory of four-webs of multidimensional surfaces on a differentiable manifold X_{2r}. (RUSSIAN) IZV. VYSSH. UCHEBN. ZAVED. MAT. **1977**, no. 11(186), 15–22. English translation: *Soviet Math. (Iz. VUZ)* **21** (1977), no. 11, 97–100.

[G 80] GOLDBERG, V. V.: On the theory of four-webs of multidimensional surfaces on a differentiable manifold X_{2r}. (Russian) *Serdica* **6** (1980), no. 2, 105–119.

[G 82a] GOLDBERG, V. V.: The solutions of the Grassmannization and algebraization problems for $(n+1)$-webs of codimension r on a differentiable manifold of dimension nr. *Tensor (N.S.)* **36** (1982), no. 1, 9–21.

[G 82b] GOLDBERG, V. V.: Grassmann and algebraic four-webs in a projective space. *Tensor (N.S.)* **38** (1982), 179–197.

[G 83] GOLDBERG, V. V.: Tissus de codimension r et de r-rang maximum. *C. R. Acad. Sci. Paris Sér. I Math* **297** (1983), no. 6, 339–342.

[G 85] GOLDBERG, V. V.: 4-tissus isoclinics exceptionnels de codimension deux et de 2-rang maximal. *C.R. Acad. Sci. Paris Ser. I Math* **301** (1985), no. 11, 593–596.

[G 86] GOLDBERG, V. V.: Isoclinic webs $W(4,2,2)$ of maximum 2-rank. *Differential Geometry*, Peniscola 1985, 168–183. Lecture Notes in Math., **1209**, Springer-Verlag, Berlin-New York, 1986.

[G 87] GOLDBERG, V. V.: Non-isoclinic 2-codimensional 4-webs of maximum 2-rank. *Proc. Amer. Math. Soc.* **100** (1987), no. 4, 701–708.

[Gr 76] GRIFFITHS, P. A.: Variations on a theorem of Abel. *Invent. Math.* **35** (1976), 321–390.

[Gr 77] GRIFFITHS, P. A.: On Abel's differential equations. Algebraic Geometry, J. J. Sylvester Sympos., Johns Hopkins Univ., Baltimore, Md., 1976, 26–51. Johns Hopkins Univ. Press, Baltimore, Md, 1977.

[L 89] LITTLE, J. B.: On webs of maximum rank. *Geom. Dedicata* **31** (1989), 19–35.

[P 68] PENROSE, R.: Structure of space-time. Battelle rencontres, 1967 Lectures in Mathematics and Physics, Chapter VII, 121–135, Benjamin, New York-Amsterdam, 1968.

[W 82] WOOD, J. A.: "An algebraization theorem for local hypersurfaces in projective space." Ph. D. Dissertation, Univ. of California, Berkeley, 1982, 87 pp.

[W 84] WOOD, J. A.: A simple criterion for local hypersurfaces to be algebraic. *Duke Math. J.* **51** (1984), no. 1, 235–237.

Vladislav V. Goldberg

Department of Mathematics
New Jersey Institute of
Technology
Newark, N.J. 07102,
U.S.A.

COLLOQUIA MATHEMATICA SOCIETATIS JÁNOS BOLYAI

56. DIFFERENTIAL GEOMETRY, EGER (HUNGARY), 1989

Bäcklund Transformations, Matrix-Riccati Systems and Isometric Immersions of Space Forms into Space Forms[*]

H. GOLLEK

§ 1. Introduction

We consider a local group action of the linear group $Gl(2N, \mathbb{R})$ on the space $M(N)$ of all matrices of type $N \times N$ given by some kind of fractional linear functions. This local action law leads to some relation between linear Pfaffian systems and certain other Pfaffian systems which we call matrix Riccati equations. If it is restricted to the pseudoorthogonal group $O(N, N)$ one obtains an action of $O(N, N)$ on the orthogonal group $O(N)$.

For an arbitrary n-dimensional simply connected manifold Y, an $\mathfrak{o}(N, N)$-valued 1-form $\tau = \tau(A_0, \varphi_0)$ on Y is associated with any pair (A_0, φ_0), where A_0 is a smooth map of Y into $O(N)$ and φ_0 is an $\mathfrak{o}(N)$-valued 1-form. The integrability condition $d\tau + \tau \wedge \tau = 0$ is equivalent to some partial differential equation for A_0 and φ_0. The whole family of equations under consideration is obtained by various choices of a symmetric $N \times N$-matrix D and a symmetric matrix valued 1-form δ involved in

---

[*] This paper is in final form and no version of it will be submitted for publication elsewhere.

the definition of τ. Denoting the set of solutions (A_0, φ_0) of this differential equation by $M(\delta, D)$, the Bäcklund transformation is obtained as a mapping B_X of $M(\delta, D)$ into itself by choosing an arbitrary matrix $X \in O(N)$ and assigning to (A_0, φ_0) a new pair (A_1, φ_1) in such a way that A_1 is the orbit of X under a map $H : Y \to O(N, N)$ obtained as solution of the linear Pfaffian system $dH = H\tau$.

The superposition formula treated in § 3 is concerned with iterations of such transformations. It gives a simple expression for the result of two iterated Bäcklund transformations and shows that two such transformations commute in a certain sense made precise in § 3.

The last three sections are concerned with geometric properties of these Bäcklund transformations. We define in § 4 a Riemannian reductive space $G_k^{N,p}$ by introducing an action of a certain subgroup $O_k(N, p)$ of $Gl(N + p, \mathbb{R})$ and an $O_k(N, p)$-invariant Riemannian metric on the Grassmannian manifold of oriented p-dimensional linear subspaces of $\mathbb{R}^{N+p}$. In the case $p = 1$, $G_k^{N,1}$ is a simply connected space form of constant curvature k. Assuming that $D = \mathrm{diag}(a, \dots, a, b, \dots, b)$ (with "a" appearing p times) is a diagonal matrix and $k = b^2 - a^2$, we associate with any pair $(A_0, \varphi_0) \in M(\delta, D)$ mappings g and G of Y into $G_k^{N,p}$ and $O_k(N, p)$, respectively, and similarly mappings f and F of Y into S^{2N-1} and $O(2N)$ with the following properties: (i) g is the composition of G with the natural projection of $O_k(N, p)$ onto $G_k^{N,p}$ and f is the composition of F with the natural projection of $O(2N)$ onto S^{2N-1} and (ii) on the open submanifold of Y consisting of those points where both g and f are regular they induce the same Riemannian metric. Then we prove in § 5 a result which is similar to the superposition formula: If (A_1, φ_1) is obtained from (A_0, φ_0) by a Bäcklund transformation and if F_1 and G_1 are the mappings associated to (A_1, φ_1), then there exist matrices $Q \in O(2N)$ and $Q_1 \in O_k(N, p)$ such that $F_1 = F A_1^T Q A_0$ and $G_1 = G A_1^T Q_1 A_0$, where the multiplication of the $O(N)$-matrices A_i with matrices of $O_k(N, p)$ and $O(2N)$ is defined by identifying $O(N)$ with certain subgroups of $O_k(N, p)$ and $O(2N)$, respectively. In the case $p = 1$, $N = n$, $Y = \mathbb{R}^n$ and where δ is a special diagonal matrix, the first of these equations implies that there exists a geodesic congruence as defined in [7] between the submanifolds of S^{2n-1} defined as images of the mappings f and f_1.

Finally, as an application, we use these transformation formulas in § 6 to construct local isometric immersions of n-dimensional space forms into the Euclidean sphere S^{2n-1}. We assume for instance $0 < b < 1$ and find the following example of an immersion $f : (u^1, u^2) \in \mathbb{R}^2 - \{u^2 = 0\} \rightarrow (x^1, \ldots, x^4) \in S^3$, whose image is a submanifold of constant curvature $k = b^2$:

$$x^1 = \sin(u^1) \tanh(bu^2),$$
$$x^2 = - \cos(u^1) \tanh(bu^2),$$
$$x^3 = - \cos\left((1 - b^2)^{1/2} u^2\right) \Big/ \cosh(bu^2),$$
$$x^4 = \sin\left((1 - b^2)^{1/2} u^2\right) \Big/ \cosh(bu^2).$$

§ 2. Matrix Riccati equations and Bäcklund transformations

Consider the general linear Group $Gl(2N, \mathbb{R})$ and the space $M(N)$ of square matrices of order N. If an element $H \in Gl(2N, \mathbb{R})$ is decomposed into 4 blocks of type $N \times N$

$$(2.1) \qquad H = \begin{pmatrix} H_{11} & H_{12} \\ H_{21} & H_{22} \end{pmatrix}$$

we define the subset $K(N) \in Gl(2N, \mathbb{R}) \times M(N)$ by the condition $(H, A_0) \in K(N)$ if and only if $A_0 H_{12} + H_{22}$ is regular.

We define the mapping $(H, A_0) \rightarrow A_0 * H$ of $K(N)$ into $M(N)$ by

$$(2.2) \qquad A_0 * H = (A_0 H_{12} + H_{22})^{-1}(A_0 H_{11} + H_{21})$$

and observe that if for a second matrix $K \in Gl(2N, \mathbb{R})$ all of the three matrices $A_0 * H$, $(A_0 * H) * K$ and $A_0 * (HK)$ are defined, then the following equation holds

$$(2.3) \qquad A_0 * (HK) = (A_0 * H) * K.$$

Therefore, if $\widetilde{V} \subset Gl(2N, \mathbb{R})$ and $V \subset M(N)$ are subgroups with $\widetilde{V} \times V \subset K(N)$ and if $A_0 * H \in V$ for any element $(H, A_0) \in \widetilde{V} \times V$, then the equation (2.2) defines a group action of $\widetilde{V}$ on V. There exist several examples of such

pairs $(\tilde{V}, V)$, for instance $\tilde{V} = O(N, N)$ — the pseudo-orthogonal group
of index N and V =the manifold underlying the orthogonal group $O(N)$.
Other examples are obtained by taking the subgroup $O(N, C)$ of $O(N, N)$
or $\tilde{V}$ =the subgroup of $Gl(2N, \mathbb{R})$ consisting of those matrices H such that
the H_{ij} in (2.1) are upper triangular matrices with positive diagonal entries.
V is $O(N)$ in the first case and the group of diagonal matrices with positive
diagonal elements in the second case. In this paper we will only be concerned
with the case of $\tilde{V} = O(N, N)$ and $V = O(N)$, except in Proposition 2.1.

The action map (2.2) transforms linear Pfaffian systems into some kind
of Riccati systems as follows: Assume that Y is an arbitrary simply connected n-dimensional smooth manifold and denote by $\tilde{C}_N^0(Y)$ and $C_N^0(Y)$
the spaces of all C^∞-mappings of Y into $\tilde{V}$ and V respectively. Similarly,
let $\tilde{C}_N^1(Y)$ and $C_N^1(Y)$ denote the spaces of all 1-forms on Y with values
in the Lie algebras $\tilde{\mathfrak{v}}$ and $\mathfrak{v}$ of $\tilde{V}$ and V respectively. We have the natural
map $H \in \tilde{C}_N^0(Y) \to \tau = H^{-1}dH \in \tilde{C}_N^1(Y)$ and, conversely, an element
$\tau \in \tilde{C}_N^1(Y)$ is integrable, i.e., of the form $\tau = H^{-1}dH$, if and only if

(2.4) $d\tau + \tau \wedge \tau = 0.$

If (2.4) is satisfied, then for any initial value $(y_0, H_0) \in Y \times \tilde{V}$ there exists
a unique map $H \in \tilde{C}_N^0(Y)$ with $\tau = H^{-1}dH$ and $H(y_0) = H_0$. Moreover, if
$X \in V$, then the mapping $A = X * H$ satisfies the following Pfaffian Riccati
system

(2.5) $dA = -A\tau_{12}A - \tau_{22}A + A\tau_{11} + \tau_{21},$

where the τ_{ij} are the 4 $N \times N$-blocks of τ.

Proposition 2.1. *Assume that Y, $\tilde{V}$ and V are as above, that D_1 and D_2
are two matrices and δ_1 and δ_2 two matrix-valued 1-forms of type $N \times N$,
that*

(2.6) $d\delta_1 = d\delta_2 = 0$ and $\delta_1 \wedge \delta_2 = \delta_2 \wedge \delta_1 = 0$

and that for any pair $(A_0, \varphi_0) \in C_N^0(Y) \times C_N^1(Y)$ the new form

(2.6a) $\tau(A_0, \varphi_0) = \tau = \begin{pmatrix} \varphi_0 & , & \delta_1 A_0^{-1} D_1 \\ D_2 A_0 \delta_2 & , & 0 \end{pmatrix}$

has values in $\tilde{\mathfrak{v}}$. Then

(i) *the integrability condition (2.4) is satisfied if*

$$(2.7) \qquad d\varphi_0 + \varphi_0 \wedge \varphi_0 + \delta_1 A_0^{-1} D_1 D_2 A_0 \wedge \delta_2 = 0,$$

$$(2.8) \qquad \delta_1 \wedge A_0^{-1} dA_0 + \varphi_0 \wedge \delta_1 = 0$$

and

$$(2.9) \qquad A_0^{-1} dA_0 \wedge \delta_2 + \delta_2 \wedge \varphi_0 = 0.$$

(ii) *Assume furthermore that τ satisfies (2.4), that $H \in \widetilde{C}_N^0(Y)$ is any solution of the system $dH = H\tau$, that $X_1 \in V$ and that $A_1 \in C_N^0(Y)$ and the matrix valued 1-form φ_1 are defined by*

$$(2.10) \qquad A_1 = X_1 * H$$

and

$$(2.11) \qquad \varphi_1 = A_0^{-1} dA_0 - \delta_2 A_1^{-1} D_2 A_0 + A_0^{-1} D_1 A_1 \delta_1.$$

Then the pair (A_1, φ_1) satisfies the equations (2.7), (2.8) and (2.9) in place of (A_0, φ_0) and D_1 with D_2 and δ_1 with δ_2 interchanged. The mapping A_1 satisfies the Pfaffian Riccati system

$$(2.12) \qquad dA_1 = -A_1 \delta_1 A_0^{-1} D_1 A_1 + A_1 \varphi_0 + D_2 A_0 \delta_2.$$

(iii) *In the case $\widetilde{V} = O(N, N)$, $V = O(N)$, $\delta_1 = \delta_2 = \delta$, $D_1 = D_2 = D$ and δ and D symmetric the 1-form τ has values in $\widetilde{\mathfrak{v}} = \mathfrak{o}(N, N)$ and φ_1 has values in $\mathfrak{v} = \mathfrak{o}(N)$.*

Remark. The assumptions (iii) will be valid throughout this paper beginning with the next section.

Proof. (i) is a direct verification as well as (iii). For instance (2.12) is a special case of (2.5). Some computations are necessary in order to prove (ii): At first, exterior multiplication of (2.11) with δ_1 from the left and with δ_2 from the right hand side yields

$$(2.13) \qquad \varphi_1 \wedge \delta_2 + \delta_2 \wedge \varphi_0 = -\delta - 2A_1^{-1} D_2 A_0 \wedge \delta_2$$

$$(2.14) \qquad \delta_1 \wedge \varphi_1 + \varphi_0 \wedge \delta_1 = \delta_1 A_0^{-1} D_1 A_1 \wedge \delta_1$$

— making use of (2.6), (2.8) and (2.9). Solving the equation (2.12) for φ_0, we obtain

$$(2.15) \qquad \varphi_0 = A_1^{-1} dA_1 + \delta_1 A_0^{-1} D_1 A_1 - A_1^{-1} D_2 A_0 \delta_2$$

and exterior multiplication of this with δ_2 from the left and with δ_1 from right hand side gives

$$(2.16) \qquad \delta_2 \wedge A_1^{-1} dA_1 + A_0^{-1} dA_0 \wedge \delta_2 = \delta_2 \wedge A_1^{-1} D_2 A_0 \delta_2,$$
$$(2.17) \qquad \delta_1 \wedge A_0^{-1} dA_0 + A_1^{-1} dA_1 \wedge \delta_1 = -\delta_1 A_0^{-1} D_1 A_1 \delta_1$$

— making again use of (2.6), (2.8) and (2.9).

Adding the equations (2.13) and (2.16), we obtain

$$(2.18) \qquad \varphi_1 \wedge \delta_2 + \delta_2 \wedge A_1^{-1} dA_1 = 0$$

and the sum of (2.14) with (2.17) is

$$(2.19) \qquad \delta_1 \wedge \varphi_1 + A_1^{-1} dA_1 \wedge \delta_1 = 0.$$

These are the equations (2.8) and (2.9) with (A_1, φ_1) in place of (A_0, φ_0) and δ_1 with δ_2 interchanged. We still have to prove the equation

$$(2.20) \qquad d\varphi_1 + \varphi_1 \wedge \varphi_1 + \delta_2 A_1^{-1} D_2 D_1 A_1 \wedge \delta_1 = 0.$$

From (2.11) we obtain, that $d\varphi_1 + \varphi_1 \wedge \varphi_1$ is the sum of the following 14 2-forms

$$\psi_1 = -A_0^{-1} dA_0 \wedge A_0^{-1} dA_0, \qquad \psi_2 = -\delta_2 \wedge A_1^{-1} dA_1 A_1^{-1} D_2 A_0,$$
$$\psi_3 = \delta_2 A_1^{-1} D_2 \wedge dA_0, \qquad \psi_4 = -A_0^{-1} dA_0 A_0^{-1} D_1 A_1 \wedge \delta_1,$$
$$\psi_5 = A_0^{-1} D_1 dA_1 \wedge \delta_1, \qquad \psi_6 = A_0^{-1} dA_0 \wedge A_0^{-1} dA_0,$$
$$\psi_7 = -A_0^{-1} dA_0 \wedge \delta_2 A_1^{-1} D_2 A_0, \qquad \psi_8 = A_0^{-1} dA_0 \wedge A_0^{-1} D_1 A_1 \wedge \delta_1,$$
$$\psi_9 = -\delta_2 A_1^{-1} D_2 A_0 \wedge A_0^{-1} dA_0, \qquad \psi_{10} = \delta_2 A_1^{-1} D_2 A_0 \wedge \delta_2 A_1^{-1} D_2 A_0,$$
$$\psi_{11} = -\delta_2 A_1^{-1} D_2 A_0 \wedge A_0^{-1} D_1 A_1 \delta_1, \qquad \psi_{12} = A_0^{-1} D_1 A_1 \delta_1 \wedge A_0^{-1} dA_0,$$
$$\psi_{13} = -A_0^{-1} D_1 A_1 \delta_1 \wedge \delta_2 A_1^{-1} D_2 A_0, \qquad \psi_{14} = A_0^{-1} D_1 A_1 \delta_1 \wedge A_0^{-1} D_1 A_1 \delta_1.$$

We have obviously $\psi_1 + \psi_6 = 0$, $\psi_3 + \psi_9 = 0$ and $\psi_4 + \psi_8 = 0$. Moreover,

$$\psi_2 + \psi_7 + \psi_{10} =$$
$$= (-\delta_2 \wedge A_1^{-1} dA_1 - A_0^{-1} dA_0 \wedge \delta_2 + \delta_2 A_1^{-1} D_2 A_0 \wedge \delta_2) A_1^{-1} D_2 A_0$$

and

$$\psi_5 + \psi_{12} + \psi_{14} =$$
$$= A_0^{-1} D_1 A_1 (A_1^{-1} dA \wedge \delta_1 + \delta_1 \wedge A_0^{-1} dA_0 + \delta_1 \wedge A^{-1} D_1 A_1 \delta_1)$$

and therefore these sums vanish by (2.16) and (2.17). The only remaining term is ψ_{11}. This proves (2.20). ∎

Example 1. Under the special assumption $N = n$, $Y = \mathbb{R}^n$, $V = O(N)$, $\widetilde{V} = O(N, N)$, $D_1 = D_2 = D = \mathrm{diag}(a, a, \ldots, a, b, \ldots, b)$, where the real numbers a and b appear p and $(n - p)$-times respectively in the diagonal, and $\delta_1 = \delta_2 = \delta = \mathrm{diag}(du^1, \ldots, du^n)$, where $u^1, \ldots, u^n$ are the canonical coordinates on $\mathbb{R}^n$, the equations (2.8) and (2.9) reduce to the same one

$$(2.21) \qquad dA_0 \wedge \delta + A_0 \delta \wedge \varphi_0 = 0.$$

If A_0 has the entries a_{ij}, then this equation implies that the entries φ_{ij} of the matrix φ_0 are of the form

$$\varphi_{ij} = -A_{ij} du^j + A_{ji} du^i$$

and that the functions A_{ij}, which are defined only for $i \neq j$, and the a_{ij} are subject to the condition

$$(2.22) \qquad \partial_i a_{lj} = a_{li} A_{ij}, \qquad i \neq j, \qquad \partial_i = \partial/\partial u^i.$$

Therefore, φ_0 is uniquely determined by A_0. The equation (2.7) can be reduced to the following system of differential equations

$$(2.23) \qquad \partial_i A_{ij} + \partial_j A_{ji} + \sum_{l \neq i, j} A_{li} A_{lj} = (a^2 - b^2) \sum_{i=1}^{p} a_{li} a_{lj},$$

$$(2.24) \qquad \partial_l A_{ji} = A_{jl} A_{li} \qquad (l \neq i \neq j \neq 1).$$

The equations (2.22), (2.23) and (2.24) are the generalized wave equation of [7] in the case $a = b$ and the generalized sine-Gordon equation in the case $p = 1$ and $a^2 - b^2 = 1$ of the paper [9]. Equation (2.12) is the Bäcklund transformation derived in [7] and [9] from pseudo-spherical and geodesic congruences.

Example 2. Proposition 2.1 provides a method for the construction of several high dimensional systems of nonlinear differential equations and Bäcklund transformations of them. If $(V, \widetilde{V})$ is a pair with $\widetilde{V} \subset Gl(2N, \mathbb{R})$

and $V \subset Gl(N, \mathbb{R})$ as defined at the beginning of this section, then the results of § 2 and the following remain valid for $(\widetilde{V}, V)$ in place of $(O(N, N), O(N))$, provided that V is a subgroup and δ_1, δ_2, D_1, D_2 are chosen in such a way that φ_1 defined by (2.11) is a 1-form with values in the Lie algebra of V and that τ has values in the Lie algebra of $\widetilde{V}$. A simple example is the following: Assume that $n = N = 2$, $Y = \mathbb{R}^2$ and that V is the group of all matrices

$$A_0 = \begin{pmatrix} 1 & a_0 \\ 0 & 1 \end{pmatrix}, \qquad a_0 \in C^\infty(\mathbb{R}^2).$$

Then φ_{12} is the only entry of φ_0 which differs from 0. Choose $\widetilde{V}$ as the group of all matrices H in $Gl(4, \mathbb{R})$ such that the blocks H_{ij} in (2.1) are upper triangular matrices with positive diagonal entries. Finally choose

$$D_1 = D_2 = D = \begin{pmatrix} r & s \\ 0 & t \end{pmatrix}, \qquad \delta_1 = \delta_2 = \delta = \begin{pmatrix} du^1 & df \\ 0 & du^2 \end{pmatrix},$$

where f is a function on $\mathbb{R}^2$ satisfying $\partial_1 f + \partial_2 f = 0$ according to (2.6). The equations (2.8) and (2.9) lead to

$$\varphi_{12} = -\partial_2 a_0 \, du^1 - \partial_1 a_0 \, du^2$$

and (2.7) is equivalent to the equation

$$\partial_1^2 a_0 - \partial_2^2 a_0 = -(r^2 \partial_2 f + t^2 \partial_1 f) + (r^2 - t^2) a_0 + (r + t) s.$$

Then equation (2.12) the provides the Bäcklund transformation of this latter equation which takes the form

$$\partial_1 a_1 = -\partial_2 a_0 - r a_1 + t a_0 - r \partial_2 f - t \partial_1 f - s,$$
$$\partial_2 a_1 = -\partial_1 a_0 - t a_1 + r a_0 + r \partial_2 f + t \partial_1 f + s.$$

§ 3. The Superposition Formula

In this section we prove a general version of the superposition formula of [7] and [9]. We shall assume that Y^n is an arbitrary simply connected manifold as in § 2 and that D_i and δ_i ($i = 1, 2$) are symmetric matrices of order N. The only additional assumption is that $D_1 = D_2 = D$ and

$\delta_1 = \delta_2 = \delta$. Then we denote by $M(\delta, D)$ the set of solutions of the equations (2.7)...(2.8). The results of § 2 show that under a definite choice of initial values $H(y_0) \in O(N, N)$ and $X_1 \in O(N)$ there is a well defined map

$$(3.1) \qquad B_{X_1, D, \delta} : M(\delta, D) \to M(\delta, D),$$

namely, $B_{X_1, D, \delta}(A_0, \varphi_0) = (A_1, \varphi_1)$, where A_1, φ_1 are given by (2.10) and (2.11). We shall assume the initial value $H(y_0) = E_{2N}$ — the identity matrix. Iterations of such maps can be considered even with different matrices D, for it is easily seen from (2.7) that $M(\delta, D_1) = M(\delta, D_2)$ whenever

$$(3.2) \qquad D_1^2 - D_2^2 = r E_N$$

with some real number r. This implies that D_1 and D_2 commute.

We shall assume during this section, that $D_1 \neq D_2$ are two symmetric matrices such that $D^2 - D_i^2 = r_i E_N$ with certain real numbers $r_1 \neq r_2$. In this case $B_{X_i, D_i, \delta}$ $(i = 1, 2)$ are two different transformations of $M(\delta, D)$.

Proposition 3.1. *If the symmetric matrices D_1 and D_2 are as above, then for any symmetric matrix valued 1-form δ and any two initial values $X_1, X_2 \in O(N)$ the two Bäcklund transformations $B_{X_i, D_i, \delta}$ commute in the following sense: For any $(A_0, \varphi_0) \in M(\delta, D)$ with initial value $X_0 = A(y_0)$ holds*

$$(3.3) \qquad B_{X_3, D_2, \delta} \circ B_{X_1, D_1, \delta}(A_0, \varphi_0) = B_{X_3, D_1, \delta} \circ B_{X_2, D_2, \delta}(A_0, \varphi_0),$$

where

$$(3.4) \qquad X_3 = (X_2^{-1} D_1 - X_1^{-1} D_2)^{-1}(X_1^{-1} D_1 - X_2^{-1} D_2) X_0.$$

Moreover, if $(A_3, \varphi_3) = B_{X_3, D_2, \delta}(B_{X_1, D_1, \delta}(A_0, \varphi_0))$ then

$$(3.5) \qquad A_3 = (A_j^{-1} D_i - A_i^{-1} D_j)^{-1}(A_i^{-1} D_i - A_j^{-1} D_j) A_0$$

where

$$(3.6) \qquad (i, j) = (1, 2) \qquad or \qquad (i, j) = (2, 1)$$

and $(A_i, \varphi_i) = B_{X_i, D_i, \delta}(A_0, \varphi_0)$.

Proof. Denote $(A_{ij}, \varphi_{ij}) = B_{X_3, D_j, \delta}(A_i, \varphi_i)$. We have $A_i = X_i * H_i$, where H_i is the solution of the system $dH_i = H_i \tau_i$ with initial value $H(y_0) = E_{2N}$, where

$$(3.7) \qquad \tau_i = \begin{pmatrix} \varphi_0 & , & \delta A_0^{-1} D_i \\ D_i A_0 \delta & , & 0 \end{pmatrix}$$

and

$$(3.8) \qquad \varphi_i = A_0^{-1} dA_0 - \delta A_i^{-1} D_i A_0 + A_0^{-1} D_i A_i \delta.$$

Furthermore, $A_{ij} = X_3 * H_{ij}$, where H_{ij} is the solution of the system $dH_{ij} = H_{ij} \tau_{ij}$ with initial value $H_{ij}(y_0) = E_{2N}$ and

$$(3.9) \qquad \tau_{ij} = \begin{pmatrix} \varphi_i & , & \delta A_i^{-1} D_j \\ D_j A_i \delta & , & 0 \end{pmatrix}.$$

Finally,

$$(3.10) \qquad \varphi_{ij} = A_i^{-1} dA_i - \delta a_{ij}^{-1} D_j A_i + A_i^{-1} D_j A_{ij} \delta.$$

The crucial part of the proof is to show that H_{ij} can be expressed in terms of matrix operations by H_j and A_i. We are going to prove that the function

$$(3.11) \qquad \widetilde{H}_{ij} = H_j \begin{pmatrix} A_i^{-1} & 0 \\ 0 & E_N \end{pmatrix} \Delta_{ij} \begin{pmatrix} A_0 & 0 \\ 0 & E_N \end{pmatrix}$$

is a solution of the system

$$(3.12) \qquad d\widetilde{H}_{ij} = \widetilde{H}_{ij} \tau_{ij}.$$

The matrix $\Delta_{ij} \in Gl(2N, \mathbb{R})$ is given here by

$$(3.13) \qquad \Delta_{ij} = \begin{pmatrix} D_i & -D_j \\ -D_j & D_i \end{pmatrix}.$$

Our assumption (3.2) implies that Δ_{ij} is invertible and

$$(3.14) \qquad \Delta_{ij}^{-1} = \frac{(-1)^j}{r} \begin{pmatrix} D_i & D_j \\ D_j & D_i \end{pmatrix}.$$

Let us prove (3.12). For the sake of a simple notation we will identify $O(N)$ with the subgroup of $O(N, N)$ consisting of all matrices of type

$$\begin{pmatrix} A & 0 \\ 0 & E_N \end{pmatrix}, \qquad A \in O(N).$$

In a similar way $\mathfrak{o}(N)$ is identified with the corresponding Lie subalgebra of $\mathfrak{o}(N, N)$. Then (3.11) is written as $\widetilde{H}_{ij} = H_j A_i^{-1} \Delta_{ij} A_0$. Differentiating this equation, we get

$$(3.15) \qquad \widetilde{H}_{ij}^{-1} d\widetilde{H}_{ij} = A_0^{-1} \Delta_{ij}^{-1} A_i \tau_j A_i^{-1} \Delta_{ij} A_0 - $$
$$- A_0^{-1} \Delta_{ij}^{-1} dA_i A_i^{-1} \Delta_{ij} A_0 + A_0^{-1} dA_0$$

and it is sufficient to show that this expression is equal to τ_{ij}. For this purpose let us introduce the matrices

$$(3.16) \qquad K_{ij} = \begin{pmatrix} A_j \delta A_0^{-1} D_j - D_j A_0 \delta A_j^{-1} & , & A_j \delta A_0^{-1} D_i \\ D_i A_0 \delta A_j^{-1} & , & 0 \end{pmatrix}$$

and

$$(3.17) \qquad L_{ij} = \begin{pmatrix} D_i A_i \delta A_0^{-1} - A_0 \delta A_i^{-1} D_i & , & A_0 \delta A_i^{-1} D_j \\ D_j A_i \delta A_0^{-1} & , & 0 \end{pmatrix}$$

Direct computations show that

$$(3.18) \qquad \tau_i = A_j^{-1} dA_j + A_j^{-1} K_{ij} A_j,$$
$$(3.19) \qquad \tau_{ij} = A_0^{-1} dA_0 + A_0^{-1} L_{ij} A_0$$

and

$$(3.20) \qquad K_{ji} = \Delta_{ij} L_{ij} \Delta_{ij}^{-1},$$

where it must be observed that we obtain for τ_i from (2.15) the equivalent expression

$$\tau_i = \begin{pmatrix} A_j^{-1} dA_j + \delta A_0^{-1} D_j A_j - A_j^{-1} D_j A_0 \delta & , & \delta A_0^{-1} D_i \\ D_i A_0 \delta & , & 0 \end{pmatrix}$$

which has to be used instead of (3.7). The three equations (3.18)...(3.20) show that (3.15) is equal to τ_{ij}.

The function $\tilde{A}_{ij} = X_j * \tilde{H}_{ij}$ is by (3.11) and (2.2) the following:

$$(3.21) \qquad \tilde{A}_{ij} = (A_j^{-1} D_i - A_i^{-1} D_j)^{-1} (A_i^{-1} D_i - A_j^{-1} D_j) A_0$$

where (i,j) is subject to (3.6). This shows $\tilde{A}_{ij} = \tilde{A}_{ji}$. Moreover, we have $\tilde{A}_{ij}(y_0) = X_3$, i.e., the initial values of $\tilde{A}_{ij}$ and A_{ij} are the same and this shows that $\tilde{A}_{ij} = A_{ij} = A_3$.

We have still to prove that $\varphi_{ij} = \varphi_{ji}$. One can derive from (3.21) and (3.14) the equivalent expression

$$(3.22) \qquad A_{ij} = (D_j A_i + D_i A_j)(D_j A_j + D_i A_i)^{-1} A_0$$

and this gives two corresponding expressions for the inverse matrix:

$$(3.23) \qquad A_{ij}^{-1} = A_0^{-1}(D_i A_i - D_j A_j)(D_i A_j - D_j A_i)^{-1} =$$
$$= A_0^{-1}(A_i^{-1} D_i + A_j^{-1} D_j)^{-1} (A_i^{-1} D_j + A_j^{-1} D_i).$$

According to (2.15) we have

$$\varphi_0 = A_i^{-1} dA_i + \delta A_0^{-1} D_i A_i - A_i^{-1} D_i A_0 \delta$$

and a similar formula with j in place of i, where the pair (i,j) is as in (3.6). Taking the difference of the two corresponding equations, we obtain

$$(3.24) \qquad 0 = A_i^{-1} dA_i - A_j^{-1} dA_j - \delta A_0^{-1}(D_i A_i - D_j A_j) +$$
$$+ (A_j^{-1} D_j - A_i^{-1} D_i) A_0 \delta.$$

On the other hand, the difference $\varphi_{ij} - \varphi_{ji}$ is by (3.10)

$$(3.25) \qquad \varphi_{ij} - \varphi_{ji} = A_i^{-1} dA_i - A_j^{-1} dA_j + \delta A_{ij}^{-1}(D_i A_j - D_j A_i) +$$
$$+ (A_j^{-1} D_i + A_i^{-1} D_j) A_{ij} \delta.$$

If we substitute now the expressions for A_{ij} and A_{ij}^{-1} taken from (3.21) and (3.23) into (3.25) we see that the right hand sides of (3.24) and (3.25) are equal. Therefore $\varphi_{ij} - \varphi_{ji} = 0$ is proved. ∎

§ 4. Associated mappings into Graßmannian manifolds

We denote by $\widehat{G}_{p,N}$ the Grassmannian manifold of oriented p-dimensional linear subspaces of $\mathbb{R}^{N+p}$. Let us define the Riemannian reductive space $G_k^{N,p} \subset \widehat{G}_{p,N}$ as follows:

Denote by $S_k^{N,p} \in Gl(N + p, \mathbb{R})$ the diagonal matrix $S_k^{N,p} = \operatorname{diag}(k^{-1},\ldots,k^{-1},1,\ldots,1)$ with the real number k^{-1} appearing p times in the diagonal, and by $O_k(N, p)$ the subgroup of $Gl(N+p, \mathbb{R})$ consisting of all matrices α such that

$$(4.1) \qquad \alpha^T S_k^{N,p} \alpha = S_k^{N,p}.$$

In the case $k = 0$ when $S_k^{N,p}$ is not defined, $O_k(N, p)$ is defined as the subgroup of those $\alpha \in Gl(N + p, \mathbb{R})$ which have a block decomposition

$$(4.2) \qquad \alpha = \begin{pmatrix} \alpha_{11} & \alpha_{12} \\ \alpha_{21} & \alpha_{22} \end{pmatrix}$$

with $\alpha_{12} = 0$, $\alpha_{11} = O(p)$ and $\alpha_{22} \in O(N)$. We define the manifold $G_k^{N,p}$ as the orbit of the subspace $\widehat{X}_0 \in \widehat{G}_{p,N}$, spanned by the first p vectors of the canonical basis of $\mathbb{R}^{N+p}$ under the action of $O_k(N, p)$. Let us denote by $\pi : O_k(N, p) \to G_k^{N,p}$ the canonical projection given by $\pi(\alpha) = \alpha \widehat{X}_0$. It assigns to any matrix α the p-dimensional subspace spanned by the first p column vectors of α with the orientation defined by the order of these vectors. As a subset of $\widehat{G}_{p,N}$ the manifold $G_k^{N,p}$ is described as follows:

(i) if $k > 0$ then $G_k^{N,p} = \widehat{G}_{p,N}$.

(ii) if $k = 0$ then $G_k^{N,p}$ consists of all p-dimensional subspaces of $\mathbb{R}^{N+p}$ which are transversal to the orthogonal complement $\widehat{X}_0^{\perp}$ of $\widehat{X}_0$. It is therefore an affine space.

(iii) if $k < 0$ then the theorem of Witt implies that $G_k^{N,p}$ consists of all p-dimensional subspaces $\widehat{X}$ of $\mathbb{R}^{N+p}$ such that the restriction of the scalar product

$$(4.3) \qquad \langle x, y \rangle = k^{-1} \sum_{i=1}^{p} x^i y^i + \sum_{k=p+1}^{N+p} x^k y^k$$

on $\mathbb{R}^{N+p}$ to $\widehat{X}$ is negative definite. The isotropy subgroup H of the element $\widehat{X}_0 \in G_k^{N,p}$ is the subgroup of those $\alpha \in O_k(N, p)$ for which

$\alpha_{12} = 0$ and $\alpha_{21} = 0$ in the block decomposition (4.2), i.e., $H = O(p) \times O(N)$.

The Lie algebra $\mathfrak{o}_k(N, p)$ of $O_k(N, p)$ splits into the direct sum

$$(4.4) \qquad\qquad \mathfrak{o}_k(N, p) = \mathfrak{h} \oplus \mathfrak{m},$$

where $\mathfrak{h}$ is the Lie algebra of H and $\mathfrak{m}$ consists of all matrices $\alpha \in \mathfrak{o}_k(N, p)$ for which $\alpha_{11} = 0$ and $\alpha_{22} = 0$ in the block decomposition (4.2).

It is easily seen that $\mathfrak{m}$ is invariant under the adjoint action of H on $\mathfrak{o}_k(N, p)$. Moreover, the quadratic form

$$(4.5) \qquad\qquad Q(\alpha) = \text{tr}(\alpha_{21}^T \alpha_{21})$$

on $\mathfrak{m}$ is positive definite and also invariant under the adjoint action of H. Therefore, $G_k^{N,p}$ is a Riemannian reductive space. The invariant metric ds^2 of $G_k^{N,p}$ is characterized by the following property: For any manifold $g = \pi \circ G : Y^n \to G_k^{N,p}$ is an immersion, the induced metric $g_*(ds^2)$ is given by

$$(4.6) \qquad\qquad g_*(ds^2) = \text{tr}(\alpha_{21}^T \alpha_{21}),$$

where in this case α_{21} denotes the corresponding block of the $\mathfrak{o}_k(N, p)$-valued 1-form $\alpha = G^{-1} dG$ on Y^n.

Let us consider for the remaining part of this section the situation of example 1 of § 2, i.e., $D_1 = D_2 = D = \text{diag}(a, \ldots, a, b, \ldots, b)$, however, the manifold Y^n and the 1-form δ on Y as well as the dimensions n and N may still be arbitrary, but Y^n simply connected and δ symmetric. Assume that

$$(4.7) \qquad\qquad k = b^2 - a^2 < 1.$$

We are going to associate with any pair $(A, \varphi) \in M(\delta, D)$ two integrable 1-forms

$$(4.8) \qquad\qquad \mu = \mu(A, \varphi) : TY^n \to \mathfrak{o}(2N)$$

and

$$(4.9) \qquad\qquad \nu = \nu(A, \varphi) : TY^n \to \mathfrak{o}_k(N, p)$$

such that the maps F_A and G_A of Y^n into $O(2N)$ and $O_k(N, p)$ obtained as solutions of the linear Pfaffian systems

$$(4.10) \qquad dG_A = G_A \nu(A, \varphi) \qquad \text{and} \qquad dF_A = F_A \mu(A, \varphi)$$

respectively have the following property: Consider the mappings $g_A = \pi \circ G_A$ and $f_A = \pi_1 \circ F_A$, where $\pi_1 : O(2N) \to G_1^{2N-p,p}$ is the canonical projection and suppose that $U \subset Y^n$ is an open subset such that the restrictions of g_A and f_A to U are immersions. Denote by ds_1^2 the invariant metric tensor of $G_1^{2N-p,p}$. Then g_A and f_A will induce the same Riemannian metric on U, i.e.,

$$(4.11) \qquad (g_A)_*(ds^2) = (f_A)_*(ds_1^2).$$

In order to define μ and ν assume

$$(4.12) \qquad \tilde{D} = \mathrm{diag}\left(1,\ldots,1,(1-k)^{1/2},\ldots,(1-k)^{1/2}\right)$$

(where the entries 1 and $(1-k)^{1/2}$ appear p and $N-p$ times respectively). Then we have

$$(4.13) \qquad \tilde{D}^2 + D^2 = (1+a^2)E_N.$$

Define the matrix $W = W(A)$ by

$$(4.14) \qquad W = \delta A^T \tilde{D}$$

and decompose W into blocks α of type $N \times p$ and β of type $N \times (N-p)$ according to

$$(4.15) \qquad W = (\alpha, \beta).$$

Then we define

$$(4.16) \qquad \mu = \begin{pmatrix} 0 & -\alpha^T & 0 \\ \alpha & \varphi & \beta \\ 0 - \beta^T & 0 \end{pmatrix} \qquad \text{and} \qquad \nu = \begin{pmatrix} 0 & -k\alpha^T \\ \alpha & \varphi \end{pmatrix}.$$

Proposition 4.1. *For any* $(A, \varphi) \in M(\delta, D)$ *the 1-forms* μ *and* ν *defined above are integrable.*

Proof. We obtain from (4.16) the following matrices for $d\mu + \mu \wedge \mu$ and $d\nu + \nu \wedge \nu$ — the "curvature tensors" of μ and ν:

$$(4.17) \qquad d\nu + \nu \wedge \nu = \begin{pmatrix} -k\alpha^T \wedge \alpha & , & -k(d\alpha^T + \alpha^T \wedge \varphi) \\ d\alpha + \varphi \wedge \alpha & , & -k\alpha \wedge \alpha^T + d\varphi + \varphi \wedge \varphi \end{pmatrix}$$

and

(4.18) $d\mu + \mu \wedge \mu =$

$$\begin{pmatrix} -\alpha^T \wedge \alpha & , & -d\alpha^T - \alpha^T \wedge \varphi & , & -\alpha^T \wedge \beta \\ d\alpha + \varphi \wedge \alpha & , & d\varphi + \varphi \wedge \varphi - \alpha \wedge \alpha^T - \beta \wedge \beta^T & , & d\beta + \varphi \wedge \beta \\ -\beta^T \wedge \alpha & , & -d\beta^T - \beta^T \wedge \varphi & , & -\beta^T \wedge \beta \end{pmatrix}.$$

We derive from (2.6), (2.7), (4.14), (4.15) and (4.16) the three following expression of $W \wedge W^T$. At first

(4.19) $W \wedge W^T = \delta A^T \tilde{D}^2 A \wedge \delta = -\delta A^T D^2 A \wedge \delta = d\varphi + \varphi \wedge \varphi.$

Secondly,

(4.20) $W \wedge W^T = \delta A^T \tilde{D}^2 A \wedge \delta = \delta A^T (\tilde{D}^2 - (1-k)E_N)A \wedge \delta =$

$$= \delta A^T \begin{pmatrix} k & 0 \\ 0 & 0 \end{pmatrix} A \wedge \delta = k\alpha^T \wedge \alpha,$$

and finally

(4.21) $W \wedge W^T = (\alpha, \beta) \begin{pmatrix} \alpha^T \\ \beta^T \end{pmatrix} = \alpha \wedge \alpha^T + \beta \wedge \beta^T.$

This proves that $d\varphi + \varphi \wedge \varphi = \alpha \wedge \alpha^T + \beta \wedge \beta^T = k\alpha \wedge \alpha^T.$

Next, (2.6) implies that $W^T \wedge W = 0$. Therefore

(4.22) $W^T \wedge W = \begin{pmatrix} \alpha^T \\ \beta^T \end{pmatrix} (\alpha, \beta) = \begin{pmatrix} \alpha^T \wedge \alpha & , & \alpha^T \wedge \beta \\ \beta^T \wedge \alpha & , & \beta^T \wedge \beta \end{pmatrix} = 0,$

i.e., $\alpha^T \wedge \alpha = \beta^T \wedge \alpha = \alpha^T \wedge \beta = \beta^T \wedge \beta = 0.$

It remains to show that $d\alpha + \varphi \wedge \alpha = 0$ and $d\beta + \varphi \wedge \beta = 0$. But these two latter equations can be fit together into the one equation $dW + \varphi \wedge W = 0$. Since $d\delta = 0$ by (2.6) and $\tilde{D}$ is invertible, this is the same as (2.8). ∎

This proposition shows the existence of the mappings G_A, F_A, g_A and f_A defined above. They are uniquely determined up to the choice of initial values. The definition (4.17) of μ and ν implies directly that the condition (4.11) is satisfied. We have

(4.23) $(g_A)_*(ds^2) = (f_A)_*(ds_1^2) = \mathrm{tr}(\alpha_1^T \alpha_1).$

In the case $n = Np$ the manifolds Y^n and $G_k^{N,p}$ have equal dimensions and if $U \subset Y^n$ is an open subset such that g_A is a diffeomorphism of U onto some open subset $\tilde{U} \subset G_k^{N,p}$ then $\tilde{U}$ is immersed isometrically into $G_1^{2N,p}$ under the mapping $f_A \circ g_A^{-1}$, for the equation (4.23) implies that f_A is regular at all points of Y^n where g_A is regular. If we consider the special case of $p = 1$, then we have $G_k^{N,1} = H_k^N$ — the simply connected complete space form of constant curvature k. Therefore, this procedure gives local isometric immersions of H_k^N into $S^{2N-1} = G_1^{2N,1}$. In the following two sections we shall develop a method for the explicit construction of such mappings F_A and G_A by Bäcklund transformations.

§ 5. The behavior of the mappings G_A and F_A under Bäcklund transformations

Assume that a pair $(A_0, \varphi_0) \in M(\delta, D)$ is transformed into a new one (A_1, φ_1) by means of a Bäcklund transformation as in (3.1), i.e.

$$(5.1) \qquad\qquad (A_1, \varphi_1) = B_{X_1, D, \delta} (A_0, \varphi_0).$$

We denote by $\mu_i = \mu(A_i, \varphi_i)$ and $\nu_i = \nu(A_i, \varphi_i)$ the corresponding 1-forms defined as in (4.17), by W_i the matrices

$$(5.2) \qquad\qquad W_i = \delta A_i^T \tilde{D}$$

and by G_i and F_i solutions of the Pfaffian systems $dG_i = G_i \nu_i$ and $dF_i = F_i \mu_i$ for $i = 0, 1$.

Proposition 5.5. *Under the assumptions made above there exist matrices $Q \in O(2N)$ and $Q_1 \in O_k(N, p)$ depending on a and b with the restriction that Q_1 is defined only for $b \neq 0$, such that*

$$(5.3) \qquad F_1 = F_0 \begin{pmatrix} E_p & 0 & 0 \\ 0 & A_1^T & 0 \\ 0 & 0 & E_{N-p} \end{pmatrix} Q \begin{pmatrix} E_p & 0 & 0 \\ 0 & A_0 & 0 \\ 0 & 0 & E_{N-p} \end{pmatrix}$$

and

$$(5.4) \qquad\qquad G_1 = G_0 \begin{pmatrix} E_p & 0 \\ 0 & A_1^T \end{pmatrix} Q_1 \begin{pmatrix} E_p & 0 \\ 0 & A_0 \end{pmatrix}.$$

Proof. Let us state at first the matrices Q and Q_1. They are

$$(5.5) \qquad Q = (1+a)^{-1/2} \begin{pmatrix} aE_p & E_p & 0 & 0 \\ -E_p & aE_p & 0 & 0 \\ 0 & 0 & bE_{N-p} & -cE_{N-p} \\ 0 & 0 & cE_{n-p} & bE_{N-p} \end{pmatrix}$$

where $c = (1-k)^{1/2}$, i.e., $c^2 + b^2 = 1 + a^2$, and

$$(5.6) \qquad Q_1 = b^{-1} \begin{pmatrix} aE_p & kE_p & 0 \\ -E_p & aE_p & 0 \\ 0 & 0 & bE_{N-p} \end{pmatrix}.$$

We introduce the notations

$$(5.7) \quad \widetilde{F}_1 = F_1 \begin{pmatrix} E_p & 0 & 0 \\ 0 & A_0^T & 0 \\ 0 & 0 & bE_{N-p} \end{pmatrix}, \qquad \widetilde{F}_0 = F_0 \begin{pmatrix} E_p & 0 & 0 \\ 0 & A_1^T & 0 \\ 0 & 0 & E_{N-p} \end{pmatrix},$$

$$(5.8) \qquad \widetilde{G}_1 = G_1 \begin{pmatrix} E_p & 0 \\ 0 & A_0^T \end{pmatrix} \quad \text{and} \quad \widetilde{G}_0 = G_0 \begin{pmatrix} E_p & 0 \\ 0 & A_1^T \end{pmatrix}.$$

Then we have to prove that

$$(5.9) \qquad \widetilde{F}_1 = \widetilde{F}_0 Q \quad \text{and} \quad \widetilde{G}_1 = \widetilde{G}_0 Q_1.$$

Introducing the 1-forms

$$(5.10) \qquad \widetilde{\mu}_i = \widetilde{F}_i^{-1} d\widetilde{F}_i, \qquad \widetilde{\nu}_i = \widetilde{G}_i^{-1} d\widetilde{G}_i \qquad (i = 0, 1),$$

the proof can be reduced to the verification of the equations

$$(5.11) \qquad \widetilde{\mu}_1 = Q^{-1} \widetilde{\mu}_0 Q \quad \text{and} \quad \widetilde{\nu}_1 = Q_1^{-1} \widetilde{\nu}_0 Q_1,$$

which are equivalent to (5.9). We infer from (5.7) and (5.10) the following special appearance of $\widetilde{\mu}_i$ and $\widetilde{\nu}_i$:

$$(5.12) \qquad \widetilde{\mu}_i = \begin{pmatrix} 0 & -\widetilde{\alpha}_i^T & 0 \\ \widetilde{\alpha}_i & \widetilde{\varphi}_i & \widetilde{\beta}_i \\ 0 & -\widetilde{\beta}_i^T & 0 \end{pmatrix}$$

and

$$(5.13) \qquad \widetilde{\nu}_i = \begin{pmatrix} 0 & k\widetilde{\alpha}_i^T \\ \widetilde{\alpha}_i & \widetilde{\varphi}_i \end{pmatrix},$$

where

$$(5.14) \qquad \widetilde{\varphi}_i = A_j \varphi_i A_j^{-1} + A_j dA_j^{-1} \qquad (i,j) = (0,1) \text{ or } = (1,0)$$

and $\widetilde{\alpha}_i$ and $\widetilde{\beta}_i$ are the blocks of type $N \times p$ and $N \times (N-p)$ respectively of the matrix valued 1-form

$$(5.15) \qquad \widetilde{W}_i = A_j W_i = (\widetilde{\alpha}_i, \widetilde{\beta}_i) = A_j \delta A_I^T \widetilde{D},$$

where again $(i,j) = (0,1)$ or $= (1,0)$. Let us state here for a later application the obvious equation

$$(5.16) \qquad \widetilde{W}_i = \widetilde{D}^{-1} \widetilde{W}_j^T \widetilde{D}.$$

In order to prove (5.11) the matrices (5.12) and (5.13) have to be made more explicit by introducing the finer block decompositions

$$(5.17) \qquad \widetilde{\mu}_i = \begin{pmatrix} 0 & -\widetilde{\alpha}_{i1}^T & -\widetilde{\alpha}_{i2}^T & 0 \\ \widetilde{\alpha}_{i1} & \widetilde{\varphi}_{i,11} & \widetilde{\varphi}_{i,12} & \widetilde{\beta}_{i1} \\ \widetilde{\alpha}_{i2} & \widetilde{\varphi}_{i,21} & \widetilde{\varphi}_{i,22} & \widetilde{\beta}_{i2} \\ 0 & -\widetilde{\beta}_{i1}^T & -\widetilde{\beta}_{i2}^T & 0 \end{pmatrix}$$

and

$$(5.18) \qquad \widetilde{\nu}_i = \begin{pmatrix} 0 & -k\widetilde{\alpha}_{i1}^T & -k\widetilde{\alpha}_{i2}^T \\ \widetilde{\alpha}_{i1} & \widetilde{\varphi}_{i,11} & \widetilde{\varphi}_{i,12} \\ \widetilde{\alpha}_{i2} & \widetilde{\varphi}_{i,21} & \widetilde{\varphi}_{i,22} \end{pmatrix}$$

which are adapted to the block structures of Q and Q_1 respectively, i.e., the matrices appearing in the diagonal of $\widetilde{\mu}_i$ have type $p \times p$, $p \times p$, $(N-p) \times (N-p)$ and $(N-p) \times (N-p)$ in this order.

Note that

$$(5.19) \qquad \widetilde{W}_i = \begin{pmatrix} \widetilde{\alpha}_{i1} & \widetilde{\beta}_{i1} \\ \widetilde{\alpha}_{i2} & \widetilde{\beta}_{i2} \end{pmatrix}.$$

Performing now the matrix multiplication we find that the first of the equations (5.11) is equivalent to the following 3 groups of equations

$$(1a) \qquad 0 = a\widetilde{\alpha}_{01}^T - a\widetilde{\alpha}_{01} + \widetilde{\varphi}_{0,11},$$
$$(1b) \qquad 0 = -ac\widetilde{\alpha}_{02} + c\widetilde{\varphi}_{0,21} + b\widetilde{\beta}_{01}^T,$$
$$(1c) \qquad 0 = ac\widetilde{\alpha}_{02}^T + c\widetilde{\varphi}_{0,12} - b\widetilde{\beta}_{01},$$
$$(1d) \qquad 0 = c^2\widetilde{\varphi}_{0,22} - bc\widetilde{\beta}_{0,2} + bc\widetilde{\beta}_{02}^T,$$

(2a) $\tilde{\alpha}_{11} = (1 + a^2)^{-1} (\tilde{\alpha}_{01}^T + a^2 \tilde{\alpha}_{01} - a\tilde{\varphi}_{0,11})$,

(2b) $\tilde{\alpha}_{12} = (1 + a^2)^{-1} (ab\tilde{\alpha}_{02} - b\tilde{\varphi}_{0,21} + c\tilde{\beta}_{01}^T)$,

(2c) $\tilde{\beta}_{11} = (1 + a^2)^{-1} (c\tilde{\alpha}_{02}^T - ac\tilde{\varphi}_{0,12} + ab\tilde{\beta}_{01})$,

(2d) $\tilde{\beta}_{12} = (1 + a^2)^{-1} (-bc\tilde{\varphi}_{0,22} + b^2 \tilde{\beta}_{02} + c^2 \tilde{\beta}_{02}^T)$,

(3a) $\tilde{\varphi}_{1,11} = (1 + a)^{-1} (-a\tilde{\alpha}_{01}^T + a\tilde{\alpha}_{01} + a^2 \tilde{\varphi}_{0,11})$,

(3b) $\tilde{\varphi}_{1,21} = (1 + a)^{-1} (b\tilde{\alpha}_{02} + ab\tilde{\varphi}_{0,21} - ac\tilde{\beta}_{01}^T)$,

(3c) $\tilde{\varphi}_{1,12} = (1 + a)^{-1} (-b\tilde{\alpha}_{02}^T + ab\tilde{\varphi}_{0,12} + ac\tilde{\beta}_{01})$,

(3d) $\tilde{\varphi}_{1,22} = (1 + a)^{-1} (b^2 \tilde{\varphi}_{0,22} + bc\tilde{\beta}_{02} - bc\tilde{\beta}_{02}^T)$.

Solving the equations (1a)...(1d) for the $\tilde{\varphi}_{0,ij}$, we get

(4a) $\tilde{\varphi}_{0,11} = a(\tilde{\alpha}_{01} - \tilde{\alpha}_{01}^T)$,   (4b) $\tilde{\varphi}_{0,21} = a\tilde{\alpha}_{02} - \dfrac{b}{c}\tilde{\beta}_{01}^T$,

(4c) $\tilde{\varphi}_{0,12} = -a\tilde{\alpha}_{02}^T + \dfrac{b}{c}\tilde{\beta}_{01}$,   (4d) $\tilde{\varphi}_{0,22} = \dfrac{b}{c}(\tilde{\beta}_{02} - \tilde{\beta}_{02}^T)$.

The equations (4a)...(4d) are equivalent to the matrix equation

(5.20) $\tilde{\varphi}_0 = \widetilde{W}_0 \tilde{D}^{-1} D - D\tilde{D}^{-1} \widetilde{W}_0^T$

and by (5.14) and (5.15) this is seen to be equivalent to (2.15).

Substituting the expressions (4a–d) for the $\tilde{\varphi}_{0,ij}$ into (2a–d) and (3a–d), we obtain

(5a) $\tilde{\alpha}_{11} = \tilde{\alpha}_{01}^T$,   (5b) $\tilde{\alpha}_{12} = \dfrac{1}{c}\tilde{\beta}_{01}^T$,

(5c) $\tilde{\beta}_{11} = c\tilde{\alpha}_{02}^T$,   (5d) $\tilde{\beta}_{12} = \tilde{\beta}_{01}^T$

and

(6a) $\tilde{\varphi}_{1,11} = a(\tilde{\alpha}_{01} - \tilde{\alpha}_{01}^T)$,

(6b) $\tilde{\varphi}_{1,21} = b\tilde{\alpha}_{02} - \dfrac{a}{c}\tilde{\beta}_{01}^T$,

(6c) $\tilde{\varphi}_{1,12} = -b\tilde{\alpha}_{02}^T + \dfrac{a}{c}\tilde{\beta}_{01}$,

(6d) $\tilde{\varphi}_{1,22} = \dfrac{b}{c}(\tilde{\beta}_{02} - \tilde{\beta}_{02}^T)$.

The equations (5a–d) are equivalent to (5.16) while the equations (6a–d) can be fit together to the matrix equation

(5.21) $\tilde{\varphi}_1 = D\widetilde{W}_0 \tilde{D}^{-1} - \tilde{D}^{-1} \widetilde{W}_0^T D$

and using (5.14) and (5.15) this is seen to be the same as (2.11).

Remark. Note that (4a), (4b), (6a), and (6d) show that

(5.22) $$\tilde{\varphi}_{1,11} = \tilde{\varphi}_{0,11} \quad \text{and} \quad \tilde{\varphi}_{1,22} = \tilde{\varphi}_{0,22}.$$

We are now going to prove the second of the equations (5.11). Performing again the matrix multiplication, we find that this equation is equivalent to the following set of 7 equations:

(7) $$0 = ak\tilde{\alpha}_{01}^T - ak\tilde{\alpha}_{01} + k\tilde{\varphi}_{0,11},$$

(8a) $$\tilde{\alpha}_{11} = b^{-2}(k\tilde{\alpha}_{01}^T + a\tilde{\alpha}_{01} - a\tilde{\varphi}_{0,11}),$$
(8b) $$\tilde{\alpha}_{12} = b^{-2}(ab\tilde{\alpha}_{02} - b\tilde{\varphi}_{0,21}),$$

(9a) $$\tilde{\varphi}_{1,11} = b^{-2}(-ak\tilde{\alpha}_{01}^T + ak\tilde{\alpha}_{01} + a^2\tilde{\varphi}_{0,11}),$$
(9b) $$\tilde{\varphi}_{1,21} = b^{-2}(bk\tilde{\alpha}_{02} + ab\tilde{\varphi}_{0,21}),$$
(9c) $$\tilde{\varphi}_{1,12} = b^{-2}(-bk\tilde{\alpha}_{02}^T + ab\tilde{\varphi}_{0,12}),$$
(9d) $$\tilde{\varphi}_{1,22} = \tilde{\varphi}_{0,22}.$$

We see that (7) is a consequence of (4a). Substituting $\tilde{\varphi}_{0,11}$ and $\tilde{\varphi}_{0,21}$ from (4a) and (4b) into (8c) and (8b) we obtain (5a) and (5b), i.e., (8a) and (8b) are valid.

Finally, substituting (4a–d) into (9a–d) respectively, we obtain the equations (6a–d). ∎

§ 6. Local isometric immersions of n-dimensional space forms into S^{2n-1} associated to 1-soliton solutions

We are going to discuss the mappings g_A and f_A constructed in § 4 for special choices of solutions of the system (2.7), (2.8), (2.9). Let us assume $p = 1$, $N = n$, $Y^n = \mathbb{R}^n$, $\delta = \text{diag}(du^1, \ldots, du^n)$, where $u^1, \ldots, u^n$ are the canonical coordinates of $\mathbb{R}^n$, and $D = \text{diag}(a, b, \ldots, b)$.

Denote again $k = b^2 - a^2$. Under these assumptions the pair $(A_0, \varphi_0) = (E_n, 0)$, where E_n denotes the identity matrix of order n, is a solution of this system. We call its image under a Bäcklund transformation $(A_1, \varphi_1) = B_{X,D,\delta}(A_0, \varphi_0)$ the 1-soliton solution.

The mappings G_{A_0} and F_{A_0} are obtained in the following way.

At first the forms μ_{A_0} and ν_{A_0} have to be established. If we denote by μ_{ij} $(i, j = 1, \ldots, 2n)$ the entries of μ_{A_0} and by ν_{ij} $(i, j = 1, \ldots, n+1)$ the entries of ν_{A_0}, then $\mu_{12} = -du^1 = -\mu_{21}$, $\mu_{i+1,i+n} = cdu^i = -\mu_{i+n,i+1}$ for $i = 2, \ldots, n$, where $c = (1-k)^{1/2}$, $\nu_{12} = -kdu^1$ and $\nu_{21} = du^1$ are the only elements of these matrices which are different from zero. As solutions of the systems $dF_{A_0} = F_{A_0}\mu_{A_0}$ and $dG_{A_0} = G_{A_0}\nu_{A_0}$ we get

$$
(6.1) \qquad F_{A_0} = \begin{pmatrix} \cos u^1 & -\sin u^1 & 0 & 0 \\ \sin u^1 & \cos u^1 & 0 & 0 \\ 0 & 0 & \cos c\overline{V} & \sin c\overline{V} \\ 0 & 0 & -\sin c\overline{V} & \cos c\overline{V} \end{pmatrix}
$$

where $\overline{V}$ is the diagonal matrix $\overline{V} = \mathrm{diag}(u^2, \ldots, u^n)$, and G_{A_0} is simply the exponential of the matrix $u^1\tilde{\nu}$, where $\tilde{\nu}$ is the coefficient of du^1 in the 1-form ν_{A_0}, i.e., the entries of $\tilde{\nu}$ are $\tilde{\nu}_{12} = -k$, $\tilde{\nu}_{21} = 1$ and $\tilde{\nu}_{ij} = 0$ for all other indices.

The matrix A_1 and the 1-form φ_1 can be computed by explicit integration of the system $dH = H\tau$. By (2.6.a) the 1-form $\tau = \tau(A_0, \varphi_0)$ is

$$
(6.2) \qquad \tau(A_0, \varphi_0) = \begin{pmatrix} 0 & \delta D \\ \delta D & 0 \end{pmatrix}
$$

and therefore, the solution H with initial value $H(0) = E_{2n}$ is

$$
(6.3) \qquad H = \begin{pmatrix} \cosh(DV) & \sinh(DV) \\ \sinh(DV) & \cosh(DV) \end{pmatrix},
$$

where $V = \mathrm{diag}(u^1, \ldots, u^n)$. If $X_1 \in O(n)$ is an arbitrary orthogonal matrix, then by (2.10) and (2.11) the general 1-soliton is given by

$$
(6.4) \qquad A_1 = (X_1 \sinh(DV) + \cosh(DV))^{-1}(X_1 \cosh(DV) + \sinh(DV))
$$

$$
(6.5) \qquad \varphi_1 = DA_1\delta - \delta A_1^T D.
$$

This shows that A_1 and φ_1 can be made explicit. The entries of A_1 are rational functions of $e^{au^1}, e^{bu^2}, \ldots, e^{bu^n}$. Therefore, the matrices F_{A_1} and G_{A_1} can also be computed explicitly by means of the rules (5.3) and (5.4), provided that $b \neq 0$.

Since $p = 1$ and $N = n$, in the case $k \geq 0$ the homogeneous space $G_k^{N,p}$ is isometric to H_k^n, the simply connected complete n-dimensional space form of constant curvature k, and to the disjoint union of two copies of H_k^n in the case $k < 0$. We will identify $G_k^{n,1}$ with the subset of $\mathbb{R}^{n+1}$ defined by the quadratic equation

$$(6.6) \qquad x^T S_k^{n,1} x = k^{-1}, \qquad x \in \mathbb{R}^{n+1}, \quad k \neq 0$$

and with the Euclidean space if $k = 0$ (compare [5]). The canonical projection π of $O_k(n,1)$ onto $G_k^{n,1}$ then assigns to any matrix of $O_k(n,1)$ its first column. The following gives the explicit view of the mappings g_{A_1} and f_{A_1}.

Proposition 6.1. *Under the assumptions made above the mappings g_{A_1} and f_{A_1} of $\mathbb{R}^n$ into H_k^n and S^{2n-1} respectively, which are associated to the 1-soliton solutions (A_1, φ_1), are the following:*

(i) $f_{A_1}(u^1, \ldots, u^n) = (x^1, \ldots, x^{2n}) \in S^{2n-1}$ *with*

$$(6.7) \qquad \begin{aligned} x^1 &= (1 + a^2)^{-1/2} (a \cos(u^1) + a_{11} \sin(u^1)), \\ x^2 &= (1 + a^2)^{-1/2} (a \sin(u^1) - a_{11} \cos(u^1)), \\ x^{2i-1} &= -(1 + a^2)^{1/2} \cos\left((1 - k)^{1/2} u^i\right) a_{1i} \\ x^{2i} &= (1 + a^2)^{-1/2} \sin\left((1 - k)^{1/2} u^i\right) a_{1i} \end{aligned} \quad\left.\vphantom{\begin{aligned} x \\ x \\ x \\ x \end{aligned}}\right\} \; i = 2, \ldots, n$$

where $a_{11}, \ldots, a_{1n}$ denote the entries for the first row of A_1.

(ii) *if $b \neq 0$ then $g_{A_1}(u^1, \ldots, u^n) = (y^0, y^1, \ldots, y^n) \in \mathbb{R}^{n+1}$ with*

$$(6.8) \qquad y^i = -b^{-1} a_{1i} \qquad \text{for} \quad i = 2, \ldots, n$$

and if $k > 0$ then

$$(6.9) \qquad \begin{aligned} y^0 &= ab^{-1} \cos(k^{1/2} u^1) + a_{11} b^{-1} k^{1/2} \sin(k^{1/2} u^1), \\ y^1 &= ab^{-1} k^{-1/2} \sin(k^{1/2} u^1) - a_{11} \cos(k^{1/2} u^1), \end{aligned}$$

if $k < 0$ then

$$(6.10) \qquad \begin{aligned} y^0 &= ab^{-1} \cosh\left((-k)^{1/2} u^1\right) - a_{11} b^{-1} (-k)^{1/2} \sinh\left((-k)^{1/2} u^1\right), \\ y^1 &= ab^{-1} (-k)^{-1/2} \sinh\left((-k)^{1/2} u^1\right) - a_{11} \cosh\left((-k)^{1/2} u^1\right), \end{aligned}$$

and finally, if $k = 0$ then

(6.11)
$$y^0 = ab^{-1}$$
$$y^1 = ab^{-1}u^1 - a_{11}b^{-1}.$$

(iii) *if $b = 0$ and $a \neq 0$ we have $k = -a^2$ and g_{A_1} is given by*

$$y^0 = 2^{-1}(p + p^{-1} + a^2 m^{-2}|a|^2 p),$$
$$y^1 = 2^{-1}a^{-1}(p - p^{-1} + a^2 m^{-1}|a|^2 p^{-1}),$$
$$y^i = x_{1i}u^i m^{-1} \qquad \text{for} \quad i = 2, \ldots, n$$

with the following meaning of symbols: (i) x_{ij} $(i, j = 1, \ldots, n)$ *are the entries of the matrix X_1 in (6.4), i.e., the initial value of A_1,* (ii) $m = x_{11}\sinh(au^1) + \cosh(au^1)$, (iii) $p = \cosh(au^1 + \rho)\cosh^{-1}\rho$ *and $\rho = -\ln(1 - x_{11})$ and finally a is the vector $a = (x_{12}u^2, x_{13}u^3, \ldots, x_{1n}u^n)^T$.*

Proof. The expressions of (i) and (ii) can be directly derived from the transformation rules (5.3) and (5.4) for the mappings F_{A_1} and G_{A_1}. Let us give a brief sketch of the proof of (iii). Since $b = 0$, one obtaines from (6.4) and (6.5) that the elements a_{1i} of A_1 are

$$a_{11} = m^{-1}(x_{11}\cosh(au^1) + \sinh(au^1))$$

and

$$a_{1i} = m^{-1}x_{1i} \qquad \text{for} \quad i = 2, \ldots, n,$$

and that the elements φ_{ij} of φ_1 are equal to zero except

$$\varphi_{1i} = -\varphi_{i1} = m^{-1}ax_{1i}du^i \qquad \text{for} \quad i = 2, \ldots, n.$$

Therefore, the matrix valued 1-form $\nu_{A_1} = \nu(A_1, \varphi_1)$ given by (4.17) can be established. One shows that the general solution of the system $dG_{A_1} = G_{A_1}\nu_{A_1}$ can be expressed by matrix exponentials, namely, if U_i is the matrix coefficient of du^i in ν_{A_1}, then the U_i depend only on the coordinate u^1 and the commutator-relations $\partial_i U_j - \partial_j U_i = [U_j, U_i]$, which are a consequence of the integrability condition of ν_{A_1} show that $[U_i, U_j] = 0$ for $i, j = 2, \ldots, n$ and $\partial_1 U_j = [U_j, U_1]$. These latter relations imply that with the notation

$$W = \int_0^{u^1} U_1(t)\, dt \qquad \text{and} \qquad V = u_2 U_2 + u_3 U_3 + \ldots + u_n U_n$$

the general solution G_{A_1} is

$$G_{A_1} = G_0 e^W e^V$$

where G_0 is an arbitrary matrix of $O_k(n, 1)$. The exponentials e^W and e^V can be computed easily. One finds for instance that $V^3 = 0$ and therefore $e^V = E_{n+1} + V + V^2/2$. In the case $G_0 = E_{n+1}$ the final result is given by the functions y^i of (iii) which make up the first column of G_{A_1}. ∎

In general, the mappings g_{A_1} are not diffeomorphisms. An exceptional case is that of $a = b = 0$. Here we have $A_1 = X_1$, i.e., A_1 is a constant mapping, and g_{A_1} is the linear mapping $(U^i) \to (x_{1i} u^i)$ of $\mathbb{R}^n$ into itself. If all the x_{1i} are different from zero then g_{A_1} is a linear isomorphism and $f_{A_1} \circ g_{A_1}^{-1}$ maps $\mathbb{R}^n$ isometrically onto a flat n-dimensional torus in S^{2n-1}.

Let us give an example of an isometric immersion of an open subset $U_1 \subset S_r^2$ of the 2-dimensional sphere of radius $r = k^{-1/2}$ into S^3, which is obtained by specializing the parameters appearing in the formulas of the mappings g_{A_1} and f_{A_1} of Proposition 6.1. We assume $n = 2$ and

(6.12)
$$X_1 = \begin{pmatrix} \cos \varphi & \sin \varphi \\ -\sin \varphi & \cos \varphi \end{pmatrix}$$

where φ is a real number $\in \mathbb{R} - \mathbb{Z}\pi$. Then (6.4) gives

(6.13)
$$a_{11} = \tanh(T + \rho) \quad \text{and} \quad a_{12} = \varepsilon / \cosh(T + \rho)$$

where

(6.14)
$$\begin{cases} T = au^1 + bu^2, & \rho = \ln|(1 + \cos \varphi)/\sin \varphi| \quad \text{and} \\ \varepsilon = \text{sign}(\sin \varphi). \end{cases}$$

If $0 < k < 1$, an isometry between $G_k^{n,1}$ and S_r^2 is given by the mapping

(6.15)
$$h : (y^0, y^1, y^2) \in G_k^{n,1} \to (z^0, z^1, z^2) = $$
$$= (k^{-1/2} y^0, y^1, y^2) \in S_r^2$$

i.e., by means of h we can identify $G_k^{n,1}$ with the standard sphere $(z^0)^2 + (z^1)^2 + (z^2)^2 = r^2 = 1/k$ of radius r in $\mathbb{R}^3$. Then the coordinates of the mapping g_{A_1} are

(6.15)
$$\begin{aligned} z^0 &= R_1 \cos(k^{1/2} u^1 + \psi_1), \\ z^1 &= R_1 \sin(k^{1/2} u^1 + \psi_1), \\ x^2 &= -a_{12}/b \end{aligned}$$

where $\psi_1 = \arg(ak^{-1/2} + ia_{11})$ and $R_1 = ((A^2 + ka_{11}^2)/kb^2)^{1/2}$.

Assuming still more special $b > 0$, $a = 0$ and $\varphi = \pi/2$, i.e., $k = b^2$, $\rho = 0$ and $T = k^{1/2}u^2$, the mappings g_{A_1} and f_{A_1} become

$$(6.16) \quad f_{A_1} : \begin{pmatrix} u^1 \\ u^2 \end{pmatrix} \to \begin{pmatrix} x^1 \\ x^2 \\ x^3 \\ x^4 \end{pmatrix} = \begin{pmatrix} \sin(u^1)\tanh(k^{1/2}u^2) \\ -\cos(u^1)\tanh(k^{1/2}u^2) \\ -\cos\left((1-k)^{1/2}u^2\right)/\cosh(k^{/2}u^2) \\ \sin\left((1-k)^{1/2}u^2\right)/\cosh(k^{/2}u^2) \end{pmatrix}$$

and

$$(6.17) \quad g_{A_1} : \begin{pmatrix} u^1 \\ u^2 \end{pmatrix} \to \begin{pmatrix} z^0 \\ z^1 \\ z^2 \end{pmatrix} = k^{1/2}\begin{pmatrix} \sin(k^{1/2}u^1)\tanh(k^{1/2}u^2) \\ -\cos(k^{1/2}u^1)\tanh(k^{1/2}u^2) \\ -1/\cosh(k^{1/2}u^2) \end{pmatrix}.$$

Let $U_1 \subset S_r^2$ be the open subset defined by the two inequalities (i) $-r < z^2 < 0$ and (ii) $z^0 \neq 0$ if $z^1 < 0$. Let $U_2 \subset \mathbb{R}^2$ one of the connected components of the set $\{(u^1, u^2) \mid u^2 \neq 0 \text{ and } k^{1/2}u^1 \neq 2p\pi \text{ for } p = 0, \pm 1, \pm 2, \dots\}$.

Corollary 6.2. *The mappings g_{A_1} and f_{A_1} are regular on the subset $u^2 \neq 0$ of $\mathbb{R}^2$ and they induce the same Riemannian metric $a_{11}^2 (du^1)^2 + a_{12}^2 (du^2)^2$ on that subset. g_{A_1} maps any of these sets U_2 diffeomorphically onto U_1. If $g_{A_1}^{-1}$ denotes the inverse diffeomorphism of the restriction of g_{A_1} to U_2, then $f_{A_1} \circ g_{A_1}^{-1}$ is an isometric immersion of U_1 into S^3.*

References

[1] A. V. BÄCKLUND, Zur Theorie der Flächentransformationen, *Math. Ann.* **19** (1882), 387–422.

[2] L. BIANCHI, Vorlesungen über Differentialgeometrie, B. G. Teubner, Leipzig, (1910).

[3] E. CARTAN, Sur les variétés de courbure constante d'un espace euclidien ou non euclidien, *Bull. Soc. Math. France* **47** (1929), 125–160 and **48** (1920), 132–208.

[4] S. HELGASON, Differential Geometry and Symmetric Spaces, Mir, Moskow 1963 (russ. transl.)

[5] S. KOBAYASHI and K. NOMIZU, Foundations of Differential Geometry, Vol. 1, New York and London, 1963.

[6] J. D. MOORE, Isometric immersions of space forms into space forms, *Pac. J. Math.*, **40**, 1972.

[7] K. TENENBLAT, Bäcklund's theorem for submanifolds of space forms and a generalized wave equation, *Bol. Soc. Bras. de Mat.* **16** (1985).

[8] K. TENENBLAT and CH. L. TERNG, Bäcklund's theorem for n-dimensional submanifolds of $\mathbb{R}^{2n-1}$, *Annals of Math.*, **111** (1980), 477–490.

[9] CH. L. TERNG, A higher dimension generalization of the sine-Gordon equation and it's soliton theory, *Annals of Math.* **111** (1980), 491–510.

Hubert Gollek

Humboldt – Univ. zu Berlin
Sektion Mathematik
PSF 1297
1086 Berlin
Germany

COLLOQUIA MATHEMATICA SOCIETATIS JÁNOS BOLYAI

56. DIFFERENTIAL GEOMETRY, EGER (HUNGARY), 1989

Abstract Poisson Structures and Star-Products

J. GRABOWSKI

1. The basic algebraic structures of classical mechanics are the algebra $V = C^\infty(N)$ of smooth functions on the phase space N under ordinary multiplication and the Lie structure on V induced by the Poisson bracket $\{\,,\,\}$ defined by the symplectic form ω on N. In canonical coordinates $(q_1, \ldots, q_n, p_1, \ldots, p_n)$ we have

$$\omega = \sum_{i=1}^{n} dq_i \wedge dp_i \quad \text{and} \quad \{f, g\} = \sum_{i=1}^{n} \left(\frac{\partial f}{\partial q_i} \frac{\partial g}{\partial p_i} - \frac{\partial g}{\partial q_i} \frac{\partial f}{\partial p_i} \right).$$

In [1] F. Bayen, M. Flato et al. have attempted to study quantization as a deformation of classical mechanics. One then does not define quantum mechanics in terms of operators but in terms of deformations of the usual multiplication and the Poisson bracket of functions on the phase space. The appropriate deformation of the associative algebra structure on V is called a *star-product*. A minimal requirement for such an approach is the existence of star products on every symplectic manifold and it was proven by M. De Wilde and P. Lecomte ([4] and [5]) on the level of formal deformations.

Since one considers nowadays more general Poisson structures than those on symplectic manifolds (e.g. the local Lie algebras of A. A. Kirillov [7] or the Jacobi structures of A. Lichnerowicz [8]), the natural question is to

The research was supported by the Alexander von Humboldt-Stiftung.

extend the result of M. De Wilde and P. Lecomte on the general case. On the other hand, the proof of M. De Wilde and P. Lecomte does not give the star-product explicitly, what is of course necessary for applications. Practically, the only well-known formula for a star-product is the one given by J. E. Moyal [8] and J. Vey [10] for the Poisson structure on a flat symplectic manifold.

In this note we would like to present a purely algebraic approach to this question, which gives us a generalization of the Vey–Moyal product and can be applied in many nonsymplectic and nonregular cases.

2.   Our object under consideration will be *an associative commutative algebra V with unit $\underline{1}$ over a field K of characteristic zero.*

We have the adjoint representation of V defined by

$$V \ni x \mapsto H_x \in \mathrm{End}_K(V), \quad \text{where } H_x(y) = xy.$$

Now, following the ideas of A. M. Vinogradov [11], define the space $\mathrm{Diff}^r(V)$ of *linear differential operators on V of order $\leq r$* recurrently:

i) $\mathrm{Diff}^0(V) = \{H_x : x \in V\}$,

ii) $\mathrm{Diff}^{r+1}(V) = \{D \in \mathrm{End}_K(V) : [H_x, D] \in \mathrm{Diff}^r(V) \text{ for all } x \in V\}$,
   where $[H_x, D] = H_x \circ D - D \circ H_x$ is the usual commutator.

One can easily prove the following:

Theorem 1. $\mathrm{Diff}^1(V) = \mathrm{Der}(V) \oplus \mathrm{Diff}^0(V)$, *where $\mathrm{Der}(V)$ is the Lie algebra of derivations of V and the splitting is given by*

$$D \mapsto (D - H_{D(\underline{1})}) + H_{D(\underline{1})},$$

i.e. derivations of V are exactly those linear differential operators D on V which vanish on constants (i.e. $D(\underline{1}) = 0$).

Since for V being the algebra $C^\infty(N)$ of smooth functions on a given manifold N the derivations are known to be vector fields on N, it is not surprising that in this case the introduced notion of linear differential operators on V coincides with the usual one. The space $\mathrm{Diff}(V) = \bigcup_{r \geq 0} \mathrm{Diff}^r(V)$ is clearly an associative subalgebra of $\mathrm{End}_K(V)$.

We define the space $\mathrm{Diff}^r_n(V)$ of n-linear differential operators of order $\leq r$ in the obvious way. For instance, a two-linear $P : V \times V \to V$ is a differential operator of order $\leq r$ if and only if it is a differential operator of order $\leq r$ with respect to each variable, i.e. if and only if $P(x, \cdot), P(\cdot, x) \in \mathrm{Diff}^r(V)$ for each $x \in V$. Set $\mathrm{Diff}_n(V) = \bigcup_{r \geq 0} \mathrm{Diff}^r_n(V)$ for the space of all n-linear differential operators.

Following the idea of A. A. Kirillov [7], define a *Jacobi structure* on V as an antisymmetric two-linear differential operator P on V satisfying the Jacobi identity, i.e. making V into a Lie algebra. If P in addition vanishes on constants (i.e. $P(\underline{1}, \cdot) = 0$), then it is called a *Poisson structure*.

We have the following algebraic version of the Kirillov's theorem for local Lie algebras.

Theorem 2 ([6]). *If V has no non-zero nilpotent elements, then every Jacobi structure on V is a differential operator of order ≤ 1.*

In particular, in virtue of Theorem 1, every Poisson structure P on such an algebra is a derivation with respect to each variable. Observe now that every such a Poisson structure defines a formal deformation of the algebra V of rank 1.

To be precise, consider the space V_h of all formal power series $\sum_{k=0}^{\infty} h^k x_k$ in h with coefficients in V.

The two-$\mathcal{K}[[h]]$-linear operator $m : V_h \times V_h \to V_h$ defined on V (naturally embedded in V_h) by

$$m(x, y) = xy + hP(x, y), \qquad x, y \in V,$$

is associative up to the rank 1, i.e.

$$m(m(x, y), z) = m(x, m(y, z)) \pmod{(h^2)},$$

since P is a two-linear derivation.

Definition. *A star-product for a Poisson structure P on an associative commutative algebra (V, m_0) with unit $\underline{1}$ is a formal deformation (V_h, m_h) of (V, m_0) (i.e. m_h is a two-$\mathcal{K}[[h]]$-linear associative product on V_h) such that*

1) $m_h(x, y) = m_0(x, y) + hP(x, y) + \sum_{k=2}^{\infty} h^k A_k(x, y)$ *for $x, y \in V$,*

2) A_k is a differential operator on V,

3) $A_k(x, y) = (-1)^k A_k(y, x)$ for $x, y \in V$,

4) A_k vanishes on constants

for $k = 2, 3, \ldots$.

Note that the assumption 4) in the above definition assures that 1 remains the unit in the algebra (V_h, m_h).

The investigation of the existence of star-products for a given Poisson structure leads to difficult questions concerning the Hochschild and Cheval- ley cohomologies (see e.g. [4], [5]). There is however one case, when we are able to write a star-product explicitely.

Assume first that P is of the form

$$P(x, y) = \sum_{i,j=1}^{n} c_{ij} D_i(x) D_j(y),$$

where $c_{ij} = -c_{ij} \in \mathcal{K}$, $D_i \in \mathrm{Der}(V)$, $i, j = 1, 2, \ldots, n$.

This is not a very restrictive assumption, since for $V = C^\infty(N)$ it is always satisfied. Note also that P can be written in such a form in many different ways and since c_{ij} are *constants*, we can not get the uniqueness in any resonable way.

Denote by $\mathcal{U}$ the algebra $\mathrm{Diff}(V)$ of all (1-linear) differential operators on V. We have a natural mapping

$$J : \mathcal{U} \otimes_\mathcal{K} \mathcal{U} \to \mathrm{Diff}_2(V) \quad \text{given by} \quad J(A \otimes B)(x, y) = A(x)B(y),$$

so P can be written as $J(\widetilde{P})$ for $\widetilde{P} = \sum_{i,j=1}^{n} c_{ij} D_i \otimes D_j$.

Since $\mathcal{U} \otimes_\mathcal{K} \mathcal{U}$ is in the obvious way an associative algebra over $\mathcal{K}$, the powers $\widetilde{P}^k$ are well-defined and we have the following.

Theorem 3 ([6]). *If a Poisson structure P on an associative algebra (V, m_0) over a field $\mathcal{K}$ can be written in the form*

$$P(x, y) = \sum_{i,j=1}^{n} c_{ij} D_i(x) D_j(y),$$

where $c_{ij} = -c_{ji} \in \mathcal{K}$, $D_i \in \mathrm{Der}(V)$, and $[D_i, D_j] = 0$, $i, j = 1, \ldots, n$, then

$$J(\exp(h\widetilde{P})) = m_0 + hP + \sum_{k=2}^{\infty} h^k J(\widetilde{P}^k)/k!,$$

where $\widetilde{P} = \sum_{i,j=1}^{n} c_{ij} D_i \otimes D_j$, is a star-product for P.

3. Let us end with few examples.

Example 1. For the flat symplectic form $\omega = \sum_{i=1}^{n} dx_i \wedge dx_{i+n}$, the corresponding Poisson structure on $V = C^{\infty}(\mathbb{R}^{2n})$ is given by $J(\widetilde{P})$, where $\widetilde{P} = \sum_{i=1}^{n}(\partial_i \otimes \partial_{i+n} - \partial_{i+n} \otimes \partial_i)$, $\partial_i = \partial/\partial x_i$, $i = 1, \ldots, 2n$.

The vector fields $\{\partial_i : i = 1, \ldots, 2n\}$ obviously commute, so we get the classical Vey–Moyal product

$$J(\exp(h\widetilde{P})) = \sum_{k=0}^{\infty} h^k P^k / k!, \qquad \text{where}$$

$$P^k(x,y) = \sum_{\substack{i_1,\ldots,i_k=1 \\ j_1,\ldots,j_k=1}}^{2n} c_{i_1 j_1} \cdots c_{i_k j_k} \partial_{i_1} \cdots \partial_{i_k}(x)\partial_{j_1} \cdots \partial_{j_k}(y),$$

$c_{ij} = \mathrm{sgn}(j-i)\delta_{|j-i|}^n$ and δ is the Kronecker's symbol.

Example 2. Let $\mathfrak{g}$ be a n-dimensional Lie algebra over $\mathcal{K}$ of length two, i.e. such that $[x,[y,z]] = 0$ for all $x, y, z \in \mathfrak{g}$ (e.g. a Heisenberg algebra). Every basis $x_1 \ldots, x_n$ of $\mathfrak{g}$ can be regarded as a system of global coordinates on the dual $\mathfrak{g}^*$ and we can define the canonical Poisson structure P on $V = C^{\infty}(\mathfrak{g}^*)$ putting $P(x_i, x_j) = [x_i, x_j]$. The Kirillov's (generalized) foliation generated by the hamiltonian vector fields consists of orbits of the coadjoint representation of the corresponding Lie group.

It is convenient to choose the basis such that $x_{r+1}, \ldots, x_n$ is a basis of $\mathfrak{g}^{(1)} = \mathrm{span}\{[x,y] : x, y \in \mathfrak{g}\}$. Since $\mathfrak{g}^{(1)}$ lies in the centre of $\mathfrak{g}$, P is of the form $J(\widetilde{P})$ for

$$\widetilde{P} = \frac{1}{2} \sum_{i,j=1}^{r} ((z_{ij}\partial_i) \otimes \partial_j - \partial_j \otimes (z_{ij}\partial_i)),$$

where $\partial_i = \partial/\partial x_i$, $z_{ij} = [x_i, x_j] \in \mathfrak{g}^{(1)}$, $i, j = 1, \ldots, r$.

The vector fields ∂_i, $z_{ij}\partial_i$, $i,j = 1,\ldots,r$, commute, so we get a star-product for P in the exponential form $J(\exp(h\widetilde{P})) = \sum\limits_{k=0}^{\infty} h^k P^k/k!$. Since $\partial_k(z_{ij}) = 0$ for $i,j,k = 1,\ldots,r$, we have

$$P^k(x,y) = \sum_{\substack{i_1,\ldots,i_k=1 \\ j_1,\ldots,j_k=1}}^{r} z_{i_1 j_1} \ldots z_{i_k j_k} \partial_{i_1} \ldots \partial_{i_k}(x)\partial_{j_1} \ldots \partial_{j_k}(y).$$

Hence it is easily seen that the star-product reduces to a star-product on the algebra $\mathcal{A}$ of polynomial functions on $\mathfrak{g}^*$. For any $a \in \mathcal{K}$ we get a new associative multiplication "$\cdot_a$" on $\mathcal{A}$ putting

$$x \cdot_a y = \sum_{k=0}^{\infty} (a/2)^k P^k(x,y)/k!,$$

since the above sum is finite for polynomials $x, y \in \mathcal{A}$.

It is interesting that the algebra $(\mathcal{A}, \cdot_a)$ is in fact the universal enveloping algebra of $\mathfrak{g}$ with the bracket $[\,,\,]_a$, where $[\,,\,]_a = a[\,,\,]$ and $[\,,\,]$ is the original bracket on $\mathfrak{g}$.

Theorem 4. *The algebra $(\mathcal{A}, \cdot_a)$ is isomorphic to the universal enveloping $(\mathcal{U}_a, *_a)$ of the Lie algebra $(\mathfrak{g}, [\,,\,]_a)$.*

Proof. For simplicity fix $a \in \mathcal{K}$ and write "$\cdot$", "$*$", etc. instead of "$\cdot_a$", "$*_a$", etc. We will also regard $\mathfrak{g}$ as naturally embedded in $\mathcal{U}$.

It is well-known (see e.g. [3]) that $\mathcal{U}$ has a basis consisting of symmetric polynomials in (non-commutative) variables $x_1,\ldots,x_n$, so we have a linear isomorphism $I : \mathcal{A} \rightarrow \mathcal{U}$ such that I maps every polynomial from $\mathcal{A}$ (with the original commutative structure) into the corresponding symmetric polynomial, i.e. $I(xy) = (x * y + y * x)/2$ for $x, y \in \mathfrak{g}$ etc. We shall show that I is in fact an isomorphism of the associative algebras $(\mathcal{A}, \cdot)$ and $(\mathcal{U}, *)$. Since symmetric polynomials are linear combinations of powers of linear polynomials (see [2]), it suffices to show that $I(x^m \cdot y^s) = x^m * y^s$ for every $x, y \in \mathfrak{g}$, $m, s = 0, 1.\ldots..$ (The powers are taken with respect to the original multiplication in $\mathcal{A}$.)

For $x, y \in \mathfrak{g}$ and $s = 0, 1,\ldots$, we have

$$(s+1)(x * y^s) = \sum_{i=0}^{s} y^{s-i} * x * y^i + ((s+1)x * y^s - \sum_{i=0}^{s} y^{s-i} * x * y^i) =$$

$$= (s+1)I(xy^s) + \sum_{i=0}^{s}[x, y^{s-i}] * y^i$$

and since $[x, y]$ lies in the center of $\mathcal{U}$ ($\mathfrak{g}$ is of length 2), the corresponding polynomials are symmetric and we have

$$x * y^s = I(xy^s) + \frac{1}{s+1} \sum_{i=0}^{s} (s-i)[x, y] * y^{s-1} = I\left(xy^s + \frac{1}{2} a P(x, y^s)\right).$$

On the other hand

$$x \cdot y^s = \sum_{k=0}^{\infty} (a/2)^k P^k(x, y^s)/k! = xy^s + (a/2)P(x, y^s),$$

since $P^k(x, \cdot) = 0$ for $x \in \mathfrak{g}$ and $k > 1$ as can be easily seen from the form of P^k.

Therefore $I(x \cdot y^s) = x * y^s$ and hence

$$I(x \cdot f) = x * I(f) \quad \text{for every} \quad x \in \mathfrak{g}, \quad f \in \mathcal{A}.$$

This implies in particular that $x^s = x^{*s}$, where the last power is taken with respect to the multiplication "$*$" and, by the associativity of all products, that $I(g \cdot f) = I(g) * I(f)$ for $f, g \in \mathcal{A}$. ∎

Example 3. Consider an interesting Poisson structure P on the unit sphere $S^2 \subset \mathbb{R}^3$ defined in global coordinates (x, y, z) in $\mathbb{R}^3$ by

$$P(x, y) = (1 - z)^2 z, \quad P(z, x) = (1 - z)^2 y, \quad P(y, z) = (1 - z)^2 x,$$

so that the corresponding two-tensor field $\widetilde{P}$ can be written in the form $\widetilde{P} = (1 - z)^z (zY \wedge X + xY \wedge Z + yX \wedge Z)$, where $X = z\partial_y - y\partial_z$, $Y = z\partial_x - x\partial_z$, $Z = x\partial_y - y\partial_x$ are vector fields on S^2. P is degenerated at $(0, 0, 1)$ and the rest of S^2 forms one 2-dimensional symplectic leaf of the corresponding Kirillov's foliation. The vector fields X, Y, Z form the Lie algebra $\mathfrak{sl}(2, \mathbb{R})$, but at some effort one can write P using only commuting vector fields.

Namely $P = J(\widetilde{P})$, where $\widetilde{P} = A \otimes B - B \otimes A$ and

$$A = -xy\partial_x + (1 - z - y^2)\partial_y + (1 - z)y\partial_z,$$
$$B = (1 - z - x^2)\partial_x - xy\partial_y + (1 - z)x\partial_z$$

are tangent to S^2 and commuting vector fields.

Hence we get a star-product for P in the form $\sum_{k=0}^{\infty} h^k P^k/k!$, where

$$P^k(u, v) = \sum_{i=0}^{k} \binom{k}{i} (-1)^i A^{k-i} B^i(u) A^i B^{k-i}(v).$$

References

[1] BAYEN, F., FLATO, M., FRONSDAL, C., LICHNEROWICZ, A. and STERN-HEIMER, D.: Deformation theory and quantization, *Ann. Phys.* **111** (1978), 61–110 and 111–151.

[2] BOURBAKI, N.: Algèbre, Ch. I, §. 8, Hermann, Paris 1971.

[3] BOURBAKI, N.: Groupes et Algèbres de Lie, Ch. II. §. 11, Hermann, Paris 1972.

[4] DE WILDE, M. and LECOMTE, P. B. A.: Existence of star-products and of formal deformations of the Poisson Lie algebra of arbitrary symplectic manifolds, *Lett. Math. Phys.* **7** (1983), 487–496.

[5] DE WILDE, M. and LECOMTE, P. B. A.: Formal deformations of the Poisson algebra of a symplectic manifold and star-products. Existence, equivalence, derivations., in Deformation Theory of Algebras and Structures and Applications, Ed. M. Hazewinkel & M. Gerstenhaber, NATO ASI Ser. C., Vol **247**, 897–960, Kluwer Acad. Publ. 1988.

[6] GRABOWSKI, J.: Abstract Jacobi and Poisson structures. Quantization and star-products, to appear in *J. Geom. Phys.*

[7] KIRILLOV, A. A.: Local Lie algebras, *Russ. Mat. Surv.* **31** (1976), 55–75.

[8] LICHNEROWICZ, A.: Les variètés de Jacobi et leurs algèbres de Lie associèes, *J. Mat. Pures. Appl.* **57** (1978), 453–488.

[9] MOYAL, J. E.: Quantum mechanics as a statistical theory, *Proc. Camb. Phil. Soc.* **45** (1949), 99–124.

[10] VEY, J.: Déformations du crochet de Poisson d'une variété sympleqtique, *Comm. Mat. Helv.* **50** (1975), 421–454.

[11] VINOGRADOV, A. M.: The logic algebra for the theory of linear differential operators, *Sov. Mat. Dokl.* **13** (1972), 1058–1062.

Janusz Grabowski

Institute of Mathematics,
Warsaw University,
Banacha 2,
00-913 Warsaw 59, Poland

COLLOQUIA MATHEMATICA SOCIETATIS JÁNOS BOLYAI
56. DIFFERENTIAL GEOMETRY, EGER (HUNGARY), 1989

On the Harmonic Rank of a Riemannian Structure

Th. HANGAN

1.

Our starting point will be the well known conjecture of Lichnerowicz according to which the harmonic Riemannian manifolds are exactly the rank one symmetric spaces. We recall that a harmonic Riemannian space (M, g) can be characterized by the property that the Laplacian Δr_m of any distance function r_m (the geodesic distance from a fixed point $m \in M$) depends only on r_m, i.e.

$$(R_1) \qquad \Delta r_m = F_m(r_m) \qquad \forall\, m \in M.$$

Should this conjecture be true, this equation would characterize the rank one symmetric spaces.

In any case, if a symmetric space satisfies R_1, it is automatically of rank one by a theorem of Ledger. We are led therefore to look for equations R_k, $k \in N$, able to distinguish the rank k symmetric spaces in the class of all Riemannian symmetric spaces. Note that in R_1 one can replace r_m by

$$\Omega_m = \frac{1}{2} r_m^2$$

the advantage being to avoid singularities. In fact, one can use instead of r_m any positive power r_m^p because one has the formula

$$\Delta r_m^p = p(r_m^{p-1}\Delta r_m + (p-1)r_m^{p-2})$$

and therefore R_1 implies an analogous equation for r_m^p.

2.

As a first candidate for R_k, consider the exterior differential equation

$$(C_k) \qquad d\Omega_m \wedge d(\Delta_m\Omega) \wedge \ldots \wedge d(\Delta^k\Omega_m) = 0 \qquad \forall\, m \in M$$

coupled with the condition

$$(\overline{C}_{k-1}) \qquad d\Omega_m \wedge d(\Delta_m\Omega) \wedge \ldots \wedge d(\Delta^{k-1}\Omega_m) \neq 0.$$

To motivate this choice, lets look at an example of rank k symmetric space: the orthogonal group $SO(2k)$ endowed with its left and right invariant metric deduced from the Killing form. One knows that every regular element $g \in SO(2k)$ belongs to a maximal torus T_g and when g describes a suitable submanifold S transversal to T_g one can introduce local coordinates in the group and compute the Laplacian Δ in a convenient way. More precisely, for any regular $g \in SO(2k)$ one can find $s \in SO(2k)$ such that

$$s^{-1}\cdot g\cdot s = \begin{pmatrix} \cos\theta_1 & -\sin\theta_1 & 0 & \cdots & & \cdots & & 0 \\ \sin\theta_1 & \cos\theta_1 & & & & & & \\ 0 & & \cos\theta_2 & -\sin\theta_2 & & & & \\ \cdot & & \sin\theta_2 & \cos\theta_2 & & & 0 & \\ \cdot & & & & \ddots & & & \\ \cdot & & & & & \cos\theta_k & -\sin\theta_k \\ 0 & \cdots & & & 0 & & \sin\theta_k & \cos\theta_k \end{pmatrix}.$$

The standard maximal torus T^k in $SO(2k)$ consist of all the matrices of this kind. When restricted to class functions, i.e. to functions on $SO(2k)$ which are invariant with respect to the adjoint action of the group onto itself, the Laplacian is represented by an elliptic operator given only in terms of the toric variables $\theta_1, \theta_2, \ldots \theta_k$. Thus, for our example, the distance of the point g from the neutral element e equals the Euclidean distance

$$(\theta_1^2 + \theta_2^2 + \ldots + \theta_k^2)^{1/2}$$

and the restriction of the Laplacian Δ to class functions is the operator

$$\tilde{\Delta} = \sum_{i=1}^{k} \frac{\partial^2}{(\partial\theta_i)^2} + \sum_{i=1}^{k} \frac{\partial G}{\partial\theta_i} \frac{\partial}{\partial\theta_i}$$

where

$$G = \ln \prod_{i<j} |\cos\theta_i - \cos\theta_j|.$$

By successive application of $\tilde{\Delta}$ to class functions one obtains no more than k independent functions and therefore the exterior differential equation C_k is satisfied; the same is true in general for all rank k symmetric spaces.

As for the condition $\overline{C}_{k-1}$, its meaning were to detect those spaces which are of rank k and not less. We conjecture that condition $\overline{C}_{k-1}$ is satisfied by all the rank k symmetric spaces. For the specific case of the group $SO(2k)$ and for small values of k this can be verified by direct computations based on the above formulas.

3.

From the joint paper [1], devoted to the study of the C_2 condition in the frame of three-dimensional Riemannian manifolds recall here the following formula which gives the second Laplacian of a power r_m^p of the distance function

$$\Delta^2 r_m^p = p\{\Delta r_m \times \Delta r_m^{p-1} + r_m^{p-1} \times \Delta^2 r_m + 2(p-1)r_m^{p-2}(\Delta r_m)' + (p-1)\Delta r_m^{p-2}\}.$$

At the right hand, beside r_m, Δr_m and $\Delta^2 r_m$ appears the radial derivative $(\Delta r_m)'$ of Δr_m. Therefore it seems that in contrast to what happens with the first Laplacian, the second Laplacian of the powers r_m^p introduces two independent functions. For this reason we was led to modify the condition C_k and to propose the following

Definition 1. *A Riemannian space* (M, g) *is said to be of harmonic rank* k *if for any system of functions of one variable* $(f_0, f_1, \ldots, f_k)$ *one has*

$$(R_k) \qquad df_0(r_m) \wedge d(\Delta f_1(r_m)) \wedge \ldots \wedge d(\Delta^k f_k(r_m)) = 0 \qquad \forall\, m \in M.$$

Moreover, there exist a system $(\overline{f}_0, \overline{f}_1, \ldots, \overline{f}_{k-1})$ *such that*

$$(\overline{R}_{k-1}) \qquad\qquad d\overline{f}_0(r_m) \wedge \ldots \wedge d(\Delta^{k-1} \overline{f}_{k-1}(r_m)) \neq 0.$$

In short, these conditions express the fact that the iterated Laplacians of radial functions produce systems with k but no more then k independent functions on some small domains of the space.

4.

We mention now the results of [1], where we have classified the Riemannian manifold (M^3, g) of harmonic rank 2. All these spaces are axially symmetric at each point. When they are homogeneous, they coincide with one of the following spaces:

i) The group $SU(2) \simeq S^3$ endowed with the Riemannian metric

$$ds^2 = \lambda^2(\omega_1^2 + \omega_2^2) + \mu^2 \omega_3^2 \qquad , \qquad \lambda, \mu \in R^+$$

where $(\omega_1, \omega_2, \omega_3)$ is the system of left invariant 1-forms on the sphere S^3 which are orthonormal for the standard Riemannian structure $(\lambda = \mu = 1)$ and satisfy the exterior differential equations

$$d\omega_1 = 2\omega_2 \wedge \omega_3, \qquad d\omega_2 = 2\omega_3 \wedge \omega_1, \qquad d\omega_3 = 2\omega_1 \wedge \omega_2.$$

Two of the Ricci principal curvatures coincide, say

$$\rho_{11} = \rho_{22} = \frac{4}{\lambda^2} - \frac{2\mu^2}{\lambda^4}$$

and the third one is

$$\rho_{33} = \frac{2\mu^2}{\lambda^4} > 0$$

so that one has

$$\rho_{11} + \rho_{33} = \rho_{22} + \rho_{33} > 0.$$

ii) The group $\widetilde{SL(2)}$ endowed with the Riemannian metric

$$ds^2 = \lambda^2(\omega_1^2 + \omega_2^2) + \mu^2\omega_3^2 \qquad \lambda, \mu \in R^+$$

where $(\omega_1, \omega_2, \omega_3)$ are left invariant 1-forms forming the dual basis of that of vector field which coincide at the neutral element of the group with the three matrices of $\mathfrak{sl}(2, R)$

$$v_1 = \begin{pmatrix} 1 & 0 \\ 0 & -1 \end{pmatrix}, \qquad v_2 = \begin{pmatrix} 0 & 1 \\ 1 & 0 \end{pmatrix}, \qquad v_3 = \begin{pmatrix} 0 & 1 \\ -1 & 0 \end{pmatrix}.$$

Here one has for the Ricci principal curvatures

$$\rho_{11} = \rho_{22} = -\frac{4}{\lambda^2} - \frac{2\mu^2}{\lambda^4}, \qquad \rho_{33} = \frac{2\mu^2}{\lambda^4} > 0$$

so that

$$\rho_{11} + \rho_{33} = \rho_{22} + \rho_{33} < o.$$

iii) The space $\mathbb{R}^3$ endowed with the Riemannian metric

$$ds^2 = dx^2 + dy^2 + \lambda^2(dz + xdy - ydx)^2$$

which is invariant under rotations around the Oz axis and also with respect to the left translations of the Heisenberg group

$$((x, y, z), (x', y', z')) \to (x + x', y + y', z + z' - xy' + x'y).$$

The spaces which are not homogeneous are warped-products $\mathbb{R} \times_f S_c$ of lines with surfaces of constant curvature S_c. Their metrics are of the type

$$ds^2 = dt^2 + f(t)^2 \cdot d\sigma_c^2$$

$d\sigma_c^2$ being the Riemannian metric of constant curvature c.

For all these spaces the two independent variables by means of which one can express all the Laplacians of radial functions $f(r_m)$ are:

— the distance to the rotation axis attached to m
— the abscis of the projection p' of the current point p on the former axis.

5.

We present now a result of geometric-algebra nature which is useful in the study of rank two pseudo-Riemannian spaces. To motivate it, observe that the first Riemannian invariants which occur when one writes the exterior differential equation

$$d\Omega_m \wedge d(\Delta\Omega_m) \wedge d(\Delta\Omega_m)' = 0$$

are:

— the fundamental form at m

$$g_m = \sum_{i,j=1}^{n} g_{ij}(m)x_i x_j,$$

— the Ricci form at m

$$\rho_m = \sum_{i,j=1}^{n} \rho_{ij}(m)x_i x_j,$$

— the cubic form

$$C_m = \sum_{i,j,k=1}^{n} (\nabla_k \rho_{ij})(m)x_i x_j x_k.$$

Here, $(x_1, \ldots, x_n)$ are normal coordinates centered at m.

These forms appear in the equation

$$dg_m \wedge d\rho_m \wedge dC_m = 0.$$

The notion of harmonic rank extends to pseudo Riemannian structures. The study of the former equation in a purely algebraic context (g_m non-definite) led us to the following

Theorem. *On the vector space $\mathbb{R}^n$, $n \geq 3$ consider g, ρ two quadratic forms and a cubic form C. Suppose moreover g non-degenerate and*

$$dg \wedge d\rho \neq 0.$$

If

$$dg \wedge d\rho \wedge dC = 0$$

and if $C \not\equiv 0$, then, there exists a linear function $\ell : \mathbb{R}^n \to \mathbb{R}$ such that

$$\rho = kg + \varepsilon \ell^2 \qquad k \in \mathbb{R}, \qquad \varepsilon = \pm 1$$

and

$$C = \ell(\alpha g + \beta \ell^2), \qquad \alpha, \beta \in \mathbb{R}.$$

If one supposes $g > 0$, the proof is straightforward due to the fact that one can diagonizale simultaneously g and ρ. But, if one renounces to this hypothesis, the proof is rather cumbersome and uses induction with respect to the dimension n.

We have the feeling that some arguments of elementary algebraic geometry could enlighten the meaning of this result.

The Riemannian and also the pseudo-Riemannian interpretation of this theorem is the following.

Let (M, g) be a pseudo-Riemannian manifold of harmonic rank 2. Then, if the Ricci tensor ρ is not cyclically parallel (in particular the space is not Einstein) its rank (as quadratic form) modulo g is one, i.e.

$$\rho = f \cdot g + \varepsilon \phi^2$$

f being a function and ϕ a 1-form. Moreover, the 1-form ϕ divides the cubic form C obtained from $\nabla \rho$ by symmetrisation and one has

$$C_m = \phi_m(\alpha(m)g_m + \beta(m)\phi_m^2) \qquad \forall\, m \in M.$$

We observe that all the spaces obtained in the classification of the three-dimensional Riemannian spaces of harmonic rank 2 are of this nature, i.e. their Ricci tensor is of rank one modulo g.

At the same time this theorem represents the rank 2 analogy of the fact that harmonic spaces are Einstein spaces i.e., their Ricci tensor is of rank zero modulo g.

We thank Renzo Caddeo for valuable discussions on this topic.

References

[1] TH. HANGAN and R. CADDEO: Une caractérisation des métriques Riemanniennes tridimensionnelles $SO(2)$-isotropes, *Rend. Sem. Fac. Sci. Univ.
Cagliari,* Vol. LVIII, 1987, pp. 61-84.

Th. Hangan

15 – Chemin des Ardennes
F–68100
Mulhouse

COLLOQUIA MATHEMATICA SOCIETATIS JÁNOS BOLYAI
56. DIFFERENTIAL GEOMETRY, EGER (HUNGARY), 1989

On a Concept of P-reducibility in Finsler Spaces

H. KAWAGUCHI

Introduction. The concept of P-reducible Finsler space was first introduced by M. Matsumoto [1] in 1977, but since then few papers have appeared on the field. Recently, S. C. Rastogi [2] has submitted to the Tensor Society his paper [2] studying P-reducibility in Finsler spaces. However, as Prof. Matsumoto indicates, some problems arise in connection with this paper so that it is not published yet. We would like to solve these problems in this publication. One of them is a concrete situation concerning the concept of P-reducibility in Finsler spaces. Now our purpose is to investigate the geometry of a subspace of the indicatrix at an arbitrary point of the Finsler space under consideration.

§1. A certain subspace of the indicatrix

Let M^n be an n-dimensional Finsler space, and I^{n-1} an indicatrix at an arbitrary fixed point x_0 of M^n. Then I^{n-1} is defined by the equation $F(x_0, X) = 1$, where F denotes the fundamental metric function of M^n and X a tangent vector of M^n at x_0. Also, I^{n-1} can be locally given by the equation $X^i = X^i(u^A)$; $i, j, \ldots = 1, 2, \ldots, n$; $A, B, \ldots = 1, 2, \ldots, n-1$ with local coordinates (u^A); I^{n-1} could be regarded as an $(n-1)$-dimensional

Riemannian space [3] by means of the induced metric

$$(1.1) \qquad g_{AB}(u^D) = g_{ij}(X^k(u^D))X_A^i X_B^j, \qquad X_A^i \stackrel{\text{def}}{=} \partial X^i/\partial u^A,$$

where $g_{ij}(X^k) \stackrel{\text{def}}{=} \frac{1}{2}(\partial^2 F^2/\partial X^i \partial X^j)$, making use of the conventional notation $g_{ij}(X^k) = g_{ij}(x_0, X^k)$. The relation $A_i(X^k)X^i = 0$ can be considered
for any X^k. The vector field A with the components A_i can be, in essential,
defined on I^{n-1} its components being given by

$$(1.2) \qquad A_A = A_i X_A^i = \partial \left(\log \sqrt{g} \right) \big/ \partial u^A, \qquad g \stackrel{\text{def}}{=} |g_{ij}|.$$

The function $\bar{g}$ is a restriction of g to I^{n-1}. A_A does not vanish if a
Finsler space does not reduce to a Riemannian one [4].

Now we are ready to give the following

Definiton. *An A-indicatrix H^{n-2} is a hypersurface of I^{n-1} defined locally
by the equation*

$$(1.3) \qquad \left(\log \sqrt{\bar{g}} \right)(u^1, u^2, \ldots, u^{n-1}) = c \qquad (c: \text{ arbitrary const.})$$

or represented in the explicit form [5]

$$(1.4) \qquad u^A = u^A(w^a); \qquad a, b, \ldots = 1, 2, \ldots, n-2$$

where (w^a) are local coordinates of H^{n-2}.

This definition leads us to the following two lemmata describing the
situation of H^{n-2} immersed in I^{n-1}:

Lemma 1. *An A-indicatrix H^{n-2} is normal to the vector field A on I^{n-1}.*

Proof. Differentiation of equation (1.3) under the condition (1.4) with
respect to the variable w^a, using equation (1.2), will furnish

$$A_A u_a^A = 0, \qquad u_a^A \stackrel{\text{def}}{=} \partial u^A/\partial w^a,$$

which completes the proof since (u_a^A) forms a frame of H^{n-2} represented on
I^{n-1}. ∎

Lemma 2. *Choosing a suitable local coordinate system* $(w^1, w^2, \ldots, w^{n-2})$ *of* H^{n-2}

$$(w_1, w_2, \ldots, w_{n-1}): \quad w_c \stackrel{\text{def}}{=} \partial/\partial w^c, \quad w_{n-1} \stackrel{\text{def}}{=} A/|A| \stackrel{\text{def}}{=} \widetilde{A}$$

forms an orthonormal frame of I^{n-1} *at a point* $u = (w^1, w^2, \ldots, w^{n-1})$, *where we put*

$$|A|^2 \stackrel{\text{def}}{=} g_{AB} A^A A^B = g^{AB} A_A A_B.$$

Proof. The general theory of hypersurfaces in Riemannian space and Lemma 1 lead us to the proof. ∎

§2. Geometry of A-indicatrices

Since in general there is a duality between a vector field and its complementary distribution, a characterization of a vector field A will be given by the geometry of A-indicatrices H^{n-2}. We may describe the following two Lemmata without proof:

Lemma 3. *An A-indicatrix H^{n-2} is an $(n-2)$-dimensional Riemann space by means of the induced Riemann metric, from equation* (1.1),

$$(2.1) \qquad g_{ab}(w^d) = g_{AB}(u^D(w^d)) u_a^A u_b^B.$$

We shall use later the following tensor on I^{n-1}

$$(2.1)' \qquad h_{AB} \stackrel{\text{def}}{=} g_{AB} - \widetilde{A}_A \widetilde{A}_B,$$

where $\widetilde{A}_A$ is a normalized vector of A_A, i.e. $\widetilde{A}_A = A_A/|A|$.

Lemma 4. *The fundamental equations of H^{n-2} in I^{n-1} are given as follows:*

[Gauss equation] $\qquad u_{a,b}^C = -(A_{A,B} u_a^A u_b^B/|A|)\widetilde{A}^C,$

[2nd fundamental tensor] $\Omega_{AB} \stackrel{\text{def}}{=} -(A_{A,B} u_a^A u_b^B/|A|), \quad \Omega_{AB} = \Omega_{BA},$

[Weingarten equation] $\qquad \widetilde{A}^A{}_{,b} = -(A_{C,B} \widetilde{A}^C u_b^B/|A|)\widetilde{A}^A,$

[Equations of Gauss and Codazzi]

$$R_{abcd} = (\Omega_{ac}\Omega_{bd} - \Omega_{ad}\Omega_{bc})R_{ABCD}\, u_a^A u_b^B u_c^C u_d^D,$$

$$\Omega_{ab,c} - \Omega_{ac,b} = R_{ABCD}\, u_a^A u_b^C u_c^D \widetilde{A}^B$$

where $(,)$ denotes the covariant differentiation, and R_{abcd} is the Riemann curvature tensor, by means of the metric (2.1).

At last in this section, we give the following theorem:

Theorem 1. *For the A-indicatrix H^{n-2} to be a congruence of the totally geodesic hypersurfaces of I^{n-1}, it is necessary and sufficient that there exist a scalar function ρ and a vector field σ_B normal to A on I^{n-1} such that*

$$A_{A,B} = \rho \widetilde{A}_A \widetilde{A}_B + \sigma_d(\widetilde{A}_A u_B^d + \widetilde{A}_B u_A^d)$$

holds where

$$\sigma_d \overset{\text{def}}{=} \sigma_B u_d^B, \qquad u_B^d = g_{BD} u_a^D g^{ad}.$$

Proof. Generally, the tensor field $A_{A,B}$ in I^{n-1} is written in the form

$$(2.2) \qquad A_{A,B} = \rho \widetilde{A}_A \widetilde{A}_B + \sigma_d(\widetilde{A}_A u_B^d + \widetilde{A}_B u_A^d) + \tau_{ab} u_A^a u_B^b,$$

where ρ, σ_B are the same as in the theorem, and τ_{ab} is a tensor field on I^{n-1} such that

$$\tau_{ab} = \tau_{AB}\, u_a^A u_b^B, \qquad \tau_{AB}\, A^A = 0.$$

Lemma 4 will readily suggest that $\tau_{AB} = A_{A,B}\, u_a^A u_b^B$ vanish if and only if H^{n-2} is totally geodesic, which completes the proof of the theorem. ∎

§3. Concrete situation of P-reducibility

In this section, we shall deal with special Finsler spaces M_0^n as follows:

(1) $|A|^2 = g_{ij} A^i A^j = g^{ij} A_i A_j$ is a function of the position on M^n,

(2) C-reducible [6]: $C_{ijk} = (n+1)^{-1}(h_{jk} C_i + h_{ki} C_j + h_{ij} C_k)$,

(3) P-reducible; $P_{ijk} = P_i h_{jk} + P_j h_{ki} + P_k h_{ij}$,
    where $P_{ijk} = C_{ijk|0}$, $P_i = (n+1)^{-1} C_{i|0}$,

(4) semi-C-reducible [7]: putting $p + q = 1$, $pq \neq 0$,

$$C_{ijk} = (p/(n+1))(h_{jk} C_i + h_{ki} C_j + h_{ij} C_k) + (q/|C|^2) C_i C_j C_k,$$

where $C_i = C_{ijk} g^{jk}$, $h_{ij} = g_{ij} - l_i l_j$, and $l_i = \partial F/\partial X^i$.

By virtue of the definition $A_{ijk} = F C_{ijk}$ and the relation $A_{ijk} l^k = 0$, the above special Finsler spaces can be characterized on I^{n-1} without loss of generality, by means of A_A, A_{ABC} and g_{AB} instead of C_i, C_{ijk} and h_{ij}, respectively, through rewriting the above conditions (1)–(4) as follows:

(1') $|A|^2 = g_{AB} A^A A^B = g^{AB} A_A A_B$ is constant,

(2') $A_{ABC} = (n+1)^{-1}(g_{BC} A_A + g_{CA} A_B + g_{AB} A_C),$

(3') $\overline{A}_{ABC} = (n+1)^{-1}(g_{BC} \overline{A}_A + g_{CA} \overline{A}_B + g_{AB} \overline{A}_C),$
    where $\overline{A}_{ABC} = C_{ijk|0} X_A^i X_B^j X_C^k$, $\overline{A}_A = C_{i|0} X_A^i$,

(4') $A_{ABC} = (p/(n+1))(g_{BC} A_A + g_{CA} A_B + g_{AB} A_C) + (q/|A|^2) A_A A_B A_C,$

where $p + q = 1$, $pq \neq 0$.

Condition (1') together with Theorem 1 has the following geometrical meaning in the implicit form:

Theorem 2. *A special Finsler space whose every A-indicatrix is a totally geodesic hypersurface of the indicatrix at every point, is a complementary specialization of the vector field A, to the one on which the length of A is a function of the position alone.*

Proof. It is sufficient to verify the theorem on an indicatrix I^{n-1}. The condition (1') is found to be equivalent to the condition $A^A A_{A,B} = 0$ which implies that there exist a tensor field τ_{ab} appearing in equation (2.2) such that

(3.1) $$A_{A,B} = \tau_{ab} u_A^a a_B^b$$

holds. Both equations (2.1) and (3.1) are specializations of the general expression (2.2) of $A_{A,B}$. This completes the proof of the theorem. ∎

In order to obtain some geometrical meanings of the conditions (2') and (4') let us consider a general form of the tensor A_{ABC} on I^{n-1} as follows:

(3.2) $A_{ABC} = \lambda \widetilde{A}_A \widetilde{A}_B \widetilde{A}_C + \mu_a (\widetilde{A}_A \widetilde{A}_B \mu_C^a + \widetilde{A}_B \widetilde{A}_C \mu_A^a + \widetilde{A}_C \widetilde{A}_A \mu_B^a) +$
$\qquad + \nu_{bc} (\widetilde{A}_A u_B^b u_C^c + \widetilde{A}_B u_C^b u_A^c + \widetilde{A}_C u_A^b u_B^c) + \eta_{abc} u_A^a u_B^b u_C^c,$

and respectively,

$$(3.2)' \qquad\qquad \lambda = |A| - \nu_b^b, \qquad \mu_a = -\eta_{ab}^b,$$

where λ, μ_a, ν_{bc}, η_{abc} are restrictions to H^{n-2} of a scalar function, a covariant vector field, a covariant symmetric tensor of order 2, and that of order 3, respectively on I^{n-1} such that

$$\mu_A A^A = 0, \quad \nu_{BC} A^C = 0, \quad \eta_{ABC} A^C = 0$$

hold.

Conversely, we have

Theorem 3. *Under the condition*

$$(3.3) \qquad\qquad \eta_{abc} = A_{ABC}\, u_a^A u_b^B u_c^C = 0,$$

the conditions $(2')$ *and* $(4')$ *are equivalent to the following ones, respectively:*

$$(3.4) \qquad\qquad \nu_{bc}\, u_B^b u_C^c = (|A|/(n+1))h_{BC},$$
$$(3.5) \qquad\qquad \nu_{bc}\, u_B^b u_C^c = (p|A|/(n+1))h_{BC}.$$

Similarly, taking into consideration a general form of the tensor A_{ABC} on I^{n-1}, we obtain a situation of P-reducibility in Finsler spaces.

References

[1] M. MATSUMOTO and H. SHIMADA, On Finsler spaces with the curvature tensors P_{ijkh} and S_{ijkh} satisfying special conditions, *Report on Math. Phys.* **12** (1977), 77–89.

[2] S. C. RASTOGI, On P-reducible Finsler spaces. (submitted to the Tensor) now on modification for publication.

[3] A. KAWAGUCHI, On the theory of non-linear connections. II. Theory of Minkowsky space and of non-linear connections in a Finsler spaces, *Tensor, N.S.* **6** (1956), 165–199.

[4] F. BRICKELL, A new proof of Deicke's theorem on homogeneous functions, *Proc. Amer. Math. Soc.* **16** (1965), 190–191.

[5] S. KOBAYASHI and K. NOMIZU, Foundations of differential geometry, Vol. I, *Interscience publishers,* 1963.

[6] M. MATSUMOTO, A theory of three-dimensional Finsler spaces in terms of scalars, *Demonst. Math.* **6** (1973), 223–251.

[7] M. MATSUMOTO and C. SHIBATA, On semi-*C*-reducibility, *T*-tensor $= 0$ and *S*4-likeness of Finsler spaces, *Journal of Mathematics of Kyoto University* **19** (1979), 301–314.

Hiroaki Kawaguchi

Sagami Institute of Technology
Tsujido, Fujisawa 251,
Japan

COLLOQUIA MATHEMATICA SOCIETATIS JÁNOS BOLYAI
56. DIFFERENTIAL GEOMETRY, EGER (HUNGARY), 1989

Finslerian Approach to Lagrange Mechanics

T. KAWAGUCHI

We attempt to make an abstract of the main results from two recent joint papers [1][1] , [2], realized together with R. Miron. *The geometrical investigation of Lagrange's dynamical systems*[2] being founded on the notion of Lagrangian gives an acceptable geometrical model for mechanics or theoretical physics. However there are many cases, as in the theory of gravitation when there is given a generalized metric tensor $g_{ij}(x, y)$ defined by the point x and the tangent vector y and we wish to study its geometrical properties. Such a tensor field, given on the total space of a vector bundle $\pi : TM \rightarrow M$, generally cannot be derived from a Lagrangian. Then the pair $M^n = (M, g_{ij}(x, y))$ was called a *generalized Lagrange space* by R. Miron. We consider a tensor field

$$g_{ij}(x, y) = \gamma_{ij}(x) + \frac{1}{c^2} y^i y^j,$$

which was used, in an approximate form by R. G. Beil [3] in electrodynamics. We prove that these spaces are not reducible to Lagrange spaces, and using the same technics as in the geometrical approach to Lagrange theory, we construct a geometrical model for gravitation and electromagnetism.

---

[1] Numbers in brackets point to the references at the end of the paper.

[2] This is called Lagrange geometry by R. Miron [4].

§1.

Let M be a differentiable manifold. (TM, π, M) its tangent bundle and (x^i, y^i), $(i, j, k = 1, \ldots, n; n = \dim M)$ the canonical coordinates of the points $u \in TM$, $\pi(u) = x$. We consider the Liouville vector field $y = y^i \frac{\partial}{\partial y^i}$ and the Riemann metric $\gamma_{ij}(x)$ on the manifold M. We put

$$(1.1) \qquad y_i = \gamma_{ij}(x)y^j, \quad \|y\|^2 = \gamma_{ij}(x)y^i y^j, \quad a_\sigma(x, y) = 1 + \frac{\sigma}{c^2}\|y\|^2,$$

where σ is a natural number.

We consider on TM the d-tensor field

$$(1.2) \qquad\qquad g_{ij}(x, y) = \gamma_{ij}(x) + \frac{1}{c^2}y_i y_j, \qquad c \in \mathbb{R}_+^*.$$

Then the pair $M^n = (M, g_{ij}(x, y))$ is a generalized Lagrange space [3]:

Theorem 1. *The generalized Lagrange space M^n is not reducible to a Lagrange space.*

Let us consider the absolute energy of the space M^n:

$$(1.3) \qquad\qquad \varepsilon = g_{ij}(x, y)y^i y^j = a_1\|y\|^2.$$

Theorem 2. *The Lagrangian $\varepsilon(x, y)$ from (1.3) is regular.*

Therefore, denoting the Christoffel symbols of the Riemann metric γ_{ij} by $\{_j{}^i{}_k\}$, we have

Theorem 3. *The canonical nonlinear connection of the space M^n has the coefficients*

$$(1.4) \qquad\qquad N^i{}_j = \{_j{}^i{}_k\}y^k.$$

Let $(\delta, \dot{\partial}_i)$ be the basis adapted to the horizontal distribution N and the vertical distribution V:

$$(1.5) \qquad\qquad \delta_i = \partial_i - N^j_i \dot{\partial}_j, \quad \partial_i = \frac{\partial}{\partial x^i}, \quad \dot{\partial}_i = \frac{\partial}{\partial y^i}.$$

Consequently, we get

Theorem 4. *The canonical metrical d-connection* $L\Gamma(N)$ *has the coeffi-cients*

(1.6)
$$\begin{cases} L^i_{jk} = \frac{1}{2}g^{ih}\left(\delta_k g_{jh} + \delta_j g_{hk} - \delta_h g_{jk}\right) = \{_j{}^i{}_k\}, \\ C^i_{jk} = \frac{1}{c^2 a_1}y^i\gamma_{jk}. \end{cases}$$

Theorem 5. *For* $n > 2$ *the Einstein equations of the space* M^n *are given by*

$$R_{ij} - \frac{1}{2}Rg_{ij} = \kappa\overset{H}{T}_{ij},$$

$$S_{ij} - \frac{1}{2}Sg_{ij} = \kappa\overset{V}{T}_{ij},$$

where κ *is a constant.*

Theorem 6. *The following properties of conservation of the energy mo-mentum tensor*

$$\overset{H}{T}{}^i{}_{j|i} = 0, \qquad \overset{V}{T}{}^i{}_j|_i = 0$$

hold good.

§2. Some Finslerian properties.

Now we suppose that there are given the gravitational tensor $\gamma_{ij}(x)$ and the vector potential $A_i(x)$ of the electromagnetic tensor field on the base manifold M. We take into account the pair $(M^n, A_i(x))$. Let $A_{ij} = \partial_i A_j - \partial_j A_i$ be the electromagnetic tensor field and denote

(2.1) $$F_{ij} = \frac{e}{mc}A_{ij}, \quad F^i{}_j = \gamma^{ih}H_{hj}, \quad F^{ij} = \gamma^{jh}F^i{}_h.$$

We admit the following postulate:

The tensor fields $g_{ij}(x, y)$ and $A_i(x)$ are related by the nonlinear con-nection $\overline{N}$:

(2.2) $$\overline{N}^i{}_j = N^i{}_j - F^i{}_j.$$

We fix this nonlinear connection for the generalized Lagrange space M^n. A first result is the following:

Theorem 7.

    I. The nonlinear connection $\overline{N}$ is globally defined on $TM \setminus \{0\}$.

    II. $\overline{N}$ is determined only by $\gamma_{ij}(x)$ and $A_i(x)$.

    III. The horizontal geodesics of the space M^n are characterized by the equations of Lorentz force:

$$\frac{d^2 x^i}{dt^2} + \{_j{}^i{}_k\} \frac{dx^j}{dt} \frac{dx^k}{dt} = F^i{}_j \frac{dx^j}{dt}.$$

Theorem 8. The transformation (2.2) of the nonlinear connection implies the transformation of the d-connection $L\Gamma(N) \to \overline{L\Gamma}(\overline{N})$ given by

$$\overline{L}^i_{jk} = L^i_{jk} + C^i_{jr} F^r{}_k, \qquad \overline{C}^i_{jk} = C^i_{jk},$$

where L^i_{jk}, C^i_{jk} are given by (1.6).

Theorem 9. The fundamental tensor field $g_{ij}(x, y)$ of the space M^n is covariant constant with respect to the d-connection $\overline{L\Gamma}(\overline{N})$.

    Denoting by ";" and "|" the h- and v-covariant derivatives with respect to $\overline{L\Gamma}(\overline{N})$, we get

Theorem 10. For $n > 2$, the Einstein equations of the metrical d-connection $\overline{L\Gamma}(\overline{N})$ are given by

$$(2.3) \quad \begin{cases} R_{ij} = \frac{1}{2} R g_{ij} + \frac{1}{c^2 a_1} g^{pq} \left(f_{ip} F_{jq} - \frac{1}{2} g^{hs} F_{rp} F_{sq} g_{ij} \right) = \kappa \overset{H}{T}_{ij}, \\ S_{ij} - \frac{1}{2} S g_{ij} = \kappa \overset{V}{T}_{ij}, \\ \overset{1}{P}_{ij} = \kappa \overset{2}{T}_{ij}, \quad \overset{2}{P}_{ij} = -\kappa \overset{3}{T}_{ij}, \end{cases}$$

where

$$\overset{1}{P}_{ij} = \frac{1}{c^2 a_1} g^{pq} F_{ip} F_{jq}, \qquad \overset{2}{P}_{ij} = \frac{1}{c^2 a_1} g^{pq} \{ \gamma_{ip} F_{jq} - \gamma_{pq} F_{ij} \}.$$

Theorem 11. The Maxwell equations with respect to $\overline{L\Gamma}(\overline{N})$ are as follows:

$$(2.4) \quad \begin{cases} F_{ij|k} + F_{jk|i} + F_{ki|j} = \frac{2}{c^2 a_1} (F_{0i} F_{jk} + F_{0j} F_{ki} + F_{0k} F_{ij}), \\ F_{ij}|_k + F_{jk}|_j + F_{ki}|_j = 0. \end{cases}$$

Some other aspects of the electromagnetic tensor field were studied, too.

References

[1] T. KAWAGUCHI and R. MIRON, On the generalized Lagrange spaces with the metric $\gamma_{ij}(x) + (1/c^2)y_i y_j$, Tensor, N.S., Vol. 48., No. 1 (1989), 52–63.

[2] T. KAWAGUCHI and R. MIRON, A Lagrangian model for gravitation and electromagnetism. Tensor, N.S., Vol. 48, No. 2 (1989), 153–168.

[3] R. G. BEIL, Electrodynamics from a metric. International Journal of Theoretical Physics, Vol. 26, No. 2 (1987), 189–197.

[4] R. MIRON and M. ANASTESIEI, Vector bundles. Lagrange spaces. Applications in relativity (in Romanian). Editura Academiei R. S. Romania, (1987).

[4] R. MIRON, A Lagrangian theory of relativity. preprint, Nr. 84, Universitatea Timişoara, (1985).

Tomoaki Kawaguchi

Institute of
Information Sciences and Electronics
University of Tsukuba
Tsukuba-shi, Japan 305

COLLOQUIA MATHEMATICA SOCIETATIS JÁNOS BOLYAI

56. DIFFERENTIAL GEOMETRY, EGER (HUNGARY), 1989

On Quasi-equivalence of two Variational Problems in Hamilton Spaces[*]

M. SZ. KIRKOVITS

1. Introduction

In the paper [3] we have defined the *Moór equivalence* of two variational problems in the *Lagrange spaces* $L^{*n} = (M, \mathcal{L}^*(x,y))$ and $L^n = (M, \mathcal{L}(x,y))$ respectively, with the relation

(1.1)
$$\mathcal{E}_i(\mathcal{L}^*(x,y)) = \lambda(x,y)\mathcal{E}_i(\mathcal{L}(x,y)); \qquad \lambda(x,y) \neq 0$$
$$\left(\mathcal{E}_i := \frac{d}{dt}\dot{\partial}_i - \partial_i; \qquad \dot{\partial}_i := \partial/\partial y^i; \qquad \partial_i := \partial/\partial x^i \right)$$

which identically holds.

We have given a necessary and sufficient condition for this equivalence, namely the transformation between the Lagrangians $\mathcal{L}^*(x,y)$ and $\mathcal{L}(x,y)$. We have shown several geometrical consequences of the equivalence-relation:

(a) L^{*n} and L^n are in *conformal* correspondence, i.e. $g_{ij}^* = \lambda(x,y)g_{ij}$.

(b) The *canonical nonlinear connections* $\overset{c}{N}{}^* \ (\overset{c}{N}{}^{*i}_{\ j}(x,y))$ and $\overset{c}{N} \ (\overset{c}{N}{}^i_{\ j}(x,y))$ are equal to each other.

---

[*] This paper is in final form and no version of it will be submitted for publication elsewhere.

Moreover, we have studied the *relations* between the corresponding *canonical metrical connection* $C\overset{*}{\Gamma}$ and $C\Gamma$. The relations between the torsion d-tensor fields of $C\overset{*}{\Gamma}$ and $C\Gamma$ have also been found.

In this paper we shall study the *quasi-equivalence* (i.e. the Rund-equivalence [11]) of two variational problems in *Hamilton spaces* using the Miron theory of Hamilton geometry [6]. He proved that the extended Legendre transformation gives rise to the fact: The Lagrange spaces L^n and the Hamilton spaces H^n are *dual* of each other.

Hence, first we shall define the quasi-equivalence of two variational problems in Lagrange spaces. Then, according to the duality existing between L^n and H^n we shall show how the relation of quasi-equivalence in Lagrange spaces is transformed into the relation of quasi-equivalence for Hamilton spaces. The main result is that *two variational problems are quasi-equivalent iff they are equivalent in the sense of Moór.*

In the last section we shall briefly study *the equivalence of two constrained variational problems in Hamilton spaces.*

2. Preliminaries

2.A. *The notion of Hamiltonian and the definition of the Hamilton space*

A function

(2.1) $\mathcal{H} : T^*M \to \mathbb{R}.$

defined on the total space T^*M of the cotangent bundle (T^*M, π^*, M) is called a *Hamiltonian.*

A Hamiltonian is called *regular* if its Hessian on $T^*M \setminus \{0\}$

(2.2)(a) $g^{ij}(x,p) = \dot{\partial}^i \dot{\partial}^j \mathcal{H}(x,p)$ $(\dot{\partial}^i := \partial/\partial p_i)$

has the property

(2.2)(b) $\mathrm{rank}\, \|g^{ij}(x,p)\| = n.$

Definition 2.1. *A Hamilton space is a pair* $H^n = (M, \mathcal{H}(x,p))$ *which consists of a* C^∞ *manifold* M *of dimension* n, *and a real and regular*

Hamiltonian $\mathcal{H}(x,p)$.

The Hamiltonian $\mathcal{H}(x,p)$ is called the *fundamental function of H^n*. $(g^{ij}(x,p))$ called *Hamilton metric*, is a symmetric, nondegenerate d-tensor field (c.f. [6]).

2.B. *Duality between L^n and H^n*

Let us consider a Lagrange space $L^n = (M, \mathcal{L}(x,y))$ having the fundamental d-tensor field called *Lagrange metric:*

(2.3) $a_{ij}(x,y) = \dot{\partial}_i \dot{\partial}_j \mathcal{L}(x,y)$ $(\dot{\partial}_i := \partial/\partial y^i)$

(c.f. [5]).

Let us compare L^n and H^n:

L^n	H^n
$1°$ $a_{ij}(x,y) = \dot{\partial}_i \dot{\partial}_j \mathcal{L}(x,y)$	$g^{ij}(x,p) = \dot{\partial}^i \dot{\partial}^j \mathcal{H}(x,p)$
Lagrange metric,	*Hamilton metric,*
(x,y) is a point of the total	(x,p) is a point of the total
space of the tangent bundle	space of the cotangent bundle
(TM, π, M)	(T^*M, π^*, M)

We consider the map $\mathfrak{L}$, i.e., the
Legendre transformation
which is a fiber preserving
local diffeomorphism:

$2°$ $(x^i, y^i) \in TM$ $\xrightarrow{\;\mathfrak{L}\;}$ $(x^i, \dot{\partial}_i \mathcal{L}) = (x^i, p_i) \in T^*M$
                 $\pi(x,y) = \pi^*(x,p) = x$ $(\forall\,(x,y) \in TM)$

Its *inverse* $\mathfrak{L}^{-1}$ is as follows:

$3°$ $(x^i, \varphi^i(x,p)) \in TM$ $\xleftarrow{\;\mathfrak{L}^{-1}\;}$ $(x^i, p_i) \in T^*M$
                 $(x^i = x^i, y^i = \varphi^i(x,p))$

$$L^n \qquad\qquad\qquad\qquad\qquad H^n$$

. . . .

Hence under the Legendre transformation $\mathfrak{L}$ we get

$4°$ $\qquad\qquad \mathcal{L}(x,y) \qquad\qquad \xrightarrow{\;\mathfrak{L}\;} \qquad \mathcal{H}(x,p) = -\mathcal{L}(x,y) + p_i y^i$

$$(y = \varphi(x,p))$$

$$(new\ Hamiltonian)$$

. . . .

In this case:

$5°$ $\qquad\qquad\qquad g^{ij}(x,p)a_{jk}(x,\varphi(x,p)) = \delta^i_k.$

. . . .

Hence we have:

$$H^n \text{ is a } dual\ space \text{ of } L^n.$$

. . . .

From $4°$ it follows

$$\partial_i \mathcal{H} = -\partial_i \mathcal{L} \qquad\qquad\qquad \dot{\partial}^i \mathcal{H} = y^i = \varphi^i(x,p).$$

. . . .

Conversely, we consider a $map\ \widetilde{\mathfrak{L}}$ with the same
properties as $\mathfrak{L}$:

$6°$ $\quad (x^i, \dot{\partial}^i \mathcal{H}) = (x^i, y^i) \in TM \quad \xleftarrow{\;\widetilde{\mathfrak{L}}\;} \qquad (x^i, p_i) \in T^*M$

. . . .

Its $inverse\ \widetilde{\mathfrak{L}}^{-1}$ is

$7°$ $\qquad\qquad (x,y \in TM) \qquad \xrightarrow{\;\widetilde{\mathfrak{L}}^{-1}\;} \qquad (x, v(x,y)) \in T^*M.$

. . . .

Under the map $\widetilde{\mathfrak{L}}$ we obtain:

$8°$ $\quad \mathcal{L}(x,y) = -\mathcal{H}(x,p) + p_i y^i \quad \xleftarrow{\;\widetilde{\mathfrak{L}}\;} \qquad\qquad \mathcal{H}(x,p)$

$$(p = v(x,y))$$

$$(new\ Lagrangian)$$

. . . .

Now:

$9°$ $\qquad\qquad\qquad g^{ij}(x, v(x,y))a_{jk}(x,y) = \delta^i_k.$

. . . .

Hence we have:

The dual of the dual of L^n is L^n.

The dual of the dual of H^n is H^n.

. . . .

$$L^n \qquad\qquad\qquad H^n$$

. . . .

Under the Legendre transformation we obtain

$10°$ $\qquad \mathcal{L}(\mathcal{E}_i(\mathcal{L})) \qquad\qquad = \qquad\qquad H_i(\mathcal{H})$

where $\mathcal{E}_i(\mathcal{L}) = 0$ are the *Euler-Lagrange equations of geodesics* in L^n and $H_i(\mathcal{H}) = 0$ are the *Hamilton equations* in H^n. Hence we have

. . . .

$11°$ $\qquad\qquad \dfrac{dx^i}{dt} = y^i; \qquad\qquad\qquad \dfrac{dx^i}{dt} = \dot{\partial}^i \mathcal{H}; \quad \dfrac{dp_i}{dt} + \partial_i \mathcal{H} = 0.$

$$\tfrac{d}{dt}(\dot{\partial}_i \mathcal{L}) - \partial_i \mathcal{L} = 0$$

In Lagrange spaces there exists a *canonical nonlinear connection* $\overset{c}{N} (\overset{c}{N}{}^i_j(x,y))$ ([5]). By Miron's Theorem ([6]) *the Legendre transformation carries the canonical nonlinear connection* $\overset{c}{N} (\overset{c}{N}{}^i_j)$ *on* TM *into the canonical nonlinear connection* $\overset{c}{N} (\overset{c}{N}{}_{ji})$ *on* T^*M *i.e. we have:*

. . . .

$$\overset{c}{N}{}_{ji}(x,p) =$$

$12°$ $\qquad \overset{c}{N}{}^i_j(x,y) = \dot{\partial}_j G^i \qquad\qquad = -\dfrac{1}{2}g_{jh}\left\{ \dfrac{1}{4}g_{ik}\dot{\partial}^k(\mathcal{H}, \dot{\partial}^h \mathcal{H}) + \right.$

$$\left. + \dot{\partial}^h \partial_i \mathcal{H} \right\},$$

where

. . . .

$13°$ $\qquad G^i = a^{ij}(y^k \dot{\partial}_j \partial_k \mathcal{L} - \partial_j \mathcal{L})$ $\qquad (\mathcal{H}, \dot{\partial}^h \mathcal{H}) :=$

$$:= \partial_i \mathcal{H} \dot{\partial}^i(\dot{\partial}^h \mathcal{H}) - \dot{\partial}^i \mathcal{H}(\partial_i \dot{\partial}^h \mathcal{H})$$
$$\text{(Poisson-bracket).}$$

Remark 2.1. We have seen that $\mathfrak{L}$ is a local diffeomorphism, however $\overset{c}{N} (\overset{c}{N}{}_{ji}(x,p))$ has global character (c.f. [6]).

Remark 2.2. The system of the Hamilton equations $H_i(\mathcal{H}) = 0$ is equivalent to the system

(2.4) $\qquad\qquad \dfrac{dx^i}{dt} = \dot{\partial}^i \mathcal{H}; \qquad\qquad \dfrac{\overset{c}{\delta} p_i}{dt} + \dfrac{\overset{c}{\delta} \mathcal{H}}{\delta x^i} = 0,$

where

$$\overset{c}{\delta}_i = \partial_i + \overset{c}{N}_{ji}(x,p)\dot{\partial}^j \quad \text{and} \quad \overset{c}{\delta} p_i = dp_i - \overset{c}{N}_{ij}dx^j.$$

These equations have an *invariant* form.

3. The Rund equivalence of two variational problems in Lagrange and Hamilton spaces

Definition 3.1. *Two variational problems in* $L^{*n} = (M, \mathcal{L}^*(x,y))$ *and* $L^n = (M, \mathcal{L}(x,y))$ *(on the same base manifold* M*) are quasi-equivalent in Rund's sense if the relations*

$$(3.1) \qquad \mathcal{E}_i(\mathcal{L}^*(x,y)) = \lambda_i^j(x,y)\mathcal{E}_j(\mathcal{L}(x,y)), \qquad \text{rank} \,\|\lambda_i^j(x,y)\| = n$$

hold identically.

3.A. *Transformation of the quasi-equivalence relation under the Legendre transformation*

Let us consider the relations of duality between L^n and H^n described in section 2. Using the results we can easily calculate the new Hamiltonians in $H^{*n} = (M, \mathcal{H}^*(x,p^*))$ and in $H^n = (M, \mathcal{H}(x,p))$:

$$(3.2) \qquad \begin{aligned} \text{(a)} \quad & \mathcal{H}^*(x,p^*) = -\mathcal{L}^*(x, \varphi^*(x,p^*)) + p_j^* \varphi^{*j}(x,p^*), \\ \text{(b)} \quad & \mathcal{H}(x,p) = -\mathcal{L}(x, \varphi(x,p)) + p_j \varphi^j(x,p). \end{aligned}$$

Differentiate the Hamiltonians mentioned above with respect to p_i^*, p_i and x^i. Since $\mathfrak{L}^{-1}(x,p^*) = \varphi^*(x,p^*) = y$ and $\mathfrak{L}^{-1}(x,p) = \varphi(x,p) = y$ we get

$$(3.3) \qquad \begin{aligned} \text{(a)} \quad & \dot{\partial}^{*i}\mathcal{H}^*(x, \psi(x,p)) = \dot{\partial}^i\mathcal{H}(x,p) \qquad (\dot{\partial}^{*i} := \partial/\partial p_i^*) \\ \text{(b)} \quad & \partial_i \mathcal{H}^* = -\partial_i \mathcal{L}^*; \qquad \partial_i \mathcal{H} = -\partial_i \mathcal{L}, \end{aligned}$$

where $\quad p_i^* = \psi_i(x,p) \quad$ and $\quad p_i = \tilde{\psi}_i(x,p^*).$

Hence we have

Theorem 3.1. *The quasi-equivalence relation* (3.1) *is transformed under the Legendre transformation into the following relation of quasi-equivalence for the Hamiltonians in Hamilton spaces*

$$\text{(a)} \qquad \dot{\partial}^{*i}\mathcal{H}^*(x,p^*) = \dot{\partial}^i\mathcal{H}(x,p)$$

(3.4)

$$\text{(b)} \qquad \frac{dp_i^*}{dt} + \partial_i\mathcal{H}^* = \lambda_i^j(x,\varphi(x,p))\left(\frac{dp_j}{dt} + \partial_j\mathcal{H}\right).$$

3.B. *Geometrical characters of the quasi-equivalence*

Proposition 3.1. *The coefficients* $\lambda_i^j(x,\varphi(x,p))$ *are the components of a d-tensor field of the type* $(1,1)$ *on* M.

Proof. Let us consider the Hamilton equations $H_i(\mathcal{H}) = 0$ in (2.4), and arbitrary fixed nonlinear connections $\overset{c}{N}{}^*_{ji}(x,p^*)$ and $\overset{c}{N}{}_{ji}(x,p)$ on $(T^*M)_1$ and on $(T^*M)_2$. Hence we can write (3.4)(b) in the following equivalent form:

$$(3.5) \qquad \frac{\overset{c}{\delta} p_i^*}{dt} + \frac{\overset{c}{\delta}\mathcal{H}^*}{\delta x^i} = \lambda_i^j(x,\varphi(x,p))\left(\frac{\overset{c}{\delta} p_j}{dt} + \frac{\overset{c}{\delta}\mathcal{H}}{\delta x^j}\right)$$

where

$$\overset{c}{\delta} p_i^* = dp_i^* - \overset{c}{N}{}^*_{ij}(x,\psi(x,p))dx^j; \qquad \overset{c}{\delta} p_i = dp_i - \overset{c}{N}{}_{ij}(x,p)dx^j,$$

$$\frac{\overset{c}{\delta}\mathcal{H}^*}{\delta x^i} = \partial_i\mathcal{H}^* + \overset{c}{N}{}^*_{ji}(x,\psi(x,p))\dot{\partial}^{*j}\mathcal{H}^*(x,\psi(x,p)),$$

$$\frac{\overset{c}{\delta}\mathcal{H}}{\delta x^i} = \partial_i\mathcal{H} + \overset{c}{N}{}_{ji}(x,p)\dot{\partial}^j\mathcal{H}(x,p).$$

We can see that the left side of (3.5) is a d-tensor field and the second factor of its right side is also a d-tensor field. Since (3.5) is an identity in (x,p) we obtain that $\lambda_i^j(x,\varphi(x,p))$ must also be a d-tensor field. ∎

Theorem 3.2. *If the relation of the quasi-equivalence in* (3.1) *holds then the fundamental tensor fields* $\overset{*}{g}{}^{ij}(x,p^*)$ *and* $g^{ij}(x,p)$ *of* H^{*n} *and* H^n, *respectively, are in the following "conformal" relation:*

$$\text{(a)} \qquad \overset{*}{g}{}^{ij}(x,\psi(x,p))\lambda_j^k(x,\varphi(x,p)) = g^{ik}(x,p)$$

(3.6)

$$\text{(b)} \qquad g^{ik}(x,\widetilde{\psi}(x,p^*))\widetilde{\lambda}_k^j(x,\varphi^*(x,p^*)) = \overset{*}{g}{}^{ij}(x,p^*).$$

Proof. Let us differentiate (3.3)(a) with respect to p_j and p_j^*. We obtain

(3.7)
(a) $\quad \dot\partial^i \dot\partial^j \mathcal{H} = \dot\partial^{*i} \dot\partial^{*k} \mathcal{H}^* \dot\partial^j p_k^* \quad (p_k^* = \psi_k(x,p))$,

(b) $\quad \dot\partial^{*i} \dot\partial^{*j} \mathcal{H}^* = \dot\partial^i \dot\partial^k \mathcal{H} \dot\partial^{*j} p_k \quad (p_k = \widetilde\psi_k(x,p^*))$.

By means of the Hamilton metric in (2.2)(a) these formulae give

(3.8)
(a) $\quad g^{ij}(x,p) = g^{ik}(x, \psi(x,p)) \dot\partial^j \psi_k$,

(b) $\quad \overset{*}{g}{}^{ij}(x,p^*) = g^{ik}(x, \widetilde\psi(x,p^*)) \dot\partial^{*j} \widetilde\psi_k$.

Now taking into account (3.4)(b) we have

$$(3.9) \quad \partial_k \psi_i \dot\partial^k \mathcal{H} + \dot\partial^k \psi_i \frac{dp_k}{dt} + \partial_i \mathcal{H}^* = \lambda_i^k(x, \varphi(x,p)) \left(\frac{dp_k}{dt} + \partial_k \mathcal{H} \right).$$

This is an identity in (x,p) and a polynomial of $\dfrac{dp_k}{dt}$, hence it necessarily follows that

(3.10)
(a) $\quad \dot\partial^k \psi_i = \lambda_i^k(x, \varphi(x,p))$,

(b) $\quad \partial_k \psi_i \dot\partial^k \mathcal{H} + \partial_i \mathcal{H}^* = \lambda_i^k \partial_k \mathcal{H}$.

After substitution of the relation (3.10)(a) in (3.8)(a) we get the statement of (3.6)(a).

On the other hand from (3.4)(b), because of rank $\|\lambda_i^j\| = n$, it follows that

$$(3.11) \quad \widetilde\lambda_j^i(x, \varphi^*(x,p^*)) \left(\frac{dp_i^*}{dt} + \partial_i \mathcal{H}^* \right) = \frac{dp_j}{dt} + \partial_j \mathcal{H} \quad (\varphi^*(x,p^*) = \varphi(x,p))$$

also holds, where $\lambda_i^k \widetilde\lambda_j^i = \delta_j^k$ and $p_k = \widetilde\psi_k(x,p^*)$.

From (3.11) it directly follows

$$(3.12) \quad \widetilde\lambda_j^i(x, \varphi^*(x,p^*)) \left(\frac{dp_i^*}{dt} + \partial_i \mathcal{H}^* \right) = \partial_i \widetilde\psi_j \dot\partial^{*i} \mathcal{H}^* + \dot\partial^{*i} \widetilde\psi_j \frac{dp_i^*}{dt} + \partial_j \mathcal{H}.$$

In the way mentioned above we get

(3.13)
(a) $\quad \dot\partial^{*i} \widetilde\psi_j = \widetilde\lambda_j^i(x, \varphi^*(x,p^*))$,

(b) $\quad \partial_i \widetilde\psi_j \dot\partial^{*i} \mathcal{H}^* + \partial_j \mathcal{H} = \widetilde\lambda_j^i \partial_i \mathcal{H}^*$.

Substituting (3.13)(a) in (3.8)(b) we get (3.6)(b). ∎

3.C. *Relation between the Rund and the Moór equivalences*

Theorem 3.3. *Two variational problems in H^{*n} and H^n are equivalent in the sense of Rund (i.e. quasi-equivalent) iff they are equivalent in the sense of Moór, i.e. iff*

$$(3.14) \qquad \lambda_i^j(x, y) = \delta_i^j \lambda(x, y).$$

Proof.

   I. Suppose that the Moór equivalence relation in (1.1) holds. Since $\mathcal{E}_j$ are the *components* of a *covariant vector* (c.f. [4]) we can write the relation (1.1) in the form

$$(3.15) \qquad \mathcal{E}_i(\mathcal{L}^*(x, y)) = \delta_i^j \lambda(x, y)\mathcal{E}_j(\mathcal{L}(x, y)).$$

Moreover, in the Definition 3.1. the matrix $(\lambda_i^j(x, y))$ is arbitrary so we can choose it as in (3.14).

   II. Now let us suppose that the Rund equivalence relation (3.1) holds. Theorem 3.2. implies

$$(3.16) \qquad \overset{*}{g}_{kj}(x, p^*) = \lambda_k^s(x, \varphi(x, p))g_{sj}(x, p).$$

Because of the symmetry of the Hamilton metric we have

$$(3.17) \qquad \lambda_k^s g_{sj} = \lambda_j^s g_{sk}.$$

This implies that

$$(3.18) \qquad g_{sj}\lambda_k^s - g_{sk}\lambda_j^s = 0$$

which is equivalent to

$$(3.19) \qquad g_{sl}(\lambda_k^s \delta_j^l - \lambda_j^s \delta_k^l) = 0.$$

In view of the properties of the Hamilton metric and the arbitrary choice of λ it results that the coefficients of g_{sl} in (3.19) must be skew-symmetric in s and l in order that the relation (3.19) should vanish. This gives for the symmetric part in s, l:

$$(3.20) \qquad \lambda_k^s \delta_j^l - \lambda_j^s \delta_k^l + \lambda_k^l \delta_j^s - \lambda_j^l \delta_k^s = 0.$$

For $l = j$ we get

$$
\begin{array}{lll}
\text{(a)} & n\lambda_k^s - \lambda_k^s + \lambda_k^s - \lambda_j^j \delta_k^s = 0, \\
\text{(b)} & n\lambda_k^s - \lambda_j^j \delta_k^s = 0.
\end{array}
$$

(3.21)   i.e.

Now putting

(3.22) $$\lambda(x, y) = \frac{1}{n}\lambda_j^j(x, y)$$

we obtain from (3.21)(b)

(3.23) $$\lambda_k^s(x, \varphi(x, p)) = \delta_k^s \lambda(x, y).$$

■

4. Some remarks about the equivalence of two constrained variational problems in Hamilton spaces

It is known from the classical calculus of variations that the variational problems subject to constraints are formulated as follows:

Given a Lagrangian $\mathcal{L}(x, y)$, it is required to find a curve $C : x^i = x^i(t)$ which satisfies the m equations of constraint $\underset{(\mu)}{\Theta}(x, y) = 0$ ($\mu = \overline{1, m}$; $m < n$), joins two fixed points $P_1(x^i(t_1))$ and $P_2(x^i(t_2))$ in M, and affords an extreme value to the integral

(4.1) $$I(C) = \int_{t_1}^{t_2} \mathcal{L}(x, y)dt; \qquad \left(y = \frac{dx}{dt} \right).$$

G. Vranceanu [14], J. Synge [13], H. Rund [12] and A. Moór [9] have studied the constrained variational problems in a geometrical manner and have given the equations of *constrained geodesics* in Riemann and Finsler spaces.

The purpose of this section is to sketch *briefly* the relation between the equations of the constrained geodesics in H^n and L^n under the following conditions:

(a)   rank $\|g_{ij}(x, y)\| = n - m$ $(g_{ij} = \dot{\partial}_i \dot{\partial}_j \mathcal{L})$

in all TM. Moreover, given a submanifold S_O in the total space T^*M of the cotangent bundle $T^*M \to M$, S_O is represented by the equations

$$(4.2) \qquad (b) \qquad\qquad \underset{(\mu)}{G}(x,p) = 0 \qquad (\mu = \overline{1,m}; \ m < n),$$

and

$$(c) \qquad\qquad \text{rank} \, \|\dot{\partial}^j \underset{(\mu)}{G}(x,p)\| = m.$$

The equations in (4.2)(b) are called Hamiltonian constraints for the variational problem.

We can see that the Lagrangian $\mathcal{L}(x,y)$ is not regular in TM. Hence, in this case we cannot use the Legendre transformation defined by R. Miron ([6]). In the forthcoming paper ([1]) Gh. Atanasiu and the authoress will construct an appropriate geometrical model for the investigation of constrained variational problems in geometrical manner. We use as a pattern the vector bundle geometry of Lagrange and Hamilton spaces.

Now we present a different method for showing the *transformation of the equations of the constrained geodesics in Hamilton space into those of Lagrange space.*

Let us define the transformation $\mathfrak{FL}$:

$$(4.3) \qquad\qquad\qquad \mathfrak{FL} : TM \to T^*M,$$

where $\mathfrak{FL}(x,y) = (x,p)$ $(x^i = x^i, \ p_i = \dot{\partial}_i\mathcal{L} \equiv \theta_i(x,y))$. We will assume that there exists an $S_1 \subset TM$ for which $\mathfrak{FL}(S_1) = S_O$ holds.

Taking into account (4.2)(b) it can be deduced that

$$(4.4) \qquad\qquad \underset{(\mu)}{G}(x^s, \dot{\partial}_s\mathcal{L}) \equiv \underset{(\mu)}{G}(x^s, \theta_s(x,y)) = O.$$

Differentiating (4.4) with respect to y^k we can derive that

$$(4.5) \qquad\quad \dot{\partial}^i \underset{(\mu)}{G}(x, \theta(x,y))\dot{\partial}_k\theta_i = O. \qquad (\dot{\partial}^i := \partial/\partial p_i \equiv \partial/\partial\theta_i)$$

Since $\dot{\partial}_k\theta_i = \dot{\partial}_k\dot{\partial}_i\mathcal{L}(x,y) = g_{ki}(x,y)$, we obtain from (4.5)

$$(4.6) \qquad\qquad\qquad \dot{\partial}^i \underset{(\mu)}{G}(x, \theta(x,y))g_{ki}(x,y) = 0.$$

Hence, by means of the conditions (4.2)(a) and (b) we have a basis for the null-vectors of g:

(4.7) $\underset{(\mu)}{b}{}^i \equiv \underset{(\mu)}{\dot{\partial}{}^i} G\,(x, \theta(x,y))$ $(i = \overline{1,n},\;\; \mu = \overline{1,m})$.

Now let us look at the function

(4.8) $\mathbf{H}(x, \theta(x,y)) = y^i \theta_i(x,y) - \mathcal{L}(x,y)$ on S_1,

from which taking the derivative with respect to y^k we obtain

(4.9)   (a) $\dot{\partial}{}^i \mathbf{H}(\dot{\partial}_k \theta_j) = \theta_k - (\dot{\partial}_k \theta_i)y^i - \dot{\partial}_k \mathcal{L} = (\dot{\partial}_k \theta_i)y^i,$

i.e.

(b) $g_{ik}(y^i - \dot{\partial}{}^i \mathbf{H}) = O.$

Therefore, $y^i - \dot{\partial}{}^i \mathbf{H}$ is a null vector of g and it can be written in the basis $\left\{ \underset{(\mu)}{b}{}^i \right\}$ in the following form:

(4.10) $y^i - \dot{\partial}{}^i \mathbf{H}(x,p) = \underset{(\mu)}{w}\,(x,y)\,\underset{(\mu)}{b}{}^i.$

This gives that

(4.11) $y^i = \dot{\partial}{}^i \mathbf{H}(x,p) + \underset{(\mu)}{w}\,(x,y)\dot{\partial}{}^i\,\underset{(\mu)}{G}\,(x,p)$

which can be considered the *inverse transformation* of $\mathfrak{FL}$. We note that it is an *implicit equation* for y^i.

Differentiating the function in (4.8) with respect to x^k we get

(4.12) $\partial_k \mathbf{H} - (y^i - \dot{\partial}{}^i \mathbf{H})\partial_k \theta_i = -\partial_k \mathcal{L}.$

Substituting (4.10) in the second term of the left side of (4.12) we obtain using (4.7) that

(4.13) $\partial_k \mathbf{H} - \underset{(\mu)}{w}\,(x,y)\dot{\partial}{}^i\,\underset{(\mu)}{G}\,(\partial_k \theta_i) = -\partial_k \mathcal{L}.$

Since in S_1 the equations $\underset{(\mu)}{G}(x, \theta(x, y)) = O$ hold, we obtain

(4.14)
$$\partial_k \underset{(\mu)}{G} = -\dot{\partial}^i \underset{(\mu)}{G} (\partial_k \theta_i).$$

We replace it in (4.13), so it follows

(4.15)
$$-\partial_k \mathcal{L} = \partial_k \mathbf{H} + \underset{(\mu)}{w}(x, y)\partial_k \underset{(\mu)}{G}.$$

Let us now consider the equations of constrained geodesics in H^n and L^n, respectively.

If a curve $C : (x^i = x^i(t), t \in (t_1, t_2))$ is a constrained geodesic of H^n then it necessarily satisfies the equations $\underset{(\mu)}{G}(x(t), p(t)) = 0$ and $H_i(\mathbf{H}) = O$ where its lifting to $T^* M$ is given by $(x(t), p(t))$. In this case the functions $p_i(t)$ defined by

(4.16)
$$p_i(t) = \theta_i\left(x(t), \frac{dx(t)}{dt}\right)$$

satisfy the following relations on S_1

(4.17)
$$\frac{dp_i}{dt} = \partial_i \mathcal{L}\left(x(t), \frac{dx(t)}{dt}\right).$$

Due to the relation (4.11) we get

(4.18)
$$\frac{dx^i}{dt} = \dot{\partial}^i \mathbf{H} + \underset{(\mu)}{w}\, \dot{\partial}^i \underset{(\mu)}{G}.$$

Moreover, if we observe the equations in (4.15) we obtain
(4.19)
$$-\partial_k \mathcal{L}\left(x(t), \frac{dx(t)}{dt}\right) = \partial_k \mathbf{H}(x(t), p(t)) + \underset{(\mu)}{w}\left(x(t), \frac{dx(t)}{dt}\right)\partial_k \underset{(\mu)}{G}(x(t), p(t)).$$

Using the equations (4.17) we get

(4.20)
$$-\frac{dp_i}{dt} = \partial_i \mathbf{H}(x(t), p(t)) + \underset{(\mu)}{w}\left(x(t), \frac{dx(t)}{dt}\right)\partial_i \underset{(\mu)}{G}(x(t), p(t)).$$

Hence we can summarize our result in the

Proposition 4.1. *If a curve C is a constrained geodesic of H^n then it is a solution of the following equations:*

(4.21) $$\frac{dx^i}{dt} = \dot\partial^i \mathbf{H} + \underset{(\mu)}{w}\, \dot\partial^i \underset{(\mu)}{G}, \qquad \frac{dp_i}{dt} + \partial_i \mathbf{H} + \underset{(\mu)}{w}\, \partial_i \underset{(\mu)}{G} = O.$$

∎

They will be called *Hamilton equations for constrained geodesics.*

The Euler–Lagrange equations of geodesics (see in $11°$ of the section 2.B.) can be written in the following "geometrical" form

(4.22) $$g_{ij}\frac{d^2 x^j}{dt^2} + G_i\left(x, \frac{dx}{dt}\right) = O,$$

where

$$G_i = y^s \partial_s \dot\partial_i \mathcal{L} - \partial_i \mathcal{L}.$$

Because of the condition (4.6), multiplying (4.22) by $\underset{(\mu)}{b}{}^i$ we get

(4.23) $$\underset{(\mu)}{\Theta} \equiv \underset{(\mu)}{b}{}^i G_i\left(x, \frac{dx}{dt}\right) = O \qquad (\mu = \overline{1,m}).$$

Hence we can see that the Hamiltonian constraints have generated new constraints in L^n. They will be called *Lagrangian constraints.*

Now we have

Theorem 4.1. *If a curve C is a constrained geodesic of H^n then it satisfies the equations $\underset{(\mu)}{\Theta} \equiv \underset{(\mu)}{b}{}^i G_i = O$ on S_1 in L^n.* ∎

Let us consider two variational problems subject to constraints in $H^{*n} = (M, \mathbf{H}^*(x, p^*))$ and $H^n = (M, \mathbf{H}(x, p))$. If the curves C^* and C are constrained geodesics in H^{*n} and H^n respectively, then they necessarily satisfy the equations
(4.24)

(a) $$\frac{dx^i}{dt} = \dot\partial^{*i} \mathbf{H}^* + \underset{(\mu)}{w}{}^* \dot\partial^{*i} \underset{(\mu)}{\overset{*}{G}}; \qquad \frac{dp_i^*}{dt} + \partial_i \mathbf{H}^* + \underset{(\mu)}{w}{}^* \partial_i \underset{(\mu)}{\overset{*}{G}} = O$$

and

(b) $$\frac{dx^i}{dt} = \dot\partial^i \mathbf{H} + \underset{(\mu)}{w}\, \dot\partial^i \underset{(\mu)}{G}; \qquad \frac{dp_i}{dt} + \partial_i \mathbf{H} + \underset{(\mu)}{w}\, \partial_i \underset{(\mu)}{G} = O.$$

By means of the above results the Hamiltonian constraints in H^{*n} and H^n generate the following Lagrangian constraints in L^{*n} and L^n, respectively:

(4.25)

(a) $$\underset{(\mu)}{\Theta^*} \equiv \underset{(\mu)}{b}{}^{*i} G_i^* = O$$

$$\left(G_i^* = y^s \partial_s \dot{\partial}_i \mathcal{L}^* - \partial_i \mathcal{L}^*, \quad \underset{(\mu)}{b}{}^{*i} = \dot{\partial}^{*i} \underset{(\mu)}{\overset{*}{G}} (x, p^*) \right);$$

(b) $$\underset{(\mu)}{\Theta} \equiv \underset{(\mu)}{b}{}^i G_i = O$$

$$\left(G_i = y^s \partial_s \dot{\partial}_i \mathcal{L} - \partial_i \mathcal{L}, \quad \underset{(\mu)}{b}{}^i = \dot{\partial}^i \underset{(\mu)}{G} (x, p) \right).$$

Now we define the Moór equivalence of two constrained variational problems in Hamilton spaces.

Definition 4.1. *Two constrained variational problems H^{*n} and H^n are equivalent in the Moór sense if the relations*

(4.26)

(a) $$\dot{\partial}^{*i} \mathbf{H}^* + \underset{(\mu)}{w}{}^* \dot{\partial}^{*i} \underset{(\mu)}{\overset{*}{G}} = \dot{\partial}^i \mathbf{H} + \underset{(\mu)}{w} \dot{\partial}^i \underset{(\mu)}{G}$$

(b) $$\frac{dp_i^*}{dt} + \partial_i \mathbf{H}^* + \underset{(\mu)}{w}{}^* \partial_i \underset{(\mu)}{\overset{*}{G}} = \delta_i^j \lambda(x, p) \left(\frac{dp_j}{dt} + \partial_j \mathbf{H} + \underset{(\mu)}{w} \partial_j \underset{(\mu)}{G} \right)$$

$$(\lambda(x, p) \neq 0)$$

hold identically in $S_O \subset T^* M$.

Using Theorem 4.1. and the relations (4.25)(a), (b) from Definition 4.1. it directly follows that the relation

(4.27) $$\underset{(\mu)}{\Theta^*} \equiv \underset{(\mu)}{b}{}^{*i} G_i^* = \lambda \underset{(\mu)}{b}{}^i G_i = \lambda \underset{(\mu)}{\Theta}$$

holds in L^{*n} and in L^n respectively.

In the paper [3] we have also proved that if two unconstrained variational problems L^{*n} and L^n are equivalent in the Moór sense then the relation

(4.28) $$\delta_i^j \lambda(x, y) G_j = G_i^*$$

holds.

Moreover, it is clear that if a curve $\mathcal{C}$ is a constrained geodesic in H^n (or in L^n) then it is necessarily an unconstrained geodesic too. Hence if the curve satisfies the equations (4.26)(a), (b) and (4.27) respectively then using (4.28) we get the following conditions for the Hamilton constraints:

(a) $\underset{(\mu)}{b}{}^{*i} = \underset{(\mu)}{b}{}^{i}$

(4.29)

i.e.

(b) $\overset{*}{\partial}{}^i \underset{(\mu)}{G}{}^{*}(x, p^*) = \overset{\cdot}{\partial}{}^i \underset{(\mu)}{G}(x, p).$

References

[1] ATANASIU, GH. and KIRKOVITS, M. SZ., On singular Lagrange spaces, *Tensor* (to appear).

[2] CARATHÈODORY, C., Variationsrechnung and partielle Differentialgleichungen erster Ordnung, Teubner, Leipzig und Berlin, 1935.

[3] KIRKOVITS, M. SZ., On equivalence of variational problems and its geometrical background in Lagrange spaces, *Anal. Sti. ale Univ. "Al I. Cuza" din Iaşi, Sect. Mat.* **35** (1989), 267–272.

[4] LOVELOCK, D. and RUND, H., Tensor, Differential forms and Variational Principles, Wiley-Interscience Publ., 1975.

[5] MIRON, R. and ANASTASIEI, M., Fibrate vectorial. Spaţii Lagrange. Aplicaţii în teoria relativitaţii, Editure Academiei, Romania, Bucureşti, 1987.

[6] MIRON, R., Hamilton geometry, Univ. Timişoara, *Sem. Mecanica, Nr. 3.,* 1987, 54.

[7] MIRON, R., Sur La Géométrie des espaces d'Hamilton, *C.R. Acad. Sci. Paris, 1988, t. 306*, Serie I. 195–198.

[8] MOÓR, A., Über äquivalente Variationsprobleme erster und zweiter Ordnung, *Journal für die reine and angewandte Mathematik*, **223** (1966) 131–137.

[9] MOÓR, A., Über nicht-holonome allgemeine metrische Linienelmentraume, *Acta Math.*, **101** (1959), 201–233.

[10] RUND, H., The Hamilton–Jacobi theory in the calculus of variations, London & New York, 1966.

[11] RUND, H., On quasi-equivalent variational problems, Coll. on the Cal. of Var. Univ. South Africa, 1966.

[12] RUND, H., Über nicht-holonome allgemeine metrische Geometrie, *Math. Nachr.* **11** (1954), 61–81.

[13] SYNGE, J. L., Geodesics in non-holonomic geometry, *Math. Ann.* **99** (1928), 738–751.

[14] VRANCEANU, G., Sur les espaces non holonomes, *Comptes Rendus* **183** (1926).

Magdalen Sz. Kirkovits

University of Forestry and
Wood Sciences
Department of Mathematics
H–9401 SOPRON
Pf. 132
Hungary

COLLOQUIA MATHEMATICA SOCIETATIS JÁNOS BOLYAI
56. DIFFERENTIAL GEOMETRY, EGER (HUNGARY), 1989

Some Gauge-natural Operators on Connections*

I. KOLÁŘ

The general concept of a natural bundle over m-manifolds was introduced by A. Nijenhuis as a modern reformulation of the classical concept of an arbitrary bundle of geometric objects, [8]. Given two natural bundles F and E over m-manifolds, the idea of a natural operator $F \to E$ reflects a geometrical construction transforming sections of FM into sections of EM for every m-dimensional manifold M, see e.g. [4]. However, a disadvantage of the theory of natural bundles and operators is that it does not cover the case of the connection bundles of arbitrary ("abstract") principal fibre bundles, which play a basic role in the gauge theories of mathematical physics. For this reason D. J. Eck modified those concepts and introduced the gauge-natural bundles and gauge-natural operators, [1].

In the present paper we solve two concrete problems about gauge-natural operators. First we determine all gauge-natural operators of the curvature type. Then we find all gauge-natural operators transforming a connection on a principal fibre bundle $P \to M$ and a linear symmetric connection on the base manifold M into a linear connection on P. Our approach is based on the methods we developped in [4] and [6]. — All manifolds and maps are assumed to be infinitely differentiable.

---

* This paper is in final form and no version of it will be submitted for publication elsewhere.

1. § Let $\mathcal{M}f_m$ be the category of m-dimensional manifolds and their local diffeormorphisms, let $\mathcal{F}M$ be the category of fibred manifolds and let B denote the base functor. Fix a Lie group G and define a category $\mathcal{P}_m(G)$, whose objects are principal G-bundles $f : P \to \overline{P}$ with the base map $Bf : BP \to B\overline{P}$ lying in $\mathcal{M}f_m$.

Definition 1. *A gauge-natural bundle over m-manifolds is a functor $F :$ $\mathcal{P}_m(G) \to FM$ such that*

(a) *every $\mathcal{P}_m(G)$-object $\pi : P \to BP$ is transformed into a fibred manifold $q_P : FP \to BP$ over BP,*

(b) *every $\mathcal{P}_m(G)$-morphism $f : P \to \overline{P}$ is transformed into an FM-morphism $Ff : FP \to F\overline{P}$ over Bf,*

(c) *for every open subset $U \subset BP$, the inclusion $i : \pi^{-1}(U) \to P$ is transformed into the inclusion $Fi : q_P^{-1}(U) \to FP$.*

If we intend to point out the structure group G, we say that F is a G-natural bundle. We remark that Eck has assumed an additional continuity condition for F, but J. Slovák has recently deduced that this condition follows from the other axioms, [6].

If two $\mathcal{P}_m(G)$-morphisms $f, g : P \to \overline{P}$ satisfy $j_y^r f = j_y^r g$ at one point $y \in P_x$ of the fibre of P over $x \in BP$, then the fact that the right translations of principal bundles are diffeomorphisms implies $j_z^r f = j_z^r g$ for every $x \in P_x$. In this case we write $j_x^r f = j_x^r g$. A gauge-natural bundle F over m-manifolds is said to be of order r, if $j_x^r f = j_x^r g$ implies $Ff|F_x P = Fg|F_x P$. Let $W^r P$ be the space of all r-jets $j_0^r \varphi$, where $\varphi : \mathbb{R}^m \times G \to P$ is a principal bundle morphism of $\mathcal{P}_m(G)$, $0 \in \mathbb{R}^m$, [3], [7]. The space $W^r P$ is a principal bundle over BP with structure group $W_m^r G$, which is the group of all r-jets $j_0^r \psi$, where $\psi : \mathbb{R}^m \times G \to \mathbb{R}^m \times G$ is a morphism of $\mathcal{P}_m(G)$ satisfying $B\psi(0) = 0$. The base functor induces a group homomorphism of $W_m^r G$ into the group G_m^r of all invertible r-jets of $\mathbb{R}^m$ into $\mathbb{R}^m$ with source and target 0. Every $\mathcal{P}_m(G)$-morphism $f : P \to \overline{P}$ is extended into a principal bundle morphism $W^r f : W^r P \to W^r \overline{P}$ defined by the jet composition $W^r f(j_0^r \varphi) = j_0^r(f \circ \varphi)$. Clearly, every smooth left action of $W_m^r G$ on a manifold S determines an r-th order G-natural bundle over m-manifolds transforming every $\mathcal{P}_m(G)$-object P into the fibre bundle associated to $W^r P$ with standard fibre S and every $\mathcal{P}_m(G)$-morphism f into $(W^r f, \mathrm{id}_S)$. The converse result was deduced by Eck, [1], who used the continuity assumption, and completed by Slovák, [6]. Hence every r-th order gauge-natural bundle is a fibre bundle associated to bundle W^r.

Let $C^\infty Y$ denote the set of all sections of a fibred manifold $Y \to M$. Given another fibred manifold $Z \to M$, an operator $A : C^\infty Y \to C^\infty Z$ is called regular, if A transforms every smoothly parametrized family of sections of Y into a smoothly parametrized family of sections of Z.

Definition 2. *Let F and E be two G-natural bundles over m-manifolds. A gauge-natural operator $A : F \to E$ is a system of regular operators $A_P : C^\infty FP \to C^\infty EP$ for every $\mathcal{P}_m(G)$-object $\pi : P \to BP$ such that for every $s \in C^\infty FP$ it holds $A_{\overline{P}}(Ff \circ s \circ Bf^{-1}) = Ef \circ A_P s \circ Bf^{-1}$ for every $\mathcal{P}_m(G)$-isomorphism $f : P \to \overline{P}$ and $A_{\pi^{-1}(U)}(s|U) = (A_P s)|U$ for every open subset $U \subset BP$.*

Obviously, the k-th jet prolongation $J^k F$ of a gauge-natural bundle of order r is a gauge-natural bundle of order $k+r$. Write $J_0^k F = J_0^k F(\mathbb{R}^m \times G)$, $E_0 = E_0(\mathbb{R}^m \times G)$. According to [4], there is a canonical bijection between the k-th order G-natural operators $F \to E$ and the $W_m^s G$-equivariant maps $J_0^k F \to E_0$, where s is maximum of the orders of $J^k F$ and E.

2. § The connection bundle $QP \to BP$ of P can be defined as the factor space $QP = J^1 P/G$. Clearly, Q is a first order gauge-natural bundle. In the case $P = \mathbb{R}^m \times G$ we shall interpret $J^1 P/G$ as the restriction of $J^1 P$ over the section (x, e) of P, $e = $ the unit of G. Write (x, y), $x = (x^i)$, $i, j, \ldots = 1, \ldots, m$, $y \in G$ for an element of $\mathbb{R}^m \times G$. Given a connection Γ on $\mathbb{R}^m \times G$, $\Gamma(0, e)$ is 1-jet $j_0^1 \psi(x)$ of a section $\psi : \mathbb{R}^m \to \mathbb{R}^m \times G$, which is identified with a map $\psi : \mathbb{R}^m \to G$, $\psi(0) = e$. Hence the standard fibre of Q is $J_0^1(\mathbb{R}^m, G)_0 = \mathfrak{g} \otimes \mathbb{R}^{m*}$. Fix a basis e_p of the Lie algebra $\mathfrak{g}$ of G, $p, q, \ldots = 1, \ldots, n = \dim G$, so that the coordinate expression of an element of $\mathfrak{g} \otimes \mathbb{R}^{m*}$ is $\Gamma_i^p e_p \otimes dx^i$. Consider a $\mathcal{P}_m(G)$-isomorphism Φ of $\mathbb{R}^m \times G$

(1) $$\overline{x} = f(x), \quad \overline{y} = \varphi(x) \cdot y, \quad f(0) = 0$$

with $\varphi : \mathbb{R}^m \to G$, where the dot denotes the multiplication in G. Then $j_0^1 \Phi \in W_m^1 G$ is characterized by

(2) $\quad a = \varphi(0) \in G$, $\quad (a_i^p) = j_0^1 \left(\varphi(x) \cdot a^{-1} \right) \in \mathfrak{g} \otimes \mathbb{R}^{m*}$, $\quad (a_j^i) = j_0^1 f \in G_m^1$.

The section ψ generating $\Gamma(0, e)$ is transformed by Φ into a section $\varphi \left(f^{-1}(\overline{x}) \right) \cdot \psi \left(f^{-1}(\overline{x}) \right)$. Hence the image of $\Gamma(0, e)$ by Φ is the 1-jet of the section

(3) $$\varphi \left(f^{-1}(\overline{x}) \right) \cdot \psi \left(f^{-1}(\overline{x}) \right) \cdot a^{-1}.$$

Let $A_q^p(a)$ be the coordinate expression of the adjoint representation of G and let tilda denote the inverse matrix. Then (3) gives the following equations of the action of $W_m^1 G$ on $\mathfrak{g} \otimes \mathbb{R}^{m*}$

$$(4) \qquad \overline{\Gamma}_i^p = A_q^p(a)\Gamma_j^q \tilde{a}_i^j + a_j^p \tilde{a}_i^j.$$

We remark that (4) expresses the well-known fact that the connection bundle $QP \to BP$ is an affine bundle with associated vector bundle $LP \otimes T^* BP$, where LP is the fibre bundle associated to P with standard fibre $\mathfrak{g}$ with respect to the adjoint action.

Let $\Gamma_{i,j}^p$ be the induced coordinates on the standard fibre $J_0^1 Q$, i.e. $\Gamma_{i,j}^p$ are the "formal" derivatives of Γ_i^p with respect to x^j, [3]. Taking the local coordinates on G determined by the basis e_p of $\mathfrak{g}$, the additional coordinates on $W_m^2 G$ are a_{ij}^p corresponding to the second partial derivetives of the map $\varphi(x) \cdot a^{-1}$ at $x = 0$ and $a_{jk}^i = \partial_{jk}^2 f^i(0)$. Let $(\tilde{a}_j^i, \tilde{a}_{jk}^i)$ be the inverse element of $(a_j^i, a_{jk}^i) \in G_m^2$. Using the general prolongation procedure of [3], we deduce from (4) that the action of $W_m^2 G$ on $J_0^1 Q$ is (4) and

$$(5) \qquad \overline{\Gamma}_{i,j}^p = A_q^p(a)\Gamma_{k,l}^q \tilde{a}_i^k \tilde{a}_j^l + B_{qr}^p(a)\Gamma_k^q a_l^r \tilde{a}_i^k \tilde{a}_j^l +$$
$$+ A_q^p(a)\Gamma_k^q \tilde{a}_{ij}^k + a_{kl}^p \tilde{a}_i^k \tilde{a}_j^l + C_{qr}^p(a)a_k^q a_l^r \tilde{a}_i^k \tilde{a}_j^l + a_k^p \tilde{a}_{ij}^k$$

where the B's and C's are some functions on G, which we shall not need.

3. § The curvature of a connection Γ on P can be considered as a section $C_P \Gamma : BP \to LP \otimes \Lambda^2 T^* BP$ and the curvature operator $C : L \otimes \Lambda^2 T^* B$ is a gauge-natural operator because of the geometric definition of the curvature. We are going to determine all gauge-natural operators $Q \to L \otimes \overset{2}{\otimes} T^* B$. (We shall see that the values of all of them lie in $L \otimes \Lambda^2 T^* B$. But this is an interesting geometric result that the antisymmetry of such operators is a consequence of their gauge-naturality.) Let $Z \subset \mathrm{Lin}(\mathfrak{g}, \mathfrak{g})$ be the subspace of all linear maps commuting with the adjoint action of G. Since every $z \in Z$ is an equivariant linear map between the standard fibres, it induces a vector bundle morphism $\overline{z}_P : LP \to LP$. Hence we can construct a modified curvature operator $C(z)_P = (\overline{z}_P \otimes \Lambda^2 T^* \, \mathrm{id}_{BP}) \circ C_P$.

Proposition 1. *All gauge-natural operators* $Q \to L \otimes \overset{2}{\otimes} T^* B$ *are the modified curvature operators* $C(z)$ *for all* $z \in Z$.

Proof. By Proposition 3 of [4], every gauge-natural operator A on the connection bundle has finite order. The r-th order gauge-natural operators correspond to the $W_m^{r+1}G$-equivariant maps $J_0^r Q \to \mathfrak{g} \otimes \overset{2}{\otimes} \mathbb{R}^{m*}$. Let $\Gamma_{i,\alpha}^p$ be the induced coordinates on $J_0^r Q$, where α is a multiindex of range m with $|\alpha| \leq r$. On $\mathfrak{g} \otimes \overset{2}{\otimes} \mathbb{R}^{m*}$ we have the canonical coordinates y_{ij}^p and the action

$$(6) \qquad\qquad \bar{y}_{ij}^p = A_q^p(a) y_{kl}^q \tilde{a}_i^k \tilde{a}_j^l.$$

Hence coordinate components of the map associated to A are some functions $f_{ij}^p(\Gamma_{k,\alpha}^q)$. If we consider the canonical inclusion of $G \times G_m^1$ into $W_m^{r+1}G$, then the transformation laws of all quantities $\Gamma_{i,\alpha}^p$ are tensorial, [3]. The equivariancy with respect to the homotheties in G_m^1 gives a homogeneity condition

$$(7) \qquad\qquad c^2 f_{ij}^p(\Gamma_{k,\alpha}^q) = f_{ij}^p(c^{1+|\alpha|}\Gamma_{k,\alpha}^q), \qquad 0 \neq c \in \mathbb{R}.$$

By the homogeneous function theorem, [6], f_{ij}^p is independent on $\Gamma_{i,\alpha}^p$ with $|\alpha| \geq 2$, linear in $\Gamma_{i,j}^p$ and quadratic in Γ_i^p. Using the invariant tensor theorem for G_m^1, [4], [6], we then get

$$(8) \qquad\qquad f_{ij}^p = a_q^p \Gamma_{i,j}^q + b_q^p \Gamma_{j,i}^q + c_{qr}^p \Gamma_i^q \Gamma_j^r$$

with real coefficients.

Considering the equivariancy of (8) with respect to the subgroup $a = e$, $a_i^p = 0$, $a_j^i = \delta_j^i$, $a_{jk}^i = 0$, we obtain $(a_q^p + b_q^p)a_{ij}^q = 0$. This gives

$$(9) \qquad\qquad b_q^p = -a_q^p.$$

The equivariancy of (8) with (9) with respect to the canonical inclusion of G into $W_m^2 G$ implies $a_q^p A_r^q(a)(\Gamma_{i,j}^r - \Gamma_{j,i}^r) = A_q^p(a)a_r^q(\Gamma_{i,j}^r - \Gamma_{j,i}^r)$. Hence a_q^p commutes with the adjoint action. Now it suffices to prove that for every a_q^p there exists at most one collection of c_{qr}^p such that (8) with (9) is a gauge-natural operator. Indeed, let $a_q^p(\Gamma_{i,j}^q - \Gamma_{j,i}^q) + \bar{c}_{qr}^p \Gamma_i^q \Gamma_j^r$ be another gauge-natural operator $\bar{A}$. Then the equivariancy of the difference $A - \bar{A}$ yields $(c_{qr}^p - \bar{c}_{qr}^p)\Gamma_i^q a_j^r = 0$, i.e. $c_{qr}^p = \bar{c}_{qr}^p$. $\blacksquare$

In the case the structure group is the general linear group $GL(n, \mathbb{R})$ in an arbitrary dimension n, the invariant tensor theorem, [4], [6], implies directly that the ad-invariant linear maps $\mathfrak{gl}(n, \mathbb{R}) \to \mathfrak{gl}(n, \mathbb{R})$ are generated by the identity and the map $A \mapsto (\text{trace } A)\,\text{id}$. Then Proposition 1 gives a two parameter family of all $GL(n, \mathbb{R})$-natural operators $Q \to L \otimes \overset{2}{\otimes} T^* B$, which we deduced by direct evaluation in [4]. (It is remarkable that the study of the case of the special structure group $GL(n, \mathbb{R})$, in which we can apply the generalized invariant tensor theorem, [4], [6], plays a useful heuristic role in the theory of gauge-natural operators.)

4. § Our next problem is to study the gauge-natural operators transforming a connection Γ on a principal fibre bundle $\pi : P \to BP$ and a linear connection Λ on the base manifold BP into a linear connection on P. In this case our operators are base-extending, see [5]. First we present a geometrical construction of such an operator N transforming (Γ, Λ) into $N_P(\Gamma, \Lambda) : TP \to J^1(TP \to P)$. Let vX be the vertical component of a vector $X \in T_y P$ and bX be its projection to the base manifold. Consider a vector field ξ on BP such that $j_x^1 \xi = \Lambda(bX)$, $x = \pi(y)$. Construct the Γ-lift $\Gamma\xi$ of ξ and the fundamental vector field $\varphi(vX)$ determined by vX. An easy calculation shows that the rule

(10) $X \mapsto j_y^1 \left(\Gamma\xi + \varphi(vX) \right)$

determines a linear connection $N(\Gamma, \Lambda)$ on P.

J. Gancarcewicz and the author discussed a similar problem in the case of a vector bundle, [2]. Their result suggests that the case the linear connection Λ is without torsion is much simplier than the general case. That is why we restrict ourselves to a symmetric Λ here. Since the difference of two linear connections on P is a tensor of $TP \otimes T^* P \otimes T^* P$, we characterize all gauge-natural operators of our type as a sum of the operator N defined by (10) and of the gauge-natural difference tensors. We recall that the fundamental vector fields on P identify the vertical tangent bundle VP of P with $P \times \mathfrak{g}$.

We construct geometrically the following 3 systems of difference tensors.

I. The connection form ω of Γ is a linear map $\omega : TP \to \mathfrak{g}$. Take any bilinear map $f_1 : \mathfrak{g} \times \mathfrak{g} \to \mathfrak{g}$ and compose $\omega \oplus \omega$ with f_1. This defines an n^3-parameter system of difference tensors $TP \otimes TP \to VP$, $n = \dim G$.

II. The curvature form $D\omega$ of ω is a bilinear map $TP \oplus TP \to \mathfrak{g}$. Take any linear map $f_2 : \mathfrak{g} \to \mathfrak{g}$ and compose $D\omega$ with f_2. This yields an n^2-parameter system of difference tensors.

III. If we apply Lemma 1 of [4] to a linear symmetric connection on BP, we find that all natural operators transforming a linear symmetric connection Λ into a tensor of $T^*BP \otimes T^*BP$ form a 2-parameter family linearly generated by both different contractions R_1 and R_2 of the curvature tensor of Λ. The tangent map of bundle projection π defines the dual injection $T^*BP \to T^*P$. Taking any fundamental vector field $\overline{Y}$ determined by a vector $Y \in \mathfrak{g}$, we obtain a $2n$-parameter system of difference tensors linearly generated by $\overline{Y} \otimes R_1$ and $\overline{Y} \otimes R_2$.

Proposition 2. *All gauge-natural operators transforming a connection on P and a linear symmetric connection on the base manifold BP into a linear connection on P form the $(n^3 + n^2 + 2n)$-parameter family generated by operator N and by the above families I., II. and III. of the difference tensors.*

The proof requires several steps (see below).

5. § Using the notation of Section 2, consider the fundamental vector fields $\overline{e}_p$ on $P = \mathbb{R}^m \times G$ corresponding to a fixed basis e_p of $\mathfrak{g}$. Define the additional coordinates X^i, E^p on TP by means of the decomposition

$$(11) \qquad X^i \frac{\partial}{\partial x^i} + E^p \overline{e}_p.$$

Let ω^p be the 1-forms on G dual to $\overline{e}_p$. Then the equations of a linear connection on $\mathbb{R}^m \times G$ are

$$(12) \qquad \begin{aligned} dX^i &= \left(D^i_{jk} X^j + D^i_{pk} E^p \right) dx^k + \left(D^i_{jq} X^j + D^i_{pq} E^p \right) \omega^q \\ dE^p &= \left(D^p_{ij} X^i + D^p_{qj} E^q \right) dx^j + \left(D^p_{ir} X^i + D^p_{qr} E^q \right) \omega^r \end{aligned}$$

where the D's are any smooth functions on $\mathbb{R}^m \times G$.

Differentiating transformation (1), we find

$$(13) \qquad \overline{X}^i = a^i_j X^j, \qquad \overline{E}^p = a^p_j X^j + A^p_q(a) E^q.$$

The same holds for dx^i and ω^p

$$(14) \qquad d\overline{x}^i = a^i_j dx^j, \qquad \overline{\omega}^p = a^p_j dx^j + A^p_q(a) \omega^q$$

while for dX^i and dE^p we obtain by the standard prolongation procedure

$$(15) \qquad d\overline{X}^i = a^i_{jk} X^j dx^k + a^i_j dX^j$$

$$(16) \qquad \begin{aligned} d\overline{E}^p &= a^p_{ij} X^i dx^j + F^p_{qr}(a) a^q_i X^i \omega^r + H^p_{qr}(a) E^q \omega^r + \\ &\quad + a^p_i dX^i + A^p_q(a) dE^q \end{aligned}$$

where the F's and H's are some smooth functions on G, which we shall not need. Let Z be the standard fibre of the bundle of all linear connections on $\mathbb{R}^m \times G$ over $0 \in \mathbb{R}^m$. Hence the elements of Z are characterized by the foot point $y \in G$ and by the D's from (12). The action of $W_m^2 G$ on Z can be deduced from (12)–(16), but we shall do it gradually in the course of the proof. The classical action of G_m^2 on the standard fibre $S = (\Lambda^i_{jk})$ of the natural bundle of all linear symmetric connections is, [4],

$$(17) \qquad \overline{\Lambda}^i_{jk} = a^i_l \Lambda^l_{mn} \tilde{a}^l_j \tilde{a}^m_k + a^i_{lm} \tilde{a}^l_j \tilde{a}^m_k .$$

The generalization of the classical Peetre theorem by J. Slovák, [6], [9], implies that every gauge-natural operator of our type has finite order. By Proposition 2 of [5], the r-th order operators of our type are in bijection with the $W_m^{r+1} G$-equivariant maps $f : J_0^r Q \times S^r \times G \to Z$ satisfying $\varphi \circ f = pr_3$, where S^r is the standard fibre of the r-th jet prolongation of the bundle of all linear symmetric connections and $\varphi : Z \to G$ is the foot point map. Thus, at the beginning the D's are any smooth functions of $\Gamma^p_{i,\alpha}, \Lambda^i_{jk,\beta}, y$ with multiindices $|\alpha| \leq r$, $|\beta| \leq r$ and they should be simplified by the equivariancy conditions.

6. § From (15) we deduce the simpliest transformation law

$$(18) \qquad \overline{D}^i_{pq} = a^i_j D^j_{rs} \tilde{A}^r_p(a) \tilde{A}^s_q(a).$$

Since the transformation laws of all Γ's and Λ's on the subgroup $G \times G_m^1 \subset W_m^{r+1} G$ are tensorial, the equivariancy with respect to the homotheties in G_m^1 yields

$$\frac{1}{c} D^i_{pq}(\Gamma^p_{i,\alpha}, \Lambda^i_{jk,\beta}, y) = D^i_{pq}(c^{1+|\alpha|} \Gamma^p_{i,\alpha}, c^{1+|\beta|} \Lambda^i_{jk,\beta}, y).$$

For $c \to 0$ this implies

$$(19) \qquad D^i_{pq} = 0.$$

Then the transformation law of D^i_{jp} is

$$(20) \qquad \overline{D}^i_{jp} = a^i_k D^k_{lq} \tilde{A}^q_p(a) \tilde{a}^l_j .$$

The homotheties in G_m^1 yield

$$D_{jp}^i(\Gamma_{i,\alpha}^p, \Lambda_{jk,\beta}^i, y) = D_{jp}^i(c^{1+|\alpha|}\,\Gamma_{i,\alpha}^p, c^{1+|\beta|}\,\Lambda_{jk,\beta}^i, y).$$

For $c \to 0$ this gives that D_{jp}^i depends on y only. Quite similarly we find that D_{pj}^i depends on y only.

The transformation law of D_{jk}^i is

(21) $\qquad a_{jk}^i + a_l^i D_{jk}^l = \overline{D}_{lm}^i a_j^l a_k^m + \overline{D}_{pl}^I a_j^p a_k^l + \overline{D}_{lp}^i a_j^l a_k^p.$

By the homogeneous function theorem, [6], the homotheties in G_m^1 yield D_{jk}^i is independent on $\Gamma_{i,\alpha}^p$ and $\Lambda_{jk,\alpha}^i$ with $|\alpha| \geq 1$ and linear in $\Gamma_i^p, \Lambda_{jk}^i$. The invariant tensor theorem then implies

$$D_{jk}^i = a_p \delta_j^i \Gamma_k^p + b_p \delta_k^i \Gamma_j^p + c\Lambda_{jk}^i + d\delta_j^i \Lambda_{lk}^l + e\delta_k^i \Lambda_{lj}^l$$

where the coefficients are smooth functions on G. Evaluating directly the equivariancy on the subgroup $a = e$, $a_j^i = \delta_j^i$, we find

(22) $\qquad D_{jp}^i = 0, \quad D_{pj}^i = 0, \quad D_{jk}^i = \Lambda_{jk}^i.$

7. § From (16) we first deduce

(23) $\qquad H_{qr}^p(a) + A_s^p(a)D_{qr}^s = \overline{D}_{st}^p A_q^s(a)A_r^t(a).$

The homotheties in G_m^1 yield D_{qr}^p depends on y only. Then (23) shows that $D_{qr}^p(a)$ is determined by $D_{qr}^p(e)$.

The transformation law of D_{iq}^p is

(24) $\qquad F_{rq}^p(a)a_i^r + A_r^p(a)D_{iq}^r = \overline{D}_{jr}^p a_i^j A_q^r(a) + \overline{D}_{rs}^p a_i^r A_q^s(a).$

The homotheties in G_m^1 yield D_{iq}^p is linear in Γ_i^p and Λ_{jk}^i. The invariant tensor theorem then implies

$$D_{iq}^p = M_{qr}^p(a)\Gamma_i^r + N_q^p(a)\Lambda_{ji}^j.$$

The equivariancy on G_m^2 gives $N_q^p(a) = 0$. Then the subgroup $a = e$, $a_j^i = \delta_j^i$ yields

(25) $\qquad F_{rq}^p(a)a_i^r - D_{rq}^p(a)a_i^r = M_{qr}^p(a)a_i^r.$

Hence $M_{qr}^p(a)$ is determined by $D_{qr}^p(a)$. Quite similarly we deduce that D_{qi}^p is linear in Γ_i^r with the coefficients determined by $D_{qr}^p(a)$. The arbitrary constants $D_{qr}^p(e)$ corresponds to the system I. of the difference tensors.

8. § The transformation law of D_{ij}^p is

$$(26) \quad a_{ij}^p + a_k^p D_{ij}^k + A_q^p(a) D_{ij}^q = \overline{D}_{kl}^p a_i^k a_j^l + \overline{D}_{qk}^p a_i^q a_j^k + \overline{D}_{kq}^p a_i^k a_j^q +$$
$$+ \overline{D}_{qr}^p a_i^q a_j^r.$$

The homotheties in G_m^1 give a homogeneity condition

$$c^2 D_{ij}^p(\Gamma_{i,\alpha}^p, \Lambda_{jk,\beta}^i, y) = D_{ij}^p(c^{1+|\alpha|}, \Gamma_{i,\alpha}^p, c^{1+|\beta|} \Lambda_{jk,\beta}^i, y).$$

This implies that D_{ij}^p is linear in $\Gamma_{i,j}^p$ and $\Lambda_{jk,l}^i$ and bilinear in Γ_i^p and Λ_{jk}^i with coefficients in y. The invariant tensor theorem yields

$$(27) \quad D_{ij}^p = a_q^p(y)\Gamma_{i,j}^q + B_q^p(y)\Gamma_{j,i}^q + c_{qr}^p(y)\Gamma_i^q\Gamma_j^r + e_q^p(y)\Gamma_i^q\Lambda_{kj}^k +$$
$$+ f_q^p(y)\Gamma_j^q\Lambda_{ki}^k + g_q^p(y)\Gamma_k^q\Lambda_{ij}^k + h_{ij}^p(y, \Lambda_{jk}^i, \Lambda_{jk,l}^i).$$

Functions h_{ij}^p for $y = e$ determine the classical natural operators with values in $T^*B \times T^*B$. By Lemma 1 of [4] they correspond to the system III. of the difference tensors. On the subgroup $a_j^i = \delta_j^i$ of G_m^2, the equivariancy of D_{ij}^p implies $e_q^p = 0$, $f_q^p = 0$, $g_q^p(y) = \delta_q^p$. In the same way as in the proof of Proposition 1 we then deduce that the remaining expression has the form $\Gamma_{i,j}^p + a_q^p(e)C_{ij}^q$, where C_{ij}^p is the coordinate expression of the curvature operator of Γ. The quantities $a_q^p(e)$ correspond to the system II. of the difference tensors, while $\Gamma_{i,j}^p$ and $\Gamma_k^p\Lambda_{ij}^k$ together with Λ_{jk}^i from (22) represent the operator N. This completes the proof of Proposition 2. ∎

References

[1] D. J. ECK, Gauge-natural bundles and generalized gauge theories, *Mem. Amer. Math. Soc.*, Vol. 33, No 247 (1981).

[2] J. GANCARCEWICZ, I. KOLÁŘ, Some gauge-natural operators on linear connections, to appear.

[3] I. KOLÁR, On the prolongations of geometric object fields, *An. Sti. Univ. "Al. I. Cuza"*, Iasi 17 (1971), 437–446.

[4] I. KOLÁR, Some natural operators in differential geometry, *Proc. "Conf. Diff. Geom. and its Applications"*, Brno 1986, Dordrecht 1987, 91–110.

[5] I. KOLÁR, Gauge-natural operators transforming connections to the tangent bundle, to appear.

[6] I. KOLÁR, P. W. MICHOR, J. SLOVÁK, Natural Operations in Differential Geometry, to appear.

[7] P. LIBERMANN, Sur les prolongements des fibrés principaux et des groupoides différentiables banachiques, *Analyse globale, Sém. Math. Supérieures,* No. 42 (Été, 1969), Presse Univ. Montreal, 1971, 7–108.

[8] A. NIJENHUIS, Natural bundles and their general properties, *Diff. Geom. in honour of K. Yano*, Kinokuniya, Tokyo 1972, 317–334.

[9] J. SLOVÁK, On the finite order of some operators, *Proc. "Conf. Diff. Geom. and its Applications"*, Brno 1986, Communications, Brno, 1987, 283–294.

Ivan Kolár

Institute of Mathematics of the ČSAV,
branch Brno
Mendelovo nám. 1
CS 66282 Brno, Czechoslovakia

COLLOQUIA MATHEMATICA SOCIETATIS JÁNOS BOLYAI
56. DIFFERENTIAL GEOMETRY, EGER (HUNGARY), 1989

Pseudoconnections on an Almost Complex Manifold[*]

Z. KOVÁCS

The following facts are well-known:

1. On an almost complex manifold (M, J) there exists a torsion-free almost complex linear connection iff the almost complex structure J is torsion-free.

2. On an almost Hermitian manifold (M, J, g) the Levi–Civita connection is almost complex iff the manifold is Kählerian.

We shall study analogous problems for pseudoconnections. The class of almost complex pseudoconnections has been studied by Italian mathematicians and they have given a necessary and sufficient condition for Problem 1 in case of the pseudoconnections ([5, Prop. 4.6]) and they have published some partial results for Problem 2 ([4]).

In this paper we shall complete the answer to Problem 2 and give some other remarks on these topics.

---

[*] This paper is in final form and no version of it will be submitted for publication elsewhere.

1. Basic notations

In this paper differentiable manifolds are of finite dimension, of Hausdorff-type, and of countable base. All manifolds, functions, and maps are C^∞. $C^\infty(M)$ denotes the algebra of real functions on the manifold M and τ_M means the tangent bundle of M with bundle projection π and $\pi^*(\tau_M)$ denotes the pullback bundle of τ_M with respect to (w.r.t.) π. The module of the sections of τ_M is denoted by $\mathfrak{X}(M)$.

For the $C^\infty(M)$-modules $M_0, M_1, \ldots, M_k$ $\mathrm{Hom}(M_1, \ldots, M_k; M_0)$ denotes the module of multilinear maps $M_1 \times \ldots \times M_k \to M_0$.

Concerning further terminology we refer to the monograph [8].

2. Preliminaries

A. Pseudoconnections ([2]) By a *pseudoconnection* on a manifold M we mean a pair (∇, A), where $A \in \mathrm{End}(\tau_M)$ is a strong bundle endomorphism of the tangent bundle of M, and $\nabla : \mathfrak{X}(M) \times \mathfrak{X}(M) \to \mathfrak{X}(M)$ is a map, which is $C^\infty(M)$-linear in its first variable, additive in its second variable, and the property

$$\nabla_X fY = (AX)fY + f\nabla_X Y \qquad (f \in C^\infty(M))$$

is also satisfied.

The covariant derivatives of the tensor fields

$$K \in \mathrm{Hom}\left(\overset{1}{\mathfrak{X}}(M), \ldots, \overset{p}{\mathfrak{X}}(M); \mathfrak{X}(M)\right)$$

and

$$F \in \mathrm{Hom}\left(\overset{1}{\mathfrak{X}}(M), \ldots, \overset{p}{\mathfrak{X}}(M); C^\infty(M)\right)$$

are given by the expressions:

$$(\nabla_X K)(Y_1, \ldots, Y_p) = \nabla_X K(Y_1, \ldots, Y_p) - \sum_{i=1}^p K(Y_1, \ldots, \nabla_X Y_i, \ldots, Y_p)$$

and

$$(\nabla_X K)(Y_1,\ldots,Y_p) = (AX)F(Y_1,\ldots,Y_p) - \sum_{i=1}^{p} F(Y_1,\ldots,\nabla_X Y_i,\ldots,Y_p)$$

respectively. On the base of these formulas a pseudoconnection (∇, A) is almost complex on an almost complex manifold (M, J) if $\nabla_X J = 0$ $(\forall X \in \mathfrak{X}(M))$ and it is metrical on a Riemannian manifold (M, g) if $\nabla_X g = 0$ $(\forall X \in \mathfrak{X}(M))$.

B. The twisted Lie-bracket and the operator d_A ([9] [5]) If $A \in \text{End}(\tau_M)$, then $L_A(X,Y) = [AX,Y] + [X,AY] - A[X,Y]$ is the *twisted Lie-bracket* of vector fields X, Y. Applying this bracket, we call the tensor $\Sigma \in \text{Hom}(\mathfrak{X}(M), \mathfrak{X}(M); \mathfrak{X}(M))$, $\Sigma(X,Y) = \nabla_X Y - \nabla_Y X - L_A(X,Y)$ the *torsion tensor* of the pseudoconnection (∇, A).

We denote by $\mathcal{D}^r(M)$ the space of all exterior differential forms of degree r on M, and let $\mathcal{D}(M) = \sum_0^{\dim M} \mathcal{D}^r(M)$. d_A denotes the antiderivation of degree 1 of $\mathcal{D}(M)$ which acts on $\mathcal{D}^0(M)$ and on $\mathcal{D}^1(M)$ respectively, in the following way:

$$(d_A f)(X) = (AX)f \qquad (f \in \mathcal{D}^0(M) \equiv C^\infty(M))$$

$$(d_A \omega)(X,Y) = \frac{1}{2}(AX\omega(Y) - AY\omega(X) - \omega(L_A(X,Y))) \quad (\omega \in \mathcal{D}^1(M)).$$

C. The Nijenhuis operator Let $K : \mathfrak{X}(M) \times \mathfrak{X}(M) \to \mathfrak{X}(M)$ be an arbitrary $\mathbf{R}$-linear map, and $C \in \text{End}(\tau_M)$. We define the map $K^C : \mathfrak{X}(M) \times \mathfrak{X}(M) \to \mathfrak{X}(M)$, as follows:

$$K^C(X,Y) = K(CX,CY) + C^2 K(X,Y) - CK(CX,Y) - CK(X,CY).$$

For $K(X,Y) = [X,Y]$ $2K^C$ is the Nijenhuis torsion N_C of C and the operator $\mathcal{N}^C : K \mapsto K^C$ is called *Nijenhuis operator*.

In the following we shall consider $C = J$ (J is a fixed almost complex structure on M.) Then $(L_A)^J \equiv L_A^J$ is a $C^\infty(M)$-bilinear map for all $A \in \text{End}(\tau_M)$ (i.e. a tensor field) and the following simple relation holds:

(1) $$L_A^J(X,Y) = N_{AJ}(JX,Y) + N_{AJ}(X,JY) + \frac{1}{2}AN_J(X,Y)$$

(where N_{AJ} means the Nijenhuis torsion of the operator pair (A, J) [10, vol. I ch.1 §3]).

3. The problem for non metrical pseudoconnections

M. Falcitelli and A. M. Pastore have proved in [6] that if the pseudoconnection (∇, A) is almost complex on an almost complex manifold (M, J) then
$\Sigma^J(X, Y) = -L_A^J(X, Y)$. (This is an equivalent formalization of Falcitelli-
Pastore's Proposition [6, Prop. 2.3], adapted to our above formalism on the
base (1).) It follows: if there exists an almost complex pseudoconnection
(∇, A) on (M, J) with vanishing torsion tensor field, then $L_A^J = 0$. The
basic result is that the reversed proposition is also true:

Theorem 1. (equivalent from of [5, Prop. 4.6]) *On an almost complex
manifold (M, J) there exists a torsion-free almost complex pseudoconnection
(∇, A) iff $L_A^J = 0$.*

Alternative proof. (Outline.) The following two Lemmas are fundamental in our proof.

Lemma 1. *If $T \in \mathrm{Hom}(\mathfrak{X}(M), \mathfrak{X}(M); \mathfrak{X}(M))$, T is antisymmetric and
$T^J = 0$, then there exists an $S \in \mathrm{Hom}(\mathfrak{X}(M), \mathfrak{X}(M); \mathfrak{X}(M))$ such that
S is symmetric and*

$$(2) \qquad T(X, JY) - JT(X, Y) = S(X, JY) - JS(X, Y).$$

Lemma 2. *If ∇^L denotes an almost complex linear connection on an
almost complex manifold (M, J) with torsion tensor $8T$ or $(X, Y) =
N_J(X, Y)$, then*

$$L_A^J(X, Y) =$$
$$= \{(\nabla^L{}_X A)Y - \nabla^L{}_Y A)X\}^J -$$
$$- \frac{1}{8}\{N_J(AX, Y) + N_J(X, AY) - AN_J(X, Y)\}^J.$$

Now, let

$$T(X, Y) =$$
$$= -\frac{1}{16}\{N_J(AX, Y) + N_J(X, AY) - AN_J(X, Y)\}+$$
$$+ \frac{1}{2}\left[(\nabla^L{}_X A)Y - (\nabla^L{}_Y A)X\right].$$

This T is antisymmetric and by reason of Lemma 2 $T^J(X,Y) = 0$, so in view of Lemma 1 there exists a symmetric S such that (2) holds. Then we construct the following (∇, A):

$$\nabla_X Y = \nabla^L_{AX} Y + T(X,Y) - S(X,Y) \quad (X,Y \in \mathfrak{X}(M)).$$

This (∇, A) is almost complex and torsion-free. ∎

Our problem is now to find an endomorphism A (or to find all endomorphisms A) which satisfy the condition $L^J_A = 0$ for a fixed J. This problem seems to be difficult in general, so we examine a special case only.

Let B the real algebra of real 2×2-type matrices and let B^t denote its sublagebra of triangular matrices. In the following $\widehat{B}$ means B or B^t. A $\widehat{B}$ structure $\widehat{\sigma}$ (σ^t or σ) on a manifold M is a morphism of real algebras $\widehat{\sigma} : \widehat{B} \to \text{End}(\tau_M)$ such that $\widehat{\sigma}(1) = id$. If the B^t structure σ^t is the restriction of a B-structure σ we say: σ is the only B structure which extends σ^t (cf. [1, Th. 2.3]).

A linear connection ∇ on M is said to be *compatible with the $\widehat{B}$ structure* $\widehat{\sigma}$ if each operator $\widehat{\sigma}(a)$ $a \in \widehat{B}$ is parallel w.r.t. ∇ (i.e. $\nabla_X \widehat{\sigma}(a) = 0$ $\forall X \in \mathfrak{X}(M)$).

A $\widehat{B}$ structure $\widehat{\sigma}$ is said to be *integrable* if each point $p \in M$ has a neighbourhood (U, φ) such that $d\varphi \circ \widehat{\sigma}(a) \circ d\varphi^{-1}$ does not depend on its variable $x \in U$.

The giving of a B^t structure σ^t (resp. a B structure σ) on M is equivalent to the giving of two operators $P, R \in \text{End}(\tau_M)$ (resp. $\widetilde{P}, Q \in \text{End}(\tau_M)$) such that $P^2 = id$, $R^2 = 0$ and $RP = -PR = R$ (resp. $\widetilde{P}^2 = Q^2 = id$ and $\widetilde{P}Q + Q\widetilde{P} = 0$). ([1, Th. 2.1]). If there is given a B^t structure σ^t which can be extended to a B structure σ, then $P = \widehat{P}$ and $J = PQ$ is an almost complex structure on M. The integrability of σ^t implies the integrability of J.

Proposition 1. *If there is given an integrable B^t structure σ^t which is extended to a B structure σ, then for every $a \in B$ there exists an almost complex torsion-free pseudoconnection $(\nabla, \sigma(a))$ w.r.t. $J = PQ$.*

Proof. In this case there exists a (unique) torsion-free and flat linear connection ∇^L on M, which is compatible with σ ([1, Th. 2.5]), consequently this ∇^L is almost complex. Then in view of Lemma 2 and integrability of J we have

$$L^J_A(X,Y) \overset{N_J=0}{=} \left\{ (\nabla^L_X \sigma^t(a)) Y - (\nabla^L_Y \sigma^t(a)) X \right\}^J.$$

Finally $\nabla^L \sigma^t = 0$, therefore $L_A^J = 0$. ∎

Remark. With the same hypothesis as in the Proposition above. The pseudoconnection $(\nabla, \sigma(a))$, where $\nabla_X Y = \nabla^L{}_{\sigma(a)X} Y$, is torsion-free and almost complex for every $a \in B$. Indeed, in the proof of Theorem 1 $T(X, Y) = S(X, Y) = 0$.

We give an example for the case $N_J \neq 0$, too.

Proposition 2. *If there is given a B^t structure σ^t which can be extended to a B structure σ and the linear connection ∇^L with torsion $\frac{1}{8} N_J$ is compatible with σ, then $L_Q^J = 0$.*

Proof. A simple but long calculation gives that

$$\{N_J(AX, Y) + N_J(X, AY) - A N_J(X, Y)\}^J =$$
$$= -2J[N_J((A \times J)X, Y) + N_J(X, (A \times J)Y) + (A \times J)N_J(X, Y)],$$

where $A \times J = AJ + JA$. Combining this fact with Lemma 2, we obtain the relation:

$$L_Q^J(X, Y) =$$
$$= \{(\nabla^L{}_X Q) Y - (\nabla^L{}_Y Q) X\}^J +$$
$$+ \frac{1}{4} J[(Q \times J)N_J(X, Y) + N_J((Q \times J)X, Y) + N_J(X(Q \times J)Y)] =$$
$$\overset{\nabla^L Q = 0}{=} \frac{1}{4} J[(Q \times J)N_J(X, Y) + N_J((Q \times J)X, Y) + N_J(X, (Q \times J)Y)].$$

But $Q(PQ) + (PQ)Q = -Q(QP) + P = -P + P = 0$. ∎

4. The problem for metrical pseudoconnections

Let G be a Hermitian metric on an almost complex manifold (M, J). Then there exists uniquely a torsion-free metric pseudoconnection (∇, A) for every A, the so called Levi–Civita pseudoconnection ([7, Prop. 4.1]). We put the question whether or not this pseudoconnection is almost complex. According to Theorem 1 $L_A^J = 0$ is necessary for this. Now we formulate the following

Theorem 2. *A necessary and sufficient condition that a torsion-free metric pseudoconnection* (∇, A) *with endomorphism* A *satisfying the condition* $L_A^J = 0$ *should be almost complex is that* $d_A \Phi = 0$, *where* $\Phi(X, Y) = g(X, JY)$ *is the fundamental 2-form.*

Proof. To prove this theorem we use the line of reasoning known from the proof of the analogous theorem for linear connections ([10, vol. II Prop. 4.2]). The difference is in the technique of computation: we use the "twisted calculus" reviewed in our preliminaries.

To prove the *sufficiency* we observe that

$$2g\left(\nabla_X Y, Z\right) =$$
$$= AX g(Y, Z) + AY g(X, Z) - AZ g(X, Y) +$$
$$+ g(L_A(X, Y), Z) + g(L_A(Z, X), Y) + g(X, L_A(Z, Y))$$

([7, (4.1)]). Taking into account this fact and the definition of d_A one can easily derive the following relation:

$$2g\left((\nabla_X J)(Y), Z\right) =$$
$$= 3d_A \Phi(X, JY, JZ) - 3d_A \Phi(X, Y, Z) + g(L_A^J(Y, Z), JX) =$$
$$= 3d_A \Phi(X, JY, JZ) - 3d_A \Phi(X, Y, Z).$$

Accordingly if $d_A \Phi = 0$, then $\nabla_X J = 0$.

The *necessity* will be based on the following

Lemma 3. $3d_A \Phi(X, Y, Z) = \sigma\left\{(\nabla_X \Phi)(Y, Z)\right\}$ *(where* σ *means the cyclic sum w.r.t.* X, Y, Z*).*

It is easy to see, that for an almost complex pseudoconnection

$$(\nabla_X \Phi)(Y, Z) = -(\nabla_X g)(JY, Z)$$

holds, i.e. for a metric almost complex pseudoconnection $\nabla_X \Phi = 0$, consequently $d_A \Phi = 0$. ∎

After these we give the following

Definition. *We shall say that the quadruple* (M, g, J, A) *is a twisted Kählerian manifold if* J *is an almost complex structure on* M, g *is a Hermitian metric,* $A \in \mathrm{End}(\tau_M)$, *and* $L_A^J = 0$, $d_A \Phi = 0$ *hold.*

Now we can state our Theorem 2 in a more attractive form: *On a twisted Kählerian manifold the Levi–Civita pseudoconnection is almost complex.*

Finally we give some examples for twisted Kählerian manifolds.

Let (M, g, J) be a manifold with a fixed almost complex structure J and with a Hermitian metric g. Let $\mathcal{H} : \pi^*(\tau_M) \to \tau_M$ be a horizontal map. This $\mathcal{H}$ determines the horizontal lift X^h of the vector field $X \in \mathfrak{X}(M)$, and the operator h of the horizontal projection. Let us denote by X^v the vertical lift of $X \in \mathfrak{X}(M)$ and by v the operator of the vertical projection w.r.t. $\mathcal{H}$. For a fixed $\mathcal{H}$ TM carries a Riemannian metric G defined by $G(X^v, Y^v) = G(X^h, Y^h) = g(X, Y) \circ \pi$, $G(H^h.Y^v) = 0$. ([3] If $\mathcal{H}$ is the horizontal map of the Levi–Civita connection on (M, g), then this G is the Sasakian metric.)

Under the hypothesis above, the relations

$$F(X^h) = X^v, \quad F(X^v) = -X^h; \quad \tilde{J}^h(X^h) = (JX)^v, \quad \tilde{J}^h(X^v) = (JX)^h$$

define two almost complex structures on TM. For these structures we have the following

Proposition 3. *If the Nijenhuis torsion $N_{\mathcal{H}}$ of $\mathcal{H}^1$ vanishes for horizontal lifts, i.e. if $v[X^h, Y^h] = 0$ $\forall X, Y \in \mathfrak{X}(M)$, then (TM, G, F, v) and $(TM, G, \tilde{J}^h, v)$ are twisted Kählerian manifolds.*

Proof. Let us denote by J the almost complex structure F or $\tilde{J}^h$. L_v^J is a tensor, so it is enough to prove its vanishing for vertical and horizontal lifts. It is easy to see that $L_v(X^\sigma, Y^\eta) = 0$ $(\sigma, \eta = h$ or $v)$:

$$L_v(X^h, Y^h) = - v[X^h, Y^h] \overset{N_{\mathcal{H}}=0}{=} 0,$$
$$L_v(X^h, Y^v) = h[X^h, Y^v] = 0 \quad ([X^h, Y^v] \text{ is vertical}),$$
$$L_v(X^v, Y^v) = v[X^v, Y^v] = 0 \quad ([X^v, Y^v] = 0).$$

J maps a (vertical or horizontal) lift of a vector field on a (vertical or horizontal) lift, so $L_v^J = 0$.

A similar method can be used to prove $d_v \Phi = 0$:

$$d_v \Phi(X^h, Y^h, Z^h) = 0,$$
$$d_v \Phi(X^v, Y^h, Z^h) = X^v G(Y^h, FZ^h) = 0 \quad (FZ^h \text{ is a vertical lift}),$$
$$d_v \Phi(X^h, Y^v, Z^v) = - Y^v G(X^h, FZ^v) + Z^v G(X^h, FY^v) = 0$$

---

[1] $\frac{1}{2} N_{\mathcal{H}}(X, Y) = [hX, hY] + h[X, Y] - h[hX, Y] - h[X, hY]$

$(FZ^v, FY^v$ are horizontal lifts and $X^v(f \circ \pi) = 0 \, \forall X \in \mathfrak{X}(M), f \in C^\infty(M))$,
$d_v \Phi(X^v, Y^v, Z^v) = X^v G(Y^v, FZ^v) - Y^v G(X^v, FZ^v) + Z^v G(X^v, FY^v) = 0$
$(FY^v$ is a horizontal lift). ∎

We remark that $N_\mathcal{H}(X^h, Y^h) = 0$ $(\forall X, Y \in \mathfrak{X}(M))$ is not an integrability condition of J, cf. [11].

References

[1] M. ANASTASIEI and I. POPOVICI, An intrinsic characterisation of Finsler connections. In *Proc. Sem. on Finsler Spaces*, 27–39, Brasov, (1980).

[2] C. DI COMITE, Pseudoconnessioni lineari su una varieta differenziabile di classe C^∞. *Ann. Mat. Pura. Appl.*, **83** (1969), 133–152.

[3] T. V. DUC, Sur la geometrie differentielle des fibres vectoriels. *Kodai Math. Sem. Rep.*, **26** (1975), 349–408.

[4] M. FALCITELLI and A. M. PASTORE, Pseudoconnessioni metriche su varieta quasi Hermitiane. *Rev. Roum. Math.*, **28** (1983), 83–832.

[5] M. FALCITELLI and A. M. PASTORE, Pseudoconnessioni su varieta quasi complesse. *Bull. Math. Roumanie*, **26** (1982), 35–44.

[6] M. FALCITELLI and A. M. PASTORE, Sulle pseudoconnessioni che conservano una struttura quasi complessa, *Rend. Mat. (Roma)*, **12** (1979), 231–247.

[7] M. FALCITELLI and A. M. PASTORE, Sulle pseudoconnessioni proiettivamente equivalenti. *Rend. Mat. (Roma)*, **13** (1980), 115–133.

[8] W. GREUB, S. HALPERIN and R. VANSTONE, *Connections, Curvature and Cohomology*. Volume I, II, Academic Press, New York, (1972, 1973).

[9] H. KIM, Curvatures and complex structures. *J. Korean Math. Soc.*, **24** (1987), 99–104.

[10] S. KOBAYASHI and K. NOMIZU, *Foundations of Differential Geometry*. Volume I, II, Interscience Publishers, New York, London, (1963, 1969).

[11] V. OPROIU, Some remarkable structures and connexions defined on the tangent bundle. *Rend. Mat. (Roma)*, **6** (1973), 503–540.

Z. Kovács

Department of Mathematics
Teacher's Training College of
Nyíregyháza
H–4400 Nyíregyháza
Sóstói út 31/b
Hungary

COLLOQUIA MATHEMATICA SOCIETATIS JÁNOS BOLYAI
56. DIFFERENTIAL GEOMETRY, EGER (HUNGARY), 1989

On Osculation of Homogeneous Connections[*]

L. KOZMA

The notion of a homogeneous connection H of a vector bundle ξ (often called as a non-linear connection) is derived from H. Friesecke's work [Frie], and was globally studied first by W. Barthel [Bart]. This means an intermediate class between general and linear connections. Its main importance was pointed out by A. Kawaguchi [Kaw], investigated Finsler structures. In a Finsler space, there exists, in general, no linear connection which would be compatible with the Finslerian metric. But the geodesics of a Finsler space correspond to a *homogeneous* connection. Naturally, the geodesics (autoparallel curves) of a linear connection are much easier to analyse. So our question is now how to approximate a homogeneous connection by a linear connection in such a way that their geodesics would be in a strong relationship. This strong relationship will now mean that the geodesics of the original homogeneous connection and those of its approximating linear connection have contact of order 2. Thus we shall call it an *osculation*.

Our plan is accomplished through combining two methods. The first one originates from A. Moór's work [Mo2] where an osculation method was given and studied for the connections of the line element bundle. These local classical investigations will be now here generalized for the recent Finsler

---

[*] This paper is in final form and no version of it will be submitted for publication elsewhere.

connection theory of M. Matsumoto [Mats]. By this osculation method we have the possibility to approximate a Finsler pair connection (H^F, H) along a fixed section $\sigma \in \mathrm{Sec}\, \xi$ by a linear one H^σ. The second construction we shall use is for Berwald connection. J. Vilms [Vil] was the first who, starting from a (homogeneous) connection H in ξ, constructed globally that linear connection H^B in the vertical bundle $V\xi$ which corresponds to the local classical connection of L. Berwald [Ber1, Ber2]. When the connection H in ξ is homogeneous, then the Finsler pair connection (H^B, H) is deflection free, and its geodesic structure can be well described. So the combination of these two steps leads us to the following: Starting from a homogeneous connection H in ξ, and osculating the deflection free Finsler pair connection (H^B, H) along $\sigma \in \mathrm{Sec}\, \xi$ we obtain a linear connection H^σ in ξ, which has the good property that the geodesics of H and H^σ have contact of order 2. (See Theorem 2.)

In the first section we recall some basic notions and notations used. In Section 2 the osculation of a Finsler pair connection is investigated. Here we point out some hereditary properties of the original and osculating connections. (For a more detailed exposition see [Koz1].) In the short third section a new, direct construction is given for the Berwald connection. In Section 4 we reach the point: the osculation of a homogeneous connection H along a fixed section $\sigma \in \mathrm{Sec}\, \xi$ is studied, and finally, as an application, we get an approximation procedure for sprays as well.

1. Bundles and connections

Manifolds, functions and maps are of class C^∞ as usual. Our basic notion is a vector bundle $\xi = (E, B, \pi, F)$ with a finite dimensional real fibre type F, a base manifold B, a total space E ($n+r$ dimensional), and a projection $\pi : E \to B$. $\mathrm{Sec}\, \xi$ denotes the set of all differentiable sections of ξ. When $f : M \to B$ is a map, $f^*(\xi)$ denotes the pull back bundle whose total space is $M \times_B E = \{(m, z) |\, m \in M,\, z \in E,\, f(m) = \pi(z)\}$ and its projection is $pr_1 : M \times_B E \to M$. Then the second projection $pr_2 : M \times_B E \to E$ is a bundle map with the property that its restriction on a fibre of $f^*(\xi)$ is a linear isomorphism.

Let $V_z E$ be the kernel of $(d\pi)_z : T_z E \to T_{\pi(z)} B$. Then the bundle $V\xi = (VE, E, \pi_V, F)$ is a subbundle of τ_E, and isomorphic to $\pi^*(\xi)$. The latter isomorphism is described as follows: $\varepsilon : \pi^*(\xi) \to V\xi$ $\varepsilon(z_1, z_2) =$ the tangent of the curve $z_1 + tz_2$ at 0, where $z_1, z_2 \in E$ with the property $\pi(z_1) = \pi(z_2)$. The map $\alpha = pr_2 \circ \varepsilon^{-1} : V\xi \to \xi$ commutes the diagram

$$\begin{array}{ccc} VE & \overset{\alpha}{\longrightarrow} & E \\ \pi_V \downarrow & & \downarrow \pi \\ E & \overset{\pi}{\longrightarrow} & B. \end{array}$$

Since α is fiberwise isomorphic, by the inverse of the restriction to the fibres we can form the *Liouville* vertical vector field $C : E \to VE$ given as $C(z) = \alpha_z^{-1}(z)$. By the identification $V\xi \cong \pi^*(\xi)$ we get accordingly a map $C : E \to E \times_B E$.

Let $\sigma \in \operatorname{Sec}\xi$ be a fixed section. Then the map $\beta_\sigma : E \to VE$ defined as $\beta_\sigma(z) = \alpha_{\sigma(\pi(z))}^{-1}(z)$ will pay an important role in the study of osculation. Accordingly again, we have $\beta_\sigma : E \to E \times_B E$ defined as $\beta_\sigma(z) = (\sigma(\pi(z)), z)$.

By *a connection* of ξ we mean a splitting $H : \pi^*(\tau_B) \to \tau_E$ of the next short exact sequence

(1) $$0 \to V\xi \overset{\iota}{\to} \tau_E \overset{\widetilde{d\pi}}{\to} \pi^*(\tau_B) \to 0$$

where $\widetilde{d\pi} : \tau_E \to \pi^*(\tau_B)$ is given by $\widetilde{d\pi}(A) = (\pi_E(A), d\pi(A))$ for $A \in TE$. H is also called a horizontal map, and its images $H_z E = \operatorname{Im} H|_{\{z\} \times T_{\pi(z)}B}$ are the horizontal subspaces which are complementary to the vertical subspaces: $\tau_E = V\xi \oplus H\xi$. Let $\varphi : I \to B$ a curve in the base space. A section $\sigma \in \operatorname{Sec}\xi$ is called *parallel* along φ if $d\sigma(\dot{\varphi})$ are horizontal vectors, where $\dot{\varphi}$ denotes the tangent curve of φ. This means that $H(\sigma \circ \varphi, \dot{\varphi}) = d\sigma(\dot{\varphi})$. From elementary calculus it follows that for a given curve φ and a vector z in $E_{\varphi(0)}$ there exists — at least locally — a section σ which is parallel along φ and $z = \sigma(\varphi(0))$. $\sigma(\varphi(t))$ is called the parallel transport of z along φ, and denoted by $P_\varphi(t, z)$. A curve φ in B is called geodesic if $\dot{\varphi}$ is paralle along φ, i.e. $H(\dot{\varphi}, \dot{\varphi}) = \ddot{\varphi}$.

Denote $\mu_t : E \to E$ the multiplication by $t \in \mathbb{R}$ in the fibres of ξ. It is said that a horizontal map satisfies the homogeneity condition [Szi2] when

(2) $$H(\mu_t(z), v) = d\mu_t(H(z, v))$$

holds for all $z \in E$, $v \in TB$ and $t \in \mathbb{R}$. If the differentiability of H is *not* assumed at the zero vectors of ξ, and H satisfies (2), we speak of a *homogeneous connection* (nonlinear connection). When H satisfies (2) and differentiable anywhere, then we get a *linear connection*. It is important that for a homogeneous connection the parallel transport exists along the entire curve φ [Bart]. In the linear case this is even linear. Homogeneous connections arise naturally in Finsler geometry. In fact, in general it is not possible to introduce a linear connection compatible (metrical) with the Finslerian metric, only a homogeneous one.

2. Osculation of Finsler type connection

Following the ideas of M. Matsumoto given for principal bundles and principal connections, we use vector bundles in this paper. Thus the pull back bundle $\pi^*(\xi) = (E \times_B E, E, pr_1, F)$ is called now a *Finsler vector bundle*.

Definition 1. *A pair (H^F, H) is called a Finsler pair connection where H^F is a connection of $\pi^*(\xi)$, and H is a connection of ξ. We speak of a linear Finsler conenction, when H^F satisfies the corresponding homogeneity condition, too:*

$$(3) \qquad\qquad d\nu_t(H^F(Z, U)) = H^F(\nu_t(Z), U)$$

for all $t \in \mathbb{R}$, $Z \in E \times_B E$, $U \in T_{pr_1(Z)} E$, where ν_t is the multiplication by $t \in \mathbb{R}$ in the fibres of $\pi^*(\xi)$.

Now we define an osculation of a Finsler pair connection along a fixed section $\sigma \in \operatorname{Sec} \xi$. In a special case — when H^F satisfies also the condition C_1 (see [Koz2, Mo1]) — we obtain the osculation method of A. Moór [Mo2]. The following statement ensures us the existence of an osculating connection H^σ along $\sigma \in \operatorname{Sec} \xi$.

Proposition 1. *Let (H^F, H) be a Finsler pair connection of ξ, and $\sigma \in \operatorname{Sec} \xi$. Then the map defined as*

$$H^\sigma : \pi^*(\tau_B) \to \tau_E$$
$$H^\sigma(z, v) = dpr_2(H^F(\beta_\sigma(z), H(\sigma(\pi(z)), v)))$$

is a horizontal map (a connection) for ξ.

Proof. We are to show that $d\pi \circ H^\sigma = pr_2$. Let $v \in TB$, and $x = \pi_B(v)$, $z \in E$. Then

$$d\pi(H^\sigma(z,v)) = d\pi(dpr_2(H^F(\beta_\sigma(z), H(\sigma(\pi(z)), v)))) =$$
$$= d\pi(dpr_1(H^F(\beta_\sigma(z), H(\sigma(\pi(z)), v)))) = d\pi(H(\sigma(\pi(z)), v)) = v.$$

for $\pi \circ pr_2 = \pi \circ pr_1$ holds and H is a horizontal map. ∎

Nomination. The connection H^σ defined in the previous proposition is called an *osculating connection* of (H^F, H) along $\sigma \in \operatorname{Sec} \xi$.

Remark. A simple, but lengthy calculation shows that the connection parameters Γ_i^α of H^σ can be derived from the connection parameters $(F_i^\alpha, C_\beta^\alpha, N_i^\alpha)$ of (H^F, H) as follows:

(4)
$$\widetilde{\Gamma}_i^\alpha = F_i^\alpha \circ \beta_\sigma - (N_i^\gamma \circ \sigma \circ \pi)(C_\gamma^\alpha \circ \beta_\sigma)$$

Proposition 2. *If (H^F, H) is a linear Finsler pair connection for ξ, then for any $\sigma \in \operatorname{Sec} \xi$ the osculating connection H^σ in ξ is also linear.*

Proof. Our assumption is (3). Notice the relationships between μ_t and $\nu_t : pr_2 \circ \nu_t = \mu_t \circ pr_2$ and $\nu_t \circ \beta_\sigma = \beta_\sigma \circ \mu_t$. Therefore for any $z \in E$ and $v \in T_{\pi(z)} B$

$$d\mu_t \circ H^\sigma(z, v) = d\mu_t \circ dpr_2 \circ H^F(\beta_\sigma(z), H(\sigma(\pi(z)), v)) =$$
$$= dpr_2 \circ d\nu_t(H^F(\beta_\sigma(z), H(\sigma(\pi(z)), v))) =$$
$$= dpr_2 \circ H^F(\nu_t(\beta_\sigma(z), H(\sigma(\pi(z)), v)) =$$
$$= dpr_2 \circ H^F(\beta_\sigma(\mu_t(z)), H(\sigma(\pi(z)), v)) = H^\sigma(\mu_t(z), v).$$

Thus H^σ really satisfies the homogeneity condition. Since each map in the representation of H^σ is of calss C^∞, H^σ is also of class C^∞, thus we get a linear connection in ξ. ∎

In order to investigate the relationships of the parallel structures on the different levels, we define a lift and sunk of sections.

Definition 2. *Let $\eta \in \operatorname{Sec} \xi$. The section $\eta^{\uparrow} \in \pi^*(\xi)$ defined as*

$$\eta^{\uparrow}(z) = (z, \eta(\pi(z))) \qquad (z \in E)$$

is called a lift of η.

Let now $\sigma \in \operatorname{Sec} \xi$ be a fixed, and $\Sigma \in \operatorname{Sec} \pi^*(\xi)$ an arbitrary section. The section $\Sigma_\sigma^\downarrow$ of ξ defined as

$$\Sigma_\sigma^\downarrow = pr_2 \circ \Sigma \circ \sigma$$

is called a sunk of Σ along $\sigma \in \operatorname{Sec} \xi$.

Theorem 1. *Let $\sigma \in \operatorname{Sec} \xi$ a parallel section along φ with respect to H.*

a) *If $\Sigma \in \operatorname{Sec} \pi^*(\xi)$ is parallel along $\sigma \circ \varphi$ with respect to H^F, then $\Sigma_\sigma^\downarrow$ is parallel along φ as well with respect to H^σ.*

b) *If $\eta \in \operatorname{Sec} \xi$ is parallel along φ with respect to H^σ, then $\eta^{\uparrow} \in \operatorname{Sec} \pi^*(\xi)$ is parallel along $\sigma \circ \varphi$ with respect to H^F.*

Proof. a) First calculate $d\Sigma_\sigma^\downarrow(\dot{\varphi})$. Using the relation $\Sigma \circ \sigma = \beta_\sigma \circ \Sigma_\sigma^\downarrow$ and the parallelity of Σ along $\sigma \circ \varphi$, we obtain

$$d\Sigma_\sigma^\downarrow(\dot{\varphi}) = dpr_2(d\Sigma(d\sigma(\dot{\varphi}))) = dpr_2(d\Sigma(\sigma \circ \varphi)^{\cdot}) =$$
$$= dpr_2(H^F(\Sigma \circ \sigma \circ \varphi, (\sigma \circ \varphi)^{\cdot})) = dpr_2(H^F(\beta_\sigma(\Sigma_\sigma^\downarrow \circ \varphi), d\sigma(\dot{\varphi}))) =$$

Considering that σ is parallel along φ we can continue as follows

$$dpr_2(H^F(\beta_\sigma(\Sigma_\sigma^\downarrow \circ \varphi), H(\sigma \circ \varphi, \dot{\varphi}))) = H^\sigma(\Sigma_\sigma^\downarrow \circ \varphi, \dot{\varphi}).$$

This proves the part a).

Before we prove the part b) note the following simple corollary. If $\eta \in \operatorname{Sec} \xi$ is parallel along φ with respect to H^σ, and $\Sigma \in \operatorname{Sec} \pi^*(\xi)$ is parallel along $\sigma \circ \varphi$ with respect to H^F, furthermore there exists such a $t_0 \in I$ that $\Sigma_\sigma^\downarrow(\varphi(t_0)) = \eta(\varphi(t_0))$, then $\Sigma_\sigma^\downarrow \circ \varphi = \eta \circ \varphi$ in some neighborhood of t_0.

b) Let $\Sigma \in \operatorname{Sec} \pi^*(\xi)$ be such a section that is parallel along $\sigma \circ \varphi$ with respect to H^F, and starts from $\beta^\sigma(\eta(\varphi(0)))$. We state that $\eta^{\uparrow} \circ \sigma \circ \varphi = \Sigma \circ \sigma \circ \varphi$. This will give our assertion b), for the parallelity of a section depends only on the value along the curve. In fact, under the previous corollary $\Sigma_\sigma^\downarrow \circ \varphi = \eta \circ \varphi$, thus considering simple properties of lifting and sinking we get

$$\eta^\uparrow \circ \sigma \circ \eta = \beta_\sigma \circ \eta \circ \varphi = \beta_\sigma \circ \Sigma_\sigma^\downarrow \circ \varphi = \Sigma \circ \sigma \circ \varphi.$$

∎

Definition 3. *A curve φ in M is called as an absolute geodesic of H^F if the curve $\Phi = (\dot\varphi, \dot\varphi)$ is horizontal for H^F.*

It is important and useful now that for a deflection free Finsler pair connection all geodesic (autoparallel) curves of H are at the same time absolute geodesic curves for H^F [Mats]. Therefore in this case we have the following important consequence.

Theorem 2. *Let (H^F, H) a deflection free Finsler pair connection for ξ, and $X \in \mathcal{X}(M)$, $x \in M$. Then the geodesic curve $\varphi \circ H$ starting from $X(x)$ has contact of second order to the geodesic curve ψ of H^X starting from $X(x)$.*

Proof. Since (H^F, H) is deflection free, φ is absolute geodesic for H^F. This means that $\dot\Phi = H^F(\Phi, \ddot\varphi)$. For $\dot\varphi = pr_2 \circ \Phi$, we obtain

$$\ddot\varphi(0) = dpr_2 \circ \dot\Phi(0) = dpr_2 \circ H^F(\Phi(0), \ddot\varphi(0)).$$

On the other hand

$$\Phi(0) = (\dot\varphi(0), \dot\varphi(0)) = (X \circ \varphi(0), \dot\varphi(0)) = \beta_X(\dot\varphi(0)),$$

thus

$$\ddot\varphi(0) = dpr_2 \circ H^F(\beta_X(\dot\varphi(0)), \ddot\varphi(0)) =$$
$$= dpr_2 \circ H^F(\beta_X(\dot\varphi(0)), H(\dot\varphi, \dot\varphi)) = H^X(\dot\varphi(0), \dot\varphi(0)) =$$
$$= H^X(X(x), X(x)) = H^X(\dot\psi(0), \dot\psi(0)) = \ddot\psi(0).$$

∎

3. A construction of Berwald's connection

Let $H : E \times_B TB \to TE$ be a horizontal map for ξ, and $v \in TB$ a fixed tangent vector, $x = \pi_B(v)$. Denote $H^v : E_x \to TE$ the map arising from H by fixing v: $H^v(z) = H(z, v)$, where $z \in E_x$. To define a connection for $V\xi$ we are to give a map $H^B : VE \times_E TE \to T(VE)$. Taking into account that $TE = HE \oplus VE$, H^B is enough to define for vertical and horizontal vectors, and then to extend it linearly for all TE.

First let $Z \in VE$ and $U \in VE$. Then we choose $H^B(Z, U)$ as an element of the induced vertical subspaces such that $d\pi_V(H^B(Z, U)) = U$. Such an element uniquely exists. Secondly let $U \in HE$ be a horizontal vector. Then the horizontal map for the Berwald connection is given as follows:

$$H^B(Z, U) = s \circ dH^v(Z),$$

where $s : TTE \to TTE$ means the canonical involution for E. Thus we really get a horizontal map, for $\operatorname{Im} s|_{T(VE)} = V(T\xi)$. The connection in $V\xi$ determined by H^B is just the *Berwald's connection*, for a dummy computation shows its local components are

$$F_i^\alpha = \frac{\partial N_i^\alpha}{\partial y^\beta} y^\beta, \qquad C_\beta^\alpha = 0.$$

Remarks.

1. It is obvious that the Berwald connection in $V\xi$ is linear, for it was defined by differential map.

2. It is also easy to prove (see [Szi1]), that H satisfies the homogeneity condition iff the Finsler pair connection (H^B, H) is deflection free.

3. Naturally, H^B gives a connection in $\pi^*(\xi) \cong V\xi$ as well.

4. Osculation of a homogeneous connection by a linear one

In this section we utilize the canonical isomorphism $\pi^*(\xi) \cong V\xi$, so we regard our osculation method described in Section 2 also valid for $V\xi$. (Other objects change accordingly, too.)

Let $\sigma \in \mathrm{Sec}\,\xi$ be an arbitrary section. Osculating the Finsler pair connection (H^B, H) formed with the Berwald connection, finally we obtain a *linear* connection in ξ derived from H.

Nomination. The osculating linear connection H^σ of the Finsler pair connection (H^B, H) formed with the Berwald connection of H is called the osculating connection of H along σ.

Remark. We can give a direct relationship between H and its osculating connection H^σ:

$$(5) \qquad H^\sigma(z, v) = d\alpha \circ s \circ dH^v \circ \beta_\sigma(z).$$

Considering the relationships of the corresponding parallel transports we may interpret the osculation as a derivation of connections.

Theorem 3. *Let H be an arbitrary connection for $\xi, \sigma \in \mathrm{Sec}\,\xi$, and φ a curve in B. Then the parallel transport P_φ^σ according to the osculating connection H^σ along σ and the derivative of the parallel transport P_φ for H at $\sigma(\varphi(0))$ have contact of first order.*

Proof. Let $x = \varphi(0) \in B$ be a starting point, and denote the tangent vector of φ at 0 by $v: v = \dot{\varphi}(0)$. We have to show now that for all $z \in E_x$ the tangent of the parallel transport curve $P_\varphi^\sigma(z)$ along φ coincides with the tangent of the curve $D\,P_\varphi|_{\sigma_0}(z)$, where $D\,P_\varphi|_{\sigma_0}$ means the derivative of the paralle transport P_φ according to H at $\sigma_0 = \sigma(\varphi(0))$.

First it is readily seen that the derivative map $D\,P_\varphi$ can be formed with the differential map of $P_\varphi(t) : E_x \to E_{\varphi(t)}$ as follows:

$$D\,P_\varphi(t)|_{\sigma_0}(z) = \alpha \circ dP_\varphi(t)\beta_{\sigma_0}(z).$$

On the other hand, note that the generalization of the equality of mixed partial derivatives for the case $F : \mathbb{R} \times M \to N$ is described in the following manner:

$$s \circ d\dot{F}(v) = (dF(v))\dot{}$$

where $s : TTN \to TTN$ is the canonical involution for N, and the dot and "d" denote the differentials with respect to $\mathbb{R}$ and M, resp. Using this for the case $F(t, z) = P_\varphi(t)(z) : E_x \to E$ we get

$$(D\,P_\varphi(t)|_{\sigma_0}(z))\dot{} = (\alpha \circ dP_\varphi(t) \circ \beta_{\sigma_0}(z))\dot{} = d\alpha \circ s \circ d\dot{P}_\varphi \circ \beta_{\sigma_0}(z).$$

$(\beta_{\sigma_0}(z)$ is substituted for v.) Applying the relationship $\dot{P}_\varphi = H^v$ between the parallel transport and the horizontal map, we continue further as

$$= d\alpha \circ s \circ dH^v \circ \beta_{\sigma_0}(z) = H^\sigma(z, v) = (P_\varphi(z))^{\cdot}$$

■

The question of geodesics arises only for tangents bundles. We are able to present a nice assertion only for homogeneous connections, namely, then (H^B, H) is deflection free. The following proposition is an easy consequence of Theorem 2.

Proposition 3. *Let H be a homogeneous connection on the manifold M, $X \in \mathcal{X}(M)$, and $x \in M$. Then the geodesic of H starting from $X(x)$ has contact of second order to the geodesic of the osculating linear connection H^X starting from $X(x)$.*

Application: An osculation of sprays by quadratic sprays

Definition 4. *A spray is a second order differential equation $S : \dot{T}M \to T\dot{T}M$ on M that satisfies the following homogeneity condition:*

$$(6) \qquad\qquad\qquad [S, C] = S,$$

($\dot{T}M$ denotes the nonzero tangent vectors.) When S is defined on the whole TM and is of class C^∞, then we speak of a quadratic spray. A curve $\varphi : I \to M$ is called a path curve when $\ddot{\varphi} = S \circ \dot{\varphi}$, i.e. $\dot{\varphi}$ is an integral curve of the vector field S.

It is well known a close relationship between sprays and connection. Ambrose, Palais, Singer [Amb] showed that the autoparallel curves of a *linear* connection H induces a *quadratic* spray $S(v) = H(v, v)$, (called as geodetic spray of H), and conversely, for a quadratic spray S there is exactly one torsion free linear connection whose autoparallel curves are just the paths of the spray S. This was generalized for sprays, and homogeneous connections by J. Grifone [Grif]. We utilize now this latter relationship to osculate a spray by a quadratic one.

Let S be a spray for the manifold M. By [Grif] there exists uniquely a torsion free compatible homogeneous connection H on M. Compatibility means here that the path curves of S and the autoparallel curves of H are identical. Regarding the osculating linear connection H^X along a vector field $X \in \mathcal{X}(M)$, and its geodetic spray S^X, we obtain an osculating quadratic spray S^X along X originating from S.

Nomination. The geodetic spray S^X of H^X is called an *osculating spray* of S along $X \in \mathcal{X}(M)$.

Remark. It can be calculated that if local parameters of a spray S are $G^i : U \subset TM \to R$, then the local parameters of S^X are

$$\widetilde{G}^j(x,y) = y^i y^k \frac{\partial^2 G^j}{\partial y^i \partial y^k}(x, X).$$

Our final assertion gives us the expected relationships of paths of S and S^X.

Proposition 4. *Let S be a spray for M, and $X \in \mathcal{X}(M)$, $x \in X$. Then a path of the spray S starting from $X(x)$ has contact of second order to the path of the osculating quadratic spray S^X starting from $X(x)$.*

Proof. It follows immediately from Theorem 2, since the paths of S and S^X are autoparallel curves of H and H^X. ∎

References

[Amb] W. AMBROSE, R. S. PALAIS and I. M. SINGER, Sprays, *An. Acad. Brasil Cien.* **32** (1960), 163–178.

[Bart] W. BARTHEL, Nichtlineare Zusammenhänge und deren Holonomiegruppen, *J. Reine Angew. Math.* **212**, (1963), 120–149.

[Ber1] L. BERWALD, Über Parallelübertragung in Räumen mit allgemeiner Massbestimmung, *Jber. Deutsch. Math. Ver.* **34** (1926), 213–220.

[Ber2] L. BERWALD, Parallelübertragung in allgemeinen Räumen, *Atti. Cong. Bologna* **4** (1931), 263–270.

[Frie] H. FRIESECKE, Vektorübertragung, Richtungsübertragung, *Metrik. Math. Ann.* **94** (1925), 101–118.

[Grif] J. GRIFONE, Structure presque tangente et connexions, I, II. *Ann. Inst. Fourier Grenoble* **22** (1,3), (1972), 287–334. and 291–338.

[Kaw] A. KAWAGUCHI, On the theory of non-linear connections I, *Tensor, N.S.* **2**, (1952), 123–142.

[Koz1] L. KOZMA, On osculation of Finsler type connections, *Acta Math. Hung.* **53** (3–4) (1989), 389–397.

[Koz2] L. KOZMA, The role of the C_1 and C_2 conditions in the theory of Finsler type connections, *Proc. Nat. Sem. on Finsler and Lagrange spaces,* Brasov, 1986, 185–200.

[Mats] M. MATSUMOTO, Foundations of Finsler Geometry and Special Finsler Spaces, Saikawa, Otsushi, Kaiseisha Press, 1986.

[Mo1] A. MOÓR, Einführung des invarianten Differentials und Integrals in allgemeinen metrischen Räumen, *Acta Math.* **86** (1959), 71–83.

[Mo2] A. MOÓR, Über oskulierende Punkträume von affinzusammenhängenden Linienelementenmannigfaltigkeiten, *Annals of Math.* **56** (1952), 397–403.

[Szi1] J. SZILASI and L. KOZMA, Remarks on Finsler type connections, Proc. Nat. Sem. on Finsler Spaces, Brasov, 1984, 185–193.

[Szi2] J. SZILASI, Horizontal map with homogeneity condition, Supl. Rend. Circ. Mat. Palermo, serie II, n.3. (1984), 307–320.

[Vil] J. VILMS, Curvature of nonlinear connections, *Proc. Amer. Math. Soc.* **19** (198), 1125–1129.

L. Kozma

Department of Mathematics
University of Debrecen
H–4010 Debrecen
Pf. 12.
Hungary

COLLOQUIA MATHEMATICA SOCIETATIS JÁNOS BOLYAI
56. DIFFERENTIAL GEOMETRY, EGER (HUNGARY), 1989

On Hypersurfaces in Hyperquadrics[*]

S. MARKVORSEN

1. Introduction

A well known construction of the geodesics in $S^n(r) = \{x \in \mathbb{R}^{n+1}; \langle P(x), P(x) \rangle = r^2\}$, where $P(x)$ denotes the position vector field, is obtained by cutting the sphere with 2-planes through the origin in $\mathbb{R}^{n+1}$.

Now consider the semi-Euclidean space $\mathbb{R}_\nu^{n+1}$ and the generalized "sphere"

$$\mathcal{S}_\nu^n(r) = \{x \in \mathbb{R}_\nu^{n+1} : |\langle P(x), P(x) \rangle| = r^2\}.$$

Following B. O'Neill [3], the connected components of $\mathcal{S}_\nu^n(r)$ are called hyperquadrics and each hyperquadric carries a fixed indicator $\varepsilon_P = \text{sign}\langle P, P \rangle$. Furthermore, every geodesic in $\mathcal{S}_\nu^n(r)$ is again obtained as the intersection of $\mathcal{S}$ with a 2-plane through the origin of $\mathbb{R}_\nu^{n+1}$ (cf. [3] p. 112, 149).

---

[*] This paper is in final form and no version of it will be submitted for publication elsewhere.

2. Example

Let Γ_κ be a geodesic of $\mathcal{S}_1^2(1) \subset \mathbb{R}_1^3$ obtained by intersection with the plane $\Pi_\kappa = \{x \in \mathbb{R}_1^3 | x^1 = \kappa \cdot x^2, \ \kappa \in [0, \infty]\}$, where x^1, x^2, x^3 are the canonical coordinate functions in $\mathbb{R}_1^3$ with x^1 in the timelike direction.

For $\kappa < 1$ we get the following coordinate functions for Γ_κ in terms of the induced arc length $s \in \mathbb{R}$:

$$\Gamma_\kappa(s) = \left(x^1(s), x^2(s), x^3(s)\right) = \left(\frac{\kappa}{\sqrt{1 - \kappa^2}} \sin(s), \frac{1}{\sqrt{1 - \kappa^2}} \sin(s), \cos(s)\right)$$

For $\kappa > 1$, the intersection $\Pi_\kappa \cap \mathcal{S}_1^2(1)$ has 4 geodesic components. Two of them (with $\varepsilon_P = -1$) can be described by

$$\Gamma_\kappa(s) = \left(x^1(s), x^2(s), x^3(s)\right) =$$
$$= \left(\pm\frac{\kappa}{\sqrt{\kappa^2 - 1}} \cosh(s), \pm\frac{1}{\sqrt{\kappa^2 - 1}} \cosh(s), \sinh(s)\right).$$

The other two (with $\varepsilon_P = +1$) are

$$\Gamma_\kappa(s) = \left(x^1(s), x^2(s), x^3(s)\right) =$$
$$= \left(\frac{\kappa}{\sqrt{\kappa^2 - 1}} \sinh(s), \frac{1}{\sqrt{\kappa^2 - 1}} \sinh(s), \pm\cosh(s)\right).$$

We observe, that for all these geodesics every single coordinate function satisfies the equation

(2.1) $\qquad \left(\varepsilon_\Gamma \cdot \frac{d^2}{ds^2}\right)\left(x^i(s)\right) = -\varepsilon_P \cdot x^i(s), \qquad i = 1, 2, 3.$

which again may be written as follows

(2.2) $\qquad \Delta_\Gamma x^i = -\varepsilon_P \cdot x^i, \qquad i = 1, 2, 3,$

where by definition $\Delta_M \phi = \operatorname{div}_M(\operatorname{grad}_M \phi)$ for any function ϕ on a semi-Riemannian manifold M.

3. A characterization of vanishing mean curvature in hyperquadrics

In [2] we showed conversely, that if a curve $\Gamma = (x^1, x^2, x^3) : \mathbb{R} \to \mathbb{R}_1^3$ satisfies the system (2.2), then Γ is a geodesic in $\mathcal{S}_1^2(1)$. In fact we proved the following result which generalizes a classical theorem of Takahashi [5].

Theorem A. *Let M^m be an immersed connected semi-Riemannian sub-manifold of $\mathbb{R}_\nu^{n+1}$. Suppose that there exists an indicator $\varepsilon \in \{-1, 1\}$, such that*

$$(*) \qquad \triangle_M x^i = -m \cdot \varepsilon \cdot x^i \quad \text{for all} \quad i = 1, 2, \ldots, n+1.$$

Then i) M^m is contained in a connected component of $\mathcal{S}_\nu^n(1)$, and $\varepsilon = \varepsilon_P$, and ii) The mean curvature $\overline{H}$ of M^m in $\mathcal{S}_\nu^n(1)$ vanishes identically. Conversely, if M^m satisfies i) and ii), then $(*)$ is also satisfied.

Remark. We note that the Laplace–Beltrami operator $\triangle_M$ is not elliptic unless the induced metric on M^m is definite. Furthermore we are not in general allowed to use the variational terms *minimal* or *maximal* for $M^m \subset \mathcal{S}_\nu^n(1)$ even if the corresponding mean curvature vanishes. In fact, the geodesics $\Gamma_{\kappa < 1}$ in section 2 neither minimize not maximize the arclength between any pair of points.

4. When one equation will do

As the main result in this note we will now show, that if we assume in advance, that M^m is a hypersurface in the unit hyperquadric (i.e. $M \subset \mathcal{S}_\nu^n(1)$ and $\dim M = m = n - 1$), then any single one of the $n+1$ equations in $(*)$ essentially implies $\overline{H} \equiv 0$. Specifically we have the following

Theorem. *Let M be a geodesically complete, connected semi-Riemannian hypersurface in $\mathcal{S}_\nu^n(1) \subset \mathbb{R}_\nu^{n+1}$ with position indicator ε_P.*

Suppose that there exists a coordinate function x^i such that

$$\triangle_M x^i = -(n-1)\varepsilon_P \cdot x^i.$$

Then either M has everywhere vanishing mean curvature or else M contains (where $\overline{H} \neq 0$) strips of generalized cylinders in $\mathcal{S}_\nu^n(1)$ with geodesic generators orthogonal to the totally geodesic hypersurface

$$V_i = \mathcal{S}_\nu^n(1) \cap \{x \in \mathbb{R}_\nu^{n+1} : x^i = 0\}.$$

Remark. It is not clear to the present author whether some principle of analytic continuation actually may sharpen this theorem to give a full analogue of the following result due to Osserman [4].

Theorem B. Let S be a surface in $\mathbb{R}^3$, and suppose that $\triangle_S x^i = 0$ for some i. Then S is either a minimal surface or else everywhere a locally cylindrical surface with its generators parallel to the x^i-axis.

5. Some preliminary identities

First we let M^m be a semi-Riemannian submanifold in any semi-Riemannian ambient space N^n and consider a smooth function $g : N \to \mathbb{R}$ which then has a smooth restriction $f = g|_M : M \to \mathbb{R}$ and a unique orthogonal decomposition of its gradient:

$$\operatorname{grad}_N(g) = \operatorname{grad}_M(f) + (\operatorname{grad}(g))^\perp.$$

Lemma B. (cf. [1]). Let $X \in TM \subset TN$. Then

$$(\operatorname{Hess}_M f)(X, X) = (\operatorname{Hess}_N g)(X, X) + \langle \operatorname{grad}_N(g), \alpha_{M \subset N}(X, X) \rangle,$$

where $\alpha_{M \subset N}$ is the second fundamental form of M in N.

Proof. By definition of the second fundamental form we have

$$\overline{D}_X(\operatorname{grad}_M f) = D_X(\operatorname{grad}_M f) + \alpha(X, \operatorname{grad}_M f),$$

where $\overline{D}$ and D are the induced Levi–Civita connections of N and M respectively (cf. [3] pp. 98 ff.). Hence we get for $Y \in \chi(M)$

$$
\begin{aligned}
(\mathrm{Hess}_M \ f)(X,Y) &= \langle D_X(\mathrm{grad}_M \ f), Y \rangle \\
&= \langle \overline{D}_X(\mathrm{grad}_M \ f) - \alpha(\mathrm{grad}_M \ f, X), Y \rangle \\
&= \langle \overline{D}_X(\mathrm{grad}_M \ f, Y \rangle \\
&= X\langle \mathrm{grad}_M \ f, Y \rangle - \langle \mathrm{grad}_M \ f, \overline{D}_X Y \rangle \\
&= \langle \overline{D}_X \ \mathrm{grad}_N \ g, Y \rangle + \left\langle (\mathrm{grad}_N \ g)^\perp, \overline{D}_X Y \right\rangle \\
&= (\mathrm{Hess}_N \ g)(X,Y) + \left\langle (\mathrm{grad}_N \ g)^\perp, \alpha(X,Y) \right\rangle \\
&= (\mathrm{Hess}_N \ g)(X,Y) + \langle \mathrm{grad}_N \ g, \alpha(X,Y) \rangle. \quad\blacksquare
\end{aligned}
$$

In our setting we let $g = x^i : \mathbb{R}^{n+1}_\nu \to \mathbb{R}$ and then

$$
h = g|_{S^n_\nu(1)} : S^n_\nu(1) \to \mathbb{R} \quad \text{and finally} \quad f = h|_M = g|_M : M^{n-1} \to \mathbb{R}.
$$

Then the lemma implies for all $X \in TM$

$$
\tag{5.1} (\mathrm{Hess}_M \ f)(X,X) = (\mathrm{Hess}_s \ h)(X,X) + \langle \mathrm{grad}_s \ h, \overline{\alpha}(X,X) \rangle
$$

$$
\tag{5.2} (\mathrm{Hess}_s \ h)(X,X) = (\mathrm{Hess}_\mathbb{R} \ g)(X,X) + \langle \mathrm{grad}_\mathbb{R} \ g, \widehat{\alpha}(X,X) \rangle,
$$

where $\widehat{\alpha}$ is the second fundamental form of $S^n_\nu(1)$ in $\mathbb{R}^{n+1}_\nu$.

Since the hyperquadrics are totally umbilical ([3] p. 111), we get

$$
\widehat{\alpha}(X,X) = -\varepsilon_P \cdot P \cdot \langle X, X \rangle,
$$

and so with $\mathrm{grad}_\mathbb{R} \ g = \mathrm{grad} \ x^i = \varepsilon_i \partial_i$ we get $\mathrm{Hess}_\mathbb{R} \ g \equiv 0$ and thus

$$
(\mathrm{Hess}_s \ h)(X,X) = -\varepsilon_P \cdot \langle X, X \rangle \cdot \langle \varepsilon_i \partial_i, P \rangle.
$$

However, by definition we also have $P = \sum_{j=1}^{n+1} x^j \partial_j$, so that $\langle \varepsilon_i \partial_i, P \rangle = x^i$.

In total we therefore get from (5.1):

$$
(\mathrm{Hess}_M \ x^i)(X,X) = -\varepsilon_P \cdot x^i \cdot \langle X, X \rangle + \langle T, \overline{\alpha}(X,X) \rangle,
$$

where we write $T = \mathrm{grad}_s \ x^i$.

From $\triangle_M x^i = \mathrm{trace}_M (\mathrm{Hess}_M \ x^i) = \sum_{j=1}^{n-1} \varepsilon_j (\mathrm{Hess}_M \ x^i)(X_j, X_j)$ for any orthonormal basis $\{X_j\}$ of TM, we finally have for every semi-Riemannian hypersurface M in $S^n_\nu(1)$:

$$
\tag{5.3} \triangle_M x^i = -(n-1)\varepsilon_P \cdot x^i + (n-1)\langle T, \overline{H} \rangle.
$$

6. Proof of the theorem

The theorem in section 4 now follows if we can show that $\langle T, \overline{H} \rangle \equiv 0$ will imply $\overline{H} \equiv 0$ or that M is cylindrical where $\overline{H}$ is not zero. We first observe that if $\overline{H}(q) \neq 0$ for some $q \in M$, then $T(q) \in T_q M$.

Thus there is a maximal neighbourhood $\mathcal{U} \subset M$ around q on which T is a vector field. Let $\Pi_i(q)$ be that 2-plane in $\mathbb{R}^{n+1}_\nu$ which contains the x^i-axis and goes through q. Then $\Pi(q) \perp T_q \mathcal{S}^n_\nu$ and since $T(q)$ is the orthogonal projection of $\varepsilon_i \partial_i$ into $T_q \mathcal{S}$, we see that $T(q)$ (when nonzero) spans the intersection $\Pi_i(q) \cap T_q M$. But this means that $\mathcal{U}$ is foliated by the integral curves $\Pi_i(\cdot) \cap \mathcal{U}$ which are geodesics in $\mathcal{S}^n_\nu(1)$ generating either a cone (over a point, where T is zero) or a cylinder over the totally geodesic hypersurface V_i. By completeness of M (and by the invariance of the property $\overline{H} \neq 0$ along generators), $\mathcal{U}$ extends along the geodesic integral curves of T to contain the cone point or to form a cylindrical strip over V_i. However, by smoothness of M, $\mathcal{U}$ cannot contain the toppoint of any cone, and we are left with the two possibilities claimed in the theorem. ∎

References

[1] K. JORGE and D. KOUTROUFIOTIS, An Estimate for the Curvature of Bounded Submanifolds, *Amer. J. Math.* **103** (1981), 711–725.

[2] S. MARKVORSEN, A Characteristic Eigenfunction for Minimal Hypersurfaces in Space Forms, *Mathematische Zeitschrift* **202** (1989), 375–382.

[3] B. O'NEILL, Semi-Riemannian Geometry, Academic Press, New York, 1983.

[4] R. OSSERMAN, Remarks on Minimal Surfaces, *Comm. Pure. Appl. Math.* **12** (1959), 233–239.

[5] T. TAKAHASHI, Minimal Immersions of Riemannian Manifolds, *J. Math. Soc. Japan* **18** (1966), 380–385.

Steen Markvorsen

Mathematical Institute
The Technical University of Denmark
Building 303
DK–2800 Lungby, Denmark.

COLLOQUIA MATHEMATICA SOCIETATIS JÁNOS BOLYAI

56. DIFFERENTIAL GEOMETRY, EGER (HUNGARY), 1989

All Unitary Representations Admit Moment Mappings

P. W. MICHOR

1. Calculus of smooth mappings

1.1. The traditional differential calculus works well for finite dimensional vector spaces and for Banach spaces. For more general locally convex spaces a whole flock of different theories were developed, each of them rather complicated and none really convincing. The main difficulty is that the composition of linear mappings stops to be jointly continuous at the level of Banach spaces, for any compatible topology. This was the original motivation for the development of a whole new field within general topology, convergence spaces.

Then in 1982, Alfred Frölicher and Andreas Kriegl presented independently the solution to the question for the right differential calculus in infinite dimensions. They joined forces in the further development of the theory and the (up to now) final outcome is the book [**F-K**].

In this section I will sketch the basic definitions and the most important results of the Frölicher–Kriegl calculus.

1.2. The c^∞-topology. Let E be a locally convex vector space. A curve $c : \mathbb{R} \to E$ is called *smooth* or C^∞ if all derivatives exist and are continuous

— this is a concept without problems. Let $C^\infty(\mathbb{R}, E)$ be the space of smooth functions. It can be shown that $C^\infty(\mathbb{R}, E)$ does not depend on the locally convex topology of E, only on its associated bornology (system of bounded sets).

The final topologies with respect to the following sets of mappings into E coincide:

(1) $C^\infty(\mathbb{R}, E)$.

(2) Lipschitz curves (so that $\left\{ \frac{c(t) - c(s)}{t - s} : t \neq s \right\}$ is bounded in E).

(3) $\{E_B \to E; B$ bounded absolutely convex in $E\}$, where E_B is the linear span of B equipped with the Minkowski functional $p_B(x) := \inf\{\lambda > 0 : x \in \lambda B\}$.

(4) Mackey-convergent sequences $x_n \to x$ (there exists a sequence $0 < \lambda_n \nearrow \infty$ with $\lambda_n(x_n - X)$ bounded).

This topology is called the c^∞-topology on E and we write $c^\infty E$ for the resulting topological space. In general (on the space $\mathcal{D}$ of test functions for example) it is finer than the given locally convex topology, it is not a vector space topology, since scalar multiplication is no longer jointly continuous. The finest among all locally convex topologies on E which are coarser then $c^\infty E$ is the bornologification of the given locally convex topology. If E is a Fréchet space, then $c^\infty E = E$.

1.3. Convenient vector spaces. Let E be a locally convex vector space. E is said to be a *convenient vector space* if one of the following equivalent (completeness) conditions is satisfied:

(1) Any Mackey–Cauchy-sequence (so that $(x_n - x_m)$ is Mackey convergent to 0) converges. This is also called c^∞-complete.

(2) If B is bounded closed absolutely convex, then E_B is a Banach space.

(3) Any Lipschitz curve in E is locally Riemannian integrable.

(4) For any $c_1 \in C^\infty(\mathbb{R}, E)$ there is $c_2 \in C^\infty(\mathbb{R}, E)$ with $c_1' = c_2$ (existence of antiderivate).

1.4. Lemma. *Let E be a locally convex space. Then the following properties are equivalent:*

(1) *E is c^∞-complete.*

(2) *If $f : \mathbb{R}^k \to E$ is scalarwise Lip^k, then f is Lip^k, for $k > 1$.*

(3) If $f : \mathbb{R} \to E$ is scalarwise C^∞ then f is differentiable at 0.

(4) If $f : \mathbb{R} \to E$ is scalarwise C^∞ then f is C^∞.

Here a mapping $f : \mathbb{R}^k \to E$ is called Lip^k if all partial derivatives up to order k exist and are Lipschitz, locally on $\mathbb{R}^n$. f scalarwise C^∞ means that $\lambda \circ f$ is C^∞ for all continuous linear functionals on E.

This lemma says that a convenient vector space one can recognize smooth curves by investigating compositions with continuous linear functionals.

1.5. Smooth mappings. Let E and F be locally convex vector spaces. A mapping $f : E \to F$ is called *smooth* or C^∞, if $f \circ c \in C^\infty(\mathbb{R}, F)$ for all $c \in C^\infty(\mathbb{R}, E)$; so $f_* : C^\infty(\mathbb{R}, E) \to C^\infty(\mathbb{R}, F)$ makes sense. Let $C^\infty(E, F)$ denote the space of all smooth mapping from E to F.

For E and F finite dimensional this gives the usual notion of smooth mappings: this has been first proved in [**Bo**]. Constant mappings are smooth. Multilinear mappings are smooth if and only if they are bounded. Therefore we denote by $L(E, F)$ the spaces of all bounded linear mappings from E to F.

1.6. Structure on $C^\infty(E, F)$. We equip the space $C^\infty(\mathbb{R}, E)$ with the bornologification of the topology of uniform convergence on compact sets, in all derivatives separately. Then we equip the space $C^\infty(E, F)$ with the bornologification of the initial topology with respect to all mappings $c^* : C^\infty(E, F) \to C^\infty(\mathbb{R}, F)$, $c^*(f) := f \circ c$, for all $c \in C^\infty(\mathbb{R}, E)$.

1.7. Lemma. *For locally convex spaces E and F we have:*

(1) *If F is convenient, then also $C^\infty(E, F)$ is convenient for any E. The space $L(E, F)$ is a closed linear subspace of $C^\infty(E, F)$, so it also convenient.*

(2) *If E is convenient, then a curve $c : \mathbb{R} \to L(E, F)$ is smooth if and only if $t \mapsto c(t)(x)$ is a smooth curve in F for all $x \in E$.*

1.8. Theorem. *The category of convenient vector spaces and smooth mappings is cartesian closed. So we have a natural bijection*

$$C^\infty(E \times F, G) \cong C^\infty(E, C^\infty(F, G)),$$

which is even a diffeomorphism.

Of course this statement is also true for c^∞-open subsets of convenient vector spaces.

1.9. Corollary. *Let all spaces be convenient vector spaces. Then the following canonical mappings are smooth:*

$$\text{ev} : C^\infty(E, F) \times E \to F, \qquad \text{ev}(f, x) = f(x)$$
$$\text{ins} : E \to C^\infty(F, E \times F), \qquad \text{ins}(x)(y) = (x, y)$$
$$(\ )^\wedge : C^\infty(E, C^\infty(F, G)) \to C^\infty(E \times F, G)$$
$$(\ )^\vee : C^\infty(E \times F, G) \to C^\infty(E, C^\infty(F, G))$$
$$\text{comp} : C^\infty(F, G) \times C^\infty(E, F) \to C^\infty(E, G)$$
$$C^\infty(\ ,\ ) : C^\infty(F, F') \times C^\infty(E', E) \to C^\infty(C^\infty(E, F), C^\infty(E', F'))$$
$$(f, g) \mapsto (h \mapsto f \circ h \circ g)$$
$$\prod : \prod C^\infty(E_i, F_i) \to C^\infty\left(\prod E_i, \prod F_i\right).$$

1.10. Theorem. *Let E and F be convenient vector spaces. Then the differential operator*

$$d : C^\infty(E, F) \to C^\infty(E, L(E, F)),$$
$$df(x)v := \lim_{t \to 0} \frac{f(x + tv) - f(x)}{t}$$

exists and is linear and bounded (smooth). Also the chain rule holds:

$$d(f \circ g)(x)v = df(g(x))\, dg(x)v.$$

1.11. Remarks. Note that the conclusion of theorem 1.8 is the starting point of the classical calculus of variations, where a smooth curve in a space of functions was assumed to be just a smooth function in one variable more.

If one wants theorem 1.8 to be true and assumes some other obvious properties, then the calculus of smooth functions is already uniquely determined.

There are, however, smooth mappings which are not continuous. This is unavoidable and not so horrible as it might appear at first sight. For example the evaluation $E \times E' \to \mathbb{R}$ is jointly continuous if and only if E is normable, but it is always smooth. Clearly smooth mappings are continuous for the c^∞-topology.

For Fréchet spaces smoothness in the sense described here coincides with the notion C_c^∞ of [**Ke**]. This is the different calculus used by [**Mic1**], [**Mil**], and [**P–S**].

A prevalent opinion in contemporary mathematics is, that for infinite dimensional calculus each serious application needs its own foundation. By a serious application one obviously means some application of a hard inverse function theorem. These theorems can be proved, if by assuming enough a priori estimates one creates enough Banach space situation for some modified iteration procedure to converge. Many authors try to build their platonic idea of an a priori estimate into their differential calculus. I think that this makes the calculus inapplicable and hides the origin of the a priori estimates. I believe, that the calculus itself should be as easy to use as possible, and that all further assumptions (which most often come from ellipticity of some nonlinear partial differential equation of geometric origin) should be treated separately, in a setting depending on the specific problem. I am sure that in this sense the Frölicher–Kriegl calculus as presented here are universally usable for most applications.

Let me point out as a final remark, that also the cartesian closed calculus for holomorphic mappings along the same lines is available in [**K–N**], and recently the cartesian closed calculus for real analytic mapping was developed in [**K–M**].

2. The moment mapping for unitary representations

The following is a review of the results obtained in [**Mic2**]. We include only one proof, the central application of the Frölicher–Kriegl calculus.

2.1. Let G be any (finite dimensional second countable) real Lie group, and let $\rho : G \to U(\mathbf{H})$ be a unitary representation on a Hilbert space $\mathbf{H}$. Then the associated mapping $\hat\rho : G \times \mathbf{H} \to \mathbf{H}$ is in general *not* jointly continuous, it is only separately continuous, so that $g \mapsto \rho(g)x, G \to \mathbf{H}$, is continuous for any $x \in \mathbf{H}$.

Definition. *A vector $x \in \mathbf{H}$ is called smooth (or real analytic) if the mapping $g \mapsto \rho(g)x$, $G \to \mathbf{H}$ is smooth (or real analytic). Let us denote by $\mathbf{H}_\infty$*

the linear subspace of all smooth vectors in $\mathbf{H}$. Then we have an embedding $j : \mathbf{H}_\infty \, C^\infty\,(G, \mathbf{H})$, given by $x \mapsto (g \mapsto \rho(g)x)$. We equip $C^\infty\,(G, \mathbf{H})$ with the compact C^∞-topology (of uniform convergence on compact subsets of G, in all derivatives separately). Then it is easily seen (and proved in [**Wa**, p. 253]) that $\mathbf{H}_\infty$ is a closed linear subspace. So with the induced topology $\mathbf{H}_\infty$ becomes a Fréchet space. Clearly $\mathbf{H}_\infty$ is also an invariant subspace, so we have a representation $\rho : G \to L(\mathbf{H}_\infty, \mathbf{H}_\infty)$. For more detailed information on $\mathbf{H}_\infty$ see [**Wa**, chapt. 4.4] or [**Kn**, chapt. III.].

2.2. Theorem. *The mapping $\hat{\rho}; G \times \mathbf{H}_\infty$ is smooth.*

Proof.   By cartesian closedness of the Frölicher–Kriegl calculus 1.8 it suffices to show that the canonically associated mapping

$$\hat{\rho}^\vee : G \to C^\infty\,(\mathbf{H}_\infty, \mathbf{H}_\infty)$$

is smooth; but it takes values in the closed subspace $L(\mathbf{H}_\infty, \mathbf{H}_\infty)$ of all bounded linear operators. So by it suffices to show that the mapping $\rho : G \to L(\mathbf{H}_\infty, \mathbf{H}_\infty)$ is smooth. But for that, since $\mathbf{H}_\infty$ is a Fréchet space, thus convenient in the sense of Frölicher–Kriegl, by 1.7(2) it suffices to show that

$$G \xrightarrow{\rho} L(\mathbf{H}_\infty, \mathbf{H}_\infty) \xrightarrow{ev_x} \mathbf{H}_\infty$$

is smooth for each $x \in \mathbf{H}_\infty$. This requirement means that $g \mapsto \rho(g)x$, $G \to \mathbf{H}_\infty$, is smooth. For this it suffices to show that

$$G \to \mathbf{H}_\infty \xrightarrow{j} C^\infty\,(G, \mathbf{H}_\infty)$$
$$g \mapsto \rho(g)x \mapsto (h \mapsto \rho(h)(g)x),$$

is smooth. But again by cartesian closedness it suffices to show that the associated mapping

$$G \times G \to \mathbf{H},$$
$$(g, h) \mapsto \rho(h)(g)x = \rho(hg)x,$$

is smooth. And this is the case since x is a smooth vector.   ■

2.3.   We now consider $\mathbf{H}_\infty$ as a "weak" symplectic Fréchet manifold, equipped with the symplectic structure Ω, the restriction of the imaginary part of the Hermitian inner product $\langle \ , \ \rangle$ on $\mathbf{H}$. Then $\Omega \in \Omega^2(\mathbf{H}_\infty)$ is a closed 2-form which is not degenerate in the sense that

$$\check{\Omega} : T\mathbf{H}_\infty = \mathbf{H}_\infty \times \mathbf{H}_\infty \to T^*\mathbf{H}_\infty = \mathbf{H}_\infty \times \mathbf{H}_\infty{}'$$

is injective (but not surjective), where $\mathbf{H}_\infty{}' = L(\mathbf{H}_\infty, \mathbb{R})$ denotes the real topological dual space. This is the meaning of "weak" above.

2.4. Review. For a finite dimensional symplectic manifold (M, Ω) we have the following exact sequence of Lie algebras:

$$0 \to H^0(M) \to C^\infty(M) \xrightarrow{\text{grad}^\Omega} \mathfrak{X}_\Omega(M) \xrightarrow{\gamma} H^1(M) \to 0$$

Here $H^*(M)$ is the real De Rham cohomology of M, the space $C^\infty(M)$ is equipped with the Poisson bracket $\{ \ , \ \}$, $\mathfrak{X}_\Omega(M)$ consists of all vector fields ξ with $\mathcal{L}_\xi \Omega = 0$ (the locally Hamiltonian vector fields), which is a Lie algebra for the Lie bracket. Also $\text{grad}^\Omega f$ is the Hamiltonian vector field for $f \in C^\infty(M)$ given by $i(\text{grad}^\Omega f)\Omega = df$, and $\gamma(\xi) = [i_\xi \Omega]$. The spaces $H^0(M)$ and $H^1(M)$ are equipped with the zero bracket.

Given a symplectic left action $\ell : G \times M \to M$ of a connected Lie group G on M, the first partial derivative of ℓ gives a mapping $\ell' : \mathfrak{g} \to \mathfrak{X}_\Omega(M)$ which sends each element X of the Lie algebra $\mathfrak{g}$ of G to the fundamental vector field. This is a Lie algebra homomorphism.

$$
\begin{array}{ccccccc}
H^0(M) & \xrightarrow{\ i\ } & C^\infty(M) & \xrightarrow{\text{grad}^\Omega} & \mathfrak{X}_\Omega(M) & \xrightarrow{\ \gamma\ } & H^1(M) \\
& & \sigma\uparrow & & \uparrow\ell' & & \\
& & \mathfrak{g} & = & \mathfrak{g} & &
\end{array}
$$

A linear lift $\sigma : \mathfrak{g} \to C^\infty(M)$ of ℓ' with $\text{grad}^\Omega \circ \sigma = \ell'$ exists if and only if $\gamma \circ \ell' = 0$ in $H^1(M)$. This lift σ may be changed to a Lie algebra homomorphism if and only if the 2-cocycle $\bar{\sigma} : \mathfrak{g} \times \mathfrak{g} \to H^0(M)$, given by $(i \circ \bar{\sigma})(X, Y) = \{\sigma(X), \sigma(Y)\} - \sigma([X, Y])$, vanishes in $H^2(\mathfrak{g}, H^0(M)))$, for if $\bar{\sigma} = \delta\alpha$ then $\sigma - i \circ \alpha$ is a Lie algebra homomorphism.

If $\sigma : \mathfrak{g} \to C^\infty(M)$ is a Lie algebra homomorphism, we may associate the *moment mapping* $\mu : M \to \mathfrak{g}' = L(\mathfrak{g}, \mathbb{R})$ to it, which is given by $\mu(x)(X) = \sigma(X)(x)$ for $x \in M$ and $X \in \mathfrak{g}$. It is G-equivariant for a suitable chosen (in general affine) action of G on $\mathfrak{g}'$. See [**We**] or [**L–M**] for all this.

2.5. We now want to carry over to the setting of 2.1 and 2.2 the procedure of 2.4. The first thing to note is that the Hamiltonian mapping grad^Ω : $C^\infty(\mathbf{H}_\infty) \to \mathfrak{X}_\Omega(\mathbf{H}_\infty)$ does not make sense in general, since $\check{\Omega} : \mathbf{H}_\infty \to \mathbf{H}_\infty{}'$ is not invertible: $\mathrm{grad}^\Omega f = \check{\Omega}^{-1}\, df$ is defined only for those $f \in C^\infty(\mathbf{H}_\infty)$ with $df(x)$ in the image of $\check{\Omega}$ for all $x \in \mathbf{H}_\infty$. A similar difficulty arises for the definition of the Poisson bracket on $C^\infty(\mathbf{H}_\infty)$.

Let $\langle x, y \rangle = \mathrm{Re}\langle x, y \rangle + \sqrt{-1}\Omega(x, y)$ be the decomposition of the Hermitian inner product into real and imaginary parts. Then $\mathrm{Re}\langle x, y \rangle = \Omega(\sqrt{-1}x, y)$, thus the real linear subspaces $\check{\Omega}(\mathbf{H}_\infty) = \Omega(\mathbf{H}_\infty, \ )$ and $\mathrm{Re}\langle \mathbf{H}_\infty, \ \rangle$ of $\mathbf{H}_\infty{}' = L(\mathbf{H}_\infty, \mathbb{R})$ coincide.

2.6. Definition. *Let $\mathbf{H}_\infty^*$ denote the real linear subspace*

$$\mathbf{H}_\infty^* = \Omega(\mathbf{H}_\infty, \ ) = \mathrm{Re}\langle \mathbf{H}_\infty, \ \rangle$$

of $\mathbf{H}_\infty{}' = L(\mathbf{H}_\infty, \mathbb{R})$, and let us call it the smooth dual of $\mathbf{H}_\infty$ in view of the embedding of test functions into distributions. We have two canonical isomorphisms $\mathbf{H}_\infty^* \cong \mathbf{H}_\infty$ induced by Ω and $\mathrm{Re}\langle \ , \ \rangle$, respectively. Both induce the same Fréchet topology on $\mathbf{H}_\infty^*$, which we fix from now on.

2.7. Definition. *Let $C_*^\infty(\mathbf{H}_\infty, \mathbb{R}) \subset C^\infty(\mathbf{H}_\infty, \mathbb{R})$ denote the linear subspace consisting of all smooth functions $f : \mathbf{H}_\infty \to \mathbb{R}$ such that each iterated derivative $d^k f(x) \in L_{sym}^k(\mathbf{H}_\infty, \mathbb{R})$ has the property that*

$$d^k f(x)(\ , y_2, \ldots y_k) \in \mathbf{H}_\infty^*$$

is actually in the smooth dual $\mathbf{H}_\infty^* \subset \mathbf{H}_\infty{}'$ for all $x, y_2, \ldots, y_k \in \mathbf{H}_\infty$, and that the mapping

$$\prod^k \mathbf{H}_\infty \to \mathbf{H}_\infty$$
$$(x, y_2, \ldots, y_k) \mapsto \check{\Omega}^{-1}(df(x)(\ , y_2, \ldots, y_k))$$

is smooth. Note that we could also have used $\mathrm{Re}\langle \ , \ \rangle$ instead of Ω. By the symmetry of higher derivatives this is then true for all entries of $d^k f(x)$, for all x.

2.8. Lemma. *For $f \in C^\infty(\mathbf{H}_\infty, \mathbb{R})$ the following assertions are equivalent:*

(1) *$df : \mathbf{H}_\infty \to \mathbf{H}_\infty{}'$ factors to a smooth mapping $\mathbf{H}_\infty \to \mathbf{H}_\infty^*$.*

(2) *f has a smooth Ω-gradient $\mathrm{grad}^\Omega f \in \mathfrak{X}(\mathbf{H}_\infty) = C^\infty(\mathbf{H}_\infty, \mathbf{H}_\infty)$ such that $df(x)y = \Omega(\mathrm{grad}^\Omega f(x), y)$.*

(3) $f \in C_*^\infty(\mathbf{H}_\infty, \mathbb{R})$.

2.9. Theorem. *The mapping* $\mathrm{grad}^\Omega : C_*^\infty(\mathbf{H}_\infty, \mathbb{R}) \to \mathfrak{X}_\Omega(\mathbf{H}_\infty)$, *given by* $\mathrm{grad}^\Omega f := \check{\Omega}^{-1} \circ df$, *is well defined; also the Poisson bracket*

$$\{\ ,\ \} : C_*^\infty(\mathbf{H}_\infty, \mathbb{R}) \times C_*^\infty(\mathbf{H}_\infty, \mathbb{R}) \to C_*^\infty(\mathbf{H}_\infty, \mathbb{R}),$$
$$\{f, g\} := i(\mathrm{grad}^\Omega f)i(\mathrm{grad}^\Omega g)\Omega = \Omega(\mathrm{grad}^\Omega g, \mathrm{grad}^\Omega f) =$$
$$= (\mathrm{grad}^\Omega f)(g) = dg(\mathrm{grad}^\Omega f)$$

is well defined and gives a Lie algebra structure to the space $C_*^\infty(\mathbf{H}_\infty, \mathbb{R})$.

We also have the following long exact sequence of Lie algebras and Lie algebra homomorphism:

$$0 \to H^0(\mathbf{H}_\infty) \to C_*^\infty(\mathbf{H}_\infty, \mathbb{R}) \xrightarrow{\mathrm{grad}^\Omega} \mathfrak{X}_\Omega(\mathbf{H}_\infty) \xrightarrow{\gamma} H^1(\mathbf{H}_\infty) = 0$$

2.10. We consider now again as in 2.1 a unitary representation $\rho : G \to U(\mathbf{H})$. By theorem 2.2 the associated mapping $\hat{\rho} : G \times \mathbf{H}_\infty \to \mathbf{H}_\infty$ is smooth, so we have the infinitesimal mapping $\rho' : \mathfrak{g} \to \mathfrak{X}(\mathbf{H}_\infty)$, given by $\rho'(X)(x) = T_e(\rho'(\ , x))$ for $X \in \mathfrak{g}$ and $x \in \mathbf{H}_\infty$. Since ρ is a unitary representation, the mapping ρ' has values in the Lie subalgebra of all linear Hamiltonian vector fields $\xi \in \mathfrak{X}(\mathbf{H}_\infty)$ which respect the symplectic form Ω, i.e. $\xi : \mathbf{H}_\infty \to \mathbf{H}_\infty$ is linear and $\mathcal{L}_\xi \Omega = 0$.

Now let us consider the mapping $= \check{\Omega} \circ \rho'(X) : \mathbf{H}_\infty \to T(\mathbf{H}_\infty) \to T^*(\mathbf{H}_\infty)$. We have $d(\check{\Omega} \circ \rho'(X)) = d(i_{\rho'(X)}\Omega) = \mathcal{L}_{\rho'(X)}\Omega = 0$, so the linear 1-form $\check{\Omega} \circ \rho'(X)$ is closed, and since $H^1(\mathbf{H}_\infty) = 0$, it is exact. So there is a function $\sigma(X) \in C^\infty(\mathbf{H}_\infty, \mathbb{R})$ with $d\sigma(X) = \check{\Omega} \circ \rho'(X)$, and $\sigma(X)$ is uniquely determined up to addition of a constant. If we require $\sigma(X)(0) = 0$ then $\sigma(X)$ is uniquely determined and is a quadratic function. In fact we have $\sigma(X)(x) = \int_{c_x} \check{\Omega} \circ \rho'(X)$, where $c_x(t) = tx$. Thus

$$\sigma(X)(x) = \int_0^1 \Omega(\rho'(X)(tx), \frac{d}{dt}tx)\,dt =$$
$$= \Omega(\rho'(X)(x), x) \int_0^1 dt$$
$$= \frac{1}{2}\Omega(\rho'(X)(x), x).$$

2.11. Lemma. *The mapping* $\rho : \mathfrak{g} \to C_*^\infty(\mathbf{H}_\infty, \mathbb{R})$, *given by* $\sigma(X)(x) = \frac{1}{2}\Omega(\rho'(X)(x), x)$ *for* $X \in \mathfrak{g}$ *and* $x \in \mathbf{H}_\infty$, *is a Lie algebra homomorphism* and $\mathrm{grad}^{\Omega} \circ \sigma = \rho'$.

For $g \in G$ we have $\rho(g)^* \sigma(X) = \sigma(X) \circ \rho(g) = \sigma(\mathrm{Ad}(g^{-1})X)$, so σ is G-equivariant.

2.12. The moment mapping. For a unitary representation $\rho : g \to U(\mathbf{H}$ we can now define the *moment mapping*

$$\mu : \mathbf{H}_{\infty} \to \mathfrak{g}' = L(\mathfrak{g}, \mathbb{R})$$

$$\mu(x)(X) := \sigma(X)(x) = \frac{1}{2}\Omega(\rho'(X)x, x),$$

for $x \in \mathbf{H}_{\infty}$ and $X \in \mathfrak{g}$.

2.13. Theorem. *The moment mapping* $\mu : \mathbf{H}_{\infty} \to \mathfrak{g}'$ *has the following properties:*

(1) $(d\mu(x)y)(X) = \Omega(\rho'(X)x, y)$ *for* $x, y \in \mathbf{H}_{\infty}$ *and* $X \in \mathfrak{g}$, *so* $\mu \in C^{\infty}_* (\mathbf{H}_{\infty}, \mathfrak{g}')$.

(2) *For* $x \in \mathbf{H}_{\infty}$ *the image of* $d\mu(x) : \mathbf{H}_{\infty} \to \mathfrak{g}'$ *is the annihilator* $\mathfrak{g}^{\Omega}_x$ *of the Lie algebra* $\mathfrak{g}_x = \{X \in \mathfrak{g} : \rho'(X)(x) = 0\}$ *of the isotropy group* $G_x = \{g \in G : \rho(g)x = x\}$ *in* $\mathfrak{g}'$.

(3) *For* $x \in \mathbf{H}_{\infty}$ *the kernel of* $d\mu(x)$ *is* $(T_x(\rho(G)x))^{\Omega} = \{y \in \mathbf{H}_{\infty} : \Omega(y, T_x(\rho(G)x)) = 0\}$, *the* Ω-*annihilator of the tangent space at* x *of the* G-*orbit through* x.

(4) *The moment mapping is equivariant:* $\mathrm{Ad}'(g) \circ \mu = \mu \circ \rho(g)$ *for all* $g \in G$, *where* $\mathrm{Ad}'(g) = \mathrm{Ad}(g^{-1})' : \mathfrak{g}' \to \mathfrak{g}'$ *is the coadjoint action.*

(5) *The pullback operator* $\mu^* : C^{\infty}(\mathfrak{g}, \mathbb{R}) \to C^{\infty}(\mathbf{H}_{\infty}, \mathbb{R})$ *actually has values in the subspace* $C^{\infty}_* (\mathbf{H}_{\infty}, \mathbb{R})$. *It also is a Lie algebra homomorphism for the Poisson brackets involved.*

2.14. Let again $\rho : G \to U(\mathbf{H})$ be a unitary representation of a Lie group G on a Hilbert space $\mathbf{H}$.

Definition. *A vector* $x \in \mathbf{H}$ *is called real analytic if the mapping* $g \mapsto \rho(g)x$, $G \to \mathbf{H}$ *is a real analytic mapping, in the real analytic structure of the Lie group* G.

We will use from now on the theory of real analytic mappings in infinite dimensions as developed in [**K-M**]. So the following conditions on $x \in \mathbf{H}$ are equivalent:

(1) x is a real analytic vector.

(2) $\mathfrak{g} \in X \to \rho(\exp X)x$ is locally near 0 given by a converging power series.

(3) For each $y \in \mathbf{H}$ the mapping $\mathfrak{g} \in X \mapsto \langle \rho(\exp X)x, y \rangle \in \mathbb{C}$ is smooth and real analytic along affine lines in $\mathfrak{g}$, locally near 0.

The only nontrivial part is (3) $\Rightarrow$ (1), and this follows from [K-M, 1.6 and 2.7] and the fact, that ρ is a representation.

Let $\mathbf{H}_\omega$ denote the vector space of all real analytic vectors in $\mathbf{H}$. Then we have a linear embedding $j : \mathbf{H}_\omega \to C^\omega(G, \mathbf{H})$ into the space of real analytic mappings, given by $x \mapsto (g \mapsto \rho(g)x)$. We equip $C^\omega(G, \mathbf{H})$ with the convenient vector space structure described in [K-M, 5.4]. Then $\mathbf{H}_\omega$ consists of all equivariant functions in $C^\omega(G, \mathbf{H})$ and is therefore a closed subspace. So it is a convenient vector space with the induced structure.

The space $\mathbf{H}_\omega$ is a dense in the Hilbert space $\mathbf{H}$ by [Wa, 4.4.5.7] and an invariant subspace, so we have a representation $\rho : G \to L(\mathbf{H}_\omega, \mathbf{H}_\omega)$.

2.15. Theorem. *The mapping $\hat{\rho} : G \times \mathbf{H}_\omega \to \mathbf{H}_\omega$ is a real analytic in the sense of* [K-M].

Proof. Similar to the proof of theorem 2.2. ∎

2.16. Again we consider now $\mathbf{H}_\omega$ as a "weak" symplectic real analytic Fréchet manifold, equipped with the symplectic structure Ω, the restriction of the imaginary part of the Hermitian inner product $\langle \ , \ \rangle$ on $\mathbf{H}$. Then again $\Omega \in \Omega^2(\mathbf{H}_\omega)$ is closed 2-form which is non degenerate in the sense that $\check{\Omega} : \mathbf{H}_\omega \to \mathbf{H}'_\omega = L(\mathbf{H}_\omega, \mathbb{R})$ is injective. Let

$$\mathbf{H}_\omega^* := \check{\Omega}(\mathbf{H}_\omega) = \Omega(\mathbf{H}_\omega, \ ) = \mathrm{Re}\langle \mathbf{H}_\omega, \ \rangle \subset \mathbf{H}'_\omega = L(\mathbf{H}_\omega, \mathbb{R})$$

again denote the *analytic dual* of $\mathbf{H}_\omega$, equipped with the topology induced by the isomorphism with $\mathbf{H}_\omega$.

2.17. Remark. All the results leading to the smooth moment mapping can now be carried over to the real analytic setting with *no* changes in the proofs. So all statements from 2.9 to 2.13 are valid in the real analytic situation. We summarize this in one more results:

2.18. Theorem. *Consider the injective linear continuous G-equivariant mapping $i : \mathbf{H}_\omega \to \mathbf{H}_\infty$. Then for the smooth moment mapping $\mu : \mathbf{H}_\infty \to$*

$\mathfrak{g}'$ *from 2.13 the composition* $\mu \circ i : \mathbf{H}_\omega \to \mathbf{H}_\infty \to \mathfrak{g}'$ *is real analytic. It is called the real analytic moment mapping.*

References

[A–K] L. AUSLANDER and B. KOSTANT, Polarization and unitary representations of solvable Lie groups, *Inventiones Math,* **14** (1971), 255–354.

[Bo] J. BOMAN, Differentiability of a function and of its compositions with functions of one variable, *Math. Scand.* **20** (1967), 249–268.

[F–K] A. FRÖLICHER and A. KRIEGL, Linear spaces and differentiation theory, *Pure and Applied Mathematics,* J. Wiley, Chichester, 1988.

[L–M] P. LIBERMANN and P. C. M. MARLE, Symplectic geometry and analytical mechanics, Mathematics and applications, D. Reidel, Dordrecht, 1987.

[Ke] H. KELLER, Differential calculus in locally convex spaces, Springer Lecture Notes 417, 1974.

[Ki1] A. A. KIRILLOV, Elements of the theory of representations, Springer-Verlag, Berlin, 1976.

[Ki2] A. A. KIRILLOV, Unitary representations of nilpotent Lie groups, *Russian Math. Surveys,* **17** (1962), 53–104.

[Kn] A. W. KNAPP, Representation theory of semisimple Lie groups, Princeton University Press, Princeton, 1986.

[Ko] B. KONSTANT, Quantization and unitary representations, in Lecture Notes in Mathematics, Vol. 170, Springer-Verlag, 1970, 87–208.

[Kr1] A. KRIEGL, Die richtigen Räumer für Analysis im Unendlich-Dimensionalen, *Monatshefte Math.* **94** (1982), 109–124.

[Kr2] A. KRIEGL, Eine kartesisch abgeschlossene Kategorie glatter Abbildungen zwischen beliebigen lokalkonvexen Vektorräumen, *Monatshefte für Math.* **95** (1983), 287–309.

[K–M] A. KRIEGL and P. MICHOR, The convenient setting for real analytic mappings, *Acta Mathematica,* (1990).

[K–N] A. KRIEGL and D. NEL, A convenient setting for holomorphy, *Cahiers Top. Géo. Diff,* **26** (1985), 273–309.

[Mic1] P. W. MICHOR, Manifolds of differentiable mappings, Shiva Mathematics Series 3, Orpington, 1980.

[Mic2] P. W. MICHOR, The moment mapping for unitary representations, J. Global Anal. Geo. **8** (1990), 299–313.

[**Mil**] J. MILNOR, Remarks on infinite dimensional Lie groups, in Relativity, Groups, and Topology II, Les Houches, 1983, B. S. DeWitt, R. Stora, Eds., Elsevier, Amsterdam, 1984.

[**P–S**] A. PRESSLEY and G. SEGAL, Loop groups, Oxford Mathematical Monographs, Oxford University Press, 1986.

[**Wa**] G. WARNER, Harmonic analysis on semisimple Lie groups, Volume I, Sringer-Verlag, New York, 1972.

[**We**] A. WEINSTEIN, Lectures on symplectic manifolds, *Regional conference series in mathematics,* **29** (1977), Amer. Math. Soc..

[**Wi**] H. WICKLICKY, Physical interpretations of the moment mapping for unitary representations, Diplomarbeit, Universität Wien, 1989.

Peter W. Michor

Institut für Mathematik,
Universität Wien
Strudlhofgasse 4
A–1090 Wien
Austria

COLLOQUIA MATHEMATICA SOCIETATIS JÁNOS BOLYAI
56. DIFFERENTIAL GEOMETRY, EGER (HUNGARY), 1989

Geodesic Mappings of Affine-connected Spaces onto Riemannian Spaces*

J. MIKES and V. BEREZOVSKI

It is known that if a space with affine connection A_n admits a geodesic mapping onto a space with affine connection $\overline{A}_n$ then in general with respect to the mapping system of coordinates $x^1, x^2, \ldots, x^n$ the objects of the connection of these spaces $\Gamma_{ij}^h(x)$ and $\overline{\Gamma}_{ij}^h(x)$ obey the following relation [1]:

(1)
$$\overline{\Gamma}_{ij}^h(x) = \Gamma_{ij}^h + \delta_{(i}^h \psi_{j)}$$

where δ_i^h — is the Kronecker symbol, $\psi_i(x)$ — is a vector and (i,j) denotes symmetrization with respect to the indices i and j.

If $\psi_i \not\equiv 0$ then the geodesic mapping is called non-trivial. By the equality (1) it is trivial to find all the spaces with affine connection which admit a non-trivial geodesic mapping (NGM) onto a given Riemannian space.

The present paper is devoted to the investigation of the geodesic mappings of affine-connected spaces A_n without torsion onto the Riemannian spaces $\overline{V}_n$. One may see that not every affine-connected space admits a non-trivial geodesic mapping onto a Riemannian space.

---

* This paper is in final form and no version of it will be submitted for publication elsewhere.

1.

Theorem 1. *The space with the affine connection A_n admits a non-trivial geodesic mapping onto a Riemannian space $\overline{V}_n$ with the metric tensor $\overline{g}_{ij}(x)$ if and only if the following set of differential equations with covariant derivatives of Cauchy type has a solution with respect to the symmetric tensor $\overline{g}_{ij}$ ($\det \|\overline{g}_{ij}\| \neq 0$), the non-zero vector $\psi_i(x)$ and the invariant $\mu(x)$:*

(2a) $$\overline{g}_{ij,k} = 2\psi_k \overline{g}_{ij} + \psi_{(i}\overline{g}_{j)k};$$

(2b) $$n\psi_{i,j} = n\psi_i\psi_j + \mu\overline{g}_{ij} + \overline{g}_{i\alpha}R^\alpha_{\beta\gamma j}\overline{g}^{\beta\gamma} - R_{ij} - \frac{2}{n+1}R^\alpha_{\alpha ij};$$

(2c) $$(n-1)\mu_{,i} = \overline{g}^{\alpha\beta}\left\{ R^\gamma_{\alpha\beta i,\gamma} - R_{\alpha i,\beta} + \right.$$
$$+ \psi_\alpha\left(4R_{\beta i} - \frac{n-5}{n+1}R^\gamma_{\gamma\beta i}\right) + 2(n-1)\psi_\gamma R^\gamma_{\alpha\beta i} +$$
$$\left. + \frac{2}{n+1}R^\gamma_{\gamma\alpha[i,\beta]} \right\}$$

where the comma denotes covariant derivative with respect to the space connection A_n, $\overline{g}^{ij}(x)$ are components of the matrix inverted to $\|\overline{g}_{ij}\|$, R^h_{ijk} and R_{ij} are respectively Riemannian and Ricci tensors of the space A_n.

Proof. Suppose that A_n admits NGM onto $\overline{V}_n$ with metric tensor $\overline{g}_{ij}(x)$. Then the connections A_n and $\overline{V}_n$ obey the relation (1) in general with respect to the mapping coordinate system. Taking into account the covariant constancy of $\overline{g}_{ij}$ in $\overline{V}_n$, the conditions are at the same time sufficient for A_n to admit NGM onto $\overline{V}_n$.

Let us consider integrability conditions of the equations (2a)

(3) $$\overline{g}_{\alpha(h}R^\alpha_{i)jk} = 2\overline{g}_{hi}\psi_{[jk]} + \overline{g}_{j(h}\psi_{i)k} - \overline{g}_{k(h}\psi_{i)j}$$

where $\psi_{ij} = \psi_{i,j} - \psi_i\psi_j$, and $[ij]$ denote alternation with respect to i and j.

Convolving (3) and $\overline{g}^{hi}$, we get $\psi_{[jk]} = \frac{1}{n+1}R^\alpha_{\alpha jk}$. Excluding $\psi_{[jk]}$ from (3), we obtain

(4) $$\overline{g}_{\alpha(h}R^\alpha_{i)jk} - \frac{2}{n+1}\overline{g}_{hi}R^\alpha_{\alpha jk} = \overline{g}_{j(h}\psi_{i)k} - \overline{g}_{k(h}\psi_{i)j}.$$

After the convolution (4) with g^{ik} one easily obtains the conditions (2b) with $\mu = \psi_{\alpha\beta}\overline{g}^{\alpha\beta}$.

Taking into account $\overline{g}^{ik}\overline{g}_{kj} = \delta^i_j$, it is not difficult to show that the equations (2a) are equivalent to the relations

(5) $$\overline{g}^{ij}_{,k} = -2\psi_k\overline{g}^{ij} - \delta^{(i}_k\psi^{j)}$$

where $\psi^i \equiv \psi_\alpha\overline{g}^{\alpha i}$.

We covariantly differentiate the conditions (2b) with respect to x^k and then alternate the result with respect to the indices j and k taking into account (2a), (2b), (5) and convolve with $\overline{g}^{ik}$, and finally we get equations (2c).

The theorem has been proved. ∎

From theorem 1 we may conclude that the set of all Riemannian spaces, the given affine-connected space A_n admits NGM onto, is dependent on $r \le r_0 = (n+1)(n+2)/2$ parameters.

Finding of all the solutions of (2) requires a consideration of their integrability conditions and differential extensions, which form a set of algebraic equations with respect to the unknown functions $\overline{g}_{ij}$, ψ_i and μ with coefficients from A_n. But this set is not linear and its solution is certainly difficult.

For equiaffine A_n when the vector ψ_i is necessarily gradient i.e. $\psi_i = \psi_{,i}$ the main equations of NGM onto Riemannian spaces $\overline{V}_n$ may be written in the following from

(6) $$a^{ij}_{,k} = \lambda^{(i}\delta^{j)}_k; \qquad n\lambda^i_{,j} = \mu\delta^i_j + a^{i\alpha}R_{\alpha j} - a^{\alpha\beta}R^i_{\alpha\beta j};$$
$$(n-1)\mu_{,i} = 2(n+1)\lambda^\alpha R_{\alpha i} + a^{\alpha\beta}(2R_{\alpha i,\beta} - R_{\alpha\beta,i})$$

where $a^{ij} = a^{ji}$, $|a^{ij}| \ne 0$,
with $a^{ij} = e^{2\psi}\overline{g}^{ij}$; $\lambda^i = -e^{2\psi}\overline{g}^{i\alpha}\psi_\alpha$.

The set of equations (6) is linear the integrability conditions and their differential extensions are a set of linear homogeneous algebraic equations with respect to the unknown functions a^{ij}, λ^i and μ.

2.

The number r of substantial parameters, the general solution of (2) depends on, we shall call the degree of mobility of a_n with respect to the geodesic mappings onto Riemannian spaces $\overline{V}_n$ (by analogy with [1]).

It is easy to prove that maximal degree of mobility $r_0 = (n+1)(n+2)/2$ with respect to the geodesic mappings onto Riemannian spaces is admitted by the projective euclidean spaces and only by them.

The following estimation is obtained for a distribution of the degrees of mobility with respect to the geodesic mappings onto Riemannian spaces.

Theorem 2. *The degree of mobility of the spaces with affine connection A_n, different from the projective-euclidean ones, with respect to the geodesic mappings onto Riemannian spaces does not exceed number $n(n-1)/2$.*

The results under discussion are generalisations of analogous theorems of N. S. Sinyukov [1] for the geodesic mappings of Riemannian spaces.

References

[1] SINYUKOV N. S., Geodesic mappings of Riemannian spaces. Moscow, Nauka, 1979.

Josef Mikes

University
ul. Petra Velikogo, 2
270057, Odessa
USSR

Vladimir Berezovski

Ped. institute
ul. K. Marxa, 2
258900, Uman
USSR

COLLOQUIA MATHEMATICA SOCIETATIS JÁNOS BOLYAI
56. DIFFERENTIAL GEOMETRY, EGER (HUNGARY), 1989

On the Generalized Hamilton Spaces*

R. MIRON

In the papers [4, 5, 6, 7] we have studied the geometry of Hamilton spaces, defined as the pairs $H^n = (M, H)$, formed by a real n-dimensional differentiable manifold M and a regular Hamiltonian $H : T^*M \setminus \{0\} \to R$. H is the fundamental function of H^n and $\gamma^{ij}(x, p) = \frac{1}{2}\frac{\partial^2 H}{\partial p_i \partial p_j}$ is the fundamental or metric d-tensor (distinguished tensor) field of H^n.

In the present paper, we consider the spaces $M^{*n} = (M, g^{ij}(x, p))$, determined by M and by a contravariant symmetric d-tensor field $g^{ij}(x, p)$ having the property rank $\|g^{ij}(x, p)\| = n$. They are called generalized Hamilton spaces. Generally, the M^{*n} are not reducible to the Hamilton spaces. Thus the methods based on symplectic geometry are not efficient for studying these spaces. It is interesting to remark that the geometry of the Hamilton spaces H^n, based on the metrical properties can be extended to the generalized Hamilton spaces M^{*n}. This is the main purpose of the paper.

We begin with the notions of non-linear connection and d-connection on the total space T^*M of the cotangent bundle of M. Then we give some

---

* This paper is in final form and no version of it will be submitted for publication elsewhere.

results in the theory of Hamilton and Cartan spaces. We introduce the notion of generalized Hamilton spaces and obtain for them some remarkable properties. Finally, we study a special class of generalized Hamilton spaces, with the fundamental d-tensor field $g^{ij}(x,p) = \gamma^{ij}(x,p) + \frac{1}{c^2}p^i p^j$, suggested by some ideas taken from the joint papers [1, 2] written with Tomoaki Kawaguchi. Also, the mentioned metric is strongly related with the Synge's metric, [9], from the relativity of the dispersive mediums.

Notations and terminology are those from our papers [3, 4, 5, 6, 7].

§ 1. The non-linear connection, d-connections

Let M be a real n-dimensional differentiable manifold and $\pi^* : T^*M \to M$ its cotangent bundle. If (x^i, p_i) are the canonical coordinates of a point $u \in T^*M$, then a transformation of coordinates on T^*M is given by

(1.1)
$$\overline{x}^i = \overline{x}^i(x^1, \ldots, x^n), \qquad \text{rank} \left\| \frac{\partial \overline{x}^i}{\partial x^j} \right\| = n,$$

$$\overline{p}_i = \frac{\partial x^j}{\partial \overline{x}^i} p_j.$$

On T^*M there are globally defined the Liouville 1-form

(1.2)
$$\widetilde{p} = p_i dx^i$$

and the symplectic structure

(1.3)
$$\theta = dp_i \wedge dx^i.$$

Let us denote by $V = \ker \pi^{*T}$ the vertical subbundle of the tangent bundle TT^*M. If the base manifold M is paracompact, then there exist horizontal subbundles N of TT^*M such that for the Whitney sum $N \oplus V$

$$TT^*M = N \oplus V$$

holds.

The fibers of N determine a distribution $N : u \in T^*M \to N_u \subset T_uT^*M$, which is supplementary to the vertical distribution $V : u \in T^*M \to V_u \subset T_uT^*M$:

(1.4)
$$T_uT^*M = N_u \oplus V_u.$$

A horizontal distribution, which verifies (1.4), is called a non-linear connection on T^*M.

On a coordinate neighbourhood $\pi^{*-1}(U) \subset T^*M$ there exists a basis adapted to the distributions N and V. This is $\left(\frac{\delta}{\delta x^i}, \frac{\partial}{\partial p_i}\right)$, where $\frac{\delta}{\delta x^i} = \frac{\partial}{\partial x^i} + N_{ji}\frac{\partial}{\partial p_j}$. The functions $N_{ji}(x,p)$ are the coefficients of the non-linear connection N.

Putting $\frac{\partial}{\partial x^i} = \partial_i$, $\frac{\delta}{\delta x^i} = \delta_i$, $\frac{\partial}{\partial p_i} = \dot{\partial}^i$ we can write

(1.5) $$\delta_i = \partial_i + N_{ji}(x,p)\dot{\partial}^j.$$

With respect to (1.1) $(\delta_i, \dot{\partial}^i)$ transform by the known laws.

Let $(dx^i, \delta p_i)$ be the dual basis of $(\delta_i, \dot{\partial}^i)$:

(1.5)' $$\delta p_i = dp_i - N_{ij}(x,p)dx^j.$$

Then

(1.6) $$\tau_{ij} = N_{ij} - N_{ji}$$

is an antisymmetric d-tensor field, globally defined on T^*M. If $\tau_{ij} = 0$ we say that N is a symmetric non-linear connection.

When $\tau_{ij} = 0$, we can write the symplectic structure θ in the form

(1.3)' $$\theta = \delta p_i \wedge dx^i.$$

This proves the geometrical character of θ.

Let us consider the d-tensor field

(1.7) $$R_{kij} = \delta_i N_{kj} - \delta_j N_{ki}.$$

A necessary and sufficient condition that the horizontal distribution N be integrable is $R_{ijk} = 0$.

If the distributions N and V satisfy (1.4) then any vector field $X \in \mathfrak{X}(T^*M)$ can be written in the form $X = X^H + X^V$, where X^H belongs to N and X^V belongs to V.

A linear connection D on T^*M is called distinguished (briefly a d-connection) if

$$\left(D_X Y^H\right)^V = 0, \qquad \left(D_X Y^V\right)^H = 0, \qquad D_X \theta = 0, \qquad \forall\, X, Y \in \Xi(T^*M).$$

We write

$$D_X^h = D_{X^H}, \qquad D_X^v = D_{X^V}$$

and say that D^h and D^v are the h- and v-covariant derivations determined by the d-connection D. Also we denote by T^h, resp. T^v the h- and vp-components of the torsion of D.

§2. Hamilton spaces

Definition 2.1. *A Hamilton space is a pair* $(M, H) = H^n$, *where* $H :$ $T^* M \to R$ *is a function (a Hamiltonian) of class* C^∞ *on* $\widetilde{T^* M} = T^* M \setminus \{0\}$, *continuous on the null section and having the metric d-tensor*

(2.1)
$$\gamma^{ij}(x, p) = \frac{1}{2} \dot{\partial}^i \dot{\partial}^j H$$

of rank n *on* $\widetilde{T^* M}$:

(2.2)
$$\mathrm{rank} \, \| \gamma^{ij}(x, p) \| = n.$$

H *is called the fundamental function, and* $\gamma^{ij}(x, p)$ *the fundamental d-tensor field of* H^n.

Let us consider $\| \gamma_{ij}(x, p) \| = \| \gamma^{ij}(x, p) \|^{-1}$.

Then

(2.3)
$$G^h = \gamma_{ij}(x, p) dx^i \otimes dx^j$$

is a symmetric tensor field of the type $(0, 2)$ and of rank n, globally defined on $\widetilde{T^* M}$. G^h is called the metric tensor of the Hamilton space H^n.

Now, it is well known that we have the

Theorem 2.1. *There exists a non-linear connection* N *on* $\widetilde{T^* M}$ *determined only by the fundamental function* H *of the Hamilton space* H^n. N *has the coefficients:*

(2.4)
$$N_{ij} = -\frac{1}{2} \gamma_{jh} \left\{ \frac{1}{4} \gamma_{ik} \dot{\partial}^k (H, \dot{\partial}^h H) + \dot{\partial}^h \partial_i H \right\}.$$

This non-linear connection is called canonical. It has the property of symmetry: $\tau_{ij} = N_{ij} - N_{ji} = 0$.

Throughout this paper we consider for the space H^n the canonical non-linear connection (2.4), and we take the basis $(\delta_i, \dot{\partial}^i)$ adapted to the distributions N and V being built by means of (2.4).

Therefore, the symplectic structure θ of the Hamilton space H^n will be given by $(1.3)'$, where δp_i is given by $(1.5)'$, (2.4).

Write

$$(2.5) \qquad\qquad \mathcal{H} = \frac{1}{2}H$$

and consider the canonical Hamiltonian vector field ξ on $\widetilde{T^*M}$

$$(2.6) \qquad\qquad i_\xi \theta = -d\mathcal{H},$$

where i denote the interior product. In the adapted basis $(\delta_i, \dot{\partial}^i)$ the vector field ξ is expressed by:

$$(2.7) \qquad\qquad \xi = (\dot{\partial}^i \mathcal{H})\delta_i - (\delta_i \mathcal{H})\dot{\partial}^i.$$

The integral curves, $t \mapsto c(t)$, of the vector field ξ are the integral curves of the Hamilton equations:

$$(2.8) \qquad\qquad \frac{dx^i}{dt} = \frac{\partial \mathcal{H}}{\partial p_i}, \qquad \frac{\delta p_i}{dt} = -\frac{\delta \mathcal{H}}{\delta x^i}.$$

The Hamilton spaces H^n have canonical d-connection determined only by the fundamental function H.

Theorem 2.1.

i) *There exists a unique d-connection D on $\widetilde{T^*M}$ with the property:*

$$D_X^h G^h = 0, \qquad D_X^v G^h = 0, \qquad T^h = T^v = 0.$$

ii) *In the adapted basis $(\delta_i, \dot{\partial}^i)$ the coefficients of D^h and D^v are given by*

$$(2.9) \qquad \begin{aligned} H^i_{jk} &= \frac{1}{2}\gamma^{ih}(\delta_j \gamma_{hk} + \delta_k \gamma_{jh} - \delta_h \gamma_{jk}) \\ C^{jk}_i &= -\frac{1}{2}\gamma_{ih}(\dot{\partial}^j \gamma^{hk} + \dot{\partial}^k \gamma^{jh} - \dot{\partial}^h \gamma^{jk}), \end{aligned}$$

respectively. ∎

§3. Cartan spaces

The first important class of the Hamilton spaces is formed by the so called Cartan spaces, [6].

Definition 3.1. *A Hamilton space $H^n = (M, H)$ for which the fundamental function $H(x, p)$ is 2-homogeneous (i.e. homogeneous of the second order) with respect to p_i on $\widetilde{T^* M}$ is called a Cartan space.*

We denote by $C^n = (M, H)$ a Cartan space.

Obviously, this definition is different from the classical definition given by Elie Cartan in his known book, [8]. The Cartan spaces appear here as the duals of Finsler spaces, [5, 6]. But the first important properties of them were studied by Cartan.

Proposition 3.1. *For the Cartan space we have:*
1° *The fundamental d-tensor field $\gamma^{ij}(x, p)$ is 0-homogeneous with respect to p_i.*
2° $\dot{\partial}^i \mathcal{H} = p^i = \gamma^{ij} p_j$.
3° C^{ijk} *is -1-homogeneous with respect to p_i and totally symmetric.*
4° $C^{ijk} p_i = 0$. ■

Let us consider the Christoffel symbols $\gamma^i_{jk}(x, p)$ of the fundamental d-tensor field $\gamma^{ij}(x, p)$ and we put

(3.1) $$\gamma^0_{ij} = \gamma^k_{ij} p_k, \qquad \gamma^0_{i0} = \gamma^0_{ij} p^j.$$

Then we have

Theorem 3.1. *The canonical non-linear connection N of the Cartan space C^n has the coefficients*

(3.2) $$N_{ij}(x, p) = \gamma^0_{ij} - \frac{1}{2} \gamma^0_{k0} \dot{\partial}^k \gamma_{ij}.$$

■

Throughout this paper we consider only this non-linear connection for the Cartan space C^n.

Then (2.9) gives us the canonical metrical d-connection of C^n.

Proposition 3.2. *The canonical metrical d-connection D of the Cartan space C^n has the properties:*

1° $\gamma^{ij}{}_{|k} = 0, \quad \gamma^{ij}|^k = 0.$

2° $T^i_{jk} = 0, \quad S^{jk}_i = 0.$

3° $C^{ijk} = -\frac{1}{4}\dot{\partial}^i\dot{\partial}^j\dot{\partial}^k H.$

4° $R_{ijk} + R_{jki} + R_{kij} = 0.$

5° $H_{|k} = \delta_k H = 0.$

6° $p_{i|j} = D_{ij} = 0, \quad p_i|^j = \delta^j_i.$

7° *The integral curves of the Hamilton equations are h-paths.*

8° *The Liouville covector p_i is parallel along the integral curves of the Hamilton equations of C^n.*

9° $R_{ijk} + R_i{}^0{}_{jk} = 0, \quad P^{jk}_i + P_i{}^{0jk} = 0, \quad S^{0jk}_i = 0,$ *where the index "0" means contraction by p_i.*

10° $S^{ijkh} = C^{rih}C^{jk}_r - C^{rik}C^{jh}_r.$ ∎

§4. Generalized Hamilton spaces

A staightforward generalization of the notion of Hamilton or Cartan space is given in our paper [5]:

Definition 4.1. *A generalized Hamilton space is a pair $M^{*n} = (M, g^{ij}(x,p))$, where $g^{ij}(x,p)$ is a symmetric d-tensor field on $\widetilde{T^*M}$ of the type $(2,0)$ and of the rank n:*

$$(4.1) \qquad \text{rank} \, \|g^{ij}(x,p)\| = n \text{ on } \widetilde{T^*M}.$$

$g^{ij}(x,p)$ *is called the fundamental (or metric) d-tensor field of the space M^{*n}.*

If there exists a Hamiltonian $H(x,p)$ such that

$$(4.2) \qquad g^{ij}(x,p) = \frac{1}{2}\dot{\partial}^i\dot{\partial}^j H,$$

then we say that M^{*n} is reducible to a Hamilton space. When $H(x,p)$ is 2-homogeneous with respect to p_i, we say that M^{*n} is reducible to a Cartan space.

A necessary and sufficient condition for M^{*n} to be reducible to a Hamilton space is that the d-tensor field $\dot{\partial}^h g^{ij}$ be totally symmetric.

In the case when M^{*n} is reducible to a Hamilton space we fix a Hamiltonian $H(x,p)$ which satisfies (4.2) and take the canonical non-linear connection N determined by H. Therefore the geometry of the space M^{*n} can be developed by means of the methods given in the paragraphs 1–4 of this paper.

If the space M^{*n} is not reducible to a Hamilton space we consider the Hamiltonian

$$(4.3) \qquad\qquad H = g^{ij}(x,p)p_i p_j.$$

We say that the space M^{*n} is weakly regular if the pair (M, H) is a Hamilton space. For this class of generalized Hamilton spaces we fix the canonical non-linear connection (2.4) determined by the Hamitonian (4.3) and apply the considerations above mentioned.

If M^{*n} is not weakly regular, then we fix an arbitrary non-linear connection N and study the pair (M^{*n}, N). Taking into account the basis $(\delta_i, \dot{\partial}^i)$ adapted to N and V we get:

Theorem 4.1.

i) *There exists a unique d-connection D with the properties:*

$$(4.4) \qquad g^{ij}{}_{|k} = 0, \qquad g^{ij}|^k = 0, \qquad T^i_{jk} = 0, \qquad S^{jk}_i = 0.$$

ii) *The coefficients of D are*

$$(4.5) \qquad \begin{aligned} H^i_{jk} &= \frac{1}{2}g^{ih}(\delta_j g_{hk} + \delta_k g_{jh} - \delta_h g_{jk}) \\ C^{jk}_i &= -\frac{1}{2}g_{ih}(\dot{\partial}^j g^{hk} + \dot{\partial}^k g^{jh} - \dot{\partial}^h g^{jk}) \end{aligned}$$

where $\|g_{ij}(x,p)\| = \|g^{ij}(x,p)\|^{-1}$. ∎

Now it is easy to study the geometrical properties of the generalized Hamilton spaces M^{*n}, [5].

Other aspects of this theory are related to the almost symplectic strucuture associated to the non-linear connection N:

$$(4.6) \qquad\qquad \theta = \delta p_i \wedge dx^i.$$

Generally θ is not integrable, [3, 5]. However, we can consider the N-lift of (M^{*n}, N). Let us take on $\widetilde{T^* M}$ the tensor fields:

(4.7)
$$G = g_{ij}(x, p)dx^i \otimes dx^j + g^{ij}(x, p)\delta p_i \otimes \delta p_j,$$
$$J = g_{ij}\dot{\partial}^i \otimes dx^j - g^{ij}(x, p)\delta_i \otimes \delta p_j.$$

We have:

Theorem 4.2. *The pair* (G, J) *is an almost Hermitian structure on* $\widetilde{T^* M}$ *having* θ *as the associated almost symplectic structure.* ■

Some important properties of the generalized Hamilton spaces M^{*n} can be studied by means of the "Almost Hermitian model" $H^{2n} = \ = (\widetilde{T^* M}, (G, J))$.

§ 5. Generalized Hamilton spaces with the metric $\gamma^{ij}(x, p) + \dfrac{1}{c^2}p^i p^j$

In what follows we study an important class of generalized Hamilton spaces suggested by physicists, [1, 2, 9]. An analogous theory for Lagrange spaces was studied by T. Kawaguchi together with the author and by J. L. Synge in the case of the relativity of dispersive mediums.

Let $\mathcal{C}^n = (M, H)$ be a Cartan space having the fundamental d-tensor field $\gamma^{ij}(x, p)$. The fundamental function $H(x, p)$ being 2-homogeneous with respect to p_i it follows that $\gamma^{ij}(x, p)$ is 0-homogeneous with respect to p_i.

On $\widetilde{T^* M}$ we consider the d-tensor field

(5.1)
$$g^{ij}(x, p) = \gamma^{ij}(x, p) + \frac{1}{c^2}p^i p^j,$$

where c is a positive real number and

(5.2)
$$p^i = \gamma^{ij}p_j = \frac{1}{2}\dot{\partial}^i H = \dot{\partial}^i \mathcal{H}.$$

We remark that the d-tensor field $g^{ij}(x, p)$ has the remarkable form:

(5.1)'
$$g^{ij}(x, p) = \dot{\partial}^i \dot{\partial}^j \mathcal{H} + \frac{1}{c^2}\dot{\partial}^i \mathcal{H}\dot{\partial}^j \mathcal{H}.$$

Writing

(5.3)
$$\|p\|^2 = \gamma^{ij} p_i p_j = H,$$
$$a = 1 + \frac{H}{c^2}$$

we get:

Proposition 5.1. *The pair $M^{*n} = (M, g^{ij}(x, p))$ is a generalized Hamilton space.*

Indeed, $g^{ij}(x, p)$ is a symmetric d-tensor field of the type $(2, 0)$ on $\widetilde{T^* M}$ satisfying (4.1). The matrix $\|g^{ij}(x, p)\|^{-1}$ has the elements

(5.4)
$$g_{ij}(x, p) = \gamma_{ij}(x, p) - \frac{1}{c^2 a} p_i p_j.$$

■

Some remarkable properties of this space M^{*n} are given in the following theorems:

Theorem 5.1. *The M^{*n} of this paragraph is not reducible to a Hamilton space.*

Proof. From (5.1), (5.2) we get

$$\dot{\partial}^k g^{ij} = \dot{\partial}^k \gamma^{ij} + \frac{1}{c^2}(\gamma^{ki} p^j + \gamma^{kj} p^i).$$

But this d-tensor is not completely symmetric. ■

Theorem 5.2. *M^{*n} is weakly regular.*

Indeed, $g^{ij}(x, p) p_i p_j = aH$ is a regular Hamiltonian, [2]. ■

Observing that the fundamental d-tensor field $g^{ij}(x, p)$ is uniquely determined by the fundamental tensors $\gamma^{ij}(x, p)$ of the Cartan space $\mathcal{C}^n$ we can take the canonical non-linear connection $\overset{\circ}{N}$ from (3.2) as the non-linear connection of the space M^{*n}. Then we fix this non-linear connection $\widetilde{T^* M}$.

In this case Theorem 4.1 is applicable and we have a canonical d-connection D metrical with respect to the fundamental d-tensor $g^{ij}(x,p)$ in (5.1).

It is convenient to express the fundamental geometrical objects of the space M^{*n} by means of the geometrical objects of the Cartan space $\mathcal{C}^n$, as in (5.1), (5.4).

Let $\overset{\circ}{D}$ be the canonical d-metrical connection of the Cartan space $\mathcal{C}^n$ and "$\overset{\circ}{|}$" and "$|$" the h- and the v-covariant derivations with respect to $\overset{\circ}{D} = (\overset{\circ}{N}, \overset{\circ}{H}, \overset{\circ}{C})$.

Applying Theorem 4.1 we can prove:

Theorem 5.3. *The canonical metrical d-connection D of the generalized Hamilton space M^{*n} has the coefficients H^i_{jk}, C^{jk}_i given by the formulas*

$$(5.5) \qquad H^i_{jk} = \overset{\circ}{H}{}^i_{jk}, \qquad C^{jk}_i = \overset{\circ}{C}{}^{jk}_i - \frac{1}{c^2 a}\gamma^{jk} p_i.$$

∎

But (5.5) is a transformation of d-connection $\overset{\circ}{D} \to D$. Using this property we can prove:

Proposition 5.2. *The canonical metrical d-connection D of the space M^{*n} determined by (5.1) has the properties:*

$1°$ $p_{i|j} = 0,$ $p_i|^j = g^{jk}\gamma_{ki}.$

$2°$ $p^i_{|j} = 0,$ $p^i|^j = \frac{1}{a}\gamma^{ij}.$

$3°$ $\gamma^{ij}{}_{|k} = 0,$ $\gamma^{ij}|^k = -\frac{1}{c^2 a}(\gamma^{ik} p^j + \gamma^{jk} p^i).$

$4°$ $H_{|k} = 0,$ $H|^k = 2p^k.$ ∎

Proposition 5.3. *The torsion of this D are*

$$T^i_{jk} = 0, \quad S^{jk}_i = 0, \quad R_{ijk} = \overset{\circ}{R}{}_{ijk}, \quad P_{jk}{}^i = \overset{\circ}{P}{}_{jk}{}^i \text{ and } C^{jk}_i \text{ from } (5.5),$$

where $\overset{\circ}{R}{}_{ijk}$, $\overset{\circ}{P}{}_{jk}{}^i$ and $\overset{\circ}{C}{}_i{}^{jk}$ are the torsions of $\overset{\circ}{D}$ of the Cartan space $\mathcal{C}^n$.
∎

Let $\overset{\circ}{R}$, $\overset{\circ}{P}$, $\overset{\circ}{S}$ be the curvature d-tensor fields of $\overset{\circ}{D}$ and R, P, S the curvature d-tensor fields of D.

Proposition 5.4. *The curvature d-tensor fields of the canonical metrical d-connection D of the space M^{*n} are related to the curvature d-tensor fields of the canonical metrical d-connection $\overset{\circ}{D}$ of the Cartan space C^n by the formulas:*

$$R_j{}^i{}_{kh} = \overset{\circ}{R}_j{}^i{}_{kh} - \frac{1}{c^2 a} p_j \gamma^{ir} \overset{\circ}{R}_{rkh},$$

$$p_j{}^i{}_k{}^h = \overset{\circ}{P}_j{}^i{}_k{}^h - \frac{1}{c^2 a} p_j \gamma^{ir} \overset{\circ}{P}_{rk}{}^h,$$

$$S_j{}^{ikh} = \overset{\circ}{S}_j{}^{ikh} - \frac{1}{c^2 a} (\gamma^{ik} \gamma^{rh} - \gamma^{ih} \gamma^{rk}).$$

■

Finally we have:

Theorem 5.4. *The almost Hermitian model $H^{2n} = (\widetilde{T^*M}, (G, J))$ of the generalized Hamilton space M^{*n}, with the metric (5.1) is not an almost Kählerian space.* ■

References

1. T. KAWAGUCHI and R. MIRON, On the generalized Lagrange spaces with the metric $\gamma_{ij}(x) + (1/c^2) y_i y_j$, *Tensor, N.S.* vol. 48, (1989), 52–63.

2. T. KAWAGUCHI and R. MIRON, A Lagrangian model for gravitation and elektromagnetism, Tensor N.S., *Tensor, N.S.* vol. 48, (1989), 153–168.

3. P. LIBERMANN et C. M. MARLE, Géométrie symplectique. Bases theoretique de la mécanique I–IV, *U.E.R. de Math.* (1986, 1987).

4. R. MIRON, Sur la géométrie des espaces d'Hamilton. *C.R. Acad. Sci. Paris,* t. 306, S.I, 1988, 195–198.

5. R. MIRON, Hamilton geometry, Seminarul de Mecanică, Nr. 3, Univ. Timişoara (1987), 54.

6. R. MIRON, The Geometry of Cartan spaces. *Prog. of Math,* vol. 22 (1& 2) (1988), 1–38.

7. R. MIRON, The geometry of Hamilton spaces. Proc. of the fifth Nat. Sem. of Finsler and Lagrange spaces, Univ. Braşov (1988), 249–278.

8. E. CARTAN, Les espaces métriques fondé sur la notion d'aire, Actualitée Sci. Ind. Herman, Paris 1933.

9. J. L. SYNGE, Relativity: The general theory, North-Holland Publ. Comp. 1960.

Radu Miron

University of Iaşi
Faculty of Mathematics
6600 Iaşi, Romania

COLLOQUIA MATHEMATICA SOCIETATIS JÁNOS BOLYAI

56. DIFFERENTIAL GEOMETRY, EGER (HUNGARY), 1989

On Isometries of Space Forms[*]

E. MOLNÁR[**]

Dedicated to my father Ernő Molnár

A *d-dimensional space form* is an orbit space $\mathcal{M}/\mathcal{G}$, where $\mathcal{M}$ is either the Euclidean ($K = 0$) or a spherical ($K > 0$) or a hyperbolic ($K < 0$) d-space of constant sectional curvature K, and $\mathcal{G}$ is an isometry group acting on $\mathcal{M}$ freely and discontinuously [19]. An isometry of $\mathcal{M}/\mathcal{G}$ can be described by an isometry φ of $\mathcal{M}$ which permutes the whole $\mathcal{G}$-orbits, i.e. $\underline{g} \mapsto \varphi^{-1}\underline{g}\varphi =: \underline{g}^\varphi$ is a metric automorphism of $\mathcal{G}$, and $\varphi \in \mathcal{N}(\mathcal{G})$ the normalizer of $\mathcal{G}$ in $\operatorname{Iso}\mathcal{M}$. Since an isometry $\underline{h}$ maps any $\mathcal{G}$-orbit onto itself iff $\underline{h} \in \mathcal{G}$, hence the isometry group of $\mathcal{M}/\mathcal{G}$ is the quotient group

$$\mathcal{N}(\mathcal{G})/\mathcal{G} =: \operatorname{Iso}\mathcal{M}/\mathcal{G}.$$

To determine isometry groups of hyperbolic space forms of finite volume, two methods will be proposed and illustrated by examples.

First we shall give a new description of the famous *Matveev–Fomenko manifold* $\overline{\mathcal{M}} = \mathcal{H}^3/\overline{G}$ conjectured as a candidate for hyperbolic manifold of minimal volume $\operatorname{Vol}\overline{\mathcal{M}} = 0{,}94272$ ($K = -1$) [7]. We shall give two presentations for the fundamental group $\overline{G}$ with the corresponding fundamental

---

[*] This paper is in final form and no version of it will be submitted for publication elsewhere.

[**] Supported by Hungarian Nat. Found. for Sci. Research Grant No. 424 (86).

domains $\overline{\mathcal{F}}_1, \overline{\mathcal{F}}_2$ (Fig. 2–3). The concave domain $\overline{\mathcal{F}}_2$ provides a simple symmetric presentation with two generators. From this we shall read off the outer automorphism group [8, 16]:

$$\operatorname{Out}\overline{G} = \operatorname{Aut}\overline{G}/\operatorname{Inn}\overline{G}.$$

The centre $\mathcal{Z}(\overline{G})$ of $\overline{G}$ is trivial now, moreover, the rigidity theorem [6, 15] guarantees that the group of automorphisms $\operatorname{Aut}\overline{G} \cong \mathcal{N}(\overline{G})$ the metric normalizer of $\overline{G}$, and the group of inner automorphisms $\operatorname{Inn}\overline{G} \cong \overline{G}$. Hence we get $\operatorname{Out}\overline{G} \cong \operatorname{Iso}\mathcal{H}^3/\overline{G}$ as a Coxeter group of order 12 with diagram

$$\bigcirc\!\!-\!\!\bigcirc\!\!\overset{3}{-\!\!-\!\!}\bigcirc$$ [2, 18]. It turns out that $\mathcal{H}^3/\overline{G}$ is a regularly minimal manifold [13], i.e. it does not cover regularly another manifold, or in other words, there does not exist a freely and discontinuously acting isometry group containing $\overline{G}$ as a proper normal subgroup (Theorem 1.1–2).

Second, we shall determine the minimal closed geodesic for any compact non-orientable hyperbolic space forms in the infinite series $\mathcal{H}/N_{tu}^1$ discovered by the author first in [9]. We start with the fundamental domain proposed in [10] (Fig. 4), and use the existence of a Coxeter supergroup $\underline{C}_t$ generated by plane reflections. The minimal closed geodesic lies on the axis of the generating screw motion $\underline{s}_u$, where u describes the rotational component of $\underline{s}_u$. This also yields that we get different (non isometric) manifolds $\mathcal{H}^3/N_{tu}^1$ for "different" (t, u) pairs. Since the fundamental domain F_{tu}^1 is just the Dirichlet–Voronoi cell $\mathcal{D}_{\underline{Q}_1}$ with a metrically well-defined point Q, hence $\operatorname{Iso}\mathcal{H}^3/\underline{N}_{tu}^1 = \operatorname{Sym}\mathcal{D}_{\underline{Q}}$ (see also [5]). The latter symmetry group of $\mathcal{D}_{\underline{Q}}$ consists of all the isometries of $\mathcal{D}_Q$ in $\mathcal{M} = \mathcal{H}^3$ such that they will preserve also the face identifications of $\mathcal{D}_Q$ by $\underline{N}_{tu}^1$ (Theorem 2.1–2).

Both methods have some disadvantages and need careful calculations which are specific in each case. We are still far from an algorithmic method, also [14] provides preparatory algorithms.

1. The Matveev–Fomenko manifold $\overline{\mathcal{M}} = \mathcal{H}^3/\overline{G}$

We take two regular simplices in $\mathcal{H}^3$ with ideal vertices on the absolut of $\mathcal{H}^3$. Figure 1.a shows their Schlegel diagrams and describes the identifying isometries of the faces:

(1)　　　$\underline{a} : a^{-1} \to a, \quad \underline{b} : b^{-1} \to b, \quad \underline{c} : c^{-1} \to c, \quad \underline{d} : d^{-1} \to d.$

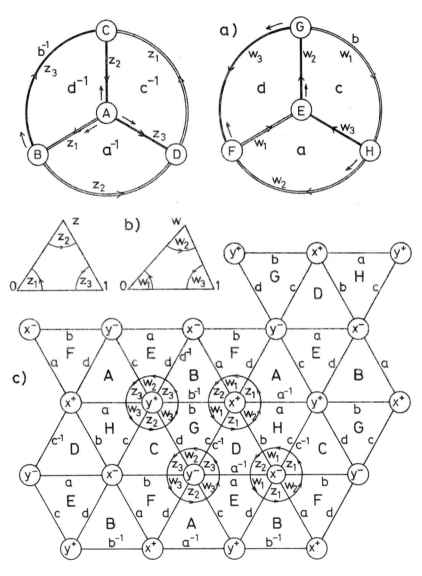

Figure 1.

The identifications induce the two equivalence classes of edges with 6 edges in each class. Since the regular ideal simplex has equal angles $\pi/3$, the simplex pair represents a non-compact hyperbolic space form of finite volume. All the ideal vertices are glued and we have described these vertex domains A, B, C, D, E, F, G, H in the Euclidean plane $\mathcal{E}_G^2$. Namely the Euclidean geometry holds on any horosphere, centred in any ideal vertex, say in G. The glued collection of the 8 vertex domains forms a fundamental domain f_G for the Euclidean translation group $\underline{T}_G \cong \mathbf{p1}$ acting freely and discontinuously on $\mathcal{E}_G^2$.

The directed edge classes x and y appear as "vertices" x^-, x^+, y^-, y^+ of f_G. Going round x^+ (or x^-) and y^+ (or y^-), we see the cyclic gluing process induced by (1):

$$(2) \qquad x \Longrightarrow: \underline{b}^{-1}\underline{a}\,\underline{d}^{-1}\underline{a}\,\underline{b}^{-1}\underline{c}; \qquad y \longrightarrow: \underline{b}^{-1}\underline{d}\,\underline{c}^{-1}\underline{a}\,\underline{c}^{-1}\underline{d}.$$

More precisely, we may assume that the two starting simplices are unified, say, at the faces d^{-1} and d. Then we get a fundamental domain $\mathcal{F}_G$ for the isometry group $\mathcal{G}$ generated by the screw motions $\underline{a}$, $\underline{b}$, $\underline{c}$ in (1). Then the glued vertex domain f_G above will be the (local) fundamental domain of the stabilizer subgroup $\underline{T}_G \subset \mathcal{G}$. $\underline{T}_G$ fixes the ideal vertex $\mathcal{G}$. Now (2) can be written because of choosing $\underline{d} = 1$:

$$(2^*) \qquad x \Longrightarrow: \underline{b}^{-1}\underline{a}^2\underline{b}^{-1}\underline{c} = 1; \qquad y \longrightarrow: \underline{b}^{-1}\underline{c}^{-1}\underline{a}\,\underline{c}^{-1} = 1.$$

The geometric meaning of the first relation, e.g., is the following.

In Fig. 1.c we see a part of $\mathcal{F}_G$ at the vertex domains G and C which are glued at $d^{-1} = d$. The face c of G is the $\underline{c}$-image of the face c^{-1} from $\mathcal{F}_G$. Hence the isometry $\underline{c}$ maps $\mathcal{F}_G$ onto $\mathcal{F}_G^{\underline{c}}$ which joins $\mathcal{F}_G$ along the face $c = (c^{-1})^{\underline{c}}$. Now we see the $\underline{c}$-image of the vertex domain D at x^+ as the part of $\mathcal{F}_G^{\underline{c}}$. Then the face $(b^{-1})^{\underline{c}}$ comes into x^+. The motion $\underline{c}^{-1}\underline{b}^{-1}\underline{c} : b^{\underline{c}} \to (b^{-1})^{\underline{c}}$ maps $\mathcal{F}_G^{\underline{c}}$ onto $\mathcal{F}_G^{\underline{b}^{-1}\underline{c}}$. This joins $\mathcal{F}_G^{\underline{c}}$ along the face $(b^{-1})^{\underline{c}} = b^{\underline{b}^{-1}\underline{c}}$. So we go round x^+ meeting the images of $\mathcal{F}_G$ as follows

$$(3) \qquad \mathcal{F}_G, \mathcal{F}_G^{\underline{c}}, \mathcal{F}_G^{\underline{b}^{-1}\underline{c}}, \mathcal{F}_G^{\underline{a}\,\underline{b}^{-1}\underline{c}}, \mathcal{F}_G^{\underline{a}\,\underline{a}\,\underline{b}^{-1}\underline{c}}, \mathcal{F}_G^{\underline{b}^{-1}\underline{a}\,\underline{a}\,\underline{c}^{-1}\underline{c}} = \mathcal{F}_G.$$

Hence we get the first relation of (2^*). Analogous algorithms have been studied in [14].

From (2^*) we read the two-generator presentation

$$(4) \qquad\qquad \mathcal{G} = (\underline{b}, \underline{c} - \underline{b}^{-1}\underline{c}\,\underline{b}\,\underline{c}^2\underline{b}\,\underline{c}\,\underline{b}^{-1}\underline{c} = 1).$$

Now, we apply the *Dehn surgery* to $\mathcal{F}_\mathcal{G}$ so that we get a new isometry group $\overline{G}$ acting freely and discontinuously on $\mathcal{H}^3$ with a compact fundamental domain [7]. The method has been developed by Thurston [16], see also [18].

We know that a complex coordinate can be introduced on the absolut of $\mathcal{H}^3$ in accordance with the Poincaré half space model. If we take the "end" $G = C = \infty$, say, then the other ideal vertices (ends) of $\mathcal{F}_\mathcal{G}$ are represented by complex numbers. Now we assume that the starting simplices are not regular, but we know, they have equal angles at the opposite edges. The angles z_1, z_1, z_3 and w_1, w_2, w_3 are indicated in Fig. 1.b. Since the geometry of any horosphere is Euclidean, we can introduce the complex parameters $\underline{z}$ and $\underline{w}$ describing the angles of the two simplices as Fig. 1 shows.

Of course this deforms also the generators in (1) also the fundamental domains $\mathcal{F}_G \to \overline{\mathcal{F}}_G$ and $f_G \to \overline{f}_G$. The picture with the images of $\overline{f}_G$ differs from the former one (Fig. 1.c), since the tiling of $\mathcal{E}_G^2$, with triangles of angles z_1, z_2, z_3 and w_1, w_2, w_3, respectively, represents a similarity group $\underline{H}_G$. To this a necessary condition is that the triangles fit at x and y. We see, e.g. at x^+, $w_1 + z_1 + w_2 + z_1 + w_1 + z_2 = 2\pi$ that means $\underline{w} \cdot \underline{z} \frac{w-1}{w} \cdot \underline{z} \cdot \underline{w} \cdot \frac{z-1}{z} = 1$. Thus we get

(5) $$\underline{z}(\underline{z} - 1)\underline{w}(\underline{w} - 1) = 1$$

as an equation for the parameters. The same guarantees $z_2 + w_3 + z_3 + w_2 + z_3 + w_3 = 2\pi$ at y^+.

The similarity group $\underline{H}_G$ of $\mathcal{E}_G^2$ has two fixed points $X_1 = G = C = \infty$ and X_2 depending on $\underline{z}(\underline{w})$. $\underline{H}_G$ extends also to a motion group in $\mathcal{H}^3$ which is generated by the identifying generators of the vertex domain $\overline{f}_G$. In Fig. 1.c and 2.a we have indicated the screw motions

(6)
$$\mu : \overline{\mathcal{F}} \supset b^{-1} \to b^{\underline{a}^{-1}\underline{b}} \subset \overline{\mathcal{F}}^{\underline{a}^{-1}\underline{b}}, \qquad \text{i.e. } \mu = \underline{b}\,\underline{a}^{-1}\underline{b} = \underline{b}\,\underline{c}^{-1}\underline{b}^{-1}\underline{c}^{-1}\underline{b};$$
$$\lambda : \quad \overline{\mathcal{F}} \supset b \to (b^{-1})^{\underline{a}^{-1}\underline{b}} \subset \overline{\mathcal{F}}^{\underline{a}^{-1}\underline{b}}, \qquad \text{i.e. } \lambda = \underline{b}^{-1}\underline{a}^{-1}\underline{c} = \underline{b}^{-1}\underline{c}^{-1}\underline{b}^{-1};$$
$$\pi : \quad b^{-1} \to b^{\underline{a}^{-1}\underline{c}^{-1}\underline{a}^{-1}\underline{b}}, \qquad \text{i.e. } \pi = \underline{b}\,\underline{a}^{-1}\underline{c}^{-1}\underline{a}^{-1}\underline{b}$$

with the screw axis $X_1 X_2$ in the half space model of $\mathcal{H}^3$ ($X_1 = \infty$).

In the case of Matveev–Fomenko manifold $\overline{\mathcal{M}} = \mathcal{H}^3/\overline{G}$ the group $\underline{H}_G$ is specified by the "surgery requirement"

(7) $$1 = \mu^5 \pi^{-2} = \mu \lambda^{-2} = \underline{b}\,\underline{c}^{-1}\underline{b}^{-1}\underline{c}^{-1}\underline{b}\,\underline{b}\,\underline{c}\,\underline{b}\,\underline{b}\,\underline{c}\,\underline{b}$$

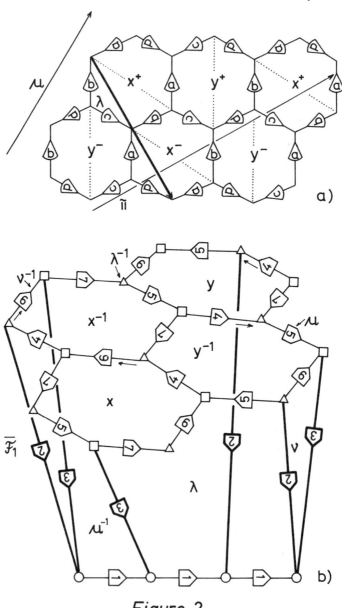

Figure 2.

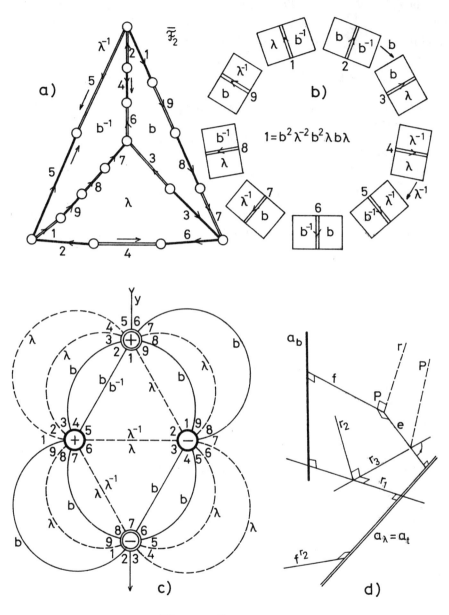

Figure 3.

where the second equation follows from the presentation (2^*) or (4). The formula (7) means that the motion group $\underline{H}_G$ is generated by the screw motion $\lambda = \underline{b}^{-1}\underline{c}^{-1}\underline{b}^{-1}$ only. We get (7) as a new defining relation for the fundamental group $\overline{G}$ of the $M - F$-manifold. We have

(8) $\overline{G} = (\underline{b}, \underline{c} - 1 = \underline{b}^{-1}\underline{c}\underline{b}\underline{c}^2\underline{b}\underline{c}\underline{b}^{-1}\underline{c} = \underline{c}^{-1}\underline{b}^{-1}\underline{c}^{1-}\underline{b}^2\underline{c}\underline{b}^2\underline{c}\underline{b}^2)$

as a presentation, or equivalently, expressing $\underline{c} = \underline{b}^{-1}\lambda^{-1}\underline{b}^{-1}$ from (6):

(9) $\overline{G} = (\lambda, \underline{b} - 1 = \lambda\underline{b}\lambda\underline{b}^2\lambda^{-2}\underline{b}^2 = \underline{b}\lambda\underline{b}\lambda^2\underline{b}^{-2}\lambda^2)$

which is the promised symmetric presentation for $\overline{G}$.

Of course, the new relation (7) gives a new restriction for the parameter $\underline{z}(\underline{w})$. From (6) we read off new angular conditions, for the homology $\underline{H}_\mu$ and $\underline{H}_\lambda$, in the language of complex numbers. The connection of the "vectors" $b^{-1} = (x^- y^-)$ and $b^{\underline{a}^{-1}\underline{b}} = (x^- y^-)^{\underline{a}^{-1}\underline{b}}$ can be expressed by multiplication by

$$\underline{H}_\mu = (1 - \underline{z}) \cdot \frac{1}{1 - \underline{w}} \cdot \frac{\underline{z}}{\underline{z} - 1} \cdot \frac{\underline{w} - 1}{\underline{w}} = \underline{z}\,\underline{w}^{-1}.$$

Similarly, $b = (x^+ y^+)$ leads to $(b^{-1})^{\underline{a}^{-1}\underline{c}} = (x^+ y^+)^{\underline{a}^{-1}\underline{c}}$ by

$$\underline{H}_\lambda = \underline{w} \cdot (1 - \underline{z})\underline{w}(1 - \underline{z}).$$

Hence (7) imply

(10) $\underline{z}\,\underline{w}^{-5}(\underline{z} - 1)^{-4} = 1$

as a new equation besides (5). The natural geometric assumptions

(11) $\operatorname{Im} \underline{z} > 0, \qquad \operatorname{Im} \underline{w} > 0$

yield a unique solution of (5) and (10). The solution determines the angles of the starting simplices and all the metric data of the generators $\underline{b}$, $\underline{c}$ or $\underline{b}$, λ. These have been implemented on computer [7].

We did not make these computations, however, we can draw a picture of a compact fundamental domain $\overline{\mathcal{F}}_1$ for $\overline{G}$ (Fig. 2.b). Moreover, we can describe the combinatorial structure of the fundamental domain $\overline{\mathcal{F}}_2$ corresponding to the presentation (9) (this is the minimal or most economical presentation in the length sum of its relators).

Fig. 2.b shows the symbolic pictures of $\overline{\mathcal{F}}_1$ as a "part of a corn-cob". Comparing with Fig. 2.a we see the procedure: The tiling with the deformed ideal simplices has been dualized into the tiling with hexagonal pyramids of ideal apices. Then the apex domains are deformed along the screw axis $X_1 X_2$ of $\underline{H}_G$. $\overline{\mathcal{F}}_1$ has 4 hexagonal faces: x^{-1} refers to the half edge class x^-, x refers to the class x^+, y^{-1} to y^-, y to y^+. We can derive the identifying generators as in (6):

$$(12) \qquad \underline{x} : x^{-1} \to x, \quad \underline{x} = \underline{b}\,\underline{a}\,\underline{b}^{-1}; \qquad \underline{y} : y^{-1} \to y, \quad \underline{y} = \underline{b}.$$

The other curved faces "connect" the former ones with the screw axis of $\underline{H}_G$. So we get the faces λ^{-1}, λ; μ^{-1}, μ and ν^{-1}, ν; determining the generators

$$(13) \qquad\qquad \lambda : \lambda^{-1} \to \lambda, \quad \mu : \mu^{-1} \to \mu, \quad \nu : \nu^{-1} \to \nu$$

by (6) and by (7). (The notations will not disturb us.)

From Fig. 2.b we can read the presentation of $\overline{G}$ again. The edge classes, labelled by $1, 2, \ldots, 7$, provides the relations:

$$(14) \qquad
\begin{aligned}
&1 : \mu\lambda^{-2}, \qquad 2 : \mu\lambda\nu^{-1}, \qquad 3 : \mu\nu^{-1}\lambda, \qquad 4 : \underline{x}\,\underline{y}^2\mu^{-1}, \\
&5 : \underline{x}\mu\underline{y}\lambda\underline{y}, \qquad 6 : \underline{x}^2\lambda^{-1}\underline{y}^{-1}\nu^{-1}, \qquad 7 : \underline{x}\mu\underline{y}^{-1}\underline{x}\lambda^{-1}.
\end{aligned}$$

These are in accordance with (6) and (12). Moreover, from (14) we can easily derive the nicest presentation (9) (with $\underline{y}$ instead of $\underline{b}$).

In Fig. 3.a we have drawn the Schlegel diagram of the minimally presenting fundamental domain $\overline{\mathcal{F}}_2$, corresponding to (9).

Fig. 3.b also indicates, how to construct the edge class $\Rightarrow\!\!\!=$ on $\overline{\mathcal{F}}_2$ from the corresponding relation (see also [12] for details of the method).

Fig. 3.c shows also the Schlegel diagram of the glued vertex domains of the unique vertex class of $\overline{\mathcal{F}}_2$. This is constructed in the analogy of the gluing algorithm described at (3). We see that $\overline{\mathcal{F}}_2$, equipped with the identifications, appears as a manifold, indeed.

Now, we turn to determination of the isometry group of $\mathcal{H}^3/\overline{G}$. From the presentation (9) we see that

$$(15) \qquad
\begin{aligned}
\rho_1 &: \underline{b}^{-1} \longleftrightarrow \underline{b}, \qquad \lambda^{-1} \longleftrightarrow \lambda; \\
\rho_2 &: \underline{b}^{-1} \longleftrightarrow \lambda^{-1}, \qquad \underline{b} \longleftrightarrow \lambda, \qquad \text{and so} \\
\rho_3 &:= \rho_1\rho_2 = \rho_2\rho_1
\end{aligned}$$

induce involutive automorphisms of the group $\overline{G}$. The rigidity theorem [6, 15] says that some involutive isometries

(16)

$$\underline{r}_1 : (\underline{b}^{-1})^{\underline{r}_1} := \underline{r}_1^{-1}\underline{b}^{-1}\underline{r}_1 = \underline{b}; \quad (\lambda^{-1})^{\underline{r}_1} := \underline{r}_1^{-1}\lambda^{-1}\underline{r}_1 = \lambda, \quad \underline{r}_1^{-1} = \underline{r}_1;$$

$$\underline{r}_2 : \underline{b}^{\underline{r}_2} := \underline{r}_2^{-1}\underline{b}\,\underline{r}_2 = \lambda, \quad \underline{r}_2^{-1} = \underline{r}_2; \quad \underline{r}_3 = \underline{r}_1\underline{r}_2 = \underline{r}_2\underline{r}_1$$

metrically realize these automorphisms, respectively. Moreover $\underline{r}_1$, $\underline{r}_2$, $\underline{r}_3$ are half-turns (line reflections) in $\mathcal{H}^3$ which map the fundamental domain $\overline{\mathcal{F}}_2$ onto itself.

Now, consider the map τ of $\overline{G}$ into itself defined by

(17)

$$\tau : (\lambda, \underline{b}) \xmapsto{\tau} (\lambda, \lambda^{-1}\underline{b}^{-1}) \quad \text{that means}$$

$$\tau : w(\lambda, \lambda^{-1}; \underline{b}, \underline{b}^{-1}) \xmapsto{\tau} w(\lambda, \lambda^{-1}; \lambda^{-1}\underline{b}^{-1}, \underline{b}\lambda),$$

where w is any finite formal word of 4 variables [2, 4]. We check that τ preserves the presentation (9) of $\overline{G}$. This also implies that τ is an automorphism. Indeed, we have by (9)

$$\lambda\underline{b}\lambda\underline{b}^2\lambda^{-2}\underline{b}^2 \xmapsto{\tau} \lambda(\lambda^{-1}\underline{b}^{-1})\lambda(\lambda^{-1}\underline{b}^{-1}\lambda^{-1}\underline{b}^{-1})\lambda^{-2}(\lambda^{-1}\underline{b}^{-1}\lambda^{-1}\underline{b}^{-1}) =$$

(18) $$= (\underline{b}^{-2}\lambda^{-1}\underline{b}^{-1}\lambda^{-1})(\lambda^{-2}\underline{b}^{-1}\lambda^{-1}\underline{b}^{-1}) = (\lambda^{-2}\underline{b}^2)(\underline{b}^{-2}\lambda^2) = 1;$$

$$\underline{b}\lambda\underline{b}\lambda^2\underline{b}^{-2}\lambda^2 \xmapsto{\tau} (\lambda^{-1}\underline{b}^{-1})\lambda(\lambda^{-1}\underline{b}^{-1})\lambda^2(\underline{b}\lambda\underline{b}\lambda)\lambda^2 = \lambda^2\underline{b}^{-2}\lambda^2\underline{b}\lambda\underline{b} = 1.$$

We see that

$$(\lambda, \underline{b}) \xmapsto{\tau} (\lambda, \lambda^{-1}\underline{b}^{-1}) \xmapsto{\tau} (\lambda, \lambda^{-1}\underline{b}\lambda) =$$

$$= (\lambda^{-1}\lambda\lambda, \lambda^{-1}\underline{b}\lambda) =: (\lambda^\lambda, \underline{b}^\lambda),$$

(19) $$(\lambda, \underline{b}) \xmapsto{\underline{r}_1} (\lambda^{-1}, \underline{b}^{-1}) \xmapsto{\underline{r}_1} (\lambda^{-1}, \underline{b}\lambda) \xmapsto{\underline{r}_1} (\lambda, \underline{b}^{-1}\lambda^{-1}) \longmapsto$$

$$\longmapsto (\lambda, \underline{b}\lambda\lambda^{-1}) = (\lambda, \underline{b}),$$

$$(\lambda, \underline{b}) \xmapsto{\underline{r}_1} (\lambda, \lambda^{-1}\underline{b}^{-1}) \xmapsto{\underline{r}_2} (b, \underline{b}^{-1}\lambda^{-1}) \xmapsto{\tau} (\lambda^{-1}\underline{b}^{-1}, \underline{b}\lambda\lambda^{-1}) \longmapsto$$

$$\xmapsto{\underline{r}_2} (\underline{b}^{-1}\lambda^{-1}, \lambda) \xmapsto{\tau} (\underline{b}\lambda\lambda^{-1}, \lambda) \xmapsto{\underline{r}_2} (\lambda, \underline{b}),$$

i.e. $\tau^2 = \lambda : \underline{x} \longmapsto \lambda^{-1}\underline{x}\lambda$ as an inner automorphism of $\overline{G}$

$$\underline{r}_1\tau\underline{r}_1\tau = 1, \quad \tau\underline{r}_2\tau\underline{r}_2\tau\underline{r}_2 = 1.$$

Again by the rigidity theorem, the automorphism τ can be realized by the isometry $\underline{t}$ with defining equations

(20) $$\underline{t} : \lambda^{\underline{t}} := \underline{t}^{-1}\lambda\underline{t} = \lambda, \quad \underline{b}^{\underline{t}} := \underline{t}^{-1}\underline{b}\,\underline{t} = \lambda^{-1}\underline{b}^{-1}.$$

Since the centre of $\overline{G}$ is trivial, e.g. by (9) as well, we get

(21) $\underline{t}^2 = \lambda,$ $\operatorname{Aut}\overline{G} \cong \mathcal{N}(\overline{G})$ and $\operatorname{Inn}\overline{G} \cong \overline{G}$

as we stated in the introduction. Moreover, we have

(22) $\operatorname{Iso}\mathcal{H}^3/\overline{G} = \mathcal{N}(\overline{G})/\overline{G} = \operatorname{Aut}\overline{G}/\operatorname{Inn}\overline{G} =: \operatorname{Out}\overline{G}$

with the Coxeter diagram ○ ○—○ .
                              $\underline{r}_1$ $\underline{r}_2$ τ

with a 3 over the edge between $\underline{r}_2$ and τ.

This last statement needs the proof that no more outer automorphism of $\overline{G}$ exist, which satisfies (9). Indeed, no other map

$$\varphi : (\lambda,\underline{b}) \overset{\varphi}{\longmapsto} \left(w_\lambda(\lambda,\lambda^{-1},\underline{b},\underline{b}^{-1}), w_{\underline{b}}(\lambda,\lambda^{-1},\underline{bb}^{-1})\right),$$

with $w(\lambda,\lambda^{-1},\underline{b},\underline{b}^{-1}) \overset{\varphi}{\longmapsto} w(w_\lambda,w_\lambda^{-1},w_{\underline{b}},w_{\underline{b}}^{-1})$ for any finite formal word w of 4 variables, preserves the relation in (9). We proceed induction by the length sum of the reduced words w_λ and $w_{\underline{b}}$. This needs lengthy but straightforward calculations, similar to those of (18), where the simple structure of relators in (9) provides a finite procedure. We do not detail these calculations here.

Now we describe the metric normalizer $\mathcal{N}(\overline{G})$ of $\overline{G}$ in $\mathcal{H}^3$. In Fig. 3.d we have drawn the screw axes a_λ and $a_{\underline{b}}$ with common normal line r_1 and symmetry line r_2 by (16).

There is a unique line f and a line reflection $\underline{f}$ with $\underline{b} = \underline{r_1 f}$ and $f \perp a_b$. Moreover,

(23) $$\lambda = \underline{b}^{r_2} = \underline{r}_1^{r_2}\underline{f}^{r_2} = \underline{r}_1\underline{f}^{r_2} = \underline{r}_1\underline{r}_2\underline{f}\,\underline{r}_2$$

holds by (16) again. The formulas (19)–(21) imply the existence of a line $e \perp a_\lambda$ and a line reflection $\underline{e}$ with $\underline{t} = \underline{e}\,\underline{r}_1$. Here

(24) $\underline{t}^2 = \lambda$ means $\underline{e}\,\underline{r}_1\underline{e}\,\underline{r}_1 = \underline{r}_1\underline{r}_2\underline{f}\,\underline{r}_2$ and

(25) $\underline{t}^{-1}\underline{b}\,\underline{t} = \underline{t}^{-2}\underline{b}^{-1}$ means $1 = \underline{t}\,\underline{b}\,\underline{t}\,\underline{b} = (\underline{e}\,\underline{r}_1\underline{r}_1\underline{f})^2 = (\underline{e}\,\underline{f})^2 =: \underline{r}^2$

with the 2-axis r. We have a point $P \in r, e, f$ and $e \perp f$, $r \perp e, f$.

(26) $(\underline{t}\,\underline{r}_2\underline{t}\,\underline{r}_2\underline{t}\,\underline{r}_2) = 1$ means $1 = (\underline{e}\,\underline{r}_1\underline{r}_2)^3 = (\underline{e}\,\underline{r}_3)^3 =: \underline{p}^3$

with a 3-rotation axis p. Finally, any relation of (9) yields

(27)
$$1 = \left[(\underline{e}\,\underline{r}_1)^3\,\underline{e}\,\underline{r}_2(\underline{e}\,\underline{r}_1)^2\,\underline{e}\,\underline{r}_2\right]^2 =: q^2$$

with the 2-axis q, which does not appear in Fig. 3.

Putting together (23)–(27), we get an economical presentation for

(28)
$$\mathcal{N}(\overline{G}) = \Big(\,\underline{e}, \underline{r}_1, \underline{r}_2 - \underline{e}^2, \underline{r}_1{}^2, \underline{r}_2{}^2, (\underline{r}_1\underline{r}_2)^2, (\underline{e}\,\underline{r}_1\underline{r}_2)^3,$$
$$\left[(\underline{e}\,\underline{r}_1)^3\,\underline{e}\,\underline{r}_2(\underline{e}\,\underline{r}_1)^2\,\underline{e}\,\underline{r}_2\right]^2\Big),$$

$\overline{G}$ is generated by $\lambda = (\underline{e}\,\underline{r}_1)^2$ and $\underline{b} = \underline{r}_2(\underline{e}\,\underline{r}_1)^2\underline{r}_2$.

We summarize the results in two theorems.

Theorem 1.1. *The Matveev–Fomenko manifold* $\mathcal{H}^3/\overline{G}$; *defined in* $\mathcal{H}^3$ *by gluing two ideal simplices with complex parameters* $\underline{z}$, $\underline{w}$, *by Fig. 1 and formulas (5), (10), (11); is a compact hyperbolic space form. The fundamental group* $\overline{G}$ *can be described by fundamental domains* $\overline{\mathcal{F}}_1$ *in Fig. 2 and* $\overline{\mathcal{F}}_2$ *in Fig. 3.* $\overline{G}$ *can be generated by two screw motions* λ *and* $\underline{b}$ *with presentation (9).* ∎

Theorem 1.2. *The isometry group of the space form* $\mathcal{H}^3/\overline{G}$ *is a Coxeter group of order 12, described at (22). The normalizer* $\mathcal{N}(\overline{G})$ *has the presentation (28).* $\mathcal{H}^3/\overline{G}$ *does not cover regularly any other space form.*

We have not proved the last statement yet. *We have to show that any group* $\overline{G}^*$, *under assumption* $\overline{G} < \overline{G}^* < \mathcal{N}(\overline{G})$, *does not act freely on the hyperbolic space.* From (22) we can see that the cosets in $\mathcal{N}(\overline{G})/\overline{G}$ are of the form $(\underline{t}\,\underline{r}_3)^n\overline{G}$ or $(\underline{t}\,\underline{r}_3)^n\underline{t}\,\overline{G}$ $n = 0, 1, 2, 3, 4, 5$. It is sufficient to show that the group extension $\langle\underline{t}\,\underline{r}_3\underline{t}\,\underline{r}_3, \overline{G}\rangle$ has a non-1 with fixed point. Namely,

(25)
$$\langle\underline{t}, \overline{G}\rangle = \langle\underline{t}\,\underline{b}, \overline{G}\rangle = \langle\underline{r}, \overline{G}\rangle,$$

(24)
$$\langle\underline{t}\,\underline{r}_3\underline{t}, \overline{G}\rangle = \langle\underline{t}\,\underline{r}_3\underline{t}\lambda^{-1}, \overline{G}\rangle = \langle\underline{t}\,\underline{r}_3\underline{t}^{-1}, \overline{G}\rangle \quad \text{furthermore}$$
$$\langle\underline{t}\,\underline{r}_3\underline{t}\,\underline{r}_3\underline{t}, \overline{G}\rangle = \langle\underline{r}_2\lambda\underline{b}, \overline{G}\rangle = \langle\underline{r}_2, \overline{G}\rangle,$$

(26)
$$\langle\underline{t}\,\underline{r}_3\underline{t}\,\underline{r}_3\underline{t}\,\underline{r}_3, \overline{G}\rangle = \langle\underline{r}_1\lambda^{-1}\underline{b}^{-1}, \overline{G}\rangle = \langle\underline{r}_1, \overline{G}\rangle$$

obviously have fixed points, $\langle\underline{t}\,\underline{r}_3, \overline{G}\rangle \supseteq \langle\underline{t}\,\underline{r}_3\underline{t}\,\underline{r}_3, \overline{G}\rangle$, and the inverse elements do the same. We have $\underline{t}\,\underline{r}_3\underline{t}\,\underline{r}_3\underline{b} = \underline{e}\,\underline{r}_2\underline{e}\,\underline{r}_2\underline{r}_2\lambda\underline{r}_2 = \underline{e}\,\underline{r}_2\underline{e}\,\underline{r}_2\underline{r}_2\underline{e}\,\underline{r}_1\underline{e}\,\underline{r}_1\underline{r}_2 = \underline{e}\,\underline{r}_3\underline{e}\,\underline{r}_3 = \underline{p}^{-1}$ as a 3-rotation by (23)–(26). ∎

We cannot exclude yet that $\mathcal{H}^3/\underline{G}$ covers another hyperbolic manifold not regularly, i.e. a freely and discontinuously acting supergroup $\underline{G}^*$ for $\underline{G}$ may exist in Iso $\mathcal{H}^3$. Then the cosets $\underline{G}^*/\underline{G}$ would provide a non-regular covering for $\mathcal{H}^3/\underline{G}^*$ by $\mathcal{H}^3/\underline{G}$. This would appear as the subdivision of a fundamental domain $\mathcal{F}_{\overline{G}}$ into the images of a $\mathcal{F}_{G^*}$ under appropriate representatives of the cosets.

I thank A. T. Fomenko, S. V. Matveev and M. I. Shtogrin for the helpful conversations during my stay in Moscow, October 1988.

2. Minimal closed geodesics and isometry groups of the manifolds $\mathcal{H}^3/\underline{N}^1_{tu}$

In Fig. 4 we have described the Schlegel diagram of the fundamental domain $\mathcal{F}^1_{tu} =: \mathcal{D}$ for the isometry group $\underline{N}^1_{tu} =: \underline{N}$, acting freely and discontinuously on $\mathcal{M} := \mathcal{H}^3$ [10]. The odd integer $q = 2t + 1$ is fixed, $t \geq 1$, $q \geq 3$. (In Fig. 4: $t = 2$, $q = 5$). In the plane m_5 of H^3 we take a regular $2q$-gon with centre Q, a vertex A, and angles π/q. In the sides of the $2q$-gon we place side faces perpendicularly to m_5. Then we take two base faces in both half spaces of m_5 symmetrically so that these base faces intersect the side faces in angles $\pi/2q$. To get compact polyhedron, we finally truncate the above solid with $4q$ triangles, each of them perpendicularly intersects one base face and 2 side faces. So we obtain the polyhedron $\mathcal{D}$ with centre Q in $\mathcal{H}^3$. To describe the metric data of $\mathcal{D}$, we have drawn in Fig. 6 the characteristic domain $QAEHLCGO =: \mathcal{F}$ cut by the symmetry planes m_1, m_2, m_5 from $\mathcal{D}$. The planes m_i, $i = 1, 2, \ldots, 6$, bound $\mathcal{F}$. Moreover, the corresponding plane reflections $\underline{m}_i$ generate the Coxeter group $\underline{C}_t$ with diagram in Fig. 6. [2, 11, 18]. This $\underline{C}_t$ will be a supergroup of index $8q$ for the group $\underline{N}$ being constructed.

We shall pair the faces of $\mathcal{D}$ by Fig. 4. Let u be a fixed integer $0 \leq u \leq t$ (in Fig. 4: $u = 1$)

1) The base faces, denoted just by $f_{s_u^{-1}}$ and f_{s_u}, are paired by the screw motion

(1) $$\underline{s}_u := (\underline{m}_1\underline{m}_2)^{q+2u} (\underline{m}_5\underline{m}_4) : f_{s_u^{-1}} \longmapsto f_{s_u}$$

with screw angle $2\pi \cdot \frac{q+2u}{2q}$ (see also Fig. 6).

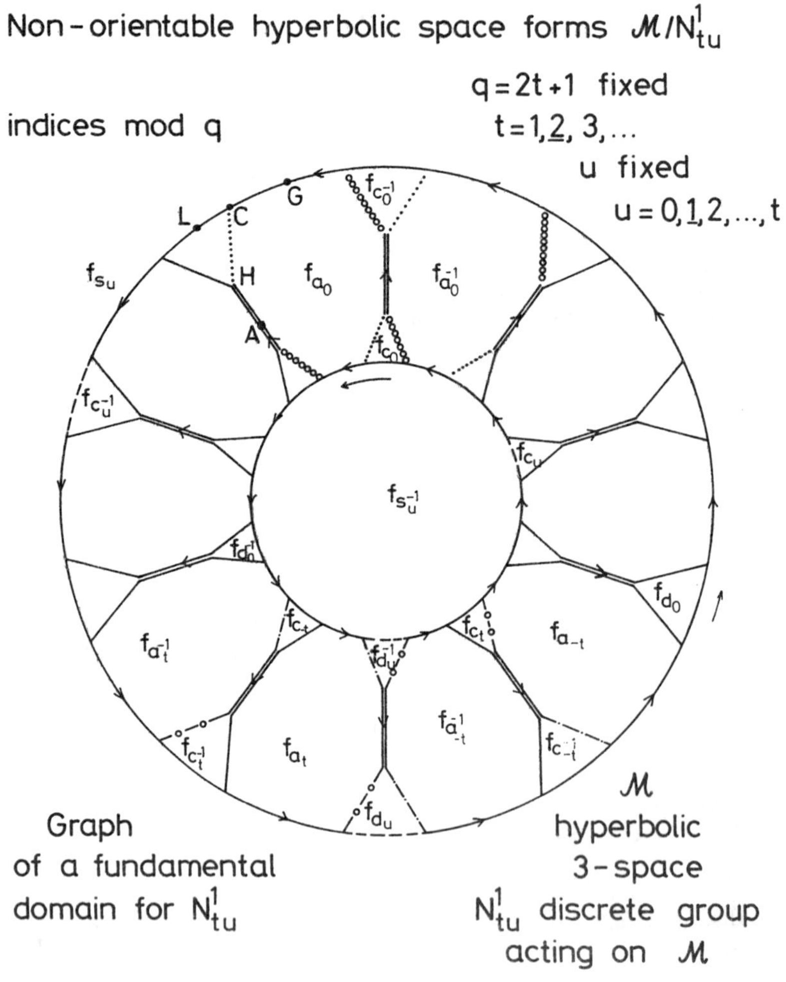

Non-orientable hyperbolic space forms $\mathcal{M}/N^1_{tu}$

indices mod q

$q = 2t + 1$ fixed
$t = 1, \underline{2}, 3, \ldots$
u fixed
$u = 0, 1, 2, \ldots, t$

Graph
of a fundamental
domain for N^1_{tu}

$\mathcal{M}$
hyperbolic
3-space
N^1_{tu} discrete group
acting on $\mathcal{M}$

Figure 4.

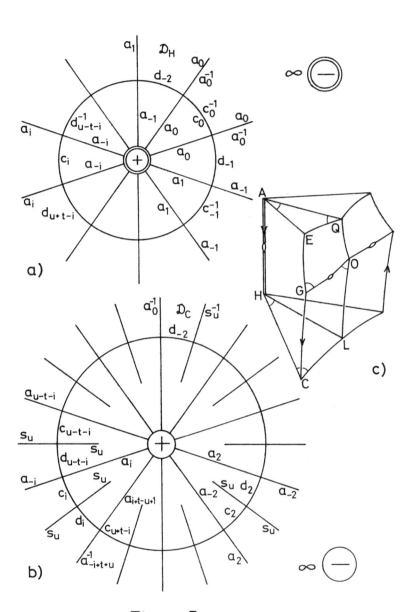

Figure 5.

2) The glide reflections

$$\underline{a}_i : f_{a_i^{-1}} \longmapsto f_{a_i} \qquad i = -t, \ldots, 0, 1, \ldots, t$$

pair the cyclically joining side faces. Fig. 6 also shows how to express $\underline{a}_i$ by the reflections $\underline{m}_1$, $\underline{m}_2$, $\underline{m}_3$. The glide planes of $\underline{a}_i$'s are perpendicular to m_5, the translational components of $\underline{a}_i$'s connect the face centres lying in m_5.

3) With the truncating face $f_{c_i^{-1}}$, surrounded by $f_{a_i^{-1}}$, f_{a_i}, f_{s_u}, we associate the truncating face f_{c_i}, surrounded by $f_{a_{-i}^{-1}}$, $f_{a_{-i}}$, $f_{s_u^{-1}}$, by the glide reflection $\underline{c}_i$ $(i = -t, \ldots, 0 \ldots, t)$.

4) The truncating face $f_{d_i^{-1}}$, surrounded by $f_{a_{i+t-u+1}^{-1}}$, $f_{a_{i+t-u}}$, $f_{s_u^{-1}}$, is paired by f_{d_i} at $f_{a_{-i+t+u+1}^{-1}}$, $f_{a_{-i+t+u}}$, f_{s_u}. The glide reflection $\underline{d}_i = \underline{s}_u \underline{c}_i \underline{s}_u$ is expressed by $\underline{c}_i$ in Fig. 6.

The indices are taken $\mod q$.

The above identifying isometries of $\mathcal{D}$ generate the group $\underline{N}$, and the angular conditions make $\mathcal{H}^3/\underline{N} \sim \mathcal{D}$ a hyperbolic space form. This space form is non-orientable because of occurring glide reflections.

To each vertex class of $\mathcal{D}$ belongs a trivial stabilizer in $\underline{N}$. This critical point of the construction is proved by gluing a local fundamental domain. We use the earlier algorithm at 1. (3). In Fig. 5.a,b we have drawn the Schlegel diagrams to the vertex classes H and C, respectively. At the edges, i.e. at the vertices of the diagrams, we look the defining relations of $\underline{N}$. We have the presentation (for fixed $t = 1, 2, \ldots$; $q = 2t + 1$; $u = 0, 1, \ldots, t$):

$$\underline{N} := \underline{N}_{tu}^1 = \left(\underline{a}_i, \underline{c}_i, \underline{d}_i; \ \underline{s}_u \quad (i = -t, \ldots, 0, \ldots, t) - \right.$$

(2) $\quad \underline{a}_{-t}^2 \cdots \underline{a}_0^2 \cdots \underline{a}_t^2; (\underline{a}_{-t} \underline{s}_u \underline{a}_{u+1} \underline{s}_u^{-1})(\underline{a}_{-t+1} \underline{s}_u \underline{a}_{u+2} \underline{s}_u^{-1}) \cdots (\underline{a}_t \underline{s}_u \underline{a}_u \underline{s}_u^{-1});$

$\quad \underline{c}_i \underline{s}_u \underline{d}_i^{-1} \underline{s}_u; \ \underline{a}_i \underline{c}_i \underline{a}_{-i} \underline{d}_{u-t-1}; \ \underline{a}_i \underline{d}_{u+t-i}^{-1} \underline{a}_{-i} \underline{c}_i^{-1} \big).$

In Fig. 6 we have indicated also the metric data of $\mathcal{F} := QAEHLCGO$ and $\mathcal{D}$ as well on the base of projective metric (see e.g. [11, 18]). It turns out that $\mathcal{D} =: \mathcal{D}_Q$ is just the Dirichlet polyhedron of the orbit $Q^{\underline{N}}$. We emphasize that $\underline{N}_{tu}^1$ is not normal subgroup of the above Coxeter group $\underline{C}_t$. That means our construction method is more general than that of F. Löbell (see [1, 3, 17]).

We summarize the construction in

Theorem 2.1. *To any odd natural number $q = 2t+1$, $t \geq 1$, and parameter $u = 0, 1, \ldots, t$ there exists a compact D-polyhedron $\mathcal{F}_{tu}^1 = \mathcal{D}_Q$ and a face pairing on $\mathcal{D}_Q$ which generate an isometry group $\underline{N}_{tu}^1$ in the hyperbolic space $\mathcal{H}^3$ so that $\mathcal{H}^3/\underline{N}_{tu}^1$ is a compact non-orientable hyperbolic space form. For different parameters (t, u) we get an infinite series of non-isometric space forms [9, 10].*

Proof of the last statement. We analyse the orbits of $\underline{N} := \underline{N}_{tu}^1$ defined above by means of the Coxeter supergroup $\underline{C}_t$.

1) We shall show that the screw motion $\underline{s}_u \in \underline{N}$ (and its $\underline{N}$-conjugates) has an extremal property. For any point X on the screw axis of $\underline{s}_u$ the equality

$$(3) \qquad \rho := \rho(X, X^{\underline{s}_u}) = \underset{X \in \mathcal{H}^3, \, \underline{h} \in \underline{N} \setminus \{1\}}{\text{minimum}} \rho(Y, Y^{\underline{h}})$$

holds for the distance ρ of any point Y from its any $\underline{N}$-image. In other words, *the screw axis of $\underline{s}_u$ provides the minimal closed geodesic* (one of them) *of the space form $\mathcal{H}^3/\underline{N}$.*

2) This minimal distance in (3) will be a monotone decreasing function of $q = 2t + 1$, so different q's yield non-isometric space forms.

3) For a fixed q the "different" u's lead to different screw angles, hence the corresponding space forms are non isometric. The screw angle shows the holonomy (rotation, of a moving frame under translation) along the screw segment as closed geodesic.

First we recall that to any screw motion $\underline{s} \in \text{Iso} \, \mathcal{H}^3$ the inequality $\rho(X, X^{\underline{s}}) \leq \rho(Y, Y^{\underline{s}})$ holds for any point X on the axis of $\underline{s}$ and for any $Y \in \mathcal{H}^3$. If $\underline{s} \in \underline{N}$, then its screw axis provides a locally minimal closed geodesic of $\mathcal{H}^3/\underline{N}$ (in a small neighbourhood of X).

To any glide reflection $\underline{g} \in \text{Iso} \, \mathcal{H}^3$ the inequality $\rho(X, X^{\underline{g}}) \leq \rho(Y, Y^{\underline{g}})$ holds for any point X on the glide line if $\underline{g}$ (in the reflection plane). If $\underline{g} \in \underline{N}$, the glide segment will be a locally minimal closed geodesic in $\mathcal{H}^3/\underline{N}$. But the Levi–Civita translation leads to on orientation reversing transformation.

Now we turn to the proof sketched before.

1) We show that

$$\rho := \underset{Y \in \mathcal{D}, \, \underline{h} \in \underline{N} \setminus \{1\}}{\text{minimum}} \rho(Y, Y^{\underline{h}}) = 2 \cdot OQ,$$

(3^*)

$$\text{where} \qquad \text{ch}^2 OQ = 1 + \frac{\sin^2(\pi/q)}{4 \cos(\pi/q)}.$$

N_{tu}^i as subgroups of Coxeter groups

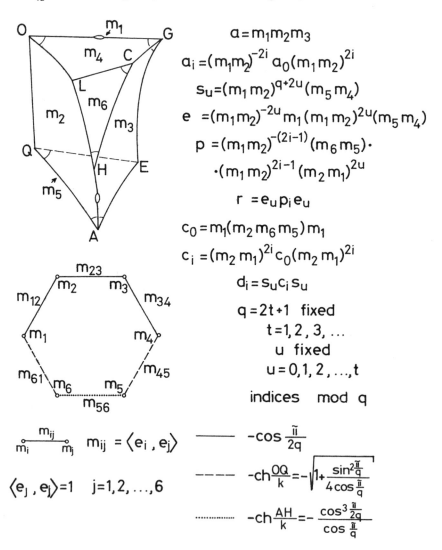

$$a = m_1 m_2 m_3$$

$$a_i = (m_1 m_2)^{-2i} a_0 (m_1 m_2)^{2i}$$

$$s_u = (m_1 m_2)^{q+2u} (m_5 m_4)$$

$$e = (m_1 m_2)^{-2u} m_1 (m_1 m_2)^{2u} (m_5 m_4)$$

$$p = (m_1 m_2)^{-(2i-1)} (m_6 m_5) \cdot$$

$$\cdot (m_1 m_2)^{2i-1} (m_2 m_1)^{2u}$$

$$r = e_u p_i e_u$$

$$c_0 = m_1 (m_2 m_6 m_5) m_1$$

$$c_i = (m_2 m_1)^{2i} c_0 (m_2 m_1)^{2i}$$

$$d_i = s_u c_i s_u$$

$$q = 2t+1 \quad \text{fixed}$$
$$t = 1, 2, 3, \ldots$$
$$u \quad \text{fixed}$$
$$u = 0, 1, 2, \ldots, t$$

indices   mod q

$$m_{ij} = \langle e_i, e_j \rangle$$

$$\langle e_j, e_j \rangle = 1 \quad j = 1, 2, \ldots, 6$$

$$-\cos \frac{\tilde{i}i}{2q}$$

$$-ch\frac{OQ}{k} = -\sqrt{1 + \frac{\sin^2\frac{\tilde{i}}{q}}{4\cos\frac{\tilde{i}}{q}}}$$

$$-ch\frac{AH}{k} = -\frac{\cos^3\frac{\tilde{i}}{2q}}{\cos\frac{\tilde{i}}{q}}$$

Figure 6.

Non-orientable hyperbolic space forms $\mathcal{M}/N_{tu}^2$

indices mod q

$q = 2t+1$ fixed

$t = 1,\underline{2},3, \ldots$

u   fixed

$u = 0,\underline{1},2,\ldots,t$

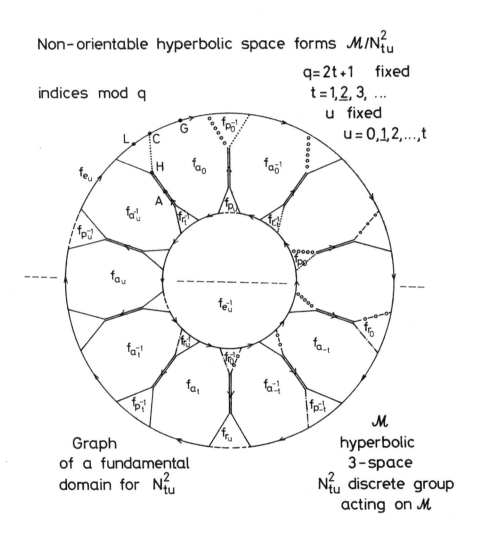

Graph
of a fundamental
domain for N_{tu}^2

$\mathcal{M}$
hyperbolic
3-space
N_{tu}^2 discrete group
acting on $\mathcal{M}$

Figure 7.

1.a) We consider the domain $\mathcal{F} := QAEHLCGO$ and its Coxeter group $\underline{C}_t > N$. First we study all the non-zero distances between any pairs of vertices, edges faces of $\mathcal{F}$ and determine the minimum of these distances. From the symmetry of the Coxeter diagram (Fig. 6) we see that the half turn, about the axis connecting the midpoints of OG and AH, maps $\mathcal{F}$ onto itself. So we can simplify the discussion, and the following three distances competite for the minimum

$$OQ = CG, \qquad AH \qquad \text{and} \qquad \rho_E = \rho_L$$

(4)
$$\operatorname{ch} AH = \frac{\cos^3(\pi/2q)}{\cos(\pi/q)}, \qquad \operatorname{ch} \rho_E = \sqrt{2} \cdot \cos(\pi/2q).$$

Here ρ_E denotes the length of the height of the triangle QEA at E. A simple comparison shows that OQ by (3^*) is the least in (4).

1.b) Now we take a locally minimal closed geodesic ℓ of $\mathcal{H}^3/\underline{N}$ which is associated with a screw motion or glide reflection denoted by $\underline{n} \in \underline{N}$. We may assume that ℓ has a point Y of $\mathcal{D}$, i.e. of a $\underline{C}_t$-image of $\mathcal{F}$, say Y just belongs to $\mathcal{F}$. The transformation $\underline{n}$ carries $Y \in \mathcal{F}$ onto $Y^{\underline{n}} \in \mathcal{F}^{\underline{n}}$.

The intersection of $\mathcal{F}$ and $\mathcal{F}^{\underline{n}}$ is empty for any pair of parameters (t, u) by the construction.

Consider the segment $YY^{\underline{n}}$ in $\mathcal{H}^3$ and note those faces $m_1, m_2, \ldots, m_6$ whose $\underline{C}_q$-images intersect the segment ℓ going from Y to $Y^{\underline{n}}$. Starting at $\mathcal{F}$ we follow the geodesic ℓ in those $\underline{C}_q$-images of $\mathcal{F}$ which join each other along whole m_i-faces, we end at $Y^{\underline{n}} \in \mathcal{F}^n$. This prescription also says how to proceed when $YY^{\underline{n}}$ goes through an edge or vertex of a $\underline{C}_q$-image of $\mathcal{F}$. We may assume $YY^{\underline{n}}$ has only Y as common point with $\mathcal{F}$, and we start with listing the faces incident to Y.

If $m_{i1}, m_{i2}, \ldots, m_{ir}$ are the listed faces of $\mathcal{F}$, that means $YY^{\underline{n}}$ passes through the $\underline{C}_q$-images

(5)
$$\mathcal{F}, \mathcal{F}^{\underline{m}_{i1}}, \mathcal{F}^{\underline{m}_{i2}\underline{m}_{i1}}, \ldots, \mathcal{F}^{\underline{m}_{ir}\cdots\underline{m}_{i2}\underline{m}_{i1}} = \mathcal{F}^{\underline{n}},$$
$$\text{hence} \quad \underline{n} = \underline{m}_{ir} \cdots \underline{m}_{i2}\underline{m}_{i1}$$

by the gluing algorithm sketched at 1.(3) (see also [2, 14]). Since we look for freely acting $\underline{n}$ which minimizes $\rho(Y, Y^{\underline{n}})$ in Y, we may assume

(6)
$$\underline{n}_z := \underline{m}_{i,r-1} \cdots \underline{m}_{i2}\underline{m}_{i1}$$

to be a transformation fixing either a face or an edge or a vertex of $\mathcal{F}$. This element of $\mathcal{F}$ is denoted by z, and $\underline{n}_z \in \underline{C}_z$ the stabilizer of z in $\underline{C}_t$. The mirror plane m_{ir}, however, does not intersect this z.

We shall discuss the cases i)–iii) in the domain $\mathcal{F}$ (in Fig. 6) and in its "neighbourhood".

i) $\underline{n}_z = \underline{m}_{i1}$, then $\underline{n} = \underline{m}_{ir}\underline{n}_z$ is a hyperbolic translation along the normal line of the mirrors $\underline{m}_{i1}$, $\underline{m}_{ir}$; $\rho(Y, Y^{\underline{n}}) = 2 \cdot \rho(m_{i1}, m_{ir})$. The preceding estimate at 1.a) guarantees (3^*).

ii) If z is an edge, then $\underline{n}_z = (\underline{m}_{i2}\underline{m}_{i1})$ is a rotation. Viewing the edges of $\mathcal{F}$, we get $\underline{n}_z$ either as a half-turn or a rotation through $\nu \cdot \frac{\pi}{q} (\leq \pi)$. In the first case $\underline{n} = \underline{m}_{ir}\underline{m}_z$ is a glide reflection and $\rho(Y, Y^{\underline{n}}) = 2 \cdot \rho(z, m_{ir}) \geq \rho$ in (3^*). In the second case the subcases

$$A) \quad z = AH \text{ and } m_{ir} = m_1; \qquad B) \quad z = OQ, \quad m_{ir} = m_3;$$
$$C) \quad z = OQ, \quad m_{ir} = m_6$$

occur, only, by the mentioned symmetry of $\mathcal{F}$.

A) We may consider

$$\underline{n} := \underline{m}_1 \ (\underline{m}_3\underline{m}); \quad m_3, m \supset AH; \quad (m_3m)\triangleleft = \nu \cdot \frac{\pi}{2q} \leq \frac{\pi}{2}.$$

The half-turn $\underline{m}_1\underline{m}_3$ has the axis EG, and we get

$$\rho(Y, Y^n) \geq 2 \cdot \rho(EG, m) \geq 2 \cdot \rho_E > 2 \cdot OQ.$$

B) $\underline{n} := \underline{m}_3 \ (\underline{m}_1\underline{m}); \quad m_1, m \supset OQ; \quad (m_1m)\triangleleft = \nu \cdot \frac{\pi}{2q} \leq \frac{\pi}{2}$

leads also to A).

C) $n := m_6 \ (\underline{m}_2\underline{m}); \quad m_2, m \supset OQ; \quad (m_2m)\triangleleft = \nu \cdot \frac{\pi}{2q} \leq \frac{\pi}{2}:$

then the half-turn $\underline{m}_6\underline{m}_2$ has the axis HL, and we get

$$\rho(Y, Y^n) \geq 2 \cdot \rho(HL, m) \geq 2 \cdot \rho(HL, m_1) \geq 2 \cdot CG = 2 \cdot OQ.$$

iii) If the fixed element z is a vertex, then $\underline{n}_z$ is a rotatory reflection. In the cases $z = E$ or L we have $\underline{n}_z$ as central inversion (point reflection). Then $\underline{n} = \underline{m}_{ir}\underline{n}_z$ is a screw motion through angle π in accordance with (3^*)

In the cases $z = O, Q, A, H, C, G$ we have

$$\underline{n}_z = (\underline{m}_{i3}\underline{m}_{i2})^\nu \qquad \underline{m}_{i1} = \underline{m}_{i1}(\underline{m}_{i3}\underline{m}_{i2})^\nu.$$

Then $\underline{n} = \underline{m}_{ir}\underline{n}_z$ is a screw motion again. We may regard only the cases $z = O, Q, A$ by the symmetry of $\mathcal{F}$. From the analogous cases we consider only $z = 0$.

First, $\underline{n} := \underline{m}_6 \; (\underline{m}_2 \underline{m})\underline{m}_4; \quad m_2, m \supset OQ; \quad (m_2 m)\sphericalangle = \nu \cdot \dfrac{\pi}{2q} \leq \dfrac{\pi}{2}.$

The half-turn $\underline{m}_6 \underline{m}_2$ has the axis $t_1 = LH$. The axis t_2 of half turn $\underline{m}\,\underline{m}_4$ is orthogonal to OQ in O. We know $\rho(Y, Y^{\underline{n}}) \geq 2 \cdot \rho(t_1, t_2)$. From the occurring axes $t_2 := OG$ provides the minimum. Since $t_1 \subset m_6$, $t_2 \subset m_1$ hold, we get

$$\rho(Y, Y^{\underline{n}}) \geq 2 \cdot \rho(t_1, t_2) \geq 2 \cdot \rho(m_6, m_1) = 2 \cdot CG.$$

Second, $\underline{n} := \underline{m}_5 \; (\underline{m}_2 \underline{m})\underline{m}_4; \quad m_1, m \supset OQ; \quad (m_2 m)\sphericalangle = \nu \cdot \dfrac{\pi}{2q} \leq \dfrac{\pi}{2}.$

The half-turn $\underline{m}_5 \underline{m}_2$ has the axis $t_1 = QA$. The axis t_2 of half-turn $\underline{m}\,\underline{m}_4$ is perpendicular to OQ in O, moreover $\rho(t_1, t_2) = OQ$.

Third, $\underline{n} := \underline{m}_3 \; (\underline{m}_1 \underline{m})\underline{m}_4; \quad m_1, m \supset OQ; \quad (m_1 m)\sphericalangle = \nu \cdot \dfrac{\pi}{2q} \leq \dfrac{\pi}{2}.$

The half-turn $\underline{m}_3 \underline{m}_1$ has the axis $t_1 = EG$. The axis t_2 of half-turn $\underline{m}\,\underline{m}_4$ is perpendicular to OQ in O. $t_2 := OL$ provides the minimum. $OL \subset m_2$ implies

$$\rho(Y, Y^{\underline{n}}) \geq 2 \cdot \rho(EG, OL) \geq 2 \cdot \rho(EG, m_2) = 2 \cdot \rho_E.$$

In cases $z = Q, A$ we prove in the same way that (3^*) is true.

2) The function ρ in (3^*) is decreasing in $q = 2t + 1$, indeed.

3) With $u = 0, 1, \ldots, t$ the screw motion $\underline{s}_u$ has essentially different screw angles. This is π if $u = 0$. Clearly, u and $-u \bmod (2t + 1)$ lead to isometric (enantiomorphic) space forms.

In addition we prove that any minimal screw motion $\underline{s}_v$ of $\underline{N}$ is $\underline{N}$-conjugate to $\underline{s}_u$ or its inverse. Otherwise we would have an isometry φ $(\notin \underline{N})$ so that $\varphi^{-1} \underline{N}^1_{tu} \varphi = \underline{N}^1_{tv}$.

In 1.b) we reviewed the possible locally minimal closed geodesics of $\mathcal{H}^3/\underline{N}$. In Fig. 6 and at (4) we have seen $OQ = CG$. To the latter one belongs the edge class⭢ in Fig. 4. The end points of these edges are $\underline{N}$-equivalent. In Fig. 5.b at the local neighbourhood of C, we also see that a glide reflection of $\underline{N}$ carries C into the other end point of ⭢. Namely, from the construction we read off that the indirect transformation

(7) $$\underline{s}_u \underline{c}_{u+t-1} = \underline{d}_{u+t-1} \underline{s}_u^{-1}$$

and its conjugates carry the end points of the class ⭢ on $\mathcal{D}$ into the corresponding starting points. The inverses of these transformations — in another order — can be expressed by

(8) $$\underline{s}_u \underline{c}_{u-2} = \underline{d}_{u-2} \underline{s}_u^{-1}.$$

Fig. 5.b also showes that the translation along the minimal closed geodesic, according to (8), reverses the orientation: the cyclic order of the faces around $\oplus$ is opposite to that around $\ominus$.

Thus we have completely proven Theorem 2.1. ∎

Theorem 2.2. *The isometry group* $\mathrm{Iso}\,\mathcal{H}^3/\underline{N}^1_{tu}$ *is just induced by the symmetries of* $\mathcal{F}^1_{tu} = \mathcal{D}_Q$ *preserving the face identifications. If* $u \neq 0$ *then only the half turn* $\underline{h}$, *with*

(9)
$$\underline{h}: f_{s_u^{-1}} \longleftrightarrow f_{s_u}, \quad f_{a_i^{-1}} \longleftrightarrow f_{a_{-i}}, \quad f_{c_i} \longleftrightarrow f_{c_i^{-1}},$$
$$f_{d_i^{-1}} \longleftrightarrow f_{d_i}, \quad i = -t, \ldots, 0, \ldots, t,$$

induces a non-trivial isometry. If $u = 0$ *then* $\mathcal{F}^1_{to}$ *allows also two extending plane reflections. One of them is* $\underline{m}_5$ *with*

(10)
$$\underline{m}_5: f_{s_u^{-1}} \longleftrightarrow f_{s_u}, \quad f_{a_i} \longleftrightarrow f_{a_i}, \quad f_{c_i} \longleftrightarrow f_{c_{-i}^{-1}},$$
$$f_{d_i^{-1}} \longleftrightarrow f_{d_{-i}},$$

the other is $\underline{h}\,\underline{m}_5 = \underline{m}_5\underline{h}$.
Any $\mathcal{H}^3/\underline{N}^1_{tu}$ *is a regularly minimal space form.*

Proof. Any element of $\mathrm{Iso}\,\mathcal{H}^3/\underline{N}^1_{tu}$ fixes the intersection of the minimal closed geodesic of $\underline{s}_u$ and the glide plane m_5 of $\underline{a}_i$'s. Thus any extending element φ of the normalizer $\mathcal{N}(\underline{N}^1_{tu})$ maps the centre Q, respectively the Dirichlet-polyhedron $\mathcal{D}_Q$, onto itself. Moreover, in the case $X \in f_{g_1^{-1}}$, $X^\varphi \in f_{g_2^{-1}}$ also $X^{g_1\varphi} = X^{\varphi g_2} \in f_{g_2}$ holds for any X and any faces f_{g_i} of $\mathcal{D}_Q$. These are in accordance with $\varphi^{-1}\underline{g}_1\varphi = \underline{g}_2$.

Then the rests are obvious, in analogous way to Sect. 1. ∎

Remarks:

1) The gluing constructions of Fig. 5 are of great importance in studying orbits of a discontinuously acting transformation groups [14]. From the construction of Fig. 5.a we can read that the edge class $\Rrightarrow$ of length $2 \cdot AH$ by (4) provides also a locally minimal closed geodesic. The glide reflection c_0 (and its $\underline{N}$-conjugates) carries the end point of $\Rrightarrow$ to its starting point.

2) The procedure in 1.b) at the proof of Th. 2.1. shows that the $\underline{C}_t$-images of $\mathcal{F} := QAEHLCGO$ (Fig. 6) can be glued also at O, G, C, H, A,

to form a new fundamental domain of $\underline{N}^1_{tu}$. Fig. 5.d indicates this at the edge AH. Of course the new face pairing will be determined by the original one, but we get a new presentation of $\underline{N}$.

3) It can also be seen that the Coxeter group $\underline{C}_t$, with the fundamental domain $\mathcal{F}$, may have other freely acting subgroups as well. The series $\underline{N}^2_{tu}$, constructed first in [10], has this property. We have only drawn it in Fig. 7 (see also Fig. 6) by giving the face pairing for it. $\mathcal{H}^3/\underline{N}^2_{tu}$ has two minimal orientation reserving closed geodesics of length $2 \cdot OQ = 2 \cdot GC$. $\mathcal{H}^3/\underline{N}^2_{tu}$ has interesting new properties which deserve further studies.

4) If $q = 4$ is even, we could give an analogous series of non-orientable compact hyperbolic space forms with another meaning of parameter u.

5) It is interesting to compare the methods in Sect. 1 and 2. Then we see their advantages and disadvantages as well. For the Matveev–Fomenko manifold $\mathcal{H}^3/\overline{G}$ the method of Sect. 2 would be just hopeless because of missing method of computation for the length of closed geodesic. In Sect. 1 the simple presentation of $\overline{G}$ and the rigidity theorem provided the isometries of $\mathcal{H}^3/\overline{G}$.

The presentation (2) of $\underline{N}^1_{tu}$ is too complicated to find outer automorphisms of $\underline{N}^1_{tu}$. The metric structure of the Coxeter supergroup $\underline{C}_t$ gave a convenient metric structure for the fundamental domain of $\underline{N}^1_{tu}$, which made possible to determine Iso $H^3/\underline{N}^1_{tu}$ for an infinite series. Now, we have also proven that the members of the series have non-isomorphic fundamental groups. Another method seems to be unsatisfactory for this decision.

References

[1] AL-JUBOURI, N. K.: On non-orientable hyperbolic 3-manifolds, *Quart. J. Math. Oxford* (2), **31** (1980), 9–18.

[2] COXETER, H. S. M. and MOSER, W. O. J: *Generators and relations for discrete groups,* Springer, Berlin–Heidelberg–New York, Fourth ed. 1980.

[3] LÖBELL, F.: Beispiele geschlossener dreidimensionaler Clifford–Kleinscher Räume negativer Krümmung, *Ber. Sachs. Akad. Wiss.* **83** (1931), 168–174.

[4] LYNDON, R. C. and SCHUPP, P. E.: *Combinatorial group theory,* Springer, Berlin–Heidelberg–New York, 1977.

[5] MAKAROVA, K. P., DAMIAN, F. L. and BALKAN, V. V.: On motion groups of 3-dimensional locally Lobachevskian manifolds, *Problems of discrete geometry, Mat. Issled.* **103**, Kishinev, Stiintsa 1988, 151–163.

[6] MARGULIS, G. A.: Isometry of closed manifolds of constant negative curvature with the same fundamental group, *Doklady Akad. Nauk SSSR* **192** (1970), 736–737 (Russian).

[7] MATVEEV, S. V. and FOMENKO, A. T.: Isoenergetic surfaces of Hamiltonian systems, account of three-dimensional manifolds in order of their complexity and computation of volumes of closed hyperbolic manifolds, *Uspehi mat. nauk* **43**, 5–22 (Russian).

[8] MEDNYKH, A. D.: Automorphism groups of three-dimensional hyperbolic manifolds, *Soviet Math. Dokl.* **32** (1985), 633–635.

[9] MOLNÁR, E.: An infinite series of compact non-orientable 3-dimensional space forms of constant negative curvature, *Annals of Glob. Anal. and Geom.* **1** (1983), 37–49, **2** (1984), 253–254.

[10] MOLNÁR, E.: Space forms and fundamental polyhedra, *Proc. of the Conf. on Diff. Geom. and Its Appl.*, Nove Mesto na Morave, Czechoslovakia, 1983, Part **1** *Differential Geometry*, 91–103 (1984).

[11] MOLNÁR, E.: Twice punctured compact Euclidean and hyperbolic manifolds and their twofold coverings, *Proc. of Coll. on Diff. Geom.*, Debrecen (Hajdúszoboszló), Hungary, 1984, *Topics in differential geometry*, ed. by J. Szenthe and L. Tamássy, *Colloq. Math. Soc. J. Bolyai*, Vol. **46**, North Holland 1987, Vol II. 883–919.

[12] MOLNÁR, E.: Minimal presentation of the 10 compact Euclidean space forms by fundamental domains, *Studia Sci. Math. Hung.* **22** (1987), 19–51.

[13] MOLNÁR, E.: Two hyperbolic football manifolds, *Diff. Geom. and Appl.*, *Proc. of the Conf.* 1988, Dubrovnik, Yugoslavia, 217–241.

[14] MOLNÁR, E.: Polyhedron complexes with simply transitive group actions and their realizations, *Acta Math. Hung.* **59** (1992), 173–214.

[15] MOSTOW, G. D.: Strong rigidity of locally symmetric spaces, *Annals of Math. Stud.* No **78**, Princeton 1973.

[16] THURSTON, W. P.: *The geometry and topology of 3-manifolds*, Princeton Univ. Lecture Notes, 1978.

[17] VESNIN, A. YU: Three-dimensional hyperbolic manifolds of Löbell type, Sib. Mat. Zh. **28** (1987), No. 5, 50–53 (Russian).

[18] VINBERG, E. B. and SHVARTSMAN, O. V.: Discrete transformation groups of spaces of constant curvature, VINITI, *Itogi Nauki i Tehniki Sovr. Probl. Math. Fund. Napr.* **29** (1988), 147–259 (Russian).

[19] WOLF, J. A: *Spaces of constant curvature*, Berkely, Univ. of California, 1972. Russian translation: Izd. "Nauka" Moscow 1982.

Emil Molnár   from September 1, 1990

Eötvös Loránd University,   *Technical University of Budapest*
Department of Geometry   *Department of Geometry*
Budapest VIII / Hungary   *Budapest XI*
Rákóczi út 5.   *Egry J. út 1.*
H–1088.   *H–1521.*

Some Non-symmetric Manifolds*

G. MOUSSONG

0. Introduction

A manifold M is called *aspherical*, if its universal cover is contractible, that is, if M is a $K(\pi, 1)$-space with $\pi = \pi_1(M)$. Examples of aspherical manifolds that most frequently occur in differential geometry are the connected complete locally symmetric Riemannian manifolds without compact de Rham factors, that is, manifolds of the form $\Gamma \backslash G/K$, where G is a connected reductive Lie group, K is a maximal compact subgroup and Γ is a torsion-free discrete subgroup in G.

It is natural to ask how much closed aspherical smooth manifolds can differ from these examples.

There exist compact Riemannian manifolds of negative curvature that do not carry a constant negative curvature. Such a manifold was first constructed by Mostow and Siu in dimension 4 in [M–S], then later Gromov and Thurston in [G–T] gave examples of manifolds of this kind in every dimension above three. These are compact aspherical manifolds that are not diffeomorphic to any locally symmetric manifold.

---

* This paper is in final form and no version of it will be submitted for publication elsewhere.

M. Davis in [D1] applied a construction using Coxeter groups to obtain, in all dimensions above three, a compact aspherical smooth manifold whose universal cover is not homeomorphic to Euclidean space. It is clear that Davis' manifolds cannot be homeomorphic to any locally symmetric manifold.

In what follows we show how Davis' technique, Gromov's theory of hyperbolic groups [G], and some results on Coxeter groups in [M] can be combined to construct five and seven dimensional compact aspherical smooth manifolds that are not homotopy equivalent to any compact locally symmetric manifold. In other words, we construct compact $K(\pi, 1)$-manifolds such that π cannot be embedded in any connected reductive Lie group as a uniform lattice.

1. Constructing aspherical manifolds

1.1. Coxeter groups

First we review some basic facts about finitely generated Coxeter groups. Details and proofs can be found in [B] or [H].

The following definition postulates the common group-theoretical properties of all properly discontinuous transformation groups on manifolds, generated by reflections. Let S be a finite set and $m : S \times S \to \mathbb{N} \cup \{\infty\}$ be a function that satisfies the conditions

$$m(s, s') = m(s', s),$$
$$m(s, s) = 1,$$
$$m(s, s') \geq 2 \quad \text{if } s \neq s',$$

for all $s, s' \in S$. Let W denote the group generated by S subject to the relations

$$(ss')^{m(s,s')} = 1, \quad (s, s') \in S \times S, \ m(s, s') < \infty.$$

The pair (W, S) is a *Coxeter system*, and W is a *Coxeter group*. Elements of S have order two in W, and the order of elements of the form ss' is $m(s, s')$.

For any subset T of S, let W_T denote the subgroup of W generated by T, then (W_T, T) is again a Coxeter system (a *subsystem* of (W, S)).

Subgroups of the form W_T are called *standard subgroups* of W. A Coxeter system is *irreducible,* if it cannot be nontrivially written as a product of subsystems, where the direct product of Coxeter systems is defined as $(W, S) \times (W', S') = (W \times W', S \times \{1\} \cup \{1\} \times S')$. Every Coxeter system can uniquely be written as the product of its maximal irreducible subsystems.

Coxeter systems are usually given by their *Coxeter graphs.* The Coxeter graph of (W, S) is a labelled graph $G(W, S)$ with vertex set S and with an edge between s and s' if $m(s, s') > 2$, labelled by $m(s, s')$ if $m(s, s') > 3$. Note that two standard subgroups generated by the subsets T and T' of S commute if and only if the subgraphs spanned by T and T' in $G(W, S)$ are independent. Maximal irreducible subsystems of (W, S) are in bijective correspondence with connected components of $G(W, S)$.

All Coxeter groups can be faithfully represented as discrete subgroups of the orthogonal group of a finite dimensional real vector space V equipped with a certain symmetric bilinear form $\langle , \rangle$. For a Coxeter system (W, S) with $|S| = n$, this form is given by the *cosine matrix,* an $n \times n$ matrix $(a_{ss'})$, where

$$a_{ss'} = \begin{cases} -\cos \frac{\pi}{m(s,s')}, & \text{if } m(s, s') < \infty \\ -1, & \text{if } m(s, s') = \infty, \end{cases}$$

and the *canonical representation* $\rho : W \to O(V, \langle , \rangle)$ of W is defined on the generators by

$$\rho(s)(x) = x - 2\langle x, v_s \rangle v_s \qquad (s \in S, \ x \in V),$$

where $\{v_s : s \in S\}$ is a basis in V in which the matrix of the form $\langle , \rangle$ is $(a_{ss'})$.

A Coxeter group is finite if and only if the bilinear form $\langle , \rangle$ is positive definite, and in this case the group is represented as a euclidean orthogonal reflection group, or in other words, as a geometric reflection group on a sphere. An irreducible Coxeter group is called *minimally infinit,* if it is infinite and all proper standard subgroups are finite. A minimally infinite irreducible Coxeter group can be realized as a geometric reflection group (with a simplex as a fundamental domain) on either a Euclidean space or a hyperbolic space (and, is called an *affine* or a *Lannér* group, respectively), according as the determinant of the cosine matrix is zero or negative. There is a well-known classification of all finite and minimally infinite irreducible Coxeter groups, see for instance the lists in [B].

1.2. Mirror structures and the universal W-complex

Given a reflection group acting on a manifold, there is a classical method of reconstructing the manifold acted on from information on a fundamental chamber, its walls, and the group. Now, following [D2], we model this situation in an abstract setting.

A *mirror structure* on a Hausdorff topological space Q is a family $\mathcal{M} = \{Q_s : s \in S\}$ of closed subsets, called *mirrors*, where S is a finite index set. For $x \in Q$ we put

$$S(x) = \{s \in S : x \in Q_s\}.$$

Suppose that (W, S) is a Coxeter system. Let $\sim$ be the equivalence relation on $W \times Q$ defined by

$$(w, x) \sim (w', x') \quad \text{if and only if} \quad x = x' \text{ and } w^{-1}w' \in W_{S(x)}.$$

The *universal space* $\mathfrak{A} = \mathfrak{A}(W, Q, \mathcal{M})$ is defined as the quotient space $(W \times Q)/\sim$, and the *universal action* of W on $\mathfrak{A}$ is given by $v[w, x] = [vw, x]$, where square brackets denote equivalence classes. W acts as a reflection group on $\mathfrak{A}$ with fundamental domain Q, where Q is identified with the subset $\{[1, x] : x \in Q\}$ of $\mathfrak{A}$. Note that the isotropy subgroup at $x \in Q$ is $W_{S(x)}$.

If Q is a simplicial complex and all mirrors are subcomplexes, then $\mathfrak{A}$ has a natural simplicial structure with Q as a subcomplex, and W acts simplicially.

Now, following [D1], for any Coxeter system (W, S) we define the *universal W-complex* $K = K(W, S)$ as follows. First, the *nerve* $N = N(W, S)$ of (W, S) is defined as the simplicial complex with vertex set S, a subset T of S spanning a simplex if and only if the subgroup W_T of W generated by T is finite. Let Q be the cone on the barycentric subdivision N' of N, and for $s \in S$, Q_s be the closed star of the vertex s in N'. Let $\mathcal{M}$ be the mirror structure $\{Q_s : s \in S\}$ on the complex Q, and finally the W-complex K is defined as the universal space $\mathfrak{A}(W, Q, \mathcal{M})$ with the universal W-action. A collection of closed stars in N' of vertices of N has a point in common if and only if these vertices form a simplex in N; therefore, the stabilizer of any $x \in Q$ is finite, and W acts properly on K. The complex K is contractible by a result of [D1]. This also follows from Remark 3.2.(i) below.

In order to obtain aspherical manifolds using this construction, one first has to note that the complex K is a PL-manifold if the nerve $N(W, S)$ of (W, S) is a PL triangulation of a sphere. (This is clear around the cone point of Q, and around a vertex which is the barycenter of, say, a top dimensional simplex spanned by $T \subset S$, the picture looks exactly as around the origin in the canonical representation of the finite subgroup W_T.) In this case W acts as a reflection group on the manifold K, that is, the fixed point sets of conjugates of elements of S are codimension one submanifolds (*reflection walls*) that separate K. Choosing a torsion free subgroup Γ of W, the quotient K/Γ is an aspherical manifold with fundamental group Γ. This is the point where one has to invoke Selberg's Lemma (cf. [S]) which says that a finitely generated matrix group is virtually torsion free. The lemma guarantees the existence of a torsion free subgroup Γ of finite index in W, and of a compact aspherical manifold with fundamental group Γ.

Suppose that Q can be given the structure of a smooth manifold with corners so that the structure of faces of Q coincides with the dual complex of the nerve N. (This is obviously the case when, for example, N is simplicially isomorphic to the boundary complex of a compact convex polytope.) Then Q carries a smooth orbifold structure (see [D1]), the group W acts smoothly on the universal orbifold cover K, and the manifold K/Γ is smooth.

2. Gromov's hyperbolic groups

2.1. Quasi-isometry

Two metric spaces (X, ρ) and (Y, σ) are said to be *quasi-isometric*, if there is a relation $\mathcal{R}$ (a *quasi-isometry*) in $X \times Y$ and there exists a positive constant C such that

for every $x \in X$ there are $x' \in X$ and $y' \in Y$ with $\rho(x, x') < C$ and $x' \; \mathcal{R} \; y'$,

for every $y \in Y$ there are $y' \in Y$ and $x' \in X$ with $\rho(y, y') < C$ and $x' \; \mathcal{R} \; y'$, and

if $x \; \mathcal{R} \; y$ and $x' \; \mathcal{R} \; y'$, then $\frac{1}{C}\rho(x, x') - C \le \sigma(y, y') \le C\rho(x, x') + C$.

Roughly speaking, quasi-isometry requires a Lipschitz-type similarity of large distances in the two metric spaces. Of course, quasi-isometric metric spaces may have very different local structures; for example, all bounded metric spaces are quasi-isometric to the one-point space.

For a less trivial example, consider a finitely generated group G. Choose a finite set F of generators in G. The distance of two elements g_1 and g_2 of G in the (left) *word metric* with respect to F is defined as the minimum length of words in elements of F and their inverses which represent $g_1^{-1}g_2$. Note that different choices of F result in quasi-isometric word metrics on G; thus, every finitely generated group is a geometric object, well-defined up to quasi-isometry. Moreover, any subgroup of finite index in G, and every finite extension of G is quasi-isometric to G.

Recall that for a finitely generated group G, the (left) *Cayley graph* of G with respect to a finite set F of generators is the graph on the vertex set G with an edge connecting g_1 and g_2 if and only if $g_1^{-1}g_2 \in F \cup F^{-1}$. If we declare all edges to have length 1 and use the shortest path metric on the graph, then the action of G on the graph by left translations is proper, isometric, and has compact quotient. The group with its word metric is obviously quasi-isometric to the Cayley graph.

It is the following generalization of this situation where the most important examples of quasi-isometric metric spaces come up. A metric space is called a *geodesic* space, if any two points can be connected with a geodesic segment, that is, with an isometric image of a segment of the real line. For example, Cayley graphs with the above metric, and all connected complete Riemannian manifolds are geodesic spaces. Consider a locally compact and complete geodesic metric space X, and suppose that a discrete group G acts properly and isometrically on X with compact quotient. Then G is finitely generated, and the inclusion of the orbit of any point in X gives a quasi-isometry between G (equipped with a word metric) and X.

2.2. Hyperbolic metric spaces and groups

M. Gromov introduced a concept of hyperbolicity for general metric spaces and finitely generated groups, and found generalizations of a host of interesting negative curvature phenomena. We review the basic definition and some important properties of hyperbolicity. The main references for this section are [G] and [G–H].

A metric space (X, ρ) is said to be *hyperbolic,* if there exists a constant C, such that for every four points x, y, z, w of X the difference of the two largest of the three real numbers

$$\rho(x, y) + \rho(z, w), \quad \rho(x, z) + \rho(y, w), \quad \text{and} \quad \rho(x, w) + \rho(y, z)$$

is less than C.

For instance, all bounded metric spaces are trivially hyperbolic. More interesting examples are the classical hyperbolic spaces (in any dimension), while Euclidean spaces of dimension greater than one are not hyperbolic. More generally, simply connected complete Riemannian manifolds with sectional curvature bounded above by a negative constant are hyperbolic. A simply connected Riemannian symmetric space of non-compact type is hyperbolic if and only if it has rank one.

The well-known procedure of adjoining a sphere at infinity to a classical hyperbolic space generalizes to a definition of the *ideal boundary* ∂X of a hyperbolic space X (see [G]). The topological space $X \cup \partial X$ is a (metrizable) compactification of X. Quasi-isometries between hyperbolic metric spaces naturally induce homeomorphisms between their ideal boundaries.

It is a highly nontrivial theorem, proved in [G], that hyperbolicity is a quasi-isometry invariant property among geodesic spaces; that is, if X and Y are quasi-isometric geodesic metric spaces and X is hyperbolic, then Y is also hyperbolic. (Here the assumption that both spaces be geodesic is essential.)

A finitely generated group G is called *hyperbolic,* if G equipped with the word metric is a hyperbolic metric space, or equivalently, if the Cayley graph of G is hyperbolic. Since the Cayley graph is geodesic, the above theorem implies that hyperbolicity of G does not depend on the choice of the system of generators. For a hyperbolic group G, the natural action of G on itself (by left translations) induces a topological action of G on ∂G. This action is minimal, that is, all orbits are dense in ∂G.

All finite groups and finite extensions of $\mathbb{Z}$ are obviously hyperbolic, while finite extensions of Abelian groups of higher rank are not. The most important examples of hyperbolic groups arise in the situation described in the last paragraph of Section 2.1: suppose that a discrete group G acts properly and (quasi-)isometrically on a hyperbolic, locally compact, complete geodesic metric space X with compact quotient, then G is hyperbolic. (For example, fundamental groups of compact Riemannian manifolds of negative sectional curvature are hyperbolic, and so are discrete groups acting properly and with compact quotient on classical hyperbolic spaces.) Further, ∂G is identified with ∂X via the quasi-isometry $g \mapsto gx$ between G and X, where all choices of $x \in X$ result in the same identification. The action of G on X and the natural action of G on itself induce the same action on $\partial X = \partial G$.

One of the main results of [M] is the following characterization of hyperbolicity among Coxeter groups.

Theorem ([M]). *Suppose that (W, S) is a Coxeter system. The group W is **not** hyperbolic if and only if either*

(i) *there is a subset T of S with $|T| \geq 3$, such that the Coxeter system (W_T, T) is affine, or*

(ii) *there is a pair of disjoint subsets T_1 and T_2 of S, such that the subgroups W_{T_1} and W_{T_2} are infinite and commute.*

(Incidentally, (i) and (ii) are the only possibilities for W to contain an Abelian subgroup of rank greater than 1.) In practice one uses the theorem to check whether a Coxeter group given by its Coxeter graph is hyperbolic as follows. First, one finds the subgraphs corresponding to the minimally infinite standard subgroups (these are all classified); then, if neither one of these fits into the list of Coxeter graphs of affine Coxeter groups, and neither two of these are independent subgraphs, then the group is hyperbolic.

3. The examples

3.1. Some hyperbolic Coxeter groups and aspherical manifolds

Consider the following Coxeter diagrams:

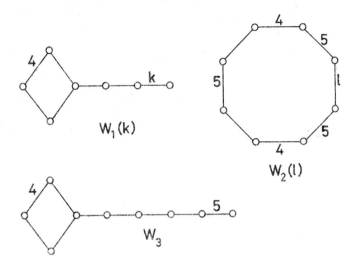

Figure 1.

Here $W_1(k)$, $W_2(l)$ and W_3 denote the corresponding Coxeter groups. Using the hyperbolicity criteria of the last section, one easily sees that W_3 is hyperbolic, and the groups in the infinite sequences $W_1(k)$ and $W_2(l)$ are also hyperbolic if $k \geq 7$ and $l \geq 4$.

The nerves of these groups are triangulations of spheres; moreover, it is not hard to check that the nerve of $W_1(k)$ for $k \geq 7$ is the boundary complex of the product of a 3-simplex and a 2-simplex, that the nerve of $W_2(l)$ for $l \geq 4$ is the boundary complex of a five dimensional cyclic polytope on eight vertices, and that the nerve of W_3 is the boundary complex of the product of a 3-simplex and a 4-simplex.

Now, applying the machinery described in the end of Section 1.2 to the groups $W_1(k)$, $W_2(l)$ and W_3, we obtain two infinite sequences $M_1(k)$ ($k \geq 7$), and $M_2(l)$ ($l \geq 4$) of five dimensional compact aspherical smooth manifolds, and M_3, a seven dimensional one.

Proposition. *The manifolds $M_1(k)$ ($k \geq 7$), $M_2(l)$ ($l \geq 4$), and M_3 are not homotopy equivalent to any compact locally symmetric Riemannian manifold.*

Proof.   Let M denote any of these manifolds, and let (W, S) be the corresponding Coxeter system. Then $M = K/\Gamma$, where K is the universal W-complex, and Γ is a torsion free subgroup of finite index in W. By passing to a smaller subgroup if necessary, we may assume that Γ is normal in W.

Suppose that M is homotopy equivalent to a compact locally symmetric manifold M'. Being a $K(\Gamma, 1)$-space, M' cannot have compact deRham factors, and since Γ is hyperbolic, M' has to have rank one. Among the non-compact rank one symmetric spaces only the real hyperbolic spaces can have odd dimension, and since M' has dimension 5 or 7, we may assume that the universal cover of M' is isometric to a real hyperbolic space $\mathbb{H}^d$ of appropriate dimension $d = 5$ or 7.

Thus, $\Gamma = \pi_1(M) = \pi_1(M')$ acts on $\mathbb{H}^d$ with compact quotient as a discrete group of isometries. Now it is a standard application of the Mostow Rigidity Theorem to embed the whole group W into the isometry group $\mathrm{Iso}\,(\mathbb{H}^d)$ of $\mathbb{H}^d$; this embedding is defined as follows. Since automorphisms of Γ are realized as conjugations by elements of $\mathrm{Iso}\,(\mathbb{H}^d)$, conjugation by an element w of W on the normal subgroup Γ is conjugation by some $w' \in \mathrm{Iso}\,(\mathbb{H}^d)$, where w' is uniquely determined since the centralizer of a cocompact subgroup in $\mathrm{Iso}\,(\mathbb{H}^d)$ is trivial. The kernel of the homomorphism $w \mapsto w'$ is the centralizer C of Γ in W. We show that C is trivial. First, C cannot contain a reflection r (a conjugate of an element of S), since r' must induce an orientation reversing homeomorphism on the sphere $\partial \mathbb{H}^d = \partial \Gamma = \partial W = \partial K$. Being disjoint from Γ, the subgroup C is finite, and since any finite subgroup of a Coxeter group is contained in a conjugate of a finite standard subgroup (cf. [D2]), the fixed point set F of C in K is non-empty and, if C is nontrivial, F is contained in some reflection wall L_r of K for some reflection $r \in W$. F is Γ-invariant, therefore L_r contains an orbit of Γ. Then the limit set of L_r in $\partial K = \partial \mathbb{H}^d$ is a proper closed subset (namely, the fixed subsphere of the sphere at infinity of $\mathbb{H}^d$ under the reflection r') containing an orbit of Γ, and this contradicts that the action of Γ on $\partial \Gamma$ is minimal.

Thus, the Coxeter group W embeds into $\mathrm{Iso}\,(\mathbb{H}^d)$ as a reflection group with bounded fundamental domain. Then the cosine matrix of (W, S) has to have precisely one negative and d positive eigenvalues, and all the remaining eigenvalues have to be 0. But, as direct calculations show, for each of the groups $W_1(k)$ $(k \geq 7)$, $W_2(l)$ $(l \geq 4)$, and W_3, the determinant of the cosine matrix is positive. This contradiction proves our proposition. ∎

3.2. Remarks

(i) In [M] it is proved that for any Coxeter group W the universal W-complex K carries a W-invariant singular metric of (piecewise constant) non-positive curvature in the sense of [G], and, in case W is hyperbolic, there exists such a metric with negative curvature. Thus, the manifolds constructed above have singular metrics of negative curvature.

(ii) The construction above cannot be carried out in arbitrarily high dimensions, since E. Vinberg's proof [V] that reflection groups with bounded fundamental polyhedron do not exist in sufficiently high dimensional hyperbolic spaces can easily be adapted to prove that if the nerve of a hyperbolic Coxeter group is a triangulation of a sphere, then the dimension of this sphere cannot be arbitrarily high.

References

[B]   N. BOURBAKI, *Groupes et Algebres de Lie*, Hermann, Paris (1968).

[D1] M. W. DAVIS, Groups generated by reflections and aspherical manifolds not covered by Euclidean space, *Ann. of Math.* **117** (1983), 293–324.

[D2] M. W. DAVIS, Some aspherical manifolds, *Duke Math. J.* **55** (1987), 105–139.

[G]   M. GROMOV, Hyperbolic groups, Essays in Group Theory, ed. by S. M. Gersten, MSRI Publ. **8** (1987), 75–263

[G–H] E. GHYS and P. DE LA HARPE (editors), *Sur les Groupes Hyperboliques d'après Mikhael Gromov*, Progress in Math. **83**, Birkhäuser, Basel, (1990).

[G–T] M. GROMOV and W. P. THURSTON, Pinching constants for hyperbolic manifolds, *Invent. Math.* **89** (1987), 1–12.

[H]   H. HILLER, *Geometry of Coxeter Groups*, Research Notes in Math. **54**, Pitman, London, (1982).

[M]   G. MOUSSONG, *Hyperbolic Coxeter groups*, Ph.D. thesis, The Ohio State Univ., Columbus, (1988).

[M–S] G. D. MOSTOW and Y.–T. SIU, A compact Kähler surface of negative curvature not covered by the ball, *Ann. of Math.* **112** (1980), 321–360.

[S]   A. SELBERG, On discontinuous groups in higher dimensional symmetric spaces, In: Int. Coll. on Function Theory, Tata Inst., Bombay (1960), 147–164.

[V] E. B. VINBERG, Absence of crystallographic reflection groups in Lobachevsky spaces of large dimension *Funktsional. Anal. i Prilozhen.* **15** (1981), 67–68 (in Russian).

Gábor Moussong

Department of Geometry
Eötvös Loránd University
Rákóczi út 5.
1088 Budapest, Hungary

COLLOQUIA MATHEMATICA SOCIETATIS JÁNOS BOLYAI

56. DIFFERENTIAL GEOMETRY, EGER (HUNGARY), 1989

The Bianchi identities and curvature tensors of Otsuki spaces*

D. F. NADJ

Introduction

The Otsuki space is a generalized affine space determined by the tensor field P_j^i and two different affine connections $''\Gamma$ and $'\Gamma$ severing for covariant derivation of co- and contravariant vectors such as

$$(0.1) \qquad \partial_k P_j^i + ''\Gamma_s{}^i{}_k P_j^s - '\Gamma_j{}^i{}_k P_s^i = 0$$

([3] (3.13)). The inverse of the tensor P_j^i is Q_j^i, and hence $P_j^i Q_r^j = P_r^j Q_j^i = \delta_r^i$ holds. The covariant differential and the basic covariant differential of a tensor X_j^i are defined in the Otsuki space by the formulas

$$(0.2) \qquad DX_j^i := P_a^i P_j^b \overline{D} X_b^a = P_a^i P_j^b X_{b|k}^a \, dx^k = \nabla_k X_j^i \, dx^k;$$

$$(0.3) \qquad X_{b|k}^a := \partial_k X_b^a + '\Gamma_s{}^a{}_k X_b^s - ''\Gamma_b{}^s{}_k X_s^a.$$

Hence

$$(0.4) \qquad \delta_{j|k}^i = '\Gamma_j{}^i{}_k - ''\Gamma_j{}^i{}_k$$

---

* This paper is in final form and no version of it will be submitted for publication elsewhere.

([3] (2.14, (3.6)–(3.8), (3.10)) i.e. the covariant derivative of the Kronecker δ-symbol is not equal to zero. This is an important characteristic of the Otsuki space.

In the Otsuki space covariant and basic covariant differentials and derivatives $'D, ''D, \ldots$ can be defined for the classical affine connections $'\Gamma$ and $''\Gamma$ resp., e.g.

$$(0.5) \qquad ''DX_j^i := P_a^i P_j^{b} {}'' \overline{D} X_b^a = P_a^i P_j^{b} {}'' \overline{\nabla}_k X_b^a \, dx^k = {}'' \nabla_k X_j^i \, dx^k,$$
$$''\overline{\nabla}_k X_b^a := \partial_k X_b^a + ''\Gamma_s{}^a{}_k X_b^s - ''\Gamma_b{}^s{}_k X_s^a$$

([3] § 3).

The coefficients of the connection $'\Gamma$ and $''\Gamma$ give two different curvature tensors, $'R_i{}^j{}_{kl}$ and $''R_i{}^j{}_{kl}$ of the form defined in the usual way.

From the different connections of Otsuki's space follows that there are different Bianchi identities of each curvature tensor which are given in §.1. In §.2. the curvature tensors are computed and it was proved that these don't have all the fine symmetry characteristics except in some special cases.

§.1. Bianchi identities

First we determine the Bianchi identities of the curvature tensor $'R_i{}^j{}_{kl}$ in the contravariant part of the connection. One of these, the Bianchi identity with respect to the basic covariant derivative, was determined in [1] (2.7) in the form

$$(1.1) \qquad 'R_p{}^a{}_{kl|m} + ''T_k{}^s{}_l\, 'R_p{}^a{}_{ms} - \delta_{p|m}^s\, 'R_s{}^a{}_{kl} + \{klm\} = 0,$$

where $\{klm\}$ denotes the sum of the expressions which we get by cyclic permutation of the indices in the foregoing expressions and $''T_j{}^i{}_k = ''\Gamma_j{}^i{}_k - ''\Gamma_k{}^i{}_j$ and $'T_j{}^i{}_k = '\Gamma_j{}^i{}_k - '\Gamma_k{}^i{}_j$ are torsions of the co- and contravariant part of the connection resp. If the connection is symmetric, i.e. $''T_j{}^i{}_k = 0$ the above formula is *not equal* to the known identity of the affine space.

The basic covariant derivatives $'\overline{\nabla}$ and $''\overline{\nabla}$ are connected with Otsuki's basic covariant derivative as follows

$$(1.2) \qquad\qquad X_{j|m}^i = '\overline{\nabla}_m X_j^i + \delta_{j|m}^s X_s^i$$

or

(1.3) $$X^i_{j|m} = {}''\overline{\nabla}_m X^i_j + \delta^i_{s|m} X^s_j.$$

Applying the formulas (1.2) and (1.3) resp. on $'R_p{}^a{}_{kl|m}$ in (1.1) and using the skew-symmetry of the curvature tensor $'R_i{}^j{}_{kl}$ in its last indices, we get

(1.4) $$'\overline{\nabla}_m\,'R_p{}^a{}_{kl} + 'T_k{}^s{}_l\,'R_p{}^a{}_{sm} + \{klm\} = 0$$

and

(1.5) $${}''\overline{\nabla}_m\,'R_p{}^a{}_{kl} + {}''T_k{}^s{}_l\,'R_p{}^a{}_{ms} + \delta^a_{s|m}\,'R_p{}^s{}_{kl} - \delta^s_{p|m}\,'R_s{}^a{}_{kl} + \{klm\} = 0.$$

Proposition 1. *The identities* (1.4) *and* (1.5) *are the Bianchi identities of the curvature tensor* $'R_i{}^j{}_{kl}$ *with respect to the contravariant and the covariant part of the connection resp.*

The contravariant part of Otsuki's connection has the characteristic of the simple affine space and if the connection $'\Gamma$ is symmetric the formula (1.4) is the known Bianchi identity.

The Bianchi identity of the curvature tensor $'R_i{}^j{}_{kl}$ of the contravariant part of the connection with respect to the covariant connection has the form (1.5). If the connection $''\Gamma$ is symmetric, i.e. $''T_j{}^i{}_k = 0$, this formula has parts which are characteristics of Otsuki spaces and this Bianchi identity is *not like* that of symmetric affine space.

In the following we observe the curvature tensor $''R_i{}^j{}_{kl}$ of the covariant part of the connection. Applying the formula (0.3) on the tensor Q^i_j and the alternation of the indices k, l we get

$$2Q^i_{j|[kl]} = -'R_p{}^i{}_{kl}Q^p_j + {}''R_j{}^p{}_{kl}Q^i_p - Q^i_{j|p}\,''T_k{}^p{}_l.$$

Since $Q^i_{j|k} = 0$ ([3] Corollary (3.6)) it follows that the curvature tensors depend on each other in the following way

(1.6) $$'R_p{}^i{}_{kl} = {}''R_j{}^s{}_{kl}Q^i_s P^j_p.$$

The Leibnitz formula is not valid in the Otsuki's space and we apply the formula (0.3) on the (1.6). Substituting $\partial_k P^i_j$ from (0.1), using (0.4)

and the cyclic permutation of the indices k, l, m with a lot of calculations it follows

$$'R_r{}^i{}_{kl|m} + \{klm\} = {}''R_j{}^s{}_{kl|m} Q_s^i P_r^j +$$
$$+ {}''R_j{}^s{}_{kl}(Q_s^i P_a^j \delta^a_{r|m} - Q_a^i P_r^j \delta^a_{s|m}) + \{klm\} = 0.$$

Substituting $'R_a{}^i{}_{kl|m}$ from (1.1), according to (1.6) after one contraction with Q_i^c and P_b^r it follows

(1.1') $''R_b{}^c{}_{kl|m} - {}''R_b{}^s{}_{kl}\delta^c_{s|m} + {}''T_k{}^s{}_l{}''R_b{}^c{}_{ms} + \{klm\} = 0.$

Applying the formulas (1.2) and (1.3) resp. on $''R_b{}^c{}_{kl|m}$ and substituting in the formula (1.1'), according to (0.4) in the same way as above we get

(1.4') $'\overline{\nabla}_m{}''R_b{}^c{}_{kl} - {}''R_b{}^s{}_{kl}\delta^c_{s|m} + {}''R_s{}^c{}_{kl}\delta^s_{b|m} +$
$$+ {}''R_b{}^c{}_{ms}(\delta^s_{[k|l]} + {}''T_k{}^s{}_l) + \{klm\} = 0$$

and

(1.5') $''\overline{\nabla}_m{}''R_b{}^c{}_{kl} + {}''T_k{}^s{}_l{}''R_b{}^c{}_{ms} + \{klm\} = 0.$

Proposition 2. *The identities* (1.1'), (1.4') *and* (1.5') *are the Bianchi identities of the curvature tensor* $''R_i{}^j{}_{kl}$ *with respect to Otsuki's basic covariant derivative and the contravariant and covariant part of the connection resp.*

The identities (1.1') and (1.4') have the parts including $\delta^i_{j|k}$ i.e. the characteristic of Otsuki space. The third identity has the form of the Bianchi identity concerning a simple affine space.

If $\delta^i_{j|k} = 0$, then the observed space is a general affine space and the above three identities are identic.

§.2. The curvature tensors of $W - O_n$ space

The Weyl–Otsuki space is an Otsuki space connected with Weyl's metric ([2]). The basic elements of that space are the tensor field P_j^i, the symmetric metric tensor g_{ij} and the vector field γ_k so that the covariant differential of the tensor g_{ij} is recurrent and $\nabla_k g_{ij} = \gamma_k g_{ij}$. We suppose that $P_a^i g_{ir} = P_r^i g_{ia}$ and $m_{ij} := g_{ab} Q_i^a Q_j^b$ hold. The covariant part of the $W - O_n$ connection is symmetric, i.e. $''T_j{}^i{}_k = 0$. It is known that for the curvature tensor $R_j{}^i{}_{kl}$ of the Riemannian space

$$R_{ijkl} = -R_{jikl} = R_{klij}$$

hold and these symmetry relations are useful in studying the validity of Schur's theorem. From the construction of the curvature tensor $''R_i{}^j{}_{kl}$ follows that $''R_i{}^j{}_{kl} = -''R_i{}^j{}_{lk}$. Let be $''R_{ijkl} = ''R_i{}^a{}_{kl} g_{aj}$. In [1] it was proved that the curvature tensor $''R_{ijkl}$ has the above symmetry characteristics only in some special cases.

In the following we compute the curvature tensor $'R_{ijkl} = 'R_i{}^a{}_{kl} g_{aj}$ of the contravariant part of the connection. The curvature tensors of the co- and contravariant part of the connection depend on each other by the relation

(2.1) $$''R_{ijkl} = 'R_{ijkl} + 2g_{sj} \delta_{i|[k|l]}^s$$

(see [1] (2.3)). Now according to [1] (4.3)

$$'R_{ijkl} + 'R_{jikl} = 2'T_{ijkl}$$

where

$$'T_{ijkl} = 2Q_r^t Q_{(i}^r m_{j)s} \gamma_{[k} \nabla_{l]} \delta_j^s + m_{ij} \delta_{[k} \gamma_{l]} - 2g_{s(i} \delta_{j)|[k|l]}^s$$

which gives

Theorem 1. *The curvature tensor $'R_{ijkl}$ is skew-symmetric in its first two indices iff $'T_{ijkl} = 0$.*

It is easy to see that the tensor $'T_{ijkl}$ is symmetric in its indices i, j and skew-symmetric in k, l.

Let us construct the tensor

$$'\overset{*}{R}_{ijkl} = 'R_{ijkl} - 'T_{ijkl}.$$

This tensor is skew-symmetric in its first two indices. Using the cyclic permutation of the indices i, k, l, adding the equations, according to the Ricci identity of the tensor $'R_{ijkl}$ we get

(2.2) $$'\overset{*}{R}_{ijkl} + 'T_i{}^a{}_{kl} g_{aj} + 'T_{ijkl} + \{ikl\} = 0.$$

D. F. NADJ

Adding to this identity a similar one obtained from the former by changing the indices $i \leftrightarrow j$, $k \leftrightarrow l$, we get the identity from which we substract the same but with changed indices i, j to k, l. Now we have

$$(2.3) \qquad '\overset{*}{R}_{ijkl} - '\overset{*}{R}_{klij} = 2'T_{i[kl]j} - i/j + A_{ijkl}$$

where

$$(2.4) \qquad A_{ijkl} = g_{ja}\,'T_{[i}{}^{a}{}_{k|l]} + g_{ia}\,'T_{[j}{}^{a}{}_{l|k]} - g_{la}\,'T_{[k}{}^{a}{}_{i|j]} - g_{ka}\,'T_{[l}{}^{a}{}_{j|i]}$$

and $[ikl]$ denote the usual alternation for i, k, l.

Lemma 1. *From the symmetry characteristics of the tensor $'T_{ijkl}$ follows that*

$$'T_{i[kl]j} - i/j = -'T_{k[ij]l} - k|l.$$

Proof.

$$'T_{i[kl]j} - 'T_{j[kl]i} = (-'T_{kijl} + 'T_{lijk} + 'T_{kjil} - 'T_{ljki}) \cdot \frac{1}{2} = -'T_{k[ij]l} - k|l.$$

∎

Lemma 2. *From the definition of the tensor A_{ijkl} it follows*

$$(2.5) \qquad A_{ijkl} = -A_{jikl} = -A_{ijlk} = -A_{klij}.$$

The proof follows according to the skew-symmetry of the torsion tensor $'T_i{}^j{}_k$.

Now we can say that $'\overset{*}{R}_{ijkl} = '\overset{*}{R}_{klij}$ iff

$$2'T_{i[kl]j} - i/j + A_{ijkl} = 0.$$

If we construct the new tensor

$$(2.6) \qquad '\widetilde{R}_{ijkl} = '\overset{*}{R}_{ijkl} - \left('T_{i[kl]j} - i/j + \frac{1}{2}A_{ijkl}\right),$$

we have

Theorem 2. *The tensor* $'\widetilde{R}_{ijkl}$ *is symmetric in its two pairs of indices.*

Proof. According to Lemma 1. and 2. with respect to the relation (2.3) follows

$$'\widetilde{R}_{ijkl} - '\widetilde{R}_{klij} = '\overset{*}{R}_{ijkl} - '\overset{*}{R}_{klij} - (2'T_{i[kl]j} - i/j + A_{ijkl}) = 0$$

which proves the statement of theorem. ■

Summarizing one can say that the observation of the $W - O_n$ space which uses the symmetry characteristics of the curvature tensors (e.g. studying the validity of Schur's theorem) can only be performed in some special cases. One of these is the case when the tensor γ is a gradient vector and $\delta^i_{j|k} = 0$ holds, but the generalized Weyl space has not the characteristics of Otsuki's space.

In the case of the Riemann–Otsuki space, i.e. if $\gamma_k = 0$, the curvature tensor $''R_{ijkl}$ has all the fine symmetry characteristics, but in the same special case the curvature tensor $'R_{ijkl}$ has not these characteristics.

References

[1] DJ. F. NADJ: On curvatures of the Weyl–Otsuki spaces, *Publ. Math. Debrecen* **28** (1981), 59–73.

[2] A. MOÓR: Otsukische Übertragung mit rekurrentem Masstensor, *Acta Sci. Math. Szeged* **40** (1978), 129–142.

[3] T. OTSUKI: On general connections I, *Math. Journal Okayama Univ.* **9** (1959–60), 99–164.

[4] L. TAMÁSSY and T. Q. BINH: On the non-existence of certain Riemannian connections with torsion and of constant curvature, *Publ. Math. Debrecen* (to appear).

Djerdji F. Nadj

University of Forestry and
Timber Industry
Dept. of Mathematics
H–9401 Sopron
Pf. 132
Hungary

COLLOQUIA MATHEMATICA SOCIETATIS JÁNOS BOLYAI
56. DIFFERENTIAL GEOMETRY, EGER (HUNGARY), 1989

On the Decomposition of Curvature Tensor Fields on Hermitian Manifolds

S. NIKČEVIĆ

1. Introduction

Before proceeding to the presentation of my results on the decomposition of curvature tensor fields on Hermitian manifolds, I would like to acquaint you in brief with the development line of the theory concerning the decomposition of curvature tensor fields under the action of some groups on some manifolds and its applications, with a view to enlightening the motivation and importance of the considered subject in differential geometry. I shall mention only some results closely connected with my considerations.

That theory was initiated by Singer and Thorpe (1969), and we have the next results in *RIEMANNIAN GEOMETRY*.

Let (V, g) be an n-dimensional real vector space with positive definite inner product. Let a tensor R of type $(1, 3)$ over V be a bilinear mapping

$$R : V \times V \to \mathrm{Hom}(V, V) : (x, y) \mapsto R(x, y).$$

We use the notation

$$R(x, y, z, w) = g(R(x, y)z, w).$$

Let us denote by $\mathcal{R}_b(V)$ and $\mathcal{R}(V)$ the subspaces of $\bigotimes^4 V^*$ consisting of all tensors having the same symmetries as the curvature tensor of Riemannian manifold, the first without, the second with first Bianchi identity, i.e. $\mathcal{R}(V)$ denotes the vector space of all curvature tensors over V. R is called a *curvature tensor* over V if it has the following properties for all $x, y, z, w \in V$:

(i) $R(x, y) = -R(y, x)$;

(ii) $R(x, y)$ is a skew-symmetric endomorphism of V, i.e.

$$R(x, y, z, w) + R(x, y, w, z) = 0;$$

(iii) $\sigma R(x, y)z = 0$ where σ denotes the cyclic sum over x, y and z. This is the first Bianchi identity.

The *Ricci tensor* $\rho(R)$ of type $(0, 2)$ associated with R is a symmetric bilinear function on $V \times V$ defined by

$$\rho(R)(x, y) = \text{trace}(z \in V \mapsto R(x, z)y \in V).$$

Then the Ricci tensor $Q = Q(R)$ of type $(1, 1)$ is given by $\rho(R)(x, y) = g(Qx, y)$ and the trace of Q is called the *scalar curvature* $\tau = \tau(R)$ of R.

In a well-known paper Singer and Thorpe [17] considered $\mathcal{R}_b(V)$ and $\mathcal{R}(V)$ (in particular for $n = 4$) and gave a geometrical useful description of the splitting of $\mathcal{R}_b(V)$ $(\mathcal{R}(V))$ into four (three) components under the action of $O(n)$:

Theorem 1.1. [17]

$$\mathcal{R}_b(V) = \mathcal{R}_1 \oplus \mathcal{R}_2 \oplus \mathcal{R}_3 \oplus \mathcal{R}_4,$$
$$\mathcal{R}(V) = \mathcal{R}_2 \oplus \mathcal{R}_3 \oplus \mathcal{R}_4,$$

(i) $R \in \mathcal{R}_1$ iff the sectional curvature R is zero.

(ii) $R \in \mathcal{R}_1 \oplus \mathcal{R}_2$ iff the curvature of R is constant.

(iii) $R \in \mathcal{R}_1 \oplus \mathcal{R}_3$ iff the Ricci tensor of R is zero.

(iv) $R \in \mathcal{R}_1 \oplus \mathcal{R}_2 \oplus \mathcal{R}_3$ iff the Ricci tensor of R is a scalar multiple of the identity.

(v) $R \in \mathcal{R}_1 \oplus \mathcal{R}_3 \oplus \mathcal{R}_4$ iff the scalar curvature of R is zero. ∎

This splitting is important because of the following conclusions:

(i). The $\mathcal{R}_i$ are the *minimal* invariant subspaces of $\mathcal{R}(V)$ under the action of $O(n)$;

(ii). The $\mathcal{R}_3$ component of curvature tensor is its *Weyl conformal curvature tensor*;

(iii). A Riemannian manifold M has curvature tensor in $\mathcal{E} = \mathcal{R}_2 \oplus \mathcal{R}_3$ if and only if M is an *Einstein space*.

Let (M, g) be a Riemannian manifold. We can get some inequalities for the quadratic invariants and by applying them we have the characterizations of some manifolds (see [1]) as follows:

(iv). $\|\rho(R)\|^2 \geq \dfrac{(\tau(R))^2}{n}$ with equality sign iff $\rho(R) = \dfrac{\tau(R)}{n} g$;
$\|R\|^2 \geq \dfrac{2}{n-1} \|\rho(R)\|^2$ with equality sign iff *sectional curvature is constant*.

This was also studied by K. Nomizu [16].

Next results we have in *KÄHLER GEOMETRY*.

Let V be a $2n$-dimensional real vector space endowed with a complex structure J compatible with positive definite inner product g, i.e.

$$J^2 = -I, \qquad g(Jx, Jy) = g(x, y).$$

Denote by $\mathcal{K}(V)$ the subspace of $\mathcal{R}(V)$ consisting of tensors satisfying the Kähler identity, i.e. $JR(x, y) = R(x, y)J$. A similar decomposition as in the previous case was given by D. L. Jonson [10], H. Mori [13] and M. Sitaramayya [18].

Theorem 1.2. [13].

$$\mathcal{K}(V) = \mathcal{K}_1(V) \oplus \mathcal{K}_b(V) \oplus \mathcal{K}_2(V),$$

where
$\mathcal{K}_1(V) = \{R \in \mathcal{K}(V) \text{ with constant holomorphic sectional curvature}\};$
$\mathcal{K}_b(V) = \{R \in \mathcal{K}(V) \text{ with vanishing Ricci tensor}\};$
$\mathcal{K}_2(V) = \text{orthogonal complement of } \mathcal{K}_b(V) \text{ in } \mathcal{K}_1^{\perp}(V).$ ∎

The splitting of $\mathcal{K}(V)$ into irreducible factors is treated for the action $U(n)$. One of the projection operators gives the *Bochner tensor*. This splitting is useful for characterizing space of constant holomorphic sectional curvature.

F. Tricerri and L. Vanhecke [21] gave the decomposition of a space of curvature tensors on a *QUATERNIONIC KÄHLER* manifold. They considered the subspace $\mathcal{K}(V)$ of $\mathcal{R}(V)$ consisting of curvature tensors satisfying

$$R(x,y,J_\alpha z, w) + R(x,y,z,J_\alpha w) = \sum_{\beta=1}^{3} \omega_{\alpha\beta}(R)(x,y)g(J_\beta z, w)$$

for all $x,y,z,w \in V$ and J_α, $\alpha = 1,2,3$ are complex structures, $\omega_{\alpha\beta}$ are 2-forms so as to satisfy $\omega_{\alpha\beta} + \omega_{\beta\alpha} = 0$.

Theorem 1.3. [21]. *The orthogonal decomposition*

$$\mathcal{K}(V) = \mathcal{K}_1(V) \oplus \mathcal{K}_2(V)$$

is unique decomposition into irreducible factors, where

$$\mathcal{K}_1(V) = \{R \in \mathcal{K}(V) \text{ with constant } Q\text{-sectional curvature}\};$$
$$\mathcal{K}_2(V) = \{R \in \mathcal{K}(V) | \tau = 0\}. \quad \blacksquare$$

The splitting of $\mathcal{K}(V)$ is under the action of $\mathrm{Sp}(n) \cdot \mathrm{Sp}(1)$.

The same authors got the results in *ALMOST HERMITIAN GEOMETRY*. They gave a complete decomposition of the space of curvature tensors $\mathcal{R}(V)$ over a Hermitian vector space into irreducible factors under the action of the unitary group.

Theorem 1.4. [22].

$$\mathcal{R}(V) = \sum_{i=1}^{10} \bigoplus W_i,$$

where the W_i are orthogonal invariants. The decomposition is irreducible for $n \geq 4$ (dim $V = 2n$); for $n = 3$, $W_6 = 0$; for $n = 2$, $W_5 = W_6 = W_{10} = 0$ and the other factors are irreducible. $\blacksquare$

They considered conformal invariants and they introduced the *general Bochner curvature tensor* as a conformal invariant tensor. They got many applications in differential geometry.

The following results are obtained in *CONTACT GEOMETRY*. First, in *ALMOST CONTACT GEOMETRY* D. Janssens and L. Vanhecke [9] got the decomposition of the vector space of AC-curvature tensors over a co-Hermitian vector space in four components under the action of group

$U(n) \times 1$; second in *NORMAL ALMOST CONTACT* case P. Matzeu [11] has some results and one part of that decomposition is the similar as the decomposition in Hermitian case.

The results in *RIEMANNIAN PROJECTIVE GEOMETRY* are given by N. Bokan [4], [5]. She gave a complete decomposition of the space of curvature tensors having the same symmetry properties as the curvature tensor associated with a symmetric connection on Riemannian manifolds. The problem is solved under the action of $SO(n)$.

Theorem 1.5. [5].

$$\mathcal{R}(V) = \sum_{i=1}^{8} \bigoplus W_i,$$

where W_i are orthogonal invariant subspaces. The decomposition is irreducible for $n > 4$ ($n = \dim M$); for $n = 2, 3, 4$ some factors are equal $\{0\}$ and the others are irreducible. ∎

The complete decomposition of this vector space under the action of $GL(n)$ was solved by R. S. Strichartz [19].

All these decompositions are in principle consequences of general theorems on group representations and they do provide good insight into some problems of differential geometry; they provide some inequalities for quadratic invariants of the curvature tensor which are very useful for characterizing some particular manifolds; quadratic invariants are used in several theories on differential geometry and some topological and algebraic studies.

Now we shall start with our results. We considered the problem of decomposition of curvature tensor fields in *HOLOMORPHICALLY PROJECTIVE GEOMETRY*. We treat that problem under the action of unitary group.

Holomorphically projective change preserves the system of holomorphically planar curves; a curve is holomorphically planar if and only if the holomorphic sections determined by its tangent vectors are parallel along the curve itself. Y. Tashiro [20] introduced the holomorphically projective tensor which is invariant under holomorphically projective changes of the connections.

The author wishes to thank N. Bokan, F. Tricerri and L. Vanhecke for their help and for their very useful suggestions.

2. Preliminaries

Let V be a $2n$-dimensional real vector space endowed with the complex structure J, compatible with the positive definite inner product g, i.e.

$$J^2 = -I, \qquad g(Jx, Jy) = g(x, y),$$

for all $x, y \in V$ and where I denotes the identity transformation of V. A tensor R of type $(1, 3)$ over V is a bilinear mapping

$$R : V \times V \to \mathrm{Hom}(V, V) : (x, y) \mapsto R(x, y).$$

R is called a *curvature tensor* over V if it has the following properties for all $x, y, z, w \in V$:

(i) $R(x, y) = -R(y, x)$;

(ii) $\sigma R(x, y)z = 0$ (the first Bianchi identity);

(iii) $JR(x, y) = R(x, y)J$ (the Kähler identity).

Let $\mathcal{R}(V)$ denote the vector space of all curvature tensors over V. This space has a natural inner product defined with that on V:

$$\langle R, \tilde{R} \rangle = \sum_{i,j,k=1}^{2n} \left(R(e_i, e_j)e_k, \tilde{R}(e_i, e_j)e_k \right),$$

where $R, \tilde{R} \in \mathcal{R}(V)$ and $\{e_i\}$ is an arbitrary orthonormal basis of V. Further, let α be the standard representation of the unitary group $U(n)$ in V. Then there is a natural induced representation $\tilde{\alpha}$ of $U(n)$ in $\mathcal{R}(V)$ given by

$$\tilde{\alpha}(a)(R)(x, y, z, w) = R\left(\alpha(a^{-1})x, \alpha(a^{-1})y, \alpha(a^{-1})z, \alpha(a^{-1})w\right)$$

for all $x, y, z, w \in V$, $R \in \mathcal{R}(V)$ and $a \in U(n)$.

We have

$$\langle \tilde{\alpha}(a)R, \tilde{\alpha}(a)\tilde{R} \rangle = \langle R, \tilde{R} \rangle, \quad a \in U(n), \quad R, \tilde{R} \in \mathcal{R}(V).$$

That implies that the orthogonal complement of an invariant subspace of $\mathcal{R}(V)$ is also invariant and the standard representation of $U(n)$ in $\mathcal{R}(V)$ is completely reducible.

N. Bokan [3] gave the partial decomposition of such space curvature tensor. We got complete decomposition of $\mathcal{R}(V)$ under the action of $U(n)$.

Why do we consider decomposition of space with those properties?

Our idea is:

(i) *to get holomorphically projective tensor as one of projection operators in splitting of $R(V)$;*

(ii) *to find the holomorphically projective invariants;*

(iii) *to get some equalities for quadratic invariants for characterizing some particular manifolds;*

(iv) *to get some topological and algebraic characteristics using that decomposition.*

I would mention that *N. Blažić and N. Bokan* got some results in the above mentioned sense [2].

We are starting with the introduction of some basic notations. There are independent traces as follows:

$$\rho_{13}(R)(x,y) = \sum_{i=1}^{2n} R(e_i, x, e_i, y),$$

$$\rho_{14}(R)(x,y) = \sum_{i=1}^{2n} R(e_i, x, y, e_i),$$

$$\tau(R) = \sum_{i,j=1}^{2n} R(e_i, e_j, e_j, e_i),$$

$$\overline{\tau}(R) = \sum_{i,j=1}^{2n} R(e_i, e_j, Je_j, e_i),$$

where $\{e_1, \ldots, e_n, Je_1, \ldots, Je_n\}$ is an arbitrary basis of V. The trace $\rho(R) = \rho_{14}(R)$ usually is called the Ricci tensor of type $(0,2)$ associated with R. The Ricci tensor $Q = Q(R)$ of type $(1,1)$ is given by $\rho(R)(x,y) = g(Qx, y)$ and the trace of Q is called the scalar curvature $\tau = \tau(R)$ of R.

In general, the traces ρ_{13} and ρ_{14} are neither symmetric nor skew-symmetric and always we have

$$\rho_{13}(Jx, Jy) = \rho_{13}(x,y);$$

ρ_{13} and ρ_{14} belong to $\mathcal{V}^2(V) = V^* \otimes V^*$, where V^* is the dual space of V. Let $\langle , \rangle$ be the inner product on $\mathcal{V}^2(V)$ given by:

$$\langle \alpha, \beta \rangle = \sum_{i,j=1}^{2n} \alpha(e_i, e_j)\beta(e_i, e_j),$$

for $\alpha, \beta \in \mathcal{V}^2(V)$. For $\alpha \in \mathcal{V}^2(V)$, we define the endomorphisms:

$$\mathcal{L}_i : \mathcal{V}^2 \to \mathcal{V}^2,$$

$i = 1, 2$ by the relations:

$$\mathcal{L}_1\alpha(x, y) = \alpha(y, x); \qquad \mathcal{L}_2\alpha(x, y) = \alpha(Jx, Jy).$$

3. The complete decomposition

In order to describe the components of decomposition of $\mathcal{R}(V)$ we start with definition.

Definition 3.1.

$$\mathcal{R}_O(V) = \{R \in \mathcal{R}(V) | \rho_{13}(R) = \rho_{14}(R) = 0\},$$
$$\mathcal{R}_O^+(V) = \{R \in \mathcal{R}_O(V) | R(Jx, Jy, z, w) = R(x, y, z, w)\},$$
$$\mathcal{R}_H(V) = \{R \in \mathcal{R}(V) | \rho_{14}(R) = 0\},$$
$$\mathcal{R}_O^\perp(V) = \text{orthogonal complement of } \mathcal{R}_O(V) \text{ in } \mathcal{R}_H(V),$$
$$\mathcal{R}_H^\perp(V) = \text{orthogonal complement of } \mathcal{R}_H(V) \text{ in } \mathcal{R}(V).$$

Before giving the fundamental theorem we shall define all the components of the decomposition.

Definition 3.2.

$$\mathcal{W}_1 \oplus \mathcal{W}_2 = \{R \in \mathcal{R}_H^\perp(V) | \mathcal{L}_i \rho_{1k}(R) = \rho_{1k}(R), \ i = 1, 2; \ k = 3, 4\},$$
$$\mathcal{W}_2 = \{R \in \mathcal{W}_1 \oplus \mathcal{W}_2 | \tau(R) = 0\},$$
$$\mathcal{W}_1 = \text{orthogonal complement of } \mathcal{W}_2 \text{ in } \mathcal{W}_1 \oplus \mathcal{W}_2,$$
$$\mathcal{W}_3 = \{R \in \mathcal{R}_H^\perp(V) | \mathcal{L}_1 \rho_{1k}(R) = \rho_{1k}(R),$$
$$\mathcal{L}_2 \rho_{14}(R) = -\rho_{14}(R), \ k = 3, 4\},$$
$$\mathcal{W}_4 \oplus \mathcal{W}_5 = \{R \in \mathcal{R}_H^\perp(V) | \mathcal{L}_i \rho_{1k}(R) = (-1)^i \rho_{1k}(R), \ i = 1, 2; \ k = 3, 4\},$$
$$\mathcal{W}_5 = \{R \in \mathcal{W}_4 \oplus \mathcal{W}_5 | \tau(R) = 0\},$$
$$\mathcal{W}_4 = \text{orthogonal complement of } \mathcal{W}_5 \text{ in } \mathcal{W}_4 \oplus \mathcal{W}_5,$$
$$\mathcal{W}_6 = \{R \in \mathcal{R}_H^\perp(V) | \mathcal{L}_1 \rho_{1k}(R) = -\rho_{1k}(R),$$
$$\mathcal{L}_2 \rho_{14}(R) = -\rho_{14}(R), \ k = 3, 4\},$$
$$\mathcal{W}_7 = \{R \in \mathcal{R}_O^\perp(V) | \mathcal{L}_1 \rho_{13}(R) = -\rho_{13}(R)\},$$
$$\mathcal{W}_8 = \{R \in \mathcal{R}_O^\perp(V) | \mathcal{L}_1 \rho_{13}(R) = \rho_{13}(R)\},$$
$$\mathcal{W}_9 = \{R \in \mathcal{R}_O(V) | R(x, y, z, w) = -R(x, y, w, z,)\},$$
$$\mathcal{W}_{10} = \{R \in \mathcal{R}_O(V) | R(x, y, z, w) = R(x, y, w, z,)\},$$
$$\mathcal{W}_{11} = \text{orthogonal complement of } \mathcal{W}_9 \oplus \mathcal{W}_{10} \text{ in } \mathcal{R}_O^+(V),$$
$$\mathcal{W}_{12} = \{R \in \mathcal{R}_O(V) | R(Jx, Jy, z, w) = -R(x, y, z, w,)\},$$

Now we obtain the next theorem.

Theorem 3.3.

3.4. $$\mathcal{R}(V) = \mathcal{W}_1 \oplus \cdots \oplus \mathcal{W}_{12},$$

where the $\mathcal{W}_i$ are the orthogonal invariant subspaces. ∎

The main result will be presented in the following theorem.

Theorem 3.5.

(i) *The decomposition 3.4. is irreducible for $n > 2$.*

(ii) *For $n = 2$, $\mathcal{W}_{11} = \mathcal{W}_{12} = \{0\}$ and the other factors are irreducible.* ∎

The proof of this theorem is based on the following facts: when ξ is a $\mathcal{G}$-concomitant between two spaces, $\mathcal{G}$ acting on these spaces, then the image for ξ of an invariant subspace is also invariant; further, the image is irreducible when the first space is irreducible (see, for example, [22]); also an invariant subspace of $\bigotimes^r V^*$ is irreducible for action of some group if and only if the space of its quadratic invariants is 1-dimensional. Using the definitions and properties of $\mathcal{W}_i$, we found that $\mathcal{W}_i$, $i = 1,\dots,8$, are the images of some invariants and irreducible subspaces of $V^* \otimes V^*$ and the restrictions of the quadratic invariants to $\mathcal{W}_i$, $i = 9,\dots,12$, are scalar multiples of $\|R\|^2$. The complete proof with details can be seen in [12].

The quadratic invariants of $R \in \mathcal{R}(V)$ are determined by their basis, given in the following theorem.

Theorem 3.6. *The vector space of all the quadratic invariants of $R \in \mathcal{R}(V)$ is spanned by the following 20 invariants:*

$$o_1(R) = \sum_{i,j,k,l} R_{ijkl} R_{ijkl}, \qquad o_2(R) = \sum_{i,j,k,l} R_{ijkl} R_{ijlk},$$

$$o_3(R) = \sum_{i,j,k,l} R_{ijkl} R_{jlki}, \qquad o_4(R) = \sum_{i,j,k,l} R_{ijkl} R_{\overline{ij}lk},$$

$$o_5(R) = \sum_{i,j,k,l} R_{ijkl} R_{ij\overline{lk}}, \qquad o_6(R) = \sum_{i,j,k,l} R_{ijki} R_{ljkl},$$

$$o_7(R) = \sum_{i,j,k,l} R_{ijki} R_{lkjl}, \qquad o_8(R) = \sum_{i,j,k,l} R_{ijki} R_{l\overline{jk}l},$$

$$o_9(R) = \sum_{i,j,k,l} R_{ijki} R_{l\overline{kj}l}, \qquad o_{10}(R) = \sum_{i,j,k,l} R_{ijki} R_{lk\overline{ji}},$$

$$o_{11}(R) = \sum_{i,j,k,l} R_{ijki} R_{ljlk}, \qquad o_{12}(R) = \sum_{i,j,k,l} R_{ijki} R_{lklj},$$

$$o_{13}(R) = \sum_{i,j,k,l} R_{ijki} R_{ljl\overline{k}}, \qquad o_{14}(R) = \sum_{i,j,k,l} R_{ijki} R_{lkl\overline{j}},$$

$$o_{15}(R) = \sum_{i,j,k,l} R_{ijik} R_{ljlk}, \qquad o_{16}(R) = \sum_{i,j,k,l} R_{ijik} R_{lklj},$$

$$o_{17}(R) = \sum_{i,j,k,l} R_{ijik} R_{lkl\overline{j}}, \qquad o_{18}(R) = \sum_{i,j,k,l} R_{ijji} R_{kllk},$$

$$o_{19}(R) = \sum_{i,j,k,l} R_{ijji} R_{kll\overline{k}}, \qquad o_{20}(R) = \sum_{i,j,k,l} R_{ij\overline{ji}} R_{kll\overline{k}},$$

where $R_{ijkl} = R(e_i, e_j, e_k, e_l)$, $R_{\overline{ij}kl} = R(Je_i, Je_j, e_k, e_l)$ *and* $\{e_1,\dots,e_n,$ $Je_1,\dots,Je_n\}$ *is an arbitrary orthonormal basis of* V. ∎

At the end of this section I would like to mention that all algebraic results, including the proofs of the theorems (3.3), (3.5) and (3.6), the algebraic characterizations of the factors, as well the dimensions of the factors, the projections and their norms can be seen in [12] and the results from next section in [15]. At the same time I want to underline that the decomposition (3.4) is not unique and I also made another decomposition [15].

Now, we can apply our algebraic results to differential geometry.

4. Hermitian manifolds with a symmetric connection

Let (M, g, J) be a Hermitian manifold with real dimension $2n$; it can be endowed with a symmetric affine connection ∇, such that $\nabla J = 0$, and curvature tensor R is given by

$$R(X, Y) = \nabla_{[X,Y]} - [\nabla_X, \nabla_Y]$$

where $X, Y \in \mathcal{X}(M)$ the algebra of C^∞ vector fields on M.

Let $T_m M$, $m \in M$ be the tangent space at m with complex structure J_m and inner product g_m. We denote by $\mathcal{R}(M)$ the vector bundle with fibre $\mathcal{R}(T_m(M))$. The decomposition (3.4) for $\mathcal{R}(T_m M)$ gives rise to a decomposition of $\mathcal{R}(M)$ in subbundles, orthogonal with respect to the fibre metric of $\mathcal{R}(M)$ induced by g. We still denote the components of the decomposition by $\mathcal{W}_i$, $i = 1, \ldots, 12$.

Theorem 4.1. $\mathcal{W}_i$, $i = 7, \ldots, 12$, are holomorphically projective invariants.
∎

Let R_i be the projection of R on the $\mathcal{W}_i(M)$.

Definition 4.2. We call

4.3. $$R_H = R_7 + \cdots + R_{12}$$

the holomorphically projective tensor.

We can express the tensor (4.3) in the following way:

$$\begin{aligned}
R_H(X, Y)Z = R(X, Y)Z &+ \frac{1}{4(n^2 - 1)}[P(X, Z)Y - P(Y, Z)X \\
&- P(X, JZ)JY + P(Y, JZ)JX] \\
&+ \frac{1}{2(n + 1)}[Q(X, Y)Z - S(X, JY)JZ]
\end{aligned}$$

where
$$P(X,Y) = [(2n-1)I + \mathcal{L}_1 - \mathcal{L}_2 - \mathcal{L}_2\mathcal{L}_1]\rho_{14}(X,Y),$$
$$Q(X,Y) = (I - \mathcal{L}_1)\rho_{14}(X,Y),$$
$$S(X,Y) = (I + \mathcal{L}_2\mathcal{L}_1)\rho_{14}(X,Y).$$

If some of the $\mathcal{W}_i$ vanish, then the corresponding manifolds have special groups of transformations and we have the following theorems.

Theorem 4.4. *Let M be a Hermitian manifold with a symmetric connection. If the homogeneous holonomy group of M has no invariant hyperplane, or, if the restricted homogeneous holonomy group has no invariant covariant vector and $\mathcal{W}_1 = \ldots = \mathcal{W}_6 = 0$, then $HP(M) = A(M)$.* ∎
Where denote by $HP(M)$ the group of all holomorphically projective transformations of M and denote by $A(M)$ the group of all affine transformations of M.

Theorem 4.5. *If a Hermitian manifold is complete with the respect to a symmetric connection and $\mathcal{W}_1 = \ldots = \mathcal{W}_6 = 0$, then $HP(M) = A(M)$.* ∎

In next two theorems we shall characterize some classes of manifolds using quadratic invariants.

Theorem 4.6. *Let M be a Hermitian manifold with symmetric connection. Then*
$$o_6(R) \geq \frac{1}{4n-3}\{o_7(R) + (n-2)[o_8(R) - o_9(R)]\}$$
with equality sign if and only if M is Ricci flat. ∎

Theorem 4.7. *Let M be a Hermitian manifold with symmetric connection. Then*
$$o_1(R) \geq \frac{1}{n^2-1}\{(4n-3)o_6(R) - o_8(R) - (n-2)[o_7(R) - o_9(R)]\}$$
with equality sign if and only if M is holomorphically projective flat, i.e. $R_H = 0$. ∎

Corollary 4.8. *Let M be a Hermitian manifold with symmetric connection and*
$$o_1(R) = \frac{1}{n^2-1}\{(4n-3)o_6(R) - o_8(R) - (n-2)[o_7(R) - o_9(R)]\}$$
then its Chern classes are powers of the first Chern class.

References

[1] M. BERGER, P. GAUDUCHON and E. MAZET, Le spectre d'une variété riemannienne, Lecture Notes in Math., vol. 194, Springer-Verlag, Berlin and New York, 1971.

[2] N. BLAŽIĆ and N. BOKAN, The Chern characteristic classes and a group of projective transformations of Hermitian manifolds, preprint 1989.

[3] N. BOKAN, Curvature tensors on Hermitian manifolds, Colloquia Mathematica Societatis János Bolyai, **46**. Topics in Diff. Geom., Debrecen (Hungary), (1984), 213–239.

[4] N. BOKAN, Curvature tensors of Riemannian manifold with connection without torsion, *Matematički vesnik,* **37** (1985), 356–364.

[5] N. BOKAN, On the complete decomposition of curvature tensors of Riemannian manifolds with symmetric connection, Rendiconti del Circolo Matematico di Palermo, Vol. 39, 1990.

[6] A. GREY and L. M. HERVELLA, The sixteen classes of almost Hermitian manifolds, *Ann. Mat. Pura Appl.* **123** (1980), 35–58.

[7] S. ISHIHARA, Holomorphically projective changes and their groups in an almost complex manifold, *Tôhoku Math. J.* **9** (1959), 273–297.

[8] N. IWAHORI, Some remarks on tensor invariants of $\mathcal{O}(n)$, $\mathcal{U}(n)$, $\mathcal{S}_p(n)$, J. *Math. Soc. Japan* **10** (1958), 145–160.

[9] D. JANSSENS and L. VANHECKE, Almost contact structures and curvature tensors, *Kodai Math. J.,* Vol. 4, No. 1, (1981), 1–27.

[10] D. L. JONSON, Sectional curvature and curvature normal forms, *Michigan Math. J.,* **27** (1980), 275–294.

[11] P. MATZEU, Decomposition of curvature tensors on almost contact manifolds, preprint 1989.

[12] P. MATZEU and S. NIKČEVIĆ, Linear algebra of curvature tensors on Hermitian manifolds, Analele ştiintifice ale Universitätii "Al. I. Cuza" Iaşi, to appear.

[13] H. MORI, On the decomposition of generalized K-curvature tensor fields, *Tôhoku Math. J.* **25** (1973), 225–235.

[14] A. NEAGU and V. OPROIU, Chern forms and H-projective curvature of complex manifolds, *Analele ştiintifice ale Universitätii "Al. I. Cuza" Iaşi,* Tomul XXIV, S. I a, f. 1, 1978, 39–46.

[15] S. NIKČEVIĆ, Induced representation of unitary group in curvature tensor fields on Hermitian manifolds, preprint 1989.

[16] K. NOMIZU, On the decomposition of generalized curvature tensor fields, *Differential geometry, in honor of K. Yano,* Kinokuniya, Tokyo, 1972, 335–345.

[17] I. M. SINGER and J. A. THORPE, The curvature of 4-dimensional Einstein spaces, *Global Analysis (papers in honor of K. Kodaira)*, Univ. of Tokyo Press, Tokyo, 1969, 355–365.

[18] M. SITARAMAYYA, Curvature tensors in Kähler manifolds, *Trans. Amer. Math. Soc.* **183** (1973), 341–353.

[19] R. S. STRICHARTZ, Linear algebra of curvature tensors and their covariant derivatives, *Can. J. Math.* **15**, 5 (1988), 1105–1143.

[20] Y. TASHIRO, On a holomorphically projective correspondence in an almost complex space, *Math J. Okayama Univ.* **6** (1957), 147–152.

[21] F. TRICERRI and L. VANHECKE, Decomposition of a space of curvature tensors on a quaternionic Kähler manifold and spectrum theory, *Simon Stevin* **53** (1979), 163–173.

[22] F. TRICERRI and L. VANHECKE, Curvature tensors on almost Hermitian manifolds, *Trans. Amer. Math. Soc.*, **267**, No. 2 (1981) 365–398.

[23] H. WEYL, Classical groups their invariants and representations, Princeton Univ. Press, Princeton, N. J., 1946.

[24] K. YANO, Differential Geometry on Complex and Almost Complex Spaces, Pergamon Press, New York 1965.

[25] K. YANO and M. KON, Structures on manifolds, Series in Pure Mathematics, Volume 3, 1984.

Stana Nikčević

University of Belgrade,
Sindjelićeva 25, 11000 Beograd,
Yugoslavia.

COLLOQUIA MATHEMATICA SOCIETATIS JÁNOS BOLYAI

56. DIFFERENTIAL GEOMETRY, EGER (HUNGARY), 1989

Isometric Immersion with Homothetical Gauss Map

S. NÖLKER

Since the very beginning of differential geometry the Gauss map has played an important role in surface theory. A natural generalization of this classical map for an isometric immersion $f : M \to \mathbb{R}^n$ of an m-dimensional Riemannian manifold into the n-dimensional Euclidean space is defined by assigning to every point p of M its tangent space $f_* T_p M$, considered as a vector subspace of $\mathbb{R}^n$. The Gauss map $g : M \to G_m(\mathbb{R}^n)$ into the Grassmannian $G_m(\mathbb{R}^n)$ obtained in this way has been extensively studied, a beautiful survey on results concerning g and on alternative definitions of the Gauss map of f can be found in [Os]. We consider the pull back b of the canonical Riemannian metric on $G_m(\mathbb{R}^n)$ via g, which is called *the third fundamental form of f*. Due to [Le] and [Ob],

$$b(v, w) = \sum_{i=1}^{m} \langle h(v, e_i), h(w, e_i) \rangle$$

for all $p \in M$, all $v, w \in T_p M$ and every orthonormal basis $(e_1, \ldots, e_m)$ of $T_p M$, where h denotes the second fundamental form of f. Under the use of the mean curvature normal H of f and of the Ricci-Tensor Ric of M the Gauss equation leads to the invariant description

$$b(v, w) = m\langle h(v, w), H \rangle - \text{Ric}(v, w).$$

It is very natural to pose the following problems:

1. *Find all f for which the Gauss map g is homothetical (i.e. b is a constant multiple of the Riemannian metric on M),*

and, more generally:

2. *Find all f for which the third fundamental form b is parallel.*

There are important examples of isometric immersions having the desired properties.

Firstly, if M is a compact connected Riemannian homogeneous space G/K such that the isotropy representation of K is irreducible, then for any $i \in \mathbb{N}_+$ the so called i-th standard immersion corresponding to the i-th positive eigenvalue of the Laplacian on M has homothetical Gauss map, see [dW], [W1], [Oh] and chapter 4, §§5–6, of [Ch] for the definition and concrete examples.

Secondly, all isometric immersion $f : M \to \mathbb{R}^n$ with parallel second fundamental form have parallel third fundamental form. Using this fact Ferus could prove a decomposition theorem for these immersions in this article [Fe1] which was an important step towards their classification in his subsequent papers [Fe2], [Fe3].

In view of the complicated geometric and algebraic structure of the above examples it seems too ambitious to try to solve the problems 1 and 2 in their general form. However, under the assumption of a flat normal bundle, we obtain complete answers by the following theorems the proofs of which will appear elsewhere:

Theorem 1. *Let M be an m-dimensional simply connected complete Riemannian manifold and $f : M \to \mathbb{R}^n$ an isometric immersion. Then:*

a) *$b = 0$ if and only if $M = \mathbb{R}^m$ and f is an isometry of $\mathbb{R}^m$ onto an m-dimensional affine subspace of $\mathbb{R}^n$.*

b) *Let f have a flat normal bundle. Then $b = \frac{1}{r^2}\langle .,.\rangle$ with $r \in \mathbb{R}_+$, if and only if there exist numbers $k_0, k_1 \in \mathbb{N}$ such that M is a Riemannian product $S^{m_1}(r) \times \ldots \times S^{m_{k_0}}(r) \times \mathbb{R} \times \ldots \mathbb{R}$ of Euclidean spheres $S^{m_i}(r) := \{p \in \mathbb{R}^{m_i+1} \mid \|p\| = r\}$ with $m_i \geq 2$ and k_1 Euclidean lines $\mathbb{R}$ and f is a Riemannian product of isometric immersions $f_1, \ldots, f_{k_0+k_1}$, where f_i is*

(i) *for $i = 1, \ldots, k_0$ a canonical imbedding of $S^{m_i}(r)$ into some Euclidean space $\mathbb{R}^{n_i}$,*

and

(ii) for $i = k_0 + 1, \ldots, k_0 + k_1$ the arclength parametrization of a curve
in some Euclidean space $\mathbb{R}^{n_i}$ which has a constant curvature $\frac{1}{r}$.

Theorem 2. *Let M be an m-dimensional simply connected complete Riemannian manifold and $f : M \to \mathbb{R}^n$ an isometric immersion. Then b is
parallel and the normal bundle of f is flat, if and only if M is a Riemannian
product of simply connected Riemannian manifolds $M_1, \ldots, M_k$ and f is a
Riemannian product of isometric immersions $f_i : M_i \to \mathbb{R}^{n_i}, i = 1, \ldots, k$,
with homothetical Gauss maps and flat normal bundles.*

Remark.

We make the assumption "M simply connected and complete" in the
theorems solely to use the global version of the de Rham decomposition
theorem. Without it the theorems hold in an appropriate "local" formula-
tion. For isometric immersions into real space forms of non-null curvature
analogous results are valid.

References

[Ch] B. Y. CHEN: Total mean curvature and submanifolds of finite type. Singa-
pure: World Scientific (1984).

[dW] M. DO CARNO and N. WALLACH: Minimal immersions of spheres into
spheres. *Ann. Math.* **95** (1971) 43–62.

[Fe1] D. FERUS: Product-Zerlegung von Immersionen mit paralleler zweiter Fun-
damentalform. *Math. Ann.* **211** (1974) 1–5.

[Fe2] D. FERUS: Immersion with parallel second fundamental form. *Math. Z.*
140 (1974) 87–93.

[Fe3] D. FERUS: Symmetric submanifolds of Euclidean space. *Math. Ann* **247**
(1980) 81–93.

[Le] K. LEICHTWEISS: Über eine Art von Krümmungsinvarianten beliebiger Un-
termannigfaltigkeiten des n-dim euklidischen Raumes. *Abh. Math. Semin.
Univ. Hamb.* **26** (1963–64) 155–190.

[Ob] M. OBATA: The Gauss map of immersions of Riemannian manifolds in
spaces of constant curvature. *J. Differ. Geom.* **2** (1968) 217–223.

[Oh] Y. OHNITA: The first standard minimal immersions of compact irreducible
symmetric spaces. In: K. Kenmotsu (ed.) Differential geometry of submanifolds. Proceedings, Kyoto 1984. *(Lecture Notes Mathematics* **Vol. 1090,
pp. 37–49**) Berlin Heidelberg New York: Springer 1984.

[Os] R. Osserman: Minimal surfaces, Gauss maps, total curvature, eigenvalue estimates, and stability. In: W. Y. Hsiang et al. (eds.) *The Chern Symposium* (1979) pp. 199–227. New York Heidelberg Berlin: Springer 1980.

[W1] N. Wallach: Minimal immersions of symmetric spaces into spheres. In: W. M. Boothby and G. L. Weiss (eds) *Symmetric Spaces*, pp. 1–40. New York: Marcel Dekker 1972.

Stefan Nölker

Mathematisches Institut
der Universität zu Köln,
Weyertal 86–90,
D–5000 Köln 41,
Germany

COLLOQUIA MATHEMATICA SOCIETATIS JÁNOS BOLYAI
56. DIFFERENTIAL GEOMETRY, EGER (HUNGARY), 1989

Codimension Reduction Problem for Real Submanifold of Complex Projective Space*

M. OKUMURA

Introduction

For a submanifold M of a Riemannian manifold $\overline{M}$, if there exists a totally geodesic submanifold M' of $\overline{M}$ such that $M \subset M'$, we say that we can reduce the codimension of the submanifold M. The codimension reduction problem was investigated by Allendoerfer [1] in the case when the ambient manifold is a Euclidean space and by Erbacher [4] in the case when the ambient manifold is a real space form. Then Cecil [3] proved a complex analogue for complex submanifold of complex projective space. When the ambient manifold is a complex manifold, the intermediate submanifold M' is requested not only to be totally geodesic, but also complex submanifold.

In the previous paper [7], the present author studied the problem for real submanifold of complex projective space. However, the conditions imposed on the submanifold in [7] are quite different from that in the theorems in [3] and [4]. In this paper, we prove a codimension reduction theorem which is more fundamental than that in [7] and may correspond to those in [3] and [4].

---

* This paper is in final form and no version of it will be submitted for publication elsewhere.

In §1, we state some formulas on real submanifold of a Kähler manifold and in §2, we define the notion of holomorphic first normal space. Making use of theory of submersion, in §3, we give relations between the second fundamental form of the real submanifold M and that of the circle bundle over M in an odd-dimensional sphere. After these preliminaries, in §4, we prove a codimension reduction theorem for real submanifold of complex projective space. In §5, we apply the theorem to CR-submanifold of complex projective space.

§1. Real submanifolds of a Kähler manifold

Let $\overline{M}$ be a real $(n+p)$-dimensional Kähler manifold with Kähler structure $(J, <,>)$, that is, J is the endomorphism of the tangent bundle $T(\overline{M})$ satisfying $J^2 = -identity$ and $<,>$ the Riemannian metric of $\overline{M}$ satisfying the Hermitian condition $< J\overline{X}, J\overline{Y} > = < \overline{X}, \overline{Y} >$ for any $\overline{X}, \overline{Y} \in T(\overline{M})$.

Let M be an immersed submanifold of $\overline{M}$ and i be the immersion. Then the tangent bundle $T(M)$ is identified with a subbundle of $T(\overline{M})$ and a Riemannian metric g of M is induced from the Riemannian metric $<,>$ of $\overline{M}$ in such a way that $g(X, Y) = < iX, iY >$ for $X, Y \in T(M)$. The normal bundle $T^{\perp}(M)$ is the subbundle of $T(\overline{M})$ consisting of all $\overline{X} \in T(\overline{M})$ which are orthogonal to $T(M)$ with respect to $<,>$. For any $X \in T(M)$ the transform JiX is written as a sum of its tangential parts iFX and the normal parts $u(X)$ in the following way:

$$(1.1) \qquad\qquad JiX = iFX + u(X).$$

Then F is an endomorphism on the tangent bundle $T(M)$ and u is a normal bundle valued 1-form on the tangent bundle. In the same way, for any $\xi \in T^{\perp}(M)$, the transform $J\xi$ is written as

$$(1.2) \qquad\qquad J\xi = -iU_{\xi} + P\xi,$$

where P defines an endomorphism on the normal bundle $T^{\perp}(M)$. It is easily verified that

$$(1.3) \qquad\qquad g(X, U_{\xi}) = < u(X), \xi >,$$

for $X \in T(M)$, $\xi \in T^{\perp}(M)$. Applying J to both members of (1.1) and (1.2), we find

(1.4) $$F^2 X = -X + U_{u(X)},$$

(1.5) $$u(FX) + Pu(X) = 0,$$

(1.6) $$FU_\xi = -U_{P\xi},$$

(1.7) $$P^2\xi = -\xi + u(U_\xi).$$

We denote by ∇ and $\overline{\nabla}$ the Riemannian connection of M and $\overline{M}$ respectively and by D the induced normal connection from $\overline{\nabla}$ to $T^\perp(M)$. Then they are related by the following Gauss and Weingarten equations:

(1.8) $$\overline{\nabla}_{iX} iY = i\nabla_X Y + h(X, Y),$$

(1.9) $$\overline{\nabla}_{iX} \xi = -iA_\xi X + D_X\xi,$$

where $h(X, Y)$ is the second fundamental form and A_ξ is a symmetric linear transformation of $T(M)$ which is called the shape operator with respect to the normal ξ. They satisfy $< h(X, Y), \xi >= g(A_\xi X, Y)$. Differentiating (1.1) and (1.2) covariantly and making use of the fact that the Riemannian connection $\overline{\nabla}$ of $\overline{M}$ leaves the almost complex structure J invariant, we have

(1.10) $$(\nabla_Y F)X = A_{u(X)} Y - U_{h(X,Y)},$$

(1.11) $$\nabla_X U_\xi = FA_\xi X - A_{P\xi} X + U_{D_X\xi},$$

(1.12) $$(D_X P)\xi = h(X, U_\xi) - u(A_\xi X),$$

where $(D_X P)$ is defined by $(D_X P)\xi = D_X(P\xi) - PD_X\xi$.

§2. Holomorphic first normal space

Let $N_0(x) = \{\xi \in T_x^\perp(M) \,|\, A_\xi = 0\}$. The first normal space $N_1(x)$ is defined to be the orthogonal complement of $N_0(x)$ in $T_x^\perp(M)$. We put $H_0(X) = JN_0(x) \cap N_0(x)$. Then $H_0(x)$ is the maximal J-invariant subspace of $N_0(x)$. Since J is isomorphism, we see that $JH_0(x) = H_0(x)$. Making use of (1.2), we can easily prove the following

Proposition 2.1. For any $\xi \in H_0(x)$, we have $A_\xi = 0$ and $U_\xi = 0$. ■

Definition. The holomorphic first normal space $H_1(x)$ is the orthogonal complement of $H_0(x)$ in $T_x^\perp(M)$.

By definition $N_1(x) \subset H_1(x)$ in $T_x^\perp(M)$. Moreover we have

Proposition 2.2. If M is a complex submanifold of a Kähler manifold, then $H_1(x) = N_1(x)$.

Proof. Since $H_1(x)$ and $N_1(x)$ are the orthogonal complements of $H_0(x)$ and $N_0(x)$ respectively, we have only to show that $H_0(x) = N_0(x)$. From (1.9), it follows that

$$(2.1) \qquad \overline{\nabla}_{iX}(J\xi) = J\overline{\nabla}_{iX}\xi = J(-iA_\xi X + D_X\xi) = -JiA_\xi X + JD_X\xi,$$

because J is covariant constant. On the other hand, M being a complex submanifold, $T_x(M)$ is J-invariant and so is $T_x^\perp(M)$, that is, for any $\xi \in T_x^\perp(M)$, $J\xi \in T_x^\perp(M)$. Hence we have

$$(2.2) \qquad\qquad\qquad \overline{\nabla}_{iX}(J\xi) = -iA_{J\xi}X + D_X(J\xi).$$

Comparing the tangential parts and the normal parts of (2.1) and (2.2), we have $A_{J\xi}X = JiA_\xi X$. Thus, if $\xi \in N_0(x)$, then $A_{J\xi} = 0$ and $\xi \in JN_0(x)$. This shows that $\xi \in N_0(x)$ implies $\xi \in H_0(x)$. This completes the proof. ∎

Proposition 2.3. Let $H(x)$ be a J-invariant subspace of $H_0(x)$ and $H_2(x)$ is the orthogonal complement of $H(x)$ in $T_x^\perp(M)$. Then $T_x(M) \oplus H_2(x)$ is a J-invariant subspace of $T_x(\overline{M})$.

Proof. Note that $T_x(\overline{M}) = T_x(M) \oplus H_2(x) \oplus H(x)$. By assumption $H(x)$ is J-invariant, we have $JH(x) = H(x)$. Thus, for any $\xi \in H(x)$ there exists $\eta \in H(x)$ such that $J\eta = \xi$. Let $Z \in T_x(M) \oplus H_2(x)$. Then for $\xi \in H(x)$, $< JZ, \xi > = < JZ, J\eta > = < Z, \eta > = 0$. This means that $JZ \in T_x(M) \oplus H_2(x)$. Hence $T_x(M) \oplus H_2(x)$ is J-invariant subspace of $T_x(\overline{M})$. ∎

§3. Real submanifolds of the complex projective space

Let the ambient manifold $\overline{M}$ be $P^{(n+p)/2}(\mathbb{C})$ with the Fubini–Study metric of constant holomorphic sectional curvature 4. The curvature tensor of $P^{(n+p)/2}(\mathbb{C})$ is given by

$$(3.1) \quad \begin{aligned} \overline{R}(\overline{X},\overline{Y})\overline{Z} =\; &< \overline{Y},\overline{Z} > \overline{X} - < \overline{X},\overline{Z} > \overline{Y} + \\ &+ < J\overline{Y},\overline{Z} > J\overline{X} - < J\overline{X},\overline{Z} > J\overline{Y} - 2 < J\overline{X},\overline{Y} > J\overline{Z}, \end{aligned}$$

from which we have the following Codazzi, Ricci equations:

$$(3.2) \qquad (\nabla_Y A_\xi)X - (\nabla_X A_\xi)Y + A_{D_X\xi}Y - A_{D_Y\xi}X =$$
$$= g(Y,U_\xi)FX - g(X,U_\xi)FY + 2g(FX,Y)U_\xi,$$

$$(3.3) \qquad R^\perp(X,Y)\xi = h(A_\xi Y,X) - h(A_\xi X,Y) + g(Y,U_\xi)u(X) -$$
$$- g(X,U_\xi)u(Y) - 2g(FX,Y)P\xi,$$

where $R^\perp(X,Y)$ is the normal curvature of M.

Now we recall that an $(n+p+1)$-dimensional sphere S^{n+p+1} of radius 1 in a Euclidean $(n+p+2)$-space is a principal circle bundle over $P^{(n+p)/2}(\mathbb{C})$. Then the Hopf-fibration $\widetilde{\pi} : S^{n+p+1} \to P^{(n+p)/2}(\mathbb{C})$ defines a submersion. The almost complex structure J of $P^{(n+p)/2}(\mathbb{C})$ is nothing but the fundamental tensor of the submersion $\widetilde{\pi}$. We construct the circle bundle over the submanifold M in such a way that the following diagram commutes:

$$
\begin{array}{ccc}
\pi^{-1}(M) & \xrightarrow{\ \widetilde{i}\ } & S^{n+p+1}\ (1) \\
{\scriptstyle \pi}\downarrow & & \downarrow{\scriptstyle \widetilde{\pi}} \\
M & \xrightarrow{\ i\ } & P^{(n+p)/2}(\mathbb{C}).
\end{array}
$$

Let V' be the unit tangent vector field to the fibre of $\pi^{-1}(M)$, then $\widetilde{i}V'$ is the unit tangent vector field to the fibre of S^{n+p+1}. We denote by g' and ∇', the Riemannian metric and the Riemannian connection of $\pi^{-1}(M)$ respectively. Also we denote by F and X^* the fundamental tensor of the submersion π and the horizontal lift of $X \in T(M)$. In the same way, ξ^* is the horizontal lift of the normal field $\xi \in T^\perp(M)$. The fundamental equations for the submersion π are the following [8, 9]:

$$(3.4) \qquad\qquad \nabla'_{X^*}Y^* = (\nabla_X Y)^* + g'(F^*X^*,Y^*)V',$$
$$(3.5) \qquad\qquad \nabla'_{X^*}V' = \nabla'_{V'}X^* = -(FX)^*.$$

The similar equations are valid for the submersion $\widetilde{\pi}$ when we replace F and V' with J and $\widetilde{i}V'$ respectively. Let $\widetilde{g}, \widetilde{\nabla}, A'$ and D' be respectively the Riemannian metric of S^{n+p+1}, the Riemannian connection for $\widetilde{g}$, the shape operator and the normal connection of $\pi^{-1}(M)$. Then, from the commutativity of the diagram, calculating $\widetilde{\nabla}_{iX^*}\xi^*$, we have

$$-\widetilde{i}A'_{\xi^*}X^* + D'_{X^*}\xi^* = (\overline{\nabla}_{iX}\xi)^* + \widetilde{g}(J(iX)^*,\xi^*)\widetilde{i}V' =$$
$$= -\widetilde{i}(A_\xi X)^* + (D_X\xi)^* + <JiX,\xi>^*\widetilde{i}V' =$$
$$= -\widetilde{i}(A_\xi X)^* - <u(X),\xi>^*\widetilde{i}V' + (D_X\xi)^* =$$
$$= -\widetilde{i}\{(A_\xi X)^* - g(U_\xi,X)^*V'\} + (D_X\xi)^*,$$

because of (1.3), (1.8), (1.9) and (3.4). Comparing the tangential parts and the normal parts of both side members of the last equation, we find

(3.6) $A'_{\xi^*} X^* = (A_\xi X)^* - g(U_\xi, X)^* V',$

(3.7) $D'_{X^*} \xi^* = (D_X \xi)^*.$

Similarly, calculating $\widetilde{\nabla}_{iV'} \xi^*$ and making use of (1.2) (1.9) and (3.5), we have

$$-i A'_{\xi^*} V' + D'_{V'} \xi^* = -(J\xi)^* = -(-iU_\xi + P\xi)^*$$

from which

(3.8) $A'_{\xi^*} V' = -U^*_\xi,$

(3.9) $D'_{V'} \xi^* = -(P\xi)^*.$

We prove the following

Lemma 3.1. *At such a point x' that $\pi(x') = x$,*

$$N'_0(x') = \{\xi^* \mid A_\xi = 0, \; U_\xi = 0\}.$$

Proof. For $\xi' \in N'_0(x')$, there exists $\xi \in T^\perp_x(M)$ such that $\xi' = \xi^*$. Since a tangent vector $X' \in T_{x'}(M')$ can be decomposed as $X' = X^* + \alpha V'$, where X^* is the horizontal lift of a tangent vector $X \in T_x(M)$, from (3.6) and (3.8), it follows that

$$A'_{\xi'} X' = A'_{\xi^*} X^* + \alpha A'_{\xi^*} V' = (A_\xi X)^* - g(U_\xi, X)^* V' - \alpha U^*_\xi.$$

If $A_\xi = 0$ and $U_\xi = 0$, then $A_{\xi'} X' = 0$, which implies that $\xi' \in N'_0(x')$.

Conversely, if $\xi' \in N'_0(x')$, it follows that

$$(A_\xi X)^* - \alpha U^*_\xi = g(U_\xi, X)^* V'.$$

In the last equation, the left hand side member is vertical. Hence $g(U_\xi, X)^* = 0$ for any $X \in T_x(M)$ and consequently $U_\xi = 0$, $A_\xi = 0$. This completes the proof. ∎

In [6] we proved that the normal connection of $\pi^{-1}(M)$ in S^{n+p+1} is flat if and only if the following two conditions are satisfied on M.

(3.10) $R^\perp(X, Y)\xi = -2g(FX, Y)P\xi,$

(3.11) $D_X P = 0,$

In this sense the normal connection of M in $P^{(n+p)/2}(\mathbb{C})$ is called lift flat normal connection if it satisfies (3.10) and (3.11).

§4. Codimension reduction problems

First of all we state the following theorem by J. Erbacher [4].

Theorem. *Let M be an n-dimensional submanifold of an $(n + p)$-dimensional Riemannian manifold $\overline{M}(\overline{c})$ of constant sectional curvature $\overline{c}$ and $N_2(x)$ be a subspace of $T_x^\perp(M)$ such that the first normal space $N_1(x) \subset N_2(x)$. If $N_2(x)$ is invariant under parallel translation with respect to the normal connection and q is the constant dimension of N_2, then there exists a totally geodesic submanifold $\widetilde{M}$ of $\overline{M}(\overline{c})$ of dimension $n + q$ such that $M \subset \widetilde{M}$.*

Applying this theorem, we now prove the following

Theorem 4.1. *Let M be an n-dimensional real submanifold of a real $(n + p)$-dimensional complex projective space $P^{(n+p)/2}(\mathbb{C})$ and $H(x)$ be a J-invariant subspace of $H_0(x)$. If the orthogonal complement $H_2(x)$ of $H(x)$ in $T_x^\perp(M)$ is invariant under parallel translation with respect to the normal connection and q is the constant dimension of H_2, then there exists a real $(n+q)$-dimensional totally geodesic complex projective subspace $P^{(n+q)/2}(\mathbb{C})$ such that $M \subset P^{(n+q)/2}(\mathbb{C})$.*

Proof. We construct the circle bundle $\pi^{-1}(M)$ over M and prove that if M satisfies the conditions of Theorem 4.1, $\pi^{-1}(M)$ satisfies the conditions of the theorem by Erbacher. Let $\xi \in H(x)$. Then, $\xi \in H_0(x)$ and by Proposition 2.1, $A_\xi = 0$ and $U_\xi = 0$, which together with (3.6) and (3.7), implies that $A'_{\xi^*} = 0$. This shows that, for a point x' such that $\pi(x') = x$, $H(x)^* = \{\xi^* \,|\, \xi \in H(x)\}$ is a subspace of $N'_0(x')$. Hence, the orthogonal complement $H_2(x)^* = \{\xi^* \,|\, \xi \in H_2(x)\}$ of $H(x)^*$ in $T_{x'}^\perp(\pi^{-1}(M))$ is a subspace of $T_{x'}^\perp(\pi^{-1}(M))$ such that $H'_1(x') \subset H_2(x)^*$. Since $H_2(x)$ is invariant under parallel translation with respect to the normal connection, so is $H(x)$. This means that for any $\xi \in H(x)$, $D_X\xi \in H(x)$. Thus, from (3.7) and (3.9), it follows that $D'_{X^*}\xi^* = (D_X\xi)^* \in H(x)^*$ and $D'_V\xi^* = -(J\xi)^* \in H(x)^*$. Hence $H(x)^*$ is invariant under parallel translation with respect to the normal connection of $\pi^{-1}(M)$. From the theorem by Erbacher, we know that there exists a totally geodesic submanifold S^{n+q+1} of S^{n+p+1} such that $\pi^{-1}(M) \subset S^{n+q+1}$. Let $U(x')$ be a neighborhood of such a point x' that $\pi(x') = x$. Then the tangent space $T_{y'}(S^{n+q+1})$ of the totally geodesic submanifold at $y' \in U(x')$ is

$T_{y'}(\pi^{-1}(M)) \oplus H_2(y)^* = (T_y(M) \oplus H_2(y))^* \oplus \{V'\}$, where $y = \pi(y')$. The geodesic γ in the direction of V' is also a geodesic of S^{n+p+1}, because S^{n+q+1} is a totally geodesic submanifold. Thus γ is a great circle on the unit sphere S^{n+q+1}. Hence, the Hopf fibration $S^{n+q+1} \to P^{(n+q)/2}(\mathbb{C})$ by γ is compatible with the Hopf fibration $\pi : S^{n+p+1} \to P^{(n+p)/2}(\mathbb{C})$ and the tangent space of the $P^{(n+q)/2}(\mathbb{C})$ at x is $T_x(M) \oplus H_2(x)$. Moreover, by Proposition 2.3, $P^{(n+q)/2}(\mathbb{C})$ is J-invariant subspace of $P^{(n+p)/2}(\mathbb{C})$. This completes the proof. ∎

For a complex submanifold M, by Proposition 2.2 we know that $H_0(x) = N_0(x)$ for any $x \in M$. Therefore we have

Corollary 4.2 [3]. *Let M be a real n-dimensional complex submanifold of $P^{(n+p)/2}(\mathbb{C})$. Suppose a J-invariant subspace of the first normal space $N_1(x)$ has constant dimension q and is parallel with respect to the normal connection. Then there is a totally geodesic real $(n+q)$-dimensional complex projective subspace $P^{(n+q)/2}(\mathbb{C})$ such that $M \subset P^{(n+q)/2}(\mathbb{C})$.*

In the following we prove that if the type number of the submanifold is greater then or equal to two and H_2 has constant dimension, then H_2 is invariant under parallel translation with respect to the normal connection. Let A_α, $\alpha = 1, \ldots, p$ be the shape operators for orthonormal normal vectors ξ_α, $\alpha = 1, \ldots, p$ respectively. The type number is defined by the largest integer r for which there are r vectors $X_1, \ldots, X_r$ such that pr vectors $A_\alpha X_i$ ($\alpha = 1, \ldots, p$, $i = 1, \ldots, r$) are linearly independent [5]. We assume that $\dim H(x) = p - q = $ const. and choose orthonormal normals $\xi_1, \ldots, \xi_p$ in a neighborhood U of x in such a way that $\xi_{q+1}, \ldots, \xi_p \in H(y)$, $y \in U$. By Proposition 2.1, for $\alpha \geq q + 1$, $A_\alpha = 0$ and $U_\alpha = 0$. On the other hand if we denote $D_X \xi_\alpha$ and $u(X)$ by $\sum_{\beta=1}^p s_{\alpha\beta}(X)\xi_\beta$ and $\sum_{\alpha=1}^p u_\alpha(X)\xi_\alpha$ respectively, the Codazzi equation (3.2) becomes

$$(4.1) \quad (\nabla_X A_\alpha)Y - (\nabla_Y A_\alpha)X = \sum_{\beta=1}^p \{s_{\alpha\beta}(X)A_\beta Y - s_{\alpha\beta}(Y)A_\beta X\} - $$
$$- u_\alpha(Y)FX + u_\alpha(X)FY - 2g(FX,Y)U_\alpha,$$

from which we have, for $\alpha \geq q + 1$,

$$(4.2) \quad \sum_{\beta=1}^q \{s_{\alpha\beta}(X)A_\beta Y - s_{\alpha\beta}(Y)A_\beta X\} = 0.$$

Since the type number is greater than or equal to two, there exist tangent vectors X_1, X_2 to M such that $A_\beta X_1$, $A_\beta X_2$, $1 \leq \beta \leq q$ are linearly independent. Then, from (4.2), we have $s_{\alpha\beta}(X_1) = s_{\alpha\beta}(X_2) = 0$ for $\alpha \geq q+1$, $\beta \leq q$. But for any Y tangent to M, we have from (4.2)

$$\sum_{\beta=1}^{q} s_{\alpha\beta}(Y) A_\beta X_1 = \sum_{\beta=1}^{q} s_{\alpha\beta}(X_1) A_\beta Y = 0.$$

Again using linearly independence of $A_\beta Y_1$, $(\beta = 1, \ldots, q)$, we conclude that $s_{\alpha\beta}(Y) = 0$, for $\alpha \geq q+1$, $\beta \leq q$ and for $\alpha \leq q$, $\beta \geq q+1$. This shows that $D_Y \xi \in H$ for $\xi \in H$. Thus H is invariant under translation with respect to the normal connection and so is H_2.

§5. CR-submanifolds

Let M be an n-dimensional real submanifold of a Kähler manifold $(\overline{M}, J, < , >)$. Suppose on M a J-invariant distribution $\Delta : x \to \Delta_x \subset T_x(M)$ is given. Moreover the complementary orthogonal distribution $\Delta^\perp : x \to \Delta_x^\perp \subset T_x(M)$ is supposed to be anti-invariant, that is, $J(\Delta_x^\perp) \subset T_x^\perp(M)$ for any $x \in M$. The submanifold M endowed with the above pair of distributions $(\Delta, \Delta^\perp)$ is called a CR-submanifold in the sense of Bejancu [2]. Particularly, if $\dim \Delta_x = 0$ for any $x \in M$, the CR-submanifold is called an anti-invariant submanifold [10]. By definition the tangent space $T_x(\overline{M})$ at $x \in M$ is decomposed as

(5.1) $$T_x(\overline{M}) = T_x(M) \oplus J\Delta_x^\perp \oplus N_x(M),$$

where $N_x(M)$ is the orthogonal complement of $J\Delta_x^\perp$ in $T_x^\perp(M)$.

Lemma 5.1. $N_x(M)$ is J-invariant, that is, $JN_x(M) = N_x(M)$.

Proof. Let $X \in T_x(M) \oplus J\Delta_x^\perp$ and $\xi \in N_x(M)$. Since X is decomposed as $X = X_1 + X_2 + JY$, where $X_1 \in \Delta_x$, $X_2, Y \in \Delta_x^\perp$, it follows that

$$< X, J\xi > = - < JX, \xi > = - < JX_1, \xi > - < JX_2, \xi > + < Y, \xi > = 0,$$

which shows that, if $\xi \in N_x(M)$, then $J\xi \in N_x(M)$. J being isomorphism of $T_x(\overline{M})$, we have $JN_x(M) = N_x(M)$.

Lemma 5.2. *Suppose that $N(M)$ is invariant under parallel translation with respect to the normal connection. Then for any $\xi \in N(M)$ and $\eta \in T^{\perp}(M)$, $A_{\xi}U_{\eta} = 0$.*

Proof. By Lemma 5.1, $N(M)$ is not only invariant under parallel translation with respect to the normal connection, but also J-invariant. Hence, $J\xi = P\xi$, $D_X\xi \in N(x)$, $JD_X\xi = PD_X\xi$ for $\xi \in N(M)$. Differentiating the first equation of those covariantly, we have

$$(5.2) \qquad \overline{\nabla}_{iX}(J\xi) = \overline{\nabla}_{iX}(P\xi) = -iA_{P\xi}X + D_X(P\xi).$$

Also we have

$$(5.3) \qquad \begin{aligned} \overline{\nabla}_{iX}(J\xi) &= J\overline{\nabla}_{iX}\xi = J(-iA_{\xi}X + D_X\xi) = \\ &= -iFA_{\xi}X + u(A_{\xi}X) + PD_X\xi. \end{aligned}$$

We note that for any $\zeta \in N(M)$, $U_{\zeta} = 0$ and therefore $u(X) \in J\Delta^{\perp}$ for $X \in T(M)$. Comparing the normal parts of (5.2) and (5.3), we find $u(A_{\xi}X) = 0$. Thus, for any $\eta \in T^{\perp}(M)$, $g(A_{\xi}U_{\eta}, X) = g(U_{\eta}, A_{\xi}X) = <u(A_{\xi}X), \eta> = 0$. This completes the proof. ∎

Theorem 5.3. *Let M be an n-dimensional anti-invariant submanifold of $P^m(\mathbb{C})$. If $N(M)$ is invariant under parallel translation with respect to the normal connection, then there exists an n-dimensional totally geodesic complex projective subspace $P^n(\mathbb{C})$ of $P^m(\mathbb{C})$ such that M is an anti-invariant subspace of $P^m(\mathbb{C})$.*

Proof. Since M is anti-invariant, the tangential parts of (5.3) vanish identically. Comparing the tangential parts of (5.2) and (5.3), we have $A_{P\xi} = 0$, for $\xi \in N(M)$. In an anti-invariant submanifold, P is an isomorphism, this implies that $A_{\xi} = 0$, for any $\xi \in N(M)$, which, together with Lemma 5.1, shows that $N(M) \subset H_0(M)$. Conversely, let $\xi \in H_0(M)$. Then for $X, Y \in T_x(M)$, we have $< \xi, X + JY > = < \xi, X > - < J\xi, Y > = 0$, because $H_0(M)$ is J-invariant. Hence ξ belongs to the orthogonal complement of $T(M) \oplus JT(M)$, that is, $\xi \in N(M)$. Thus we proved that $N(M) = H_0(M)$ and consequently $JT_x(M)$ is the holomorphic first normal space. Applying Theorem 4.1, we know that there exists a real $2n$-dimensional, totally geodesic, complex projective subspace $P^n(\mathbb{C})$ of $P^m(\mathbb{C})$ such that M is an anti-invariant submanifold of $P^n(\mathbb{C})$. ∎

Lemma 5.4. Let M be a CR-submanifold of $P^{(n+p)/2}(\mathbb{C})$ with lift flat normal connection. Then for $\xi \in N_x(M)$, $\eta \in T_x^{\perp}(M)$, $A_\xi A_\eta = A_\eta A_\xi$.

Proof. Since the normal connection is lift flat, from (3.3) and (3.10), it follows that

$$h(A_\xi Y, X) - h(A_\xi X, Y) + g(Y, U_\xi)u(X) - g(X, U_\xi)u(Y) = 0.$$

Particularly, if $\xi \in N_x(M)$, $U_\xi = 0$ and we have for $\xi \in N_x(M)$,

(5.4) $$h(A_\xi Y, X) = h(A_\xi X, Y).$$

Let $\xi \in N_x(M)$, $\eta \in T_x^{\perp}(M)$, then, by (5.4),

$$\begin{aligned}
g\left((A_\eta A_\xi - A_\xi A_\eta)X, Y\right) &= g(A_\eta A_\xi X, Y) - g(A_\eta A_\xi Y, X) = \\
&= <h(A_\xi X, Y), \eta> - <h(A_\xi Y, X), \eta> = \\
&= <h(A_\xi X, Y) - h(A_\xi Y, X), \eta> = 0,
\end{aligned}$$

which implies that $A_\eta A_\xi = A_\xi A_\eta$. ∎

Theorem 5.5. Suppose that the normal connection of a CR-submanifold M of $P^{(n+p)/2}(\mathbb{C})$ is lift flat and $N(M)$ is invariant under the parallel translation with respect to the normal connection. Then there exists a totally geodesic complex projective subspace $P^{(n+q)/2}(\mathbb{C})$ such that M is a CR-submanifold of the complex subspace.

Proof. By means of Theorem 4.1, we have only to prove that $N(M) = H_0(M)$. We choose orthonormal normal fields $\xi_1, \ldots, \xi_p$ in such a way that $\xi_1, \ldots, \xi_q \in J\Delta^{\perp}$, $\xi_{q+1}, \ldots, \xi_p \in N(M)$ and denote by A_α the shape operator for ξ_α. Since $N(M)$ is not only invariant under parallel translation with respect to the normal connection, but also J-invariant, it follows that

(5.5) $$D_X \xi_\lambda = \sum_{\mu=q+1}^{p} s_{\lambda\mu}(X)\xi_\mu, \lambda \geq q+1,$$

(5.6) $$J\xi_\lambda = P\xi_\lambda = \sum_{\mu=q+1}^{p} P_{\lambda\mu}\xi_\mu, \lambda \geq q+1,$$

from which, we have

(5.7) $$JD_X \xi_\lambda = \sum_{\mu,\nu=q+1}^{p} s_{\lambda\mu}(X)P_{\mu\nu}\xi_\nu, \lambda \geq q+1.$$

On the other hand, comparing the tangential parts of (5.2) and (5.3), and making use of (5.6), we find

$$(5.8) \qquad FA_\lambda X = \sum_{\mu=q+1}^{p} P_{\lambda\mu} A_\mu X, \lambda \geq q+1.$$

Substituting X for $A_\lambda X$ in (5.8) and summing over $\lambda = q+1, \ldots, p$, we have

$$(5.9) \qquad F \sum_{\lambda=q+1}^{p} A_\lambda^2 X = \sum_{\lambda,\mu=q+1}^{p} P_{\lambda\mu} A_\mu A_\lambda X = 0,$$

because $P_{\lambda\mu}$ is skew-symmetric with respect to λ, μ and by Lemma 5.4. Thus we have

$$F^2 \sum_{\lambda=q+1}^{p} A_\lambda^2 X = - \sum_{\lambda=q+1}^{p} A_\lambda^2 X + \sum_{\lambda=1}^{p} U_{u(A_\alpha^2 X)} = 0.$$

However, as we have seen in the proof of Lemma 5.2, $u(A_\alpha^2 X) = u(A_\alpha A_\alpha X) = 0$. Hence, $\sum_{\lambda=q+1}^{p} A_\lambda^2 X = 0$, which shows that $A_\lambda = 0$ for $\lambda \geq q+1$ and that $N_x(M)$ is J-invariant subspace of $N_0(M)$. Since $H_0(x)$ is J-invariant subspace of $N_0(x)$, $N_x(M) \subset H_0(M)$. Let $\xi \in H_0(M)$ and $\eta \in J\Delta^\perp$. Then there exists $X \in \Delta^\perp \subset T(M)$ such that $\eta = JX$. Therefore, because of the fact that $H_0(M)$ is J-invariant, we have

$$< \xi, \eta > = < \xi, JX > = - < J\xi, X > = 0.$$

This means that $\xi \in N(M)$. Hence it follows that $N(M) = H_0(M)$. This completes the proof. ∎

References

[1] C. B. ALLENDOERFER, Rigidity for spaces of class greater than one, *Amer. J. of Math.* **61** (1939), 633–644.

[2] A. BEJANCU, CR-submanifolds of a Kähler manifold I, *Proc. of the Amer. Math. Soc.* **69** (1978), 135–142.

[3] T. E. CECIL, Geometric applications of critical point theory to submanifolds of complex projective space, *Nagoya Math. J.* **55** (1974), 5–31.

[4] J. ERBACHER, Reduction of the codimension of an isometric immersion, *J. of Differential Geom.* **5** (1971), 333–340.

[5] S. KOBAYASHI and K. NOMIZU, "Foundations of differential geometry vol. II", Interscience, New York, 1969.

[6] M. OKUMURA, Normal curvature and real submanifold of the complex projective space, *Geometriae Dedicata* **7** (1978), 509–517.

[7] M. OKUMURA, Reducing the codimension of a submanifold of a complex projective space, *Geometriae Dedicata* **13** (1982), 277–289.

[8] B. O'NEILL, The fundamental equations of a submersion, *Michigan Math. J.* **13** (1966), 459–469.

[9] K. YANO and S. ISHIHARA, Fibred space with invariant Riemannian metric, *Kōdai Math. Sem. Rep.* **19** (1967), 317–360.

[10] K. YANO and M. KON, "Anti-invariant submanifold", Marcel Dekker, New York, 1976.

Masafumi Okumura

Department of Mathematics
Saitama University
Urawa
Japan

On the Holonomy Group of the Normal Connection

C. OLMOS [1] [2]

1. Introduction

Given a Riemannian manifold the representation on the tangent space of the restricted holonomy group is a direct product of irreducible representations and a trivial one. Each one of the non-trivial factors is either the isotropy representation of a simple Riemannian symmetric space or it is the representation of a compact Lie group which acts transitively on the unit sphere. This fact is a consequence of Berger's list [B] and the proof depends strongly on the classification of compact Lie groups.

Later Simons [S] gave a purely algebraic proof of Berger's results. He defined a holonomy system to be a triple $[\mathbb{V}, R, \Phi]$, where $\mathbb{V}$ is an Euclidean space, R is an algebraic Riemannian curvature tensor on $\mathbb{V}$ and Φ is a connected compact Lie group acting effectively on $\mathbb{V}$ by isometries, such that $R_{x,y}$ belongs to the Lie algebra of Φ for all $x, y \in \mathbb{V}$. He proved that if $[\mathbb{V}, R, \Phi]$ is an irreducible holonomy system such that Φ doesn't act

[1] supported by I.C.T.P., Trieste (Italy) — Univ. Nac. Cordoba, Cordoba (Argentina).

[2] The author wants to thank Prof. A. Salam of the I.C.T.P. for hospitality.

transitively on the unit sphere of $\mathbb{V}$, then $[\mathbb{V}, R, \Phi]$ must be symmetric. In the case that R has non-zero scalar curvature he was able to prove that there always exists a non-zero algebraic Riemannian curvature tensor R' on $\mathbb{V}$ such that $[\mathbb{V}, R', \Phi]$ is symmetric. So, in this case, the representation of Φ on $\mathbb{V}$ is always the isotropy representation of a simple Riemannian symmetric space (see [S]).

In [O] it was proved that the non-trivial part of the representation on the normal space of the normal holonomy group of any submanifold of a space of constant curvature, is always the isotropy representation of a semisimple Riemannian symmetric space (briefly s-representation). This result is far from being expected and its proof is easily derived from Simons' methods after defining an appropriate algebraic Riemannian curvature tensor on the normal bundle (with non-positive sectional curvatures) which gives the same geometric information as the usual normal curvature tensor.

An application of the above results, done independently by Heintze, Thorbergsson and the author, is that the submanifolds of the Euclidean space with constant principal curvatures (isoparametric in the sense of [St]) are exactly the isoparametric ones (see [T]) and its focal manifolds (see [H–O–T]).

In this note we are mostly concerned with the natural problem that arises from the results in [O]: which irreducible s-representations can occur as normal holonomy representations of submanifolds of the sphere. We show that almost all s-representations of rank ≥ 2 do actually occur in this way. For it, we study the normal holonomy group of (singular) orbits of s-representations (the so called imbedded R-spaces).

2. Preliminaries

Let M^n be a differentiable manifold and let $i : M^n \rightarrow Q^n$ be an isometric immersion, where $(Q^n, \langle,\rangle)$ is a Riemannian manifold of constant curvature.

Denote by

$$\nu(M) = \{(p, \xi) : p \in M, \, \xi \in (i_{*_p}(T_pM))^{\perp}\}$$

the normal bundle over M induced by i. For the sake of simplicity we shall also denote by $\langle,\rangle$ the induced metric on the fibres of $\nu(M)$ as well as the induced Riemannian metric on M. The connection in $\nu(M)$ will be the usual $\nabla^{\perp}$. As usual A will be the shape operator of i and $R^{\perp}$ the curvature tensor of the normal connection. $C^{\infty}(\nu(M))$ denotes the set of C^{∞} sections from M into $\nu(M)$.

Define $\mathcal{R}^{\perp} : C^{\infty}(\nu, M))^3 \to C^{\infty}(\nu(M))$ by

$$\mathcal{R}_p^{\perp}(\xi_1, \xi_2)\xi_3 = \sum_{j=1}^{n} R_p^{\perp}(A_{\xi_1}(e_j), A_{\xi_2}(e_j))\xi_3$$

where $p \in M$, $\xi_1, \xi_2, \xi_3 \in \nu(M)_p$ and $\{e_1, \ldots, e_n\}$ is an arbitrary orthonormal basis of $T_p M$.

Using the Ricci identity (i.e. $\langle R^{\perp}(X,Y)\xi, \eta \rangle = \langle [A_\xi, A_\eta](X), Y\rangle$), it is easy to obtain (see [O])

Lemma 2.1. *For all $p \in M$, $\xi_1, \xi_2, \xi_3, \xi_4 \in \nu(M)_p$ it is verified:*

i) $\mathcal{R}^{\perp}(\xi_1, \xi_2) = -\mathcal{R}_p^{\perp}(\xi_2, \xi_1)$

ii) $\langle \mathcal{R}_p^{\perp}(\xi_1, \xi_2)\xi_3, \xi_4\rangle = -\langle \xi_3, \mathcal{R}_p^{\perp}(\xi_1, \xi_2)\xi_4\rangle$

iii) $\mathcal{R}_p^{\perp}(\xi_1, \xi_2)\xi_3 + \mathcal{R}_p^{\perp}(\xi_2, \xi_3)\xi_1 + \mathcal{R}_p^{\perp}(\xi_3, \xi_1)\xi_2 = 0$

iv) $\langle \mathcal{R}_p^{\perp}(\xi_1, \xi_2)\xi_3, \xi_4\rangle = \langle \mathcal{R}_p^{\perp}(\xi_3, \xi_4)\xi_1, \xi_2\rangle =$
$= \frac{1}{2}$ trace $([A_{\xi_1}, A_{\xi_2}] \circ [A_{\xi_3}, A_{\xi_4}])$.

If $\mathbb{V}$ is an Euclidean space, we denote by $\mathcal{A}(\mathbb{V})$ the vector space of skew-symmetric endomorphisms of $\mathbb{V}$.

Let $p \in M$ and let $\tilde{A}p : \Lambda^2(\nu(M)_p) \to \mathcal{A}(T_p M)$ be defined by $\tilde{A}_p(\xi_1 \wedge \xi_2) = [A_{\xi_1}, A_{\xi_2}]$. Let $\tilde{R}_p^{\perp} : \Lambda^2(T_p M) \to \mathcal{A}(\nu(M)_p)$ be defined by $\tilde{R}_p^{\perp}(X \wedge Y) = R^{\perp}(X, Y)$ and let $\tilde{\mathcal{R}}^{\perp} : \Lambda^2(\nu(M)_p) \to \mathcal{A}(\nu(M)_p)$ be defined by $\tilde{\mathcal{R}}^{\perp}(\xi_1 \wedge \xi_2) = \mathcal{R}^{\perp}(\xi_1, \xi_2)$.

An easy calculation shows that $\tilde{\mathcal{R}}^{\perp} = -\tilde{R}_p^{\perp} \circ j_p^{-1} \circ \tilde{A}_p$, where $j_p : \Lambda^2(T_p M) \to \mathcal{A}(T_p M)$ is the usual isomorphism, i.e.,

$$\langle j_p(X \wedge Y)Z, W\rangle = \langle X, Z\rangle\langle Y, W\rangle - \langle X, W\rangle\langle Y, Z\rangle.$$

An also easy calsulation shows that

$$\ker \tilde{R}_p^{\perp} = (j_p^{-1} \circ \tilde{A}_p(\Lambda^2(\nu(M)_p)))^{\perp},$$

where the orthogonal complement is taken with the usual inner product on
$\Lambda^2(T_pM)$.

It is now clear the following (see [O])

Proposition 2.2. *For all $p \in M$ the linear span of the set $\{R_p^{\perp}(X, Y) :$
$X, Y \in T_pM\}$ in $\mathcal{A}(\nu(M)_p)$ coincides with the linear span of the set
$\{\mathcal{R}_p^{\perp}(\xi_1, \xi_2) : \xi_1, \xi_2 \in \nu(M)_p\}$.*

Observe that the above proposition and the theorem of Ambrose–Singer
imply that $\mathcal{R}^{\perp}$ carries all the normal holonomy information.

Using the tensor $\mathcal{R}^{\perp}$ and a slight modification of Simons' methods, it
is proved in [O] the following

Theorem 2.3. *Let M^n be an immersed submanifold of a Riemannian
manifold Q^N of constant curvature. Let $p \in M$ and let Φ^* be the restricted
holonomy group of the normal connection at p. Then Φ^* is compact, there
exists a unique (up to order) orthogonal decomposisiton of the normal space
at p $\nu(M)_p = \mathbb{V}_0 \oplus \ldots \oplus \mathbb{V}_k$ into Φ^*-invariant subspaces and there exist
$\Phi_0, \ldots, \Phi_k$ normal Lie subgroups of Φ^* such that:*

   i) $\Phi^* = \Phi_0 \times \ldots \times \Phi_k$ (direct product).
   ii) Φ_i acts trivially on $\mathbb{V}_j$ if $i \neq j$.
   *iii) $\Phi_0 = \{1\}$ and if $i \geq 1$ Φ_i acts irreducible on $\mathbb{V}_i$ as the isotropy
        representation of a simple Riemannian symmetric space.*

Isotropy representations of semisimple Riemannian symmetric spaces
will be called *s-representations*.

Normal holonomy group of R-spaces

We are now concerned with the natural problem of which irreducible *s*-
representations can occur as normal holonomy representations of submanifolds of the sphere. As we shall see later the orbits of *s*-representations,
considered as submanifolds of the sphere, will produce many examples. In

this way we shall see that almost all irreducible s-representations of rank ≥ 2 can occur as normal holonomy representations.

Let us recall that the orbits of the s-representations turn out to be the so called canonically imbedded R-spaces.

For our purposes we shall need the following result in [O–S], which generalizes the results of Ferus [F] on symmetric R-spaces and those in [P–T 1] about homogeneous isoparametric submanifolds.

Theorem 3.1. *(Olmos–Sanchez). Let M^n be an imbedded compact Riemannian submanifold of $\mathbb{R}^N$. The following statements are equivalent:*

i) *M^n admits a linear metric connection ∇^c such that $\nabla^c(\nabla - \nabla^c) = 0$ and $\nabla^c \alpha = 0$, where ∇ is the Levi–Civita connection and α is the second fundamental form.*

ii) *M^n is a homogeneous weakly isoparametric submanifold of $\mathbb{R}^N$.*

iii) *M^n is an orbit the isotropy representation of a semisimple Riemannian symmetric space.*

Let us say that in the above theorem

$$(\nabla^c_X \alpha)(Y, Z) = \nabla^\perp_X \alpha(Y, Z) - \alpha(\nabla^c_X Y, Z) - \alpha(X, \nabla^c_X Z)$$

and that M^n is said to be a homogeneous weakly isoparametric submanifold of $\mathbb{R}^N$ if given $p, q \in M$ and c curve in M from p to q then there exists an isometry g of $\mathbb{R}^N$ such that:

a) $g(M) = M$.

b) $g(p) = q$.

c) $g_{*p}|_{(T_p M)^\perp}$ coincides with the $\nabla^\perp$-parallel displacement along c.

Note that in this case the tensor of Lemma 2.1 is parallel with respect to the normal connection.

Let (K, ρ) be an irreducible s-representation, where $\rho : K \to SO(N)$ is a Lie group monomorphism (K connected). Let $0 \neq u \in \mathbb{R}^N$, let $M'^n = \rho(K)u$ and let $K_u = \{k \in K : \rho(k)u = u\}$. If M'^n is not a sphere (i.e. (K, ρ) is of rank ≥ 2) then $\rho(K)_0 = \{g \in I(\mathbb{R}^N) : g(M') = M'\}_0$, where the subscript 0 denotes the connected component of the identity and $I(\ )$ denotes the full isometry group.

The above equality is due to [S. th. 6] and [S, cor. pp. 232].

Decompose orthogonally $\nu(M')_u = \mathbb{V}_0 \oplus \mathbb{V}_1$ where $\mathbb{V}_0$ is the set of fixed points of the restricted normal holonomy group Φ_u^*.

Let $H = \{\rho(k)_{*_u}|_{(T_u M')^{\perp}} : k \in K_u\}$. Then by Theorem 3.1 and the previous paragraph we have that $H_0 \supset \Phi_u^*$. As it was seen in [O–S] H_0 must fix $\mathbb{V}_o$ point-wise and $L(H_0) \oplus \mathbb{V}_1$ is a semisimple Lie algebra, where

$$[\xi, \eta] = \mathcal{R}_u^{\perp}(\xi, \eta) \text{ if } \xi, \eta \in \mathbb{V}_1,$$
$$[X, \xi] = -[\xi, X] = X(\xi) \text{ if } X \in L(H_0), \ \xi \in \mathbb{V}_1,$$
$$[X, Y] = X \circ Y - Y \circ X \text{ if } X, Y \in L(H_0).$$

Thus $L(H_0) = \langle\{\mathcal{R}_u^{\perp}(\xi, \eta) : \xi, \eta \in \mathbb{V}_1\}\rangle$ and therefore $H_0 = \Phi_u^*$ (of course this is also true if M' is a sphere).

Then we have the following

Proposition 3.2. *Let $(K, \rho, \mathbb{R}^N)$ be an s-representation, let $u \in \mathbb{R}^N$, let $M = \rho(K)u$ be an orbit and let $(K_u)_0$ be the identity component of the isotropy subgroup of K at u. Then the image of the representation of $(K_u)_0$ on the normal space of M at u coincides with the restricted normal holonomy group at u.*

Let $(K, \rho, \mathbb{R}^N)$ be an irreducible s-representation and let $v \in \mathbb{R}^N$ be a principal orbit such that $\widetilde{M}^n = \rho(K)v$ (i.e. of highest dimension). Then $\widetilde{M}$ is a homogeneous isoparametric submanifold of $\mathbb{R}^N$ and we can associate to it a marked Dynkin diagram (see [T], [P–T 2], [H–P–T]).

Let $\alpha_1, \ldots, \alpha_k$ be the simple roots of such a diagram and let $n_1, \ldots, n_k \in \nu(\widetilde{M})_v$ be the corresponding principal normal curvatures $(K = N - n)$. We must have that n_i is proportional to α_i and that $\langle -v, n_i \rangle = 1$, for all i.

Let $i \in \{i_0, \ldots, i_k\}$ be fixed and let $w \in (T_v\widetilde{M})^{\perp}$ be defined by $w = n_{i_0} - pr_{i_0}(n_{i_0})$, where pr_{i_0} denotes the orthogonal projection into $\langle\{n_1, \ldots, n_k\} \setminus \{n_{i_0}\}\rangle$.

Let $z = w - v$. Then $\langle z, n_i \rangle = 1$ if $i \neq i_0$ and $\langle z, n_{i_0}\rangle \neq 1$. Let $\widetilde{z}$ be the $\nabla^{\perp}$-parallel normal vector field on $\widetilde{M}$ defined by $\widetilde{z}(v) = z$ and let $\phi : \widetilde{M} \rightarrow \mathbb{R}^N$ be defined by $\phi(p) = p + \widetilde{z}(p)$. Let $M = \phi(\widetilde{M}) = \rho(K)w$. Then M is a focal manifold of $\widetilde{M}$ (see [P–T 1]).

We have

(1)
$$(T_w M)^\perp = (T_v \widetilde{M})^\perp \oplus \mathcal{H}_v$$

where $\mathcal{H}_v = \ker(\mathrm{Id} - \widetilde{A}_z)$ and $\widetilde{A}$ is the shape operator of $\widetilde{M}$.

Let $\mathbb{V} = v + (T_v \widetilde{M})^\perp \oplus \mathcal{H}_v = (T_v \widetilde{M})^\perp \oplus \mathcal{H}_v$. Then, by the Slice Theorem of [H–P–T], $\overline{M} = (\widetilde{M} \cap \mathbb{V})_v$ is an isoparametric submanifold of $\mathbb{V}$ contained in $v + \mathbb{V} \cap \langle w \rangle^\perp$ and not contained in a smaller affine subspace. Its marked Dynkin diagram is obtained by taking out the root α_{i_0} from the marked Dynkin diagram associated to $\widetilde{M}$.

By (1) we have that $\overline{M}$ is an isoparametric submanifold of the normal space $\nu(M)_w$. It is not hard to see that $\overline{M} = \rho((K_w)_0)v$ and hence, by Proposition 3.2, $\overline{M} = w + \Phi_w^* \cdot (v - w)$.

If Φ_w^* is the restricted normal holonomy group of M at w, considered as a submanifold of the $|w|$-sphere S, then $\widetilde{\Phi}_w^*$ can be identified with $\Phi_w^*|_{\nu(M)_w \cap \langle w \rangle^\perp}$. It is now clear that $\widetilde{\Phi}_w^*$ acts irreducible on the normal space $\widetilde{\nu}(M)_w$ of M at w (in S). Moreover, the isoparametric foliation of $\widetilde{\nu}(M)_w$ induced by $\widetilde{\Phi}_w^*$ has the same marked Dynkin diagram as $\overline{M}$.

In this way we can obtain any proper connected subdiagram of the Dynkin diagrams of the homogeneous isoparametric submanifolds.

Irreducible isoparametric submanifolds of rank ≥ 2 are determined by its marked Dynkin diagrams, with the exception of the following two cases (see [H–P–T]):

i) The families of uniform multiplicity 2 associated to the adjoint representation on its Lie algebra of $Sp(r)$ and $SO(2r + 1)$ give the same marked Dynkin diagram of type B_r $(r \geq 2)$.

ii) The families of uniform multiplicity 1 associated with $Sp(r)$ and $SO(2r + 1)$ have the same marked Dynkin diagram of type B_r $(r \geq 2)$.

By considering both cases separately and by the table in [P–T 2, pp. 175] we get the following

Theorem 3.3. *All irreducible s-representations of rank ≥ 2 can arise from normal holonomy represesentations of submanifolds of the sphere with the*

possible exception of the isotropy representation of the following symmetric spaces

$$(E_6, \mathrm{Spin}(10) \times U(1)/\mathbb{Z}_2)$$
$$(G_2 \times G_2, \mathrm{diag}(G_2 \times G_2)$$
$$(G_2, SO(4))$$
$$(F_4 \times F_4, \mathrm{diag}(F_4 \times F_4)$$
$$(F_4, Sp(3) \times Sp(1)/\mathbb{Z}_2)$$
$$(E_6, SU(6) \times Sp(1)/\mathbb{Z}_2)$$
$$(E_7, Sp(12) \times Sp(1)/\mathbb{Z}_2)$$
$$(E_8, E_7 \times Sp(1)/\mathbb{Z}_2)$$
$$(E_8 \times E_8, \mathrm{diag}(E_8 \times E_8))$$
$$(E_8, \mathrm{Spin}(16)/\mathbb{Z}_2).$$

References

[B] M. BERGER, Sur les groupes d'holonomie homogene des varietes a connexion affine et des varietes Riemanniennes, *Bull. Soc. Math. France* **83** (1955), 279–330.

[F] D. FERUS, Symmetric submanifolds of Euclidean space, *Math. Ann.* **247** (1980), 81–93.

[H–O–T] HEINTZE, C. OLMOS and THORBERGSSON. Work in preparation.

[H–P–T] W. J. HSIANG, R. P. PALAIS and C. L. TERNG, The topology of isoparametric submanifolds, *J. Diff. Geom.* **27** (1988), 423–460.

[O] C. OLMOS, The normal holonomy group, To appear in *Proc. Amer. Math. Soc.*

[O–S] C. OLMOS and SANCHEZ, A geometric characterization of R-spaces, Preprint.

[P–T 1] R. S. PALAIS and C. L. TERNG, A general theory of canonical forms, *Trans. Amer. Math. Soc.* **300** (1987), 771–789.

[P–T 2] R. S. PALAIS and C. L. TERNG, Critical point theory and submanifold geometry, Springer-Verlag Lectures Notes in Math. **1353** (1988).

[S] J. SIMONS, On the transitivity of holonomy systems, *Annals of Mathematics* **76 n.2** (1962), 213–234.

[St] W. STRÜBING, Isoparametric submanifolds, *Geom. Dedicata* **20 n.3** (1986), 367–387.

[T] C. L. TERNG, Isoparametric submanifolds and their Coxeter groups, J. Diff. Geom. **21** (1985), 79–107.

Carlos Olmos

Mathematical Section
I. C. T. P.
P. O. Box 586
I–34100 Trieste, Italy

COLLOQUIA MATHEMATICA SOCIETATIS JÁNOS BOLYAI
56. DIFFERENTIAL GEOMETRY, EGER (HUNGARY), 1989

The Group of Isometries of a Compact Riemannian Homogeneous Space[*]

A. L. ONISHCHIK

Let G be a compact connected Lie group acting transitively and effectively on a manifold M, let $H = G_o$ be the stabilizer of a point $o \in M$. It is well known that there exist Riemannian metrics on M invariant with respect to G. They correspond bijectively to those Euclidean metrics on the tangent space $T_o(M)$ at o which are invariant with respect to the isotropy representation Iso : $H \to GL(T_o(M))$. If we fix a G-invariant Riemannian metric γ on M, then the group of all isometries $I(M, \gamma) = I(M)$ of γ is a compact Lie group containing G as a Lie subgroup. We consider here the problem of determining the group $I(M)$, assuming G to be simple. In studying the identity component $I(M)^0$, we use the old results of [7] about inclusions between compact connected transitive Lie transformation groups. To describe the full group $I(M)$ some general considerations concerning autosimilitudes of an action are used.

The results obtained here include the following special cases which were studied till now: M is an irreducible Riemannian symmetric space (Cartan [1], Wolf [10]); H is maximal in G (Onishchik [7]); M is isotropy

---

[*] This paper is in final form and no version of it will be submitted for publication elsewhere.

irreducible (Wolf [11]); $H = \{e\}$ i.e., M is a simple compact Lie group endowed with a left-invariant Riemannian metric (Ochiai–Takahashi [6], D'Atri–Ziller [2]). The main result is Theorem 6 describing the group $I(M)$ for the natural Riemannian metric on any simply connected homogeneous space M of a simple compact Lie group.

As a rule, we denote Lie groups by Roman capitals and their tangent Lie algebras by the corresponding Gothic small letters. By G^0 the connected identity component of a Lie group G is denoted. All manifolds and actions are supposed to be of class C^∞, and Diff M is the group of diffeomorphisms of a manifold M onto itself. Lie subgroups are supposed to be virtual (i.e., not necessarily closed).

1. Automorphisms and autosimilitudes of a transitive action

An action of a Lie group G on a manifold M is, as usually, a homomorphism $t : G \to \text{Diff } M$ such that the mapping $(g, x) \mapsto t(g)(x)$ of $G \times M$ into M is smooth. A G-manifold is a manifold with an action of G. Actions of a fixed Lie group G are objects of a category, whose morphisms are equivariant mappings, i.e., smooth mappings $f : M \to M'$ of G-manifolds such that $f(gx) = gf(x)$, $g \in G$, $x \in M$. Isomorphisms of this category are equivariant diffeomorphisms.

Now we consider a larger category, whose objects are actions of various Lie groups. Its morphisms are defined as follows. Let M be a G-manifold and M' a G'-manifold. A smooth mapping $f : M \to M'$ is called a morphism if there exists a homomorphism $\varphi : G \to G'$ such that $f(gx) = \varphi(g)f(x)$, $g \in G$, $x \in M$. A morphism f is called a similitude if f is a diffeomorphism and f^{-1} is a morphism. If the actions of G on M and of G' on M' are effective and $f : M \to M'$ is a similitude, then the corresponding homomorphism $\varphi : G \to G'$ is a uniquely determined isomorphism of Lie groups.

Two actions are called isomorphic (similar) if there exists an isomorphism (a similitude) between them. Note that two similar actions of the same Lie group need not be isomorphic.

Now we introduce two important groups associated with a G-manifold M. Let $\operatorname{Aut}_G M$ be the group of automorphisms (i.e., isomorphisms onto itself) of M and $\operatorname{Sim}_G M$ the group of autosimilitudes (i.e., similitudes onto itself) of M. Clearly, $\operatorname{Aut}_G M$ is a subgroup of $\operatorname{Sim}_G M$. Since $u(gx) = (ugu^{-1})(ux)$ for any $u, g \in G$, $x \in M$, we see that $t(u) \in \operatorname{Sim}_G M$ for each $u \in G$. One verifies easily that $t(G)$ is a normal subgroup of $\operatorname{Sim}_G M$.

Suppose that the action $t : G \to \operatorname{Diff} M$ is transitive, fix a point $o \in M$ and denote $H = G_o$. It is well known that the mapping $gH \mapsto go$, $g \in G$, is an isomorphism of the G-manifold G/H on M, where G acts on the coset manifold G/H by left translations. We identify M with G/H via this isomorphism. Then the automorphisms of the G-manifold M are expressed as "right translations"

$$gH \mapsto gnH, \qquad n \in N_G(H),$$

where $N_G(H)$ is the normalizer of H in G. Hence,

$$\operatorname{Aut}_G M \simeq N_G(H)/H.$$

The action is called asystatic if $\operatorname{Aut}_G M$ is discrete.

Now we describe the group $\operatorname{Sim}_G M$ in terms of the pair (G, H). Let $\operatorname{Aut}(G, H)$ be the group of all automorphisms of G mapping H onto itself. Each $\alpha \in \operatorname{Aut}(G, H)$ gives rise to the autosimilitude s_α of $M = G/H$ by

$$s_\alpha(go) = \alpha(g)o, \qquad g \in G.$$

Then $A = \{s_\alpha : \alpha \in \operatorname{Aut}(G, H)\}$ is a subgroup of $\operatorname{Sim}_G M$ fixing the point o.

Proposition 1. *Assume the action t to be transitive and effective. Then the subgroup A coincides with the stabilizer $(\operatorname{Sim}_G M)_o$, and $\operatorname{Sim}_G M = t(G)A$.*

Proof. For any $s \in (\operatorname{Sim}_G M)_o$ we have by definition $s(go) = \varphi(g)s(o) = \varphi(g)o$ for a certain $\varphi \in \operatorname{Aut} G$. If $g \in H$, then $\varphi(g)o = s(o) = o$, whence $\varphi \in \operatorname{Aut}(G, H)$. The second assertion is a consequence of the transitivity of t. $\blacksquare$

As an example, consider the action ℓ of a Lie group G onto itself by left translations. Recall that the holomorph of G is, by definition, the group of diffeomorphisms of G onto itself, generated by left translations and automorphisms (it contains, clearly, all right translations).

Corollary 1. *For the action ℓ of a Lie group G on $M = G$, the group $\mathrm{Sim}_G M$ coincides with the holomorph $\mathrm{Hol}\, G = \ell(G)\,\mathrm{Aut}\, G$.* ∎

Corollary 2. *Let G be a connected non-abelian simple Lie group. Consider the action of the group $G \times G$ on $M = G$ by two-sided translations. Then $\mathrm{Sim}_{G\times G} M = \mathrm{Hol}\, G \rtimes \langle s \rangle$, where s is the inversion $g \mapsto g^{-1}$ of G.*

Proof. Each $\alpha \in \mathrm{Aut}\, G$ gives rise to the automorphism $\alpha \times \alpha$ of $G \times G$ and, clearly, $\alpha = s_{\alpha \times \alpha}$. Denoting by σ the automorphism $(u, v) \mapsto (v, u)$ of $G \times G$, we get $s = s_\sigma$. Therefore, it is sufficient to prove that $\mathrm{Aut}(G \times G, G_d) = \mathrm{Aut}\, G \times \langle \sigma \rangle$, where $G_d = (G \times G)_e$ is the diagonal of $G \times G$ and $\mathrm{Aut}\, G$ is imbedded into $\mathrm{Aut}(G \times G)$ by $\alpha \mapsto \alpha \times \alpha$. But this follows easily from the uniqueness of the decomposition of $G \times G$ into simple factors. ∎

Now we consider the case when G is semisimple. For an effective action t of G on M, we identify $t(G)$ with G.

Theorem 1. *Let $M = G/H$ where G is a connected semisimple Lie group acting on M effectively. Then $\mathrm{Sim}_G M$ has a finite number of components, and $(\mathrm{Sim}_G M)^0 = G(\mathrm{Aut}_G M)^0$ (locally direct product). If G is compact, then $\mathrm{Sim}_G M$ is compact too, and the natural homomorphism $\mathrm{Aut}(G, H) \to A$ is an isomorphism.*

Proof. By Proposition 1, $(\mathrm{Sim}_G M)^0 = G\, A^0$. Since G is semisimple, all the automorphisms from A^0 are interior. Thus, each s_α, $\alpha \in A^0$, has the form

$$s_\alpha(go) = (ngn^{-1})o, \qquad n \in N_G(H).$$

This implies that $A^0 \subset \mathrm{Aut}_G M$. Clearly, G and $\mathrm{Aut}_G M$ commute elementwise, and $G \cap \mathrm{Aut}_G M = Z(G)$ is finite. Therefore, $(\mathrm{Sim}_G M)^0 = G(\mathrm{Aut}_G M)^0$ (locally direct product). Now, $\mathrm{Sim}_G M/(\mathrm{Sim}_G M)^0 = = GA/GA^0 \simeq A/G \cap A$ is a homomorphic image of the finite group A/A^0 and, hence, a finite group. If G is compact, then so is $\mathrm{Aut}_G M$ and, hence, $(\mathrm{Sim}_G M)^0$ and $\mathrm{Sim}_G M$.

To prove the last assertion of the theorem, consider the tangent algebras $\mathfrak{g} \supset \mathfrak{h}$ of G and H. If $\alpha \in \mathrm{Aut}(G, H)$ is such that $s_\alpha = id$, then $d\alpha(x + \mathfrak{h}) =$

$x + \mathfrak{h}$ for any $x \in \mathfrak{g}$. Choose a scalar product in $\mathfrak{g}$ which is invariant with respect to Aut $\mathfrak{g}$. Then the orthogonal complement $\mathfrak{m} = \mathfrak{h}^{\perp}$ is invariant under $d\alpha$, and so is the subspace $\mathfrak{i} = \mathfrak{m} + [\mathfrak{m}, \mathfrak{m}] + [\mathfrak{m}[\mathfrak{m}, \mathfrak{m}]] + \ldots$ which is an ideal of $\mathfrak{g}$. The ideal $\mathfrak{i}^{\perp}$ is contained in $\mathfrak{h}$, whence $\mathfrak{i}^{\perp} = 0$ due to the effectivity. Therefore, $\mathfrak{i} = \mathfrak{g}$ and $\alpha = id$. ■

We apply these results to the study of enlargements of transitive actions. Let G be a Lie subgroup of a Lie group G'. Then for any action $t' : G' \to$ Diff M, its restriction $t = t'|G$ is an action of G on M. We call t' an enlargement of the action t; it is said to be proper if $G \neq G'$.

Let $t : G \to$ Diff M be an action of a Lie group G. Denote by $N(t)$ the natural action of the group $\mathrm{Sim}_G M$ on M. Assuming t to be effective and identifying G with $t(G)$, we may consider $N(t)$ as an enlargement of t. We call this enlargement the natural one. It is clear from the definitions that the natural enlargement is the largest one among those enlargements which contain G as a normal subgroup.

Using Proposition 1 and Theorem 1, we get

Proposition 2. *Let G be a normal Lie subgroup of a Lie group G', $t' : G' \to$ Diff M an effective enlargement of an action $t = t'|G$. Then G' is identified with a subgroup of $\mathrm{Sim}_G M$ of the form $G' = GB$, where B is a subgroup of A, the action t' being the restriction of $N(t)$. If G is connected and semisimple, then $G'^0 = GC$ (locally direct product), where C is a connected Lie subgroup of $(\mathrm{Aut}_G M)^0$.* ■

A transitive action of a connected Lie group G is called irreducible if no proper connected virtual normal Lie subgroups of G act transitively. Proposition 2 permits us to describe transitive actions in terms of irreducible ones.

Since the identity component G^0 of G is normal in G and acts on any connected homogeneous space of G transitively, we get

Corollary. *Let t be a transitive effective action of a Lie group G on a connected manifold M. Identifying G with $t(G)$, we have $G \subset \mathrm{Sim}_{G^0} M$ and $G = G^0 B$ where B is a subgroup of $A = (\mathrm{Sim}_{G^0} M)_0$.* ■

2. Compact enlargements of a transitive simple compact transformation group

A factorization of a Lie group G is any triple (G, A, B) where A, B are some Lie subgroups of G and $G = AB$ (i.e., any $g \in G$ has the form $g = ab$ with $a \in A$, $b \in B$). The factorization (G, A, B) is called trivial if $A = G$ or $B = G$. There exists an intimate relation between factorizations and enlargements of transitive actions: if (G, A, B) is a factorization, then A acts on G/B transitively and $G/B = A/A \cap B$. Conversely, if a subgroup $A \subset G$ acts on G/B transitively, then $G = AB$.

We note the following properties of factorizations (see Onishchik [7, 8]):

i) If the triple (G, A, B) is a factorization, then so is (G, B, A).

ii) If (G, A, B) is a factorization, then so are (G, gAg^{-1}, hBh^{-1}) for any $g, h \in G$.

iii) If (G, A, B) is a factorization, then so is (G^0, A^0, B^0).

iv) If (G, A, B) is a factorization, G' a connected Lie group, A', B' its Lie subgroups such that the triple (G', A', B') is locally isomorphic to (G, A, B), then (G', A', B') is a factorization.

In the class of all factorizations (G, A, B) of connected Lie groups G, consider the equivalence relation generated by relations that follow from the properties i)–iv). In [7] all factorizations of connected compact Lie group, up to this equivalence, were described. The main point is the description of factorizations of simple compact Lie groups which we shall recall here.

To make this description more simple, let us give the following definition. A factorization (G, A, B) with connected G, A, B is called irreducible if neither A nor B may be replaced in it by its proper normal Lie subgroup. For an arbitrary factorization $(G, A < B)$ with connected G, A, B, denote by A_0 and B_0 some minimal connected normal Lie subgroups of A and B respectively such that $G = A_0 B_0$. Clearly, the factorization (G, A_0, B_0) is irreducible, and $A \subset N_G(A_0)^0$, $B \subset N_G(B_0)^0$. Thus, it is sufficient to describe the irreducible factorizations (G, A, B) and to indicate the subgroups $N_G(A)^0$, $N_G(B^0)$. The result is as follows.

Theorem 2 (Onishchik [7]). *Any non-trivial irreducible factorization* (G, A, B), *where G, A, B are connected, G compact and simple, is equiv-*

alent to one of factorizations from Table 1 (the group $N_G(A^0)$ or $N_G(B^0)$ not indicated there coincides with A or B respectively). ∎

Table 1

G	A	B	$N_G(B)^0$	$A \cap B$	G	A	B	$A \cap B$
SU_{2n} $n \geq 2$	Sp_n	SU_{2n-1}	$SU_{2n-1}U_1$	Sp_{n-1}	SO_7	SO_6	G_2	SU_3
SO_{2n} $n \geq 4$	SO_{2n-1}	SU_n	SU_nU_1	SU_{n-1}	SO_8	SO_7	Spin_7	G_2
SO_{4n} $n \geq 2$	SO_{4n-1}	Sp_n	Sp_nSp_1	Sp_{n-1}	SO_{16}	SO_{15}	Spin_9	Spin_7

Now we go over to the description of connected compact enlargements of simple compact transitive groups. This is equivalent to the description of factorizations of connected compact Lie groups such that one factor is simple. We present three types of enlargements of this sort.

Type I. Let (G', G, H') be a non-trivial factorization, G' and G connected, simple and compact, and denote $H = G \cap H'$. Then we have an enlargement of the standard action of G on $G/H = G'/H'$ to the standard action of G'.

Type II. Let us identify a connected simple compact Lie group G with the diagonal subgroup $G_d \subset G \times G$. Suppose a non-trivial factorization (G, A', A'') be given and denote $H = A' \cap A''$. Then

$$G \times G = G(A' \times A''), \qquad G \cap (A' \times A'') = H,$$

whence

$$G/H = (G \times G)/(A' \times A'') = G/A' \times G/A''.$$

Thus, we have an enlargement of the standard action of G on G/H to an action of $G \times G$.

Type III. Consider an enlargement of type II. It may happen that the action of G on one of the manifolds G/A', G/A'' (but not on both!) admits an enlargement $G \subset G'$ of type I. As a result, we get an enlargement of the action of G to $G' \times G$.

A homogeneous space G/H (or the corresponding transitive action) of a connected simple compact Lie group G is called exceptional if the action of G admits an enlargement of type I or II. The exceptional homogeneous spaces may be listed easily with the help of Theorem 2.

The following theorem describes all effective compact connected enlargements of transitive actions of simple compact Lie groups.

Theorem 3 (Onishchik [7]). *Let $t : G \to \mathrm{Diff}\, M$ be a transitive action of a connected simple compact Lie group G, and let $t' : G' \to \mathrm{Diff}\, M$ be its effective proper enlargement, G' also being connected and compact. If t' is irreducible, then t' is of type I, II or III. In general case, one of the following possibilities holds:*

i) $G' = GC$ *(locally direct product), where $C \subset (\mathrm{Aut}_G\, M)^0$;*

ii) $G' = G^\sim C$ *(locally direct product), where $G^\sim$ acts as an enlargement of type I, II or III of t and $C \subset (\mathrm{Aut}_{G^\sim}\, M)^0$.* ∎

Corollary. *If a non-exceptional transitive action of a connected simple compact Lie group G is asystatic, then this action does not admit any effective proper compact connected enlargement. The exceptional asystatic homogeneous spaces are presented (up to local similarity) in Table 2. Every exceptional asystatic homogeneous space G/H admits a unique proper effective connected compact enlargement G'/H' indicated in the same Table.* ∎

Now we present some results concerning not necessarily connected compact enlargements. For simplicity, we consider only the non-exceptional homogeneous spaces.

Proposition 3. *Let $M = G/H$ be a non-exceptional homogeneous space of a connected simple compact Lie group G, $H^0 \neq \{e\}$. Then any effective action of a compact Lie group G' on M enlarging the action t of G is a restriction of the natural enlargement $N(t)$, and $G' = GB$ where $B \subset A \simeq \mathrm{Aut}(G, H)$.*

Table 2

G/H	G'/H'	Type
SO_{2n-1}/U_{n-1} $n \geq 4$	SO_{2n}/U_n	I
$SO_{4n-1}/Sp_{n-1}Sp_1$ $n \geq 2$	$SO_{4n}/Sp_n Sp_1$	I
$SO_{4n-1}/Sp_{n-1}U_1$ $n \geq 2$	$SO_{4n}/Sp_n U_1$	I
$Sp_n/Sp_{n-1}U_1 = \mathbb{C}P^{2n-1}$ $n \geq 2$	$SU_{2n}/SU_{2n-1}U_1$	I
$\mathrm{Spin}_7/G_2 = S^7$	SO_8/SO_7	I
$\mathrm{Spin}_9/\mathrm{Spin}_7 = S^{15}$	SO_{16}/SO_{15}	I
SO_{15}/Spin_7	SO_{16}/Spin_9	I
$G_2/SU_3 = S^6$	SO_7/SO_6	I
SO_{16}/Spin_7	$SO_{16}/SO_{15} \times SO_{16}/\mathrm{Spin}_9$	II
$\mathrm{Spin}_8/G_2 = S^7 \times S^7$	$SO_8/SO_7 \times \mathrm{Spin}_8/\mathrm{Spin}_7$	II

Proof. By Theorem 3, $G'^0 = GC$ (locally direct product), C being a connected Lie subgroup of $(\mathrm{Aut}_G M)^0 \simeq N_G(H)^0/H \cap N_G(H)^0$. Since $H^0 \neq \{e\}$, $\dim C \leq \dim(\mathrm{Aut}_G M)^0 < \dim G$. Therefore, G is normal in G' and we may apply Proposition 2 and Theorem 1. ∎

Proposition 4. Let $M = G$ be a connected simple compact non-abelian Lie group. Suppose that an effective action of a compact Lie group G' on M enlarging the action ℓ of G is given. Then $G' \subset \mathrm{Hol}\, G \rtimes \langle s \rangle$ where s is the inversion $g \mapsto g^{-1}$ of G.

Proof. Since the action ℓ is non-exceptional, by Theorem 3 $G'^0 = \ell(G)C$ (locally direct product) where C is a connected Lie subgroup of the group of right translations $r(G)$. If $C \neq r(G)$, then $\ell(G)$ is normal in G', and $G' \subset \mathrm{Sim}_\alpha M = \mathrm{Hol}\, G$ (see Corollary 1 of Proposition 1). If $C = r(G)$, then, by Corollary of Proposition 2, $G' \subset \text{Sim}_{G \times G} M$, and Corollary 2 of Proposition 1 may be applied.

3. Groups of isometries of Riemannian homogeneous spaces of simple compact Lie groups

Let $M = G/H$ be a homogeneous space of a connected simple non-abelian compact Lie group G endowed with a G-invariant Riemannian metric γ. We may suppose that G acts effectively. Then the group of isometries $I(M) = I(M, \gamma)$ is compact (but not necessarily connected), and its natural action on M is an enlargement of that of G. In this section, the main results of the paper (Theorems 4, 5 and 6) are formulated. Theorems 4 and 5 are easy consequences of the above results. The proof of Theorem 6 is much more complicated and will be given in sections 4 and 5.

Theorem 4. *If G/H is non-exceptional, then*

$$I(M)^0 = GC \qquad \text{(locally direct product)},$$

where C is a connected Lie subgroup of $\text{Aut}_G M$. If we exclude the case when $M = G$ with the action ℓ and γ is bi-invariant, then $I(M) \subset \text{Sim}_G M$, and

$$I(M) = GB,$$

where B is a Lie subgroup of $A \simeq \text{Aut}(G, H)$.

Proof. Use Theorem 3, Propositions 3 and 4. ∎

Corollary (D'Atri–Ziller [2]). *Let γ be a left-invariant (but not bi-invariant) Riemannian metric on a connected simple non-abelian compact Lie group G. Then $I(G, \gamma) \subset \text{Hol}\, G$, i.e.,*

$$I(G, \gamma) = \ell(G)B,$$

where B is a Lie subgroup of $\text{Aut}\, G$. ∎

We note that in [2] the restriction on γ not to be bi-invariant is missing and that for bi-invariant γ the assertion is false (see Theorem 6).

Theorem 5. *Let $M = G/H$ be asystatic. If it is non-exceptional then $I(M)^0 = G$. If it is exceptional, then $I(M)^0$ coincides with G or with the enlargement indicated in Table 2.*

Proof. Use Theorem 4 and Corollary of Theorem 3. ∎

Let G be a connected simple compact Lie group. Then there exists on the Lie algebra $\mathfrak{g}$ a unique (up to a positive factor) $\operatorname{Ad} G$-invariant scalar product $\langle , \rangle$. We may define it by

$$\langle x, y \rangle = -k_{\mathfrak{g}}(x, y), \qquad x, \in \mathfrak{g},$$

where $k_{\mathfrak{g}}$ is the Killing form of $\mathfrak{g}$. This scalar product is, clearly, invariant with respect to $\operatorname{Aut} \mathfrak{g}$.

Let H be a closed Lie subgroup of G. Then $\mathfrak{g} = \mathfrak{h} \oplus \mathfrak{m}$ where $\mathfrak{m} = \mathfrak{h}^{\perp}$. The subspace $\mathfrak{m}$ is $\operatorname{Ad} H$-invariant and is naturally isomorphic, as a H-module, to the tangent space $T_o(G/H)$ at the point $o = H$, regarded as a H-module via the isotropy representation. Restricting the scalar product onto $\mathfrak{m}$, we get an invariant Riemannian metric γ_0 on $M = G/H$ called the natural Riemannian metric. Denote $I(M) = I(M, \gamma_0)$. The main result of this paper is the following

Theorem 6. *Let $M = G/H$ be a simply connected homogeneous space of a connected simple compact Lie group G endowed with the natural Riemannian metric. Then*

$$I(M)^0 = G(\operatorname{Aut}_G M)^0 \quad \text{(locally direct product)},$$
$$I(M) = \operatorname{Sim}_G M = GA, \qquad A \simeq \operatorname{Aut}(G, H),$$

except of the following cases:
 a) $M = G_2/SU_3 = S^6$, $I(M) = O_7$;
 b) $M = \operatorname{Spin}_7/G_2 = S^7$, $I(M) = O_8$;
 c) $M = \operatorname{Spin}_8/G_2 = S^7 \times S^7$, $I(M) = (O_8 \times O_8) \rtimes \langle s \rangle$, s being the transposition of factors;
 d) $M = G$ with the action ℓ, γ_0 is the bi-invariant Riemannian metric on G, $I(M) = (\operatorname{Hol} G) \rtimes \langle s \rangle$, $s : g \mapsto g^{-1}$ for $g \in G$.

Before proving this theorem, we deduce from it a known consequence. A homogeneous space is called isotropy irreducible if its isotropy representation is irreducible (over reals). The classification of isotropy irreducible homogeneous spaces of compact Lie groups was obtained by Manturov [5] and Wolf [11] independently (we note that in each of these two classifications some cases are missing, but the union of two lists seems to contain all the cases). Since any two bilinear invariant of an irreducible representation differ by a scalar factor only, so are any two invariant Riemannian metrics on an isotropy irreducible homogeneous space. Thus, all these metrics are natural.

Corollary (Wolf [11]). *Let $M = G/H$ be an isotropy irreducible simply connected homogeneous space of a connected simple non-abelian compact Lie group G, endowed with an invariant Riemannian metric. Then*

$$I(M)^0 = G,$$
$$I(M) = \operatorname{Sim}_G M = GA, \quad where \quad A \simeq \operatorname{Aut}(G, H),$$

except of the cases a) and b) of Theorem 6.

Proof. We note that the action of G on M is asystatic. In fact, denote by $\mathfrak{n}$ the subalgebra of $\mathfrak{g}$ corresponding to the subgroup $N_G(H)$ of G. Then $\mathfrak{n}/\mathfrak{h}$ is a subspace of $\mathfrak{g}/\mathfrak{h} \simeq T_o(M)$ invariant with respect to Iso. If $\mathfrak{n} = \mathfrak{g}$, then H is normal in G and, hence, $H = \{e\}$. This is impossible, since Iso is irreducible and $\dim G > 1$.

If M is non-exceptional, then $I(M)^0 = G$ by Theorem 5. Applying Theorem 6, we see that among four excepted cases only the spaces a) and b) are isotropy irreducible. ∎

4. Some necessary lemmas

In this section, we denote by G a connected simple non-abelian compact Lie group, acting on a manifold $M = G/H$ transitively and effectively.

Lemma 1. *If M is endowed with the natural G-invariant Riemannian metric, then $\operatorname{Sim}_G M \subset I(M)$.*

Proof. Clearly, it is sufficient to prove that $(\operatorname{Sim}_G M)_o$ leaves invariant the scalar product in $T_o(M) \simeq \mathfrak{m}$. By Proposition 1 $(\operatorname{Sim}_G M)_o$ consists of transformations s_α, $\alpha \in \operatorname{Aut}(G, H)$. Evidently, $(ds_\alpha)_o$ is identified with the restriction of $d\alpha$ on $\mathfrak{m}$. Our assertion follows from the invariance of the scalar product in $\mathfrak{g}$ with respect to $\operatorname{Aut}\mathfrak{g}$. ∎

Now, let an effective action of a connected compact semisimple Lie group G' on $M = G/H$ enlarging the action of G be given. Denote $H' = G'_o$. The corresponding Lie algebras $\mathfrak{g}$, $\mathfrak{h}$, $\mathfrak{g}'$, $\mathfrak{h}'$ satisfy

$$\mathfrak{g}' = \mathfrak{g} + \mathfrak{h}', \qquad \mathfrak{g} \cap \mathfrak{h}' = \mathfrak{h}.$$

Since G is simple, we have $k_{\mathfrak{g}'}|\mathfrak{g} = c k_{\mathfrak{g}}$ with $c > 0$. Since we consider invariant Riemannian metrics on M up to scalar factors, we may use the scalar product

$$\langle u, v\rangle = -k_{\mathfrak{g}'}(u, v)$$

on $\mathfrak{g}'$ and $\mathfrak{g}$. Denote $\mathfrak{m}' = \mathfrak{h}'^{\perp}$, $\mathfrak{m} = \mathfrak{h}^{\perp}$ (orthogonal complements in $\mathfrak{g}'$ and $\mathfrak{g}$ respectively) and let $\pi_0 : \mathfrak{g}' \to \mathfrak{h}'$, $\pi_1 : \mathfrak{g}' \to \mathfrak{m}'$ be the orthogonal projections. Then $\pi_1 : \mathfrak{m} \to \mathfrak{m}'$ is an isomorphism of vector spaces. If we identify, in the natural way, $T_o(M)$ with $\mathfrak{m}$ and $\mathfrak{m}'$, then π_1 becomes the identity mapping id. Denote by $\langle,\rangle'$ the scalar product in $\mathfrak{m}'$ induced by the given scalar product in $\mathfrak{m}$, i.e., satisfying

$$\langle \pi_1(u), \pi_1(v)\rangle' = \langle u, v\rangle, \qquad u, v \in \mathfrak{m}.$$

Lemma 2. *Let $\mathfrak{m}' = \mathfrak{m}'_1 \oplus \ldots \oplus \mathfrak{m}'_r$ be the decomposition of the H'-module $\mathfrak{m}'$ into irreducible components. The natural G-invariant metric γ_0 on M is G'-invariant if and only if the following conditions is satisfied:*
  a) For any i, there exists $c_i > 0$ such that $\langle u, v\rangle' = c_i\langle u, v\rangle$ for $u, v \in \mathfrak{m}'_i$. If this is true, the following condition is also satisfied:
  b) If $\mathfrak{m}'_i \cap \mathfrak{g} \neq 0$ for a certain $i = 1, \ldots, r$, then $\mathfrak{m}' \subset \mathfrak{g}$.
  If G' is simple and G'/H' isotropy irreducible, then G'-invariance of the natural Riemannian metric on G/H implies that either $\mathfrak{m}' \cap \mathfrak{g} = 0$ or $G' = G$.

Proof. Clearly, the metric γ_0 is G'-invariant if and only if

(1) $\qquad ((\operatorname{Ad}h)u, (\operatorname{Ad}h)v)' = (u, v)', \qquad h \in H', \quad u, v \in \mathfrak{m}'.$

Since $\operatorname{Ad}H'$ acts in each $\mathfrak{m}'_i$ irreducibly and preserves $\langle,\rangle$, the equation (1) implies a). The converse is clear.

Suppose a) to be true and denote $m_i = \pi^{-1}(m_i') \cap m$. Then $\langle u, u \rangle = c_i \langle \pi_1(u), \pi_1(u) \rangle$ for any $u \in m_i$. Hence,

$$\langle \pi_0(u), \pi_0(u) \rangle = \langle u, v \rangle - c_i \langle \pi_1(u), \pi_1(u) \rangle = (1 - c_i) \langle \pi_1(u), \pi_1(u) \rangle$$

for $u \in m_i$. By substituting a non-zero $u \in m_i' \cap g = m_i' \cap m$, we see that $c_i = 1$. Therefore, $\pi_0(u) = 0$ for $u \in m_i$. Thus, $m_i \subset m'$ and $m_i = m_i'$.

To prove the last assertion, we note that if $m' \subset g$, then the ideal $i' = m' + [m', m'] + [m', [m', m']] + \dots$ of g' lies in g. Since g' is simple, $i' = g' = g$. ∎

Let (G, A', A'') be a non-trivial factorization of G, and $H = A' \cap A''$. As we have seen in Section 2, the standard action of G on $M = G/H$ admits the enlargement of type II, identifying M with $(G \times G)/(A' \times A'') = M' \times M''$ where $M' = G/A'$, $M'' = G/A''$. If we denote $o = H \in M$, $o' = A' \in M'$, $o'' = A'' \in M''$, then the orbits $A''(o) = A'/H$ and $A'(o) = A'/H$ are identified with $M' \times \{o''\}$ and $\{o'\} \times M''$ respectively.

Lemma 3. *If γ is a G-invariant Riemannian metric on M, then its restrictions γ' and γ'' onto $M' \times \{o''\}$ and $\{o'\} \times M''$ are invariant, under A'' and A'. If, in addition, γ is $(G \times G)$-invariant, then γ' and γ'' are G-invariant. If $\gamma = \gamma_0$ is the natural G-invariant metric and A', A'' simple, then γ' and γ'' are the natural A''- and A'-invariant Riemannian metrics respectively.*

Proof. The first and the second assertions are evident. To prove the third one, we fix an invariant scalar product on g and note that its restrictions onto a' and a'' are (unique up to positive factors) invariant scalar products on these simple Lie algebras. Denote by m_1 and m_2 the orthogonal complements to $\mathfrak{h}$ in a'' and a' respectively. Then $m = m_1 + m_2$ is orthogonal to $\mathfrak{h}$ and $g = a' + a'' = m + \mathfrak{h}$. Therefore, $m = \mathfrak{h}^\perp$ in g. Our assertion follows easily from the inclusions $m_1 \subset m$, $m_2 \subset m$. ∎

5. Proof of Theorem 6

Denote by G' the smallest connected normal Lie subgroup of $I(M)$ containing G. Then $I(M) \subset \text{Sim}_{G'} M$. If $G' = G$, then Lemma 1 implies that

$I(M) = \mathrm{Sim}_G M$. By Theorem 1, we have to prove that $G' = G$ except of the cases a), b), c), d). We divide the proof into two parts: (i) G' simple, and (ii) G' not simple.

(i) If G' is simple and $G' \neq G$, then, by Theorem 3, the enlargement $G \subset G'$ is of type I. If we want to verify that the condition a) or b) of Lemma 2 is not satisfied, then it is sufficient to do this for all maximal H' such that $G' = GH'$ (see Theorem 2 to determine all these subgroups). We also note that for a classical linear Lie algebra $\mathfrak{g}'$ we may set

$$\langle u, v \rangle = -\mathrm{tr}(uv), \qquad u, v \in \mathfrak{g}'.$$

Now we go over to the case-by-case verification. As usually, we denote by E_{ij} the matrix whose entries are zero except of the entry 1 located in the i-th row and in the j-th column.

1) $G' = SU_{2n}$, $G = Sp_n$, $H' = SU_{2n-1}U_1$, $H = Sp_{n-1}U_1$, $n \geq 2$. The H'-module $\mathfrak{m}'$ is irreducible. We have $E_{1,n+1} + E_{n+1,1} \in \mathfrak{m}' \cap \mathfrak{g}$. Hence, by Lemma 2, the natural G-invariant Riemannian metric is not G'-invariant.

2) $G' = SU_{2n}$, $G = SU_{2n-1}$, $H' = Sp_n$, $H = Sp_{n-1}$, $n \geq 2$. The H'-module $\mathfrak{m}'$ is irreducible and consists of all matrices of the form

$$\begin{pmatrix} \overline{X} & \overline{Y} \\ Y & -X \end{pmatrix}, \quad X, Y \in M_n(\mathbb{C}), \quad \overline{X}^\top = -X, \quad \mathrm{tr}\, X = 0, \quad \overline{Y}^\top = -Y.$$

For $n > 2$ it is clear that $\mathfrak{m}' \cap \mathfrak{g} \neq 0$, and we are able to apply Lemma 2. If $n = 2$, then the considered factorization is equivalent to (SO_6, SU_3, SO_5) which we are going to study next.

3) $G' = SO_{2n}$, $G = SU_n$, $H' = SO_{2n-1}$, $H = SU_{n-1}$, $n \geq 3$. The H'-module $\mathfrak{m}'$ is irreducible and consists of all matrices of the form

$$\begin{pmatrix} 0 & x_1 & \cdots & x_{2n-1} \\ -x_1 & & & \\ \vdots & & 0 & \\ -x_{2n-1} & & & \end{pmatrix}, \qquad x_i \in \mathbb{R}.$$

Clearly, $x = E_{1n} - E_{n1} + E_{n+1,2n} - E_{2n,n+1} \in \mathfrak{m}$, $x' = \pi_1(x) = E_{1n} - E_{n1}$, $\langle x', x' \rangle' = 4$ and $\langle x', x' \rangle = 2$. On the other hand, $y = (n-1)E_{1,n+1} - E_{2,n+2} - \ldots - E_{n,2n} - (n-1)E_{n+1,1} + E_{n+1,2} + \ldots, + E_{2n,2} \in \mathfrak{m}$, $y' = \pi_1(y) = (n-1)(E_{1,n+1} - E_{n+1}, 1)$, $\langle y', y' \rangle' = 2n(n-1)$ and $\langle y', y' \rangle = 2(n-1)^2$. Thus, the condition a) of Lemma 2 is not satisfied.

4) $G' = SO_{2n}$, $G = SO_{2n-1}$, $H' = U_n$, $H = U_{n-1}$, $n \geq 3$. The H'-module $\mathfrak{m}'$ is irreducible and consists of all matrices of the form

$$\begin{pmatrix} X & Y \\ Y & -X \end{pmatrix}, \quad X, Y \in M_n(\mathbb{R}), \quad X^\top = -X, \quad Y^\top = -Y.$$

Clearly, $\mathfrak{m}' \cap \mathfrak{g} \neq 0$, and thus Lemma 2 is applicable.

5) $G' = SO_{4n}$, $G = Sp_n$, $H' = SO_{4n-1}$, $H = Sp_{n-1}$, $n \geq 2$. As was shown before (case 1)), the Sp_n-invariant natural Riemannian metric on S^{4n-1} is not preserved by the action of the subgroup $SU_{2n} \subset SO_{4n}$. Hence, it is not SO_{4n}-invariant.

6) $G' = SO_{4n}$, $G = SO_{4n-1}$, $H' = Sp_n Sp_1$, $H = Sp_{n-1} Sp_1$, $n \geq 2$. The H'-module $\mathfrak{m}'$ is irreducible (this case is missing in [11]). One verifies that $E_{1,n+1} - E_{n+1,1} - E_{3n,n+1} + E_{3n+1,3n} \in \mathfrak{m}' \cap \mathfrak{g}$, and thus Lemma 2 is applicable.

7) $G' = SO_8$, $G = \mathrm{Spin}_7$, $H' = SO_7$, $H = G_2$. In this case $G/H = S^7$ is isotropy irreducible and, hence, the natural Riemannian metric is the standard metric on S^7. Thus, $I(M) = O_8$.

8) $G' = SO_7$, $G = G_2$, $H' = SO_6$, $H = SU_3$. In this case $G/H = S^6$ also is isotropy irreducible and, hence, $I(M) = O_7$.

9) $G' = SO_7$, $G = G_2$, $H' = SO_5 \times SO_2$, $H = SU_2 U_1$. The H'-module $\mathfrak{m}'$ is irreducible. From the description of the subalgebra $G_2 \subset \mathfrak{so}_7$ given e.g. in Postnikov [9], Lecture 14, it is easy to conclude that $\mathfrak{m}' \cap \mathfrak{g} \neq 0$. Thus, Lemma 2 is applicable.

10) $G' = \mathrm{Spin}_7$, $G = SU_4$ or Sp_2, $H' = G_2$, $H = SU_3$ or Sp_1. here $M = S^7$. As was shown before (case 1), the natural Sp_2-invariant Riemannian metric on S^7 is not SU_4-invariant and, hence, not Spin_7-invariant. Now suppose that the natural SU_4-invariant Riemannian metric is Spin_7-invariant. Since the homogeneous space Spin_7 / G_2 is isotropy irreducible, this metric would coincide with the standard metric on S^7. But this is false, as we have seen before (case 3)).

11) $G' = SO_{16}$, $G = \mathrm{Spin}_9$, $H' = SO_{15}$, $H \simeq \mathrm{Spin}_7$. The H'-module $\mathfrak{m}'$ is irreducible and is described in 3) ($n = 8$). The group Spin_9 is contained, as usually, in the Clifford algebra $C(\mathbb{R}^9)$ associated with the negative definite quadratic form $-\sum_{i=1}^{9} y_i^2$, where y_i are coordinates in the standard basis $v_1, \ldots, v_9$ of $\mathbb{R}^9$, and is imbedded in SO_{16} with the help of the spinor representation μ. The tangent algebra $\mathfrak{spin}_9$ of Spin_9 is spanned by the elements

$v_i v_j$ $(i < j)$ of $C(\mathbb{R}^9)$. The group Spin_8 may be regarded as a subgroup of Spin_9 contained in $C(\mathbb{R}^8)$ where $\mathbb{R}^8 = \langle v_1, \dots, v_8 \rangle$. The restriction $\mu| \, \mathrm{Spin}_8$ is the spinor representation of Spin_8; it splits into two irreducible components λ and ρ of dimension 8 (the semi-spinor representations). We may assume that the corresponding invariant subspaces of $\mathbb{R}^{16} = \langle e_1, \dots, e_{16} \rangle$ are $\langle e_1, \dots, e_8 \rangle$ and $\langle e_9, \dots, e_{16} \rangle$, H' being the stabilizer of e_1. Then H is contained in Spin_8 and coincides with the stabilizer $\lambda^{-1}(SO_7)$ of e_1.

Take $x = v_1 v_9 \in \mathfrak{spin}_9$. Clearly, $x \in \mathfrak{spin}_8^\perp$ and, hence, $x \in \mathfrak{m}$. One verifies that

$$d\mu(x) = E_{19} - E_{91} + \sum_{k=2}^{8} (-E_{k,k+8} + E_{k+8,k}).$$

Therefore, $(\pi_1(\mu(x)), \pi_1(\mu(x)))' = 16$, $(\pi_1(\mu(x)), \pi_1(\mu(x))) = 2$. Now, for $y = v_1 v_2 + v_3 v_4 + v_5 v_6 - v_7 v_8 \in \mathfrak{spin}_8$ we have $d\lambda(y) = 4(-E_{01} + E_{10})$, $d\mu(y) = 4(-E_{01} + E_{10} + E_{09} - E_{90})$. Hence, $y \in \mathfrak{m}$, $(\pi_1(d\mu(y)), \pi_1(d\mu(y)))' = 64$, $(\pi_1(d\mu(y)), \pi_1(d\mu(y)))' = 32$. One sees that the condition a) of Lemma 2 is not satisfied.

12) $G' = SO_{16}$, $G = SO_{15}$, $H' = \mathrm{Spin}_9$, $H \simeq \mathrm{Spin}_7$. The H'-module $\mathfrak{m}'$ is irreducible (this case is missing in [5] and in the resulting list of [11]). By Lemma 2, it is sufficient to prove that $\mathfrak{m}' \cap \mathfrak{g} \neq 0$.

We have a decomposition $\mathfrak{so}_{16} = (\mathfrak{so}_8 \times \mathfrak{so}_8) \oplus \mathfrak{p}$ where $\mathfrak{so}_8 \times \mathfrak{so}_8$ consists of operators preserving the direct decomposition of $\mathbb{R}^{16}$ defined in 11), and $\mathfrak{p}$ of operators transposing the summands. As we have seen $d\mu(\mathfrak{spin}_8) \subset \mathfrak{so}_8 \times \mathfrak{so}_8$. Clearly, $\mathfrak{spin}_9 = \mathfrak{spin}_8 \oplus \mathfrak{q}$ where $\mathfrak{q} = \langle v_i v_9 : i = 1, \dots, 8 \rangle$. One verifies that $d\mu(\mathfrak{q}) \subset \mathfrak{p}$. By dimension argument, there exists a non-zero $x \in \mathfrak{p} \cap \mathfrak{so}_{15}$ which is orthogonal to $d\mu(\mathfrak{q})$. Clearly, $x \in \mathfrak{m}'$.

II. Suppose that G' is not simple. Consider for G' the possibilities i) and ii) of Theorem 3.

In the case i) $G' = GC$ (locally direct product) where $C \subset (\mathrm{Aut}_G M)^0$. If $C \not\subset G$, then G is normal in $I(M)$ which gives the contradiction. Thus, $C \simeq G$ and $H = \{e\}$, i.e., $M = G$ with the action ℓ and γ_0 is the bi-invariant metric on G. Using Lemma 1, we see that $I(M) = \mathrm{Sim}_{G'} M$, and Corollary 2 of Proposition 1 applies.

In the case ii) G' is an enlargement of G of type II or III. Suppose it to be an enlargement of type II associated with a certain non-trivial factorization (G, A', A''), $A' \cap A'' = H$. As Theorem 2 shows, there exist simple factors $A_0' \subset A'$, $A_0'' \subset A''$ such that $G = A_0' A_0''$ (in fact, at least one of the subgroups A', A'' is simple). Then, by Lemma 3, the natural Riemannian metric on $M = G/H$ induces the A_0''- and A_0'-invariant natural Riemannian metrics γ' on $M' = G/A_0'$ and γ'' on $M'' = G/A_0''$ respectively. By the same Lemma, γ' and γ'' are G-invariant. Since the factorization is non-trivial, the part I of the proof implies that our factorization is equivalent to $(SO_8, SO_7, \text{Spin}_7)$. This gives us the case c) of the theorem. Clearly, the natural metric on $\text{Spin}_8 / G_2 = S^7 \times S^7$ is isometric to the product of two standard metrics on S^7.

The group G' cannot be an enlargement of type III. In fact, in this case it would contain an enlargement $G^\sim$ of type II. Hence, $M = S^7 \times S^7$ and $G^\sim = SO_8 \times SO_8$, but the action of SO_8 on S^7 admits no proper connected compact enlargements. ■

6. Some remarks concerning a conjecture of Wu Yi Hsiang

Let $M = G/H$ be a homogeneous space of a connected simple compact Lie group G. Then the number $\dim I(M, \gamma)$ depends upon the Riemannian metric γ chosen on M. In Hsiang [3] it was conjectured that the natural G-invariant metric γ_0 on M is the "most symmetric" one among all Riemannian metrics γ, i.e.,

$$\dim I(M, \gamma_0) > \dim I(M, \gamma)$$

for any Riemannian metric γ non-isometric with $c\gamma_0$, $c > 0$. This conjecture was disproved by Lukesh [4] constructing uncountably many homothetically distinct Riemannian metrics on $V_{n,2} = SO_n/SO_{n-2}$ whose connected group of isometries is $SO_n \times SO_2 = (\text{Sim}_{SO_n} V_{n,2})^0$. He proposed the following revised form of the conjecture:

$$\dim I(M, \gamma_0) \geq \dim I(M, \gamma)$$

for any Riemannian metric γ on M. The following two examples show that this weakened conjecture is false, too.

Example 1. Let M be $Sp_n/Sp_{n-1} = S^{4n-1}$, $n > 1$, with the natural Riemannian metric γ_0. By Theorem 6, $I(M, \gamma_0) = Sp_n Sp_1$. Hence,

$\dim I(M, \gamma_o) < \dim O_{4n}$ where O_{4n} is the group of isometries of the standard metric on S^{4n-1}.

Example 2. One can construct a Riemannian metric γ on S^{4n-1} such that $I(S^{4n-1}, \gamma) = O_{4n-1}$. Since $\dim O_{4n-1} > \dim I(M, \gamma_0)$ for $n \geq 2$, where γ_0 is the natural metric from Example 1, the metric γ on Sp_n / Sp_{n-1} contraversing the conjecture may be chosen non-homogeneous.

References

[1] CARTAN E., Sur certaines formes riemanniennes rémarquables des géométries à groupe fondamentale simple, *Ann. Sc. Ec. Norm. Sup.*, **44** (1927), 345–467.

[2] D'ATRI J. E. and ZILLER W., Naturally reductive metrics and Einstein metrics on compact Lie groups, *Mem. Amer. Math. Soc.*, **18**, No. 215 (1979), 1–72.

[3] HSIANG W. Y., In: Mostert P., Ed. Proc. Conf. Transformation groups, New Orleans, 1967. Springer, N.Y., 1968.

[4] LUKESH G. W., Variations of metrics on homogeneous manifolds, *J. Math. Soc. Japan*, **31** (1979), 655–667.

[5] MANTUROV O. V., Homogeneous Riemannian spaces with irreducible rotation groups, *Trudy Semin. vekttenz. anal.*, **13** (1966), 68–146 (Russian).

[6] OCHIAI T. and TAKAHASHI T., The group of isometries of a left invariant Riemannian metric on a Lie group, *Math. Ann.*, **223** (1976), 91–96.

[7] ONISHCHIK A. L., Inclusion relations between transitive compact transformation groups, *Trudy Mosk. mat. Obshch.*, **11** (1962), 199–242 (Russian). English translation in: *Amer. Math. Soc. Transl.*, **50** (1966), 5–58.

[8] ONISHCHIK A. L., Factorizations of reductive Lie groups, *Mat. Sb.*, **80** (1969), 555-599 (Russian). English translation in: *Math. USSR – Sbornik*, **9** (1969), 525–554.

[9] POSTNIKOV M. M., Lie groups and Lie algebras, Nauka, M., 1982 (Russian).

[10] WOLF J. A., Locally symmetric homogeneous spaces, *Comm. Math. Helv.*, **37** (1962), 65–101.

[11] WOLF J. A., The geometry and structure of isotropy irreducible homogeneous spaces, *Acta Math.*, **120** (1968), 59–148.

A. L. Onishchik

Yaroslavl University
150 000 Yaroslavl
USSR

A Quasi-umbilical Hypersurface of a Conformally Recurrent Space[*]

M. PRVANOVIĆ

1. Introduction

Let $(\overline{M}, \overline{g})$ be an $(n+1)$-dimensional $(n \geq 4)$ Riemannian space covered by a system of coordinate neighborhoods (U, y^α). Let (M, g) be a hypersurface of $\overline{M}$, defined in a local coordinate system by means of the system of parametric equations $y^\alpha = y^\alpha(x^i)$, where g is the metric induced on M by $(\overline{M}, \overline{g})$. Here and in the sequal, Greek indices take the values $1, 2, \ldots, n+1$ and Latin indices — the values $1, 2, \ldots, n$. Let N^α be a local unit normal to (M, g). Then

$$(1.1) \qquad g_{ij} = \overline{g}_{\alpha\beta} B_i^\alpha B_j^\beta,$$

$$(1.2) \qquad \overline{g}_{\alpha\beta} N^\alpha B_j^\beta = 0, \qquad \overline{g}_{\alpha\beta} N^\alpha N^\beta = \varepsilon, \quad \varepsilon = \pm 1,$$

and

$$(1.3) \qquad B_i^\alpha B_j^\beta g^{ij} = \overline{g}^{\alpha\beta} - \varepsilon N^\alpha N^\beta,$$

---

[*] This paper is in final form and no version of it will be submitted for publication elsewhere.

where $B_i^\alpha = \frac{\partial y^\alpha}{\partial \chi^i}$.

We denote by $\overline{R}_{\alpha\beta\gamma\delta}$, $\overline{R}_{\alpha\beta}$ and $\overline{R}$ the curvature tensor, the Ricci tensor and the scalar curvature of $(\overline{M},\overline{g})$, and by R_{ijkl}, R_{ij} and R — the corresponding objects of (M,g), while h_{ij} are the components of the second fundamental tensor. Then the Gauss and Codazzi equations for the (M,g) can be written in the forms respectively (c.f. [1], p. 149)

$$\overline{R}_{\alpha\beta\gamma\delta}\, B_i^\alpha B_j^\beta B_k^\gamma B_l^\delta = R_{ijkl} - \varepsilon(h_{il}h_{jk} - h_{ik}h_{jl}),$$
$$\overline{R}_{\alpha\beta\gamma\delta}\, N^\alpha B_j^\beta B_k^\gamma B_l^\delta = \nabla_l h_{jk} - \nabla_k h_{jl},$$

where ∇ is the operator of the van der Wearden–Bartolotti covariant derivative. Also (c.f. [1], pp. 147–148)

$$(1.4) \qquad \nabla_r B_j^\beta = \varepsilon h_{rj} N^\beta, \qquad \nabla_r N^\alpha = -h_{ra} g^{at} B_t^\alpha.$$

The hypersurface (M,g) is said to be quasi-umbilically immersed in $(\overline{M},\overline{g})$ if [2]:

$$(1.5) \qquad h_{ij} = \alpha g_{ij} + \beta v_i v_j,$$

where α and β are some functions and v is a vector field. If $\beta = 0$, (M,g) is a umbilical hypersurface of $(\overline{M},\overline{g})$.

Totally umbilical subspace of Riemannian spaces with certain conditions imposed on the Weyl conformal curvature tensor of an ambient space were investigated by many authors. For example, Olszak proved [3] that totally umbilical subspace of a conformally recurrent space is conformally recurrent and found [4] the relation $HC_{ijkl} = 0$, where $H = \overline{g}_{\alpha\beta}H^\alpha H^\beta$, H^α is the mean curvature vector field and C_{ijkl} is the Weyl conformal curvature tensor of the subspace. The above results were generalized in [5] in the sense that the ambient space is conformally birecurrent.

In the present paper we also generalize Olszak's results investigating a quasi-umbilical hypersurface of a conformally recurrent space.

2. Some preliminary results

For a quasi-umbilical hypersurface, the Gauss and Codazzi equations take the forms

(2.1)
$$\overline{R}_{\alpha\beta\gamma\delta} B_i^\alpha B_j^\beta B_k^\gamma B_l^\delta = R_{ijkl} - \varepsilon\alpha^2 (g_{il}g_{jk} - g_{ik}g_{jl}) -$$
$$- \varepsilon\alpha\beta(g_{il}v_j v_k + g_{jk}v_i v_l - g_{ik}v_j v_l - g_{jl}v_i v_l),$$

(2.2)
$$\overline{R}_{\alpha\beta\gamma\delta} N^\alpha B_j^\beta B_k^\gamma B_l^\delta = \alpha_l g_{jk} - \alpha_k g_{jl} + (\beta_l v_k - \beta_k v_l)v_j +$$
$$+ \beta[v_k \nabla_l v_j - v_l \nabla_k v_j + v_j(\nabla_l v_k - \nabla_k v_l)],$$

where $\alpha_i = \frac{\partial\alpha}{\partial x^i}$, $\beta_i = \frac{\partial\beta}{\partial x^i}$, while the equalities (1.4) become

(2.3)
$$\nabla_r B_j^\alpha = \varepsilon(\alpha g_{rj} + \beta v_r v_j)N^\beta,$$

(2.4)
$$\nabla_r N^\alpha = -\alpha B_r^\alpha - \beta v_r v^t B_t^\alpha.$$

Transvecting (2.1) with g^{il} and using (1.3), we get

(2.5)
$$\overline{R}_{\beta\gamma} B_j^\beta B_k^\gamma = \varepsilon\overline{R}_{\alpha\beta\gamma\delta} N^\alpha B_j^\beta B_k^\gamma N^\delta + R_{jk} -$$
$$- \varepsilon[\alpha^2(n-1) + \alpha\beta v_a v^a]g_{jk} - \varepsilon(n-2)\alpha\beta v_j v_k.$$

Transvecting (2.5) with g^{jk} and using (1.3), we find

(2.6)
$$\overline{R}_{\beta\gamma} N^\beta N^\gamma = \frac{1}{2}\varepsilon(\overline{R} - R) + \frac{n(n-1)}{2}\alpha^2 + (n-1)\alpha\beta v_a v^a.$$

At last, transvecting (2.2) with g^{jk}, we obtain

(2.7)
$$\overline{R}_{\alpha\delta} N^\alpha B_l^\delta = (n-1)\alpha_l + \beta_l v_a v^a + \beta(2v^a \nabla_l v_a - v^a \nabla_a v_l) -$$
$$- v_l(\beta_a v^a + \beta\nabla_a v^a).$$

Now, let us consider the conformal curvature tensor of the ambient space

$$\overline{C}_{\alpha\beta\gamma\delta} = \overline{R}_{\alpha\beta\gamma\delta} - \frac{1}{n-1}(\overline{R}_{\alpha\delta}\overline{g}_{\beta\gamma} - \overline{R}_{\alpha\gamma}\overline{g}_{\beta\delta} + \overline{R}_{\beta\gamma}\overline{g}_{\alpha\delta} - \overline{R}_{\beta\delta}\overline{g}_{\alpha\gamma}) +$$
$$+ \frac{\overline{R}}{n(n-1)}(\overline{g}_{\alpha\delta}\overline{g}_{\beta\gamma} - \overline{g}_{\alpha\gamma}\overline{g}_{\beta\delta}).$$

In view of (1.1) and (1.2), we have

$$\overline{C}_{\alpha\beta\gamma\delta} N^\alpha B_j^\beta B_k^\gamma B_l^\delta = \overline{R}_{\alpha\beta\gamma\delta} N^\alpha B_j^\beta B_k^\gamma B_l^\delta -$$
$$- \frac{1}{n-1}(g_{jk}\overline{R}_{\alpha\delta} N^\alpha B_l^\delta - g_{jl}\overline{R}_{\alpha\gamma} N^\alpha B_k^\gamma),$$

from which, using (2.2) and (2.7), we get

(2.8)
$$\overline{C}_{\alpha\beta\gamma\delta} N^\alpha B_j^\beta B_k^\gamma B_l^\delta =$$
$$= v_j(\beta_l v_k - \beta_k v_l) + \beta[v_k \nabla_l v_j - v_l \nabla_k v_j + v_j(\nabla_l v_k - \nabla_k v_l)]-$$
$$- \frac{1}{n-1} g_{jk}[\beta_l v_a v^a + \beta(2v^a \nabla_l v_a - v^a \nabla_a v_l) - v_l(\beta_a v^a + \beta \nabla_a v^a)]+$$
$$+ \frac{1}{n-1} g_{jl}[\beta_k v_a v^a + \beta(2v^a \nabla_k v_a - v^a \nabla_a v_k) - v_k(\beta_a v^a + \beta \nabla_a v^a)].$$

Also

$$\overline{C}_{\alpha\beta\gamma\delta} N^\alpha B_j^\beta B_k^\gamma N^\delta = \overline{R}_{\alpha\beta\gamma\delta} N^\alpha B_j^\beta B_k^\gamma N^\delta - \frac{\varepsilon}{n-1}\overline{R}_{\beta\gamma} B_j^\beta B_k^\gamma -$$
$$\frac{1}{n-1} g_{jk}\overline{R}_{\alpha\delta} N^\alpha N^\delta + \frac{R}{n(n-1)}\varepsilon g_{jk}.$$

In view of (2.6), this can be rewritten in the form

$$\overline{C}_{\alpha\beta\gamma\delta} N^\alpha B_j^\beta B_k^\gamma N^\delta = \overline{R}_{\alpha\beta\gamma\delta} N^\alpha B_j^\beta B_k^\gamma N^\delta - \frac{\varepsilon}{n-1}\overline{R}_{\beta\gamma} B_j^\beta B_k^\gamma +$$
$$g_{jk}\left[-\frac{n-2}{2n(n-1)}\varepsilon\overline{R} + \frac{\varepsilon}{2(n-1)}R - \frac{n}{2}\alpha^2 - \alpha\beta v_a v^a\right].$$

Finally, substituting $\overline{R}_{\alpha\beta\gamma\delta} N^\alpha B_j^\beta B_k^\gamma N^\delta$ from (2.5), we get

$$\overline{C}_{\alpha\beta\gamma\delta} N^\alpha B_j^\beta B_k^\gamma N^\delta = \varepsilon\frac{n-2}{n-1}\overline{R}_{\beta\gamma} B_j^\beta B_k^\gamma - \varepsilon R_{jk} + (n-2)\alpha\beta v_j v_k +$$
$$+ g_{jk}\left[-\frac{n-2}{2n(n-1)}\varepsilon\overline{R} + \frac{\varepsilon}{2(n-1)}R + \frac{n-2}{2}\alpha^2\right].$$

Let us put

(2.9)
$$\overline{C}_{\alpha\beta\gamma\delta} N^\alpha B_j^\beta B_k^\gamma N^\delta = \varepsilon Q_{jk}.$$

Then the preceding relation can be rewritten as follows

$$
\frac{1}{n-1}\overline{R}_{\alpha\beta}B_j^\beta B_k^\gamma = \frac{1}{n-2}Q_{jk} + \frac{1}{n-2}R_{jk} - \varepsilon\alpha\beta v_j v_k +
$$

(2.10)

$$
+ g_{jk}\left[\frac{1}{2n(n-1)}\overline{R} - \frac{1}{2(n-1)(n-2)}R - \frac{\varepsilon}{2}\alpha^2\right].
$$

Now, we can compute

$$
\overline{C}_{\alpha\beta\gamma\delta}B_i^\alpha B_j^\beta B_k^\gamma B_l^\delta = \overline{R}_{\alpha\beta\gamma\delta}B_i^\alpha B_j^\beta B_k^\gamma B_l^\delta -
$$

$$
-\frac{1}{n-1}(\overline{R}_{\alpha\delta}B_i^\alpha B_l^\delta g_{jk} - \overline{R}_{\alpha\gamma}B_i^\alpha B_k^\gamma g_{jl} + \overline{R}_{\beta\gamma}B_j^\beta B_k^\gamma g_{il} - \overline{R}_{\beta\delta}B_j^\beta B_l^\delta g_{ik}) +
$$

$$
+ \frac{\overline{R}}{n(n-1)}(g_{il}g_{jk} - g_{ik}g_{jl}).
$$

Using (1.1), (2.1) and (2.10), we have

$$
\overline{C}_{\alpha\beta\gamma\delta}B_i^\alpha B_j^\beta B_k^\gamma B_l^\delta = R_{ijkl} - \frac{1}{n-2}(g_{jk}R_{il} - g_{jl}R_{ik} +
$$

$$
+ g_{il}R_{jk} - g_{ik}R_{jl}) + \frac{R}{(n-1)(n-2)}(g_{jk}g_{il} - g_{jl}g_{ik}) -
$$

$$
-\frac{1}{n-2}(g_{jk}Q_{il} - g_{jl}Q_{ik} + g_{il}Q_{jk} - g_{ik}Q_{jl}),
$$

that is

$$
(2.11)\quad \overline{C}_{\alpha\beta\gamma\delta}B_i^\alpha B_j^\beta B_k^\gamma B_l^\delta = C_{ijkl} - \frac{1}{n-2}(q_{jk}Q_{il} - g_{jl}Q_{ik} + g_{il}Q_{jk} - q_{ik}Q_{jl}),
$$

where

$$
C_{ijkl} = R_{ijkl} - \frac{1}{n-2}(g_{jk}R_{il} - g_{jl}R_{ik} + g_{il}R_{jk} - g_{ik}R_{jl}) +
$$

$$
+ \frac{R}{(n-1)(n-2)}(g_{jk}g_{il} - g_{ik}g_{jl})
$$

is the conformal curvature tensor of the hypersurface (M, g).

It is known ([2], p. 154, Corollary (2.1)) that

The conformally flat hypersurface of a conformally flat space is quasi-umbilical.

Now, we can prove the converse theorem [6]:

Theorem. *Quasi-umbilical hypersurface of a conformally flat space is conformally flat.*

Proof. First, we note that the conformal curvature tensor satisfies the conditions

(2.12)
$$C_{ijkl} = -C_{jikl} = -C_{ijlk} = C_{klij},$$
$$C^a{}_{akl} = C^a{}_{kal} = C^a{}_{kla} = 0.$$

Therefore, transvecting (2.9) with g^{jk} and using (1.3), we have

(2.13) $$Q_{jk}g^{jk} = \varepsilon \overline{C}_{\alpha\beta\gamma\delta} N^\alpha N^\delta (\overline{g}^{\beta\gamma} - \varepsilon N^\beta N^\gamma) = 0.$$

Now, let us suppose that the ambient space is conformally flat. Then (2.11) reduces to

(2.14) $$C_{ijkl} = \frac{1}{n-2}(g_{jk}Q_{il} - g_{jl}Q_{ik} + g_{il}Q_{jk} - g_{ik}Q_{jl}).$$

Transvecting (2.14) with g^{il}, we find $Q_{jk} = 0$, and (2.14) reduces to $C_{ijkl} = 0$. This completes the proof of the theorem. ∎

Specially, if the ambient space is an $(n+1)$-dimensional space form, it can be proved ([2], p. 165, equ. (4.6)) that

(2.15) $$\nabla_j v_i = (v^k \beta_k)v_j v_i + \beta(\nabla_k v_j)v^k v_i - \beta(\nabla_k v_i)v^k v_j - \beta_j v_i.$$

If $v_a v^a \neq 0$, we may assume, without loss of generality, that $v_a v^a = 1$. Then we have

$$(\nabla_k v_i)v^k = \frac{\beta}{1+\beta}(\nabla_k v_j)v^k v^j v_i.$$

Substituting this into (2.15), we find

$$\nabla_j v_i = (v^k \beta_k)v_j v_i - \beta_j v_i,$$

which can be rewritten in the form

$$\nabla_j v_i = t_j v_i, \quad \text{where } t_j = v^k \beta_k v_j - \beta_j.$$

3. A quasi-umbilical hypersurface satisfying some additional conditions

In the following, we suppose that v in (1.5) is non-null torse-forming vector field, i.e. we suppose $v_a v^a \neq 0$ and

(3.1) $$\nabla_j v_i = t_j v_i$$

where t is a vector field. Then (2.8) can be written in the form

(3.2) $$\overline{C}_{\alpha\beta\gamma\delta} = N^\alpha B_j^\beta B_k^\gamma B_l^\delta = T_l v_j v_k - T_k v_j v_l - \frac{1}{n-1} g_{jk} M_l + \frac{1}{n-1} g_{jl} M_k,$$

where

(3.3) $$T_l = \beta_l + 2\beta t_l, \qquad M_l = v_a v^a T_l - T_a v^a v_l.$$

Applying the operator ∇_r to (2.11), using (2.3) and taking into account the properties (2.12) of the conformal curvature tensor, we find

$$\nabla_\rho \overline{C}_{\alpha\beta\gamma\delta} B_i^\alpha B_j^\beta B_k^\gamma B_l^\delta B_r^\rho +$$
$$+ \varepsilon(\alpha g_{ri} + \beta v_r v_i) \overline{C}_{\alpha\beta\gamma\delta} N^\alpha B_j^\beta B_k^\gamma B_l^\delta -$$
$$- \varepsilon(\alpha g_{rj} + \beta v_r v_j) \overline{C}_{\beta\alpha\gamma\delta} N^\beta B_i^\alpha B_k^\gamma B_l^\delta +$$
$$+ \varepsilon(\alpha g_{rk} + \beta v_r v_k) \overline{C}_{\gamma\delta\alpha\beta} N^\gamma B_l^\delta B_i^\alpha B_j^\beta -$$
$$- \varepsilon(\alpha g_{rl} + \beta v_r v_l) \overline{C}_{\delta\gamma\alpha\beta} N^\delta B_k^\gamma B_i^\alpha B_j^\beta =$$
$$= \nabla_r C_{ijkl} - \frac{1}{n-2}(g_{jk}\nabla_r Q_{il} - g_{jl}\nabla_r Q_{ik} + g_{il}\nabla_r Q_{jk} - g_{ik}\nabla_r Q_{jl}).$$

Substituting (3.2), we find
(3.4)
$$\nabla_\rho \overline{C}_{\alpha\beta\gamma\delta} B_i^\alpha B_j^\beta B_k^\gamma B_l^\delta B_r^\rho +$$
$$+ \varepsilon(\alpha g_{ri} + \beta v_r v_i)\left(T_l v_j v_k - T_k v_j v_l - \frac{1}{n-1}g_{jk}M_l + \frac{1}{n-1}g_{jl}M_k\right) -$$
$$- \varepsilon(\alpha g_{rj} + \beta v_r v_j)\left(T_l v_i v_k - T_k v_i v_l - \frac{1}{n-1}g_{ik}M_l + \frac{1}{n-1}g_{il}M_k\right) +$$
$$+ \varepsilon(\alpha g_{rk} + \beta v_r v_k)\left(T_j v_l v_i - T_i v_l v_j - \frac{1}{n-1}g_{li}M_j + \frac{1}{n-1}g_{lj}M_i\right) -$$
$$- \varepsilon(\alpha g_{rl} + \beta v_r v_l)\left(T_j v_k v_i - T_i v_k v_j - \frac{1}{n-1}g_{ki}M_j + \frac{1}{n-1}g_{kj}M_i\right) =$$
$$= \nabla_r C_{ijkl} - \frac{1}{n-2}(g_{jk}\nabla_r Q_{il} - g_{jl}\nabla_r Q_{ik} + g_{il}\nabla_r Q_{jk} - g_{ik}\nabla_r Q_{jl}).$$

Now, let us suppose that the ambient space is conformally recurrent, i.e.

(3.5) $$\nabla_\rho \overline{C}_{\alpha\beta\gamma\delta} = a_\rho \overline{C}_{\alpha\beta\gamma\delta}.$$

Then, in view of (2.11),

$$\nabla_\rho \overline{C}_{\alpha\beta\gamma\delta} B_i^\alpha B_j^\beta B_k^\gamma B_l^\delta B_r^\rho =$$

$$a_r \left[C_{ijkl} - \frac{1}{n-2}(g_{jk}Q_{il} - g_{jl}Q_{ik} + g_{il}Q_{jk} - g_{ik}Q_{jl}) \right],$$

where

$$a_r = a_\rho B_r^\rho.$$

Substituting the last relation into (3.4), we find

(3.6)
$$a_r \left[C_{ijkl} - \frac{1}{n-2}(g_{jk}Q_{il} - g_{jl}Q_{ik} + g_{il}Q_{jk} - g_{ik}Q_{jl}) \right] +$$

$$+ \varepsilon(\alpha g_{ri} + \beta v_r v_i)\left(T_l v_j v_k - T_k v_j v_l - \frac{1}{n-1}g_{jk}M_l + \frac{1}{n-1}g_{jl}M_k \right) -$$

$$- \varepsilon(\alpha g_{rj} + \beta v_r v_j)\left(T_l v_i v_k - T_k v_i v_l - \frac{1}{n-1}g_{ik}M_l + \frac{1}{n-1}g_{il}M_k \right) +$$

$$+ \varepsilon(\alpha g_{rk} + \beta v_r v_k)\left(T_j v_l v_i - T_i v_l v_j - \frac{1}{n-1}g_{li}M_j + \frac{1}{n-1}g_{lj}M_i \right) -$$

$$- \varepsilon(\alpha g_{rl} + \beta v_r v_l)\left(T_j v_k v_i - T_i v_k v_j - \frac{1}{n-1}g_{ki}M_j + \frac{1}{n-1}g_{kj}M_i \right) =$$

$$= \nabla_r C_{ijkl} - \frac{1}{n-2}(g_{jk}\nabla_r Q_{il} - g_{jl}\nabla_r Q_{ik} + g_{il}\nabla_r Q_{jk} - g_{ik}\nabla_r Q_{jl}).$$

Transvecting (3.6) with g^{jk} and using (2.12), (2.13) and (3.3), we get

$$\nabla_r Q_{il} = a_r Q_{il} + \frac{n-2}{n-1}\varepsilon\beta(v_l M_i + v_i M_l)v_r +$$

$$+ \varepsilon\alpha \left[(T_i v_l + T_l v_i)v_r - 2T_r v_i v_l + \right.$$

$$\left. + \frac{2}{n-1}g_{li}M_r - \frac{1}{n-1}(g_{rl}M_i + g_{ri}M_l) \right].$$

Substituting this relation into (3.6), we find

(3.7)
$$\nabla_r C_{ijkl} = a_r C_{ijkl} + \frac{4\varepsilon}{(n-1)(n-2)}\alpha M_r(g_{jk}g_{il} - g_{jl}g_{ik}) -$$
$$- \frac{2\varepsilon}{n-2}\alpha T_r(g_{jk}v_i v_l - g_{jl}v_i v_k + g_{il}v_j v_k - g_{ik}v_j v_l) +$$
$$+ \frac{\varepsilon}{n-2}\alpha v_r[g_{jk}(T_i v_l + T_l v_i) - g_{jl}(T_i v_k + T_k v_i) +$$
$$+ g_{il}(T_j v_k + T_k v_j) - g_{ik}(T_j v_l + T_l v_j)] +$$
$$+ \varepsilon\alpha[g_{ri}v_j(T_l v_k - T_k v_l) + g_{rj}v_i(T_k v_l - T_l v_k) +$$
$$+ g_{rk}v_l(T_j v_i - T_i v_j) + g_{rl}v_k(T_i v_j - T_j v_i)] +$$
$$+ \frac{\varepsilon}{n-2}\alpha[(g_{rk}g_{lj} - g_{rl}g_{kj})M_i + (g_{rj}g_{ik} - g_{ri}g_{jk})M_l +$$
$$+ (g_{ri}g_{jl} - g_{rj}g_{il})M_k + (g_{rl}g_{ik} - g_{rk}g_{il})M_j].$$

If $\alpha = 0$, (3.7) reduces to

(3.8)
$$\nabla_r C_{ijkl} = a_r C_{ijkl},$$

i.e. the hypersurface is conformally recurrent or (in the case $a_r = 0$) conformally symmetric.

If $\alpha \neq 0$, the condition (3.8) is satisfied if and only if

$$\frac{4}{n-1}(g_{jk}g_{il} - g_{jl}g_{ik})M_r -$$
$$- 2T_r(g_{jk}v_i v_l - g_{jl}v_i v_k + g_{il}v_j v_k - g_{ik}v_j v_l) +$$
$$+ v_r[g_{jk}(T_i v_l + T_l v_i) - g_{jl}(T_i v_k + T_k v_i) +$$
$$+ g_{il}(T_j v_k + T_k v_j) - g_{ik}(T_j v_l + T_l v_j)] +$$
$$+ (g_{rk}g_{lj} - g_{rl}g_{jk})M_i + (g_{rj}g_{ik} - g_{ri}g_{jk})M_l +$$
$$+ (g_{ri}g_{lj} - g_{rj}g_{il})M_k + (g_{rl}g_{ik} - g_{rk}g_{il})M_j +$$
$$+ (n-2)[g_{ri}v_j(T_l v_k - T_k v_l) + g_{rj}v_i(T_k v_l - T_l v_k) +$$
$$+ g_{rk}v_l(T_j v_i - T_i v_j) + g_{rl}v_k(T_i v_j - T_j v_i)] = 0.$$

Transvecting this relation with g^{ir}, we find

$$(n+1)(n-3)\left[-\frac{1}{n-1}(g_{jk}M_l - g_{jl}M_k) + T_l v_j v_k - T_k v_l v_j\right] = 0,$$

or

$$-\frac{1}{n-1}(g_{jk}M_l - g_{jl}M_k) + T_l v_j v_k - T_k v_l v_j = 0$$

because of $n > 3$.

Transvecting the last relation with v^l and taking into account $M_a v^a = 0$, we get

$$-\frac{n-2}{n-1}v_j M_k = 0,$$

from which

$$T_k v_a v^a - v_k T_a v^a = 0.$$

In view of $v_a v^a \neq 0$, the preceding relation can be rewritten as follows

(3.9) $T_k = f v_k,$

where f is a scalar function.

Conversely, if (3.9) holds good, (3.7) reduces to (3.8). Thus we have:

Theorem. *Let $(\overline{M}, \overline{g})$ be a conformally recurrent Riemannian space with a_ρ as a recurrence vector field. Let (M, g) be its quasi-umbilical hypersurface satisfying (3.1). Then*
   if $\alpha = 0$, (M, g) is conformally recurrent space with the recurrence vector $a_r = a_\rho B_r^\rho$;
   if $\alpha \neq 0$, in order that the (M, g) is a conformally recurrent space with the recurrence vector a_r, it is necessary and sufficient that (3.9) holds good.
■

Let us consider the case $\alpha \neq 0$, $T_k = f v_k$ more deeply. In this case, (3.2) reduces to

(3.10) $\overline{C}_{\alpha\beta\gamma\delta} N^\alpha B_j^\beta B_k^\gamma B_l^\delta = 0.$

Applying the operator ∇_r to (3.10) and using (2.3) and (2.4), we get

$$\nabla_r \overline{C}_{\alpha\beta\gamma\delta} N^\alpha B_j^\beta B_k^\gamma B_l^\delta -$$
$$- \alpha \overline{C}_{\alpha\beta\gamma\delta} B_r^\alpha B_j^\beta B_k^\gamma B_l^\delta - \beta v_r v^t \overline{C}_{\alpha\beta\gamma\delta} B_t^\alpha B_j^\beta B_k^\gamma B_l^\delta -$$
$$- \varepsilon(\alpha g_{rk} + \beta v_r v_k)\overline{C}_{\alpha\beta\delta\gamma} N^\alpha B_j^\beta B_l^\delta N^\gamma +$$
$$+ \varepsilon(\alpha g_{rl} + \beta v_r v_l)\overline{C}_{\alpha\beta\gamma\delta} N^\alpha B_j^\beta B_k^\gamma N^\delta = 0.$$

Substituting (2.9) and (2.11), we find

$$\nabla_\rho \overline{C}_{\alpha\beta\gamma\delta} N^\alpha B_j^\beta B_k^\gamma B_l^\delta B_r^\rho -$$

$$- \alpha \left[C_{rjkl} - \frac{1}{n-2}(g_{jk}Q_{rl} - g_{jl}Q_{rk} + g_{rl}Q_{jk} - g_{rk}Q_{jl}) \right] -$$

$$- \beta v_r v^t \left[C_{tjkl} - \frac{1}{n-2}(g_{jk}Q_{tl} - g_{jl}Q_{tk} + g_{tl}Q_{jk} - g_{tk}Q_{jl}) \right] -$$

$$- (\alpha g_{rk} + \beta v_r v_k)Q_{jl} + (\alpha g_{rl} + \beta v_r v_l)Q_{jk} = 0,$$

which reduces to

$$- \alpha C_{rjkl} + \frac{\alpha}{n-2}(g_{jk}Q_{rl} - g_{jl}Q_{rk} + g_{rl}Q_{jk} - g_{rk}Q_{jl}) -$$

(3.11) $$- \beta v_r v^t C_{tjkl} + \frac{\beta}{n-2}v_r(g_{jk}v^t Q_{tl} - g_{jl}v^t Q_{tk} + v_l Q_{jk} - v_k Q_{jl}) -$$

$$- (\alpha g_{rk} + \beta v_r v^k)Q_{jl} + (\alpha g_{rl} + \beta v_r v_l)Q_{jk} = 0$$

because of (3.5) and (3.10).

Transvecting (3.11) with v^r, we have

$$(\alpha + \beta v_a v^a) \left[\frac{1}{n-2}(g_{jk}v^r Q_{rl} - g_{jl}v^r Q_{rk} + v_l Q_{jk} - v_k Q_{jl}) - \right.$$

$$\left. - v^t C_{tjkl} - v_k Q_{jl} + v_l Q_{jk} \right] = 0.$$

Thus, if $\alpha + \beta v_a v^a \neq 0$, we obtain

$$v^t C_{tjkl} + v_k Q_{jl} - v_l Q_{jk} = \frac{1}{n-2}(g_{jk}v^r Q_{rl} - g_{jl}v^r Q_{rk} + v_l Q_{jk} - v_k Q_{jl}).$$

Substituting this into (3.11), we get

(3.12) $$\alpha \left[-C_{rjkl} + \frac{1}{n-2}(g_{jk}Q_{rl} - g_{jl}Q_{rk} + g_{rl}Q_{jk} - g_{rk}Q_{jl}) - \right.$$

$$\left. - g_{rk}Q_{jl} + g_{rl}Q_{jk} \right] = 0.$$

Transvecting (3.12) with g^{rk} and taking into account (2.12) and (2.13), we obtain

$$\alpha Q_{jl} = 0$$

Therefore, (3.12) reduces to

$$\alpha C_{rjkl} = 0.$$

Thus the following theorem is true.

Theorem. *Let (M,g) be a quasi-umbilical hypersurface of a conformally recurrent space and let it satisfy (3.1), (3.9), $\alpha \neq 0$ and $\alpha + \beta v_a v^a \neq 0$. Then (M,g) is conformally flat.*

References

[1] L. P. EISENHART, Riemannian geometry, Princeton, 1949.

[2] BANG-YEN CHEN, Geometry of submanifolds, Marcel Dekker, New York, 1973.

[3] Z. OLSZAK, On totally umbilical surfaces immersed in Riemannian conformally recurrent and conformally symmetric spaces, *Demonstratio Math.* **8** (1975), 303–311.

[4] Z. OLSZAK, On totally umbilical surfaces in some Riemannian spaces, *Colloqu. Math.* **37** (1977), 105–111.

[5] R. DECZCZ, S. EWERT-KRZEMINIEWSKI and J. POLICHT, On totally umbilical submanifolds of conformally birecurrented manifolds, *Colloqu. Math.* **55** (1988), 79–96.

[6] S. NISHIKAWA and Y. MADEA, Conformally flat hypersurfaces in a conformally flat Riemannian manifold, *Tôhoku Math. Journ.* **26** (1974), 159–168.

Mileva Prvanović

Institute of Mathematics
University of Novi Sad
Novi Sad, Yugoslavia

COLLOQUIA MATHEMATICA SOCIETATIS JÁNOS BOLYAI
56. DIFFERENTIAL GEOMETRY, EGER (HUNGARY), 1989

The Reconstruction Problem at Prequantic Level*

M. PUTA

1. Introduction

Let M be the phase space of a classical mechanical system and G a Lie group of symmetries on M. It is well known that the reconstruction theory gives us a technique for reconstructing the dynamics on M from that on the reduced phase space [3]. More precisely, this is done by using a canonical connection on the principal G_μ-bundle $J^{-1}(\mu) \to M_\mu$, where J is the momentum map of the free action of G on M, G_μ is the coadjoint isotropy group at μ and M_μ is the corresponding reduced phase space.

The goal of this paper is to extend this technique at prequantic level from the geometric quantization point of view. We shall consider only the particular case of the cotangent bundle in view of some technical difficulties which will be pointed out in the last section.

---

* This paper is in final form and no version of it will be submitted for publication elsewhere.

2. Geometric prequantization

Let (M, ω) be a symplectic finite dimensional quantizable manifold, i.e. there exists an open cover $\{U_i \,|\, i \in I\}$ of M together with a collection $\{\theta_i, u_{ij} \,|\, i, j \in I\}$ such that θ_i is a symplectic potential of ω defined on U_i $(d\theta_i = \omega)$, $\theta_j = \theta_i + du_{ji}$ on $U_i \cap U_j$ and $u_{kj} + u_{ji} + u_{ik} = 2\pi n_{ijk}\hbar$ on $U_i \cap U_j \cap U_k$, where $n_{ijk} \in \mathbb{Z}$ and $\hbar$ is the Planck constant divided by 2π. Then there exists a complex line bundle $p : L \to M$ over M called the prequantum bundle which has the $\exp(iu_{ji}/\hbar)$ as gauge transformations, i.e. if $\Psi_i : p^{-1}(U_i) \to U_i \times \mathbb{C}$ is a local trivialization of L, then

$$\Psi_j \cdot \Psi_i^{-1}(m, z) = (m, z \cdot \exp(i \cdot u_{ji}/\hbar));$$

an Hermitian structure $(\cdot, \cdot)$, i.e. a complex inner product on the fibres of L given by: $(l_1, l_2) = z_1 \bar{z}_2$, if $l_1, l_2 \in p^{-1}(m)$ and $\Psi_i(l_j) = (m, z_j)$, $j = 1, 2$; and a connection ∇ whose curvature is $(1/\hbar)\omega$. Moreover the connection ∇ and the Hermitian structure $(\cdot, \cdot)$ are compatible, i.e.

$$X(s, t) = (\nabla_X s, t) + (s, \nabla_X t)$$

for each $X \in \mathfrak{X}_{\mathbb{C}}(M)$ and $s, t \in \Gamma(L)$.

If we take into account the fact that a section $s \in \Gamma(L)$ of L can be identified with a collection of functions $s_i : U_i \to \mathbb{C}$ such that $s_j = \exp(iu_{ji}/\hbar)s_i$ on $U_i \cap U_j$, when we put $\psi_i \cdot s(m) = (m, s_i(m))$, then ∇ is given by

$$\begin{cases} \nabla_X s = t, \\ t_j = X(s_j) - \dfrac{1}{\hbar}\theta_j(X)s_j. \end{cases}$$

In this picture the Hilbert representation space of geometric prequatization is constructed from the smooth sections of the prequantum bundle $p : L \to M$ and for each $f \in C_{\mathbb{R}}^\infty(M)$ the prequantum operator δ_f is given by

$$\delta_f(s) = -i\hbar\nabla_{X_f} s + f \cdot s.$$

We shall finish this section with an alternative description of the space of smooth sections of $p : L \to M$ and of the prequantum operator δ_f which will be very useful in all that follows. Consider the bundle $p : L^* \to M$

over M, where L^* equals L without the zero-section, (equivalently L^* is the principal fibre bundle associated to L with multiplicative structure group, $\mathbb{C}^* = \mathbb{C} \setminus \{0\}$, or equivalently L^* is the $\mathbb{C}^*$ bundle over M with the same gauge transformations as L). Then L^* has local charts $U_i \times \mathbb{C}^*$ and transition functions:

$$(U_i \cap U_j) \times \mathbb{C}^* \subset U_i \times \mathbb{C}^* \to (U_i \cap U_j) \times \mathbb{C}^* \subset U_j \times \mathbb{C}^*$$
$$(m, z) \mapsto (m, z \cdot \exp(iu_{ji}/\hbar)).$$

On L^* we can consider also the connection 1-form α, i.e. the connection 1-form in the principal fibre bundle $p; L^* \to M$ associated with the connection ∇ in the prequantum bundle $p : L \to M$. In general the connection form α takes values in the Lie algebra of the structure group G, here $G = \mathbb{C}^*$, hence α takes complex values. On $U_j \times \mathbb{C}^*$ it is defined by:

$$\alpha_j = \frac{1}{\hbar}\theta_j + i\frac{dz}{z}.$$

For each $f \in C_{\mathbb{R}}^{\infty}(M)$, let V_f be the real vector field on L^* defined by:

$$\begin{cases} p_*(V_f) = X_f \\ \alpha(V_f) = -i(f \circ p), \end{cases}$$

or equivalently V_f is the "lift" of X_f with respect to the connection 1-form α. On a local chart $U_j \times \mathbb{C}^*$, V_f is given by:

$$(V_f)_j = X_f + \frac{1}{\hbar}\theta_j(X_f)(Z - \overline{Z}),$$

where the vector fields Z and $\overline{Z}$ are defined on the local charts $U_i \times \mathbb{C}^*$, by:

$$Z = z\frac{\partial}{\partial z} \quad ; \quad \overline{Z} = \overline{z}\frac{\partial}{\partial \overline{z}},$$

and we note that these definitions are correct, i.e. they coincide on the intersection of two charts.

Let $\mathcal{K}$ be the space of complex valued functions on L^* satisfying the relation:

$$\varphi(zl) = z^{-1}\varphi(l),$$

for each $l \in L^*$ and $z \in \mathbb{C}^*$. Then we can define a map $\lambda : \Gamma(L) \to \mathcal{K}$, $s \mapsto \lambda_z$, by the following process: if $l \in p^{-1}(m) \in L^*$, then there exists a unique element $\lambda_s(l) \in \mathbb{C}$ such that $s(m) = \lambda_s(l) \cdot l \in L$, $(\lambda_s(l)$ is the component of $s(m)$ in the basis $\{l\}$ in $p^{-1}(m))$; this function λ_s obviously satisfies $\lambda_s(zl) = z^{-1}\lambda_s(l)$ because

$$\lambda_s(zl)zl = s(m) = \lambda_s(zl) \cdot z = \lambda_s(l) \cdot l.$$

Furthermore, the map λ is obviously injective, but also surjective because of the inverse construction:

$$s(m) \overset{\text{def}}{=} \lambda_s(l) \cdot l, \qquad m = p(l).$$

Thus $\Gamma(L)$ can be identified with $\mathcal{K}$.

With all of these in mind we have:

Theorem 2.1. ([2]) *The map $f \in C_{\mathbb{R}}^{\infty}(M) \mapsto V_f \in \mathfrak{X}(L^*)$ is an injective one and:*

(i) $\lambda_{\delta_f s} = -i\hbar V_f(\lambda_s)$, *for each $f \in C_{\mathbb{R}}^{\infty}(M)$ and $s \in \Gamma(L)$;*

(ii) $V_{\{f,g\}} = [V_f, V_g]$, *for each $f, g \in C_{\mathbb{R}}^{\infty}(M)$.*

Remark 2.1. The above theorem gives us an elegant description of geometric prequantization in terms of $\mathcal{K}$ and V_f. More precisely the Hilbert representation space can be obtained from $\mathcal{K}$ via its identification with $\Gamma(L)$ and for each $f \in C_{\mathbb{R}}^{\infty}(M)$ the prequantum operator Q_f is given by:

$$Q_h = -i\hbar V_f.$$

3. The reconstruction problem for the cotangent bundle

Let M be a smooth manifold and G a Lie group which acts freely and properly on M by transformations $\emptyset_g : M \to M$ and define the lifted action to the cotangent bundle $(\emptyset_g)_* : T^*M \to T^*M$ by pushing forward one form:

$$(\emptyset_g)_*(\beta)(v) = \beta(T\emptyset_g^{-1}(v)),$$

where $\beta \in T_m^*M$ and $v \in T_{\emptyset_g(m)}M$. The lifted action preserves the canonical 1-form θ on T^*M and the momentum mapping for this lifted action is given by:

$$J(\beta_m)(\xi) = \beta_m(\xi_M(m)), \qquad \beta_m \in T_m^*M,$$

where ξ_M is the infinitesimal generator of the action of $\varnothing$ on M, i.e.

$$\xi_M(m) = \left.\frac{d}{dt}\right|_{t=0} \varnothing_{\exp(t\xi)}(m).$$

Assume that $\mu \in \mathcal{G}^*$ is a regular value of J and that the isotropy group $G_\mu = \{g \in G | \, J(\varnothing_g(m)) = J(m) = \mu\} = \{g \in G | \, \cdot \, Ad^*_{g-1}(J(m)) = \mu\} = \{g \in G | \, Ad^*_{g-1}(\mu) = \mu\}$ acts freely and properly on $J^{-1}(\mu)$. Then the manifold $(T^*M)_\mu \overset{\text{def}}{=} J^{-1}(\mu)/G_\mu$ turns out to be symplectic. More precisely there exists a unique symplectic structure ω_μ on $(T^*M)_\mu$ consistent with the structure ω on T^*M, that is:

$$\pi^*_\mu \omega_\mu = i^*_\mu \omega,$$

where $i_\mu : J^{-1}(\mu) \hookrightarrow T^*M$ is the inclusion and $\pi_\mu : J^{-1}(\mu) \rightarrow (T^*M)_\mu$ is the projection.

Let us assume that $M \rightarrow M/G_\mu$ is a principal G_μ-bundle and $\gamma \in \Omega^1(M, \mathcal{G})$ is a connection 1-form on M. Then it is a straightforward computation to show that $\widetilde{\gamma} = (\pi \circ i)^*(\gamma)$ is a connection 1-form on the principal G_μ-bundle $\pi_\mu : J^{-1}(\mu) \rightarrow (T^*M)_\mu$, induced by the pull-back — see the following diagram:

$$
\begin{array}{ccc}
M & \longrightarrow & M/G_\mu \\
{\scriptstyle \pi}\uparrow & & \\
T^*M & \overset{i_\mu}{\hookleftarrow} \; J^{-1}(\mu) & \overset{\pi_\mu}{\longrightarrow} \; (^*M)_\mu.
\end{array}
$$

Now we have:

Theorem 3.1. ([3]) *Let H be a G-invariant Hamiltonian on T^*M inducing the Hamiltonian H_μ on $(T^*M)_\mu$. Then the flow of the Hamiltonian vector field X_H can be determined from the flow of the reduced Hamiltonian vector field X_{H_μ} in the following way:*

(i) *If $c_\mu(t)$ is an integral curve of X_{H_μ}, $c_\mu(0) = [p_0]$, let $d(t)$ be its horizontal lift in $J^{-1}(\mu)$ with respect to the connection 1-form $\widetilde{\gamma}$, and let $\xi(t) \in \mathcal{G}$ be given by:*

$$\xi(t) \overset{\text{def}}{=} \gamma(\pi(c(t))(FH(d(t))),$$

where FH denotes the fiber-derivate of H.

(ii) *Let $g(t) \in G$ be the unique solution of the equation:*

$$\dot{g}(t) = T_e L_{g(t)}\left(\xi(t)\right),$$

with the initial condition $g(0) = e$.

(iii) *Then $c(t) \overset{\text{def}}{=} \emptyset_{g(t)} d(t)$ is the integral curve of X_H with initial condition $c(0) = p_0$.*

In other words the above theorem tells us that the Hamiltonian vector field X_H is completely determined by the Hamiltonian vector field X_{H_μ} via the connection 1-forms γ and $\widetilde{\gamma}$.

4. The reconstruction problem at prequantic level

Let M be a smooth manifold and G a Lie group which acts freely and properly on M. If $\mu \in \mathcal{G}^*$ is a regular value of the momentum map J corresponding to the lifted action of G on T^*M (see §3), we shall suppose in the sequel that the following conditions are satisfied:

 (i) the isotropy group G_μ acts freely and properly on $J^{-1}(\mu)$;

 (ii) the reduced phase space $((T^*M)_\mu, \omega_\mu)$ is a quantizable one;

(iii) $M \to M/G_\mu$ is a principal G_μ-bundle.

Let $p : L \to T^*M$ [resp. $p_\mu : L_\mu \to (T^*M)_\mu$] be the prequantum bundle on T^*M [resp. $(T^*M)_\mu$] α [resp. α_μ] the corresponding connection 1-forms and Q [resp. Q^μ] the prequantum operators given by geometric quantization (see §2). Then we can prove the main result of this paper:

Theorem 4.1. *Let H be a G-invariant smooth function on T^*M and H_μ its restriction to $(T^*M)_\mu$. Then the prequantum operator Q_H on the extended phase space T^*M can be determined from the prequantum operator Q_{H_μ} on the reduced phase space $(T^*M)_\mu$ via the connection 1-form α^μ, $\widetilde{\gamma}$ and γ.*

Proof. The proof will be performed in three steps:

Step 1. Let $Q^\mu_{H_\mu}$ be the prequantum operator which is canonically associated to the smooth function H_μ. Using the results of §2 it is easy to see that $Q^\mu_{H_\mu}$ is the "lift" of the Hamiltonian vector field $-i\hbar X_{H_\mu}$, with respect to the connection 1-form α^μ, or equivalently:

$$\begin{cases} (p_\mu)_*(Q^\mu_{H_\mu}) = -i\hbar X_{H_\mu}, \\ \alpha^\mu(Q^\mu_{H_\mu}) = \hbar\,(p \circ H_\mu). \end{cases}$$

Step 2. Let $-i\hbar X_H$ be the smooth vector field on T^*M which is canonically associated to $-i\hbar X_{H_\mu}$ via Theorem 3.1.

Step 3. Let Q_H be the "lift" of $-i\hbar X_H$ with respect to the connection 1-form α. Then Q_H is the desired prequantum operator. ∎

Example 4.1. Let us consider the particular case of the two coupled rigid bodies in the plane moving under the influence of a potential depending only on their relative position. In this case the configuration space of the mechanical system is the 2-dimensional torus T^2, the phase space is its cotangent bundle T^*T^2 and the Lie group G is $S^1 = \{x \in \mathbb{C} | |z| = 1\}$. It acts on the torus T^2 in a canonical way:

$$(\theta(\theta_1, \theta_2)) \mapsto (\theta + \theta_1, \theta + \theta_2),$$

and the corresponding lifted action to T^*T^2 has the moment map:

$$J(\alpha, \psi, p_\varphi, p_\psi) = p_\varphi$$

where $\varphi = \frac{\theta_1 + \theta_2}{\sqrt{2}}$, $\psi = \frac{\theta_1 - \theta_2}{\sqrt{2}}$ and p_φ and p_ψ are the corresponding moments. Then it is not hard to see that the reduced phase space $(T^*T^2)_\mu$ is diffeomorphic to T^*S^1 with the canonical symplectic structure $dp_\psi \wedge d\psi$. Therefore $(T*T^2)_\mu$ is a quantizable manifold and also the conditions (i) and (ii) from the beginning of this section are satisfied.

I want to finish my considerations with the observation that in this paper I have considered only the particular case when the phase space of the mechanical system is a cotangent bundle, and this because it is only for these mechanical systems, to my knowledge that there exist sufficient conditions for the reduced phase space to be quantizable. For more details on this problem see [1], [4], [5].

References

[1] M. J. GOTAY, Constraints, Reduction and Quantization, *J. Math. Phys.* **27** (1986), 2051–2066.

[2] B. KOSTANT, Quantization and unitary representation, *Lect. Notes in Math.*, vol. 170 (1970), 87–208.

[3] J. MARSDEN, R. MONTGOMERY and T. RAŢIU, Reduction, symmetry and Berry's phase in mechanics (to appear).

[4] M. PUTA, On the reduced phase space of a cotangent bundle, *Lett. Math. Phys.* **8** (1984), 189–194.

[5] M. PUTA, Geometric quantization of the reduced phase space of the cotangent bundle, Proceedings of the Conference on Differential Geometry and its Application, August 24–30, 1986, Brno, Edited by D. Krupka, A. Svec, J. E. Purkyne, University Brno, Czechoslovakia, 1987.

Mircea Puta

Seminarul de Geometrie–Topologie
University of Timişoara
1900 Timişoara
Romania

On the Fixed Point Set of an Isometry

J. SZENTHE*

Let $M = K/H$ be a homogeneous Riemannian manifold and $\phi : M \to M$ an isometry. Then the components of the fixed point set $F(\phi; M)$ are totally geodesic submanifolds of M ([K] pp. 59–60) and since any totally geodesic submanifold L of a homogeneous Riemannian manifold is itself a homogeneous manifold ([KN] vol. II pp. 59–60), the components of $F(\phi; M)$ are homogeneous manifolds too. The above observation is the starting point for those results which are presented below. In fact, if L is a totally geodesic submanifold of a homogeneous Riemannian manifold $M = K/H$, then in general there is no subgroup $G \subset K$ such that L as a homogeneous manifold is an orbit of G. Actually, if a geodesic γ of M is not the orbit of a subgroup of K, then $L = \gamma$ is an example. Thus the following definition is introduced: Let ϕ be an isometry of a homogeneous Riemannian manifold $M = K/H$, if $G \subset K$ is a subgroup such that every component of $F(\phi; M)$ is an orbit of G then the subgroup is said to be *associated* with the isometry. At first some results are given concerning the existence of the associated subgroups. Assuming that the components of $F(\phi; M)$ are orbits of a subgroup $G \subset K$, the question arises, whether these orbits are distinguished among the other orbits of G. It is shown that the components of $F(\phi; M)$ as orbits of G are isolated in their strata provided that $M = K/H$ is an irreducible Riemannian symmetric space of compact

* Supported by Hungarian Nat. Found. for Sci. Research Grant No. 424(86).

type and ϕ is an involutive isometry commuting with the symmetry of M at $o = H \in K/H$.

1. The construction of a subgroup associated with an isometry

The problem to construct a subgroup $G \subset K$ associated with an isometry ϕ of a homogeneous Riemannian manifold $M = K/H$ is considered subsequently. The construction is based on the assumption that the isometry ϕ is induced by an automorphism ρ of the group K in a sense which will be explained in what follows now.

Let K be a Lie group which is not necessarily connected, and $H \subset K$ a closed subgroup and $M = K/H$ the uniquely defined real analytic manifold whose elements are the left cosets of H and such that both the canonical left action

$$\kappa : K \times M \to M$$

and the canonical projection $\pi : K \to K/H$ are real analytic. Let there be a Riemannian metric on M which is invariant under κ; in other words such that the action κ is isometric. Consider now an isometry $\phi : M \to M$ of the homogeneous Riemannian manifold which leaves the point $o = H \in K/H = M$ fixed. Let there be a Lie group automorphism ρ of K such that $\rho(H) = H$ and the diagram

$$
\begin{array}{ccc}
K & \xrightarrow{\rho} & K \\
\pi \downarrow & & \downarrow \pi \\
M & \xrightarrow{\phi} & M
\end{array}
$$

is commutative; then the isometry ϕ is said to be *induced* by the automorphism ρ. In the special case when K is equal to $I(M)$ the full isometry group of M, then any isometry ϕ leaving the point o fixed is induced by an inner automorphism. In fact, now there is an $f \in K = I(M)$ such that $\phi = \kappa_f$ holds, where $\kappa_f(x) = \kappa(f, x)$, $x \in M$. But then one has

$$faH = faf^{-1}(fH) = (faf^{-1})H, \qquad \kappa(f, aH) = \pi \circ ad(f)a, \qquad aH \in M$$

consequently, $\phi \circ \pi = \kappa_f \circ \pi = \pi \circ ad(f) = \pi \circ \rho$ where $\rho = ad(f)$ holds.

Assume now that the isometry ϕ is induced by an automorphism ρ and consider the subgroup $F(\rho; K) \subset K$; let G be the identity component of $F(\rho; K)$. If in particular K is the full isometry group of M and ρ is the inner automorphism defined by ϕ then G is the identity component of the centralizer of ϕ in K. Consider now again the case of a general K and the fixed point set $F(\phi; M)$ of an isometry ϕ. Then $F(\phi; M)$ is invariant under the canonical action of G on M. In fact, left $aH \in F(\phi; M)$ and $g \in G$, then

$$\phi(gaH) = \phi \circ \pi(ga) = \pi \circ \rho(ga) = \pi(g\rho(a)) = g\pi \circ \rho(a) =$$
$$= g\phi \circ \pi(a) = g\phi(aH) = gaH$$

holds. Consequently, $F(\phi; M)$ is the union of orbits of G. Moreover, the inclusion

$$\pi(F(\rho; K)) \subset F(\phi; M)$$

is valid too; namely, if $a \in F(\rho; K)$ then $\phi \circ \pi(a) = \pi \circ \rho(a) = \pi(a)$ holds.

The fact that the identity component G of $F(\rho; K)$ concerns the problem of associated subgroup, is shown by the following Proposition and by some facts in its proof.

Proposition 1. *Let K be compact and $\phi : M \to M$ an isometry of the Riemannian manifold $M = K/H$ leaving o fixed and induced by an automorphism $\rho : K \to K$. If*

$$\pi(F(\rho; K)) = F(\phi; M)$$

holds then G is associated with the isometry ϕ.

Proof. Consider the decomposition $F(\rho; K) = \cup\{a_iG; i = 0, 1, \ldots\}$ into left cosets. Then $\pi(a_iG)$ is an orbit of G for every $i = 0, 1, \ldots$; in fact, G is an invariant subgroup of $F(\rho; K)$, therefore

$$G\pi(a_i) = Ga_iH = a_iGH = \pi(a_iG)$$

holds. Moreover, if $\pi(a_iG) \cap \pi(a_iG)$ is not empty then $\pi(a_iG) = \pi(a_jG)$ holds; consequently, any component of $F(\phi; M)$ is equal to some $\pi(a_iG)$. Thus the subgroup G is associated with the isometry ϕ. ∎

As the following example shows the crucial assumption of the preceding Proposition 1 is not fulfilled even in simple cases.

Example. Let K be a compact connected semisimple Lie group $T \subset K$ a maximal torus and $M = K/T$. Then the Killing form of the Lie algebra $\mathfrak{k}$ of K induces an invariant Riemannian metric on M. Let now $f \in T$ be a regular element of K, then $F(ad(f); K) = T$. On the other hand let $\phi = \kappa_f$ then

$$F(\phi; M) = \{aT : faT = aT, \ aT \in M\}$$

this means that $aT \in F(\phi; M)$ if and only if $a^{-1}fa \in T$; but this holds if and only if a is in the normalizer $N(T)$ of T, since f is regular. Consequently,

$$F(\phi; M) = N(T)/T;$$

thus $F(\phi; M)$ is given by the Weyl group of K, therefore it is not a connected subset of M in general. But $\pi(F(\rho; K)) = \pi(T) = o \in M$ is connected.

Another sufficient condition for G to be associated with ϕ is given by the following result.

Theorem 1. Let K be compact $M = K/H$ and ϕ an isometry of M leaving o fixed and induced by an automorphism $\rho : K \to K$. Then the subgroup $G \subset K$ the identity component of $F(\rho; K)$ is associated with ϕ provided that H is connected and ρ is involutive.

Proof. Considering former observations and the preceding Proposition 1 it is sufficient to show that the inclusion

$$F(\phi; M) \subset \pi(F(\rho; K))$$

is valid. In fact, assume that $\phi(gH) = gH$ holds for a $gH \in K/H$. Then $\rho(g) = gh$ holds with some $h \in H$ where $h \neq e$ can be assumed without loss of generality; consequently, $g = \rho^2(g) = \rho(gh) = gh\rho(h)$ is valid and therefore $\rho(h) = h^{-1}$. Since K is compact and H is connected, the exponential map of H is surjective ([Hel] pp 135) consequently, there is an $X \in \mathfrak{h}$ with $h = \exp X$. Put $\widetilde{h} = \exp \frac{1}{2}X$; then $\widetilde{h}^2 = h$ and $\rho(\widetilde{h}) = \widetilde{h}^{-1}$ since the Lie algebra isomorphism $T\rho : \mathfrak{k} \to \mathfrak{k}$ induced by ρ is involutive and therefore $T\rho(X) = -X$ must be valid. Thus $\rho(g\widetilde{h}) = gh\rho(\widetilde{h}) = g\widetilde{h}^2\widetilde{h}^{-1} = g\widetilde{h}$ is valid. But $g\widetilde{h} \in gH$ and $g\widetilde{h} \in F(\rho; K)$; consequently $gH \in \pi(F(\rho; K))$ is valid. Therefore the assumption of the preceding Proposition 1 holds and consequently G is associated with ϕ. ∎

In the special case when the Riemannian homogeneous manifold $M = K/H$ is a Riemannian symmetric space, then any component of $F(\phi; M)$ as a totally geodesic submanifold is obviously the orbit of some subgroup of K, whose Lie algebra can be obtained by making use of the Lie triple system associated with the totally geodesic submanifold ([Hel] pp 224–226). As the following Proposition shows in this case the subgroup G is such that all the components of $F(\phi; M)$ are orbits of G, which means that the subgroup G is associated with ϕ.

Proposition 2. *Let $M = K/H$ be a Riemannian symmetric space of the compact type and $\phi : M \to M$ an involutive isometry of M. If K is the full isometry group $I(M)$ and $f \in K = I(M)$ defined by $\phi = \kappa_f$ then G the identity component of $F(\rho; K)$ is associated with ϕ where $\rho = ad(f)$ is the inner automorphism defined by f.*

Proof. In fact, $F(\phi; M)$ is the union of orbits of G by a preceding observation. Consider a component C of $F(\phi; M)$ and a point $x \in C$. Then $G(x) \subset C$. Assume that $C \neq G(x)$. Then there is a geodesic segment γ starting at x, included in C and orthogonal to $G(x)$. Since C is totally geodesic, there is an isometry τ mapping C onto itself and x onto the other endpoint of γ. Then $\tau^{-1} \circ \phi \circ \tau$ leaves C pontwise fixed. Moreover, $\tau^{-1} \circ \phi \circ \tau$ is involutive. If U is a tubular neighborhood of C, the $\tau^{-1} \circ \phi \circ \tau$ leaves only the points of C fixed in U. Therefore

$$\tau^{-1} \circ \phi \circ \tau \lceil U = \phi \lceil U$$

but then $\tau^{-1} \circ \phi \circ \tau = \phi$. Consequently, $\tau \in G$ must be valid, since γ can be taken arbitrary small. Thus a contradiction is obtained. ∎

In the special case when $M = K/H$ is a Riemannian symmetric space the preceding observation have bearings on some important known results. Namely, consider a Riemannian symmetric space $M = K/H$ and the canonically associated involutive automorphism $\sigma : K \to K$. Let $\phi : M \to M$ be an isometry which is induced by an involutive automorphism $\rho : K \to K$. Consider now the decomposition $\mathfrak{k} = \mathfrak{m} \oplus \mathfrak{h}$ of the Lie algebra $\mathfrak{k}$ of K into the -1 and $+1$ eigenspaces of $T\sigma$. Then $\mathfrak{h}$ is the Lie algebra of H, therefore $T\rho$ maps $\mathfrak{h}$ onto itself and since $T\rho$ is an automorphism it maps $\mathfrak{m}$ onto itself too. Consider also $\mathfrak{k} = \mathfrak{p} \oplus \mathfrak{g}$ the decomposition into the -1 and $+1$ eigenspaces of $T\rho$. Then $\mathfrak{k} = (\mathfrak{m} \cap \mathfrak{p}) \oplus (\mathfrak{m} \cap \mathfrak{g}) \oplus (\mathfrak{h} \cap \mathfrak{p}) \oplus (\mathfrak{h} \cap \mathfrak{g})$ is a decomposition into common eigenspaces of $T\sigma$ and $T\rho$. Therefore the automorphisms

σ and ρ commute. Geometric considerations based on the existence of an involutive automorphism commuting with the canonical automorphism of a symmetric space were made by R. Hermann [Her] in studying group actions with totally geodesic orbits. Later on D. S. P. Leung studied the so-called reflective submanifolds of Riemannian symmetric spaces in a series of papers [L1–3] where one of his constructions was based on the existence of an involutive automorphism commuting with the canonical one. Recently B. Y. Chen and T. Nagano have studied pairs of totally geodesic submanifolds of Riemannian symmetric spaces [CN1-3] where the automorphism ρ appears in a special case.

2. The components of the fixed point set are isolated orbits

Let $\alpha : G \times M \rightarrow M$ be an isometric action of a compact connected Lie group on a complete Riemannian manifold M. The orbits $G(x)$, $G(y)$ of $x, y \in M$ are said to be of the *same type* if the isotropy subgroups G_x, $G_y \subset G$ are conjugate in G. The union of those orbits of G which have the same type as the orbit $G(x)$ is called the *stratum* of $G(x)$. The orbit $G(x)$ is said to be *isolated in its stratum* if there is a neighborhood U of $G(x)$ in M which intersects the stratum of $G(x)$ exactly in $G(x)$. It will be shown subsequently, that if ϕ is an isometry of an irreducible Riemannian symmetric space $M = K/H$ of the compact type induced by an involutive automorphism of K then the components of $F(\phi; M)$ as orbits of the associated subgroup G are isolated in their strata. A preparatory lemma is presented to this end first.

Lemma 1. *Let $\mathfrak{k}$ be a compact semisimple Lie algebra without exceptional factors $\sigma : \mathfrak{k} \rightarrow \mathfrak{k}$ an involutive automorphism such that $(\mathfrak{k}, \sigma)$ is an irreducible orthogonal symmetric Lie algebra and let $\mathfrak{k} = \mathfrak{m} \oplus \mathfrak{h}$ be the decomposition into -1 and $+1$ eigenspaces of σ. Let ρ be another involutive automorphism commuting with σ and $\mathfrak{k} = \mathfrak{p} \oplus \mathfrak{g}$ the corresponding decomposition into -1 and $+1$ eigenspaces of ρ. If $\mathfrak{b} \subset \mathfrak{m} \cap \mathfrak{p}$ is a subspace such that $[\mathfrak{h} \cap \mathfrak{g}, \mathfrak{b}] = \{0\}$ holds then $\mathfrak{b} = \{0\}$ must be valid as well.*

Proof. Consider first the case when the irreducible orthogonal symmetric Lie algebra (k, σ) is of type I. Fix the interior product $\langle,\rangle$ on $\mathfrak{k}$ which is given by the negative of the Killing form. Let now $\mathfrak{b} \subset \mathfrak{m} \cap \mathfrak{p}$ be a subspace such that $[\mathfrak{h} \cap \mathfrak{g}, \mathfrak{b}] = \{0\}$ holds. Assume that $\mathfrak{b} \neq \{0\}$ is valid. Then the inclusion $[\mathfrak{m} \cap \mathfrak{p}, \mathfrak{b}] \subset \mathfrak{h} \cap \mathfrak{g}$ holds. Moreover, the validity of $\langle [\mathfrak{m} \cap \mathfrak{p}, \mathfrak{b}], \mathfrak{h} \cap \mathfrak{g} \rangle = \langle \mathfrak{m} \cap \mathfrak{p}, [\mathfrak{h} \cap \mathfrak{g}, \mathfrak{b}] \rangle = \{0\}$ yields that even $[\mathfrak{m} \cap \mathfrak{p}, \mathfrak{b}] = \{0\}$ is true. But then the equality $[(\mathfrak{m} \cap \mathfrak{p}) \oplus (\mathfrak{h} \cap \mathfrak{g}), \mathfrak{b}] = \{0\}$ is valid as well. Put now $\mathfrak{g}' = (\mathfrak{m} \cap \mathfrak{p}) \oplus (\mathfrak{h} \cap \mathfrak{g})$ and let $\mathfrak{a}$ be the centralizer of $\mathfrak{b}$ in $\mathfrak{k}$. Then $\mathfrak{a}$ is nontrivial since $\{0\} \neq \mathfrak{h} \cap \mathfrak{g} \subset \mathfrak{a}$. By the Jacobi identity the following is obtained

$$[[\mathfrak{a}, \mathfrak{a}], \mathfrak{b}] + [[\mathfrak{b}, \mathfrak{a}], \mathfrak{a}] + [[\mathfrak{a}, \mathfrak{b}], \mathfrak{a}] = \{0\}, \quad [[\mathfrak{a}, \mathfrak{g}'], \mathfrak{b}] + [[\mathfrak{b}, \mathfrak{a}], \mathfrak{g}'] + [[\mathfrak{g}', \mathfrak{b}]\mathfrak{a}] = \{0\}$$

where of course a somewhat loose but condensed notation is applied. Therefore, $[\mathfrak{a}, \mathfrak{a}] \subset \mathfrak{a}$ and $[\mathfrak{a}, \mathfrak{g}'] \subset \mathfrak{a}$ hold. But then $\mathfrak{a} + \mathfrak{g}' \subset \mathfrak{k}$ is a subalgebra and therefore $\mathfrak{g}'$ is not maximal in $\mathfrak{k}$ since $\mathfrak{b} \neq \{0\}$ implies that $\{0\} \neq \mathfrak{a} \neq (\mathfrak{m} \cap \mathfrak{g}) \oplus (\mathfrak{h} \cap \mathfrak{p})$ holds. On the other hand the validity of

$$[\mathfrak{g}', (\mathfrak{m} \cap \mathfrak{g}) \oplus (\mathfrak{h} \cap \mathfrak{p})] \subset (\mathfrak{m} \cap \mathfrak{g}) \oplus (\mathfrak{h} \cap \mathfrak{p})$$

implies that $(\mathfrak{k}, \mathfrak{g}')$ is a symmetric pair. Since $(\mathfrak{k}, \sigma)$ is of type I, the Lie algebra $\mathfrak{k}$ is simple ([Hel] pp. 379–380). But then $(\mathfrak{k}, \mathfrak{g}')$ is irreducible as well and consequently $\mathfrak{g}'$ is a maximal subalgebra in $\mathfrak{k}$ ([Hel] pp. 377–388) in contradiction with the observation above that $\mathfrak{g}'$ is not maximal in $\mathfrak{k}$. Therefore $\mathfrak{b} = \{0\}$ follows.

Consider secondly the case when the orthogonal irreducible symmetric Lie algebra $(\mathfrak{k}, \sigma)$ is of type of II. Then $\mathfrak{k} = \mathfrak{k}' \oplus \mathfrak{k}'$ where $\mathfrak{k}'$ is a compact simple Lie algebra and σ is given as follows $\sigma : (X, Y) \mapsto (Y, X)$ for $(X, Y) \in \mathfrak{k}' \oplus \mathfrak{k}'$ ([Hel pp. 379–380). Let now $\mathfrak{k} = \mathfrak{m} \oplus \mathfrak{h}$ be the decomposition into -1 and $+1$ eigenspaces of σ. Then

$$\mathfrak{m} = \{(X, -X) : X \in \mathfrak{k}'\}, \quad \mathfrak{h} = \{(Z, Z) : Z \in \mathfrak{k}'\} = \Delta(\mathfrak{k}' \oplus \mathfrak{k}').$$

Consider now another involutive automorphism $\rho : \mathfrak{k} \to \mathfrak{k}$ commuting with σ. Then there is an involutive automorphism $\rho' : \mathfrak{k}' \to \mathfrak{k}'$ such that the following two cases can occur:

$$
\begin{aligned}
1st &: \rho(X, Y) = (\rho'(X'), \rho'(Y')), & (X, Y) \in \mathfrak{k}', \oplus \mathfrak{k}' \\
2nd &: \rho(X, Y) = (\rho'(Y'), \rho'(X')), & (X', Y') \in \mathfrak{k}' \oplus \mathfrak{k}'
\end{aligned}
$$

[L2]. Let now $\mathfrak{k} = \mathfrak{p} \oplus \mathfrak{g}$ and $\mathfrak{k}' = \mathfrak{p}' \oplus \mathfrak{g}'$ be the decompositions into -1 and $+1$ eigenspaces of ρ and ρ'. Then in the first case

$$\mathfrak{p} = \{(U', V') : U', V' \in \mathfrak{p}'\}, \quad \mathfrak{g} = \{(X', Y') : X', Y' \in \mathfrak{g}'\}$$

$$\mathfrak{m} \cap \mathfrak{p} = \{(U', -U') : U' \in \mathfrak{p}'\}, \qquad \mathfrak{h} \cap \mathfrak{g} = \{(X', X') : X' \in \mathfrak{g}\}.$$

Assume now that there is a subspace $\mathfrak{b} \subset \mathfrak{m} \cap \mathfrak{p}$ such that $[\mathfrak{h} \cap \mathfrak{g}, \mathfrak{b}] = \{0\}$. Then $\mathfrak{b} = \{U', -U') : U' \in \mathfrak{b}' \subset \mathfrak{p}'\}$ where $\mathfrak{b}'$ is a subspace which centralizes $\mathfrak{g}'$ in $\mathfrak{k}'$. But $\mathfrak{k}'$ is simple and therefore the restricted adjoint representation ad: $\mathfrak{g}' \times \mathfrak{p}' \to \mathfrak{p}'$ is irreducible. Therefore, $\mathfrak{b}' = \{0\}$ must be valid. In the second case

$$\mathfrak{p} = \{(U', -\rho'(U') : U' \in \mathfrak{k}'\}, \qquad \mathfrak{g} = \{(X', \rho(X')) : X' \in \mathfrak{k}'\},$$

$$\mathfrak{m} \cap \mathfrak{p} = \{(U', U') : U' \in \mathfrak{g}'\}, \qquad \mathfrak{h} \cap \mathfrak{g} = \{(X', X') : X' \in \mathfrak{g}'\}.$$

Therefore, if there is a subspace $\mathfrak{b} \subset \mathfrak{m} \cap \mathfrak{p}$ such that $[\mathfrak{h} \cap \mathfrak{g}, \mathfrak{b}] = \{0\}$, then $\mathfrak{b} = \{(U', -U') : U' \in \mathfrak{b}'\}$ where $\mathfrak{b}' \subset \mathfrak{g}'$ is a subspace which centralizes $\mathfrak{g}'$. Since $\mathfrak{k}$ has no exceptional factors, $\mathfrak{k}'$ is not exceptional. Consequently, $\mathfrak{g}'$ is a semisimple according to the list of irreducible symmetric Lie algebras [Hel] pp. 515–518), but then $\mathfrak{b}' = \{0\}$ must be valid. ∎

As the above proof shows, the assumption that $\mathfrak{k}$ has no exceptional factors pertains only the 2nd case of the type II irreducible Lie algebras.

The subsequent arguments need also the fact that the associated subgroup does not depend on the coset space representation of the homogeneous Riemannian manifold under suitable assumptions. In fact, consider a homogeneous Riemannian manifold $M = K/H$ and fix a point $x = aH \in M$. Then the isotropy subgroup of the canonical action $\kappa : K \times M \to M$ at x is $H_x = aHa^{-1}$ and there is a bijection $\iota : K/H \to K/H_x$ given by

$$\iota : gH \mapsto ga^{-1}H_x, \qquad gH \in K/H$$

which is equivariant for κ and the canonical left action κ_x of K on K/H_x; moreover, there is a unique Riemannian metric on K/H_v such that ι is isometric.

Lemma 2. Let H be connected and the isometry $\phi : K/H \to K/H$ leaving $o = H$ fixed be induced by an involutive automorphism $\rho : K \to K$. If ϕ leaves also $x = aH \in K/H$ fixed then the isometry

$$\phi_x = \iota \circ \phi \circ \iota^{-1} : K/H \to K/H_x$$

is also induced by ρ and if $G \subset K$ is associated with ϕ then it is also associated with ϕ_x.

Proof. Considering Theorem 1 there is no loss of generality by assuming $\rho(a) = a$. Thus the map $\phi_x = \iota \circ \phi \circ \iota^{-1}$ is explicitly given as follows

$$gH_x \mapsto gaH \mapsto \phi(gaH) = \pi \circ \rho(ga) = \rho(g)aH \mapsto \rho(g)H_x$$

which means that $\phi_x \circ \pi_x = \pi_x \circ \rho$ with the canonical projection $\pi_x : K \to K/H_x$. The fact that G is associated with ϕ_x too follows now from the observations that ι is equivariant for κ and κ_x and that ι maps $F(\phi; K/H)$ onto $F(\phi_x; K/H_x)$. ∎

The following theorem yields the second main result by showing that the components of the fixed point set as orbits of the associated subgroup are distinguished among the other orbits of this subgroup.

Theorem 2. *Let $M = K/H$ be an irreducible Riemannian symmetric space of compact type where K has no exponential factors, H is connected and $\Phi : M \to M$ an isometry induced by an involutive automorphism $\rho : K \to K$ commuting with the canonical automorphism σ of M. Then the identity component G of $F(\rho; K)$ is associated with ϕ and the components of $F(\phi; M)$ as orbits of G are isolated in their strata.*

Proof. The fact that G is associated with ϕ follows by Theorem 1. Consider an invariant tubular neighborhood U of $G(o)$ in M. If $N_oG(o) \subset T_oM$ is the normal space to $G(o)$ at o then the submanifold

$$S_o = U \cap exp_o(N_oG(o))$$

is a slice of the action $\alpha : G \times M \to M$ of G at o. If there is another orbit of G in U which has the same type as $G(o)$ then it intersects S_o in a point x such that $G_x = G_o$. Moreover, if $x = exp_o(v)$ with $v \in N_oG(o)$ then $v = T_o\alpha_g(v)$ holds for $g \in G_o$ since exp_o is equivariant for the actions

$$T_o\alpha_g : T_oM \to T_oM, \qquad \alpha_g : M \to M, \qquad g \in G_0.$$

Therefore, $G(x)$ has the same type as $G(o)$ if and only if $T_o\alpha_g(v) = v$ for $g \in G_o$.

Let now $\mathfrak{m}, \mathfrak{h} \subset \mathfrak{k}$ and $\mathfrak{p}, \mathfrak{g} \subset \mathfrak{k}$ be the -1 and $+1$ eigenspaces of $T\sigma$ and $T\rho$. Since σ and ρ commute the decomposition $\mathfrak{k} = (\mathfrak{m} \cap \mathfrak{p}) \oplus (\mathfrak{m} \cap \mathfrak{g}) \oplus (\mathfrak{h} \cap \mathfrak{p}) \oplus (\mathfrak{h} \cap \mathfrak{g})$ holds and the canonical identification $T_oM = \mathfrak{m}$ yields the identifications $T_oG(o) = \mathfrak{m} \cap \mathfrak{g}$, $N_oG(o) = \mathfrak{m} \cap \mathfrak{p}$. Moreover, the above identifications yield

$$T_o \alpha_g(v) = Ad(g)X, \qquad v = X \in T_oM, g \in G_o.$$

Therefore, $x = exp_o(v)$ has a G orbit of the same type as $G(o)$ if and only if

$$[Z, X] = 0, \qquad Z \in \mathfrak{g}_o = \mathfrak{g} \cap \mathfrak{h}$$

is valid where $\mathfrak{g}_o$ is the Lie algebra of G_o and $X = v$. But the set of such $X \in \mathfrak{m} \cap \mathfrak{p}$ is equal to $\{0\}$ by the preceding Lemma 1.

The validity of the assertion of the theorem for the other components of $F(\phi; M)$ follows obviously by taking into account the preceding Lemma 2 as well. ■

References

[CN1] B. Y. CHEN and T. NAGANO, Totally geodesic submanifolds of symmetric spaces I, *Duke Math. J.*, **44** (1977), 745–755.

[CN2] B. Y. CHEN and T. NAGANO, Totally geodesic submanifolds of symmetric spaces II, *Duke Math. J.*, **45** (1978), 405–425.

[CN3] B. Y. CHEN and T. NAGANO, Un invariant geometrique Riemannien, *C. R. Acad. Sci. Paris,* **295** (1982), 389–391.

[Hel] S. HELGASON, Differential Geometry, Lie Groups, and Symmetric Spaces, New York, 1978.

[Her] R. HERMANN, Totally geodesic orbits of isometries, *Nederl. Akad. Wetensch. Proc. Ser A* **65** (1962), 291–298.

[K] S. KOBAYASHI, Transformation Groups in Differential Geometry, Berlin, 1972.

[KN] S. KOBAYASHI and K. NOMIZU, Foundations of Differential Geometry I–II. New York, 1963–69.

[L1] D. S. P. LEUNG, The reflection principle for minimal submanifolds of Riemannian symmetric spaces, *J. Differential Geometry* **8** (1973), 153–160.

[L2] D. S. P. LEUNG, On the classification of reflective submanifolds of Riemannian symmetric spaces, *Indiana Univ. Math. J.* **24** (1974), 327–339.

[L3] D. S. P. LEUNG, Reflective submanifolds III, *J. Differential Geometry* **14** (1979), 167–177.

J. Szenthe

Department of Geometry
Eötvös University
Budapest, Rákóczi út 5.
H-1088, Hungary

On Finsler Connections*

J. SZILASI

1. In this paper we shall discuss some new basic ideas concerning Finsler connections. Our approach uses two key notions: *horizontal structure* and (general) *pseudoconnection*. We hope that this new approach will result in a real conceptual and technical simplification of M. Matsumoto's well-known theory [9]. Also, it may be helpful in better understanding the "pure essence" of Finsler connections.

Our ideas also have natural relations with R. Miron's recent theory [10]. Miron's theory is more general than ours in the sense that it works on the whole tangent bundle of the total space of a vector bundle, while we limit ourselves to the vertical subbundle. However, this restriction is perfectly adequate with the demands of classical Finsler geometry.

We shall always be working in the category of the finite dimensional, second countable, smooth manifolds. We are concerned with global properties, but for the sake of a comparison with other (mainly the classical) treatments, we shall systematically give coordinate expressions. As for the notations and basic conventions we refer to the monographs [4], [5] and the paper [14].

---

* This paper is in final form and no version of it will be submitted for publication elsewhere.

2. To begin with, let us consider a fibered space $\xi = (E, \pi, M)$. (Then the only requirement is that $\pi : E \longrightarrow M$ is a surjective submersion.) As in the case of vector bundles, one can construct the vertical subbundle $V\xi$ of the tangent bundle τ_E of the total space E of ξ. Assigning a Whitney-complement $H\xi$ to $V\xi$, we say that a *horizontal structure* H has been given in ξ. This will be written as follows:

$$(2.1) \qquad\qquad H : \quad \tau_E = V\xi \oplus H\xi.$$

H naturally determines the *vertical projector* $v : \quad \tau_E \longrightarrow V\xi$ and the *horizontal projector* $h : \quad \tau_E \longrightarrow H\xi$. The decomposition (2.1) also induces a direct decomposition

$$\mathfrak{X}(E) = \mathfrak{X}_V E \oplus \mathfrak{X}_H E$$

of the module of the vector fields on E onto the submodule of the vertical vector fields and that of the horizontal ones. For the sake of simplicity, the projectors $\mathfrak{X}(E) \longrightarrow \mathfrak{X}_V E$ and $\mathfrak{X}(E) \longrightarrow \mathfrak{X}_H(E)$ also will be denoted by v and h, resp. Every horizontal structure (2.1) determines a unique mapping

$$\ell^H : \quad \mathfrak{X}(M) \longrightarrow \mathfrak{X}_H E, \quad X \longmapsto X^H$$

where X and X^H are π-related. If $A^1_P(M, \tau_E)$ denotes the graded Lie algebra of the *projectable vector 1-forms on M* [8], then ℓ^H can be interpreted as such an element of $A^1_P(M, \tau_E)$ which is projectable onto the identity (1,1) tensor field on M. Conversely, if $\Gamma \in A^1_P(M, \tau_E)$ and it is projectable onto $1_{\mathfrak{X}^1_1(M)}$, then Γ determines a unique horizontal structure in ξ. Such a vector 1-form Γ will be mentioned as a *lifting form* in the sequel.

In particular, let $\xi = (E, \pi, M)$ be a *vector bundle*. We shall denote by Z the Liouville vector field, that is the vertical vector field generated by the homothetic transformations of E.

(2.2) Definition. *A horizontal structure H in the vector bundle ξ is said to be satisfying the homogeneity condition* (HC) *if the Lie derivative $\mathcal{L}_Z v$ of the vertical projector vanishes.*

For other formulations of (HC) and a discussion see [13]. Here we only mention the useful criterion

$$(2.3) \qquad\qquad (HC) \Longleftrightarrow \forall X \in \mathfrak{X}(M): \quad [X^H, Z] = 0.$$

3. Suppose that H is a horizontal structure in the fibered space ξ. Let λ be a vertical vector 1-form on M, symbolically $\lambda \in A^1(M, V\xi)$ (cf. [3]). In this case the pair (H, λ) — or (ℓ^H, λ) — is said to be an *affine structure* in ξ. Obviously, $\ell^H - \lambda \in A^1_P(M, \tau_E)$; this projectable vector 1-form is said to be the *affine lifting form* of the affine structure (H, λ). In the special case $\xi := \tau_M$ there exists a canonical vertical vector 1-form, the vertical lift

$$\ell^v : \quad \mathfrak{X}(M) \longrightarrow \mathfrak{X}_V TM, \quad X \longmapsto X^v.$$

Then an affine structure (H, ℓ^v) is nothing but an immediate (nonlinear) generalization of the classical affine structures (cf. [12] 2.108).

We shall see soon that our general concept of an affine structure proves to be useful in the theory of Finsler connections, too.

4. We briefly recall the definition of the (generalized) pseudoconnections, which was proposed by the present author and Z. Kovács in [14]. — Let ξ and $\tilde{\xi}$ be *vector bundles* over the same manifold M. A pair (∇, A) is called a *pseudoconnection* in ξ with respect to (w.r.t.) $\tilde{\xi}$ if $A : \tilde{\xi} \longrightarrow \tau_M$ is an M-morphism and

$$\nabla : \quad \operatorname{Sec} \tilde{\xi} \times \operatorname{Sec} \xi \longrightarrow \operatorname{Sec} \xi, \quad (\tilde{\sigma}, \sigma) \longmapsto \nabla_{\tilde{\sigma}} \sigma$$

is a mapping which is $C^\infty(M)$-linear in its first variable, additive in its second variable and has the following characteristic property:

$$\forall \tilde{\sigma} \in \operatorname{Sec} \tilde{\xi}, \quad \sigma \in \operatorname{Sec} \xi, \quad f \in C^\infty(M): \quad \nabla_{\tilde{\sigma}} f\sigma = [(A \circ \tilde{\sigma})f]\sigma + f\nabla_{\tilde{\sigma}} \sigma.$$

A trivial but surprisingly important example of pseudoconnections is given in the following

(4.1) Lemma. *Let ι denote the identity morphism of $V\xi$. — There exists a unique pseudoconnection (∇^i, ι) (briefly ∇^i) in $V\xi$ w.r.t. itself (briefly in $V\xi$) satisfying the following condition:*

$$\forall \sigma \in \operatorname{Sec} \xi, \quad X \in \mathfrak{X}_V E : \quad \nabla^i_X \sigma^v = 0,$$

where σ^v is the vertical lift of the section σ (cf. [14], Lemma 3). ∎

For a *proof* see again [14], Lemma 3 and 4. (Another, coordinate-free, proof has been communicated to the author by Z. Kovács. His reasoning is based on elementary homological algebra.) We mention in advance that the trivial pseudoconnection ∇^i will play the same role in our theory as the "flat vertical connection" ([9], p. 57) in Matsumoto's theory.

5. Now we are in a position to formulate some of the indispensable definitions concerning Finsler connections. But what is a Finsler connection? We begin with an answer to this question. — Suppose that $\xi = (E, \pi, M)$ is a fixed vector bundle of rank r with n-dimensional base manifold M.

(5.1) Definition. *A Finsler connection is a linear connection in the vertical bundle $V\xi$, i.e. a linear connection of type*

$$\mathfrak{X}(E) \times \mathfrak{X}_V E \longrightarrow \mathfrak{X}_V E, \quad (X, Y) \longmapsto \nabla_X Y$$

(cf. [6], [11]). If H is a horizontal structure in ξ and ∇ is a Finsler connection, then the pair (∇, H) is said to be a Matsumoto pair.

Let $\left(U, (u^i)_{i=1}^n\right)$ be a chart in M. Consider and fix the vector bundle chart $\left(\pi^{-1}(U); (x^i)_{i=1}^n, (y^\alpha)_{\alpha=1}^r\right)$ — where $x^i := u^i \circ \pi$ — for ξ as described in [14], p. 1168. If H is a horizontal structure in ξ, then there exists a unique family of functions

$$N_i^\alpha : \quad \pi^{-1}(U) \longrightarrow \mathbb{R} \quad (1 \le i \le n, 1 \le \alpha \le r)$$

such that

$$\forall i \in \{1, \dots, n\} : \quad \left(\frac{\partial}{\partial u^i}\right)^H = \frac{\partial}{\partial x^i} - N_i^\alpha \frac{\partial}{\partial y^\alpha}.$$

The functions N_i^α are called the (local) *parameters* of H w.r.t. the fixed vector bundle chart. It is easy to see that the lifts

$$\frac{\delta}{\delta x^i} := \left(\frac{\partial}{\partial u^i}\right)^H, \quad 1 \le i \le n$$

constitute a local basis for the horizontal subbundle $H\xi$. — Returning to the Finsler connection ∇, its parameters over $\pi^{-1}(U)$ are defined by the relations

$$(5.2) \qquad \nabla_{\frac{\partial}{\partial x^i}} \frac{\partial}{\partial y^\alpha} = \Gamma_{i\alpha}^\beta \frac{\partial}{\partial y^\beta}, \quad \nabla_{\frac{\partial}{\partial y^\alpha}} \frac{\partial}{\partial y^\beta} = C_{\alpha\beta}^\gamma \frac{\partial}{\partial y^\gamma}$$

$$(1 \le i \le n; \quad 1 \le \alpha, \beta, \gamma \le r)$$

as usual in case of linear connections (our notation is the traditional one).

(5.3) Definition. *A Finsler connection ∇ is said to be regular if the kernel of the mapping $L: X \in \mathfrak{X}(E) \longmapsto L(X) := \nabla_X Z$ is a horizontal structure in ξ, i.e. if $\mathfrak{X}(E) = \mathfrak{X}_V E \oplus \operatorname{Ker} L$ (cf. [6], def. 1.6.1).*

(5.4) Proposition. *A Finsler connection ∇ is regular iff*

$$\det \left(\delta_\alpha^\beta + y^\lambda C_{\alpha\lambda}^\beta \right) \neq 0. \tag{5.5}$$

Proof. $\mathfrak{X}(E) = \mathfrak{X}_V E \oplus \operatorname{Ker} L \iff L \upharpoonright \mathfrak{X}_V E$ is an isomorphism $\iff$ $\left(L \left(\frac{\partial}{\partial y^\alpha} \right) \right)_{\alpha=1}^n$ is a local basis of $\mathfrak{X}_V E \iff$ (5.5), because

$$\forall \alpha \in \{1, \ldots, r\}: \quad L \left(\frac{\partial}{\partial y^\alpha} \right) = \nabla_{\frac{\partial}{\partial y^\alpha}} y^\lambda \frac{\partial}{\partial y^\lambda} = \frac{\partial}{\partial y^\alpha} +$$

$$+ y^\lambda C_{\alpha\lambda}^\beta \frac{\partial}{\partial y^\beta} = \left(\delta_\alpha^\beta + y^\lambda C_{\alpha\lambda}^\beta \right) \frac{\partial}{\partial y^\beta}.$$

∎

(5.5) Corollary. *If ∇ is a regular Finsler connection, then the parameters of the horizontal structure $\operatorname{Ker} L$ are the unique functions L_i^α satisfying the equations*

$$y^\beta \Gamma_{i\beta}^\alpha - L_i^\lambda \left(\delta_\lambda^\alpha + y^\beta C_{\lambda\beta}^\alpha \right) = 0.$$

Proof. If $\frac{\delta}{\delta x^i} = \frac{\partial}{\partial x^i} - L_i^\alpha \frac{\partial}{\partial y^\alpha}$ $(1 \leq i \leq n)$ is a local basis for $\operatorname{Ker} L$, then

$$\forall i \in \{1, \ldots, n\}: \quad 0 = L \left(\frac{\delta}{\delta x^i} \right) = \nabla_{\frac{\partial}{\partial x^i}} - L_i^\lambda \frac{\partial}{\partial y^\lambda} y^\beta \frac{\partial}{\partial y^\beta} =$$

$$= \nabla_{\frac{\partial}{\partial x^i}} y^\beta \frac{\partial}{\partial y^\beta} - L_i^\alpha \nabla_{\frac{\partial}{\partial y^\alpha}} y^\beta \frac{\partial}{\partial y^\beta} = \left(y^\beta \Gamma_{i\beta}^\alpha - L_i^\lambda \left(\delta_i^\alpha + y^\beta C_{\lambda\beta}^\alpha \right) \right) \frac{\partial}{\partial y^\alpha},$$

so (5.6) holds. At the same time, (5.5) guarantees that the functions L_i^α are uniquely determined by (5.6). ∎

6. Recall that any horizontal structure H in the vector bundle ξ gives rise to a splitting $\mathcal{H}: \pi^* \tau_M \longrightarrow \tau_E$ of the canonical short exact sequence

$$0 \longrightarrow V\xi \xrightarrow{\lambda} \tau_E \xrightarrow{\mu} \pi^* \tau_M \longrightarrow 0$$

(λ is the natural inclusion, $\mu: a \in T_z E \longmapsto (z, T_z \pi(a))$) such that $H\xi = \operatorname{Im} \mathcal{H}$ — and conversely.

(6.1) Definition. *Let H be a horizontal structure in the vector bundle ξ and consider the splitting $\mathcal{H}$ derived by H. — A triplet (∇^h, ∇^v, H) is said to be a* Cartan triad *if $(\nabla^h, \mathcal{H})$ is a pseudoconnection in $V\xi$ w.r.t. $\pi^* \tau_M$ and (∇^v, ι) is a pseudoconnection in $V\xi$ w.r.t. itself. ∇^h and ∇^v are called the* h-connection term *and the* v-connection term *of the triad, resp.*

We represent a Cartan triad locally by the family

$$\left(F_{i\alpha}^{\beta},\ C_{\alpha\beta}^{\gamma},\ N_i^{\alpha} \right)$$

where the functions N_i^{α} are the parameters of H, the $C_{\alpha\beta}^{\gamma}$-s are defined by (5.2) (replacing ∇ with ∇^v) and the functions $F_{i\alpha}^{\beta}$ are determined by the relations

$$(6.2) \qquad\qquad \nabla^h_{\widehat{\frac{\partial}{\partial u^i}}} \frac{\partial}{\partial y^{\alpha}} = F_{i\alpha}^{\beta} \frac{\partial}{\partial y^{\beta}}.$$

In (6.2) $\widehat{\frac{\partial}{\partial u^i}}$: $z \in \pi^{-1}(U) \longmapsto \left(z, \frac{\partial}{\partial u^i} \circ \pi(z) \right) \in E \times_M TM$. More generally, we shall need the following convention: if $X \in \mathfrak{X}(M)$, $\widehat{X}$ denotes the section

$$z \in E \longmapsto \widehat{X}(z) := (z, X \circ \pi(z)) \in E \times_M TM.$$

The next result (cf. [14], Theorem) provides a simple but important connection between Cartan triads and Matsumoto pairs.

(6.3) Theorem. *Let (∇^h, ∇^v, H) be a Cartan triad. If the mapping $\nabla :\ \mathfrak{X}(E) \times \mathfrak{X}_V E \longrightarrow \mathfrak{X}_V E$ is given by the formula*

$$\nabla_X Y = \nabla^v_{vX} Y + \nabla^h_{\mu \circ X} Y,$$

then ∇ is a Finsler connection, consequently (∇, H) is a Matsumoto pair *which is said to be* associated *with (∇^h, ∇^v, H).*

Conversely, *suppose that (∇, H) is a Matsumoto pair and form the mappings*

$$\nabla^h :\ \operatorname{Sec} \pi^* \tau_M \times \mathfrak{X}_V E \longrightarrow \mathfrak{X}_V E, \quad (\widetilde{X}, Y) \longmapsto \nabla_{\mathcal{H} \circ \widetilde{X}} Y,$$

$$\nabla^v :\ \mathfrak{X}_V E \times \mathfrak{X}_V E \longrightarrow \mathfrak{X}_V E, \quad (X, Y) \longmapsto \nabla_X Y.$$

Then the triad (∇^h, ∇^v, H) is a Cartan triad. ∎

The (not too hard) *proof* follows the same line as the proof of the above cited theorem of [14], so we omit it. Similarly, a quite straightforward calculation yields the

(6.4) Proposition. If (∇, H) is a Matsumoto pair with parameters $\left(\Gamma^\alpha_{i\beta}, C^\alpha_{\beta\gamma}, N^\alpha_i\right)$ — see (5.2) — and (∇^h, ∇^v, H) is the associated Cartan triad of (∇, H) with parameters $\left(F^\alpha_{i\beta}, \tilde{C}^\alpha_{\beta\gamma}, N^\alpha_i\right)$, then

$$F^\alpha_{i\beta} = \Gamma^\alpha_{i\beta} - N^\lambda_i C^\alpha_{\lambda\beta}, \qquad \tilde{C}^\alpha_{\beta\gamma} = C^\alpha_{\beta\gamma}$$

$$(1 \le i \le n; \quad 1 \le \alpha, \beta, \gamma \le r).$$

∎

(6.5) Definition. A Cartan triad (∇^h, ∇^v, H) is said to be satifying the C_2-condition if

$$\forall X \in \mathfrak{X}_V E : \quad \nabla^v_X Z = \nabla^i_X Z$$

(cf. [7] and [9] def. 13.1).

(6.6) Proposition. If the Cartan triad (∇^h, ∇^v, H) satisfies the C_2-condition, then the connection term ∇ in its associated Matsumoto pair is regular.

Proof. Due to the proof of (5.4) it is enough to show that $L\left(\frac{\partial}{\partial y^\alpha}\right)$, $1 \le \alpha \le r$ is a local basis of $\mathfrak{X}_V E$. But this follows immediately by (C_2), because

$$L\left(\frac{\partial}{\partial y^\alpha}\right) := \nabla_{\frac{\partial}{\partial y^\alpha}} Z \overset{(6.3)}{=} \nabla^v_{\frac{\partial}{\partial y^\alpha}} Z + \nabla^h_{\mu \circ \frac{\partial}{\partial y^\alpha}} Z \overset{(C_2)}{=} \nabla^i_{\frac{\partial}{\partial y^\alpha}} y^\beta \frac{\partial}{\partial y^\beta} =$$

$$= \frac{\partial}{\partial y^\alpha} + y^\beta \nabla^i_{\frac{\partial}{\partial y^\alpha}} \frac{\partial}{\partial y^\beta} \overset{(4.1)}{=} \frac{\partial}{\partial y^\alpha}.$$

∎

7. In this section we introduce and briefly discuss a further important notion: the notion of *deflection*.

(7.1) Proposition and Definition. (cf. [9], p. 67)
(a) Let (∇^h, ∇^v, H) be a Cartan triad. The mapping

$$D: \quad \mathfrak{X}(M) \longrightarrow \mathfrak{X}_V E, \quad X \longmapsto \nabla^h_{\overset{h}{X}} Z$$

is a vertical vector 1-form (i.e. $D \in A^1(M, V\xi)$) whose coordinate expression is

$$D = \left(y^\lambda F_{i\lambda}^\alpha - N_i^\alpha\right) \frac{\partial}{\partial y^\alpha} \otimes dx^i.$$

D is said to be the *deflection* of the given Cartan triad.

(b) The deflection of a Matsumoto pair (∇, H) is the $(1,1)$ tensor field

$$\tilde{D}: \quad X \in \mathfrak{X}(E) \longmapsto \tilde{D}(X) := \nabla_{hX} Z.$$

It appears in coordinates thus:

$$\tilde{D} = y^\lambda \left(\Gamma_{i\lambda}^\alpha - N_i^\nu C_{\nu\lambda}^\alpha\right) \frac{\partial}{\partial y^\alpha} \otimes dx^i.$$

∎

(7.2) Proposition. If ∇ is a regular Finsler connection, then the Matsumoto pair $(\nabla, \operatorname{Ker} L)$ and the Cartan triad associated with $(\nabla, \operatorname{Ker} L)$ have no deflection.

Proof. From the meaning of the horizontal structure $\operatorname{Ker} L$, $\forall X \in \mathfrak{X}(E)$: $hX \in \operatorname{Ker} L$, hence

$$\tilde{D}(X) := \nabla_{hX} Z \overset{(5.3)}{=} L(hX) = 0.$$

Let us denote by $\mathcal{L}$ the splitting belonging to $\operatorname{Ker} L$. Taking into account (6.3), the deflection of $(\nabla^h, \nabla^v, \operatorname{Ker} L)$ is the mapping

$$D: \quad X \in \mathfrak{X}(M) \longmapsto \nabla_{\widehat{X}}^h Z = \nabla_{\mathcal{L} \circ \widehat{X}} Z.$$

Here $\mathcal{L} \circ \widehat{X} \in \operatorname{Ker} L$, thus

$$0 = L(\mathcal{L} \circ \widehat{X}) := \nabla_{\mathcal{L} \circ \widehat{X}} Z \Longleftrightarrow D = 0.$$

∎

From (7.1) we immediately get the

(7.3) Proposition. *If D is the deflection of the Cartan triad (∇^h, ∇^v, H), then (H, D) is an affine structure in the vector bundle ξ whose affine lifting form has the coordinate expression*

$$(7.4) \qquad \ell^H - D = \left(\frac{\partial}{\partial x^i} - y^\beta F^\alpha_{i\beta} \right) \frac{\partial}{\partial y^\alpha} \otimes dx^i.$$

∎

Comparing (7.4) with the formula (8.17') of [9], we can conclude that the quite natural affine structure (H, D) plays the same role in our theory as Matsumoto's somewhat mysterious "non-linear connection associated with a V-connection" in his theory. (See also Matsumoto's instructive Remark 8.2 in [9].)

8.

(8.1) Definition. *The mapping*

$$P^1 : \; \operatorname{Sec} \pi^* \tau_M \times \mathfrak{X}_V E \longrightarrow \mathfrak{X}_V E, \quad (\tilde{X}, Y) \longmapsto v[\mathcal{H} \circ \tilde{X}, Y] - \nabla^h_{\tilde{X}} Y$$

is said to be the P^1-torsion of the Cartan triad (∇^h, ∇^v, H).

Note that if $X \in \mathfrak{X}(M)$, then $\mathcal{H} \circ \hat{X} = \ell^H(X) = X^H$, hence

$$\forall X \in \mathfrak{X}(M), \; Y \in \mathfrak{X}_V E : \quad P^1(\hat{X}, Y) = [X^H, Y] - \nabla^h_{\hat{X}} Y.$$

Thus $\forall i \in \{1, \ldots, n\}, \; \alpha \in \{1, \ldots, r\}$:

$$P^1 \left(\widehat{\frac{\partial}{\partial u^i}}, \frac{\partial}{\partial y^\alpha} \right) = \left(\frac{\partial N^\beta_i}{\partial y^\alpha} - F^\beta_{i\alpha} \right) \frac{\partial}{\partial y^\beta} ;$$

the functions

$$P^\beta_{i\alpha} := \frac{\partial N^\beta_i}{\partial y^\alpha} - F^\beta_{i\alpha}$$

are said to be the *coordinate functions* of the P^1-torsion.

(8.2) Proposition. *Under the homogeneity condition (HC),*
$\forall X \in \mathfrak{X}(M) : \; P^1(\hat{X}, Z) = -D(X)$; *if, moreover, the deflection vanishes, we get the relation*

$$y^\beta P^\alpha_{i\beta} = 0.$$

Proof. From the preceding remark

$$P^1(\widehat{X}, Z) = [X^H, Z] - \nabla^h_{\widehat{X}} Z \overset{(2.3)}{=} -\nabla^h_{\widehat{X}} Z \overset{(7.1)}{=} -D(X);$$

$$y^\beta P^\alpha_{i\beta} = y^\beta \frac{\partial N^\alpha_i}{\partial y^\beta} - y^\beta F^\alpha_{i\beta} \overset{(HC)}{=} N^\alpha_i - y^\beta F^\alpha_{i\beta} \overset{(7.1)}{=} 0,$$

if D vanishes. ∎

With the help of the P^1-torsion and the trivial pseudoconnection ∇^i (lemma (4.1)) one can construct some important special Cartan triads (cf. [8], p. 120).

(8.3) Definition. *A Cartan triad (∇^h, ∇^v, H) is said to be a* *Hashiguchi triad if $P^1 = 0$,* *a Rund triad if $\nabla^v = \nabla^i$,* *a Berwald triad if it is a Rund triad and a Hashiguchi triad at the same* *time.*

So we have the following "commutative diagram":

$$
\begin{array}{ccc}
\dfrac{(\nabla^h, \nabla^v, H)}{\text{Cartan triad}} & \xrightarrow{\ \nabla^v := \nabla^i\ } & \text{Rund triad} \\[2ex]
\Big\downarrow{\scriptstyle P^1 = 0} & & \Big\downarrow{\scriptstyle P^1 = 0} \\[2ex]
\text{Hashiguchi triad} & \xrightarrow{\ \nabla^v := \nabla^i\ } & \text{Berwald triad}
\end{array}
$$

(8.4) Definition. *If (∇, H) is a Matsumoto pair and its associated Cartan triad is the Berwaldian one, then the Finsler connection ∇ is said to be the Berwald connection.*

From our previous results one can deduce with a little extra work the

(8.5) Theorem. *Any horizontal structure $H : \tau_E = V\xi \oplus H\xi$ in a vector bundle ξ determines a unique Berwald triad and — consequently — a unique Berwald connection ∇^B. H satisfies the homogeneity condition (HC) iff ∇^B has no deflection.* ∎

(Further details — from a slightly different point of view — can be found in [14] and [15].)

9. In this concluding section we shall have a look at a very special classical situation. — Suppose that (M, L) is a *Lagrange space* in Miron's sense [10], that is $L : TM \longrightarrow \mathbb{R}$ is a regular Lagrange function:

$$\det \left(\frac{\partial^2 L}{\partial y^i \partial y^i} \right) \neq 0.$$

Let X_L be the Lagrange vector field for L. It can be characterized by the well-known relation $i_{X_L} \omega_L = -dE$ ($E := Z(L) - L$, ω_L is the Lagrangian 2-form; cf. e.g. [1], def. 3.5.11). The coordinate expression of X_L is

$$X_L = y^i \frac{\partial}{\partial x^i} + G^i \frac{\partial}{\partial y^i},$$

where

$$G^i = g^{ij} \left(\frac{\partial^2 L}{\partial x^k \partial y^j} \cdot y^k - \frac{\partial L}{\partial x^j} \right), \quad (g^{ij}) := \left(\frac{\partial^2 L}{\partial y^i \partial y^j} \right)^{-1}.$$

Since the 25's of this century it has been a well-established folklore that the family

$$(G^i_j) := \left(\frac{1}{2} \frac{\partial G^i}{\partial y^j} \right)$$

determines — in our terminology — a horizontal structure and hence — owing to (8.5) — a Berwald connection in τ_M. This connection is called the *classical Berwald connection* of the Lagrange space (M, L) (cf. [9], Teorema 4.1). Utilizing now a clever observation of M. Crampin [2] we obtain the following elegant result:

(9.1) Theorem. *Let (M, L) be a Lagrange space and let X_L be the Lagrange vector field for L. The mapping*

$$\ell^B : \quad X \in \mathfrak{X}(M) \longmapsto \frac{1}{2} \left(X^c + [X^v, X_L] \right)$$

(X^c and X^v are the obvious complete and vertical lifts of X) is a lifting form in the sense of section 2. The horizontal structure determined by ℓ^B induces just the classical Berwald connection.

Proof. Calculating by brute force, we get the coordinate expression

$$\ell^B \left(\frac{\partial}{\partial u^i} \right) = \frac{1}{2} \left(\left(\frac{\partial}{\partial u^i} \right)^c + \left[\frac{\partial}{\partial y^i}, y^j \frac{\partial}{\partial x^j} + G^j \frac{\partial}{\partial y^j} \right] \right) =$$

$$= \frac{1}{2} \left(\frac{\partial}{\partial x^i} + \frac{\partial G^j}{\partial y^i} \frac{\partial}{\partial y^j} \right)$$

for ℓ^B, from which our assertions follow easily. ∎

So at least three things of a seemengly quite different kind, namely horizontal structures, Berwald connection and Lagrangian mechanics have met in a fascinating manner.

References

[1] R. ABRAHAM and J. E. MARSDEN, Foundations of Mechanics, 2nd ed., Benjamin / Cummings, Reading, Massachusets, 1978.

[2] M. CRAMPIN, On horizontal distribution on the tangent bundle of a differentiable manifold, *J. London Math. Soc. (2)*, **3** (1971), 178–182.

[3] M. CRAMPIN, Generalized Bianchi identities for horizontal distributions, *Math. Proc. Camb. Phil. Soc.* **94** (1983), 125–132.

[4] J. DIEUDONNÉ, Treatise on Analysis, Vol. 3, Academic Press, New York, 1972.

[5] W. GREUB, S. HALPERIN and R. VANSTONE, Connections, Curvature, and Cohomology, Vols. 1, 2, Academic Press, New York – London, 1972, 1973.

[6] B. T. M. HASSAN, The theory of geodesics in Finsler spaces, Ph. D. Thesis, Southampton, 1967.

[7] L. KOZMA, The role of the C_1 and C_2 conditions in the theory of Finsler connections, *The Proceedings of the National Seminar on Finsler Spaces, Brasov (1986)*, 185–200.

[8] L. MANGIAROTTI and M. MODUGNO, Graded Lie algebras and connections on a fibered space, *J. Math. pures et appl.*, **63** (1984), 111–120.

[9] M. MATSUMOTO, Foundations of Finsler geometry and special Finsler spaces, Kaiseisha Press, 1986.

[10] R. MIRON and M. ANASTESIEI, Vector bundles, Lagrangian geometry, applications in relativity (Roumanian), Ed. Academiei R.S.R., 1987.

[11] V. OPROIU, Some properties of the tangent bundle related to the Finsler geometry, *The Proceedings of the National Seminar on Finsler spaces, Brasov (1980)*, 195–207.

[12] W. POOR, Differential Geometric Structures, Mc Graw – Hill, New York, 1981.

[13] J. SZILASI, Horizontal maps with homogeneity condition, *Proc. of the 11th Winter School, Suppl. Rend. Palermo (1984)*, 307–320.

[14] J. SZILASI and Z. KOVÁCS, Pseudoconnections and Finsler-type connections, *Colloquia Math. Soc. János Bolyai 46, Topics in Differential Geometry*, 1165–1184.

[15] J. SZILASI and L. KOZMA, Remarks on Finsler-type connections, *The Proceedings of the National Seminar on Finsler Spaces, Brasov (1984)*, 181–195.

J. Szilasi

Department of Mathematics
University of Debrecen
H–4010 Debrecen, Pf. 12
Hungary

COLLOQUIA MATHEMATICA SOCIETATIS JÁNOS BOLYAI
56. DIFFERENTIAL GEOMETRY, EGER (HUNGARY), 1989

On Weakly Symmetric and Weakly Projective Symmetric Riemannian Manifolds*

L. TAMÁSSY and T. Q. BINH

1. Introduction

The notions of pseudo symmetric (*p.s.*) resp. pseudo projective symmetric (*p.p.s.*) Riemannian manifolds have recently been introduced by M. C. Chaki [1], resp. M. C. Chaki and S. K. Saha [2]. Such spaces are investigated also in the paper of M. Tarafdar [3].

We recall the definition of these spaces. Let (M, g) be an n-dimensional Riemannian manifold, and U a tensor field of type $(1, 3)$ on it. Let $X, Y, Z, V \in \mathfrak{X}(M)$ be tangent vector fields and α a 1-form on M. Let us consider the relation

$$
\begin{aligned}
(\nabla_X U)(Y, Z, V) = {} & 2\alpha(X)U(Y, Z, V) + \alpha(Y)U(X, Z, V) + \\
& + \alpha(Z)U(Y, X, V) + \alpha(V)U(Y, Z, X) + \langle U(Y, Z, V), X \rangle A,
\end{aligned}
\tag{1}
$$

where $A \in \mathfrak{X}(M)$ is a vector field defined by

$$
\langle X, A \rangle = \alpha(X) \qquad \forall X,
$$

---

* This paper is in final form and no version of it will be submitted for publication elsewhere.

$\langle , \rangle$ means the inner product according to the metric g, and ∇ denotes the Levi–Civita connection of (M, g). If (1) holds for $U \equiv R$ (the curvature tensor of (M, g)), then the space is called pseudo symmetric (see [1]); and if (1) holds for $U \equiv W$ (the conformal curvature tensor of (M, g)), the space is called pseudo projective symmetric (see [2]). α is the associated 1-form. — In case of the vanishing of α the pseudo symmetric Riemannian manifold (M, g) is symmetric, for in this case $\nabla_X R = 0$.

As well known

$$(2) \qquad W(Y, Z, V) = R(Y, Z, V) - \frac{1}{n-1}[S(Z, V)Y - S(Y, V)Z],$$

where S is the Ricci tensor of (M, g). Using the relation

$$(\nabla_X P)(Y, Z, V) = \nabla_X (P(Y, Z, V)) - P(\nabla_X Y, Z, V) -$$
$$- P(Y, \nabla_X Z, V) - P(Y, Z, \nabla_X V)$$
$$P \equiv S(Z, V)Y - S(Y, V)Z,$$

the covariant derivation of (2) yields (see also [2] (1.12))

$$(3) \qquad (\nabla_X W)(Y, Z, V) = (\nabla_X R)(Y, Z, V) -$$
$$- \frac{1}{n-1} \{ [(\nabla_X S)(Z, V)] Y - [(\nabla_X S)(Y, V)] Z \}.$$

We can put the question, which Riemannian manifolds are $p.s.$ and also $p.p.s.$ at the same time with the same associated 1-form α. One can find a necessary condition for this in the paper [2] of Chaki and Saha, namely the vanishing of the scalar curvature r of (M, g). They give also a sufficient condition which consists of the existence of an absolute parallel unit (or not vanishing) vector field $v : \nabla_X v = 0, \forall X$.

In this paper we go further in the weakening of the condition of symmetry by introducing the notions of weakly symmetric ($w.s.$), and weakly projective symmetric($w.p.s.$) Riemannian manifold, and we find *necessary and sufficient* conditions for a $w.s.$ (M, g) to be $w.p.s.$, or conversely, with the same associated 1-forms. Special cases will also be investigated. A corollary of these yields a necessary and sufficient condition for a $p.s.$ space to be $p.p.s.$ too. — All vector and tensor fields, forms, etc. are supposed to be of class C^∞.

2. Weakly symmetric spaces

Definition 1. *A Riemannian manifold* (M, g) *is called weakly symmetric (w.s.) if there exist 1-forms* $\alpha, \beta, \gamma, \sigma$ *and a vector field* F *such that*

(4)
$$
\begin{aligned}
(\nabla_X R)(Y, Z, V) = {} & \alpha(X)R(Y, Z, V) + \beta(Y)R(X, Z, V) + \\
& + \gamma(Z)R(Y, X, V) + \sigma(V)R(Y, Z, X) + \\
& + \langle R(Y, Z, V), X \rangle F,
\end{aligned}
$$

where R *is the curvature tensor of* (M, g).

Definiton 2. *A Riemannian manifold* (M, g) *is called weakly projective symmetric (w.p.s.) if there exist 1-forms* $\alpha, \beta, \gamma, \sigma$ *and a vector field* F *such that*

(5)
$$
\begin{aligned}
(\nabla_X W)(Y, Z, V) = {} & \alpha(X)W(Y, Z, V) + \beta(Y)W(X, Z, V) + \\
& + \gamma(Z)W(Y, X, V) + \sigma(V)W(Y, Z, X) + \\
& + \langle W(Y, Z, V), X \rangle F,
\end{aligned}
$$

where W *is the projective curvature tensor of* (M, g).

$\alpha, \beta, \gamma, \sigma$ *are called in both cases the associated 1-forms and* F *the associated vector field.*

We want to derive conditions on $\alpha, \beta, \gamma, \sigma$ and the Ricci tensor S of (M, g) from the fact that a *w.s.* space is also *w.p.s.* with the same associated 1-forms $\alpha, \beta, \gamma, \sigma$ and associated vector field F, and $n \geq 4$. Thus we assume both relations (4) and (5) with the same associated 1-forms and associated vector field. Substituting the value of W from (2) in the right hand side of (5), we obtain

$$
\begin{aligned}
(\nabla_X W)(Y, Z, V) = {} & (\nabla_X R)(Y, Z, V) - \\
& - \frac{1}{n-1}\{\alpha(X)[S(Z, V)Y - S(Y, V)Z] + \\
& + \beta(Y)[S(Z, V)X - S(X, V)Z] + \gamma(Z)[S(X, V)Y - S(Y, V)X] + \\
& + \sigma(V)[S(Z, X)Y - S(Y, X)Z] + \langle S(Z, V)Y - S(Y, V)Z, X \rangle F\}.
\end{aligned}
$$

This and (3) are two expressions for $\nabla_X W$. Equating the right hand sides, we get

(6)
$$[\beta(Y)S(Z,V) - \gamma(Z)S(Y,V)]X + [\alpha(X)S(Z,V) +$$
$$+ \gamma(Z)S(X,V) + \sigma(V)S(Z,X) - (\nabla_X S)(Z,V)]Y -$$
$$- [\alpha(X)S(Y,V) + \beta(Y)S(X,V) + \sigma(V)S(Y,X) -$$
$$- (\nabla_X S)(Y,V)]Z + \langle S(Z,V)Y - S(Y,V)Z, X \rangle F = 0$$
$$\forall X, Y, Z, V \in \mathfrak{X}(M).$$

Let p be an arbitrary element of M and U_p a coordinate neighbourhood of p. The following considerations will be performed in this neighbourhood. Because of $n \geq 4$ there exists an X which does not lie in the subspace spanned by Y, Z and F, i.e. $X \notin \{Y, Z, F\}$. Thus we obtain from (6)

(7) $$\beta(Y)S(Z,V) - \gamma(Z)S(Y,V) = 0 \qquad \forall Y, Z, V \in \mathfrak{X}(U_p).$$

Similarly we can choose a Y such that $Y \notin \{X, Z, F\}$ and also a Z such that $Z \notin \{X, Y, F\}$. In these cases we obtain from (6)

(8) $$\alpha(X)S(Z,V) + \gamma(Z)S(X,V) + \sigma(V)S(Z,X) = (\nabla_X S)(Z,V)$$

and

(9)
$$\alpha(X)S(Y,V) + \beta(Y)S(X,V) + \sigma(V)S(Y,X) = (\nabla_X S)(Y,V)$$
$$\cdot \quad \forall X, Y, V \in \mathfrak{X}(U_p),$$

(7), (8) and (9) are true in any coordinate neighbourhood U_p $p \in M$. Thus they hold for any $X, Y, Z, V \in \mathfrak{X}(M)$.

By repeated application of (7) and the well known symmetry of the Ricci tensor S we obtain

(10)
$$\gamma(Z)S(Y,V) = \beta(Y)S(Z,V) = \beta(Y)S(V,Z) = \gamma(V)S(Y,Z) =$$
$$= \gamma(V)S(Z,Y) = \beta(Z)S(V,Y) = \beta(Z)S(Y,V) =$$
$$= \gamma(Y)S(Z,V) = \gamma(Y)S(V,Z) = \beta(V)S(Y,Z)$$
$$\forall Y, Z, V \in \mathfrak{X}(M).$$

This yields the following relations

(11) $$\beta(Y)S(Z,V) = \gamma(Y)S(Z,V),$$

(12) $$\beta(Y)S(Z,V) = \beta(Z)S(Y,V) = \beta(V)S(Y,Z),$$

(13) $$\gamma(Y)S(Z,V) = \gamma(Z)S(Y,V) = \gamma(V)S(Y,Z).$$

(12) means that βS is totally symmetric, i.e. $(\beta S)(Y, Z, V) \equiv \beta(Y) S(Z, V)$ is independent of the order of the vectors Y, Z, V. (13) means the same for γS. (11) results in $\beta = \gamma$ at those points p only, where there exist Z and V such that $S(Z, V)|_p \neq 0$.

Reforming the second term of (8) according to (13) and in view of the symmetry of S, we have

$$\alpha(X)S(Z, V) + \gamma(X)S(Z, V) + \sigma(V)S(Z, X) = (\nabla_X S)(Z, V).$$

Interchanging in this Z and V, and in view of the symmetry of S, we obtain

$$\sigma(V)S(Z, X) = \sigma(Z)S(V, X),$$

and again by repeated application of this and of the symmetry of S we obtain

$$(14) \qquad \sigma(Y)S(Z, V) = \sigma(Z)S(Y, V) = \sigma(V)S(Y, Z) \qquad \forall Y, Z, V,$$

i.e. σS too is totally symmetric. Finally, (8) can be transformed by the use of (13) and (14) into

$$(15) \qquad [\alpha(X) + \gamma(X) + \sigma(X)]S(Z, V) = (\nabla_X S)(Z, V).$$

These give

Proposition 1. *If an (M, g) $(n \geq 4)$ is weakly symmetric and also weakly projective symmetric with the same associated 1-forms $\alpha, \beta, \gamma, \sigma$ and associated vector field F (where the vanishing of F is also allowed), then:*

a) $\beta S = \gamma S$,

b) βS, c) γS, d) σS are totally symmetric and (15) holds.

3. The case $F = 0$

The above results hold, of course, in case of $F = 0$ too. We show that in this case the converse of the statement of Proposition 1 is also true.

Theorem 1. *A Riemannian manifold (M, g) $(n \geq 3)$ is weakly symmetric and also weakly projective symmetric with the same associated 1-forms $\alpha, \beta, \gamma, \sigma$ and associated vector field $F \neq 0$ iff $\beta S = \gamma S$; βS, γS and σS are totally symmetric, and (15) holds.*

Proof. The necessity has been proved above. (It is easy to see that now, because of $F = 0$, $n \geq 3$ suffices).

Conversely, we assume (5) with $F = 0$; (15); a) $\beta S = \gamma S$; and that b) βS, c) γS, d) σS are totally symmetric, and then we prove (4). Taking into account (2) and $F = 0$ in (5) we get

$$
\begin{aligned}
(\nabla_X W)(Y, Z, V) = & \{ \alpha(X) R(Y, Z, V) + \beta(Y) R(X, Z, V) + \\
& + \gamma(Z) R(Y, X, V) + \sigma(V) R(Y, Z, X) \} - \\
& - \frac{1}{n-1} \{ \alpha(X)[\overset{1}{S(\breve{Z}, V)}Y - \overset{2}{S(\breve{Y}, V)}Z] + \\
& + \beta(Y)[\overset{3}{S(\breve{Z}, V)}X - \overset{4}{S(\breve{X}, V)}Z] + \gamma(Z)[\overset{5}{S(\breve{X}, V)}Y - \overset{3}{S(\breve{Y}, V)}X] + \\
& + \sigma(V)[\overset{6}{S(\breve{Z}, X)}Y - \overset{7}{S(\breve{Y}, X)}Z] \}.
\end{aligned}
$$

(16)

Making use of (15) in (3) there results

$$
\begin{aligned}
(\nabla_X W)(Y, Z, V) = & (\nabla_X R)(Y, Z, V) - \\
& - \frac{1}{n-1} \{ [\overset{1}{\alpha(\breve{X})} + \overset{5}{\gamma(\breve{X})} + \overset{6}{\sigma(\breve{X})}] S(Z, V) Y - \\
& - [\overset{2}{\alpha(\breve{X})} + \overset{4}{\gamma(\breve{X})} + \overset{7}{\sigma(\breve{X})}] S(Y, V) Z \}.
\end{aligned}
$$

(17)

But the value of the second curly bracket of (16) equals the curly bracket of (17): $\underset{\smile}{1} = \underset{\smile}{1}$, $\underset{\smile}{2} = \underset{\smile}{2}$, ..., $\underset{\smile}{7} = \underset{\smile}{7}$ because of a), b), c), d). Then a comparison of the right hand sides of (16) and (17) yields (4). ∎

4. The case $F \neq 0$

Theorem 2. *A Riemannian manifold (M, g) $(n \geq 4)$ is weakly symmetric and also weakly projective symmetric with the same associated 1-forms $\alpha, \beta, \gamma, \sigma$ and associated vector field $F \neq 0$ iff the Ricci tensor S vanishes.*

Proof. A) The necessity of $S = 0$. — In this case we have the relations (6), (7), (8), (9), and from these

$$\langle S(Z, V)Y - S(Y, V)Z, X \rangle F = 0.$$

Since $F \neq 0$, we obtain

$$\langle S(Z, V)Y - S(Y, V)Z, X \rangle = 0 \qquad \forall\, X, Y, Z \in \mathfrak{X}(U_p).$$

Let now $Y \perp X$ and $Z = X$, $X \neq 0$. Thus $S(Y, V)\langle X, X \rangle = 0$ and hence

$$S(Y, V) = 0 \qquad \forall\, Y, V \in \mathfrak{X}(U_p).$$

But this is true for any $p \in M$, which results in $S = 0$ on M.

We remark that in this case (i.e. $S = 0$) the conditions of Proposition 1 and of Theorem 1, namely a), b), c), d) and (15) are trivially satisfied.

B) The sufficiency of $S = 0$ is nearly trivial. In this case (2) reduces to $W = R$ and then (4) coincides with (5). ∎

We remark that in case of $\alpha = 2\beta = 2\gamma = 2\sigma$, and $\langle X, F \rangle = \alpha(X)$ $\forall\, X$, $\alpha \neq 0$ a weakly symmetric space is a pseudo-symmetric space, and a *w.p.s.* space is a *p.p.s.* space in the sense of Chaki and Saha. Hence, as a consequence of Theorem 2, we obtain the following

Corollary. *A Riemannian manifold (M, g) $(n \geq 4)$ is pseudo-symmetric and also projective pseudo-symmetric in the sense of Chaki and Saha with the same associated 1-form $\alpha \neq 0$ iff the Ricci tensor S vanishes.*

References

[1] M. C. CHAKI, On pseudo symmetric manifolds, *An. Sti. Int. Univ. "Al. I. Cuza" Iasi* **33** (1987), 53–58.

[2] M. C. CHAKI and S. K. SAHA, On pseudo projective symmetric manifolds, *Bull. Inst. Math. Acad. Scinica* **17** (1989), 59–65.

[3] M. TARAFDAR, On pseudo-symmetric and pseudo-Ricci-symmetric Sasakian manifolds, *Per. Math. Hung.,* to appear.

L. Tamássy

Department of Mathematics
University of Debrecen
H–4010 Debrecen
Pf. 12.
Hungary

T. Q. Binh

Department of Mathematics
University of Debrecen
H–4010 Debrecen
Pf. 12.
Hungary

COLLOQUIA MATHEMATICA SOCIETATIS JÁNOS BOLYAI
56. DIFFERENTIAL GEOMETRY, EGER (HUNGARY), 1989

Secondary Characteristic Classes: a Survey[*]

I. VAISMAN

The secondary (or exotic) characteristic classes are global invariants of geometric structures which are derived from the curvature of adequate connections. They have been discovered around 1970 in foliation theory (Godbillon–Vey [GV], Bernshtein–Rozenfeld [BR 1, 2], Bott–Haefliger [BH], [Ha 2], [Bt 3], Kamber–Tondeur [KT 1, 2, 3], D. Lehmann [Le 1, 2], etc.). This development was in connection with the study of Haefliger's classifying spaces e.g., [Ha 1], [Bt 3], and that of the cohomology of the Lie algebras of vector fields due to Gelfand and Fuks [GF], [Fu 1, 2], [Go], [Ms]. At the same time, a different approach to secondary classes appeared in conformal geometry (Chern–Simons [CS 1, 2]). Since then, the secondary classes have been the object of an extensive research and literature.

The present survey is not one of approaches and theorems in the theory of secondary classes, but rather one of the *geometric phenomena* encountered in this theory: *existence, rigidity, variation, residues*, etc. and of structures where such classes appear. Moreover, it is intended for nonspecialists. All these imply the following features: i) the survey refers to basic facts only, and there is no claim of completeness neither in the text nor in the references; ii) only one approach to the theory, that of D. Lehmann [Le 2], has been chosen from the many which exist in the literature; iii) while some of the

---

[*] This paper is in final form and no version of it will be submitted for publication elsewhere.

characteristic classes encountered here have been studied by the author [Va 1–6], this exposition is not intended to focus on personal results.

1. The general theory

The theory of real valued characteristic classes is "easy". All you have to do is: Take connections θ, θ_0, θ_1, etc. on a principal G-fibre bundle $P \to M$, take invariant polynomials f of degree k on the corresponding local curvature forms Θ, and notice the formulas (e.g., [KN])

$$(1.1) \qquad df(\Theta) = 0 \text{ (Bianchi)} ,$$

$$(1.2) \qquad f(\Theta_1) - f(\Theta_0) = d\triangle(\theta_0, \theta_1)f \quad \text{(Stokes)} ,$$

where, if Θ_t is the curvature of $(1 - t)\theta_0 + t\theta_1$, then

$$(1.3) \qquad \triangle(\theta_0, \theta_1)f = k \int_0^1 f\left(\theta_1 - \theta_0, \Theta_t^{(k-1)}\right) dt.$$

Then, look for a cohomological interpretation of these formulas via de Rham's theorem. Hence, let us start looking for it!

In topology, the *primary characteristic classes* of a principal bundle $P \to M$ with structure group G are those which belong to $\mathrm{im}(\phi^* : H^*(B_G) \to H^*(M))$, where $\phi : M \to B_G$ is the *classifying map* of P.

In the sequel, we stay in the C^∞-category, and consider only real or complex cohomology. Then it turns out that the characteristic classes are determined by the curvature of the connections of P. Namely, take $I^k(G)$ to be the space of ad G-invariant polynomials of degree k on the Lie algebra $\mathfrak{G}$ of G, and $I(G) = \oplus_{k \geq 0} I^k(G)$, with the symmetric product. Define the homomorphism $\rho(\theta) : I(G) \to \wedge M$ by the evaluation of $f \in I^k(G)$ on the curvature Θ of a connection θ on P. Then the Bianchi identity implies $d \circ \rho = 0$, which is exactly formula (1.1), and we obtain the *Chern–Weil homomorphism*

$$w(P) = \rho^* : I(G) \to H^*(M, \mathbb{R}),$$

which is independent of θ because of (1.2).

Thereby, the functorial character of $w(P)$ is ensured, and an inductive limit procedure yields a universal Chern–Weil homomorphism $w(G)$: $I(G) \to H^*(B_a, \mathbb{R})$ such that $\phi^* \circ w(G) = w(P)$. If G is compact, $w(G)$ is an isomorphism, and $\operatorname{im}(w(P))$ contains precisely the characteristic classes of P [Bo].

Now, following D. Lehmann [Le 2], let J_0, J_1 be two homogeneous ideals of $I(G)$, and $I^+(G) = \oplus_{k \geq 1} I^k(G)$. Define the algebra

$$(1.4) \qquad W(J_0, J_1) = \left[I(G)/_{J_0}\right] \otimes \left[I(G)/_{J_1}\right] \otimes \wedge(I^+(G)),$$

with the graduation

$$\operatorname{grade} [f]_{J_0} = \operatorname{grade} [f]_{J_1} = 2k, \qquad \operatorname{grade} \widehat{f} = 2k - 1,$$

where $f \in I^k(G)$, and it defines $[f]_{J_0}$, $[f]_{J_1}$, $\widehat{f}$ in the three factors of (1.4), and, finally, with the differential

$$d[f]_{J_0} = d[f]_{J-1} = 0, \qquad d\widehat{f} = [f]_{J_1} - [f]_{J_0}.$$

Then, for two connections θ_0, θ_1 on the G-principal bundle P such that $J_a \subseteq \ker \rho(\theta_a)$ $(a = 0, 1)$ (J_a-*connections*) we get a homomorphism of differential graded algebras

$$(1.5) \qquad \qquad \rho'(\theta_0, \theta_1) : W(J_0, J_1) \to \wedge M$$

by putting

$$\rho'([f]_{J_0}) = \rho(\theta_0)f, \qquad \rho'([f]_{J_1}) = \rho(\theta_1)f, \qquad \rho'\widehat{f} = \triangle(\theta_0, \theta_1)f.$$

(1.1) and (1.2) ensure the commutation of ρ' with the differentials, hence the existence of a *secondary Chern–Weil homomorphism*

$$(1.6) \qquad \qquad \rho'^* (\theta_0, \theta_1) : H^*(W(J_0, J_1)) \to H^*(M, \mathbb{R}).$$

It is clear that $\operatorname{im} \rho'^*$ contains the primary characteristic classes of P. If it also contains other classes, the latter are called *secondary characteristic classes of the triple* (P, θ_0, θ_1).

Remark. One can prove that the same characteristic classes are obtained if $W(J_0, J_1)$ is replaced by the subalgebra where the third factor is taken to be $\wedge(A)$ instead of $\wedge(I^+(G))$, A being a system of generators of $I^+(G)$ [Le 2].

The main geometric phenomena of the theory of secondary classes are:

a) *Existence of J-connections for certain ideals J. This amounts to vanishing phenomena of some primary classes.*

b) *Existence of non-vanishing secondary characteristic classes.*

c) *Rigidity and variation of secondary classes by change of connections.*

d) *The residual role and character of secondary classes.*

e) *Special features: localization, classifying space, etc.*

Let us give some explanations for the subjects c), d) and e).

Let $\phi_a (a = 0, 1)$ be connections on $P \times I \to M \times I (I = \{s/\ 0 \leq s \leq 1\})$, and $\theta_{a,s} = \phi_a|_{s=const.}$. Then, if we apply Stokes' theorem for integrals along the fibers of $M \times I \to M$ to the differential form $\rho'(\phi_0, \phi_1)(w)$, where $[w] \in W(J_0, J_1)$, we get

$$(1.7) \qquad \rho'(\theta_{0,1}, \theta_{1,1})(w) - \rho'(\theta_{0,0}, \theta_{1,0})(w) =$$
$$= d(\Gamma(\phi_0, \phi_1)(w)) + \Gamma(\phi_0, \phi_1)(dw),$$

where for any $[w] \in W(J_0, J_1)$ one has

$$(1.8) \qquad \Gamma(\phi_0, \phi_1)(w) = \int_{s=0}^{s=1} \rho'(\phi_0, \phi_1)(w).$$

The connections ϕ_a are said to be a *link* between $\theta_{a,0}$ and $\theta_{a,1}$, and if ϕ_a are J_a-connections $(\theta_{a,0}, \theta_{a,1})$ are said to be J_a- *homotopic*. Now, if $[w]$ is a cocycle, (1.7) yields the

Rigidity Theorem. *The secondary characteristic classes of a triple* (P, θ_0, θ_1) *do not change if* θ_a *are replaced by* J_a*-homotopic connections* $\theta'_a (a = 0, 1)$.

Hence, the secondary classes are obstructions to the existence of a connection θ of P which is simultaneously J_a-homotopic with $\theta_a (a = 0, 1)$. Indeed, if such θ exists we may take $\theta_0 = \theta_1 = \theta$, and all the secondary classes vanish because of (1.3). Particularly, if $J_0 = J_1 = 0$ no secondary classes appear, which stresses again the role of the vanishing phenomena.

On the other hand, (1.7) also indicates the possibility of a *continuous variation* of the secondary classes. Indeed, $\theta_0(\tau)$, $\theta_1(\tau)$ ($\tau \in \mathbb{R}$) may depend continuously on τ without being J_a-homotopic. Examples will be given in Section 2.

Our next subject is *residues*. These have been studied by many authors ([Bt 1], [BB], [LP 2], [He 2, 3], [Ni], [Le 3, 4], [Va 4], etc.), and we mention, particularly, the basic paper of Baum and Bott [BB]. A *clean* approach to the theory has been developed recently by D. Lehmann [Le 4] as follows.

Let us take M^m compact and oriented, and a certain *singular set* $S \subset V \subset \overline{V} \subset U$, where U, V are open neighborhoods which retract to S, and $\partial\overline{V} \cap S = \phi$. Then, $H^*(M, \mathbb{R})=$ the cohomology of the *Mayer–Vietoris cochain complex* [BT]

$$(1.9) \qquad L^*(M, S) = \wedge^*(M \setminus S) \oplus \wedge^* U \oplus \wedge^{*-1}(U \cap (M \setminus S))$$

with the differential

$$D(a, b, \xi) = (da, \ db, \ -d\xi \ + \ b \ - \ a).$$

Moreover, the subcomplex

$$(1.10) \qquad L_0^*(U, S) = \wedge^* U \oplus \wedge^{*-1}(U \cap (M \setminus S))$$

determines precisely the relative cohomology $H^*(U, U \setminus S; \mathbb{R})$, and the latter is sent to the homology $H_{m-*}(S, \mathbb{R})$ by the Lefschetz duality [Sp]

$$(1.11) \ ((b, \xi) \in L_0^*(U, S)) \mapsto \left\{ (\gamma \in \wedge^{m-*} U) \mapsto \left(\int_V b \wedge \gamma - \int_{\partial\overline{V}} \xi \wedge \gamma \right) \right\}.$$

Now, let us consider the principal G-bundle $P \to M$, and take: i) an arbitrary connection θ_0 on P, ii) a J-connection θ_1 on $P|_{M \setminus S}$, where J is a certain ideal of $I(G)$. Then

$$(f \in J \cap I^k(G)) \mapsto (\rho(\theta_0)f, \ \rho'(\theta_0, \theta_1)f)$$

induces a cohomology homomorphism $J \ \to \ L_0^*(M, S)$ whose image in $H^*(M, \mathbb{R})$ consists exactly of the primary characteristic classes [Le 4] of P associated with the elements of J. Furthermore, the restriction to U

(where only a local connection θ_0 on $p|_U$ is enough) followed by Lefschetz duality (1.11) yields *residues* along S such that the *sum of residues* along all the components of S dualizes back to the respective characteristic classes. The significance of these procedure follows from the fact that the residues represent a *local information* around S, while the summation of this local information yields a *global invariant* (the characteristic class). A typical example is Hopf's theorem which says that the Euler characteristic of M is equal to the sum of indices of singularities of a vector field on M.

Furthermore, if for two ideals J_a ($a = 0, 1$) there are connections θ_0 on $P|_U$, θ_0' on $P|_{M \setminus S}$, *linked* by a J_0-homotopy ϕ_0 along $U \cap (M \setminus S)$, and a global J_1-connection θ_1 of P, it is possible to construct a homomorphism of differential graded algebras $W(J_0, J_1) \to L^*(M, S)$ by

$$(1.12) \qquad [w] \in W(J_0, J_1)) \mapsto (\rho'(\theta_c', \theta_1)w, \ \rho'(\theta_0, \theta_1)w, \ \Gamma(\phi_0, \theta_1)w).$$

In case that θ_0, θ_0' are restrictions of some global J_0-connection $\widetilde{\theta}_0$, (1.12) sends the cohomology class of $[w]$ to the secondary characteristic class $[\rho'^*(\widetilde{\theta}_0, \theta_1)w]$, if $H^*(M, \mathbb{R})$ is seen as $H^*(L(M, S))$ [Le 4]. Finally, if $\widetilde{\theta}_0$ and θ_1 are J_0-homotopic over $M \setminus S$, we may take $\theta_0' = \theta_1$, and (1.12) sends $[w]$ to a class in $H^*(U, U \setminus S; \mathbb{R})$ which has residues along S.

Of course, for the subject *special features* we do not have, a general theory. However, we shall indicate here the important idea of *localization* due to G. Duminy and S. Hurder [Hu 3], [HH], which has been studied only in foliation theory but, perhaps, it might have some other applications as well, if seen as follows.

For $f \in I^k(G)$, define

$$(1.13) \qquad\qquad \wedge_f(M) = \{\lambda \in \wedge M / (\rho(\theta_a)f) \wedge \lambda = 0\}$$

where θ_a are J_a-connections ($a = 0, 1$), and denote by $H_f^*(M)$ the cohomology of $(\wedge_f M, d)$. Then f defines an operator

$$(1.14) \qquad \widetilde{f} : H_f^*(M) \to H^{*+2k-1}(M, \mathbb{R}); \qquad \widetilde{f}([\lambda]) = [(\rho'\widehat{f}) \wedge \lambda],$$

with ρ' of (1.5), which is called the *Weil operator* associated with f. Now, let us assume that M is compact. Topologize conveniently the cohomology spaces of (1.14) such as to make $\widetilde{f}$ continuous, then take the continuous linear functional $\widetilde{f}' = \int \circ \widetilde{f}$ on $H_f^{m-2k+1}(M)$. It will determine $\widetilde{f}$ itself, and, thereby, the secondary classes which contain the factor $\widehat{f}$. Finally, if in the definition of $\widetilde{f'}$ we use the integral over some measurable sets only we obtain the announced *localization* of the secondary characteristic classes. This leads to vanishing theorems for secondary classes if M is a union of measurable sets where the localized classes vanish [Hu 3], [HH].

2. The applications

Remember that we agreed to discuss basics only. If G is a linear group, the algebra $I(G)$ is generated by the *Chern polynomials* on complex square matrices A:

$$(2.1) \qquad c_k(A) = \frac{i^k}{2^k \pi^k} tr \wedge^k A = \frac{i^k}{2^k \pi^k} \delta^{j_1 \ldots j_k}_{i_1 \ldots i_k} a^{i_1}_{j_1} \ldots a^{i_k}_{j_k},$$

as shown in the following table [Bt 3], [KN], [KT 3], etc.

Table I: Important algebras $I(G)$

No.	Group G	Algebra $I(G)$	
1	$Gl(n, \mathbb{C})$ $(= L_n\mathbb{C})$	$\mathbb{C}[\ldots, c_k, \ldots]$	$(k = 1, \ldots, n)$
2	$Gl(n, \mathbb{R})$ $(= L_n\mathbb{R})$	$\mathbb{R}[\ldots, i^k c_k, \ldots]$	$(k = 1, \ldots, n)$
3	$U(n)$	$\mathbb{R}[\ldots, c_k, \ldots]$	$(k = 1, \ldots, n)$
4	$O(n)$	$\mathbb{R}[\ldots, (-1)^h c_{2h}, \ldots]$	$(h = 1, \ldots, [n/2])$
5	$Sp(n)$	$\mathbb{R}[\ldots, c_{2h}, \ldots]$	$(h = 1, \ldots, n)$

Correspondingly, we have the following most important vanishing phenomena

Table II: Vanishing Phenomena

No.	Type of connection	Vanishing Ideal
1	Flat Connection	$J = I^+(G)$
2	Hermitian Connection	$J = \{c_k - \bar{c}_k\} \subset I(L_n\mathbb{C}) \otimes I(L_n\mathbb{C})$
3	Volume Preserving Connection	$J = \{ic_1\} \subset IL_n\mathbb{R}$
4	Leafwise Flat Connection for a Foliation $\mathcal{F}$ [Bt3]	$J = \{f \in I(G)/\deg f > \operatorname{codim}\mathcal{F}\}$
5	Projectable Connection for a Foliation $\mathcal{F}$ [Mo]	$J = \left\{f \in I(G)/\deg f > \dfrac{1}{2}\operatorname{codim}\mathcal{F}\right\}$
6	Orthogonal Connection	$J_{\text{odd}} = \{., ic_{2h-1}, .\} \subset IL_n\mathbb{R}$
7	Orthogonal Connection	$J_{\text{odd}} = \{., c_{2h-1}, .\} \subset I(U(n))$
8	Quaternionic Metric Connection [KN], [Va 5]	$J_{\text{odd}} = \{., c_{2h-1}, .\} \subset I(U(2n))$
9	Holomorphic Connection over Compact Kähler Manifold [At]	$J = I^+(G)$

As a consequence of some of these phenomena we can now discuss the most important secondary characteristic classes.

a) Flat and foliated bundles

Secondary characteristic classes of flat bundles are of an independent interest [KT 3]. However, in this restricted survey, we only mention them as a particular case of the foliated bundles, where the foliation has codimension zero.

The foliated bundles are the field where most of the work on secondary classes was done. If $\mathcal{F}$ is a foliation of codimension q on the manifold M, there are adapted coordinates such that $\mathcal{F}$ is defined locally by $x^a = 0$

$(a = 1, \ldots, q)$, and the geometric objects which depend locally on x^a alone are called either *foliate* or *projectable* (they project onto the space of leaves). Particularly, if the transition functions of a bundle are projectable the bundle is said to be *foliate*, and it is the pull back of a bundle over the space of leaves. By a partition of unity argument, we see that such a bundle has leafwise flat connections i.e., such that the curvature forms have a factor dx^a. This explains the vanishing phenomena no. 4 of Table II.

Foliate bundles *may* also have *projectable connections* (which is not always true [Mo]). For instance, this is the case for the *transversal bundle* $\nu\mathcal{F} = TM/T\mathcal{F}$ of a *Riemannian foliation* i.e., a foliation endowed with a projectable Riemannian metric of $\nu\mathcal{F}$. Then, the curvature contains two factors dx^a, and we have the vanishing phenomena no. 5 of Table II.

The following table summarizes some of the secondary characteristic classes of the transversal bundle $\nu\mathcal{F}$ of a foliation.

Table III: Secondary classes of $\nu\mathcal{F}$

No.	Geometric Situation	Ideals	Subalgebra of $W(J_0, J_1)$ which yields the Secondary Classes
1	Real Foliation of Codim. $= q$	$J_0 = \{\deg > q\}$ (θ_0 leaf. flat) $J_1 = J_{\text{odd}}$ (θ_1 orthogonal)	$WO_q = \mathbb{R}[., i^k c_k, .]/\{\deg > q\}^{\otimes}$ $\otimes \wedge(., i^{2h-1}\widehat{c}_{2h-1}, .)$ (Ref. [Bt3])
2*	Holomorph. Foliation of Codim. $= q$	$J_0 = \{\deg > 2q\}$ (θ_0 leaf.' flat) $J_1 = \{c_k - \bar{c}_k\}$ (θ_1 Hermitian)	$WU_q = \mathbb{C}[., c_k, .]/\{\deg > 2q\}^{\otimes}$ $\otimes \mathbb{C}[., \bar{c}_k, .]/\{\deg > 2q\}^{\otimes}$ $\otimes \wedge_{\mathbb{C}}(., \widehat{c}_k, .)$ (Ref. [Pi])
3**	Riemannian Foliation of Codim. $= q$	$J_0 = \{\deg > q/2\}$ (θ_0 projectable and orthogonal) $J_1 = I^+\,\mathrm{Gl}(q, \mathbb{R})$ θ_1 flat; if any)	$RW_q = \mathbb{R}[., c_{2h}, .]/\{\deg > q/2\}^{\otimes}$ $\otimes \wedge(., \widehat{c}_{2h}, .)'$ (Ref. [LP1])
4	Codim. q Foliation with $\nu\mathcal{F}$ 2-foliated	$J_0 = \{\deg > q\}$ (θ_0 leaf. flat 1) $\theta_1 = \{\deg > q\}$ (θ_1 leaf. flat 2)	$W = \mathbb{R}[., i^k c_k, .]/\{\deg > q\}^{\otimes}$ $\otimes \mathbb{R}[., i^k c'_k, .]/\{\deg > q\}^{\otimes}$ $\otimes \wedge(., i^k \widehat{c}_k, .)$ (Ref. [Va 6])

* This case is not a direct result of the general theory. It is derived by means of convenient tensorisations with the complex conjugate algebras, and we have to take $\widehat{dc}_k = c_k - \bar{c}_k$.

** In this case we use $I(O(q))$. See [LP1] for using $I(SO(q))$ i.e., using also the Euler class of $\nu\mathcal{F}$.

Among the many other secondary classes, not quoted in this table one has classes of Kähler foliations [MM], of conformal and projective foliations [Ya], of flags of foliations [Fe], [Wo], of subfoliations [CG], [CM], [Do], etc.

In order to illustrate the typical phenomena of the theory of secondary characteristic classes, we quote the following results.

i) The cohomology of WO_q has been completely calculated. It is generated by the classes of those of the monomials $i^u c_{i_1} \ldots c_{i_s} \widehat{c}_{j_1} \ldots \widehat{c}_{j_t}$ ($u = \sum_{h=1}^{s} i_h + \sum_{k=1}^{t} j_k$) which are cocycles, and, in fact, a basis has been determined (the *Vey basis* [Go]). Of a particular interest is the class $gv(\mathcal{F}) = \rho'^* (i^{q+1} [c_1^q \widehat{c}_1])$ called the *Godbillon–Vey class* [GV], [Pi]. Examples where this class is non zero are known (see later on Thurston's example). Recently, it has been proven that the Godbillon–Vey class of a foliation of codimension 1 is functorial by C^1-diffeomorphisms [Ra], but not by homeomorphisms [Gh]. A Riemannian representation of the Godbillon–Vey class using the second fundamental form of the leaves was given in [RW] and [KY]. A generalized Godbillon–Vey class is encountered in case 4 of Table III namely, the class defined by the cocycle $\gamma = \sum_{h=0}^{q} i^{q+1} (c_1)^h (c_1')^{q-h} \widehat{c}_1$ [Va 6].

ii) Concerning variation and rigidity, let us first quote briefly an example of a family $\mathcal{F}_s$ ($s \in \mathbb{R}$) of foliations of codimension 1 of S^3 such that $gv(\mathcal{F}_s)$ varies continuously. The example is due to Thurston [Th]. Let P be a convex k-gon in the hyperbolic plane $\mathbb{H}$, P' its reflection in one of its sides. $Q = P \cup P'$, $\overline{Q} = Q\backslash$ disks around the vertices, $\pi : (Sl(2,\mathbb{R}) =$ the circle bundle over $\mathbb{H}) \to \mathbb{H}$, and $\pi^{-1}(\overline{Q}) \approx \overline{Q} \times SO(2)$. Here, identifications by the $\gamma_i \in Sl(2,\mathbb{R})$ which send the sides of P to the corresponding sides of P' yield a manifold $M \approx S^3\backslash(k$ solid tori), foliated by the upper triangular matrices of $Sl(2,\mathbb{R})$. Now, if the "holes" of M are filled in with *Reeb components* [Re], one gets the announced foliations $\mathcal{F}$, and computations give $\int_{S^3} gv(\mathcal{F}) = 4\pi$ (area of P). Since area of P = arbitrary number s, we are done.

A systematic study of rigidity and variation has been done by several authors [GFF], [He 1], [He 3], [Ts], etc., and it uses *derivatives* of the secondary classes and the theory of residues. Of course, the general homotopy argument of Lehmann [Le 2] is also used. The results show that there exist secondary classes with linearly independent continuous variations but that one also has the following

Rigidity Theorem [He 1]. *The secondary characteristic classes defined by the monomials $i^u c_{i_1} \ldots c_{i_s} \widehat{c}_{j_1} \ldots \widehat{c}_{j_t}$, where $\sum_{h=1}^{s} i_h + i_0 > q + 1$, q is the codimension of $\mathcal{F}$, and $i_0 = i_1$ if factors c appear, $i_0 = \infty$ otherwise, do not change if $\mathcal{F}$ is subject to a smooth deformation by foliations $\mathcal{F}_s (s \in \mathbb{R})$ of codimension q.*

iii) Foliation theory is also the field where many applications of residues theory can be made. The corresponding basic paper is that of Baum and Bott [BB]. There, one studies a complex analytic manifold M, with a holomorphic foliation of codimension q on $M \setminus S$, where S is a singularity set. Then, as described in Section 1, for $f \in I(U(q))$, $\deg f > q$, the primary characteristic class associated with f is represented in $H^*(U, U \setminus S)$, U being a neighborhood of S, and it dualizes to a residue in $H_{m-*}(S, \mathbb{R})$. In certain cases, the residue is expressible as a Grothendieck residue of several complex variables [Hr]. Similar residues computations have been developed for Riemannian and projective foliations [LP 2], [Ni], and in the study of the variation of secondary classes [He 3].

iv) The construction of the classifying space B_G of a Lie group G has been extended to groupoids, and, particularly, a classifying space $B\Gamma_q$ of the groupoid Γ_q of germs of diffeomorphisms of $\mathbb{R}^q$ was obtained [Ha 1], [Bt 3] together with a universal Γ_q-structure $\pi : E\Gamma_q \to B\Gamma_q$. Now, Haefliger showed that the Γ_q-structures can be seen as *foliations with singularities*, hence foliations $\mathcal{F}$ are given by certain classifying maps $\varphi : M \to B\Gamma_q$ (e.g., [Ha 1], [Bt 3], [Rn]). Like in the case of the primary classes, one gets a universal homomorphism

$$(2.2) \qquad w : H^*(WO_q) \to H^*(B\Gamma_q; \mathbb{R})$$

such that $\varphi^* \circ w = \rho'^* = $ the secondary Chern–Weil homomorphism.

Thereby, the secondary characteristic classes will offer information about the topology of $B\Gamma_q$ ([Ha 1], [Bt 3], [Hu 1], etc.). Particularly, one sees that these spaces may have *uncountable* generated homotopy. For instance, in [Hu 1] it is proven that *there exist epimorphisms* $\pi_n(B\Gamma_q) \to \mathbb{R}^{\nu_n}$, *where* $\{\nu_n\}$ *is a sequence which possesses a subsequence tending to infinity.*

v) Another *special feature* is Hurder's localization theory. Namely [Hu 3], [HH], one looks at the algebra

$$(2.3) \qquad \wedge_{\mathcal{F}} M = \{\Phi \in \wedge M / \forall U, \Phi|_U \text{ factorizes by } dx^1 \wedge \ldots \wedge dx^q\},$$

where $x^a = 0 \ (a = 1, \ldots, q)$ defines $\mathcal{F}$ over U. Then the Weil operator

$$(2.4) \qquad \widetilde{f} : H^{m-2k+1}(\wedge_{\mathcal{F}} M) \to H^m(M, \mathbb{R})$$

is defined for all $f \in J^k_{odd}$, and it localizes by composition with integration (see Section 1).

It is particularly interesting to integrate over sets $B \in \mathcal{B}(\mathcal{F})$ =the σ-algebra of Lebesgues measurable saturated subsets of M, thereby defining the *Weil measure* associated with f [HH]. To get the flavor of the applications we quote

Theorem [HH]. *Assume that $M = U_{u=1}^{\infty} B_u$ with $B_u \in \mathcal{B}(\mathcal{F})$, and such that $\mathcal{F}|_{B_u}$ $(u = 1, 2, \ldots)$ have compact leaves. Then the Godbillon–Vey class $gv(\mathcal{F}) = 0$.*

The same result holds if $\mathcal{F}|_{B_u}$ has some convenient *transverse measure*, if the leaves of $\mathcal{F}|_{B_u}$ have *subexponential growth*, etc. [HH].

vi) Finally let us just mention the quite remarkable fact that the secondary classes of a Riemannian foliation $\mathcal{F}$, seen from the viewpoint of Chern and Simons [CS 2], enter in *index theorems for transversal elliptic operators* [Lz].

b) Lagrangian and quaternionic bundles

Here, we shall review some personal results [Va 1, 2, 4, 5].

Let $\pi : E \to M$ be a real (complex) vector bundle of rank $2n$, endowed with a symplectic structure ω. Then E has a corresponding well defined homotopy class of $U(n)$-reductions ($Sp(n)$-reductions) of its structure group, and the Chern classes of these reductions provide us with well defined Chern classes of E.

Furthermore, in the real case, if one has also a Lagrangian subbundle L of E we shall say that (E, L) is a *Lagrangian bundle*, and we notice that giving L is equivalent to a further reduction of the structure group of E to $O(n)$. In the complex case, because of the $Sp(n)$-reductions, we shall also refer to (E, ω) as a *quaternionic bundle*.

The vanishing phenomena no. 7 and 8 of Table II show that orthogonal connections of a Langrangian bundle (quaternionic metric connections of a quaternionic bundle) are J_{odd}-connections, and this leads to residues and to secondary characteristic classes.

For instance, if a real symplectic vector bundle has a Lagrangian subbundle L defined outside a *singular set* S of its basis M, residues

$$\mathrm{res}_{c_{2h-1}}(L, S) \in H_{m-4h+2}(S, \mathbb{R})$$

will appear. As an application we quote

Proposition [Va 4]. *Let M be a $(4h - 2)$-dimensional compact almost symplectic manifold. Let L be a Lagrangian distribution over $M \setminus S$, where S consists of isolated points S_k. Then the Euler characteristic $\chi(M)$ is equal to the sum of the residues of c_{2h-1} at S_k.*

Now, let (E, L_a) $((E < \omega_a))$ $(a = 1, 2)$ be two Lagrangian (quaternionic) bundles with the same real (complex) bundle E. Then we have two families of orthogonal (metric quaternionic) connections, and, following the general construction of Section 1, it is possible to derive secondary characteristic classes from the algebra

$$(2.5) \qquad WL_n = \mathbb{R}[., c_{2h}, .] \otimes \mathbb{R}[., c'_{2h}, .] \otimes \wedge(., \widehat{c}_k, .),$$

where all the generators are copies of the Chern polynomials. Like in foliation theory, one sees that the same classes are provided by $\widehat{W}L_n$ obtained by replacing the third factor of WL_n by $\wedge(., \widehat{c}_{2h-1}, .)$ [Va 2]. Since the differential of $\widehat{W}L_n$ is zero, the secondary characteristic classes under discussion are algebraic combinations of the Pontrjagin classes of L_a (the Chern classes of E) and of the classes $\mu_h \in H^{4h-3}(M, \mathbb{R})$ defined by the generators $\widehat{c}_{2h-1}$. The latter are called the *Maslov classes* of (L_1, L_2) $((\omega_1, \omega_2))$ since, up to a constant factor, $\mu_1(L_1, L_2)$ is the class discovered by Maslov in the mathematical analysis of quantum physics [Ma]. (See also [Au] and its references for a survey of various other approaches to Maslov classes.)

The most important facts about Maslov classes are contained in the following theorem [Va 2, 4, 5]

Theorem. *i) The Maslov classes do not depend on the choice of the $O(n)$-connections (the $S_p(n)$-connections). ii) The Maslov classes are obstructions to the transversality of the Lagrangian subbundles L_1, L_2 (to the homotopy of the complex symplectic structures ω_1, ω_2 via similar structures). iii) If M is compact and oriented, the Maslov classes can be expressed as sums of residues along nice singularity sets.*

The most interesting case is obtained if one takes M to be a Lagrangian submanifold of a cotangent bundle T^*N, endowed with the canonical symplectic structure. Then, let $E = TT^*N|_M$, $L_1 = $ (the tangent bundle of the fibers of $T^*N|_M$, $L_2 = TM$. In this case it is possible to compute differential forms which represent the Maslov classes, if an auxiliary Riemannian metric of N is chosen. For instance, if N is locally flat, so is T^*N, and μ_h is represented by

$$(2.6) \qquad m_h = \nu_h \delta^{j_1 \cdots j_{2h-1}}_{i_1 \cdots i_{2h-1}} \, b^{i_1}_{j_1} \wedge b^{k_2}_{j_2} \wedge b^{i_2}_{k_2} \wedge \ldots \wedge b^{k_{2h-1}}_{j_{2h-1}} \wedge b^{i_{2h-1}}_{k_{2h-1}},$$

where $\nu_h =$const., and (b^j_i) is the matrix of 1-forms which represents the second fundamental form of M in T^*N [Va 2], [MN].

Similar computations can be done for certain quaternionic structures of cotangent bundles of Kähler manifolds [Va 5].

c) Holomorphic bundles

Here we give just a very brief quotation from [Va 3]. Let P be a holomorphic C^∞-trivial G-principal bundle, over the m-dimensional basis M, endowed with a holomorphic connection θ. Using θ and C^∞-trivial connections θ_0 on P, we arrive at secondary characteristic classes derived from the algebra

$$(2.7) \qquad Wh(m, G) = I_{\mathbb{C}}(G)|_{\{\deg > (m+1)/2\}} \otimes \wedge I^+(G, \mathbb{C}).$$

These classes are independent on the connections θ and θ_0. On the contrary, if one factorizes by $\{\deg > (m/2)\}$, some of the classes may vary continuously.

If the basis M is compact Kähler, the Atiyah vanishing theorem (no. 9, Table II), associates secondary classes to all the elements of $\wedge I^+(G, \mathbb{C})$.

References

[At] M. F. ATIYAH, Complex analytic connections in fibre bundles, *Transactions American Mat. Soc.*, **85** (1957) 181–207.

[Au] M. AUDIN, Classes caractéristiques Lagrangiennes, *Algebraic Topology, Barcelona 1986, Lecture Notes in Math.*, **1298**, Springer–Verlag, Berlin, New York, (1987) 1–16.

[BB] P. BAUM and R. BOTT, Singularities of holomorphic foliations, *J. Diff. Geometry*, **7** (1972) 279–342.

[BR 1] I. N. BERNSHTEIN and B. I. ROZENFELD, Characteristic classes of foliations, *Funkt. Anal. i Priloz̆*, **6–1** (1972) 68–69, *Func. Anal. and Appl.*, **6** (1972) 60–61.

[BR 2] I. N. BERNSHTEIN and B. I. ROZENFELD, Homogeneous spaces of infinite dimensional Lie algebras and characteristic classes of foliations, *Uspehi. Math. Nauk*, **28(4)** (1973) 103–138, *Russian Math. Surveys*, **28(4)** 107–142.

[Bo] A. BOREL, Sur la cohomologie des espaces fibrés principaux et des espaces homogènes de groupes de Lie compactes, *Ann. Math.*, **57** (1953) 115–207.

[Bt 1] R. BOTT, A residue formula for holomorphic vector fields, *J. Diff. Geometry*, **1** (1967) 311–330.

[Bt 2] R. BOTT, On a topological obstruction to integrability, *Global Analysis, Berkeley 1968, Proc. Sympos. Pure Math.*, **16** (1970) 127–131.

[Bt 3] R. BOTT, Lectures on characteristic classes and foliations, *Lectures on algebraic and differential topology, Lecture Notes in Math.*, **279** Springer-Verlag, Berlin, New York, (1972) 1–94.

[BH] R. BOTT and A. HAEFLIGER, On characteristic classes of Γ-foliations, *Bull. American Math. Soc.*, **78** (1972) 1039–1044.

[BT] R. BOTT and L. W. TU, *Differential forms in algebraic topology*, Springer–Verlag, Berlin, New York, (1982).

[CS 1] S. S. CHERN and J. SIMONS, Some cohomology classes in principal fiber bundles and their applications to Riemannian geometry, *Proc. Nat. Acad. Sci. U.S.A.*, **68** (1971) 791–794.

[CS 2] S. S. CHERN and J. SIMONS, Characteristic forms and geometric invariants, *Ann. Math.*, **99** (1974) 48–69.

[CG] L. A. CORDERO and P. M. GADEA, Exotic characteristic classes and subfoliations, *Ann. Inst. Fourier, Grenoble*, **26** (1976) 225–237. (Erratum, Idem 27 (1977)).

[CM] L. A. CORDERO and X. M. MASA, Characteristic classes of subfoliations, *Ann. Inst. Fourier, Grenoble*, **31** (1981) 61–86.

[Do] D. DOMINGUEZ, Classes caractéristiques nontriviales de sousfeuilletages localelement homogènes, *Geometriae Dedicata*, **28** (1988) 229–249.

[Fe] B. L. FEIGIN, Characteristic classes of flags of foliations, *Funct. Anal. i Priloz̆*, **9(4)** (1975) 49–56., *Funct. Anal and Appl.*, **9** (1975) 312–317.

[Fu 1] D. B. FUKS, Characteristic classes of foliations, *Uspehi Math. Nauk.*, **28(2)** (1973) 3–17, *Russian Math. Surveys*, **28(2)** (1973) 1–16.

[Fu 2] D. B. FUKS, Cohomology of infinite dimensional Lie algebras and characteristic classes of foliations, *Itogi Nauki, Matematica*, **10** (1978) 179–286, *J. Soviet. Math.*, **11** (1979) 922–980.

[GFF] I. M. GELFAND, B. L. FEIGIN and D. B. FUKS, Cohomologies of the Lie algebras of formal vector fields with coefficients in its adjoint space and variations of characteristic classes of foliations, *Funkt. Anal. i Priloz̆*, **8(2)** (1974) 13–29, *Funct. Anal. and Appl.*, **8** (1974) 99–112.

[GF] I. M. GELFAND and D. B. FUKS, Cohomologies of the Lie algebra of tangential vector fields of a smooth manifold, *Funkt. Anal. i Priloz̆*, **3(3)** (1969) 32–52, *Funct. Anal. and Appl.*, **3** (1969) 194–210.

[Gh] E. GHYS, Sur l'invariance topologique de la classe de Godbillon–Vey, *Ann. Inst. Fourier, Grenoble*, **37** (1987) 59–76.

[Go] C. GODBILLON, Cohomologies d'algèbres de Lie de champs de vecteurs formels, *Séminaire Bourbaki 1972/73, exp. 421, Lecture Notes in Math.*, **383**, Springer–Verlag, Berlin, New York, (1974) 69–87.

[GV] C. GODBILLON and J. VEY, Un invariant des feuilletages de codimension 1, C. R. Acad. Sci. Paris 273 (1971) 92–95.

[Ha 1] A. HAEFLIGER, Homotopy and integrability, *Manifolds, Amsterdam, 1970, Lecture Notes in Math.*, **197**, Springer–Verlag, Berlin, New York, (1971) 133–163.

[Ha 2] A. HAEFLIGER, Sur les classes charactéristiques des feuilletages, *Séminaire Bourbaki 1971/72, Exp. 412, Lecture Notes in Math.*, **317**, Springer–Verlag, Berlin, New York, (1973) 239–260.

[Hr] R. HARTSHORNE, *Residues and duality, Lecture Notes in Math.*, **20**, Springer–Verlag, Berlin, New York, (1966).

[He 1] J. L. HEITSCH, Deformations of secondary characteristic classes, *Topology*, **12** (1973) 381–388.

[He 2] J. L. HEITSCH, Residues and characteristic classes of foliations, *Bull. American Math. Soc.*, **83** (1977) 397–399.

[He 3] J. L. HEITSCH, Independent variation of secondary characteristic classes, *Ann. Math.*, **108** (1978) 421–460.

[HH] J. L. HEITSCH and S. HURDER, Secondary classes, Weil measures and the geometry of foliations, *J. Diff. Geometry*, **20** (1984) 291–309.

[Hu 1] S. HURDER, Dual homotopy invariants of *G*-foliations, *Topology*, **20** (1981) 365–387.

[Hu 2] S. HURDER, On the homotopy and cohomology of the classifying space of Riemannian foliations, *Proc. American Math. Soc.*, **81** (1981) 485–489.

[Hu 3] S. HURDER, Global invariants for measured foliations, *Transactions American Math. Soc.*, **280** (1983) 367–391.

[KT 1] F. W. KAMBER and PH. TONDEUR, Characteristic classes of modules over a sheaf of Lie algebras, *Notices American Math. Soc.*, **19**, (1972) A401.

[KT 2] F. W. KAMBER and PH. TONDEUR, Characteristic invariants of foliated bundles, *Manuscripta Math.*, **11** (1974) 51–89.

[KT 3] F. W. KAMBER and PH. TONDEUR, *Foliated bundles and characteristic classes, Lecture Notes in Math.*, **493**, Springer–Verlag, Berlin, New York, (1975).

[KY] H. KITAHARA and S. YOROZU, Godbillon–Vey invariant and its differential geometric interpretation, *Ann. Sci. Kanazawa Univ.*, **12** (1975) 41–51.

[KN] S. KOBAYASHI and K. NOMIZU, *Foundations of differential geometry*, **I, II** Intersci. Publ., New York, 1963, 1969.

[La] H. B. LAWSON, JR., *The quantitative theory of foliations*, Conf. Board of Math. Sci., Providence, R. I., 1977.

[Lz] C. LAZAROV, An index theorem for foliations, *Illinois J. of Math.*, **30** (1986) 101–121.

[LP 1] C. LAZAROV and J. PASTERNACK, Secondary characteristic classes for Riemannian foliations, *J. Diff. Geometry*, **11** (1976) 365–385.

[LP 2] C. LAZAROV and J. PASTERNACK, Residues and characteristic classes for Riemannian foliations, *J. Diff. Geometry*, **11** (1976) 599–612.

[Le 1] D. LEHMANN, J-homotopie dans les espaces de connexions et classes exotiques de Chern–Simons, *C. R. Acad. Sci. Paris*, **275** (1972) 835–838.

[Le 2] D. LEHMANN, Classes caractéristiques exotiques et $\mathcal{J}$-connexité des espaces de connexions, *Ann. Inst. Fourier, Grenoble*, **24** (1974) 267–306.

[Le 3] D. LEHMANN, Résidus des connexions à singularités et classes caractéristiques, *Ann. Ins. Fourier, Grenoble*, **31** (1981) 83–98.

[Le 4] D. LEHMANN, Classes caractéristiques résiduelles, *Publ. IRMA, Lille*, **17**, No. VII, 1989.

[Ma] V. P. MASLOV, *Théorie des perturbations et méthodes asymptotiques*, Moskow Univ. Publ. House, Moskwa 1965, Dunod, Paris, 1972.

[MM] T. MATSUOKA and S. MORITA, On characteristic classes of Kähler foliations, *Osaka J. Math.*, **16** (1979) 539–550.

[Mo] P. MOLINO, Propriétés cohomologiques et propriétés topologiques des feuilletages à connexion transverse projetable, *Topology*, **12** (1973) 317–325.

[MN] J. M. MORVAN and L. NIGLIO, Classes caractéristiques des couples de sous-fibrés Lagrangiens, *Ann. Inst. Fourier, Grenoble*, **36** (1986) 193–209.

[Ms] M. A. MOSTOW, Variations, characteristic classes, and the obstruction to mapping smooth to continuous cohomology, *Transactions American Math. Soc.*, **240** (1978) 163–182.

[Ni] S. NISHIKAWA, Residues and characteristic classes for projective foliations, *Japan J. Math.*, **7** (1981) 45–108.

[Pi] H. V. PITTIE, *Characteristic classes of foliations, Research Notes in Math.*, **10**, Pitman Publ., London, 1976.

[Ra] G. RABY, Invariance des classes de Godbillon–Vey par C^1-difféomorphismes, *Ann. Inst. Fourier, Grenoble*, **38** (1988) 205–212.

[Re] G. REEB, Sur certaines propriétés topologiques des variétées feuilletées, Hermann, Paris, 1952.

[Rn] B. L. REINHART, Differential geometry of foliations, *Ergebnisse der Math.*, **99**, Springer–Verlag, Berlin, New York, 1983.

[RW] B. L. REINHART and J. W. WOOD, A metric formula for the Godbillon-Vey invariant for foliations, *Proc. American Mat. Soc.*, **38** (1973) 427–430.

[Sp] E. H. SPANIER, *Algebraic Topology*, Mc. Graw–Hill, New York, 1966.

[Th] W. THURSTON, Noncobordant foliations of S^3, *Bull. American Math. Soc.*, **78** (1972) 511–514.

[Ts] T. TSUJISHITA, Characteristic classes of families of foliations, *Foliations, Tokyo 1983, Adv. Studies in Pure Math.*, **5** (1985) 195–210.

[Va 1] I. VAISMAN, Lagrangian foliations and characteristic classes, *Differential Geometry (L. A. Cordero, ed.), Research Notes in Math.*, **131**, Pitman Publ., London, (1985) 245–256.

[Va 2] I. VAISMAN, *Symplectic geometry and secondary characteristic classes, Progress in Math.*, **72**, Birkhäuser, Boston, 1987.

[Va 3] I. VAISMAN, Holomorphic structures and connections on differentiable fibre bundles, *Manuscripta Math.*, **62** (1988) 33–63.

[Va 4] I. VAISMAN, Residues of Chern and Maslov classes, Geometry and Topology of Submanifolds II, World Sci. Publ. Co., London, 1990, 370–385.

[Va 5] I. VAISMAN, Exotic classes of quaternionic bundles, *Israel J. of Math.*, **69** (1990), 46–58.

[Va 6] I. VAISMAN, Foliated partial holomorphic structures on principal bundles, *Monatshefte für Math.* (to appear).

[Wo] R. WOLAK, Characteristic classes of almost-flag structures, *Geometriae Dedicata*, **24** (1987) 207–220.

[Ya] K. YAMATO, On exotic characteristic classes of conformal and projective foliations, *Osaka Math. J.*, **16** (1979) 589–604.

Izu Vaisman

Department of Mathematics
University of Haifa
Israel

COLLOQUIA MATHEMATICA SOCIETATIS JÁNOS BOLYAI
56. DIFFERENTIAL GEOMETRY, EGER (HUNGARY), 1989

Isoparametric Submanifolds of General Riemannian Manifolds*

L. VERHÓCZKI**

1) Introduction

In the present paper we study isoparametric submanifolds of Riemannian manifolds. A submanifold is called isoparametric if the induced connection of its normal bundle is flat and the principal curvatures of this submanifold with respect to any parallel normal vector field are constant. At first the notion of isoparametric submanifolds was introduced only in spaces of constant curvature. In the last ten years such submanifolds have been extensively studied (see [2], [5], [7] and [10]). However, L. S. Palais and C. L. Terng have pointed out that general Riemannian manifolds may also have isoparametric submanifolds. They have shown that principal orbits of an isometric action admitting sections are submanifolds of this type in the ambient space (see [6]).

The purpose of this paper is to discuss properties of isoparametric submanifolds omitting the assumption that the ambient manifold is a space

---

* This paper is in final form and no version of it will be submitted for publication elsewhere.

** Supported by Hungarian Nat. Found. for Sci. Research Grant No. 424(86).

of constant curvature.

In chapter 2 at first we show an example for isoparametric submanifolds by deforming the metric of a product Riemannian manifold. Furthermore, we show that a submanifold with flat normal bundle and with parallel second fundamental form is isoparametric.

It is known that shape operators of an isoparametric submanifold M commute if the ambient manifold is a space of constant curvature. Hence, we can take involutive distributions on M such that they are invariant with respect to any shape operator of M and we can study integral submanifolds of these distributions which are totally geodesic in M (see [10]). However, this is not valid in general case. Therefore considering an isoparametric submanifold M, we shall fix a parallel normal vector field on M and shall regard the so-called eigendistributions of the shape operator with respect to this normal vector field. We shall give a sufficient and necessary condition for such involutive eigendistributions to have totally geodesic integral submanifolds in M.

In chapter 3 homogeneous isoparametric submanifolds will be considered. It is known that the parallel manifolds to an isoparametric hypersurface are also isoparametric if the ambient manifold is a space of constant curvature (see [11; p. 102]). In this paper we shall consider an isometric action on a Riemannian symmetric space which admits sections. We shall prove that if there is a principal orbit which is an umbilical submanifold, then the principal curvatures and the eigendistributions on a parallel submanifold to it are determined by the curvature tensor of the ambient manifold (see Theorem 2).

The notation and the notions for submanifolds of Riemannian manifolds used by us can be found in [1]. Throughout this paper N denotes a complete connected Riemannian manifold of dimension $m + k$ with the metric tensor $\langle \, , \, \rangle$ and with the Levi–Civita connection $\widetilde{\nabla}$. Let M be a connected m-dimensional submanifold of N and let ∇ be the induced connection on M. Considering three vector fields X, Y, ξ on M, where X, Y are tangential to M and ξ is normal to M, the Gauss and the Weingarten equations are given by

$$\widetilde{\nabla}_X Y = \nabla_X Y + h(X, Y),$$
$$\widetilde{\nabla}_X \xi = -A_\xi(X) + D_X \xi$$

where h denotes the second fundamental form of M and A_ξ denotes the shape operator with respect to ξ. D is the induced normal connection in the normal bundle of M denoted by $\nu(M)$. Considering a point p of M, the eigenvalues of the linear operator $A_{\xi(p)}$ are said to be principal curvatures of M at p with respect to the normal vector $\xi(p)$. A differentiable section ξ of the vector bundle $\nu(M)$ is said to be parallel (more precisely D-parallel) if $D\xi = 0$. It is well-known that the normal connection is flat if and only if on a suitable neighborhood of any point of M there exist parallel normal vector fields $\eta_1, \ldots, \eta_k$ which form a local orthonormal frame in $\nu(M)$.

We shall use the definition of isoparametric submanifolds given by Palais and Terng in [6].

Definition 1. *A connected submanifold M is said to be an isoparametric submanifold of N if the following conditions are satisfied:*

a) *The normal connection in $\nu(M)$ is flat.*

b) *The principal curvatures of M with respect to any parallel normal vector field defined on a connected open domain in M are constant on the domain.*

Obviously, if M is an isoparametric submanifold of N and ξ is a parallel normal vector field on M, then the eigenvalues of A_ξ have constant multiplicities on M.

In chapter 3 the following notion will be used.

Definition 2. *Let M be an isoparametric submanifold in N and let ξ be a parallel normal vector field on M, such that $\xi(q)$ is not a focal point of M for any $q \in M$. The submanifold $Exp(\xi)$ is said to be the parallel manifold to M defined by ξ, where Exp denotes the exponential map of $\nu(M)$ into N.*

2) Properties of isoparametric submanifolds in general Riemannian manifolds

In this chapter at first we give an example for isoparametric submanifolds by suitable deforming the metric tensor of a product Riemannian manifold.

These isoparametric submanifolds and ones of spaces of constant curvatures have similar properties.

Let L_0, $L_1, \ldots, L_r$ be connected differentiable manifolds and let us consider the product manifold $N = L_0 \times L_1 \times \ldots \times L_r$. Let $\pi_s : N \to L_s$ denote the natural projection of N onto L_s $(s = 0, \ldots, r)$.

Considering at every point p of N the tangent subspace $(E_s)_p$ corresponding to the leaf given by L_s, we obtain an involutive distribution E_s on N. Obviously, the tangent bundle of N is the Whitney sum of the vector bundles $E_0, \ldots, E_r$ over N.

Let X_s be a differentiable vector field on L_s. Clearly, there exists exactly one vector field $\overline{X}_s$ on N belonging to E_s such that $T\pi_s \circ \overline{X}_s = X_s \circ \pi_s$ holds where $T\pi_s$ denotes the tangent linear map of π_s. Later this assignment $X_s \mapsto \overline{X}_s$ will be used $(s = 0, \ldots, r)$.

Let $g_s : \mathfrak{A}L_s \times \mathfrak{A}L_s \to \mathfrak{F}(L_s)$ be a Riemannian metric on the manifold L_s where $\mathfrak{F}(L_s)$ denotes the ring of differentiable functions on L_s and $\mathfrak{A}L_s$ denotes the space of smooth vector fields on L_s. Let us regard the tensor field $\widetilde{g}_s : \mathfrak{A}N \times \mathfrak{A}N \to \mathfrak{F}(N)$ defined by $\widetilde{g}_s(\widetilde{X}, \widetilde{Y}) = g_s(T\pi_s \circ \widetilde{X}, T\pi_s \circ \widetilde{Y})$ for any vector fields $\widetilde{X}, \widetilde{Y}$ on N. Furthermore, let $f_1, \ldots, f_r$ be real-valued positive differentiable functions on L_0. Let us consider the Riemannian metric

$$\langle \, , \, \rangle = \widetilde{g}_0 + \sum_{s=1}^{r}(f_s \circ \pi_0)\widetilde{g}_s$$

on the manifold N.

Let q be an arbitrary point of L_0. Introduce the notations $M = L_1 \times \ldots \times L_r$ and $M_q = \{q\} \times M$.

Proposition 1. M_q *is an isoparametric submanifold of the Riemannian manifold N endowed with the metric $\langle \, , \, \rangle$.*

Proof. Let $\iota : M_q \to N$ denote the inclusion map of M_q into N. In order to determine the Levi–Civita connection on N the following well-known equality will be used (see e.g. [3; p. 83])

$$\langle \widetilde{\nabla}_{\widetilde{X}} \widetilde{Y}, \widetilde{Z} \rangle = \frac{1}{2}\{\widetilde{X}\langle \widetilde{Y}, \widetilde{Z}\rangle + \widetilde{Y}\langle \widetilde{Z}, \widetilde{X}\rangle - \widetilde{Z}\langle \widetilde{X}, \widetilde{Y}\rangle +$$
$$+ \langle \widetilde{Z}, [\widetilde{X}, \widetilde{Y}]\rangle + \langle \widetilde{Y}, [\widetilde{Z}, \widetilde{X}]\rangle - \langle \widetilde{X}, [\widetilde{Y}, \widetilde{Z}]\rangle\}$$

where $\widetilde{X}$, $\widetilde{Y}$, $\widetilde{Z}$ are arbitrary vector fields on N and $[\widetilde{X},\widetilde{Y}]$ denotes the Lie bracket of $\widetilde{X}$ and $\widetilde{Y}$. Clearly, for any vector fields X_s, Y_s on L_s and X_j on L_j $(s \neq j)$ the following equalities hold

$$[\overline{X}_s, \overline{Y}_s] = \overline{[X_s, Y_s]}, \qquad [\overline{X}_s, \overline{X}_j] = 0.$$

Let us consider two vector fields X_0, Y_0 on L_0. Clearly, we have $\langle \overline{X}_0 \circ \iota, \overline{Y}_0 \circ \iota \rangle = g_0(X_0(q), Y_0(q))$. Since the distributions $E_0, \ldots, E_r$ are orthogonal to one another, $\overline{X}_0 \circ \iota$ and $\overline{Y}_0 \circ \iota$ normal vector fields on M_q. Supposing that $s \neq 0$, the equalities given above imply

$$\langle \widetilde{\nabla}_{\overline{X}_s} \overline{X}_0, \overline{Y}_0 \rangle = \frac{1}{2} \overline{X}_s \langle \overline{X}_0, \overline{Y}_0 \rangle = \frac{1}{2} \overline{X}_s (g_0(X_0, Y_0) \circ \pi_0) = 0.$$

Hence, $D(\overline{X}_0 \circ \iota) = 0$, that is, $X_0 \circ \iota$ is a parallel normal vector field on M_q. Since the normal bundle $\nu(M_q)$ has parallel orthonormal frames, $\nu(M_q)$ is flat.

It remained to prove that the principal curvatures of M_q with respect to $\overline{X}_0 \circ \iota$ are constant.
Similarly, we have

$$\langle \widetilde{\nabla}_{\overline{X}_s} \overline{X}_0, \overline{Y}_s \rangle = \frac{1}{2} \overline{X}_0 \langle \overline{X}_s, \overline{Y}_s \rangle = \frac{1}{2} \overline{X}_0 (f_s \circ \pi_0)(g_s(X_s, Y_s) \circ \pi_s)$$

$$= \frac{1}{2(f_s \circ \pi_0)}((X_0 f_s) \circ \pi_0)\langle \overline{X}_s, \overline{Y}_s \rangle$$

and $\langle \widetilde{\nabla}_{\overline{X}_x} \overline{X}_0, \overline{X}_j \rangle = 0$ provided $s \neq j$.
It follows from this that $\widetilde{\nabla}_{\overline{X}_s} \overline{X}_0 = F_s \overline{X}_s$ where $F_s = \frac{1}{2(f_s \circ \pi_0)}(X_0 f_s) \circ \pi_0$.
Therefore the shape operator $A_{\overline{X}_0 \circ \iota}$ has constant eigenvalues

$\lambda_s = -\frac{1}{2f_s(q)} df_s(X_0(q))$ where df_s denotes the differential of f_s $(s = 1, \ldots, r)$, which completes the proof. ∎

Remark. M_q is a totally geodesic submanifold in N if and only if q is a critical point of every function f_s $(s = 1, \ldots, r)$.
Remark. Let p_j be an arbitrary point of L_j. By the equalities written above it can be seen that the submanifold
$\{q\} \times \{p_1\} \times \ldots \times L_s \times \ldots \times \{p_r\}$ is totally geodesic in M_q and totally umbilical in N.

After having given an example we show that a connected submanifold having flat normal bundle with parallel second fundamental form is isoparametric. To obtain this fact we need the following simple lemma.

Lemma 1. *Let M be a connected submanifold of a Riemannian manifold N such that there exists a parallel normal vector field ξ on M. If the shape operator A_ξ is parallel then the principal curvatures of M with respect to ξ are constant on M.*

Proof. Let us suppose that A_ξ is parallel, that is, ∇A_ξ vanishes. Let us consider a unit tangential vector field X on a connected domain U of M such that $A_\xi(X) = \lambda X$ holds for a differentiable function λ on U. Regarding a tangential vector v at a point p of U, by definition we have

$$0 = (\nabla_v A_\xi)X(p) = v(\lambda)X(p) + \lambda(p)\nabla_v X - A_\xi(\nabla_v X).$$

Since X is a unit vector field and A_ξ is a self-adjoint map, $\nabla_v X$ and $A_\xi(\nabla_v X)$ are orthogonal to $X(p)$. Therefore $v(\lambda) = 0$ holds which implies that λ is a constant function on U.

Considering the eigenvalues of A_ξ with the suitable multiplicities at every point of M, we obtain m continuous functions on M. It can be easily proved that the set of those points in M where these functions are differentiable is an open dense subset of M. Hence, applying the fact obtained above, it follows from this that the principal curvatures of M with respect to ξ are constant with constant multiplicities. ∎

Regarding the second fundamental form h of a submanifold M as a section of the product bundle $T^*M \otimes T^*M \otimes \nu(M)$, where T^*M denotes the cotangent bundle of M, the covariant derivative of h denoted by $\overline{\nabla}h$ can be defined. h is said to be parallel if $\overline{\nabla}h = 0$.

Lemma 2. *Let M be a connected submanifold of a Riemannian manifold N such that the normal connection in $\nu(M)$ is flat. Then h is parallel if and only if $\nabla A_\xi = 0$ holds for any parallel normal vector field ξ on a domain of M.*

Proof. Let $\eta_1, \ldots, \eta_k$ be parallel normal vector fields defined on a connected domain U of M such that they form a local orthonormal frame in $\nu(M)$. Let X, Y, Z be arbitrary tangent vector fields on U. By definition we have

$$(\overline{\nabla}_X h)(Y, Z) = D_X h(Y, Z) - h(\nabla_X Y, Z) - h(Y, \nabla_X Z).$$

We shall frequently use the following well-known equality

$$(1) \qquad\qquad h(Y, Z) = \sum_{i=1}^{k} \langle A_{\eta_i}(Y), Z \rangle \eta_i.$$

This implies

$$(2) \qquad (\overline{\nabla}_X h)(Y, Z) = D_X \left(\sum_{i=1}^{k} \langle A_{\eta_i}(Y), Z \rangle \eta_i \right) -$$

$$- \sum_{i=1}^{k} (\langle A_{\eta_i}(\nabla_X Y), Z \rangle + \langle A_{\eta_i}(Y), \nabla_X Z \rangle) \eta_i.$$

Applying the equality $X \langle Y, Z \rangle = \langle \nabla_X Y, Z \rangle + \langle Y, \nabla_X Z \rangle$, we obtain

$$(3) \qquad (\overline{\nabla}_X h)(Y, Z) = \sum_{i=1}^{k} \langle A_{\eta_i}(Y), Z \rangle (D_X \eta_i) +$$

$$+ \sum_{i=1}^{k} \langle (\nabla_X A_{\eta_i})(Y), Z \rangle \eta_i.$$

Since η_i $(i = 1, \ldots, k)$ is parallel, by (3) $\overline{\nabla} h = 0$ holds if and only if ∇A_{η_i} vanishes for any index i.

Obviously a normal vector field ξ defined on U is parallel in $\nu(M)$ if and only if there exist real numbers $c_1, \ldots, c_k$ such that $\xi = \sum_{i=1}^{k} c_i \eta_i$ is true. Hence our lemma is proved. ∎

The previous lemmas immediately imply the following assertion.

Corollary 1. *Let M be a connected submanifold of a complete Riemannian manifold N such that the normal connection in $\nu(M)$ is flat and M has parallel second fundamental form. Then M is an isoparametric submanifold.*

It is known that the shape operators of a submanifold having flat normal bundle commute if the ambient manifold is a space of constant curvature. Hence, in this case if M is an isoparametric submanifold, then we can take the so-called eigendistributions on M which are invariant distributions with respect to any shape operator. It is known, that these distributions are involutive and their integral submanifolds are totally geodesic (see e.g. [10]). However, the above facts do not hold in general case. Therefore eigendistributions of the shape operator with respect to a fixed parallel normal vector field will be studied.

Hereafter M will denote an isoparametric submanifold of a Riemannian manifold N. (We shall omit the assumption that N is a space of constant curvature.) Let ξ be a parallel normal unit vector field on a connected open domain U in M, and let $\lambda_1, \ldots, \lambda_r$ ($\lambda_s < \lambda_{s+1}$) be the constant principal curvatures of M with respect to ξ ($r \leq m$). Considering a point p of U, the tangent space $T_p M$ is a direct sum of eigensubspaces of $A_{\xi(p)}$, more precisely $T_p M = \sum_{s=1}^{r} (E_s)_p$ where $(E_s)_p$ denotes the eigensubspace belonging to the eigenvalue λ_s ($s = 1, \ldots, r$). Regarding these subspaces at every point of U, we obtain differentiable distributions $E_1, \ldots, E_r$ on U which will be called eigendistributions with respect to A_ξ. Obviously, they are orthogonal to one another. Concerning these distributions the following theorem is true.

Theorem 1. If E_s is an involutive distribution, then its integral submanifolds are totally geodesic in M if and only if the following condition (A) is satisfied:

(A) For any tangential vector fields X, Y, Z on U such that X and Y belong to E_s the vector field $\widetilde{R}(X, Z)Y$ is normal to ξ, where $\widetilde{R}$ denotes the Riemannian curvature tensor in N.

Proof. Let us suppose that E_s is involutive and the condition (A) is satisfied. Let X, Y be two tangential vector fields on U belonging to E_s. We shall prove that $\nabla_X Y$ also belongs to the eigendistribution E_s.

Let Z be a vector field on U such that Z belongs to another distribution E_j different from E_s. Since the condition (A) is satisfied, we have

$$(4) \qquad \langle \widetilde{R}(X, Z)Y, \xi \rangle = 0.$$

Let us consider parallel normal vector fields $\eta_1, \ldots, \eta_k$ on U such that they form an orthonormal frame in $\nu(M)$ and $\eta_k = \xi$. Then using (2) and the equality $\widetilde{\nabla}_Z \xi = -\lambda_j Z$, we have

$$(\overline{\nabla}_Z h)(X, Y) = Z(\lambda_s \langle X, Y \rangle)\xi - \lambda_s(\langle \nabla_Z X, Y \rangle + \langle X, \nabla_Z Y \rangle)\xi +$$
$$+ \sum_{i=1}^{k-1} Z \langle A_{\eta_i}(X), Y \rangle \eta_i -$$
$$- \sum_{i=1}^{k-1} (\langle \nabla_Z X, A_{\eta_i}(Y) \rangle + \langle A_{\eta_i}(X), \nabla_Z Y \rangle)\eta_i.$$

Since the metric tensor on M is parallel, the equality given above implies

$$(5) \qquad \langle (\overline{\nabla}_Z h)(X, Y), \xi \rangle = 0.$$

Let us regard the well-known Codazzi equality (see [1; p. 3])

$$\widetilde{R}(X,Z)Y^{\perp} = (\overline{\nabla}_X h)(Z,Y) - (\overline{\nabla}_Z h)(X,Y)$$

where the expression on the left hand side denotes the component of $\widetilde{R}(X,Z)Y$ normal to M. Applying (4) and (5), this implies

(6) $$\langle (\overline{\nabla}_X h)(Z,Y), \xi \rangle = 0.$$

On the other hand, using again (2) and the equality $A_\xi(Z) = \lambda_j Z$ it can be easily seen that

$$\langle (\overline{\nabla}_X h)(Z,Y), \xi \rangle = -\lambda_j \langle \nabla_X Y, Z \rangle - \lambda_s \langle Y, \nabla_X Z \rangle$$

holds. As the metric tensor on M is parallel and $\langle Y, Z \rangle = 0$, (6) implies

$$(\lambda_j - \lambda_s)\langle \nabla_X Y, Z \rangle = 0.$$

Since $\lambda_j - \lambda_s \neq 0$, $\nabla_X Y$ is normal to Z, that is, $\nabla_X Y$ is a vector field belonging to E_s. This immediately implies that if L_s is an integral submanifold of E_s, then its second fundamental form in M vanishes everywhere, that is, L_s is totally geodesic in M.

Conversely, let us assume that E_s is involutive and its integral submanifolds are totally geodesic in M. Considering the equalities written above, we obtain that the condition (A) is satisfied if Z is a normal vector field to E_s. We have only to show that the equality (4) is also true in that case when Z belongs to the distribution E_s.

Let X, Z be vector fields on U belonging to E_s. Obviously, $\widetilde{\nabla}_X \xi = -\lambda_s X$ and $\widetilde{\nabla}_Z \xi = -\lambda_s Z$ hold. Since h is symmetric in its variables, we have

$$\widetilde{R}(X,Z)\xi = \lambda_s(\widetilde{\nabla}_Z X - \widetilde{\nabla}_X Z) - \widetilde{\nabla}_{[X,Z]} \xi.$$

By our assumption $[X,Y]$ also belongs to E_s, that is, $\widetilde{\nabla}_{[X,Z]} \xi = -\lambda_s [X,Z]$ is true. The above equalities imply

(7) $$\widetilde{R}(X,Z)\xi = 0.$$

Applying (7) and the symmetric property of $\widetilde{R}$, we obtain that the equality (4) holds which completes the proof. ∎

Remark. Obviously, the condition (A) is satisfied if the ambient manifold N is a space of constant curvature.

Remark. Concerning Theorem 1 it can be seen that E_s is involutive if and only if

$$(\nabla_X A_\xi)(Y) - (\nabla_Y A_\xi)(X) = 0$$

holds for any vector fields X, Y belonging to E_s. Therefore by Lemma 2 if M is an isoparametric submanifold with parallel second fundamental form, then the eigendistributions given above are integrable.

3) Homogeneous isoparametric submanifolds in symmetric spaces

In this chapter we deal with those isometric actions on Riemannian manifolds which admit sections (or in other words admit orthogonally transversal submanifolds). These actions are important for us since the principal orbits are isoparametric submanifolds of the ambient manifold. In order to get more details on these actions see [6], [8], and [9]. Notions and basic facts concerning differentiable actions used by us can be found in [4].

Let G be a compact connected Lie group acting isometrically on a connected complete Riemannian manifold N and let $\alpha : G \times N \to N$ denote this action.

A closed connected submanifold is said to be a section of α if it meets all the orbits orthogonally. Obviously, if α admits a section, then for any point of N there exists a section passing through it. Considering a point p in N, $G(p)$ will denote the orbit of p which is a compact connected submanifold of N, and G_p will denote the isotropy group of p. For an element g of G let α_g be the isometry defined by $\alpha_g(p) = \alpha(g,p)$ where $p \in N$. $T_p G(p)$ denotes the tangent space of the submanifold $G(p)$ and $\nu_p G(p)$ denotes its orthogonal complement.

It is known that if $G(p)$ is a principal orbit, then the representation of the group G_p on $\nu_p G(p)$ is trivial. Therefore any vector u included in $\nu_p G(p)$ uniquely determines a section ξ of the normal bundle $\nu G(p)$ such that $\xi(p) = u$ and for every $g \in G$

$$T\alpha_g \circ \xi = \xi \circ \alpha_g$$

holds. If the action admits sections, then it can be shown that ξ is a parallel normal vector field (see [6]).

Later on we use the following result proved by Palais and Terng (see [6; p. 788]).

Proposition 2. *If $\alpha : G \times N \to N$ is an isometric action admitting sections, then the principal orbits are isoparametric submanifolds in N.*

Remark. Examples for isometric actions admitting orthogonally transversal submanifolds (sections) on compact symmetric spaces have been given by J. Szenthe in [9].

Henceforth N denotes a Riemannian symmetric space and $\alpha : G \times N \to N$ denotes an isometric action admitting sections. It is known that all the sections are totally geodesic submanifolds of N (see [8; p. 283]). Let p be a point of N such that $G(p)$ is a principal orbit of dimension m.

Let u be a unit vector in $\nu_p G(p)$ and consider the self-adjoint map $\widetilde{R}_u : T_p N \to T_p N$ defined by $\widetilde{R}_u(v) = \widetilde{R}(v, u)u$ for $v \in T_p N$ where $\widetilde{R}$ denotes the curvature tensor of N. Since $\nu_p G(p)$ is the tangent space of a totally geodesic submanifold at p, it can be easily seen that $\nu_p G(p)$ and $T_p G(p)$ are invariant subspaces of $\widetilde{R}_u$. The restriction of $\widetilde{R}_u$ onto the subspace $T_p G(p)$ will be denoted by R_u. Hence, we can take an orthonormal basis $v_1, \ldots, v_m$ of $T_p G(p)$ consisting of characteristic vectors of R_u. Suppose that $R_u(v_i) = \mu_i v_i$ holds ($i = 1, \ldots, m$). Let ξ be the parallel normal vector field on $G(p)$ having the property $\xi(p) = u$. According to the facts written above, ξ defines a differentiable section of the product bundle $T^* G(p) \otimes T G(p)$ denoted by R_ξ. Obviously, for any $q \in G(p)$ $R_{\xi(q)}$ has the same eigenvalues $\mu_1, \ldots, \mu_r$ ($\mu_{s-1} < \mu_s$; $s = 2, \ldots, r$) as R_u has. Considering the eigensubspaces $F_1(q), \ldots, F_r(q)$ of $R_{\xi(q)}$ at any point q of $G(p)$, we obtain r smooth distributions $F_1, \ldots, F_r$ on $G(p)$ which will be called the eigendistributions of R_ξ.

Let Exp denote the exponential map of $\nu G(p)$ onto N. Since $G(p)$ is compact, there exists a positive number ε such that the restriction of Exp onto the tubular neighborhood of $G(p)$ in $\nu G(p)$ with radius ε, denoted by $\nu^\varepsilon G(p)$, is a diffeomorphism. Clearly we have

$$Exp \circ T\alpha_g|_{\nu^\varepsilon G(p)} = \alpha_g \circ Exp|_{\nu^\varepsilon G(p)}$$

for any $g \in G$. Let us regard the geodesic $\gamma : (-\varepsilon, \varepsilon) \to N$ defined by $\gamma(t) = Exp(tu)$. Using the equality written above, we obtain that $G(\gamma(t)) = Exp(t\xi)$ holds for every $t \in (-\varepsilon, \varepsilon)$, that is, $G(\gamma(t))$ is a parallel manifold to $G(p)$ defined by $t\xi$. (See Definition 2.)

Let $\eta_1, \ldots, \eta_k$ ($\xi = \eta_k$) be parallel normal vector fields on $G(p)$ such that they form an orthonormal frame in $\nu G(p)$.

Let us take a local coordinate system $(x^1, \ldots, x^m)$ in $G(p)$ on a connected open neighborhood U of p such that $\frac{\partial}{\partial x^i}(p) = v_i$ is true ($i = 1, \ldots, m$). We can take a Fermi coordinate system $(z^1, \ldots, z^{m+k})$ on $Exp(\nu^\varepsilon(U))$

$$z^i = x^i \circ \pi \circ Exp^{-1} \qquad (i = 1, \ldots, m)$$
$$z^{m+j} = \langle \eta_j, Exp^{-1} \rangle \qquad (j = 1, \ldots, k)$$

where $\pi : \nu G(p) \to G(p)$ denotes the natural projection of $\nu G(p)$ onto $G(p)$.

Considering the principal orbit $Exp(t\xi)$, it is clear that the curves $\alpha_g \circ \gamma|[0, t]$ ($g \in G$) are minimizing geodesics joining the submanifolds $G(p)$ and $Exp(t\xi)$. Let us regard the vector fields $\widehat{\eta}_1, \ldots, \widehat{\eta}_{k-1}, \widehat{\xi}$ on $G(\gamma(t))$ which are obtained by parallel translation of the vector fields $\eta_1, \ldots, \eta_{k-1}, \xi$ along $\alpha_g \circ \gamma$ ($g \in G$). Recall that as $G(p)$ is a principal orbit, the isotropy group G_p acts trivially on $\nu_p G(p)$ and $\eta_j \circ \alpha_g = T\alpha_g \circ \eta_j$ is true for any $g \in G$. Let us consider the parallel vector field X_{m+j} along γ defined by $X_{m+j}(p) = \eta_j(p)$. The above facts immediately imply that $T\alpha_g \circ X_{m+j}$ is a parallel vector field along $\alpha_g \circ \gamma$ such that $T\alpha_g \circ X_{m+j}(0) = \eta_j \circ \alpha_g(p)$. Therefore we can easily see the following lemma.

Lemma 3. *The vector fields* $\widehat{\eta}_1, \ldots, \widehat{\eta}_{k-1}, \widehat{\xi}$ *form a parallel orthonormal frame in* $\nu Exp(t\xi)$. *Furthermore,* $\widehat{\xi} = \frac{\partial}{\partial z^{m+k}} \circ \iota$ *holds where* ι *denotes the natural inclusion map of* $Exp(t\xi)$ *into* N.

Let us consider the vector fields $Y_1, \ldots, Y_m$ along γ defined by $Y_i = \frac{\partial}{\partial z^i} \circ \gamma$ ($i = 1, \ldots, m$). Fixing a curve $\sigma : (-\delta, \delta) \to G(p)$ such that the tangent vector of σ at 0 coincides with $v_i = \frac{\partial}{\partial z^i}(p)$, take the geodesic variation Γ of γ defined by $\Gamma(t, \tau) = Exp(t(\xi \circ \sigma)(\tau))$. Since Y_i is the transversal vector field of this geodesic variation along γ, we obtain that Y_i is a Jacobi vector field. We shall use the following equality.

Lemma 4. $\widetilde{\nabla}_u Y_i = -A_u(v_i)$.

Proof. The equalities given below verify our lemma

$$\tilde{\nabla}_u Y_i = \tilde{\nabla}_u \frac{\partial}{\partial z^i} = \tilde{\nabla}_{v_i} \frac{\partial}{\partial z^{m+k}} + \left[\frac{\partial}{\partial z^i}, \frac{\partial}{\partial z^{m+k}} \right] (p) =$$

$$= \tilde{\nabla}_{v_i} \xi = -A_u(v_i) + D_{v_i}\xi = -A_u(v_i).$$

∎

Let us consider the parallel vector fields $X_1, \ldots, X_m$ along γ having the property $X_i(0) = v_i$ $(i = 1, \ldots, m)$. We have chosen the basis $v_1, \ldots, v_m$ of $T_p G(p)$ so that $R_u(v_i) = \mu_i v_i$ holds. Hence, it can be seen that if v_i is a characteristic vector of A_u, then the vector field Y_i is parallel to X_i. More precisely, using Lemma 4 and the well-known property of Jacobi vector fields in symmetric spaces (see [3; p. 132]), the following statement can be easily proved.

Lemma 5. *Let us suppose that $G(p)$ is an umbilical submanifold with respect to ξ, that is, $A_\xi = -\kappa id_{TG(p)}$ holds for a number κ. Then $Y_i(t) = c_i(t)X_i(t)$ is true where*

$$c_i(t) = \begin{cases} \cos(\sqrt{\mu_i} t) + \frac{\kappa}{\sqrt{\mu_i}} \sin(\sqrt{\mu_i} t) & \text{if } \mu_i > 0, \\ 1 + \kappa t & \text{if } \mu_i = 0, \\ \mathrm{ch}(\sqrt{-\mu_i} t) + \frac{\kappa}{\sqrt{-\mu_i}} \mathrm{sh}(\sqrt{-\mu_i} t) & \text{if } \mu_i < 0. \end{cases}$$

The previous assertion imply the following theorem concerning parallel manifolds to isoparametric submanifolds.

Theorem 2. *Let $\alpha : G \times N \to N$ be an isometric action admitting sections on a Riemannian symmetric space N and let ξ be a parallel normal unit vector field on a principal orbit $G(p)$. If $G(p)$ is an umbilical submanifold with respect to ξ, then there exists a positive number ε such that for any $t \in (-\varepsilon, \varepsilon)$ the eigendistributions of $A_{\hat{\xi}}$ on the parallel manifold $Exp(t\xi)$ to $G(p)$ coincide with ones of $R_{\hat{\xi}}$. ($\hat{\xi}$ denotes the normal vector field on $Exp(t\xi)$ obtained by parallel translation of ξ.)*

Proof. Let us consider the eigendistributions $F_1, \ldots, F_r$ of R_ξ on $G(p)$. It can be easily seen that the distributions $\hat{F}_1, \ldots, \hat{F}_r$ on $Exp(t\xi)$ obtained by parallel translation of $F_1, \ldots, F_r$ along geodesics $\alpha_g \circ \gamma |[0, t]$ $(g \in G)$ are just eigendistributions of $R_{\hat{\xi}}$ (with the same eigenvalues).

Hence the subspace $\widehat{F}_s(\gamma(t))$ is spanned by some suitable vectors from $X_1(t), \ldots, X_m(t)$.

Assume that $A_\xi = -\kappa id_{TG(p)}$ holds for a number κ. In this case we can use Lemma 5. Denoting the tangent vector of γ at t by $\dot\gamma(t)$ and the point $\gamma(t)$ by q, we have

$$\widetilde{\nabla}_{Y_i(t)} \widehat{\xi} = \widetilde{\nabla}_{\frac{\partial}{\partial z^i}(q)} \frac{\partial}{\partial z^{m+k}} = \widetilde{\nabla}_{\widehat{\xi}(q)} \frac{\partial}{\partial z^i} + \left[\frac{\partial}{\partial z^i}, \frac{\partial}{\partial z^{m+k}} \right](q) =$$

$$= \widetilde{\nabla}_{\dot\gamma(t)} Y_i = \widetilde{\nabla}_{\dot\gamma(t)} (c_i X_i) = \frac{dc_i}{dt} X_i(t)$$

where $\frac{dc_i}{dt}$ denotes the derivative of c_i at t. This immediately implies

(8) $$A_{\widehat{\xi}}(X_i(t)) = -\frac{1}{c_i(t)} \frac{dc_i}{dt} X_i(t).$$

The equality (8) shows that the theorem is true. ∎

Notice that using (8), the principal curvatures of a parallel manifold can be simply determined.

Preserving the notation used in Theorem 2, by Lemma 5 and the equality (8) we obtain the following assertion.

Corollary 2. *If A_ξ vanishes, then the principal curvatures of the submanifold $Exp(t\xi)$ with respect to $\widehat{\xi}$ are*

$$\lambda_i = \begin{cases} \sqrt{\mu_i}\, \mathrm{tg}(\sqrt{\mu_i}\,t) & \text{if } \mu_i > 0, \\ 0 & \text{if } \mu_i = 0, \\ -\sqrt{-\mu_i}\, \mathrm{th}(\sqrt{-\mu_i}\,t) & \text{if } \mu_i < 0. \end{cases}$$

Considering Theorem 2, it seems to be interesting to seek such isometric actions which have a totally geodesic principal orbit. Obviously, on spheres natural examples can be given for isometric actions having this property.

References

[1] B. Y. CHEN, Geometry of submanifolds and its applications, Tokyo, 1981.

[2] D. FERUS, H. KARCHER und H. F. MÜNZNER, Cliffordalgebren und neue isoparametrische Hyperflächen, *Math. Z.* **177** (1982) 479–502.

[3] D. GROMOLL, W. KLINGENBERG und W. MEYER, Riemannsche Geometrie im Grossen, Heidelberg, 1968.

[4] K.JÄNICH, Differenzierbare *G*-Mannigfaltigkeiten, Berlin, 1968.

[5] H. F. MÜNZNER, Isoparametrische Hyperflächen in Sphären I, II, *Math. Ann.* **251** (1980) 57–71, **256** (l981) 215–232.

[6] L. S. PALAIS and C. L. TERNG, A general theory of canonical forms, *Trans. Amer. Math. Soc.* **300** (1987), 771–789.

[7] W. STRÜBING, Isoparametric submanifolds, *Geometriae Dedicata* **20** (1986) 367–387.

[8] J. SZENTHE, Orthogonally transversal submanifolds and the generalization of the Weyl group, *Period. Math. Hungar.* **15** (1984) 281–299.

[9] J. SZENTHE, Some isometric actions with orthogonally transversal submanifolds on Riemannian symmetric spaces, *Studia Scien. Math. Hung.* **21** (1986) 175–179.

[10] C. L. TERNG, Isoparametric submanifolds and their Coxeter groups, *J. Diff. Geom.* **21** (1985), 79–107.

[11] C. L. TERNG, Submanifolds with flat normal bundle, *Math. Ann.* **277** (1987) 95–111.

L. Verhóczki

Department of Geometry
Technical University of Budapest
1521 Budapest, Hungary

COLLOQUIA MATHEMATICA SOCIETATIS JÁNOS BOLYAI
56. DIFFERENTIAL GEOMETRY, EGER (HUNGARY), 1989

Some Remarks on Parallel Immersions[*]

B. WEGNER

The aim of this short note is to give a comment on the work of H. R. Farran and S. A. Robertson on parallel immersions [2] and to present some some more precise results in the case of diagonal products. The main tool for this purpose is an equivalent interpretation of parallelism of immersions in terms of the geometry of the normal bundle, which previously has been used in the study of transnormal submanifolds [5].

In the first part it is shown that there is one gap in the considerations on parallel rank in [2]. Four different kinds of parallel ranks for immersions can be introduced. Two of them have been identified in [2] by some inaccuracy. Furthermore, the question posed at the end of the corresponding section in [2] is answered positively by quotation of another work and by an explicit example.

The second part discusses the parallel rank for diagonal products of immersions. Theorem 9 of [2] is given in a more precise form by determining the parallel rank of the diagonal product of two similar immersions explicitly. In contrast to the statement made there, Theorem 10 of [2] only presents

---

[*] This paper is in final form and no version of it will be submitted for publication elsewhere.

a necessary condition for the diagonal product of two pairs of parallel immersions to be parallel again. The precise necessary and sufficient condition for this property is more restrictive as will be proved below. Furthermore, detailed calculations for rotational surfaces will show that in general there will be no simple relation between the parallel rank of a diagonal product and the parallel ranks of its factors.

1. Parallel rank

Let $f : M \to E^n$ be a sufficiently differentiable immersion of an m-manifold M into Euclidean n-space, $n - m > 0$. As usual we denote by $T_x f : T_x M \to T_{f(x)} E^n$ the differential of f at $x \in M$, by $L_{f,x} = T_x f(T_x M)$ its image as a subspace of the vector space $\mathbb{R}^n$ underlying E^n and by $L_{f,x}{}^\perp$ its orthogonal complement. The corresponding affine subspaces give the tangent space $\tau_f(x) = \{f(x)\} + L_{f,x}$ and the normal space $\nu_f(x) = \{f(x)\} + L_{f,x}{}^\perp$ to f at x.

According to [2] two immersions $f, g : M \to E^n$ are called *parallel* if for every $x \in M$ $\nu_f(x) = \nu_g(x)$. The fundamental tool for the following considerations is

Lemma 1. *The two immersions f and g of M into E^n are parallel, if and only if the vector field given by $\xi(x) = g(x) - f(x)$ is a parallel section of the normal bundle of f.*

Before proving this assertion some further notations should be introduced. Each $v \in \mathbb{R}^n$ has a unique decomposition $v = \mathrm{prn}_{f,x} v + \mathrm{prt}_{f,x} v$ into a normal part $\mathrm{prn}_{f,x} v \in L_{f,x}{}^\perp$ and tangential part $\mathrm{prt}_{f,x} v \in L_{f,x}$ with respect to f at x. A normal vector field ξ of f is a called parallel in the normal bundle of f, if $\mathrm{prn}_f \overline{\nabla}_u \xi = 0$ for all tangent vectors u of f, $\overline{\nabla}_u$ denoting the directional derivative in E^n in direction of $T_x f(u)$.

Proof of Lemma 1: i) Assume that f and g are parallel. Then $\xi(x) = g(x) - f(x)$ is a section of the normal bundle of f and $L_{f,x}{}^\perp = L_{g,x}{}^\perp$ for all $x \in M$. Hence $L_{f,x} = L_{g,x}$ for all $x \in M$ and $\overline{\nabla}_u \xi = \overline{\nabla}_u g - \overline{\nabla}_u f = (T_x g)(u) - (T_x f)(u) \in L_{f,x}$ i.e., $\mathrm{prn}_f \overline{\nabla}_u \xi = 0$.

ii) Let $\xi(x) = g(x) - f(x)$ determine a normal section of f such that $\mathrm{prn}_f \, \overline{\nabla}_u \xi = 0$ for all tangent vectors u of M. Then

$$\overline{\nabla}_u g = \overline{\nabla}_u \xi + \overline{\nabla}_u f \in L_{f,x}$$

for all $u \in T_x M$, implying $L_{g,x} = L_{f,x}$ because f and g are immersions. But ξ was a normal section of f and therefore $\nu_{g,x} = \nu_{f,x}$ for all $x \in M$. ∎

Using the considerations made in [2] or Lemma 1 and standard results from differential geometry, we see that the following notions are well-defined.

Definition. Let $f : M \to E^n$ be a differentiable immersion.
a) The *parallel rank* of f is given by $\rho(f) = \dim A(f)$, where $A(f)$ is the affine space spanned by the immersions of M into E^n which are parallel to f.
b) The *covering parallel rank* of f is given by $\widetilde{\rho}(f) = \rho(f \circ \gamma)$, γ denoting the universal covering of M.
c) The *uniform local parallel rank* of f is given by

$$\rho_{\mathrm{loc}}(f) = \min\{\rho_{\mathrm{loc}}(f, x) | x \in M\},$$

where
d) $\rho_{\mathrm{loc}}(f, x) = \max\{\rho(f|U)|U \text{ open neighborhood of } x \text{ in } M\}$ denotes the *local parallel rank* of f at x.

Proposition 1. For every differentiable immersion $f : M \to E^n$ the following inequalities are satisfied

$$\rho_{\mathrm{loc}}(f, x) \geq \rho_{\mathrm{loc}}(f) \geq \widetilde{\rho}(f) \geq \rho(f).$$

There are examples for which the strict inequalities are valid.

Proof. The proposed inequalities follow directly from the definitions given above.
a) An example for which $\widetilde{\rho}(f) > \rho(f)$ is given in the original paper [2] or in the work of F. J. Craveiro de Carvalho and S. A. Robertson [1].
b) As an example for which $\rho_{\mathrm{loc}}(f) > \widetilde{\rho}(f)$ a plane curve with unbounded curvature may serve (see also [6]).
c) Examples for which $\rho_{\mathrm{loc}}(f, x) > \rho_{\mathrm{loc}}(f)$ can be found by including patches of 2-submanifolds in E^4 having parallel rank 2 which obviously exist and

those having parallel rank 0 (which will be shown to exist) in a common 2-submanifold.

d) Examples for which $\widetilde{\rho}(f) = \rho_{\text{loc}}(f) = \rho_{\text{loc}}(f, x) = 0$ for all $x \in M$ are given at the end of this paper or according to Lemma 1 by simply connected pieces of 2-submanifolds in E^4 with non-flat normal connection (see [3] for example). ∎

The final part of the preceding proof answers a question raised in [2].

2. Diagonal products of immersions

Let $f : M \rightarrow E^n$ and $\widetilde{f} : M \rightarrow E^{\widetilde{n}}$ be differentiable immersions. The diagonal product

$$f \triangle \widetilde{f} : M \rightarrow E^{n+\widetilde{n}}$$

is given by $(f \triangle \widetilde{f})(x) = (f(x), \widetilde{f}(x))$. According to Lemma 1 the parallel ranks of $f, \widetilde{f}$ and $f \triangle \widetilde{f}$ are given by the maximal numbers of linearly independent parallel sections in the corresponding normal bundles. The following calculations will show how the computation of the parallel rank of $f \triangle \widetilde{f}$ depends on the first and second fundamental tensors (see [4]) of f and $\widetilde{f}$.

Let $\Pi : E^{n+\widetilde{n}} \rightarrow E^n$ and $\widetilde{\Pi} : E^{n+\widetilde{n}} \rightarrow E^{\widetilde{n}}$ denote the orthogonal projections onto the factors of $E^{n+\widetilde{n}} = E^n \times E^{\widetilde{n}}$. Then $\Pi \circ (f \triangle \widetilde{f}) = f$ and $\widetilde{\Pi} \circ (f \triangle \widetilde{f}) = \widetilde{f}$. Every vector field $\xi_\triangle$ along $f \triangle \widetilde{f}$ in $E^{n+\widetilde{n}}$ decomposes into two projections ξ and $\widetilde{\xi}$ along f and $\widetilde{f}$ respectively such that $(\xi_\triangle (f \triangle \widetilde{f})(x)) = (\xi(f(x)), \widetilde{\xi}(\widetilde{f}(x)))$.

The decomposition of ξ into tangential and normal parts with respect to f,

$$\xi = \eta + \mu, \quad \eta = \text{prt}_f \, \xi, \quad \mu = \text{prn}_f \, \xi$$

defines a tangent vector field τ for M by $Tf(\tau) = \eta$. Let G denote the Riemannian metric which is induced by f on M, and consider the corresponding quantities for $\widetilde{f}$. Then the condition that $\xi_\triangle$ is a normal vector field along $f \triangle \widetilde{f}$ is equivalent to the equation

$$G(\tau, u) + \widetilde{G}(\widetilde{\tau}, u) = 0 \quad \text{for all } u \in TM.$$

The equation

$$\overline{\nabla}_u \xi_\triangle = \frac{d}{dt} \xi((f \triangle \widetilde{f})(\alpha(t))) \Big|_{t=0}$$

for some differentiable curve α in M with $\dot{\alpha}(0) = u$ implies

$$\overline{\nabla}_u \xi_\triangle = \left(\overline{\nabla}_u^n \xi, \widetilde{\overline{\nabla}_u^n \xi} \right).$$

This shows that $\mathrm{prn}_{f \triangle \widetilde{f}} \overline{\nabla}_u \xi_\triangle = 0$ is equivalent to

$$\overline{\nabla}_u^n \xi = T f(v) \text{ and } \widetilde{\overline{\nabla}_u^n \xi} = T \widetilde{f}(v)$$

for some tangent vector v to M.

In terms of the connections $\nabla, \widetilde{\nabla}$ induced by the metrics $G, \widetilde{G}$ on M, the corresponding normal connections $\nabla^{\mathrm{nor}}, \widetilde{\nabla}^{\mathrm{nor}}$ in the normal bundles of f and $\widetilde{f}$, the second fundamental tensors $V, \widetilde{V}$,

$$V(u, w) = \mathrm{prn}_f \overline{\nabla}_u^n T f(w)$$

(w suitably extended to a tangent vector field), and the shape operators (Weingarten maps) $A_\mu, \widetilde{A}_{\widetilde{\mu}}$ in direction of the normal vectors $\mu, \widetilde{\mu}$, given by

$$G(A_\mu(u), w) = \langle V(u, w), \mu \rangle_n,$$

$\langle, \rangle_n$ denoting the Euclidean scalar product of E^n, these equations can be interpreted for the normal and tangential parts of ξ and $\widetilde{\xi}$. This gives

Lemma 2. *With the notations introduced above the condition for $\xi_\triangle$ to be parallel in the normal bundle of $f \triangle \widetilde{f}$ is equivalent to the following system of equations:*

(A)	$G(\tau, u) + \widetilde{G}(\widetilde{\tau}, u) = 0,$
(B)	$V(\tau, u) + \nabla_u^{\mathrm{nor}} \mu = 0,$
$(\widetilde{B})$	$\widetilde{V}(\widetilde{\tau}, u) + \widetilde{\nabla}_u^{\mathrm{nor}} \widetilde{\mu} = 0,$
(C)	$\nabla_u \tau - A_\mu(u) = \widetilde{\nabla}_u \widetilde{\tau} - \widetilde{A}_{\widetilde{\mu}}(u),$

for all tangent vectors u of M.

With these preparations it is easy to prove the correct version of Theorem 10 in [2].

Finally two examples should be considered which show that there is no simple general relation between the parallel ranks of two immersions and that of their diagonal product. The differential equations contained in Lemma 2 seem to be non-integrable in the general case. The fundamental equations of surface theory will not be of much help for this purpose.

Since the constructions in the examples are based on surfaces of revolution, it is useful to recall the fundamental geometric quantities for these surfaces: Let $f(u^1, u^2) = (\alpha(u^2) \sin u^1, \alpha(u^2) \cos u^1, \beta(u^2))$ be a parametrized surface of revolution in space such that the profile curve is parametrized by arc length, i.e. $(\alpha'(u^2))^2 + (\beta'(u^2))^2 = 1$. Then the notions used in Lemma 2 have representations with respect to the parameter tangents $\partial f / \partial u^1$ and $\partial f / \partial u^2$ as follows: $G_{ij} = G_i \delta_{ij}$ with $G_1 = \alpha^2(u^2)$, $G_2 = 1$,

$$N = \frac{1}{|\alpha(u^2)|} \left(\frac{\partial f}{\partial u^1} \times \frac{\partial f}{\partial u^2} \right), \quad L_{ij} := \left\langle V \left(\frac{\partial f}{\partial u^i}, \frac{\partial f}{\partial u^j} \right), N \right\rangle = L_i \delta_{ij}$$

with $L_1 = \alpha(u^2) \beta'(u^2)$ and $L_2 = \kappa(u^2)$ (the curvature of the profile), while the matrix representation of A_N with respect to $\partial f / \partial u^1$ and $\partial f / \partial u^2$ has coefficients $L_i^j = \kappa_i \delta_{ij}$ with $\kappa_1 = \beta'(u^2)/\alpha(u^2)$ and $\kappa_2 = \kappa(u^2)$. At most two of the Christoffel symbols of the induced connection do not vanish:

$$\Gamma_{11}^2 = -\alpha'(u^2)\alpha(u^2) \text{ and } \Gamma_{21}^1 = \Gamma_{12}^1 = \alpha'(u^2)/\alpha(u^2).$$

Every section ξ of the normal bundle of f is defined by a function λ such that $\xi = \lambda N$. Since N is parallel in the normal bundle of f, we get $\nabla_u^{\text{nor}} \xi = (\nabla_u \lambda) N$.

Example 1. Let f and $\widetilde{f}$ be parametrizations of two straight circular cylinders with radius 1 and orthogonally intersecting axes:

$$f(u^1, u^2) = (\sin u^1, \cos u^1, u^2),$$
$$\widetilde{f}(u^1, u^2) = (u^1, \sin u^2, \cos u^2).$$

Then $G_{ij} = \widetilde{G}_{ij} = \delta_{ij}$, $L_{11} = L_1^1 = 1 = \widetilde{L}_2^2 = \widetilde{L}_{22}$, the other coefficients of L and $\widetilde{L}$ vanish, $\Gamma_{ij}^k = \widetilde{\Gamma}_{ij}^k = 0$. Hence for $\mu = \lambda N$ and $\widetilde{\mu} = \widetilde{\lambda} \widetilde{N}$ the equations in Lemma 2 are equivalent to

(A_3) $\tau = -\widetilde{\tau},$

(B_3)
$$\frac{\partial \lambda}{\partial u^2} = 0 \quad \text{and} \quad \tau^1 = -\frac{\partial \lambda}{\partial u^1},$$

$(\widetilde{B}_3)$
$$\frac{\partial \widetilde{\lambda}}{\partial u^1} = 0 \quad \text{and} \quad \widetilde{\tau}^2 = -\frac{\partial \widetilde{\lambda}}{\partial u^2},$$

(C_3)
$$\frac{\partial(\tau^1 - \widetilde{\tau}^1)}{\partial u^1} - \lambda = 0, \quad \frac{\partial(\tau^1 - \widetilde{\tau}^1)}{\partial u^2} = 0,$$
$$\frac{\partial(\tau^2 - \widetilde{\tau}^2)}{\partial u^2} - \widetilde{\lambda} = 0, \quad \frac{\partial(\tau^2 - \widetilde{\tau}^2)}{\partial u^1} = 0.$$

This implies that τ^1 is a function of u^1 only and satisfies the differential equation $2(\tau^1)'' + \tau^1 = 0$. Hence

$$\tau^1 = c_1^1 \sin \frac{1}{\sqrt{2}} u^1 + c_2^1 \cos \frac{1}{\sqrt{2}} u^1$$

for arbitrary real constants c_1^1, c_2^1. Similarly we get

$$\tau^2 = c_1^2 \sin \frac{1}{\sqrt{2}} u^2 + c_2^2 \cos \frac{1}{\sqrt{2}} u^2.$$

Clearly the other data can be determined uniquely from τ^1 and τ^2 using the equations given above.

According to Lemma 1 and Lemma 2 this shows $\widetilde{\rho}(f \triangle \widetilde{f}) = 4$. Clearly the conditions of Theorem 10 in [2] are satisfied for this example. But no parallel normal section can be found for $f \triangle \widetilde{f}$ which has projections along f and $\widetilde{f}$ being purely normal to f and $\widetilde{f}$.

Example 2. Let f parametrize a circular cone and $\widetilde{f}$ a unit sphere as rotational surfaces:

$$f(u^1, u^2) = (au^2 \sin u^1, au^2 \cos u^1, bu^2), \quad a^2 + b^2 = 1, \quad a \neq 0 \neq b,$$
$$\widetilde{f}(u^1, u^2) = (\cos u^2 \sin u^1, \cos u^2 \cos u^1, \sin u^2),$$

(u^1, u^2) varying in a suitable plane domain D. Then we have:

$$G_{ij} = G_i \delta_{ij}, \quad G_1 = a^2 (u^2)^2, \quad G_2 = 1,$$
$$L_{ij} = L_i \delta_{ij}, \quad L_1 = abu^2, \quad L_2 = 0,$$
$$L_i^j = \kappa_i \delta_i^j, \quad \kappa_1 = \frac{b}{au^2}, \quad \kappa_2 = 0, \quad \Gamma_{11}^2 = -a^2 u^2,$$
$$\Gamma_{21}^1 = \Gamma_{12}^1 = \frac{1}{u^2}, \quad \Gamma_{ij}^k = 0 \text{ else};$$

$$\widetilde{G}_{ij} = \widetilde{G}_i \delta_{ij}, \quad \widetilde{G}_1 = \cos^2 u^2, \quad \widetilde{G}_2 = 1,$$

$$\widetilde{L}_{ij} = \widetilde{L}_i \delta_{ij}, \quad \widetilde{L}_i = \cos^2 u^2, \quad \widetilde{L}_2 = 1,$$

$$\widetilde{L}_i^j = \delta_i^j, \quad \widetilde{\Gamma}_{11}^2 = \sin u^2 \cos u^2, \quad \widetilde{\Gamma}_{21}^1 = \widetilde{\Gamma}_{12}^1 = -\frac{\sin u^2}{\cos u^2},$$

$$\widetilde{\Gamma}_{ij}^k = 0 \text{ else.}$$

Hence with the usual notations the equations of Lemma 1 obtain the form

(A_4)
$$a^2 (u^2)^2 \tau^1 + \cos^2 u^2 \widetilde{\tau}^1 = 0, \quad \tau^2 + \widetilde{\tau}^2 = 0,$$

(B_4)
$$abu^2 \tau^1 + \frac{\partial \lambda}{\partial u^1} = 0, \quad \frac{\partial \lambda}{\partial u^2} = 0,$$

$(\widetilde{B}_4)$
$$\cos^2 u^2 \widetilde{\tau}^1 + \frac{\partial \widetilde{\lambda}}{\partial u^1} = 0, \quad \widetilde{\tau}^2 + \frac{\partial \widetilde{\lambda}}{\partial u^2} = 0,$$

(C_4)
$$\frac{\partial \tau^1}{\partial u^1} + \frac{1}{u^2} \tau^2 - \frac{b}{au^2} \lambda = \frac{\partial \widetilde{\tau}^1}{\partial u^1} - \frac{\sin u^2}{\cos u^2} \widetilde{\tau}^2 - \widetilde{\lambda},$$

$$\frac{\partial \tau^1}{\partial u^2} + \frac{1}{u^2} \tau^1 = \frac{\partial \widetilde{\tau}^1}{\partial u^2} - \frac{\sin u^2}{\cos u^2} \widetilde{\tau}^1,$$

$$\frac{\partial \tau^2}{\partial u^1} - a^2 u^2 \tau^1 = \frac{\partial \widetilde{\tau}^2}{\partial u^1} + \sin u^2 \cos u^2 \widetilde{\tau}^1,$$

$$\frac{\partial \tau^2}{\partial u^2} = \frac{\partial \widetilde{\tau}^2}{\partial u^2} - \widetilde{\lambda}.$$

From (B_4) we get that λ and $u^2 \tau^1$ are functions of u^1 only. Using (A_4) we derive from $(\widetilde{B}_4)$ and the last equation of (C_4):

$$-a^2 u^2 (u^2 \tau^1) + \frac{\partial \widetilde{\lambda}}{\partial u^1} = 0, \quad -\tau^2 + \frac{\partial \widetilde{\lambda}}{\partial u^2} = 0, \quad 2\frac{\partial \tau^2}{\partial u^2} + \widetilde{\lambda} = 0.$$

The integrability conditions for the first two equations imply that $\partial \tau^2 / \partial u^1$ is a function of u^1 only, because $u^2 \tau^1$ is a function of u^1 only. But then differentiation of the last equation gives $\partial \widetilde{\lambda} / \partial u^1 = 0$.

Hence $\widetilde{\lambda}$ and τ^2 are functions of u^2 only, $\tau^1 = 0$, $\lambda = c$ for some $c \in \mathbb{R}$, $\widetilde{\tau}^1 = 0$, $\widetilde{\tau}^2$ is a function of u^2 only. Furthermore $\widetilde{\lambda}$ has to satisfy the differential equation $2\widetilde{\lambda}'' + \widetilde{\lambda} = 0$, i.e. $\widetilde{\lambda}$ has the form

$$\widetilde{\lambda}(u^2) = A \sin \frac{1}{\sqrt{2}} u^2 + B \cos \frac{1}{\sqrt{2}} u^2$$

with $A, B \in \mathbb{R}$. Hence

$$\tau^2 = A \frac{1}{\sqrt{2}} \cos \frac{1}{\sqrt{2}} u^2 - B \frac{1}{\sqrt{2}} \sin \frac{1}{\sqrt{2}} u^2 = -\widetilde{\tau}^2.$$

Then the first equation of (C_4) implies

$$\frac{1}{\sqrt{2}} \left(\frac{1}{u^2} + \frac{\sin u^2}{\cos u^2} \right) \left(A \cos \frac{1}{\sqrt{2}} u^2 - B \sin \frac{1}{\sqrt{2}} u^2 \right) +$$

$$+ A \sin \frac{1}{\sqrt{2}} u^2 + B \cos \frac{1}{\sqrt{2}} u^2 - \frac{b}{au^2} C = 0$$

for all u^2. This only can be satisfied for $A = B = C = 0$.

The calculations made above show that $\widetilde{\rho}(f \triangle \widetilde{f}) = 0$ though $\widetilde{\rho}(f) = 1 = \widetilde{\rho}(\widetilde{f})$. In particular the question posed in [2] at the and of section 2 is answered positively by an explicit example having codimension 4.

References

[1] F. J. CRAVEIRO DE CARVALHO and S. A. ROBERTSON: Self-parallel curves, *Math. Scand.* **65** (1989), 67–74.

[2] H. R. FARRAN and S. A. ROBERTSON: Parallel immersions in Euclidean space, *J. London Math. Soc* **(2) 35** (1987), 527–538.

[3] D. FERUS, A. C. ASPERTI and L. RODRIGUEZ: Surfaces with non-zero normal curvature tensor, *Atti Acad. Lincei*, 8. Ser, vol. 73 (1982), 109–115.

[4] N. J. HICKS: Notes on differential geometry, van Nostrand (Princeton, 1965).

[5] B. WEGNER: Einige Bemerkungen zur Geometrie transnormaler Mannigfaltigkeiten, *J. Diff. Geom.* **16** (1981), 93–100.

[6] B. WEGNER: Cylinders of constant width. *Geometry, Proc. Congr.*, Thessaloniki / Greece 1987 (1988), 231–240.

Bernd Wegner

Fachbereich Mathematik
TU Berlin
Straße des 17. Juni 136
D–1000 Berlin 12
Germany

List of Participants

D. V. ALEKSEEVSKY

ul. Antonova, d. 2
kv. 99
117279 Moscow
USSR

S. I. ANDERSSON

Research Group of Global Analysis
Chalmers Univ. of Techn.
S–41296 Göteborg
Sweden

A. ASADA

Dept. of Math.
Sinsyu University
Matumoto
Nagano pref.
Japan

GH. ATANASIU

Dept. of Math.
Univ. of Braşov
R–2200 Braşov
Romania

S. BÁCSÓ

Dept. of Math.
Univ. of Debrecen
H–4010 Debrecen
Pf. 12
Hungary

C. BAIKOUSSIS

Dept. of Math.
Univ. of Ioannina
G–45110 Ioannina
Greece

T. F. BANCHOFF

Mathematics Dept.
Brown University
Providence
RI 02912
USA

J. K. BEEM

Mathematics Dept.
Univ. of Missouri
Columbia
MO 65211
USA

C. L. BEJAN

Seminarul Matematic
Universitatea "Al. I. Cuza"
R–6600 Iaşi
Romania

A. BEJANCU

Dept. of Math.
Polytechnic Inst. of Iaşi
C.P. 17, Iaşi 1
R–6600 Iaşi
Romania

K. BÉLTEKY

Dept. of Math.
Univ. of Debrecen
H–4010 Debrecen
Pf. 12
Hungary

T. Q. BINH

Dept. of Math.
Univ. of Debrecen
H–4010 Debrecen
Pf. 12
Hungary

N. BLAŽIĆ

Faculty of Math.
Univ. of Belgrade
Studentski trg. 16
YU–11000 Belgrade
Yugoslavia

N. BOKAN

Faculty of Math.
Univ. of Belgrade
Studentski trg. 16
YU–11000 Belgrade
Yugoslavia

K. BUCHNER

Institute of Math.
Techn. Univ. of München
Arcisstr. 21
D–8000 München 2
Germany

K. BUREŠ

Faculty of Math. and Anal.
Charles University
Sokolovská 83
CS–18600 Prague
Czechoslovakia

I. ČOMIĆ

Faculty of Technical Sciences
YU–21000 Novi Sad
Yugoslavia

A. CRUMEYROLLE

Univ. Paul Sabatier
Secreteriat Mathématiques
118, route de Narbonne
F–31062 Toulouse
France

B. CSIKÓS

Dept. of Geometry
Eötvös University
H–1088 Budapest
Rákóczi út 5
Hungary

A. DEKRÉT

Dept. of Math. and Phys.
VSLD, Maxova 24
CS–96053 Zvolen
Czechoslovakia

M. FALCITELLI

Dept. of Math.
Camput University
Via G. Fortunato
I–70125 Bari
Italy

V. T. FOMENKO

ul. Vichnevaja 56
kv. 42
347937 Taganrog
USSR

S. FORMELLA

ul. Klemensiewicza 10/4
PL–70028 Szczecin
Poland

M. FRANCAVIGLIA

Inst. of Math. and Phys.
Via C. Alberto 10
I–10123 Torino
Italy

S. GARBIERO

Dept. of Math.
Univ. of Lecce
Via Arnesano
I–73100 Lecce
Italy

P. B. GILKEY

Mathematics Dept.
Univ. of Oregon
Eugene, OR 97403
USA

V. V. GOLDBERG

Mathematics Dept.
New Jersey Institute of Technology
Newark, NJ 07102
USA

H. GOLLEK

Humboldt Universität zu Berlin
Sektion Mathematik
Berlin
Germany

E. M. GORELIK

ul. Taskentnaja 4–2–7
109444 Moscow
USSR

J. GRABOWSKI

Mathematisches Institut
Auf der Morgenstelle 10
D–7400 Tübingen
Germany

A. GYARMATI

Dept. of Math.
Univ. of Debrecen
H–4010 Debrecen
Pf. 12
Hungary

TH. HANGAN

15–Chemin des Ardennes
F–68100 Mulhouse
France

F. Hehl

Inst. Theor. Phys.
Universität Köln
Zülpicher Str. 77
D–5000 Köln 41
Germany

M. Husty

Inst. für Angewandte Geom.
Montanuniversität Leoben
A–8700 Leoben
Austria

T. Kántor

Dept. of Math.
Univ. of Debrecen
H–4010 Debrecen
Pf. 12
Hungary

H. Kawaguchi

Sagami Inst. of Technology
Tsujido Nishi-kaigan 1–1–25
Fujisawa 251
Japan

T. Kawaguchi

Institute of Information Sciences
Univ. of Tsukuba
Tsukuba-shi 305
Japan

M. Sz. Kirkovits

Dept. of Math.
Univ. of Forestry
H–9401 Sopron
Pf. 132
Hungary

B. Kis

Dept. of Math.
Univ. of Debrecen
H–4010 Debrecen
Pf. 12
Hungary

F. C. Klepp

Dept. of Math.
Polytechnical Inst. Timişoara
R–1900 Timişoara
Romania

I. Kolář

Math. Inst. of CSAV
Branch Brno
Mendelová nám. 1
CS–66282 Brno
Czechoslovakia

J. J. Konderak

Institute of Math.
Univ. of Salerno
I–84100 Salerno
Italy

Z. Kovács

Dept. of Math.
Teacher's Training College of
Nyíregyháza
H–4400 Nyíregyháza
Sóstói út 31/b
Hungary

O. Kowalski

MFF UK
Sokolovská 83
CS–18600 Prague
Czechoslovakia

J. KOZMA

Bolyai Institute
Univ. of Szeged
H–6720 Szeged
Aradi vértanuk t. 1
Hungary

L. KOZMA

Dept. of Math.
Univ. of Debrecen
H–4010 Debrecen
Pf. 12
Hungary

A. KURUSA

Bolyai Institute
Univ. of Szeged
H–6720 Szeged
Aradi vértanuk t. 1
Hungary

N. LAUTERSZTAJN-SKARBINSKI

Dept. of Struct. Mech.
Chalmers Univ. of Techn.
S–41296 Göteborg
Sweden

P. LECOMTE

Inst. de Mathematique
Universite de Liege
Des Tilleuls 15
B–4000 Liege
Belgium

K. LEICHTWEISS

Math. Inst.
Universität Stuttgart
Pfaffenwaldring 57
D–7000 Stuttgart
Germany

L. LEJEUNE

78 bis rue Thiers
F–59650 Willeneuve d'Ascq.
France

P. LIBERMANN

116 Av. du Général Leclerc
F–75014 Paris
France

K. MARATHE

Mathematics Dept.
Brooklyn College of CUNY
Brooklyn, NY 11210
USA

R. MARINOSCI

Dipartimento di Mat.
Universitá di Lecce
I–73100 Lecce
Italy

S. MARKVORSEN

Math. Institute
Technical Univ. of Denmark
DK–2800 Lungby
Denmark

P. W. MICHOR

Inst. für Mathematik
Universität Wien
Strudlhofgasse 4
A–1090 Wien
Austria

K. MIKAMI
Dept. of Math.
Akita University
Akita 010
Japan

J. MIKES
ul. Viljamsa 54/287
270015 Odessa
USSR

R. MIRON
Faculty of Math.
Univ. of Iaşi "Al. I. Cuza"
R–6600 Iaşi
Romania

E. MOLNÁR
Dept. of Geometry
Eötvös University
H–1088 Budapest
Rákóczi út 5
Hungary

G. MOUSSONG
Dept. of Geometry
Eötvös University
H–1088 Budapest
Rákóczi út 5
Hungary

D. F. NADJ
Dept. of Math.
Univ. of Forestry
H–9401 Sopron
Pf. 132
Hungary

P. T. NAGY
Bolyai Institute
Univ. of Szeged
H–6720 Szeged
Aradi vértanuk t. 1
Hungary

S. NIKČEVIĆ
Faculty of Math.
Univ. of Belgrade
Studentski trg. 16
YU–11000 Belgrade
Yugoslavia

J. NIKIĆ
Institute of Math.
Univ. of Novi Sad
YU–21000 Novi Sad
Yugoslavia

S. NÖLKER
Math. Institut
Universität Köln
Weyertal 86–90
D–5000 Köln
Germany

M. OKUMURA
Dept. of Math.
Saitama University
Urawa
Japan

C. OLMOS
Math. Section
I. C. T. P.
P. Box 586
I–34100 Trieste
Italy

A. ONISHCHIK

nab. Kolomenskaja 10–153
115142 Moscow
USSR

P. E. PARKER

Mathematics Dept.
Wichita State Univ.
Wichita
KS 67208
USA

J. F. POMMARET

ENPC/CERMA
La courtine
F–93167 Noisy-le-Grand
France

M. PRVANOVIĆ

Institute of Math.
Univ. of Novi Sad
YU–21000 Novi Sad
Yugoslavia

N. RAHMANI

Laboratoire de Matématiques et Informatique
4, rue des Freres Lumiere
F–68093 Mulhouse
France

T. RAPCSÁK

MTA SZTAKI
H–1111 Budapest
Kende u. 13
Hungary

M. RAVA

Inst. of Theor. Physics, C.T.H.
S–41296 Göteborg
Sweden

C. ROGER

Universite de Metz
U. F. R. de Mathematiques
F–57045 Metz Cedex 01
France

H. SHIMADA

Hokkaido Tokai University
Minamiku
Minamisawa 5–1
Sapporo 005
Japan

N. S. SINYUKOV

ul. Engels. H. 1–A, F. 4
270014 Odessa
USSR

GY. SOÓS

Dept. of Geometry
Eötvös University
H–1088 Budapest
Rákóczi út 5
Hungary

G. STANILOV

ul. Trepetlika n. 16–20/b
BG–1407 Sofia
Bulgaria

726

M. Stojanović
Gospodara Vučića 176
YU–11000 Belgrade
Yugoslavia

J. Szenthe
Dept. of Geometry
Eötvös University
H–1088 Budapest
Rákóczi út 5.
Hungary

J. Szilasi
Dept. of Math.
Univ. of Debrecen
H–4010 Debrecen
Pf. 12
Hungary

A. Szűcs
Dept. of Analysis
Eötvös University
H–1088 Budapest
Múzeum krt. 6–8
Hungary

L. Tamássy
Dept. of Math.
Univ. of Debrecen
H–4010 Debrecen
Pf. 12
Hungary

L. Vaisman
Dept. of Math.
Univ. of Haifa
31999 Haifa
Israel

L. Vanhecke
Dept. of Math.
Catholic Univ. of Leuven
Celestijnenlaan200 B
B–3030 Leuven
Belgium

L. Verhóczki
Dept. of Geometry
Technical Univ. of Budapest
H–1521 Budapest
Stoczek u. 4
Hungary

A. M. Vinogradov
Dept. of Geometry
Faculty of Mech. and Math.
119899 Moscow
USSR

B. Wegner
Fachbereich Mathematik
TU Berlin
Str. des 17, Juni 136
D–1000 Berlin
Germany

T. Yamada
Asahikawa National College of Techn.
2502 chōme Sunkōdai
Asahikawa 070
Japan

H. Yasuda
Asahikawa Medical College
4–5 Nishikagura
Asahikawa 078
Japan

Z. ZEKANOWSKI

ul. Marszalkowska 140 m. 88
PL–00061 Warszawa
Poland

V. ZOLLER

Research Inst. for Telecomm.
H–1026 Budapest
Gábor A. u. 65
Hungary

N. ŽUKOVA

ul. Brinskoga 1–90
603163 Gorky
USSR